AQUEOUS TWO-PHASE PARTITIONING

AQUEOUS TWO-PHASE PARTITIONING

Physical Chemistry and Bioanalytical Applications

Boris Y. Zaslavsky
KV Pharmaceutical Company
St. Louis, Missouri

Marcel Dekker, Inc. New York • Basel • Hong Kong

Library of Congress Cataloging-in-Publication Data

Zaslavsky, Boris Y.
Aqueous two-phase partitioning : physical chemistry and bioanalytical applications / Boris Y. Zaslavsky.
p. cm.
Includes bibliographical references and index.
ISBN 0-8247-9461-3 (acid-free paper)
1. Liquid chromatography. 2. Biomolecules—Separation. I. Title.
QP519.9.L55Z37 1994
574.19'285—dc20 94-22868
CIP

The publisher offers discounts on this book when ordered in bulk quantities. For more information, write to Special Sales/Professional Marketing at the address below.

This book is printed on acid-free paper.

MARCEL DEKKER, INC.
270 Madison Avenue, New York, New York 10016

Current printing (last digit):
10 9 8 7 6 5 4 3 2 1

PRINTED IN THE UNITED STATES OF AMERICA

To my mother

Irina S. Spector

PREFACE

Separation of chemical compounds by extraction or partition in immiscible (usually organic or water-organic) solvents was pioneered by Lyman C. Craig [1], mainly using partition in the countercurrent mode. For the development of the partition chromatography, Martin and Synge [2] were awarded a Nobel prize in 1952. Partition of substances in water-organic solvent systems, octanol-water, in particular, is one of the corner stones of studies of quantitative structure-activity relationships (QSAR) in drug design, medicinal chemistry, toxicology, etc. The use of organic solvents made these techniques, however, generally unsuitable for separation or analysis of biological materials.

Per-Åke Albertsson in Sweden in the mid-1950s showed that the partition technique may be used for separation of biological materials (from proteins to cells) provided aqueous two-phase systems are employed [3]. These systems, composed of two immiscible aqueous phases, occur in aqueous mixtures of different water-soluble polymers such as dextran and poly(ethylene glycol), or a single polymer and a specific salt (e.g., poly(ethylene glycol) and ammonium sulfate). An aqueous two-phase system contains mainly water, with the first polymer predominating in one phase and the second polymer (or salt) predominating in the other phase. Since the solvent in both phases is water, the phases provide a suitable environment for biological macromolecules, cells, viruses, and so on.

When a mixture of, for example, proteins, is added to an aqueous two-phase system, each protein distributes uniquely between the two phases. Protein partitioning depends on the specific features of the protein and partition conditions (composition of the system, pH, etc.). Under appropriate conditions the target protein may be concentrated in the upper phase, while all the other proteins partition into the lower phase resulting in the target protein isolation.

Partitioning of biomaterials in aqueous polymer two-phase systems is widely recognized today as a highly efficient separation technique. Readily scaled-up extraction in these systems has gained increasing attention as the separation method of choice in biotechnology. Versatility of the technique capable of separating cells and viruses as well as proteins and other soluble materials is an additional important advantage of the method.

Much empirical research has been devoted to the separation of cells, viruses, subcellular particles, and biopolymers both on the laboratory and the industrial scale [3-8]. Nevertheless, the mechanism governing aqueous two-phase partition is largely unknown [3]. Hence the optimal choice of partition conditions for the separation of particular biological products remains mostly a trial-and-error practice.

The rational design of optimal partition conditions depends on understanding the forces governing the partition behavior of solutes in an aqueous two-phase system. This understanding may hardly be gained from the empirical research using biological macromolecules which are too complicated themselves and hence leave too much room for uncertain speculations.

Recently numerous attempts to understand aqueous two-phase systems using the principles of polymer chemistry have been undertaken [7,8]. The possibility of applying the principles developed for nonpolar organic systems to highly polar aqueous systems is rather uncertain, however.

An alternative approach [9] is based on the analogy between aqueous two-phase systems and systems formed by water and immiscible organic solvents. This approach, used throughout this book, provides a better insight into the basic physicochemical principles of partitioning of solutes in aqueous two-phase systems.

The most fascinating aspect of the partition studies, in my view, is that they cross many fundamental fields of biomedical research beyond the scope of bioseparation - from drug design to organization of metabolic pathways. This aspect in regard to the information provided by the solute partition behavior in aqueous two-phase systems is the focus of this book. Unavoidably many of the issues in the book are discussed briefly. The purpose in many cases is to raise the questions rather than to provide the answers. I hope that these questions may provoke a reader to look at the known facts from a new perspective and inspire interest in aqueous two-phase partitioning, which I found to be very intellectually rewarding.

The three main issues in physical chemistry of the aqueous two-phase partition technique and its analytical applications are: the mechanism of phase separation in the aqueous mixtures of polymers or a single polymer and a salt; the physicochemical properties of the two phases governing partitioning of a solute and the ways to manipulate these properties; and, finally, the information provided by the partition behavior of a solute in a given aqueous two-phase system. This book is composed of three parts, each devoted to one of these issues.

The first part of the book describes the phenomenon of phase separation of two polymers in an aqueous medium and the role of the solvent in this phenomenon. Following the first two chapters on the specific features of water and aqueous solutions, phase separation in aqueous mixtures of polymers and a single polymer and an inorganic salt is considered to be due to the appearance of two immiscible aqueous media with different solvent properties. The phases of an aqueous two-phase system are suggested to be viewed as two different solvents of the same aqueous nature.

The second part deals with the physicochemical properties of the aqueous polymer phases governing the solute partitioning in comparison with those of water-organic solvent systems. The model of partitioning of solutes in aqueous two-phase systems, accounting for all the experimental data available, is advanced. According to this model, partitioning of a solute between the two phases of an aqueous two-phase system is governed by the difference in the total relative strength of the solute-water interactions in the phases.

The final part deals with the unique information provided by the partition behavior of a solute in aqueous two-phase systems. It is shown that the information in question is related to the solute relative hydrophobicity. The relative hydrophobicity of a solute as measured by the solute partitioning in an aqueous two-phase system may be used as a general descriptor of a biomolecule structure. Applications of this descriptor in quantitative structure-activity relationships (QSARs) for drugs and biological solutes are discussed. The relative hydrophobicity of a polar solute is shown to depend on the composition of an aqueous medium. The role of the relative hydrophobicity of biological solutes in their functioning *in vivo* is considered. It is suggested that partition behavior of solutes in aqueous two-phase systems simulates their behavior *in vivo*. This consideration leads to some new ideas about the role of variability of the water state in control of proteins functioning, in arrangement of blood-tissue barriers, in detoxification effects of some water-soluble polymers, etc. Several new applications of the aqueous two-phase partition technique in medical diagnostics, quality control of recombinant proteins, characterization of biopolymers in medicine, biology, and biotechnology are described. Finally, the concepts developed throughout the book are applied to the data on separation of biomolecules in aqueous two-phase systems, and the approach to the separation method development is discussed. The last chapter presents over 150 phase diagrams with polymer and salt composition of the phases for various aqueous polymer two-phase systems.

Most of the concepts discussed throughout the book originate from research performed by my co-workers and me at the Nesmeyanov's Institute of Organoelement Compounds, Russian Academy of Sciences, Moscow, Russia. I would like to express my gratitude to Anna A. Borovskaya, Nelli D. Gulaeva, Dr. Natalia M. Mestechkina, and Dr. Larisa M. Miheeva not only for technical skills and many helpful suggestions but firstly and most of all for the friendship and moral support during many years of our work together.

Many stimulating and fruitful hours of discussions with Dr. Michael A. Chlenov and Dr. Victor Y. Levin are greatly appreciated.

I started to work on this book in New York City where I arrived with my family in 1991 from Moscow, Russia. I would like to thank many wonderful friends who helped us to adjust and made the hard transition period easier and more tolerable, particularly, Dimitri and Sophie Stein, Gene and Gloria Sosin, Brigite Sauget, Ruth Polack, and Dr. Olaf Andersen of Cornell University Medical College.

Many wonderful people have been a source of inspiration and encouragement during the development of this book. My father, Dr. Yurii S. Zaslavsky, has always given me an example of true devotion to research and scientific integrity. I am greatly indebted to Dr. James S. Clegg of California University, Dr. David Kessel of Wayne State University School of Medicine, and Dr. Carel Van Oss of New York State University at Buffalo for critical reading and editing parts of the manuscript. The encouragement and moral support of Dr. Carl Djerassi of Stanford University is sincerely appreciated. Thanks are also due to the authors and publishers of copyrighted materials. I am also thankful to Anita Lehkwani, Eric Stannard and other members of the staff of Marcel Dekker, Inc. for their help in publishing this work. Without the patience of my wife, Ira, and son, Alex, this book would not have been written.

Boris Y. Zaslavsky

REFERENCES

1. L. C. Craig and D. Craig, In: Techniques of Organic Chemistry, (A. Weissberger, ed.), Vol. #3, Interscience, New York, 1956.
2. A. J. P. Martin and R. L. M. Synge, Biochem. J., 35, 1358 (1941).
3. P.Å. Albertsson, Partition of Cell Particles and Macromolecules, 3rd. ed., Wiley, New York, 1986.

4. Partitioning in Aqueous Two-Phase Systems: Theory, Methods, Uses, and Applications to Biotechnology (H. Walter, D.E. Brooks, and D. Fisher, eds.), Academic Press, Orlando, Florida, 1985.
5. Separations Using Aqueous Phase Systems: Applications in Cell Biology and Biotechnology (D. Fisher and I. A. Sutherland, eds.), Plenum Press, New York, 1989.
6. W. Muller, Liquid-Liquid Partition Chromatography of Biopolymers, GIT Verlag, Darmstadt, 1988.
7. H. Walter, G. Johansson, and D.E. Brooks, Anal.Biochem., Partitioning in Aqueous Two-Phase Systems: Recent Results, 197, 1-18 (1991).
8. Methods in Enzymology, Aqueous Two-Phase Systems Vol. 228 (H. Walter and G. Johansson, eds.), Academic Press, Orlando, Florida, 1994.
9. B.Yu. Zaslavsky, Anal.Chem., Bioanalytical Applications of Partitioning in Aqueous Polymer Two-Phase Systems, 64, 765A-773A (1992).

CONTENTS

PART 1.

PHASE SEPARATION IN AQUEOUS POLYMER SYSTEMS

A vast variety of methods currently exists for separation and characterization of biological materials. Numerous chromatographic and extraction techniques have been described in the literature. The special feature of the technique of partition in aqueous polymer two-phase systems (distinguishing it from other bioanalytical and separation methods) is the fact that the solvent in both phases of a system is of the same aqueous nature. As shown throughout this book, this feature not only provides the surroundment suitable for biological materials in both phases. It is also fundamentally important for analytical applications of the technique.

The occurrence of immiscible aqueous phases is not as readily understandable as that of common two-phase systems composed of solvents of different chemical nature. Two-phase systems commonly used for liquid-liquid extraction and partition chromatography are formed by partially immiscible solvents such as water and ether or water and octanol. Occurrence of these systems may be viewed as the result of limited mutual solubility of solvents originated from their different chemical nature. Addition of inorganic salts, temperature changes, etc., may cause phase separation in mixtures of water with water-soluble organic solvents such as iso-propanol or ethylene glycol, for example. This phenomenon may also be regarded as resulting from changes in the mutual solubility of two solvents under influence of external factors.

Phase separation in mixtures of two polymers or of a single polymer and a salt in water as an only solvent is not readily understood. This phenomenon is, however, crucial for gaining an insight into forces governing partitioning of solutes in aqueous two-phase systems. Hence, the first part of this book is devoted to phase separation in aqueous polymer systems. Since any aqueous polymer two-phase systems include (a) water and (b) at least one macromolecular compound, properties of water and aqueous polymer solutions are discussed in the first two chapters. These properties are discussed from the viewpoint of effects of macromolecular and low molecular weight compounds on the structure and solvent features of aqueous media. Chapter 3 deals with a possible mechanism of phase separation in aqueous polymer systems.

CHAPTER 1.

WATER IN THE PRESENCE OF ADDITIVES

The properties of liquid water in relation to its structure and/ or thermodynamic state in pure water and in the presence of additives of low molecular weight are considered in this chapter. The purpose is to substantiate the fundamental role of water in specific features of aqueous two-phase systems discussed below.

In a sense aqueous polymer solutions are similar to those of some polar organic solvents. Organic solvents such as alcohols, acetone, acetonitrile, etc., are fully miscible with water (i.e., there is no real solubility limit) like certain water-soluble polymers such as polyethylene glycol, polyvinylpyrrolidone, etc. Phase separation in water-organic solvent mixtures can be induced, for example, by temperature change or salt addition exactly as in aqueous polymer solutions. Therefore the properties of water in the mixtures are worthy of consideration. Effects of inorganic salts on the water properties will also be considered as these effects are important for phase separation and for the properties of the phases in aqueous two-phase systems with salt additives.

The literature on the properties of water and aqueous solutions is enormous (see, e.g., [1-8]) and hence it is unavoidable to be selective in the choice of references. The selection was often based not only on the merits of the publications but also on my personal views as well as my unfortunate ignorance due to the vast number of publications and limited expertise in the corresponding fields. The physical assumptions underlying some of the most important concepts on the structure of water and its change under influence of various solutes are described below, together with the implications for the subject of the present book.

1.1. PROPERTIES OF LIQUID WATER

Water has many peculiar physical properties such as contraction on melting, density maximum at 4°C, isothermal compressibility minimum (at 46°C) in the normal liquid range, etc. All these peculiarities are due to the features of the water molecule.

The isolated water molecule is symmetrical and triangular, the O-H bond length being 0.957 Å and the H-O-H angle 104.5°. The molecule is highly polar, with the permanent electric dipole moment of 1.83 D (debye units). The intermolecular interactions in ice and in liquid water increase the effective dipole moment per molecule; the estimates given in literature [6, p.8] amount to 3.6 D in ice, 2.45 D in liquid water at 0°C and 2.37 D at 83°C.

One of the essential features of the water molecule important for the properties of liquid water is its ability to participate in hydrogen bonding (H-bonding). This ability is due to the specific distribution of electron density in the molecule.

The association of water molecules in the vapor phase produces dimers with a linear hydrogen bond (H-bond) as found by molecular beam microwave spectroscopy [9]. According to the quantum mechanical calculations for the water dimer, moderate rotations of the water molecules about their oxygen atoms are of small energy cost, provided the donated hydrogen remains essentially on the symmetry axis. Motions that move the donated hydrogen off axis increase the dimer energy significantly (see, e.g., in [5]). The calculations also indicate that after forming a first H-bond, the charge distribution within the molecules' pair is altered in such a way that the hydrogen acceptor molecule becomes potentially an even better hydrogen donor than before. In other words, the water molecule is capable of forming a stronger second H-bond due to existence of the first H-bond. Similarly, the proton donor acquires an enhanced ability to accept a hydrogen due to the H-bond that it has already formed. This particular feature of H-bonding in water is called "cooperativity" and it is fundamentally important for the structure of liquid water.

According to the definition, the term of structure covers the distribution of the distances and angles between the particles in a system under examination. The structure of a liquid in a three-dimensional space fluctuates in time. Therefore the question of the time over which a measurement of the structure-related properties of the liquid can be made by a given experimental method is of a critical importance (for review see refs.1,5,7). The diffusional correlation time, i.e. the time between the changes of position for a water molecule, is about 10^{-5} sec in ice and ca. 10^{-11} sec in pure liquid water. Hence it is clear that ice is a much more suitable subject for such "standard" structure-exploring methods as the X-ray and neutron diffraction techniques. These methods have been essential for analysis of the structures of different forms of ice (see, e.g. in [1,2]).

Each water molecule in the ice crystal has four nearest neighbors at a distance of 2.76 Å to which it is H-bonded. The molecule donates its two hydrogens to two of the four neighbor molecules and accepts hydrogens from the other two neighbors. These four H-bonds are spatially arranged with local tetrahedral symmetry. That means that the oxygen atoms of the neighbor molecules occupy the vertices of a regular tetrahedron surrounding the oxygen atom of the central molecule. The bond angle of the isolated water molecule (104.5°) is only slightly different from the ideal tetrahedral angle (109.5°) corresponding to the strictly linear H-bonding in the ice crystal. The O-H equilibrium distance within an individual water molecule is 0.99 Å, so that the O...H distance in the H-bonded interconnections is 1.77 Å. There are 12 second nearest neighbors, each at 4.5 Å from oxygen atom of the water molecule taken as center. The arrangement is a relatively open one, with interstices of considerable size.

The open space in the tetrahedrally structured ice phase or liquid water seems to explain a wide variety between the estimates of the size of a water molecule given in the literature. It is reported [6, p.58] to be ca. 18 cm^3/mole as deduced from the normal liquid density at atmospheric pressure and 4°C. From space-filling models, the actual volume of a static water molecule is estimated by Conway [6, p.58] as 7.5 cm^3/mole, and the spherical rotation envelope of such an individual molecule is reported by the same author as about 22.5 m^3/ mole.

The unfilled space within ordinary ice crystals allows ice to respond to increased pressure by formation of higher density structures. The water molecules in most of the known ice forms retain the basic tetrahedral H-bonded structure with nothing resembling close packing of spheres. The arrangement of the four nearest neighbors around the one water molecule taken as center may be a highly distorted tetrahedron, with H-bonded neighbors being at distances of 2.75-2.87 Å and the nonbonded second neighbor molecules at distance as small as 3.24 Å [7a]. In some forms of ice there is a single network of H-bonds connecting the water molecules, which is essentially tetrahedral though distorted. In the other ice forms there are even two interpenetrating but not interconnected networks of H-bonded water molecules. Each network fills empty space in the other. Each of these networks is perfectly tetrahedral, but each oxygen atom has eight nearest neighbors instead of four. Each water molecule is H-bonded to four of these eight neighbors, and is in repulsive (i.e., nonbonded) contact with the other four molecules. The O...O nearest neighbor distance, owing to these repulsive interactions is 2.86 Å, or 0.11 Å larger than in the ordinary ice crystal. In view of the subject of this book this case may be considered as an example of two coexisting but immiscible water structures in the specific ice form.

The heat of ice melting amounts to merely 13% percent of the sublimation energy of ice. That implies that the majority of H-bonds is not destroyed upon melting. Direct structural information available for the crystalline state of water cannot be obtained for liquid water as in the liquid state only the short-range order may exist long enough. As the short-range order in water must be governed by the features of the water molecule per se, it is to be assumed that the short-range order in ice and in liquid water are very similar. The experimental observations support this assumption.

X-ray diffraction analysis of liquid water gives a radial distribution function for water molecules in a coordinate system in which the one water molecule is taken as center. The analysis shows a very sharp peak in the distribution function at a distance of 2.84 Å at 4°C and 2.94 Å at 200°C. This peak characterizes the average distance at which the nearest neighbor molecules are most likely to be located. The above values compared to the one given above for the ordinary ice crystal (2.76 Å) indicate that the nearest neighbors are farther apart in liquid water than they are in ordinary ice. The intensity of the peak in the distribution function indicates the number of the nearest neighbor molecules varying from 4.5 to 5.0 [5] depending on what maximum distance is selected for counting nearest neighbors. For simple liquids this number is usually relatively large (for close-packed spheres there would be 12 nearest neighbors) as compared to the 4.5-5.0 value for liquid water. This is an unambiguous evidence that the local order about a given water molecule in liquid water remains close to tetrahedral. A broad second peak in the distribution function is observed at a distance of about 4.5 Å which corresponds to the oxygen-oxygen distance range where successive bonds at the tetrahedral angle would require the second nearest neighbor molecules to be located. The number of the second nearest neighbors is found to be 12. Beyond about 8 Å the random distribution of molecules is observed. The neutron scattering technique provided the information completely consistent with that obtained by the X-ray diffraction method. Hence, it is universally accepted that each molecule in liquid water tends to coordinate in a tetrahedral grouping with its nearest neighbors. The tetrahedra, however, are distorted compared to those in ice, and the orientations of nearest neighbor molecules are much more variable than in ice.

Heat capacity of liquid water is practically twice that of ice. This fact cannot be attributed to the difference between the energy states of individual molecules in the liquid and crystalline states and must be explained by a temperature dependence of the energy of intermolecular interactions. The reason may be either a decrease of the number of H-bonds with increasing temperature or variable intermolecular arrangement of molecules, the average state gradually shifting to higher energy as the temperature is increased.

The essential feature of the H-bond responsible for the intermolecular arrangement of water molecules is the angular dependence of the bond energy. Pople [10] was the first to establish the importance of this dependence for the water structure. Pople indicated that "bent" H-bonds, i.e. the ones nonlinear in the O-H...O direction (up to about 25-30° out of line), are associated with an attractive potential energy dependent on the O-H...O angle. The angular dependence of the H-bond energy was computed by Popkie et al.[11] for water dimer in vapor phase. In liquid water this dependence is affected by the existence or absence of the second H-bonds formed by the participating molecules and by other closely adjacent non-bonded water molecules. This function is not available as yet but the view [12] that the energy levels of water molecules (or those of H-bonds) in liquid water should be described by an essentially continuous broad distribution is consistent with numerous experimental observations.

The very broad distribution of vibrational bands in liquid water observed in infrared and Raman spectra [13] implies a substantial variety of energies (or distances and bond angles) for the molecules. Any attempt to specify how many H-bonds a given water molecule forms with its neighbors must involve an element of arbitrary definition [5,7] independent of dealing in terms of interactions energies or in terms of geometry (i.e., distances and bond angles). There is no universally agreed method of estimating the H-bond energy but the estimate made from the experimentally determined energies of sublimation and vaporization of ice -3.57 kcal/mole of bonds [1] is in agreement with the estimate of -4.0 kcal/mole obtained by theoretical calculations [14].

The structure of liquid water could be defined if the following four factors were determined: i) the positional correlation between a given molecule and its first, second, third, etc. nearest neighbors; ii) the coordination number, i.e. the mean number of nearest neighbors to a given molecule; iii) the angular correlation between the nearest neighbors; and iv) the number of broken H-bonds and the extent to which the existent H-bonds are bent while still providing a given intermolecular arrangement. It was noted above that while the first two factors are determined experimentally by means of the X-ray and neutron scattering techniques, the latter two present unresolved problems so far.

The most powerful methods available at present for studying the structure of liquid water seem to be advanced computer simulations of water structure by the Monte Carlo and molecular dynamics methods (for review see, e.g., [5,7,15]). The most important of the two appears to be the molecular dynamics technique. This technique allows one to calculate not only the radial distribution function but also the kinetic properties (e.g., self-diffusion coefficient, rates of rotational relaxation) of molecules in liquid water. In this type of calculations, an assembly of molecules is considered with each molecule experiencing forces and torques due to other molecules out to some cut-off radius depending on the limitations of the computer used. The model of a water molecule is usually represented by a rigid tetrahedron of the appropriate size, with the oxygen at the center, and two positive and two negative charges at the corners, and an interaction potential between the molecules is assumed. Pair-wise molecular interactions are commonly treated in terms of a '6-12' Lennard-Jones potential energy function with an attractive (negative) term varying as the reciprocal 6th power of the intermolecular distance and a repulsive (positive) term varying as the reciprocal 12th power of the distance. The limitations of the technique clearly follow from the rigidity of the water molecule model (though it is possible to introduce some additive parameters to account for molecular vibrations into the model) and from the additivity principle used to describe the total interaction for the system as a sum of potentials between individual pairs of molecules [7]. In attempt to obviate the latter limitation different effective pair potential functions have been used to account for the cooperative interactions involved in a condensed assembly of molecules in liquid water (for review see [7,15].

In spite of the above limitations, the computer simulation studies of water give an excellent fit to the experimentally observed radial distribution function. The calculated heat capacity and internal energy are in good agreement with the experimental values for liquid water. The liquid water density maximum and isothermal compressibility minimum phenomena are also accounted for by the results of the simulation studies.

Analysis of the distribution of molecules engaged in various numbers of H-bonds [14] used the interaction energy necessary for H-bond formation defined as about -4 kcal/mole. The result was [14] a relatively broad distribution of the fractions of molecules connected by 0, 1, 2, 3, or 4 H-bonds with the mean of the distribution shifting to lower number of H-bonds with increasing temperature. Geiger et al.[16] used the computer simulation approach in terms of the so-called percolation theory and showed that a large single space-filling random network of H-bonds exists in liquid water around room temperature and below. This network has a local preference for tetrahedral geometry, but it contains a large proportion of strained, distorted and broken bonds. The liquid water as a whole is communally connected, the strained bonds presumably playing a fundamental role in the ability of the liquid to alter the network topology under influence of solutes and other factors [4,5,7,12]. This concept of the liquid water structure is in agreement with the general features of the infrared and Raman spectra [13,17,18], ^{1}H-NMR spectra [19] of liquid water, etc.

Essentially all the experimental techniques used to study liquid water and aqueous solutions measure average properties over all molecules of a given type in the system. Hence, the interpretation of the results usually depends on the particular conceptual model of the water structure used by the investigator. It is from model-dependent interpretation that the controversy between the investigators comes, not from the actual experimental observations upon which most scientists do agree. A lot of different models are used in the literature to interpret the experimental observations on aqueous systems. Some of the most popular models should be briefly outlined here.

There are different classifications of numerous models of the water structure proposed by different authors [1,2,6-8,12,20]. Until recently the models have been usually classified into three major categories (see, e.g., ref.7): mixture, interstitial, and continuum models, the second category being dispelled by the recent experimental findings [6]. According to the more recent classification given by Angell and Rodgers [20] in relation to the vibrational spectra of liquid water, the main types of models include: a) simple, unrepentant, two-state models that imply two distinct species, ideally mixing; b) generalized two-state models, according to which there exist two classes of molecules, strongly H-bonded and weakly H-bonded, respectively; c) quasi-lattice with broken bonds, i.e. an effective two-state model with bond states replacing molecular states; d) continuum models with preferential exchange of strong bonds for weak bonds with increasing temperature (leaving intermediate strength bonds constant in population); and e) continuum models with continuous bond weakening on increases of temperature.

The a- and b-type models represent the aforementioned mixture and interstitial models according to the other classifications. An example of the interstitial model is the one proposed by Samoilov [21], according to which an ice-like H-bonded structure coexists with nonbonded "interstitial" single molecules located in the lattice cavities. In mixture models the water molecules are divided into two, or slightly more number of classes, e.g., H-bonded and not H-bonded, or those included in associated clusters and not included. One of the most well-known models of this type is the "flickering cluster" model advanced by Frank and Wen [22] and elaborated in a detailed statistical mechanical treatment by Nemethy and Scheraga [23]. For review of the models the reader is referred to [2,7]. In view of the results of the computer simulation studies of the water structure it has been numerously emphasized in the literature (see, e.g., [7,12]) that simple two-state or two-component mixture models are basically inadequate. More elaborate mixture models, e.g., the one advanced by Scheraga and his colleagues [24,25], appear to be more adequate. These models are essentially incompatible with a simple two-state concept of water structure and have much in common with a continuum point of view.

The continuum models are based on the assumption that the energy levels of water molecules (or those of intermolecular H-bonds) in liquid water can be described by an essentially continuous distribution function (see [2,4,6, 7,12]). The models of this type usually suffered from structural underdefinition

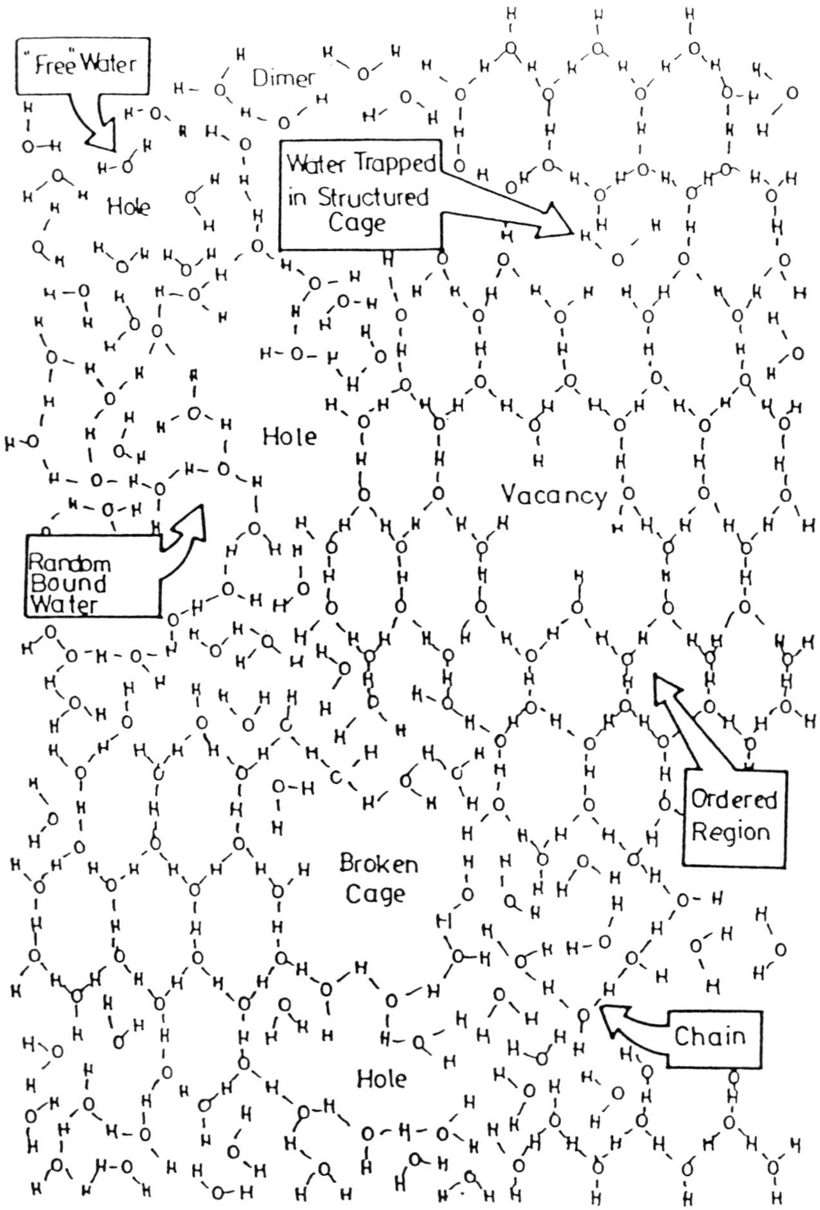

Figure 1.1. Schematic illustration of structure of liquid water. (From R. Kaliszan, Quantitative Structure - Chromatographic Retention Relationships, Wiley-Interscience, New York, 1987, p.20. Reprinted by permission of Wiley-Interscience.)

in the past. They are gaining in importance recently due to the results of the computer simulation studies which seem to give better insight into the actual behavior and properties of liquid water than have done treatments based on arbitrary models of water structure.

The most adequate tentative concept of liquid water structure seems to be the one according to which the main pattern of water structure is an uniform continuous four-coordinated irregular network of H-bonds with considerable fluctuations of such features as the length of the oxygen-oxygen distance, bond angle, and the bonds' energy parameters [26,27]. For fluid properties, discrete units of H-bonded polygons, $(H_2O)_n$, which can move independently of each other must exist. The smallest structural unit among a numerous models suggested in the literature seems to be a polycyclic, cubic-shaped octamer capable of dissociation into two cyclic tetramers [28]. Other possible discrete entities constantly formed and braked in liquid water may be cyclic pentamers $(H_2O)_5$ along with their sandwich dimers, the decameric $(H_2O)_{10}$. Presence of much larger polymeric aggregates seems also to be likely [29,30]. It was reported recently that the most stable gas-phase cluster of water $H_3O^+(H_2O)_{20}$ consists of 20 water molecules arranged as a pentagonal dodecahedron with an H_3O^+ trapped inside the cage-like structure [31].

The above model illustrated schematically in Figure 1.1 is in accordance with the most recent experimental evidence and it is used below as the basis for consideration of the specific solvent features of water, the mechanism of phase separation in aqueous polymer systems, and the properties of the phases in these systems.

1.2. SOLVENT PROPERTIES OF MIXTURES OF WATER WITH ORGANIC SOLVENTS

The solvent properties term covers the ability of the solvent molecules to participate in all possible, specific and unspecific, intermolecular interactions between the solvent and solute molecules. The possibility to use the physical characteristics of a solvent, such as its dielectric constant, dipole moment, refractive index, surface tension, etc. as a measure of its solvent properties is a much debated question [33,34]. It was argued [33], particularly, that the majority of the physical properties of a solvent represents the intermolecular interactions occurring between the molecules in the pure solvent, but not the ability of the solvent to interact with a solute.

Using the following short summary of the intermolecular interactions in liquids given by Buckingham [35], the intermolecular force of the interaction of two spherical molecules is equal to $-dU(R)/dR$ where $U(R)$ is the interaction energy of the pair of molecules at the distance R. There is also a torque component depending on the angle between the molecules. In solution, in the presence of other molecules, it is convenient to consider a potential of average force A(R) which is a Helmholtz free energy and is the mean interaction energy of the two molecules at the distance R, averaged over all configurations of all other molecules in the system. $A(R) = U(R) - T \cdot S(R)$ is the sum of an energy U(R) and an entropic contribution $-T \cdot S(R)$ both being dependent on the temperature T. The entropic contribution results from the change in order in the environment resulting from the interaction of the pair. Hydrophobic effect that plays an important role in the properties of aqueous solvents depends on S(R), for the decrease in entropy on forming a cage of water molecules around a nonpolar molecule is clearly less when the two nonpolar molecules are close together than when they are well separated.

One of the most important properties of a solvent clearly is its ability to dissolve various solutes. The phenomenon of the relative insolubility of nonpolar compounds in water by comparison to their solubility in non-aqueous solvents is known as the hydrophobic effect. There is no need to go into the details of this phenomenon here as it has been discussed at length in numerous books and reviews (see, for example, in [2,3,5]). Hydrophobic effect is observed in aqueous solutions as well as in pure water but the strength of the effect depends on the solution composition.

In order to quantify the propensity of an aqueous solution to provoke a hydrophobic effect on a given solute the following approach has been proposed [36]. The difference in the Gibbs free energy, ΔG_{tr}^{o} associated with transferring of a given solute from pure water to an aqueous solution or to an organic solvent is used as a measure of the so-called solvophobic effect. This term was employed by Abraham et al. [36] to stress that the effect is measured not only in pure water but in water-organic solvent mixtures of various compositions. The solubilities of nonpolar solutes, such as argon, alkanes up to n-octane, and some cycloalkanes and other alkane-like compounds, in pure water and in water-organic solvent mixtures were determined. From these the corresponding ΔG_{tr}^{o} values were calculated using the molar concentration scale. It was established [36] that the data obtained are described as:

$$\Delta G_{tr}^{o}(\text{water} \rightarrow \text{solvent}) = M_s \cdot R_T + D_s \tag{1.1}$$

where R_T is parameter characteristic of the solute and related to the solute size (details see in [37,38]); M_s and D_s are parameters characteristic of the solvent (water-organic solvent mixture of a given composition is considered as a particular solvent).

The M_s-values were determined and used as a measure of a solvent solvophobic effect in reference to hydrophobic effect in pure water, since $M_s = 0$ for water by definition. After that a scale of solvophobic power, Sp_s was constructed by defining $Sp_s = 0$ for the most nonpolar solvent n-hexadecane [36] so that

$$Sp_s = 1 - M_s/M_{hexadecane} = 1 - M_s/4.2024 \quad (1.2)$$

as the M_s-value of 4.2024 as determined for hexadecane from Equation 1.1.

The Sp_s values were reported [36] for mixtures of water with several organic solvents. It should be stressed that due to the reasons considered by Abraham et al.[36] the solvophobic power parameter Sp_s is not exactly a measure of the relative strength of the hydrophobic effect manifestation in a given solvent medium. The concentration effects of polar organic solvents on the Sp_s values for the corresponding water-solvent mixtures are, however, worthy of notice (see Figure 1.2).

It appears from the data [36] presented in Fig. 1.2 that each curve may be represented by the combination of three connected straight lines corresponding to the three different composition ranges of a given mixture. The ranges are those of the water-rich mixtures, the organic solvent-rich mixtures, and the

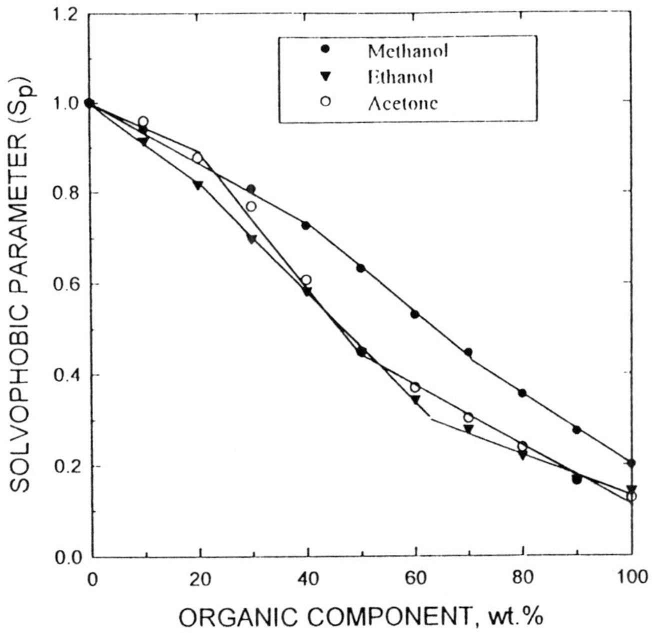

Figure 1.2. Solvophobic parameter Sp_s of water-organic solvent mixtures as a function of the concentration of the organic component (Data from M. H. Abraham, P. L. Grellier, R. A. McGill, J.Chem.Soc., Perkin Trans. II, 339 (1988), by permission of the Royal Society of Chemistry).

mixtures of intermediate compositions. It may be assumed that the composition of the water-rich mixture corresponding to the first linear fragment of each curve plotted in Fig. 1.2 should be considered as that of an aqueous solution of a given organic solvent. The latter linear fragment seems to correspond to the solutions of water in the solvent, while the intermediate one represents the so-called water-organic solvent mixtures.

Actually the typical dependence of a physicochemical property of a liquid mixture is generally described by a curve characterized as [39,40]:

$$\Delta = \Delta_D \cdot \ln(c_p/c^* + 1) + \Delta^o \tag{1.3}$$

where Δ is the quantitative measure of the property examined; c_p is the molar concentration of the more polar component; Δ^o is the value of the property Δ in the pure component with lower polarity; the pure component with lower polarity; Δ_D and c^* are the empirical parameters of the equation. It can be seen from Equation 1.3, however, that the Δ - c_p dependence may be approximated by the linear function at least at low c_p values.

It should be noted that different physicochemical properties of water mixtures with various polar additives, e.g., overall solvent polarity, acidity, viscosity, permittivity [41], seem to follow the similar pattern. It follows that among mixtures of water with a polar solvent only the water-rich mixtures of the composition range described by the linear physical property - concentration relationship should be placed into the group of aqueous solutions of the solvent.

From now on the term of aqueous solution for the simplicity sake will be used for the water/additive mixtures the physical or physicochemical properties of which fit linear relationship between the property and the additive concentration including the property value characteristic of pure water. The mixtures fallen into this category will be viewed as those composed essentially of the water H-bonds network more or less altered by an additive present in the mixture. The composition range corresponding to this category may vary depending on the type of additive and on the physicochemical property under examination (see, e.g., in [42]). Different properties of the mixtures may be differently sensitive to an alteration induced by a given additive on the water structure and/or thermodynamic state. Some of the properties may be even totally insensitive to the additive presence up to a certain critical concentration. In this case the linear relationship may start from this critical concentration value. The aqueous solutions below and at the critical concentration of the additive may be viewed as similar to pure water in regard to the property analyzed.

It follows from the above definition that essentially all the physicochemical properties of aqueous solutions of a given additive are to be interrelated. Note that all the properties of pure water are governed by the two also

Table 1.1
Free Energy of Transfer of a CH_2 Group from Aqueous Medium into Organic Solvents [a]

Organic solvent	$-\Delta G(CH_2)_{tr}$, cal/mole CH_2		$\Delta[\Delta G(CH_2)_{tr}]$ cal/mole CH_2
	Ref.[43] [b]	Ref.[38] [c]	
Hexadecane	761 ± 65	920 ± 10	159 ± 75
Cyclohexane	1127 ± 43	940 ± 10	-187 ± 53
Hexane	1010 ± 31	920 ± 10	- 90 ± 41
Benzene	842 ± 66	920 ± 10	78 ± 76
1-Octanol	727 ± 17	860 ± 10	133 ± 27
1-Butanol	542 ± 58	840 ± 10	298 ± 68

[a] $\Delta[\Delta G(CH_2)_{tr}] = [\Delta G(CH_2)_{tr}]_1 - [\Delta G(CH_2)_{tr}]_2$ where indexes 1 and 2 denote the $\Delta G(CH_2)_{tr}$ values determined from the solubility measurements [38] and from the partition measurements [43], respectively;
[b] Values determined from partition experiments [43];
[c] Values determined from the solubility measurements [38].

interrelated features, namely the structure of the H-bonds network and the distribution of the H-bonds energies. Presence of an additive affects any property of water in the solution by altering these features. Hence the aforementioned interrelationship seems to be reasonable. (The properties absent in pure water and imposed on the medium by an additive are beyond the scope of the present consideration.)

Different effects of chemically similar additives on the solvent properties of water are illustrated by the values of the free energy of transfer of a methylene group, methylene group, $\Delta G(CH_2)_{tr}$, from an organic solvent to water given in Table 1.1. These values were determined from the solubilities of a series of n-alkanes in pure water and in several pure organic solvents [38] and from the partition coefficients of homologous series of solutes in water-organic solvent two-phase systems [43]. (Solubility measurements and partition technique are discussed in detail below.)

The data given in Table 1.1 suggest that the difference between the $\Delta G(CH_2)_{tr}$ values determined by the two techniques increases with increasing solubility of organic solvent in water. The solubility of water in organic solvents increases in the same series but this may be neglected for simplicity sake as the solvent properties of pure octanol or hydrocarbon and water-saturated octanol or hydrocarbon were shown to be essentially identical [38]. Hence it may be assumed that the difference observed between the $\Delta G(CH_2)_{tr}$ values under comparison represents the difference between the affinity for a CH_2 group of pure water and that of the aqueous medium altered by the presence of a given organic additive. The difference induced by octanol indicates particularly that the affinity for a CH_2 group of the octanol-saturated aqueous phase exceeds that of pure water by 133 cal/mole CH_2, i.e. the phase is more hydrophobic than pure water. The differences observed for hexane and cyclohexane imply that the effect of an additive on the aqueous medium in regard to its affinity for a CH_2 group should not be attributed just to nonpolar character of the additive. It should be particularly noted that any direct interaction between a solute, such as inert gases, alkanes and alkane-like compounds, and an additive present in the aqueous phase is highly unlikely in the case under consideration. An alteration of the solvent properties of the aqueous solution of an additive in reference to those of pure water appears to be due to the effect of the additive on the structure and/or thermodynamic state of water in the solution.

At least two types of parameters of a solute are known to be responsible for its solubility in a solvent: structural parameters, accounting for the geometry of the solute molecule, and interaction parameters which account for the various interactions energies between the solute and solvent molecules.

The solution process is commonly imagined as taking place in three hypothetical steps: 1) removal of the solute molecule from its original environment; 2) formation of a cavity in the solvent to accommodate the solute molecule; and 3) interactions of the solute molecule placed into the cavity with its new solvent environment. For simplicity, it may be assumed that a solute molecule is removed from a hypothetical gas phase, i.e. the first step may be neglected.

The free energy change associated with transferring a solute from a hypothetical gas phase into solution may be described as suggested by Sinanoglu [44]:

$$\Delta G_{tr} = \Delta G_{cav} + \Delta G_{int} + RT \cdot \ln(RT/PV_s) \qquad (1.4)$$

where ΔG_{cav} is the free energy change for formation of a cavity in the solvent to accommodate the solute molecule; ΔG_{int} is the free energy change due to the solute-solvent interactions; the last term measures the entropy effect, in a rough approximation taking care of the "free volume" of the solute in the solvent of the average molecular volume V_s; R, T and P have the standard meaning.

According to Sinanoglu [44], the free energy change for cavity formation comes from the surface energy of the cavity related to the energy required to separate the solvent molecules from one another and to the cavity size. It can be expressed in simplified form as [44]:

$$\Delta G_{cav} = c \cdot (V_1)^{2/3} \cdot \gamma_s \tag{1.5}$$

where V_1 is the solute molar volume; γ_s is the solvent surface tension; and c is the constant accounting for correction of the macroscopic surface tension of the solvent to molecular dimensions.

The second term in Equation 1.4, which represents the interaction of the solute molecule with the solvent molecules, is assumed [44] to be the sum of a van der Waals term, ΔG(VdW), including all the nonbonded, nonelectrostatic solute-solvent interactions, and an electrostatic free energy term, ΔG(e.s.), covering electrostatic solute-solvent interactions.

The van der Waals term is given as [44]:

$$\Delta G(\text{VdW}) = \mathbf{a} \cdot \Delta_1 \cdot D_1 \cdot D_s \tag{1.6}$$

where $\Delta_1 = 1.35 \cdot I_1 \cdot I_s/(I_1 + I_s)$ with the ionization potentials I_s and I_1 of solvent molecule and solute molecule, respectively; D_1 and D_s are the corresponding Clausius-Mosotti functions for pure solute and pure solvent given by $D_s = (n_s^2 - 1)/(n_s^2 + 2)$ and $D_1 = (n_1^2 - 1)/(n_1^2 + 2)$ where n_s and n_1 are the refractive indexes of pure solvent and pure solute, respectively; **a** is the constant specific for the solute-solvent pair examined.

The electrostatic free energy term is expressed as [44]:

$$\Delta G(\text{e.s.}) = b \cdot \mu_1^2 \cdot (\varepsilon_s - 1)/V_1 \cdot \alpha_1 \tag{1.7}$$

where μ_1 is the solute dipole moment; ε_s the solvent dielectric constant; α_1 the solute polarizability; b constant; V_1 as defined above.

From the model suggested by Sinanoglu [44] Equation 1.4 can be expressed in simplified form as:

$$\Delta G_{tr} = a \cdot \Delta_1 \cdot D_1 \cdot D_s - b \cdot \mu_1^2 \cdot (\varepsilon_s - 1)/V_1 \cdot \alpha_1 + c \cdot (V_1)^{2/3} \cdot \gamma_s + RT \cdot \ln(RT/PV_s) \tag{1.4a}$$

It follows from Equation 1.4a that ΔG_{tr} exhibits a linear dependence on the solvent surface tension, a slight dependence on the solvent molar volume which is often negligible [44] and a more complicated dependence on the solute (or additive) molecular volume.

According to this equation there are mainly two usually opposing solvent effects. These effects may be classified as solvation or "inverse volume" forces and surface forces. "Solvophobic forces" (with "hydrophobic" as an especially strong particular case) arise from the surface forces and act not only on nonpolar, hydrocarbon-like solutes but also on polar solutes [44]. It is just that if the solutes are polar, they may act in a way to counteract some of the surface effect. For small solutes solvation or 'inverse volume' forces may dominate [44] but with large solutes surface forces are supposed to take over.

The above model describes the process of solution of a substance in a given solvent. It follows from the above that the energy of the solvent in a given solution is changed in reference to the pure solvent due to formation of cavities and the solvent engagement in the interactions with the additive. In the case of water as the solvent this means that the intermolecular arrangement of water molecules and the distribution of the H-bonds energies in the medium are changed in the presence of an additive in reference to those in pure water. Various additives are known to alter different properties of water, such as surface tension [36,45], dielectric constant [46], etc. All these observations are consistent with the above assumption. It can be concluded that the solvent properties of an aqueous medium in the presence of a given additive are changed in reference to those of pure water due to the effect of the additive on the structure and/or thermodynamic state of water.

The solvent properties of the aqueous solution of a given additive are displayed with respect to a solute which means that we deal not just with a binary "water + additive" system but with the ternary "water + additive + solute" system. The latter type of aqueous systems is much more complicated than the binary systems. The most important question in studies of these systems is whether there is a direct solute-additive interaction. The possibility to provide the answer usually depends on the particular experimental technique employed.

Numerous physical properties of any component of an aqueous ternary system may be used to monitor the interactions in such a system. If optical, NMR spectroscopic, etc., properties of water are examined the conclusion is usually drawn in relation to the water structure in the system (see, e.g., in [47] and the references cited therein). An illustrative example is offered by the data [48,49] that organic additives, such as dimethyl sulphoxide and 1,4-dioxane, or acetonitrile and 1,4-dioxane, affect each other's capability of modifying the structure and/or the energy distribution of the water-water H-bonds when water is present in a great excess.

If the physical properties of a solute in an aqueous ternary system are monitored to examine the interactions in the system, the interpretation of the experimental observations depends on the answer to the question of preferential solvation of the solute either by water, or by additive, or by both (see, e.g., in [50] and references cited therein). To illustrate the common difficulties encountered in the studies of ternary aqueous systems those typical for the solvatochromic studies of water-organic solvent mixtures may be considered.

The solvatochromic method used to study the solvent properties of liquids is based on the solvent effects on electronic spectra of certain dye probes

[33,34]. The phenomenon of changes of position and shape of absorption band(s) in the ultraviolet(UV)-visible spectra of some dyes resulting from the solvent change is called solvatochromism. The intramolecularly ionic dyes in which an electron-donating group is linked by a conjugated system to an electron-accepting group are those the most usually employed as solvatochromic probes. The electronic transition of these dyes is usually associated with a charge transfer between these two groups, producing an excited state with a dipole moment different from that in the electronic ground state (for details see [33,51]).

To illustrate the effects under discussion it may be mentioned that the solvatochromic absorption band of a pyridinium-N-phenoxide betaine dye is situated at 795 nm in 1,4-dioxane, at 622 nm in acetonitrile, and at 515 nm in methanol [33].

The molar transition energy of a dye is expressed as:

$$E_T/\text{kcal/mole} = h \cdot c \cdot N_A \cdot \nu_{max} = 2.859 \cdot 10^{-3} \cdot \nu_{max}/\text{cm}^{-1} \qquad (1.8)$$

where ν_{max} is the wavelength of the solvatochromic absorption band of a given dye; h is the Planck constant; c the speed of light in vacuum; and N_A the Avogadro's number.

Different dyes are used to characterize the solvent properties of organic liquids (for review see [33,34]). It has been shown that the solvent properties of liquids may be quantified not only by the characteristics of the dyes electron spectra but also by those of IR, ESR, NMR, and fluorescence spectra and NMR coupling constants of various probes, etc., (see [34] and references cited therein).

The effects of different solvents on a given dye probe or a set of probes may be described in terms of the multiparameter model as (Equation 1 in ref.[34]):

$$XYZ = XYZ^o + \text{cavity term} + \text{dipolar term} + \text{H-bonding term} \qquad (1.9a)$$

where a cavity term represents the free energy or enthalpy input required to create a suitably sized cavity in the solvent to accommodate the solute molecule; a dipolar term represents the exoergic effects of solute-solvent dipole-dipole, dipole-induced dipole, and mutually induced dipole interactions (it should be noted that recently Yalkowsky et al.[52] indicated that this term actually represents just the dipole moment of the solvent); and an H-bonding term represents the exoergic effects of H-bonding (or Lewis acid-base) complexation between the solute and the solvent.

Using the subscript 's' to denote the solvent and the subscript '1' to denote the solute Equation 1.9a becomes (Equation 2 in [34]):

$$XYZ = XYZ^o + A \cdot (\delta_H^2)_s \cdot V_1/100 + B \cdot \pi^*_s \cdot \pi^*_1 + C \cdot \alpha_s \cdot (\beta_m)_1 + D \cdot \beta_s \cdot (\alpha_m)_1 \qquad (1.9b)$$

where δ_H is the Hildebrand solubility parameter, the δ_H^2 represents the solvent contribution to the cavity term; V_1 is the solute molar volume; π^* a measure of solute or solvent dipolarity/polarizability; α is a measure of solute or solvent H-bond donor acidity; β is a measure of solute or solvent H-bond acceptor basicity; A, B, C, and D are constants.

When effects of different solvents on a single probe are studied, Equation (1.9b) is expressed in terms of the so-called solvatochromic parameters of the solvents (Equation 3 in [34]):

$$XYZ = XYZ^o + h \cdot (\delta_H^2)_s + S \cdot \pi^*_s + a \cdot \alpha_s + b \cdot \beta_s \qquad (1.9c)$$

where h, S, a, and b are the coefficients specific for a given probe; all the other terms as defined above.

Experimental procedure to determine the π^*_s values of solvents which are neither H-bond acceptors nor donors is no problem as $\alpha_s = \beta_s = 0$ in Equation (1.9b). The π^*_s values of solvents which are H-bond acceptors but not donors ($\beta_s \neq 0$, $\alpha_s = 0$) are estimated using the non-H-bond donor indicator solutes, and those of amphiprotic solvents, such as water, alcohols, etc., (α_s and β_s both $\neq 0$) are ascertained using the selected solvatochromic probes which are incapable of H-bonding [51]. It should be emphasized here that the aforementioned conclusion by Yalkowsky et al.[52] questions the validity of this procedure. The appropriately selected sets of probes are employed to estimate the α_s or the β_s solvent parameters (for details see [34,51] and the references cited therein).

The solvatochromic method was numerously used to study the solvent properties of mixtures of water with various organic solvents [53-57]. The molar transition energy for the solvatochromic band of the aforementioned pyridinium-N-phenoxide betaine dye, $E_T(30)$, (the number in parentheses refers to the indicator designation by Dimroth et al., see in [33,55]) was particularly used as a measure of the overall polarity of these mixtures. The "polarity" term used by the authors [33,53,55] differs from the above dipolarity/polarizability term advanced by Kamlet et al.[51]. The polarity term used in regard to the $E_T(30)$ parameter includes the ability of a solvent to participate in the H-bond interactions.

Cheong et al.[56] reported that it was shown by the Kamlet's group that the $E_T(30)$ polarity of a solvent may be described by the following combination of the Kamlet-Taft dipolarity parameter, π^*_s, and the solvent H-bond donor acidity parameter, α_s:

$$E_T(30) = 31.00 + 13.43 \cdot \pi^*_s + 15.06 \cdot \alpha_s \qquad (1.10a)$$

$$N = 40;\ r^2 = 0.984;\ s = 1.65$$

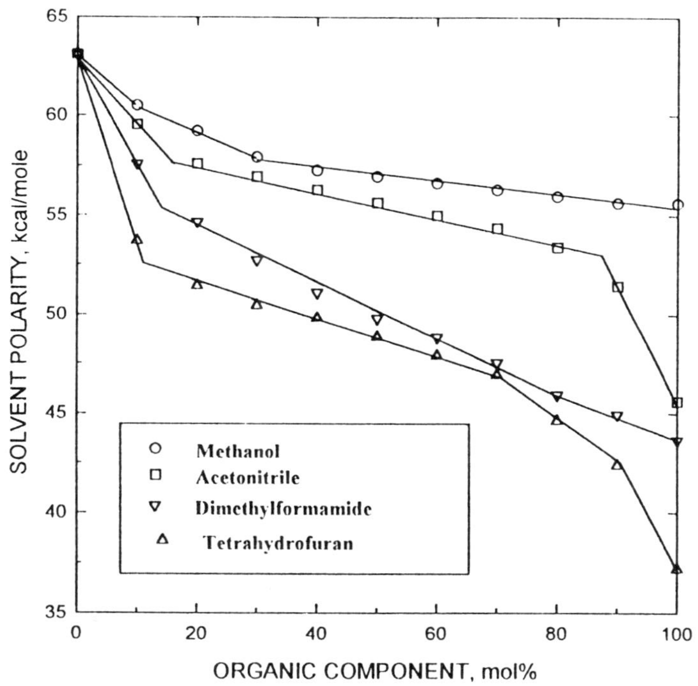

Figure 1.3. Solvent polarity of water-organic solvent mixtures as a function of the concentration of the organic component. Calculated from data in [54].

where $E_T(30)$ is expressed in kcal/mole; N is the number of solvents examined; r^2 the correlation coefficient; s the standard deviation. This equation is claimed [56] to hold for 40 pure solvents including aliphatic alcohols and other H-bond donors.

On the other hand, according to Krygowski et al.[54] the same solvent polarity parameter in the case of water-organic solvent mixtures is related to the H-bond acceptor basicity parameter β_s of the mixture as:

$$E_T(30) = b + a \cdot \beta_s \quad (1.10b)$$

where a and b are empirical coefficients (b-value is close to the $E_T(30)$-value for pure water; a-value is negative and depends on the particular organic solvent studied in mixture with water).

Equation 1.10b holds for entire mixture composition range only for mixtures of water with solvents of high permittivity ($\varepsilon_s > 30$), such as alcohols and dimethyl sulfoxide [54]. For the mixtures of water with acetonitrile, dimethyl formamide, acetone, and dioxane this equation is applicable only for a limited solvent concentration range not exceeding 20-30 mol% of the solvent. Thus, Equation 1.10a seems to be preferable to Equation 1.10b.

Equation 1.10a has been employed [56] to study mixtures of water with acetonitrile, tetrahydrofuran, methanol, and iso-propanol. The overall solvent dipolarity π^*_s-values were estimated for the mixtures of various composition using the solvatochromic probes incapable of H-bonding. The H-bond donor acidity α_s values for the mixtures were calculated [56] according to Equation 1.10a from the π^*_s values and those of $E_T(30)$ measured by Krygowski et al.[54]. Both solvatochromic parameters π^*_s and a_s values appear to be essentially linearly related to the concentration of an organic solvent in aqueous solution up to 20-30 mol% of the additive.

The variation in overall solvent polarity $E_T(30)$ as a function of the organic additive concentration in a mixture with water is generally nonlinear as shown in Figure 1.3.

Results of the studies of the overall polarity of water-organic solvent mixtures (see in [53-56] and the references cited therein) usually agree with the well-known structural model for aqueous solutions of low molecular weight nonionic additives [58,59].

According to this model [58], addition of initial amount of an additive results in replacement of non-bonded single water molecules located in the cavities of the H-bonds network of water by the additive molecules. The structural equilibrium in water is shifted increasing the fraction of H-bonded network molecules. At the composition of about 0.2 mole fraction of an additive the water network becomes "saturated" with the additive. Further addition of the additive results in the formation of two "microphases" [58]: a highly structured microphase consisting predominantly of water and a relatively disordered microphase containing mostly the additive molecules.

According to the other model [59], the additive molecules do not replace the single water molecules but are located in the zones of weak H-bonds. The additive molecules change the H-bonds angles inducing the overall strengthening of the existent H-bonds network. An increase of the additive concentration results in the distortion or rapture of the H-bonds. Whether water-water H-bonds are distorted rather than broken in the presence of an additive is open to discussion.

The most current view seems to favor the idea of a distorted network, but deciding at what point a distorted bond is turned into a broken one may be largely a matter of opinion. It seems also impossible at present to suggest at what degree of distortion of the water H-bonds network a water-organic solvent mixture should be viewed as the solution of water in the solvent and not as that of the solvent in water.

It seems reasonable to assume, however, that the critical value in the mixture composition corresponds to the point of the departure of the initial fragment of the measured quantity-concentration curve from a straight line. It was mentioned above that this critical point depends both on the organic additive and the physicochemical property examined. In the solvatochromic polarity measurements this critical value appears to vary in the range of 0.15-0.3 mole fraction of the additive (in the case of polar additives). Until this value is reached the binary water-organic solvent mixture may be viewed as an essentially aqueous solution. The polarity of this water-rich composition range is, however, the least studied due to the low solubility of the most solvatochromic dyes in pure water.

The water-soluble solvatochromic, carboxylate-substituted anionic pyridinium-N-phenoxide betaine dye has been synthesized recently by Reichardt et al.[60]. This dye was used to study the polarity of aqueous solutions of inorganic salts, polymers and phases of aqueous polymer two-phase systems (see below). Polarity of water-ethylene glycol, water-ethanol, and water-urea mixtures were measured with this dye (designated as the betaine dye 6 by Reichardt et al. [60]). The results partially published in [61] showed that the $E_T(6)$ polarity value is linearly dependent on the additive up to about 52% wt. ethylene glycol, 34% wt. ethanol, and 47% wt. urea.

The $E_T(6)$-value can be used as the polarity measure due to the existence of interrelationship between the $E_T(6)$ and $E_T(30)$ values according to [60]:

$$E_T(6) = (0.932 \pm 0.014) \cdot E_T(30) + (3.335 \pm 0.685) \qquad (1.11)$$

$$N = 22;\ r^2 = 0.998;\ s = 0.388$$

where both E_T values are expressed in kcal/mole (22 solvents for which the E_T values have been compared include 13 H-bond donor solvents and 9 solvents incapable of H-bonding). Hence the polarity scale of $E_T(30)$ values can be extended to include aqueous solutions (see below). It should be noted, however, that the above water-soluble betaine dye 6 may be used in aqueous medium in an alkaline range of pH only. At the pH values of less than 8.5 the intensity of the solvatochromic absorption band of the dye 6 is greatly reduced due to protonation at the phenolic oxygen atom of the dye anion [33,60].

The results of the solvatochromic studies of water-organic solvent mixtures imply that all the solvent features (dipolarity, H-bond donor acidity, H-bond acceptor basicity) of aqueous solutions are related to the structure of water in these solutions. This conclusion is consistent with numerous other experimental observations some of which are discussed below.

The aforementioned experimental finding by Cheong et al. [56] that the solvatochromic H-bond donor acidity α_s of the medium decreases in aqueous solutions of organic additives, such as acetonitrile, methanol, tetrahydrofuran, etc., with increasing concentration of the additive agrees with the above structural models. These findings are also consistent with the known fact that the water structuring is accompanied by a decrease of the water acidity. It is known particularly that pH value of pure water at 0°C is 7.47 and in the region of -35°C it is ca.8.4 [61] as compared to 7.0 of pure water at 25°C.

Gordon [63] used the ^{1}H-chemical shift of chloroform as a probe to study the H-bond acceptor basicity of water in the presence of the water-structure-promoting additives. It was shown [63] that the basicity of the aqueous medium increases with increasing water structure. The Hammett acidity function [64] was used as a measure of acidity of aqueous solutions of various organic additives in the presence of a strong inorganic acid (see in [65] and the references cited therein). It was shown [65] that the additives enhancing water structure decrease the H-bond donor acidity of aqueous medium in agreement with the aforementioned results of the solvatochromic observations [56].

It was noted above that the data obtained in the studies of ternary aqueous systems may be viewed as those indicative of the probe-additive interactions. The study performed by Dawber et al.[57] offers a typical example. The molar transition energy for the solvatochromic absorption band, E_T, of a betaine dye was studied as a function of composition in several binary organic solvent mixtures and in those of water with alcohols, tetrahydrofuran, and acetone. The E_T value was interpreted in terms of the solvation of the dye by the one or the other component of a given binary solvent mixture [57]. The assumption used was that if the solvation of the dye is non-specific, a linear relationship between the E_T of the mixture E_T/mix/, and mole fractions of the two components of the system would be:

$$E_T/mix/ = E_T^o/1/\cdot X_1 + E_T^o/2/\cdot X_2 \qquad (1.12)$$

where E_T^o/1/ and E_T^o/2/ are the values for the pure solvents 1 and 2; X_1 and X_2 represent the mole fractions of the solvents 1 and 2 in the mixture examined. Here are Dawber et al.[57] own comments: "Certainly changes in solvent liquid structure do occur in mixed solvents, but in this work it is assumed that the major reason for the deviation of E_T/mix/ from a linear function of mole fraction is due to preferential solvation phenomena".

Thus, the essentially identical data obtained in [53-56] and in [57] are interpreted in different terms; in those of the overall solvent structure in the former case and in terms of preferential solvation in the latter case with an arbitrary choice of terms in both cases. It should be noted that the former view is more relevant to the subject of the present book (see below). Each view, however, may be correct depending on the system and the properties of the system under examination. The only way to get out of this difficulty seems to make an allowance for the experimental data obtained on the mixture examined with an experimental technique sensitive to the water features only. As an example, the conclusion [57] that in the water-tetrahydrofuran and water-acetone mixtures the deviation of E_T/mix/ value from Equation 1.12 in the water-rich composition ranges are due merely to the preferential solvation of the dye by the organic component neglects that the water structure in the aqueous solutions of the above organic additives is significantly altered as shown, e.g., by the measurements of argon solubility in these solutions [66].

The question of the probable solute-additive interactions in an aqueous solution in regard to the structure and/or thermodynamic state of water in very complex. The additive may incorporate in the hydration shell of the solute and may alter the solute-water interactions even in the absence of a direct solute-additive interactions. In particular, Kuharski and Rossky [67,68], in a molecular dynamics study, simulated the incorporation of urea in the hydration shell of a non-polar solute. These authors have shown that interactions between water-water and water-urea molecules in contact with the solute are somewhat stronger than the same interactions in the bulk solution, although the enhancement was found to be smaller than in pure water. Numerous experimental observations may be explained by that an additive is built-in into the structure of the hydration shell of a solute (see, e.g., in [69]).

Essentially all the experimental information reported in the enormous literature on the subject implies that the presence of an additive in water alters the solvent properties of the aqueous medium in the solution relatively to those in pure water. It seems quite possible for the simplisity sake to attribute these alterations to a change in the water structure. Note once again that by the term of water structure is meant not only the space arrangement of water molecules but also the distribution of the energies of interactions between the molecules.

Most of the solvation studies seems to be based on the concept that the intermolecular arrangement and the distribution of the intermolecular interactions' energies in a solvent are completely allowed for by the aforementioned cavity term (see Eqations 1.5 and 1.9a). The interaction term is generally assumed to depend on the properties of a solute and on those of the pure solvent. This is a very rough approximation on numerous occasions leading to erroneous interpretation of experimental observations on aqueous systems.

The amount of non-H-bonded water molecules, thermodynamic activity of water, the reorientational motion of water molecules, their self-diffusion coefficient, etc., depend on the composition of an aqueous medium. It follows that the average ability of water molecules (non-H-bonded as well as H-bonded ones) to participate in the interactions with a given solute molecule should be a function of the solution composition. This assumption agrees with that the electrostatic free energy term in Equation 1.5a includes particularly the solvent dielectric constant (see Equation 1.7) which is known to depend on the solution composition. The ionization potential of water molecule in liquid water appears to be a function of the water structure as follows from the aforementioned temperature dependence of pH of pure liquid water. Hence this parameter must also depend on the composition of an aqueous medium as shown by the effects of different additives on the basicity and acidity of the medium (see above). The ionization potential of a solvent is present in the van der Waals term (see Equation 1.6). Thus, it follows that the free energy of the solute-solvent interactions in an aqueous medium should depend on the composition of the medium.

This conclusion seems to be trivial. It is important, however, in view of the generally accepted idea that when the interactions of a solute with water are examined the main variables are the nature and shape (size) of the solute molecule, and the aqueous medium engaged in the interactions is invariable. This is undoubtedly correct when the interaction is considered as taking place between the individual isolated molecules, e.g., in the hypothetical gas phase. Actually, the water molecules coming into direct contact with the solute molecule in a liquid aqueous medium are not the separate water molecules but the ones of varied initial energies depending on their intermolecular arrangement and distribution of the water-water H-bonds energies. Since the latter are functions of the medium composition, the solute-water interactions in the presence of a given additive should depend on the effect of the additive on the structure and/or thermodynamic state of water [70]. Hence the term "water" when applied to an aqueous solution covers the aqueous component of the system the variable features of which must be taken into account. This concept [70] is not particularly new. It is clearly in line with the earlier considerations by Tanford [71], Beall [12] and others which will be discussed below.

It should be noted that up till now I used the term "structure of water" combining it with the "and/or thermodynamic state of water". The reason is that in spite of the numerous papers, reviews, and books written on the subject, the very basic questions of how to define the "structure of water" and which experiment should be performed to measure the "structural changes" in water remain unanswered.

The ambiguity of the meaning of the above term as conceived by different authors results in a wide-spread disagreement among investigators using different experimental techniques. To give one example out of many, the structure of water in aqueous solutions of low molecular weight carbohydrates is concluded to be slightly disturbed merely in the neighborhood of a solute molecule by Franks [72] and Kiyosawa [73] from the measurements of the osmotic pressure in the solutions. At the same time Ueberreiter from the viscosity measurements [74] concluded that glucose at the concentration up to about 5 mol% disrupts the water structure and at the concentrations exceeding 5-6 mol% it builds a new water structure specific for the aqueous solution of this carbohydrate. According to Ueberreiter [74], ribose displays the qualitatively similar effect. Boyer et al.[75] reported an enhancement of the water structure in aqueous solutions of 0.2 M HCl with increasing concentration of a carbohydrate, such as glucose, fructose, sorbitol, etc. Miyajima et al.[76], measuring the excess partial molar entropy of water for glucose, mannose, and galactose, reported the ordering of water molecules induced by the dissolution of these monosaccharides. The final conclusion seems currently to be unattainable.

One of the attempts to quantify the structural changes in the solvent occurring in aqueous solutions is worthy of notice to illustrate the most typical difficulty. Ben-Naim [3] proposed to estimate the structural alteration in water by the change in the average number of H-bonds occurring in the solvent. This approach has led to many interesting findings described in [3]. This measure, however, is similar to the average value specific for a set of values described by a broad non-Gaussian distribution function. In other words, it is highly likely that the one and the same value of the average number of H-bonds may correspond to quite different water structures. Essentially all the other approaches to estimation of the water structure and its alterations seem to be open to the same objection.

In order to illustrate the typical conflict of opinions with respect to the influence of additives on the water structure, a part of the typical discussion [4, pp.163-168] should be cited:

"H.L. Friedman - When you put a solute particle in the water it only affects water in its neighborhood and yet the thermodynamic theory pretends that the effect is everywhere. This can't be. The range of any interaction is finite. It's only the water molecules in the neighborhood of the solute which will be affected.

M.J. Blandamer - This is the conflict between the statistical and classical thermodynamics. It is possible to discuss the factors which control the limiting partial molar volume of an alcohol in water. One measures the volume of a real solution, alcohol and water. In understanding the partial molar volumes in a real solution, one considers interactions between an alcohol molecule and other molecules, water and alcohol within the system.

M.L. Friedman - What you have here is a thermodynamic coefficient, but for a molecular interpretation one needs a local coefficient, something telling you how far out from the solute particle the solvent is affected.

M.J. Blandamer - In a sense that is what the activity coefficient does by describing the interaction of, say, a solute molecule with the total system.

B. Everett - Surely if one introduces a solute molecule into the water structure which changes the chemical potential of the structure, then to maintain a uniform chemical potential throughout the structure this change of chemical potential must be propagated throughout the system. In that sense the influence of the added solute is long ranged and limited only by the boundaries of the phase."

It seems there is a very long way to go before we establish the proper characteristics of the structure of water in an aqueous solution. From now on the term "structure of water" will be used to cover both intermolecular arrangement and distribution of the water-water H-bonds energies existent in a given aqueous medium.

1.3. SOLVENT PROPERTIES OF AQUEOUS ELECTROLYTE SOLUTIONS

The solvent properties of aqueous electrolyte solutions differ from those of aqueous solutions of non-ionic additives due to the presence of the electrostatic forces. These forces are the most long-range intermolecular interactions. The ion-water interactions, also called ionic hydration, have been discussed at length in many reviews and books (see, e.g., [6,7,77-79] and will be dealt with here very briefly. The purpose of the present consideration is to give an outline of the current concepts on the effects of ionic additives on the properties of water relevant to the problems of phase separation and partitioning of solutes in aqueous polymer two-phase systems.

The water molecules located in the vicinity of an ion tend to orient with the negative oxygen end of the molecule inward or outward according to the sign of the charge on the ion. This kind of centra-symmetrical ordering is incompatible with the tetrahedral arrangement of water molecules in pure water. Hence when the two types of ordering come into conflict some disordering of water may be expected. The structure of water in the vicinity of an ion may be more stable than that in pure water. For instance, the translational motion of the water molecules in the ion-affected zone may be decreased in reference to that in pure water (the phenomenon called the positive hydration by Samoilov [21]). At the same time the intermolecular arrangement of water molecules in this zone may be disordered compared to that in pure water. The generally accepted picture of an ion in water is the inner hydration sphere (or the first hydration layer) of the oriented, immobilized, and closely packed water molecules, and the transition zone (or outer hydration sphere) in which the preferred arrangement of water molecules corresponds neither to that of the inner hydration sphere nor to that of bulk water.

The bulk water in an aqueous electrolyte solution is the part of water not included in both inner and outer hydration spheres of the ions. The identity of the bulk water with pure water is still a much debated question [2,4,6,7, 77,78]. If the bulk water is assumed to be identical with pure water, an aqueous electrolyte solution may be described in terms of the two-state mixture model. The water molecules are divided into those affected by the ions present and those unaffected and undistinguishable from the molecules in pure water. The hydration number then may be defined as the amount of the water molecules associated with or affected by the ion. The hydration number has been used in same cases as a measure of the water-perturbing efficiency of an ion. It should be noted that the hydration number measure is applied in the literature not only to the ionic but also to non-ionic solutes. Various experimental techniques differ in their sensitivity to the effect of a given additive on the features of its aqueous solution. That causes a significant disparity between the hydration number values for a given additive obtained by different methods (for details see, e.g., [6, pp.582-606]). The main concept of the hydration number measure is that the effective amount of the solvent in solution is less than the total amount of the solvent as some water molecules are tied to the ions (or non-ionic additive molecules) and are not free to act as the constituents of the solvent medium. The phenomenon of negative hydration, in which some of the water is "released" by the ions and is more free to move is neglected by the concept.

Essentially the same model of ion hydration is considered in terms of the mean distance of separation between hydration spheres around each ion decreasing with increasing salt concentration. For example [80], for solutions containing a 1:1 salt at 10^{-3} mole/dm^3 the mean separation distance is about $94 \cdot 10^{-10}$ m. The hydration spheres of the ions in these dilute solutions are supposed not to interact. With increasing salt concentration the distance decreases and at 10^{-1} mole/dm^3 it is about $20 \cdot 10^{-10}$ m. Simultaneously the thermodynamic properties of aqueous salt solution reflect the growing role of contributions from overlap of the hydration spheres around the ions [80].

According to the concept developed by Krestov [77], the ionic hydration alters the structure of water in both hydration spheres of the ions and bulk water in the solution. The results of the studies of the solubility of inert gases in aqueous electrolyte solutions clearly support this concept. For example, the solubility of argon in aqueous solution of $[Co(NH_3)_5Cl]Cl_2$ at the salt concentration up to $5 \cdot 10^{-4}$ mol% exceeds its solubility in pure water, but it decreases with further increasing salt concentration [79].

According to Pan [81], ionic hydration should be regarded as a result of the difference between water-water interactions in pure water, and ion-water and water-water interactions in solution. If the interactions in solution are stronger than those in pure water, positive hydration will result. Otherwise negative hydration will occur. It should be stressed here that this definition seems to cover any kind of hydration independent of the ionic or non-ionic type of additive being hydrated. The particular specific feature of ionic hydration which is the impact of the electrostatic field of an ion on the water dipoles in the neighborhood is ignored.

Actually, ionic additives to aqueous polymer systems, with the exception of the ionic solutes being partitioned in the systems, are usually employed at concentrations of about 0.1 mole/kg and higher, i.e. when overlap of the hydration spheres of ions would occur. Therefore the question of the properties of bulk water in reference to those of pure water in our case seems to be just of purely academic interest. The effects of ionic additives on the structure of water in the hydration spheres is more relevant to the major subjects discussed further on.

The ions with respect to the effect on water are divided by different authors into different groups. According to Samoilov [21], the ions are divided into those reducing the translational motion of the nearest water molecules in reference to that of molecules in pure water (positively hydrated ions), and into those increasing the motion (negatively hydrated ions). According to the other type of classification, the distinction is drawn between the ions strongly interacting with neighboring water molecules, the interaction being electrostrictive in some cases and H-bonding in others, and the hydrophobic ions enhancing the water-water interactions within their hydration spheres. For details on these and other types of ions classification the reader is referred to the book by Conway [6].

Results of the studies of the effects of inorganic salts on the structure of water obtained by different methods generally agree in that the effects of cations are relatively small in comparison with those of anions, especially the large, polyatomic ones. This fact is generally believed to be due to the orientation of water molecules at anions radially through the O-H bond direction, leaving three other structure-sensitive H-bonding vectors for interaction with water. At cations, the water orientation at ions is probably with the dipole axis colinear with the center of the ion (less structure-sensitive) [6, pp.312-341].

As a general rule, two solute ions will attract each other if their structural effects or tendencies to orient water molecules are compatible, but they will repel each other if their effects are incompatible [80]. The pairwise interaction parameters estimated by Antonini et al. [82] point to a strong repulsion between hydration spheres around K^+ and F^- ions or Na^+ and Br^- ions indicating that the structures of water in these spheres are incompatible. It should be noted here that the incompatibility of the water structures in the hydration shells of certain inorganic ions and non-ionic macromolecules, such as poly(ethylene glycol), polyvinylpyrrolidone, etc., is supposed [83,84] to be the cause of a reduced salt concentration in the water zone surrounding the polymer chain (in more details see below).

About 60 years ago Bernal and Fowler introduced still widely used though oversimplified terms "structure-breaking" and "structure-making" to describe the effects of different ions on the structure of water. By these terms it is meant that the effect of a structure-breaking ion on water is qualitatively similar to that of an increase in temperature, while a structure-making ion produces an opposite effect like that of a decrease in temperature. The structure-affecting properties of the ions are displayed in such properties of their aqueous solutions as the viscosity (structure-breaking ions reduce it), the rate of exchange of water molecules between the hydration shell and bulk water (structure-breaking ions decrease its energy of activation), the longitudinal relaxation rate of the water molecules, measured by NMR (structure-breaking ions increase it), etc. [6,7, 77,78]. According to various measures of the effects of ions on the structure of water (see, e.g., in [85]), the water structure-making category of ions includes cations, such as Li^+, Na^+, NH_4^+, Ca^{2+}, Mg^{2+}, etc., and anions, such as F^-, SO_4^{2-}, CO_3^{2-}, PO_4^{3-}, CH_3COO^-, etc., and the structure-breaking ions are K^+, Rb^+, Cs^+, Cl^-, Br^-, I^-, SCN^-, NO_3^-, ClO_4^-, etc. It has been emphasized by Blandamer particularly [80] that the term "electrostrictive structure breaker" is more appropriate for ions, such as Cl^-, Br^-, I^-, etc., as their mechanism of water structure breaking is quite different from, e.g., the structure breaking action of urea. Similarly the ions, such as Li^+ and F^-, should be called "electrostrictive structure makers" because the mechanism of water structure-making by ions is different from that for non-polar compounds. It is recognized [86] that the terminology of structure making and breaking is too limited for adequate description of the molecular reality. Actually, what appears to happen is a reorganization of water molecules or H-bonds that can be perceived either as structure making or breaking depending on the particular experimental method used to study the process. For the sake of simplicity, the former terms will be employed here with due regard for their limitations as well as for the type of additive (and the mechanism of its action on the water structure) considered.

In studies of the effects of inorganic salts on the solvent properties of an aqueous medium the influence of salts on the water solubilities of non-polar and polar solutes have been extensively investigated. An increase in the solubility of a given solute in an aqueous medium containing an ionic additive in reference to its solubility in pure water is generally attributed to the salting-in action of the additive. The opposite effect is attributed to the salting-out action of the additive. The effects of added salts on the water solubility of a given solute may be summarized using the following empirical equation:

$$\ln(S_o/S) = k_S \cdot C_s \tag{1.13}$$

where S_o is the saturation solubility of a solute in the salt-free solution and S is its solubility in the presence of salt at concentration C_s; k_S is the Setschenov coefficient. A positive value for k_S means that the solute is "salted-out" by the salt additive, and a negative k_S value means the salting-in action of the additive in relation to the solute.

Thermodynamically, the solubility equilibrium between a pure solute phase and its saturated solution in a solvent is determined by the equality of chemical potentials of the solute in its own phase (μ^o) and in the solvent. It follows that if the solubility of the solute is sufficiently small, i.e. the solute-solute and ion-solute interactions are assumed to be absent,

$$S_o/S = f \tag{1.14}$$

where f is the activity coefficient of the solute in the salt-containing solution (the activity coefficient of the solute in the salt-free solution f_o can be taken as 1 if S_o is small).

The empirical Setschenov Equation 1.13 takes the form:

$$\ln f = k_S \cdot C_s = \ln(S_o/S) \tag{1.13a}$$

The RT·lnf term can be identified with a non-ideal free energy contribution for the solute in the solution due to the presence of the salt since

$$\mu = \mu^o + RT \cdot \ln(f \cdot S) = \mu^o + RT \cdot \ln S + RT \cdot \ln f \tag{1.15}$$

where μ is the chemical potential of the solute in the salt-containing solution; μ^o, S, and f as defined above; (μ^o + RTlnS) term is the "ideal solution" value of the chemical potential of the solute (see below).

Qualitatively salting-in and salting-out effects may be explained as follows [6]. A given volume of the solution experiences a polarization energy in the ionic field. The polarization per unit volume experienced by water dipoles is usually much greater than that experienced by non-ionic solute molecules, so, near an ion, the concentration of the latter is relatively reduced. Integrated over the whole solution for all ions, this corresponds to a decrease in solubility or a salting-out effect. Salting-in effects, which are rarer, will tend to arise if an ion can more strongly polarize the solute than the solvent. These effects are equivalent to changes of the activity coefficient of the solute as indicated by Equation 1.13a.

Numerous attempts have been made to account for the solubilities of non-ionic solutes in salt solutions using different theoretical models (details and a fairly extensive set of key references may be found in the book by Conway [6, pp.444-465]). The effectiveness of salting-out and salting-in action varies considerably with salt but the order is usually the same for different solutes. The problem of understanding solubility patterns is closely linked with that of understanding ion-solvent interactions. For example, large organic ions often salt-in non-polar solutes, a phenomenon termed hydrotropism. The situation becomes more complex when the solute is amphiphilic, i.e. it contains both hydrophobic and hydrophilic centers.

According to Melander and Horvath [87], the salting-out and salting-in effects of salts on amphiphilic solutes, such as proteins, may be described in terms of the aforementioned model by Sinanoglu [44]. The free energy change for transfer of a solute from a hypothetical gas phase into solution is described by the above Equation 1.4. It is assumed [87] that when only the salt concentration in the aqueous solvent is changed (provided there are no specific solute-ion interactions), the energetics of the transfer process is affected only by changes in the cavity term, ΔG_{cav}, and in the electrostatic free energy term, $\Delta G(e.s.)$ (see Equations 1.5 and 1.7). The static dielectric constant of solution, ε_s, entered into the expression for the $\Delta G(e.s.)$ term (Equation 1.7) is known to depend on the salt concentration (up to about 1.0 M) as ([88], Equation 6.2):

$$\varepsilon_s = \varepsilon^o{}_w + \delta \cdot C_s \qquad (1.16)$$

where $\varepsilon^o{}_w$ represents the static dielectric constant of pure water; δ is a negative quantity known as the dielectric decrement; C_s as defined above.

Combining Equations 1.7 and 1.16 we obtain

$$\Delta G(e.s.) = -b \cdot \mu_1{}^2 \cdot (\varepsilon^o{}_w + \delta \cdot C_s - 1)/V_1 \cdot \alpha_1 = -B \cdot \mu_1{}^2 \cdot (Q + C_s \cdot \delta) \qquad (1.17)$$

where $B = b/V_1 \cdot \alpha_1$ and $Q = (\varepsilon^o{}_w - 1)$, both being constants for a given solute.

The cavity term Equation 1.5 was modified [87] on the base of the known fact [89] that the surface tension of many inorganic salt solutions γ_s can be approximately described as:

$$\gamma_s = \gamma^o{}_w + \sigma \cdot C_s \qquad (1.18)$$

where $\gamma^o{}_w$ is the surface tension of pure water; C_s is the salt concentration (expressed in terms of molality, since molality is independent of temperature); σ is a constant called [87] the molal surface tension increment of a salt.

Under the above assumptions [87] Equation 1.4 can be expressed in simplified form as a sum of the two groups of terms:

$$\Delta G_{tr}{}^o = [a \cdot \Delta_1 \cdot D_1 \cdot D_s - B \cdot \mu_1{}^2 \cdot Q + c \cdot (V_1)^{2/3} \cdot \gamma^o{}_w +$$

$$+ RT \cdot \ln(RT/PV_s)] + (c \cdot V_1{}^{2/3} \cdot \sigma - Q \cdot \mu_1{}^2 \cdot \delta) \cdot C_s \qquad (1.19)$$

where all the terms are as defined above (see Equations 1.4 - 1.7 and 1.16 - 1.18).

Equation 1.19 obviously describes the solubility of a solute in the salt-containing solution, the solubility in the salt-free solution being represented by the first group of terms.

Hence Equation 1.13a can be expressed as:

$$\ln f = k_S \cdot C_s = (1/RT) \cdot (c \cdot V_1{}^{2/3} \cdot \sigma - Q \cdot \mu_1{}^2 \cdot \delta) \cdot C_s \qquad (1.20)$$

thus,

Table 1.2

Molal Surface Tension Increments of Some Inorganic Salts in Water at 25°C

Salt	$\sigma \cdot 10^3$ dyn·g/cm·mole	Salt	$\sigma \cdot 10^3$ dyn·g/cm·mole
KSCN	0.45 [a]	NaSCN	0.57-0.41 [b]
LiI	0.79	KI	0.84-1.08 [b]
NH_4NO_3	0.85	NaI	1.02-1.20 [b]
$NaNO_3$	1.06	NH_4Br	1.14
$LiNO_3$	1.16	LiBr	1.26
KBr	1.31	NaBr	1.28-1.32 [b]
CsI	1.39	NH_4Cl	1.04-1.39 [b]
$KClO_4$	1.40	LiCl	1.63-1.67 [b]
CsCl	1.53	NaCl	1.64-1.86 [b]
KCl	1.46-1.88 [b]	LiF	2.0
Na_2HPO_4	2.02	$(NH_4)_2SO_4$	2.16-2.30 [b]
K_2SO_4	2.58	Na_3PO_4	2.66
Na_2SO_4	2.42-2.73 [b]	Li_2SO_4	2.78
NaH_2PO_4			

[a] In the concentration range up to 0.5 mole/kg only;

[b] Values cover the range of different values reported in [87,89,90].

$$RT \cdot k_S = c \cdot V_1^{2/3} \cdot \sigma - Q \cdot \mu_1^{2} \cdot \delta \tag{1.21}$$

As the dielectric decrement term δ is of negative value for all the inorganic salts studied so far [88], Equation 1.21 implies that the Setschenov coefficient value k_S must be of positive value, i.e., any salt must manifest the salting-out action in relation to any amphiphilic solute. Since this conclusion is clearly wrong, it means that either the model or the parameters employed are inappropriate. The latter is argued by Bull and Breese [90] to be the case. The molal surface tension increments of inorganic salts are always positive (see Table 1.2), while the increments of the salting-in salts, such as KSCN, NaSCN, $KClO_4$, etc., should have been negative. That seems to be the case for the interfacial tension changes produced by salts in two-phase systems composed of hydrocarbon, such as decane, and aqueous salt solution [90]. The molal interfacial tension increments, however, have been reported only for very limited number of salts [90]. Hence the surface tension increment proposed [87] as a quantitative measure of the water structure-affecting capability of a salt remains to be used. The molal surface tension increment values for the most commonly used inorganic salts are listed in Table 1.2.

It was noted above that the mechanism of action of ionic additives on the water structure is different from that of non-ionic additives. At the same time the total result of the actions of both ionic and non-ionic additives is the change of the structure of water in a given solution in reference to that in pure water. In both cases the solvent properties of the aqueous medium are altered. The acidity of the medium is affected by inorganic salts as well as by non-ionic organic additives [65]. The overall polarity of the medium measured by the solvatochromic betaine dye 6 is also altered by both types of additives [55,61]. Partitioning of solutes in water-organic solvent two-phase systems is affected by the presence of non-ionic additives in the aqueous phase (see Table 1.1) as well as by that of salt additives [91,92].

The data reported in [91,92] imply that ionic additives affect the relative affinity of the aqueous phase for a CH_2 group similarly to the aforementioned non-ionic organic additives. The free energy of transfer of a CH_2 group from octanol phase to aqueous phase, $\Delta G_{tr}(CH_2)$, can be estimated as 598 cal/mole CH_2 in the presence of 0.11 M sodium phosphate buffer, pH 7.4 [92], and 718 cal/mole CH_2 in the presence of 0.01 M acetate buffer, pH 4.0 [91] as compared to 727 cal/mole CH_2 in the case of the salt-free aqueous phase. The increase in the affinity of the aqueous medium for a CH_2 group in the presence of the acetate and phosphate buffers salts is likely to be due to the water structure-making actions of these salts.

Comparison between the effects of different additives on the water structure indicates that while there is a clear distinction between those of inorganic salts and non-ionic organic additives, there is no such distinction between the effects of non-ionic nonpolar and hydrophilic additives. In practice there is rather a gradual change in the hydration properties and respectively in the water structure-affecting properties proceeding from hydrocarbon through ethers, ketones, etc., to amides, alcohols, etc. The introduction of an ionic or hydrophilic non-ionic group into a nonpolar solute molecule dramatically changes its effect on the water structure consistent with the disruption of non-polar hydration due to strong solute-solvent H-bonding or ion-dipole interactions. These effects seem to be manifested considerably more in the case of macromolecular solutes.

REFERENCES:

1. D. Eisenberg, W. Kauzmann, The Structure and Properties of Water, Clarendon Press, London, 1969.
2. Water. A Comprehensive Treatise. (editor F. Franks), Plenum Press, New York, Vols.1-6, 1972-1979.
3. A. Ben-Naim, Hydrophobic Interactions, Plenum Press, New York, 1980.
4. Water and Aqueous Solutions (eds. G. W. Neilson, J. E. Enderby), Adam Higler, 1986.
5. F. H. Stillinger, Science, 209, 451 (1980).
6. B. E. Conway, Ionic Hydration in Chemistry and Biophysics, Elsevier, Amsterdam, 1981.
7. J. T. Edsall, H. A. McKenzie, Adv.Biophys.: a) (part 1) 10, 137 (1978); b) (part 2) 16, 53 (1983).
8. Water and Ions in Biological Systems (eds. A. Pullman, V. Vasilescu, L. Packer), Plenum Press, New York, 1985.
9. T. R. Dyke, K. M. Mack, J. S. Muenter, J.Chem.Phys., 66, 498 (1977).
10. J. A. Pople, Proc.Roy.Soc.(London), A205, 163 (1951).
11. H. Popkie, H. Kistenmacher, E. Clementie, J.Chem.Phys., 59, 1325 (1973).
12. P. T. Beall, Cryobiology, 20, 324 (1983).
13. B. Z. Gorbunov, Yu. I. Naberukhin, Zh.Strukt.Khimia (Rus), 16, 703 (1975).
14. F. H. Stillinger, A. Rahman, J.Chem.Phys., 60, 1545 (1974).
15. R. R. Dogonadze, A. Kornyshev, J. Ulstrup - In: The Chemical Physics of Solvation. Part A. Theory of Solvation. (eds. R. R. Dogonadze, E. Kalman, A. Kornyshev, J. Ulstrup), Elsevier, Amsterdam, 1985, pp.3-35.

16. A. Geiger, F. H. Stillinger, A. Rahman, J.Chem.Phys., 70, 4185 (1979).
17. V. I. Yashkichev, Adv.Mol.Relaxation and Interaction Processes, 24, 157 (1982).
18. V. I. Yashkichev, Zh.Phys.Khimia (Rus), 60, 267 (1986).
19. I. M. Svischev, V. V. Goncharov, Yu.A.Buslaev, Doklady Acad.Nauk USSR (Rus), 298, 1430 (1988).
20. C. A. Angell, V. Rodgers, J.Chem.Phys., 80, 6245 (1984).
21. O. Ya. Samoilov, Structure of Aqueous Electrolyte Solutions and the Hydration of Ions, Consultants Bureau, New York, 1965.
22. H. S. Frank, W. Y. Wen, Discuss.Fraday Soc., 24, 137 (1957).
23. G.Nemethy, H.A.Scheraga, J.Chem.Phys., 41, 680 (1964).
24. A.T.Hagler, H.A.Scheraga, G.Nemethy, J.Chem.Phys., 76, 3229 (1973).
25. J.C.Owicki, B.R.Lentz, A.T.Hagler, H.A.Scheraga, J.Phys. Chem., 79, 2352 (1976).
26. Yu. I. Naberukhin, Zh.Struct.Khimia (Rus), 25, 60 (1984).
27. A. R. Henn and W. Kauzmann, J.Phys.Chem., 93, 3770 (1989).
28. S. W. Benson, E. D. Siebert, J.Am.Chem.Soc., 114, 4269 (1992).
29. A. W. Castleman Jr., X. Yang, J.Am.Chem.Soc.,111, 6845 (1989)
30. A. W. Castleman Jr., X. Yang, J.Phys.Chem., 94, 8500 (1990)
31. A. W. Castleman, Jr., X. Yang, J.Chem.Phys., 94, 3268 (1991).
32. R. Kaliszan, Quantitative Structure - Chromatographic Retention Relationships, Wiley-Interscience, New York, 1987, p.20.
33. C. Reichardt, In: Molecular Interactions (eds. H. Ratajczak, W. J. Orville-Thomas), Vol.3, Wiley, Chichester, 1982, pp. 241-282.
34. M. J. Kamlet, R. W. Taft, Acta Chem.Scand., B39, 611 (1985).
35. A. D. Buckingham, In: Organic Liquids: Structure, Dynamics, and Chemical Properties (eds. A. D. Buckingham, E. Lippert, S. Bratos), Wiley, New York, 1978, pp.327-336.
36. M. H. Abraham, P. L. Grellier, R. A. McGill, J.Chem.Soc., Perkin Trans. II, 339 (1988).
37. M. H. Abraham, J.Am.Chem.Soc., 101, 5477 (1979).
38. M. H. Abraham, J.Am.Chem.Soc., 104, 2085 (1982).
39. H. Langhals, Angew.Chem. Int.Ed.Engl., 21, 724 (1982).
40. H. Langhals, Tetrahedron Letts., 27, 339 (1986).
41. M. Tabellout, P. Lanceleur, J. Emery, D. Hayward, R. A. Pethrick, J.Chem.Soc. Faraday Trans., 86, 1493 (1990).
42. E. D. Katz, K. Ogan, R. P. W. Scott, J.Chromatogr., 352, 67 (1986).
43. B. Y. Zaslavsky, L. M. Miheeva, S. V. Rogozhin, J.Chromatogr., 216, 103 (1981).

44. O. Sinanoglu, In: Molecular Interactions (eds. H. Ratajczak, W. J. Orville-Thomas), Vol.3, Wiley, Chichester, 1982, pp.283-342.
45. R. R. Salem, Zh.Phys.Khimia (Rus), 62, 1582 (1988).
46. P. S. Yastremsky, V. S. Harkin, V. S. Goncharov, A. K. Lyaschenko, Zh. Phys. Khimia (Rus), 57, 91 (1983).
47. W. A. P. Luck, Pure & Appl. Chem., 59, 1215 (1987).
48. V. Zelano, P. Mirty, Z.Phys.Chem. (Leipzig), 267, 857 (1986).
49. P. Mirty, V. Zelano, J.Chem.Soc., Faraday Trans.I, 84, 29 (1988).
50. P. Chatterjee, S. Bagchi, J.Phys.Chem., 95, 3311 (1991).
51. M. J. Kamlet, J. L. Abboud, R. W. Taft, J.Am.Chem.Soc., 99, 6027 (1977).
52. S. H. Yalkowsky, R. Pinal, S. Banerjee, J.Pharm.Sci., 77, 74 (1988).
53. S. Balakrishnan, A. J. Easteal, Aust.J.Chem., 34, 943 (1981).
54. T. M. Krygowski, P. K. Wrona, U. Zielkowska, C. Reichardt, Tetrahedron, 41, 4519 (1985).
55. B. P. Johnson, B. Gabrielsen, M. Matulenko, J. G. Dorsey, C. Reichardt, Anal.Letts., 19, 939 (1986).
56. W. J. Cheong, P. W. Carr, Anal.Chem., 60, 820 (1988).
57. J. G. Dawber, J. Ward, R. A. Williams, J.Chem.Soc., Faraday Trans.I, 84, 713 (1988).
58. Yu. I. Naberukhin, V. A. Rogov, Uspehi Khimia (Rus), 40, 207 (1971).
59. B. Z. Gorbunov, Yu. I. Naberukhin, V. E. Slivkov, Zh.Strukt.Khimia (Rus), 15, 403 (1974).
60. C. Reichardt, E. Harbusch-Gornert, G. Schafer, Liebigs Ann. Chem., 839 (1988).
61. B. Y. Zaslavsky, L. M. Miheeva, E. A. Masimov, S. Djafarov, C. Reichardt, J.Chem.Soc., Faraday Trans.I, 86, 519 (1990).
62. M. J. Taylor, Cryo-Letters, 2, 231 (1981).
63. J. E. Gordon, J.Am.Chem.Soc., 94, 650 (1972).
64. L. P. Hammett, Physical Organic Chemistry, 2nd ed., McGraw-Hill Book Co., New York, 1970.
65. A. S. Chernyak, M. L. Schepotko, A. K. Lyaschenko, D. B. Poblinkov, Doklady Acad.Nauk USSR (Rus), 254, 377 (1980).
66. V. I. Vinogradova, G. A. Krestov, In: Current Problems of Solution Chemistry (ed. B. D. Beresin) (Rus), Nauka, Moscow, 1986, pp.39-42.
67. A. Kuharski, P. J. Rossky, J.Am.Chem.Soc., 106, 5786 (1986).
68. A. Kuharski, P. J. Rossky, J.Am.Chem.Soc., 106, 5794 (1986).
69. P. M. Brandts, W. J. Gelsema, C. L. De Ligny, J.Chromatogr.,437, 337 (1988).

70. B. Y. Zaslavsky, E. A. Masimov, Topics Current Chem., 146, 171 (1988).
71. C. Tanford, Science, 200, 1012 (1978).
72. F. Franks, Pure & Appl.Chem., 59, 1189 (1987).
73. K. Kiyosawa, Bull.Chem.Soc.Jpn., 61, 633 (1988).
74. K. Ueberreiter, Colloid Polymer Sci., 260, 37 (1982).
75. J. P. H. Boyer, R. J. P. Corriu, R. J. M. Perz, C. G. Reye, Tetrahedron, 31, 2075 (1975).
76. K. Miyajima, M. Sawada, M. Nakagaki, Bull.Chem.Soc.Jpn., 56, 1620 (1983).
77. G. A. Krestov, Thermodynamics of Ionic Processes in Solutions (Rus), Khimia, Leningrad, 1984.
78. Y. Marcus, Ion Solvation, Wiley, Chichester, 1985.
79. V. K. Abrosimov, In: Current Problems of Solution Chemistry (ed.B. D. Beresin) (Rus), Nauka, Moscow, 1986, pp.97-156.
80. M. J. Blandamer, Adv.Phys.Org.Chem., 14, 204 (1977).
81. C. Pan, J.Chem.Soc., Faraday Trans.I, 84, 1341 (1988).
82. A. C. R. Antonini, M. J. Blandamer, J. Burgess, A. W. Hakin, N. D. Hall, A. H. Blandamer, J.Chem.Soc., Faraday Trans.I, 84, 1889- (1988).
83. E. Florin, R. Kjellander, J. C. Eriksson, J.Chem.Soc., Faraday Trans.I, 80, 2889 (1984).
84. M. J. Garvey, I. D. Robb, J.Chem.Soc., Faraday Trans.I, 75, 993 (1979).
85. Y. Marcus, Pure & Appl.Chem., 59, 1093 (1987).
86. N. Muller, Acc.Chem.Res., 23, 23 (1990).
87. W. Melander, C. Horvath, Arch.Biochem.Biophys., 183, 200 (1977).
88. J. B. Hasted, Aqueous Dielectrics, Chapman & Hall, London, 1973, pp.136-175.
89. N. L. Jarvis, M. A. Scheiman, J.Phys.Chem., 72, 74 (1968).
90. H. B. Bull, K. Breese, Arch.Biochem.Biophys., 202, 116 (1980).
91. P. H. Wang, E. J. Lien, J.Pharm.Sci., 69, 662 (1980).
92. B. Y. Zaslavsky, L. M. Miheeva, S. V. Rogozhin, J.Chromatogr., 212, 13 (1981).

CHAPTER 2.
AQUEOUS POLYMER SOLUTIONS

As shown above, the solvent properties of aqueous solutions of various additives differ from those of pure liquid water depending on the type and concentration of the additive. The reason seems to be that an additive molecule perturbs the local water structure which may affect the total aqueous solvent medium. It appears intuitively clear and is supported by experimental evidence (see below) that all other features being identical, the more the size of the additive molecule, the larger water perturbing effect it may display.

To give just one example, it is known that if there is no chemical reaction, two solutions of small molecules in a common solvent are always miscible, provided they are far enough from saturation. In other words, an unsaturated solution of a low molecular weight solute dissolves another solute in a quantity comparable with the amount that can be dissolved in the pure solvent. This is not at all true, however, for solutions of macromolecules. A polymer solution that is still far from its saturation point is, in general, almost totally impenetrable to another high polymer. Possible reasons for this and other properties of aqueous polymer solutions are discussed here.

2.1. THERMODYNAMICS OF POLYMER SOLUTIONS

No general theory of solutions has been developed as yet. In respect to the type of solvent, solutions are usually classified into polar and non-polar, aqueous and non-aqueous solutions, etc. In regard to the type of solutes, solutions are often divided into three groups: solutions of electrolytes, solutions of non-electrolytes and solutions of polymers.

According to the second law of thermodynamics, two components will mix, e.g., to form a solution, if the Gibbs free energy of mixing, ΔG^{mix}, is negative:

$$\Delta G^{mix} = \Delta H^{mix} - T \cdot \Delta S^{mix} < 0 \tag{2.1}$$

where ΔH^{mix} and ΔS^{mix} are the enthalpy and entropy of mixing, respectively; T is the temperature.

The simplest model for a liquid solution is one in which intermolecular forces can be ignored. This occurs in ideal solutions which are analogous to perfect gases. The molecules of the solute are assumed to interact with each other exactly as they do with the solvent in which they are dissolved. In contrast to the absence of forces in the ideal gas, there is a uniformity of forces in the ideal solution. Due to this uniformity of interactions there is nothing to prevent the arrangement of molecules in an ideal solution in a completely random fashion.

Though the ideal solution is a theoretical concept, mixtures consisting of non-polar and/or low polarity molecules often have properties that agree closely to ideal conditions. Liquids that mix to form ideal solutions do so without energy change and are miscible in all proportions. There is no volume or enthalpy change on mixing, and the components obey Raoult's law. Concentration and activity are equal in ideal solutions.

The entropy of mixing of an ideal solution can be calculated by considering the number of possible configurations in a random distribution:

$$\Delta S^{mix} = -R \cdot (X_1 \cdot \ln X_1 + X_2 \cdot \ln x_2) \tag{2.2}$$

where X_1 and X_2 are mole fractions of solvent and solute, respectively.

This expression applies to spherical molecules of similar shape and size. When these properties are significantly different for solute and solvent, the distribution will be influenced in favor of the non-random most economical way in which the molecules can fit together.

A preference of solute for solvent, or solvent for solvent, or any other orderly pattern in the molecular array of a real solution, eliminates the system from ideal status. Hildebrand suggested that when deviations from Raoult's law are not too great, thermal movement in the mixture is sufficient to keep the molecules randomly dispersed. He designated these regular solutions [1].

Since entropy is a measure of disorder, it can be used as a criterion of ideality by specifying that the entropy of mixing in an ideal solution is equal or greater than that involved in any other solution process. A regular solution can be defined as one involving no entropy change when a small amount of one of its components is transferred to it from an ideal solution of the same composition, the total volume remaining unchanged.

The quasi-crystalline approach, one of many theoretical concepts of liquid structure [2], is based on the assumption that the structure of liquid near the melting point is not greatly different from that of the corresponding solid. This follows from the very low expansion in pacing from solid to liquid. The crystalline state is an ordered structure. Each molecule occupies a specific position and its only motion is to vibrate about this position. The lattice points are highly specific and can be occupied by only particular species of molecule.

The quasi-crystalline approach was successfully used by Flory and Huggins [3] to describe polymer solutions. The following equation for large linear molecules with segments occupying different sites in a lattice of small molecules was derived [3]:

$$\Delta S^{mix} = -R(X_1 \cdot \ln\phi_1 + X_2 \cdot \ln\phi_2) \tag{2.3}$$

where ϕ_1 and ϕ_2 are the volume fraction of the corresponding components of the mixture.

Differentiation gives the partial entropy of mixing, e.g.,

$$\Delta S_2^{mix} = -R \cdot \ln\phi_2 + \phi_1 \cdot (1 - V_2/V_1) \tag{2.4}$$

where V_1 and V_2 are the molar volumes of the corresponding components of the mixture.

The entropies of mixing of real solutions can vary considerably from the ideal value. If the solute complexes with the solvent, the entropy of mixing will be lower than ideal since the intermolecular arrangement of the solution will be more ordered than the isolated individual components. However, as the solution becomes more dilute, the entropy of solution will approach ideal, as the solute molecules will be so far apart that even if complexation did not occur, each molecule would be surrounded entirely by solvent molecules.

It is clear that marked differences in molar volumes between solute and solvent could give rise to low entropies of mixing. If, e.g., a low molecular weight solute were dissolved in a high molecular weight solvent, solute molecules would be expected to locate themselves preferentially in the interstices between the solvent molecules, with a resulting increase in order and decrease in entropy. This is, in fact, suggested by the Flory-Huggins Equation 2.4, which includes the ratio of solute and solvent molar volumes [3].

Solution stability may be better illustrated as shown by Patterson [5] through transforming Equation 2.3 to express the combinatorial ΔS^{mix} per volume of solution V_t:

$$\Delta S^{mix}/R \cdot V_t = -[(\phi_1/V_1)\cdot \ln\phi_1 + (\phi_2/V_2)\cdot \ln\phi_2] \qquad (2.3a)$$

where V_1 and V_2 are molar volumes of the pure components 1 and 2, respectively.

Three main thermodynamic effects contribute to the enthalpy and entropy of mixing. These contributions correspond to: the combinatorial entropy of mixing [3] expressed by Equation 2.3; the intermolecular interactions in the mixture; and the so-called free volume effect.

The molecules of the two components of a non-polar polymer solution usually interact through dispersion forces. According to the solubility parameter theory [1], the dispersion force or random dipole-induced dipole interaction always leads to a positive contribution to ΔG^{mix}. It therefore cannot be the cause for the solubility of a given solute in a given solvent. A negative contribution to ΔG^{mix} arises in the case of "specific interactions", e.g., charge transfer, H-bonding, etc. For either dispersion forces or specific interactions, the main effect is on the enthalpy of mixing, ΔH^{mix}, positive and unfavorable to mixing for dispersion forces and negative and favorable in the case of specific interactions.

The second main contribution to ΔG^{mix} corresponds to the molecular interactions between the components. Since the pure solvent and pure liquid polymer have been taken as reference standard states for the Flory-Huggins treatment of the polymer solution [3], the parameter of importance is clearly the difference between the total interaction energy in the solution as compared with that for the pure (liquid) components. The forces between the molecules considered in the theory [3] (dispersion forces only) decrease rapidly with the distance of separation. Hence the consideration was restricted to the energies developed by first neighbor molecules, or segments of the polymer chain, in the solution. Interactions between molecules (segments) which are not immediate neighbors are assumed to make no contribution to the total interaction energy.

According to the lattice model with each cell able to accommodate either a solvent molecule or a polymer segment, three types of first neighbor contact represented by the self-explanatory symbols, 1-1, 2-2, and 1-2, may occur [1]. The energy change for the formation of an unlike contact pair Δw_{12} is:

$$\Delta w_{12} = w_{12} - 0.5\cdot(w_{11} + w_{22}) \qquad (2.5)$$

The enthalpy of mixing ΔH^{mix} is described as:

$$\Delta H^{mix} = RT\cdot V_t\cdot(z\cdot\Delta w_{12}/kTV_p)\cdot\phi_1\cdot\phi_2 = RT\cdot(V_t\cdot\chi_{12}/V_1)\cdot\phi_1\cdot\phi_2 \qquad (2.6)$$

where V_t is the volume of solution; z is the lattice coordination number; V_p is the molar volume of a polymer segment; V_1, ϕ_1, ϕ_2, and Δw_{12} are as defined above.

Parameter χ_{12} is as defined by Flory [3] a dimensionless quantity which characterizes the interaction energy per solvent molecule divided by kT:

$$\chi_{12} = z \cdot \Delta w_{12}/kT \tag{2.7}$$

According to Flory, the quantity $kT \cdot \chi_{12}$ represents the difference in energy of a solvent molecule immersed in the pure polymer compared with one surrounded by molecules of its own kind, i.e. in the pure solvent.

Combining Equation 2.6 with Equation 2.3a for ΔS^{mix} in accordance with Equation 2.1, ΔG^{mix} becomes:

$$\Delta G^{mix}/RT \cdot V_t = \underbrace{[(\phi_1 \cdot \ln\phi_1)/V_1 + (\phi_2 \cdot \ln\phi_2)/V_2]}_{\text{combinatorial}} + \underbrace{\chi_{12} \cdot \phi_1 \cdot \phi_2/V_1}_{\text{interactional}} \tag{2.8}$$

The first two terms are small and the third term with χ_{12} positive may cause phase separation ($\Delta G^{mix} > 0$) in the polymer solution at a small value of the polymer volume fraction ϕ_2 (see below).

The original Flory-Huggins theory [3] ignored the role of the aforementioned free volume effect. The Prigogine-Flory theory developed later [5,6] has shown that χ-parameter includes both interactional and free volume effects with different temperature dependence. Since in this part of the discussion we are not concerned with predicting the temperature dependence of phase separation in polymer solution, the free volume effect will not be considered here. For detailed treatment of the Prigogine-Flory theory the reader is referred to [4-6]. Somewhat more detailed consideration of the theory as applied to mixtures of two polymers in a common solvent will be given below in the next chapter.

The interaction parameter χ_{12} in the polymer solution theory is used as a measure of the polymer-solvent interactions. The advantages and limitations of this measure will be discussed below.

The other way to describe the interactions in polymer solutions is based on the virial expansion approach. The chemical potential of the solvent μ_1 in solution is related to its concentration as:

$$\mu_1 - \mu^o_1 = RT \cdot \ln(f_1 \cdot X_1) \tag{2.9}$$

where μ^o_1 is the chemical potential of the pure solvent; f_1 is the coefficient of activity of the solvent; X_1 is mole fraction of the solvent.

The virial expression is a power series in the concentration of the solute that expresses the chemical potential of the solvent. In ideal solution $f_1 = 1$, hence Equation 2.9 takes the form:

$$\mu_1 - \mu^o{}_1 = RT \cdot \ln x_1$$

Replacing $X_1 = 1 - X_2$ for $(1 - X_2)$, where X_2 - mole fraction of the solute, leads to

$$\mu_1 - \mu^o{}_1 = RT \cdot \ln(1 - X_2) \tag{2.9a}$$

Substituting a convenient power series expression for $\ln(1 - X_2)$ when X_2 is small compared to 1, into Equation 2.9a, gives

$$\mu_1 - \mu^o{}_1 = -RT \cdot [X_2 + 1/2 \cdot X_2^2 + 1/3 \cdot X_2^3 + ...] \tag{2.10}$$

Neglecting all powers of X_2 higher than one, which is an acceptable procedure when $X_2 << 1$, one obtains

$$\mu_1 - \mu^o{}_1 = -RT \cdot X_2 \tag{2.10a}$$

which is the basis for useful equations describing the behavior of macromolecules.

Expressing the solute concentration c_2 (supposedly dilute) in weight/volume units, $X_2 = c_2 \cdot V^o{}_1/M_2$, where M_2 is the solute molecular weight, and $V^o{}_1$ is the partial molar volume of solvent in the solution. Therefore, when c_2 is small, Equation 2.10a becomes

$$\mu_1 - \mu^o{}_1 = -RT \cdot V^o{}_1 \cdot c_2/M_2 \tag{2.10b}$$

which is the simplest approximation for the solvent chemical potential, expressed as a linear function of solute concentration.

When the solution is not within the ideal limit ($f_1 = 1$), the coefficients of the power series expansion depend on the activity coefficient f, so they are written as unknown quantities:

$$\mu_1 - \mu^o{}_1 = -RT \cdot V^o{}_1 \cdot [c_2/M_2 + B \cdot c_2^2 + C \cdot c_2^3 + ...] \tag{2.11}$$

where B, the coefficient of c_2^2, is called the second virial coefficient, C is the third virial coefficient, etc. Equation 2.11 is the virial expansion of the solvent chemical potential.

A number of physical techniques allow determination of the chemical potential difference between pure solvent and solvent in solution. Among these are measurements of the vapor pressure, freezing point depression, boiling point elevation, and osmotic pressure.

When the solution is ideal ($f_1 = 1$), it follows from Equation 2.10 that $B_{ideal} = V^o{}_1/2M_2^2$ (see, for example, in [7]). In the case of $B > B_{ideal}$ in real solution, the first and second terms in the virial expansion have the same sign which means that the solute-solvent interactions are favorable, more so than solute-solute or solvent-solvent interactions. Solvents for which $B > 0$ for a particular macromolecule are called good solvents for this macromolecule. When the value of $B < 0$, the solvent is called a poor solvent for a given macromolecule. Precipitation of the polymer occurs relatively easily from poor solvents.

The quasi-crystalline concept used in the lattice solution model involves the formation of a cavity in the solvent to accommodate a solute molecule. The energy required for the cavity formation is assessed in terms of the energy required to separate the solvent molecules from one another and to the cavity size (see above, Equation 1.5). Girifalco and Good [8] applied the regular solution theory to interfaces assuming that the free energy of adhesion, ΔG_{12}, between phases 1 and 2 would be the geometric mean of the free energies of cohesion of the separate phases, ΔG_{11} and ΔG_{22}. This treatment leads to:

$$\Delta G_{11} = -2\gamma_1 \tag{2.12}$$

$$\Delta G_{12} = \gamma_{12} - \gamma_1 - \gamma_2 \tag{2.13}$$

where γ is the surface (or interfacial) free energy per unit area or surface (or interfacial) tension.

It should be noted that adhesion (Equation 2.13) is usually considered in thermodynamics as the process of forming a 1-2 interface from a 1-vacuum surface and a 2-vacuum surface. Consequently, the sign of ΔG_{12} is the opposite of that in the conventionally-defined energy of mixing.

Good and Girifalco [9] showed that the interfacial tension between non-polar phases, 1 and 2, is described as:

$$\gamma_{12} = \left(\sqrt{\gamma_1} + \sqrt{\gamma_2}\right)^2 \tag{2.14}$$

This model, however, is applicable only to non-polar systems [10]. A new theoretical model applicable to systems with H-bonds and other polar interactions was suggested recently by van Oss et al. [10-12]. This model was used, in particular, to explain phase separation in aqueous polymer systems (see below).

According to the model [10-12], there are two major contributions into the surface tension: a) non-polar interactions or Lifshitz-van der Waals (LW) interactions, and b) polar interactions including all electron acceptor-electron donor or Lewis acid-base (AB) interactions. Non-polar (LW) interactions are described by Equations 2.12 - 2.14. In contrast to LW interactions, polar AB interactions are essentially asymmetrical and it is taken into account by the model [10-12]. The electron acceptor and electron donor parameters of the polar component (γ_i^{AB}) of the surface tension of compound i are expressed as, respectively, γ_i^+ and γ_i^-. It was shown [10,13,14] that the polar component of the free energy of interaction between materials 1 and 2 (adhesion) can be expressed as:

$$\Delta G_{12}^{AB} = -2(\sqrt{\gamma_1^+}\cdot\sqrt{\gamma_2^-} + \sqrt{\gamma_1^-}\cdot\sqrt{\gamma_2^+}) \quad (2.15)$$

and the polar component of the free energy of cohesion of any compound i is

$$\Delta G_{ii}^{AB} = -4\cdot\sqrt{\gamma_i^+}\cdot\sqrt{\gamma_i^-} \quad (2.16)$$

so that

$$\gamma_i^{AB} = 2\cdot\sqrt{\gamma_i^+}\cdot\sqrt{\gamma_i^-} \quad (2.17)$$

Applying Equation 2.13 to the polar component of the free energy of interaction between components 1 and 2, and combining Equations 2.15 and 2.17, the following expression for the polar component of the interfacial tension between materials 1 and 2 was obtained (see, e.g., in [14]):

$$\gamma_{12}^{AB} = 2\cdot(\sqrt{\gamma_1^+} - \sqrt{\gamma_2^+})\cdot(\sqrt{\gamma_1^-} - \sqrt{\gamma_2^-}) \quad (2.18)$$

When Equation 2.18 is compared with equation 2.14, it becomes clear that while γ_{12}^{LW} cannot be less than zero, γ_{12}^{AB} can readily be negative, i.e., when

$$\gamma_1^+ > \gamma_2^+ \quad \text{and} \quad \gamma_1^- < \gamma_2^-$$

or when

$$\gamma_1^+ < \gamma_2^+ \quad \text{and} \quad \gamma_1^- > \gamma_2^-.$$

This approach has the advantage that the interfacial tensions are readily measurable. Van Oss et al.[10-15] showed that γ^{LW}, γ^+, and γ^- parameters can be experimentally determined for different polymers and solvents, and $\gamma_{12} = \gamma_{12}^{AB} + \gamma_{12}^{LW}$ can be calculated.

To describe the polymer solubility or miscibility in a given solvent van Oss et al. [12,16] used parameter ΔG_{121}. This is the interfacial free energy change (per minimum effective surface area of contact between two polymer molecules 1) associated with bringing together two polymer molecules 1 initially present in solvent 2 with an effectively infinite layer of phase 2 separating two surfaces of phase 1 [12]. This parameter is related to the interfacial tension as

$$\Delta G_{121} = -2\gamma_{12} \quad (2.19)$$

If $\Delta G_{121} > 0$, then molecules of 1 will repel each other in solvent 2, and polymer 1 will disperse or dissolve in 2. If $\Delta G_{121} < 0$, molecules of 1 in solution, at equilibrium, will attract each other, and so will tend to precipitate per kinetic unit in the system will cause segments of the polymer molecule that meet in a dilute solution to separate again [12]. Hence, according to van Oss et al. [12,16], $\Delta G_{121} > -1.5 \cdot kT$ is a general condition for solubility or dispersibility of the polymer 1 in solvent 2.

According to van Oss et al. [12], an important difference between polar and non-polar systems is that γ_{12} can be negative in polar systems, while the lowest value of γ_{12} in apolar ones is zero. Thus, ΔG_{121} is always negative or zero in non-polar systems. Hence, to achieve solubility in apolar systems the value of ΔG_{121} must be in the range from zero to $-1.5 \cdot kT$, but in polar systems it may be of any value above $-1.5 \cdot kT$ including all the positive values of ΔG_{121} that can exist.

It was also shown [16] that the interfacial free energy ΔG_{121} (per contactable surface area, S_c, and expressed in units of kT) of a polymer 1, dissolved in a solvent 2, is related to the Flory-Huggins χ_{12} parameter as:

$$\Delta G_{121} = -\chi_{12} \cdot kT/S_c \qquad (2.20)$$

or

$$\chi_{12} = 2S_c \cdot \gamma_{12}/kT \qquad (2.21)$$

Using Equation 2.21 van Oss et al. [16] calculated χ_{12}-values for the solutions of poly(ethylene glycol) in water at different polymer concentrations from the surface tension data for poly(ethylene glycol) and water. These values are discussed below.

All three measures of the polymer-solvent interactions described above (Flory-Huggins χ_{12} parameter, second virial coefficient B, and the interfacial free energy ΔG_{121}) seem to be complimentary and cover a wide range of solutes and solvents. The experimental values of these parameters reported in the literature for polymer solutions will be discussed below in regard to phase separation in aqueous polymer systems. These measures based on thermodynamic considerations, however, cannot answer the most important question: What is the molecular mechanism behind the experimentally observed features of aqueous polymer solutions? The experimental observations discussed below indicate, first of all, that many properties of aqueous solutions of macromolecules are generally different from those of aqueous solutions of low molecular weight solutes at the comparable concentrations.

2.2. PROPERTIES OF WATER NEAR INTERFACES

Solutions of macromolecules may be considered as one-phase with two homogeneously mixed components or as two-phase systems with the large macromolecule constituting a separate phase. The choice depends on how the sys-

2.2. PROPERTIES OF WATER NEAR INTERFACES

Solutions of macromolecules may be considered as one-phase with two homogeneously mixed components or as two-phase systems with the large macromolecule constituting a separate phase. The choice depends on how the system behaves in regard to the property under examination, and what is convenient for the analysis [7]. If a given polymer in solution is considered as a separate phase, an existence of the interface is implied.

The creation of an interface, such as an air-water interface, in a bulk aqueous phase gives rise to asymmetry in the bonding of the water molecules. Hence it is obvious that the properties of the water at the air-water (or solid surface-water) interface are different from those of the bulk structure.

It has been numerously observed that water and aqueous solutions adjacent to most (solid) interfaces possess significantly different properties from those of the respective bulk systems. The term vicinal water was suggested by Drost-Hansen (17-33) for such interfacial water. According to the definition by Drost-Hansen [29], vicinal water is water the structure of which is modified by proximity to an interface but excluding chemically "bound" water directly on the surface (the water of primary hydration).

Comparison of some properties of bulk water with those of vicinal water given in [17] is presented in Table 2.1. Since the properties of vicinal water differ from those of bulk aqueous systems, it was concluded [17-33] that the structures of vicinal water and of bulk water are different. According to Drost-Hansen, the modification of the water structure extends over considerable distances, as much as 30 to 300 molecular diameters, or 100 to 1000 Å [32]. The evidence for extensive structure modification stems partially from the following measurements: (a) viscosity [26]; (b) ultrasonic absorption and velocities [23,24]; (c) conductance data [33]; (d) ultra-slow mechanical relaxation (shear) [27,33]; (e) dielectric data [34, 35], and other experimental observations reviewed in, e.g., [17,22,29,33].

The properties of vicinal water exhibit thermal anomalies, displayed as rather abrupt changes in the temperature coefficient of the above properties over at least four temperature intervals, particularly near 14-16, 29-32, 44-46, and 59-62°C. For example, in contrast to air/water surface tension measurements, those made by the capillary rise method (i.e., the glass/ water interfacial tension) at different temperatures have shown several thermal anomalies (see, e.g., in [29]). These anomalies appear as inflection points in the surface tension versus temperature near the above intervals. Since the surface (interfacial) tension is a free energy (see above), the temperature derivative is an entropy of surface formation. Drost-Hansen reported [30] that for water in narrow glass capillaries there is an increase in the entropy of about a factor of 2 over a temperature interval of 3-4°C near 30°C. This effect is considered by Drost-Hansen

Table 2.1.

Comparison of Some Properties of Bulk and Vicinal Water

Property	Bulk water	Vicinal water	Ref.
Density, g/cm^3	1.00	0.97	[18,19]
Specific heat capacity, cal/K·g	1.00	1.25 ± 0.05	[20,21]
Thermal expansion coefficient at 25°C, °C^{-1}	$250 \cdot 10^{-6}$	$(300\text{-}700) \cdot 10^{-6}$	[22]
Adiabatic compressibility coefficient, Atm^{-1}	$45 \cdot 10^{-6}$	$(60\text{-}100) \cdot 10^{-6}$	[23]
Excess sound adsorption (α/v^2), cm^{-1}·sec^2	$7 \cdot 10^{-17}$	ca. $35 \cdot 10^{-17}$	[24]
Heat conductivity, cal·sec^{-1}·cm^{-3}·°C	$1.4 \cdot 10^{-3}$	ca. $(1\text{-}5) \cdot 10^{-2}$	[25]
Viscosity (cP), Energy of activation	0.89	2-10	[26]
Ionic conductance (KCl), kcal/mole	ca. 4	5-8	[27]
Dielectric relaxation	$19 \cdot 10^9$	$2 \cdot 10^9$	[28]

[From J. S. Clegg, W. Drost-Hansen, In: The Biochemistry and Molecular Biology of Fishes (Eds. P. W. Hochachka, M. T. P. Mommsen), Vol.1, Elsevier Science Publishers, Amsterdam, 1991, pp. 1-23. Reprinted by permission of Elsevier Science Publishers.]

and Etzler [31] as manifestation of specific, relatively long-range and temperature-dependent vicinal restructuring of water, originating from the glass surface/water interface.

The estimates of the distance over which solid surfaces affect water structure differ between different authors. As indicated above, Drost-Hansen and his colleagues estimate this distance up to 1000 Å [17], while Clifford [36] argues that the likely limit is about 100 Å. Rand and Parsegian [37] conclude that water is affected by phospholipid bilayers over 10-30 Å from the bilayer surfaces, and Robb [38] estimates the distance over which an amphiphilic polymer may affect the water structure as 20-30 Å. The question clearly remains open but the differences between these estimates do not contradict the concept of the influence of solid surface-water interface on the water structure.

While the vicinal water hypothesis by Drost-Hansen appears to be sound and confirmed by numerous experimental evidence, the so-called "paradoxical effect" (see, for example, in [17]) seems to be doubtful. The "effect" in question is that the vicinal water is induced by any kind of solid (or macromolecular) surface independent of the specific chemical nature of the surface. The likely reason for this concept may be the insufficient sensitivity of the current methods employed for the analysis of the properties of vicinal water.

Anomalous properties of water in macromolecular gels reported recently by several authors [39-43] may be attributed to the effect of the gel matrix on the structure of water confined to the gel pores. Van Steveninck et al. [39] examined elution characteristics of low-molecular weight solutes on gel filtration columns using Sephadex G-10 and G-25, and Bio-Gel P-6 and showed that the results obtained could be explained only by anomalous solvent behavior of internal gel water. Janado et al.[40] analyzed the elution behavior of sodium dodecyl sulfate on Bio-Gel P-2 and concluded that the preferential partition of the surfactant in the Bio-Gel phase is primarily due to the specific nature of water in the gel matrix. This conclusion was also supported by the observations [41] that internal water of swollen Sephadex gels could dissolve significant amounts of water-insoluble dyes. The conclusion that the structures of water inside and outside the gel matrix are different [39-41] is supported by Wiggins et al.[42,43] explaining the observations in somewhat different terms.

Wiggins et al.[42,43]. examined distribution of different ions between water in swollen Sephadex, and Bio-Gel gels and water in external solution. The results indicated particularly that highly hydrated ions, such as Na^+ and H^+ are accumulated in the gel water, while less highly hydrated ions, such as K^+ and NH_4^+ accumulate in the external water. This was explained by the authors [42,43] as being due to the difference in the density and thermodynamic activity coefficients of water in gel and in the external solution.

The vast amount of additional evidence for the long-ranged influence of macromolecular surfaces on the structure and solvent properties of water also exists in the literature on the problem of water in biological systems. In view of the complexity of the object the evidence is usually less direct and will be briefly commented on in the other section (see Chapter 8).

2.3. SOLVENT PROPERTIES OF AQUEOUS POLYMER SOLUTIONS

There is an enormous literature on aqueous polymer solutions. (For complete bibliography the reader is referred to the books by Molyneux [44,45].) The problem of the structure of water in the polymer solutions is as yet in an embryonic state. The structure of water is of paramount importance in the occurrence and properties of the phases of aqueous polymer two-phase systems. Hence the experimental results from the literature are discussed below with regard to the influence of polymers on the structure of water.

Essentially all the non-ionic water-soluble polymers can be viewed as composed of two kinds of groups, those participating in hydrophobic hydration and groups capable of H-bonding and dipole-dipole interactions with water molecules. For example, polyvinylpyrrolidone (PVP) can be viewed as composed of two kind of groups, dipolar imide group on the pyrrolidone rings and hydrophobic groups as methine and methylene in the backbone and the methylene in the same ring. Similarly, poly(ethylene glycol) (PEG), the simple linear polymer, possesses a hydrophobic region (-CH_2-CH_2-) and a single H-bonding ether oxygen per monomer unit. Even in the case of polysaccharide dextran the chemically similar hydroxyl groups in the glucose monomer unit are known to differ in respect of the hydration interactions. H-bonding of equatorial hydroxyl groups with water molecules appears to be relatively more energetically favorable than that of axial hydroxyl groups [46].

Thus, the local water structure around a given macromolecule usually consists of water molecules H-bonded to the polar hydrophilic site and the hydrophobic hydration water structure around the non-polar site. The interplay of these effects results in the specific structuredness of water molecules around the macromolecule which overcomes the surface forces (see above) sufficiently to stabilize the macromolecule in aqueous solution. This outline of a number of possible contributing factors seems to be as far as one can go at present in discussing the effects manifested by macromolecules on their local water environment. Further discussion requires information on the conformation of polymer, spatial compatibility of the polymer groups topology with that of water molecules, etc. The detailed liquid structures of water near the polymer chain cannot be elucidated by the experimental methods available at present. The point to be emphasized is that as the sizes of the local water structures in the polymer solutions are very large, it is possible to regard them as regions of water with the particular structures [47].

It should also be stressed that the presence of hydrophilic highly polar groups in the water-soluble polymers clearly differentiate them from non-polar solutes. The latter ones according to Hvidt [48] are assumed to fluctuate between solvated (hydrated) and nonsolvated states, and the probability of finding them in either state depends on the concentration. Due to the presence of hydrophilic polar groups water-soluble polymers are usually strongly hydrated and the fluctuation between hydrated and dehydrated states for these polymers in aqueous medium is highly unlikely. This point is important in view of some theoretical considerations of phase separation in aqueous polymer systems to be discussed below.

It is well known that the solute-solute interactions may be transmitted by and through intervening solvent molecules. An example of the importance of such long-range effects in aqueous solutions of amphiphilic nonelectrolytes may be illustrated by the volumetric properties of monohydric alcohols in water. The partial molar volume of t-butyl alcohol differs from the standard state volume even in very dilute solutions of ca. 10^{-3} mole fraction of alcohol, i.e. when the solute molecules are on average 60 Å apart [49]. That means that the two relatively small non-ionic alcohol molecules in water at this distance still "feel" each other's presence through the intervening water. It is generally believed that, e.g., PEG and PVP macromolecules manifest much stronger effects on the water structure than that displayed by t-butanol molecule. The strong water structure-making effects of dextran [50], PEG and PVP [51] have been revealed by different experimental techniques. Therefore it can be assumed that the two PEG (or Dex, PVP, etc.) macromolecules in water should "feel" each other's presence through the intervening solvent at the distance much larger than that found for t-butanol molecules.

The difference between the physicochemical properties of aqueous solutions of low molecular weight non-ionic solutes and those of nonionic macromolecules is mainly due to the difference in the sizes of the molecules' hydration shells. This conclusion is consistent with the aforementioned large deviations of behavior of aqueous polymer solutions from ideality even in very dilute solutions and numerous experimental findings to be considered below. It was indicated above that two solute species will generally attract or repel each other in water depending on the compatibility of the structures of water in their hydration shells [49,52]. Non-ionic macromolecules are known to form aggregates or association complexes in aqueous solutions. Hydrophobic interactions, van der Waals forces and H-bonding are generally assumed to be responsible for the association in aqueous macromolecular solutions but the nature of the aggregation process remains an open question as yet. An examination of the behavior of aqueous PEG solutions enables one to consider the question of the compatibility of hydration shells' structures of the macromolecules of this typical water-soluble polymer.

It is commonly known that there is a good structural fit between the water and the PEG macromolecule. A stabilized local water structure presumably due partially to the orientational polarization of water molecules H-bonded with the macromolecule and partially to the enhanced water-water interactions in the vicinity of the polymer hydrophobic sites is confirmed by numerous experimental data (see, e.g., in [44,45]). The ether oxygen of the monomer unit of PEG macromolecule is generally accepted to be strongly hydrated, with two or three water molecules H-bonded to it. According to Mank et al. [53] PEG macromolecule affects much more water molecules than it forms H-bonds with, exerting a structure-making action upon 16 water molecules per monomer unit. If this is the case all the water would be specifically "structured" in solutions of PEG that were greater than about 13 %wt.

At elevated temperatures aqueous PEG solutions are known to phase separate into two coexisting PEG-rich and PEG-poor aqueous phases [54]. According to Kjellander et al.[55], this is caused by the disruption of the specific structure of the PEG hydration shell due to increased thermal motion. At higher temperatures when the structure of water essentially disappears, PEG and water become completely miscible again.

The overlap of the PEG hydration shells may occur either at some distance of the macromolecules apart or in a close polymer-polymer contact. The difference between these two situations is of no particular importance in the theoretical model developed by Kjellander et al.[55].

The extent of the hydration of PEG in aqueous solutions remains uncertain despite numerous investigations. Each monomer group is supposed to be associated with a definite number of water molecules. The estimates of this number vary from less than one [56,57] to more than five [58], with a value of two, which is attributed to H-bonding of the water molecules to the ether oxygen, being most frequently quoted [59,60]. The difference between the total amount of water molecules affected by PEG macromolecule (16 per monomer unit, according to [53]) and that of H-bonded water molecules (about 2 per monomer unit) is 14 water molecules per monomer unit of PEG. It is likely that these 14 water molecules are under different influence of the monomer unit. As a first approximation let us assume that there are two kinds of molecules among those under the influence, namely, those "strongly" and "weakly" affected by the unit. In this case the distance at which the hydration shells of PEG macromolecules overlap is likely to be governed by the amount of the "strongly" affected water molecules. The amount of the "weakly" affected water molecules is likely to determine the concentration of water in the PEG-poor phase when phase separation occurs [54]. It is impossible to estimate the amounts of these two fractions of water from the compositions of the coexisting phases, as phase separation occurs at elevated temperature when the situation is clearly different from that at the ambient temperature [53]. In any case, the obvious conclusion to be drawn from the above is that the structures of water in the hydration shells of the PEG macromolecules are compatible with each other.

It should be mentioned here that instead of hydration shell term the above vicinal water term may be used to the same purpose. The question of terminology is always open to discussion, and it may be suggested, e.g., to use the terms bounded water layer and perturbed water zone to describe the situation with water in the vicinity of a macromolecule.

The so-called lower critical temperature demixing is very common in aqueous solutions of non-ionic polymers, such as PEG, polyvinyl alcohol (PVA), polymethacrylamide, etc. [61]. Thus, the association complexes well known to be present in aqueous polymer solutions can be viewed as those resulting from the overlap of the compatible hydration shells of macromolecules or from a "fusion" of the specifically structured water regions with the corresponding change in the amounts of the "weakly" and "strongly" affected water molecules in the solution. In this case an increase in the polymer concentration leads to a gradual decrease in the amount of "weakly" affected water molecules, on the one hand, and to a gradual increase in the strength of the effect experienced by these molecules, on the other hand. It seems impossible to divide contributions of these two factors at present. It is clear, however, that the water-structure-perturbing factors, such as urea, inorganic salts, etc., should affect the above "fusion" process underlying phase separation or association phenomena as the experimental data show they do [62-66].

The concept of a long-range restructuring of water in aqueous solutions of water-soluble polymers is supported particularly by the results reported recently by Scherbakov and Monin [67]. The authors [67] showed that: a) the macroscopic orientation in aqueous solutions of PEG-400, PEG-600 and PEG-1500 over the polymer concentration range from 2 to about 20 %(w/w) is clearly displayed in ^{17}O, ^{2}H and proton resonance spectra; and b) the solutions are oriented in the external constant magnetic field of 304 and 875 gauss as indicated by the time-dependent change of the solution electrical resistance. The results obtained enabled the authors [67] to suggest that aqueous PEG solutions should be viewed as lyotropic liquid crystalline systems [68]. It was found also by Scherbakov and Monin [67] that the activity coefficient of Na^+ cation in the 0.14 M NaCl aqueous solution in the presence of PEG increases with increasing polymer concentration. The effect decreases with increase of the PEG molecular weight. In view of the data reported by Florin [69] and Breen et al. [70] (also see the references cited in [70]), it seems possible to agree with the opinion [67] that the influence of PEG on the Na^+ activity coefficient is due to the polymer effect on the water structure.

The nuclear magnetic relaxation rates of ions, such as Li^+, Na^+, Rb^+, Cs^+, Cl^-, Br^-, and I^- were studied in aqueous PEG solutions [69,70]. The enhancement of the ion relaxation rate is proportional to the PEG concentration up to about 10-15 %wt., while in more concentrated polymer solution the ion relaxation is no longer proportional to the PEG concentration. The NMR and neutron diffraction analysis shows that no direct ion-PEG interaction occurs (see in [70]). The polymer effect on the relative molal increase of the ion relaxation is increased in the series: $Li^+ < Na^+ < Rb^+ < Cs^+ < Cl^- < Br^- < I^-$ [69, 70], i.e. the enhancement of the ion relaxation rate is largest for ions with a large hydration shell. The results [69,70] were interpreted in terms of polymer perturbation of the dynamics and preferential orientation of water molecules in the ion hydration shell. It may be concluded from these results [69,70] that the presence of PEG affects the ionic hydration interactions due to the polymer influence on the water structure.

It was mentioned above that various additives affect the acidity and basicity of water according to their action on the water structure [71-73] (see in Chapter 1). In view of the polymer influence on the activity coefficients of different ions in aqueous solutions (see, e.g., in [38,42,43,67]) the water-structure-perturbing non-ionic polymers may be expected to affect the acid-base equilibrium in water. The experimental evidence was obtained in the study [74] of the acid-base equilibria of sulphonephthalein dyes in aqueous solutions of PEG, Ficoll, dextran, and some other non-ionic polymers. The results of studying acid-base and tautomeric equilibria of fluorescein and eosin in solutions of PEG and Ficoll showed [74] that the equilibrium constants for the dyes change in reference to pure water. The changes determined are similar to those observed for the dyes in water-organic solvent mixtures. These changes were argued [74] to be due to the polymer effects on the water structure, leading to modified hydration and thermodynamic activity of both H^+ and OH^- ions.

The changes in the equilibria observed in [74] may also be partially due to the polymer effects on the dielectric properties of the aqueous medium [58,75-79].

The complex permittivity measurements is a method providing direct information about the orientational mobility of water, i.e., about the structure of water in aqueous solutions [80]. The advantage of this method is that there is essentially no contribution of the solute molecules into the complex permittivity of water at the frequencies in the 1 - 40 GHz range [58,75-80].

Measurements of complex permittivity of aqueous solutions of PVP [75], PEG, PVA and poly(vinyl methyl ether) each of various molecular weights show that the water reorientation time in the solution differs in a characteristic manner from what one expects on the basis of the data for aqueous solutions of small organic molecules [76]. The data on the permittivity of aqueous solutions of various homologues of PEG [76,78,79] indicate that the wavelength of dielectric relaxation and the microviscosity of water are significantly increased in solutions of PEG with a degree of polymerization exceeding 12 monomer units compared to those observed in solutions of PEGs with lower polymerization degree. The effect of PEG of a high polymerization degree on

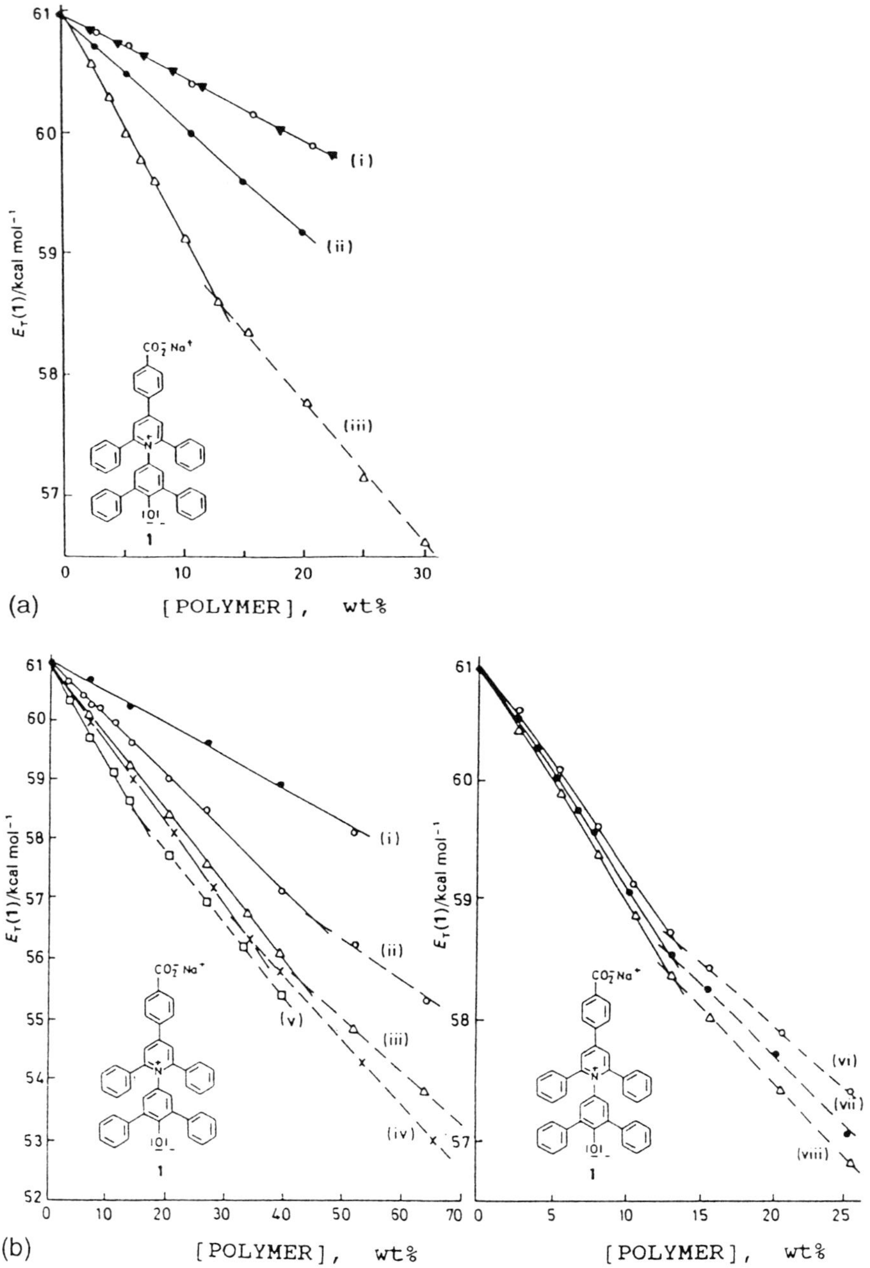

Figure 2.1. Solvent polarity of aqueous medium (measured with the betaine dye 1 of the structure shown) in aqueous polymer solutions, pH 9.3 as a function of the polymer concentration: a: (i) dextran; (ii) Ficoll; (iii) PEG-6,000; b: (i) ethylene glycol; (ii) diethylene glycol; (iii) PEG-200; (iv) PEG-300; (v) PEG-600; (vi) PEG-2,000; (vii) PEG-6,000; (viii) PEG-20,000. Data from [82].

temperature. The static dielectric constants, ε_s, and the logarithm of the dielectric relaxation time of water, τ, in aqueous solutions of PEG, dextran, PVP, and Ficoll [77] are linearly dependent on the polymers concentrations:

$$\Delta = A(\Delta) + B(\Delta)_i \cdot C(P)_i \tag{2.22}$$

where Δ is the property of the aqueous medium under study; $C(P)_i$ is the polymer concentration in wt.%; $A(\Delta)$ and $B(\Delta)_i$ are constants, $A(\Delta)$ is equal to Δ in the polymer-free medium, and $B(\Delta)_i$ depends on the polymer type and molecular weight; subscript 'i' denotes the polymer under study.

The absolute $B(\varepsilon_s)$-values [77] decrease in the order: PVP (M_w = 12,700) > PEG (M_w = ~8,000) > Ficoll (M_w = 400,000) > dextran (M_w = 57,200) which obviously disagrees with the order of the size (or molecular weight) of the polymers. It follows that the ability of water molecules to orient in the applied field decreases in line with the structuring effect of the polymer on water but not according to the particle volume effect. The observed order of the B(t)-values in the same polymer solutions [77], slightly different from that for the $B(\varepsilon_s)$-values, is in agreement with the order of the relative hydrophobicity of the polymers (see below). This implies that the polymer effect on the dielectric orientational mobility of water is governed by the relative immobilizing structuring influence of a given polymer on water molecules.

The first data on the polymer effect upon the solvent polarity of the aqueous medium have been reported by Arnold et al.[81]. The authors used the shift of the maximum of fluorescence and change in the quantum yield of 1-anilinonaphthalene-8-sulfonate (ANS) employed as a probe to measure the changes in the solvent polarity of the aqueous medium with increasing concentration of PEG. The results [81] indicated that PEG reduces the polarity of the medium. This conclusion was confirmed by the observation that the affinity of nonpolar compound pyrene for the aqueous medium increases with increasing PEG concentration. More recently, the overall solvent polarity of aqueous alkaline solutions of PEGs of different molecular weights, dextran, and Ficoll was estimated [82] using the solvatochromic anionic betaine dye described above (see Chapter 1). The molar transition energy for the solvatochromic absorption band of the dye, E_T, was used as a measure of the solvent polarity of the aqueous medium in the polymer solutions (see Equation 1.8).

Figure 2.1a shows the data obtained for the aqueous solutions of dextran (Dex), Ficoll (Fic), and PEG-6000 at different polymer concentrations. The initial concentration effects of the polymers on the solvent polarity parameter is described by Equation 2.22 [82]. The polymer concentrations up to which Equation 2.22 was found to be valid [82] amount to about 13 %wt. for PEGs of 2,000 to 20,000 molecular weight, increasing with decrease in the polymer molecular weight (see Figure 2.1b).

An important question considered in [82] is if the effects observed result from a direct polymer-dye interaction, or they are due to a change in the solvent properties induced by the presence of the polymer. As it is impossible to give an unambiguous answer to this question, the authors [82] used the following indirect arguments.

According to Mank et al.[53], PEG macromolecule exerts a structure-making action upon 16 water molecules per monomer unit, implying that all the water must be specifically "structured" in solutions of PEG at the polymer concentration of 13 wt% and more. The agreement with the "critical" values of the polymer concentrations up to which Equation 2.22 was found to be valid [82] is obvious (see in Fig. 2.1). The other argument follows from the consideration [82] of the water zone in which the dye molecules are likely to be located. Location of the dye in the inner hydration zone of PEG macromolecule cannot be differentiated from the direct PEG-dye interaction. If the dye is located in the bulk unperturbed water zone, no solvatochromic effect would be observed. The DSC [83] and NMR [53] measurements indicate that no water remains unbound in aqueous solutions of PEGs at more than 52 wt%. The linear relationship between E_T and C(PEG) is observed below this concentration [82]. Hence it was concluded [82] that the contribution of the dye molecules located in the inner hydration zone to the solvatochromic effect observed is small or non-existent. Therefore it was suggested that the dye molecules are located mainly in two water zones - in the bulk water (if it exists) and in the outer hydration zone, i.e. the zone of non-bound water perturbed by the presence of PEG macromolecules.

As a final argument in favor of the absence of direct dye-polymer interactions, the quantitatively similar results were obtained in the studies of the solvatochromic effects of several other dyes in the presence of the same polymers [74,82]. Since it is highly unlikely that the different dyes interact with the polymers in identical fashion, that confirms the assumption [82] that the polymers examined change the solvent polarity of water or, in other words, solvation (hydration) capability of water.

It follows for the experimental data described above that the presence of polymer additives in water alters essentially all the solvent properties of aqueous medium. Hence it was suggested (see, e.g., in [84]) that aqueous polymer solutions may be viewed as different solvents of the same aqueous nature.

The approach to the experimental study of the influence of water-soluble polymers on the relative hydrophobic character and the relative ionic hydration power of water in aqueous polymer solutions has been suggested by Zaslavsky et al.[85]. It should be noted that from now on under the relative hydrophobic character of a solvent term is meant the thermodynamic affinity of the solvent medium for non-polar groups and solutes, namely, for a CH_2 group,

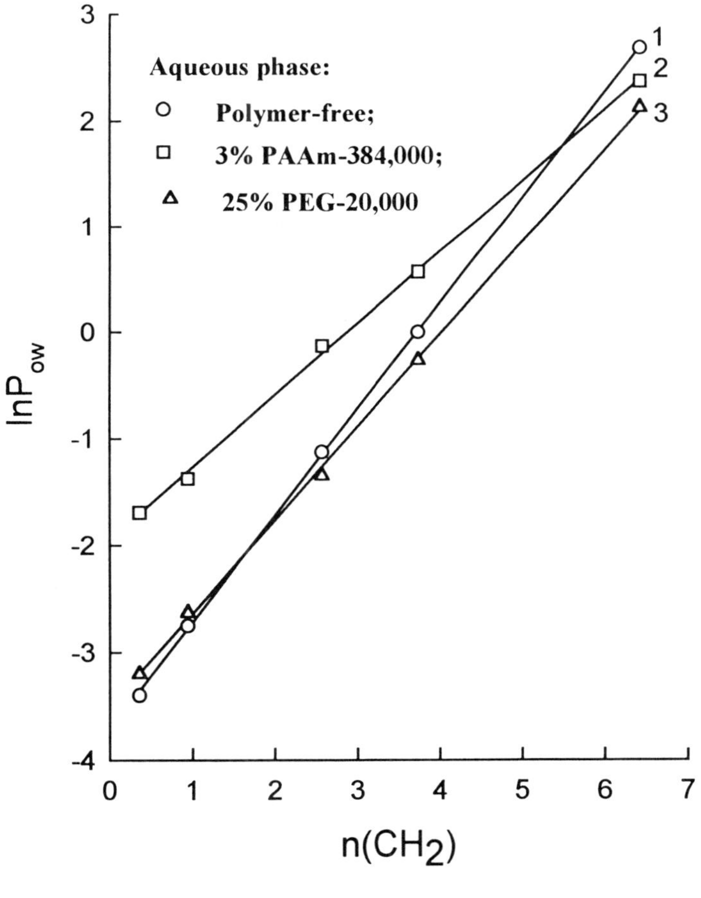

Figure 2.2 Logarithm of the partition coefficient of dinitrophenylated (DNP-) amino acid as a function of the aliphatic side-chain length of the DNP-amino acid in the following two-phase systems: (1) polymer-free aqueous salt solution-octanol; (2) 3%wt. polyacrylamide in aqueous salt solution - octanol; (3) 25 %wt. PEG-20,000 in aqueous salt solution - octanol. Salt composition of aqueous phase - 0.15 M NaCl in 0.01 M sodium phosphate buffer, pH 7.4. Data from [86].

in reference to that of a solvent (pure water, octanol-saturated water, etc.) chosen as a reference solvent. The relative ionic hydration power of a solvent medium means the ability of the solvent to participate in the electrostatic interactions, e.g., ion-dipole interactions, with an ionic solute relative to that of pure (or octanol-saturated) water.

The approach [85] is based on the use of the partition technique. Partitioning of a homologous series of solutes with different aliphatic chain lengths in water-organic solvent and in aqueous polymer solution - organic solvent two-phase systems is examined. The results obtained are used to calculate the free energies of transfer of a CH_2 group and of ionic polar group from organic solvent-saturated water to the solvent-saturated aqueous solution of the polymer under study [85-88].

To illustrate the technique under consideration several typical relationships between the logarithm of the partition coefficient and the equivalent number of methylene groups [89] in the aliphatic chain of the compounds being partitioned in a given two-phase system formed by octanol and aqueous solution of a given polymer are shown in Figure 2.2.

The relationships observed are described as:

$$\ln P_{ow} = A + E \cdot n(CH_2) \tag{2.23}$$

where P_{ow} is the solute distribution coefficient defined as the ratio of the concentration of the solute in the organic phase to the solute concentration in the aqueous phase; $n(CH_2)$ is the equivalent number of CH_2 groups in the solute aliphatic chain [89] (see below); E is the slope of a linear plot of $\ln P_{OW}$ versus $n(CH_2)$ and it represents an average $\ln P_{ow}$ increment per methylene group; A is the intercept of the $\ln P_{ow}$ plot at $n(CH_2) = 0$, i.e., it represents the contribution of a polar group present in the solutes being partitioned into the $\ln P_{ow}$ value.

Note that [90]: a) E value depends on the properties of the coexisting phases but is independent of the particular homologous series of solutes employed; and b) A value depends on properties of the phases and on the nature of an ionic or non-ionic polar group present in the solutes being partitioned.

Parameter E, characterizing the CH_2 increment into $\ln P_{ow}$ is related to the free energy of transfer of a methylene group from the aqueous to nonaqueous phase:

$$\Delta G(CH_2)_{w(P)\to o} = -RT \cdot E \tag{2.24}$$

where $\Delta G(CH_2)_{w(P)\to o}$ is the free energy of transfer of a CH_2 group from organic solvent-saturated aqueous phase to the water-saturated organic phase; E as defined above; R universal gas constant; T temperature.

It should be noted that the similar expression for parameter A includes the difference in the states of the ionized and non-ionized species of the solute being partitioned in aqueous and organic phases. That is one of the major drawbacks of parameter A as a measure of the relative ionic hydration power of the aqueous medium. An additional assumption used in the technique [85-88] is that the solubility of octanol in water altered by the presence of the polymer additive does not affect the affinity of the aqueous phase for the groups in question. Octanol was chosen as the reference organic solvent and the aqueous solution of 0.15 M NaCl in 0.01 M sodium phosphate buffer, pH 7.4, was used as the reference aqueous medium [85-88]. It should be noted that the refraction measurements of the organic phase implied that the composition of the octanol phase is not changed by the presence of the polymers in the aqueous phase within the experimental error limits.

It is possible to calculate the free energy of transfer of a CH_2 group between the phases of the hypothetical two-phase system formed by an aqueous polymer solution and the same but polymer-free solution, $\Delta g(CH_2)$, according to:

$$\Delta g(CH_2) = \Delta G(CH_2)_{w(P)\rightarrow o} - \Delta G(CH_2)_{w\rightarrow o} \qquad (2.25)$$

where $\Delta G(CH_2)_{w(P)\rightarrow o}$ is the free energy of transfer of a CH_2 group from aqueous polymer-containing phase to octanol phase; $\Delta G(CH_2)_{w\rightarrow o}$ is the free energy of transfer of a CH_2 group from aqueous polymer-free phase to octanol phase (-618 cal/mole of CH_2 [85] when the ionic composition of the aqueous phase is as indicated above).

The $\Delta g(CH_2)$ used as a measure of the relative hydrophobic character of a polymer aqueous solution [85-88] was found to depend on the polymer concentration as shown in Fig.2.3 for polyvinylpyrrolidones (PVPs) of different molecular weights. Typical data presented in Figure 2.3 indicate that the relative hydrophobic character of an aqueous polymer solution depends on the molecular weight and concentration of the polymer. It also depends on the type of polymer.

The general trend in the solutions of the polymers examined up to the present [85-88] seems to be a decrease in the relative ionic hydration power of water in parallel with an increase of the relative hydrophobic character of the aqueous medium under increasing concentration of a non-ionic polymer additive. The molecular weight of a polymer affects its influence depending on the type of the polymer. In the case of polyvinyl alcohol (PVA) [87] the molecular weight of the polymer in the range from $2 \cdot 10^4$ to $1 \cdot 10^5$ does not affect the polymer influence on the water features under study. The effects of polyacrylamide and PEG increase [86], while that of PVP decreases [88] with increasing polymer molecular weight. The influence of PVA is affected by the per-

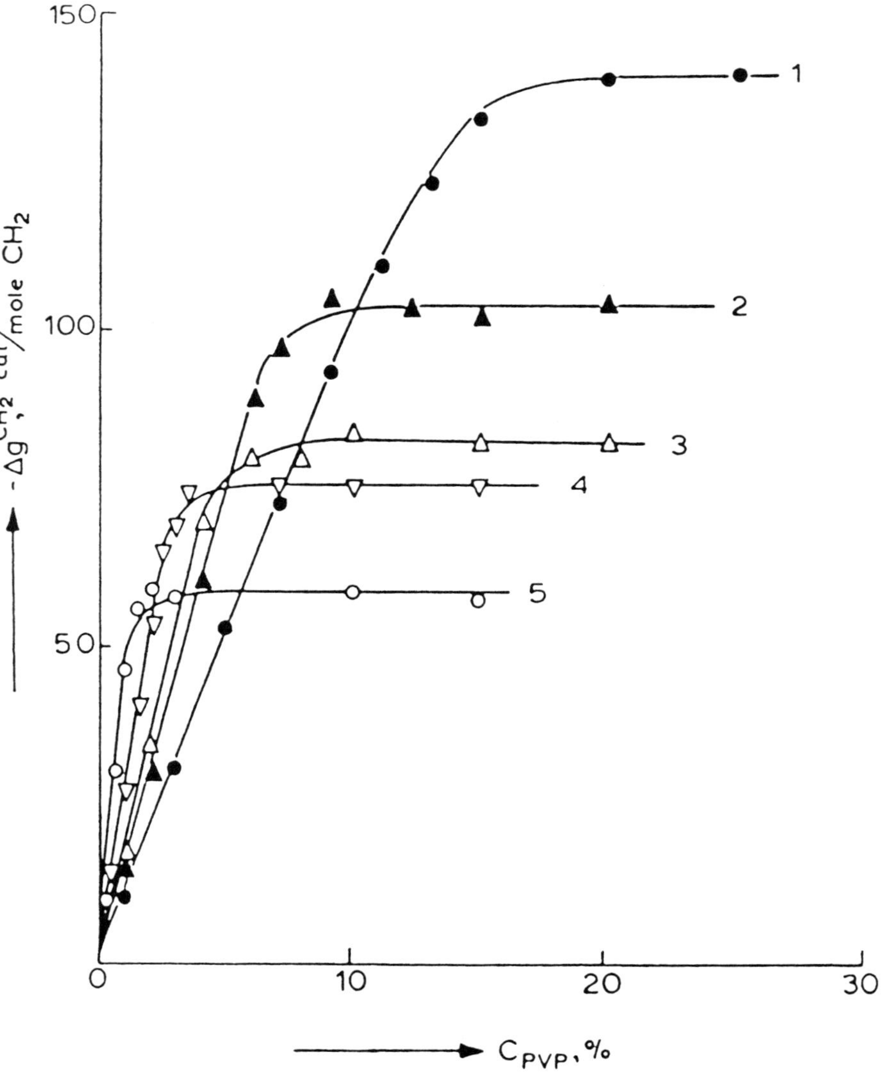

Figure 2.3. Free energy of transfer of a CH_2 group from polymer-free aqueous medium to solution of polyvinylpyrrolidone (PVP) as a function of the polymer concentration, C_{PVP}. Salt composition of the media - 0.15 M NaCl in 0.01 M sodium phosphate buffer, pH 7.4. Molecular weight, M_w, of the polymers: 1 - $5 \cdot 10^3$; 2 - $12 \cdot 10^3$; 3 - $17 \cdot 10^3$; 4 - $5 \cdot 10^4$; 5 - $1.8 \cdot 10^5$. Data from [88].

centage of acetate groups in the polymer (varied from 1% to 18%) [87] which clearly indicates that the effects under discussion are due to the specific polymer-water interactions.

As may be seen from Figure 2.3, the relative hydrophobic character of the aqueous polymer solutions attains a limit denoted as lim[$\Delta g(CH_2)$] specific for a given polymer. It seems that a given polymer may change the aqueous medium affinity for a non-polar solute or group, i.e., the relative hydrophobic character of the medium, only up to a certain limit. The polymer concentration at which this limit is achieved usually increases with increasing polymer molecular weight [85-88], while the lim[$\Delta g(CH_2)$]-value itself seems to be related to the polymer molecular weight in a more complicated manner.

The finding [85-88] that the relative affinity of the aqueous medium for a methylene group is usually increased in the presence of non-ionic polymers seems to be due to the structuring influence of these polymers on water. It was noted above that the low solubility of non-polar solutes in water is mainly due to the requirements of rearrangement of water molecules and enhancement of the water structure in the solute vicinity. It seems reasonable to imagine that the structuring action of a non-ionic polymer additive leads to such an intermolecular arrangement of water molecules in the solvent medium that the above demands are partially met. Hence the medium accommodates non-polar solutes or groups more readily than the polymer-free aqueous medium.

It follows from the essentially all the aforementioned experimental data that the aqueous solution of a polymer additive may be viewed as a particular solvent of the aqueous nature. This conclusion seems to be consistent with the notion [91] that pure water at different temperatures should be regarded as different solvents with their specific features. It should be emphasized that there is no single criterion for comparison of the polymer-containing aqueous solvents. An illustrative example offer the aqueous solutions of dextran which differ from pure water in the dielectric relaxation mobility of water molecules, the overall solvent polarity, etc. (see above) but manifest the relative affinity for a methylene group equal to that of the polymer-free aqueous medium. Hence, it should be kept in mind that a quantitative agreement between some characteristics' values for aqueous solutions of two different polymer additives is not to be taken as an evidence for the similar structures of water in these solutions.

It is possible, however, to construct a solvent scale combining organic solvents, water, and aqueous polymer solutions according to one or the other physicochemical feature of the solvent media.

One of the possibilities is provided by the above measurements of the solvent polarity in aqueous polymer solutions [82]. To scale the solvent polarities for different solvents it is necessary to combine the values determined with the use of different betaine dyes [92,93]. For instance, the carboxylate-substituted anionic betaine dye 1 has too low solubility in non-polar solvents, and its tert-butyl-substituted analog 2 is usually used in these cases [92,93]. The $E_T(2)$ values of dye 2 and those of dye 1 can be converted to the E_T of the nonsubstituted "parent" dye 30 [92] by linear transformation:

$$E_T(30) = [E_T(j) - a]/b \qquad (2.26)$$

where j denotes the betaine dye employed, and a and b are constants depending on the particular dye used.

To convert the $E_T(30)$ to a dimensionless unit, Reichardt [92] suggested the normalized E_T^N scale. This scale uses the polarity extremes of

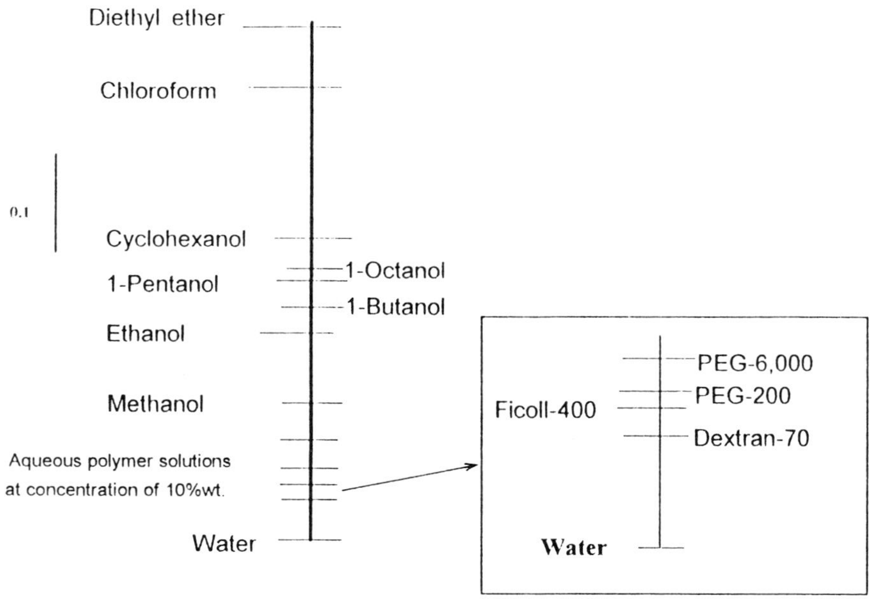

Figure 2.4. Relative solvent polarity scale for aqueous medium in polymer solutions at concentrations of 10%wt. (box on the right) and some organic solvents. Calculated from data in [82].

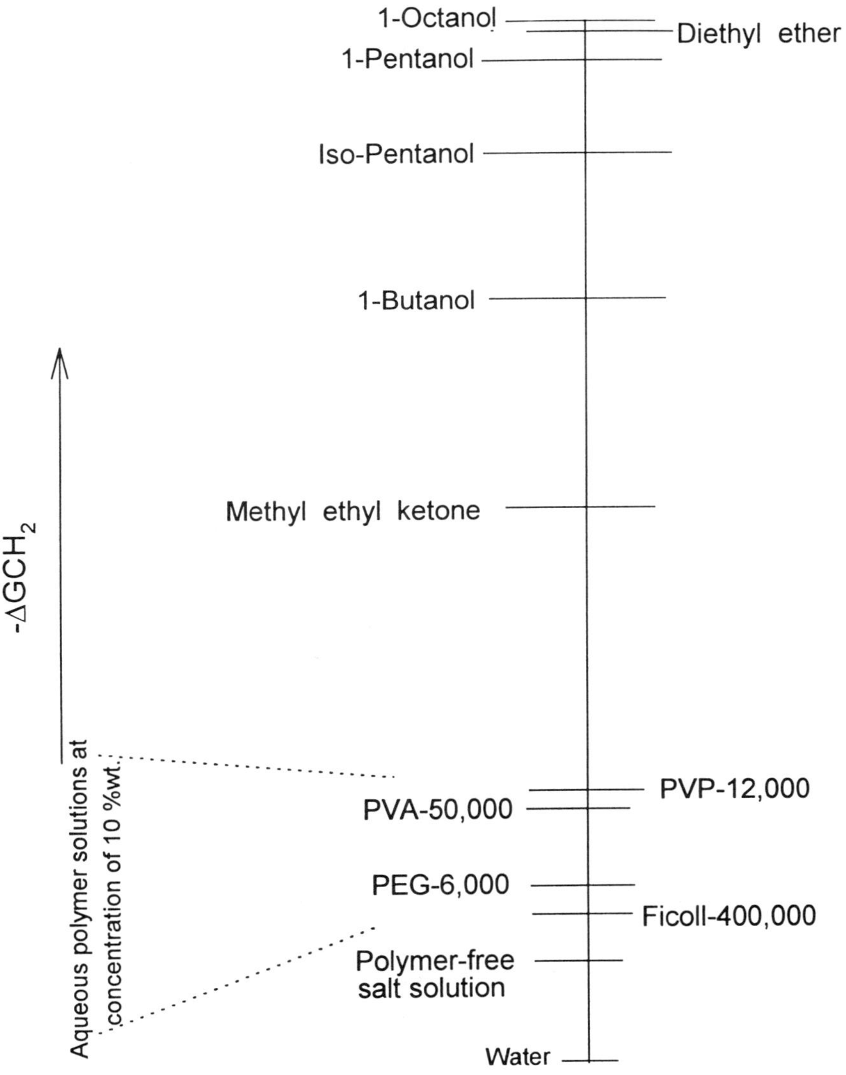

Figure 2.5. Relative solvent hydrophobicity scale for aqueous medium in polymer solutions at concentrations of 10%wt. and some organic solvents. Calculated from data in [85-88,94].

water and tetramethylsilane (TMS) as reference solvents:

$$E_T^N = [E_T(\text{test solvent}) - E_T(\text{TMS})]/[E_T(\text{water}) - E_T(\text{TMS})]$$

$$= [E_T(\text{test solvent}) - 30.7]/32.4 \qquad (2.27)$$

Therefore, non-polar solvents have an E_T^N value close to zero, and polar solvents have an E_T^N close to 1. Equation 2.27 was used to calculate the positions of the aqueous polymer solutions studied in [82] and organic solvents on the scale, as presented in Figure 2.4.

As may be seen in Fig. 2.4., the positions of the aqueous polymer solutions on the dimensionless solvent polarity scale are close to that of pure water. An enlarged view of the scale for the aqueous solutions of several polymers at the concentration of 10 wt% on the right indicates the positions of the solutions. These positions, i.e., the solvent polarity of the aqueous polymer solutions, depend on the type, molecular weight, and concentration of a polymer.

The relative hydrophobic character, i.e., the relative affinity of the solvent medium for a non-polar CH_2 group, was also used to scale the solvent properties of different organic solvents [94,95] and various aqueous polymer solutions [84-88,96]. The free energy of transfer of a CH_2 group from pure water to a given organic solvent or to an aqueous polymer solution was used to calculate the positions of these solutions and organic solvents on the scale, as shown in Figure 2.5.

As indicated above, the relative hydrophobic character of an aqueous polymer solution depends on the polymer concentration (similar to the solvent polarity of the solution). To compare the scale under consideration with the one in Fig. 2.4 the positions of the polymer solutions were calculated for those at the polymer concentration of 10 wt%. The enlarged view of the scale for the solutions in question is given on the right. It can be seen that (1) on both scales the positions of the polymer solutions are different but close to that of pure water in comparison with those of even highly polar organic solvents, such as methanol, ethanol, etc., and (2) the relative hydrophobic character of the aqueous polymer solutions (at 10 wt% concentration) is increased in the order of decreasing the solvent polarity of the medium in the same solutions.

Taking into account all the above experimental facts, it seems reasonable to suggest that aqueous polymer solutions should be considered as different solvents of the same aqueous nature, particularly, in view of their often observed partial immiscibility - the phenomenon discussed in the next section.

2.4. SUMMARY

Being dependent on relatively weak forces, the water structure is easily perturbed by temperature and solutes of various kinds. The spatial arrangement of water molecules and the distribution of the energies of interactions between water molecules (covered by the term of the water structure) respond very differently to the type and concentration of soluble additive. The influence of additives on dielectric permittivity and ionization of water, its solvent polarity and viscosity, heat capacity and thermodynamic activity, etc., imply the long range effects of additives, and enables one to regard aqueous solutions of different additives or even those of the same additive at different concentrations as different water-like solvents of different solvent features. This view seems to be of particular importance for aqueous polymer solutions which according to some experimental evidence may be regarded as lyotropic liquid crystalline systems.

Finally, to quote F.Franks [97, pp.13-14]: "Molecular interactions in aqueous solution are seen to be long range effects, modulated by the hydration shells of the two interaction partners. At the present time there are no theories which would make possible a detailed specification of these hydration shells. On the other hand, any theory which purports to account for interactions between molecules without consideration of the molecular nature of the solvent is suspect."

Bearing this in mind, it is possible to consider the numerous experimental facts and theoretical treatments of phase separation in aqueous polymer systems.

REFERENCES:

1. J. H. Hildebrand and R. L. Scott, The Solubility of Non-Electrolytes, 3rd ed., Dover Publications, New York, 1964.
2. K. C. James, Solubility and Related Properties, Marcel Dekker, New York, 1986.
3. P. J. Flory, Principles of Polymer Chemistry, Cornell University Press, Ithaca, New York, 1953.
4. D. Patterson, Polymer Eng.Sci., 22 (2), 64 (1982).
5. I. Prigogine, The Molecular Theory of Solutions, North-Holland, Amsterdam, 1967.
6. P. J. Flory, Disc.Faraday Soc., 49, 7 (1970).
7. D. Eisenberg, D. Crothers, Physical Chemistry with Applications to the Life Sciences, Benjamin/Cummings Publ.Co., London, 1979.

8. L. A. Girifalco and R. J. Good, J.Phys.Chem., 61, 904 (1957).
9. R. J. Good and L. A. Girifalco, J.Phys.Chem., 64, 561 (1960).
10. C. J. van Oss, M. K. Chaudhury, R. J. Good, Adv. Colloid Interface Sci., 28, 35 (1987).
11. C. J. van Oss, R. J. Good, M. K. Chaudhury, Langmuir, 4, 884 (1988).
12. C. J. van Oss and R. J. Good, J.Macromol.Sci.-Chem., A26, 1183 (1989).
13. C. J. van Oss, R. J. Good, M. K. Chaudhury, Sep.Sci. Technol., 22, 1515 (1987).
14. C. J. van Oss, M. K. Chaudhury, R. J. Good, Chem.Rev., 88, 927 (1988).
15. C. J. van Oss, R. J. Good, H. J. Busscher, J.Dispersion Sci. Technol., 11, 75 (1990).
16. C. J. van Oss, K Arnold, R. J. Good, K. Gawrisch, S. Ohki, J.Macromol.Sci.-Chem., A27, 563 (1990).
17. W. Drost-Hansen, In: Bennett, Bryant, Hulbert (eds.), Microstructure of Fine-Grained Sediments: From Mud to Shales, Springer Verlag, Berlin, 1990, p.259.
18. F. M. Etzler and D. M. Fagundus, J.Colloid Interface Sci., 93, 585 (1983).
19. F. M. Etzler and D. M. Fagundus, J.Colloid Interface Sci., 115, 513 (1987).
20. C. V. Braun and W. Drost-Hansen, In: M. Kerker (ed.), Colloid and Interface Science, Academic Press, New York, v.3, 1976, pp.533-541.
21. F. M. Etzler, Langmuir, 4, 878 (1988).
22. W. Drost-Hansen, In: F. Franks and S. Mathias (eds.), Biophysics of Water, Wiley, New York, 1982, pp.163-169.
23. W. Drost-Hansen, L. Singleton, In: Foundations of Medical Cell Biology, (E. E. Bittar, ed.), JAI Press, Greenwich, Conn, Vol. 3A, 1991, pp.157-180.
24. J. S. Clegg, W. Drost-Hansen, In: The Biochemistry and Molecular Biology of Fishes (Eds. P. W. Hochachka, M. T. P. Mommsen), Vol.1, Elsevier Science Publishers, Amsterdam, 1991, pp. 1-23.
25. W. Drost-Hansen, In: H. D. Brown (ed.), Chemistry of the Cell Interfaces, Academic Press, New York, Vol.B, 1971, pp.1-184.
26. G. Peschel and K. H. Adlfinger, J. Colloid Interface Sci., 34, 505 (1970).

27. J. A. Schufle, C. T. Huang, W. Drost-Hansen, J. Colloid Interface Sci., 54, 184 (1976).
28. W. Alpers and H. Huhnerfuss, J.Phys.Chem., 87, 5251 (1983).
29. W. Drost-Hansen, Ind.Eng.Chem., 61, 10 (1969).
30. W. Drost-Hansen, Annals New York Acad. Sci., 125, 471 (1965).
31. F. M. Etzler and W. Drost-Hansen, Croatica Chem.Acta, 56, 563 (1983).
32. F. M. Etzler and W. Drost-Hansen, In: M. Blank (ed.), Bioelectrochemistry: Ions, Surfaces, Membranes, Advances Chem.Series, # 188, Am.Chem.Soc., Washington, 1980, pp.485-497.
33. W. Drost-Hansen, J.Geophys.Res., 77, 5132 (1972).
34. C. Ballario, A. Bonincontro, C. Cametti, J. Colloid Interface Sci., 54, 415 (1975).
35. J. S. Clegg and W. Drost-Hansen, In: L. Taylor, A. Y. Cheung (eds.), The Physical Basis of Electromagnetic Interactions with Biological Systems, Institute for Physical Sciences and Technology and School of Medicine: University of Maryland, 1977, p.121.
36. J. Clifford, In: F. Franks (ed.), Water - A Comprehensive Treatise, Vol.5, Plenum, New York, 1975, pp.75-132.
37. R. P. Rand and V. A. Parsegian, Biochim.Biophys.Acta, 988, 351 (1989).
38. I. D. Robb, In: Chemistry and Technology of Water-Soluble Polymers (ed. C. A. Finch), Plenum Press, New York, 1983, pp.193-202.
39. J. Van Steveninck, M. Paardekooper, T. Dubbelman, E. Ben-Hur, Biochim.Biophys.Acta, 1115, 96 (1991).
40. M. Janado, R. Nakayama, Y. Yano, H. Nakamori, J.Biochem., 86, 795 (1979).
41. M. Janado, K. Takanaka, H. Nakamori, Y. Yano, J.Biochem., 87, 57 (1980).
42. P. M. Wiggins, R. T. van Ryn, J.Biophys., 58, 585 (1990).
43. P. M. Wiggins, R. T. van Ryn, D. G. C. Ormrod, Biophys.J., 60, 8 (1991).
44. P. Molyneux, Water-Soluble Synthetic Polymers: Properties and Behavior, Vols.1 & 2, CRC Press, Boca Raton, Florida, 1983.
45. P. Molyneux, Water-Soluble Synthetic Polymers. Update 1, Macrophile Associates, London, 1987.
46. F. Franks, Pure & Appl.Chem., 59, 1189 (1987).
47. K. Gekko, In: Ions and Molecules in Solution (eds. N. Tanaka, H. Ohtaki, R. Tamamushi), Elsevier, Amsterdam, 1983, pp.339-358.
48. A. Hvidt, Ann. Rev. Biophys. Bioeng., 12, 1 (1983).

49. M. J. Blandamer, Adv.Phys.Org.Chem., 14, 204 (1977).
50. M. Aizawa, S. Suzuki, T. Kuoka, N. Nakajima, Y. Iwao, Bull. Chem.Soc.Jpn., 49, 2061 (1976).
51. P. Molyneux, In: Water. A Comprehensive Treatise., F. Franks (ed.), Plenum Press, New York, Vol.4, 1975, pp.569-801.
52. A. C. R. Antonini, M. J. Blandamer, J. Burgess, A. W. Hakin, N. D. Hall, A. H. Blandamer, J.Chem.Soc., Faraday Trans.I, 84, 1889 (1988).
53. V. V. Mank, I. M. Solomentseva, A. A. Baran, O. D. Kurilenko, Ukrainian Chim.Zh.(Rus), 40, 28 (1974).
54. S. Saeki, N. Kuwahara, M. Nakata, M. Kaneko, Polymer, 17, 685 (1976).
55. R. Kjellander, E. Florin, J.Chem.Soc., Faraday Trans.I, 77, 2053 (1981).
56. F. Bordi, C. Cametti, A. DiBiasio, J.Phys.Chem., 92, 4772 (1978).
57. M. J. Hey, S. M. Ilett, J.Chem.Soc. Faraday Trans., 87, 3671 (1991).
58. U. Kaatze, O. Gottman, R. Podbielski, R. Pottel, U. Terveer, J.Phys.Chem., 82, 112 (1978).
59. T. de Vringer, J. G. H. Joosten, H. E. Junginger, Colloid Polymer Sci., 264, 623 (1986).
60. Y. Miyazaki, H. Matsuura, Bull.Chem.Soc.Jpn., 64, 288 (1991).
61. W. Burchard, In: Chemistry and Technology of Water-Soluble Polymers (editor C. A. Finch), Plenum Press, New York, 1983, pp.125-142.
62. S. Saeki, N. Kuwahara, M. Nakata, M. Kaneko, Polymer, 18, 1027 (1977).
63. B. Y. Zaslavsky, T. O. Bagirov, A. A. Borovskaya, N. D. Gulaeva, L. M. Miheeva, A. U. Mahmudov, M. N. Rodnikova, Polymer, 30, 2104 (1989).
64. E. Florin, R. Kjellander, J. C. Eriksson, J.Chem.Soc., Faraday Trans.I, 80, 2889 (1984).
65. M. J. Garvey, I. D. Robb, J.Chem.Soc., Faraday Trans.I, 75, 993 (1979).
66. A. Guner, O. Guven, Makromol.Chem., 179, 2789 (1978).
67. V. A. Scherbakov, Y. G. Monin, Doklady Acad.Nauk USSR (Rus), 300, 1412 (1989).
68. A. A. Vedenov, Physics of Solutions, Nauka, Moscow, 1984.
69. E. Florin, Macromolecules, 18, 360 (1985).
70. J. Breen, L. Huis, J. de Bleijser, J. C. Leyte, Ber.Bunsenges. Phys.Chem., 92, 160 (1988).

71. M. J. Taylor, Cryo-Letters, 2, 231 (1981).
72. J. E. Gordon, J.Am.Chem.Soc., 94, 650 (1972).
73. A. S. Chernyak, M. L. Schepotko, A. K. Lyaschenko, D. B.Poblinkov, Doklady Acad.Nauk USSR (Rus), 254, 377 (1980).
74. B. Y. Zaslavsky, L. M. Miheeva, N. D. Gulaeva, A. A. Borovskaya, M. I. Rubtsov, L. L. Lukatskaya, N. O. Mchedlov-Petrossyan, J.Chem.Soc. Faraday Trans., 87, 931 (1991).
75. U. Kaatze, Adv. Mol. Relaxation Processes, 7, 71 (1975).
76. G. Masszi, L. Koszorus, T. Lakatos, Acta Biochim. Biophys. Hung., 21, 263 (1986).
77. B. Y. Zaslavsky, L. M. Miheeva, M. N. Rodnikova, G. V. Spivak, V. S. Harkin, A. U. Mahmudov, J.Chem.Soc., Faraday Trans.I, 85, 2857 (1989).
78. A. Herrmann, L. Pratsch, K. Arnold, C. Lassmann, Biochim. Biophys.Acta, 738, 87 (1983).
79. K. Arnold, A. Herrmann, L. Pratsch, K. Gawrisch, Biochim. Biophys. Acta, 815, 515 (1985).
80. J. B. Hasted, Aqueous Dielectrics, Chapman & Hall, London, 1973, pp.136-175.
81. K. Arnold, L. Pratsch, K. Gawrisch, Biochim.Biophys. Acta, 728, 121 (1983).
82. B. Y. Zaslavsky, L. M. Miheeva, E. A. Masimov, S. F. Djafarov, C. Reichardt, J.Chem.Soc. Faraday Trans., 86, 519 (1990).
83. C. P. S. Tilcock, D. Fisher, Biochim.Biophys.Acta, 577, 53 (1979)
84. B. Y. Zaslavsky, Anal.Chem., 64, 765A (1992).
85. B. Y. Zaslavsky, E. A. Masimov, L. M. Miheeva, S. V. Rogozhin, D. P. Hasaev, Doklady Acad.Nauk USSR (Rus), 261, 669 (1981).
86. E. A. Masimov, B. Y. Zaslavsky, A. A. Gasanov, S. V. Rogozhin, J.Chromatogr., 284, 337 (1984).
87. E. A. Masimov, B. Y. Zaslavsky, A. A. Gasanov, Y. A. Davidovich, S. V. Rogozhin, J.Chromatogr., 284, 349 (1984).
88. B. Y. Zaslavsky, E. A. Masimov, A. A. Gasanov, S. V. Rogozhin, J.Chromatogr., 294, 261 (1984).
89. B. Y. Zaslavsky, N. M. Mestechkina, L. M. Miheeva, S. V. Rogozhin, J.Chromatogr., 240, 21 (1982).
90. B. Y. Zaslavsky, E. A. Masimov, Topics Current Chem., 146, 171 (1988).
91. V. K. Abrosimov, In: Current Problems of Solution Chemistry (ed. B. D. Beresin) (Rus), Nauka, Moscow, 1986, pp.97-156.
92. C. Reichardt, Solvents and Solvent Effects in Organic Chemistry, 2nd ed., Verlag Chemie, New York, 1986.

93. S. J. Gluck, M. P. Wingeier, J.Chromatogr., 547, 69 (1991).
94. S. S. Davis, T. Higuchi, J. H. Rytting, Adv.Pharm.Sci. (H. S. Bean, ed.) Acad.Press, 73 (1974).
95. R. F. Rekker, The Hydrophobic Fragmental Constant: Its Derivation and Application. A Means of Characterizing Membrane Systems, Amsterdam, Elsevier, 1977.
96. B. Y. Zaslavsky, L. M. Miheeva, S. V. Rogozhin, J.Chromatogr., 212, 13 (1981).
97. F. Franks, Biophysics and Biochemistry at Low Temperatures, University Press, Cambridge, 1985

CHAPTER 3.

PHASE SEPARATION IN AQUEOUS POLYMER SYSTEMS: EXPERIMENTAL FACTS AND THEORETICAL MODELS

When two particular chemically different polymers (e.g., dextran (Dex) and poly(ethylene glycol) (PEG)) or one polymer and a specific salt (e.g., PEG and sodium phosphate) are mixed at certain concentrations in an aqueous solution, the solution separates into two immiscible phases [1]. One phase is rich in one polymer, and the second phase is rich in the other polymer (or salt) with water as a solvent in both phases. Pairs of polymers capable of phase separation in water are listed in Table 3.1, and new polymer combinations are constantly being introduced.

Phase separation may also occur in aqueous and nonaqueous single polymer solutions above or below a certain temperature, in polymer mixtures in nonaqueous solvents, and in polymer mixtures without any solvent at all [2-6]. The incompatibility of certain high polymers toward each other, clearly important for polymer composite materials, can be easily recognized, e.g., by that films obtained from mixtures of such polymers are not homogeneous but turbid or opaque and possess mechanical properties inferior to those of films obtained from the separate constituents. This phenomenon has been recognized as a characteristic property of macromolecules and has received wide attraction of polymer chemists. Since most of the synthetic polymers are soluble in organic

Table 3.1.

Polymer Systems Capable of Phase Separation in Aqueous Media [a]

Component 1	Component 2
1. Nonionic polymer (P) - Nonionic polymer (Q) - Water	
Polypropylene glycol	Methoxypolyethylene glycol
	Polyethylene glycol
	Polyvinyl alcohol
	Hydroxypropyldextran
	Dextran
Polyethylene glycol	Polyvinyl alcohol
	Polyvinylpyrrolidone
	Dextran
	Arabinogalactan [b]
	Hydroxypropyl starch [c]
	Ficoll
Polyvinyl alcohol	Methylcellulose
	Hydroxypropyldextran
	Dextran
Polyvinylpyrrolidone	Methylcellulose
	Maltodextrin [d]
	Dextran
Methylcellulose	Hydroxypropyldextran
	Dextran
Ethylhydroxyethylcellulose	Dextran
Hydroxypropyldextran	Dextran
Ficoll	Dextran
2. Polyelectrolyte (P) - Nonionic polymer (Q) - Water	
Na dextran sulfate	Polypropylene glycol
	Methoxypolyethylene glycol NaCl
	Polyethylene glycol NaCl
	Polyvinyl alcohol NaCl
	Polyvinylpyrrolidone NaCl
	Methylcellulose NaCl
	Ethylhydroxyethylcellulose NaCl
	Hydroxypropyldextran NaCl
	Dextran NaCl
DEAE dextran HCl	Polypropylene glycol NaCl
	Polyethylene glycol Li_2SO_4
	Polyvinyl alcohol
	Methylcellulose

Component 1	Component 2
Casein [e]	Dextran
	Pectin
	Ficoll
	Amilopectin
Na carboxymethyldextran	Methoxypolyethylene glycol NaCl
	Polyethylene glycol NaCl
	Polyvinyl alcohol NaCl
	Polyvinylpyrrolidone NaCl
	Methylcellulose NaCl
	Ethylhydroxyethylcellulose NaCl
	Hydroxypropyldextran NaCl
Na carboxymethylcellulose	Polypropylene glycol NaCl
	Methoxypolyethylene glycol NaCl
	Polyethylene glycol NaCl
	Polyvinyl alcohol NaCl
	Polyvinylpyrrolidone NaCl
	Methylcellulose NaCl
	Ethylhydroxyethylcellulose NaCl
	Hydroxypropyldextran NaCl
3. Polyelectrolyte (P) - Polyelectrolyte (Q) - Water	
Na dextran sulfate	Na carboxymethyldextran
	DEAE dextran HCl NaCl
	Na carboxymethylcellulose
Na carboxymethyldextran	Na carboxymethylcellulose
Casein [e]	Sodium alginate, 0.1 M NaOH
	Na carboxymethylcellulose, 0.1M NaOH
Ovalbumin (pH 6.6) [e]	Soybean globulins
	Ovalbumin thermotropic aggregates
	Casein
4. Polymer (P) - Low Molecular Weight Component (Q) - Water	
Polypropylene glycol	Potassium phosphate
	Glycerol
Methoxypolyethylene glycol	Potassium phosphate
Polyethylene glycol	Inorganic salts, e.g., K^+ (Na^+, Li^+, $(NH_4)^+$, etc.) PO_4^{3-}, SO_4^{2-}, etc.[f]
	Glucose, maltose, cellobiose, iso-maltose, maltotriose, iso-maltotriose, ß-cyclodextrin [g]
Polyvinylpyrrolidone	Butylcellosolve
	Potassium phosphate

Table 3.1 Continued.

Component 1	Component 2
Polyvinyl alcohol	Butylcellosolve
Dextran	Butylcellosolve
	Propyl alcohol, Iso-propyl alcohol
Na dextran sulfate	Sodium chloride (0°C)

[a] All pairs except noticed specifically are taken from [1];
[b] Zaslavsky et al., unpublished data; [c] from [7]; [d] from [8]; [e] from [9]; [f] from [10].

solvents, phase separation in polymer mixtures in a common solvent has been studied mostly in nonaqueous systems.

The main types of these systems according to the classification given by Patterson [5] include: a) polymer + polymer; b) polymer + solvent (polymer solution); c) two highly compatible polymers + solvent; and d) two miscible polymers + solvent. The recent theoretical views on the systems of the a-type are outlined in [6]. The concepts developed for the non-polar solvent-containing polymer systems will be briefly described below as many attempts (to be discussed later) to use these concepts for aqueous polymer systems have been reported recently.

The aim of the present chapter is to discuss the likely mechanisms of phase separation in aqueous polymer systems. I believe that the water structure governs phase separation in these systems. The common way to elucidate the role of water in any physico-chemical process is to monitor the effects of water structure-perturbing factors (temperature, urea, inorganic salts, etc.) on the process in question. Therefore the influence of these factors on phase separation in aqueous polymer systems will be considered in detail. Phase diagrams used for description of two-phase systems are outlined first. After that aqueous systems containing a single polymer are considered and the theoretical models of phase separation in such systems are discussed. Aqueous mixtures of two different polymers and the current theoretical treatments of phase separation in these systems are then considered.

3.1. PHASE DIAGRAMS

The composition of aqueous polymer two-phase systems may be conveniently represented on a triangular phase diagram as illustrated in Figure 3.1. Polymers, Ficoll and dextran, and solvent, water, are represented by the apexes of the equilateral triangle in the diagram. Points within the triangular area represent mixtures of the three components. The percentage of each component in a mixture is read along a line perpendicular to the side opposite the corresponding apex. The sum of the coordinates of any point on the diagram is always 100%. The composition is usually specified in weight percents but any convenient concentration unit may be used.

Most aqueous polymer systems to be discussed below conform to phase diagram of the type illustrated in Fig. 3.1. This is a 'simple' type consisting of a single binodial curve.

All mixtures of the compositions represented by points under the binodial line give rise to phase separation. The mixtures represented by points above the binodial line give an apparently homogeneous one-phase system. To understand the diagram, consider what happens when amounts p of Ficoll, q of dextran and s of water are mixed. The total composition of the mixture is represented by the point A under the binodial line, i.e., the mixture separates

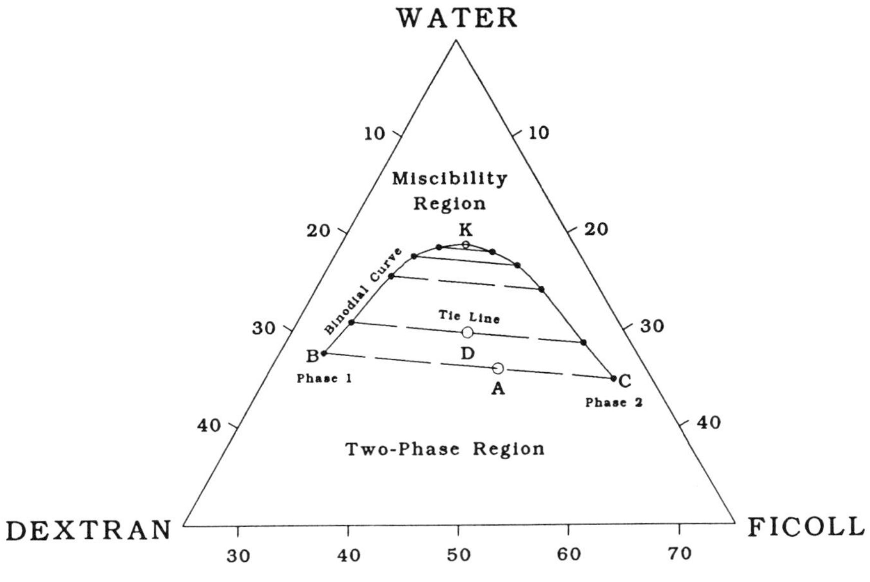

Figure 3.1. Triangular phase diagram for dextran-70 — Ficoll-400 — water system.

into two phases. The compositions of these two phases are represented by the points B and C called nodes and located at the binodial line. The line joining the points B and C representing the compositions of the coexisting phases is called a tie line. The point A representing the total mixture must be positioned on the same tie line as the nodes B and C characterizing the compositions of the coexisting phases originated from this mixture.

It should be particularly noted that mixtures of different total compositions represented by different points on the same tie line give rise to two-phase systems with the identical compositions though different volumes of the coexisting phases. This is similar to a pair of two immiscible solvents when any

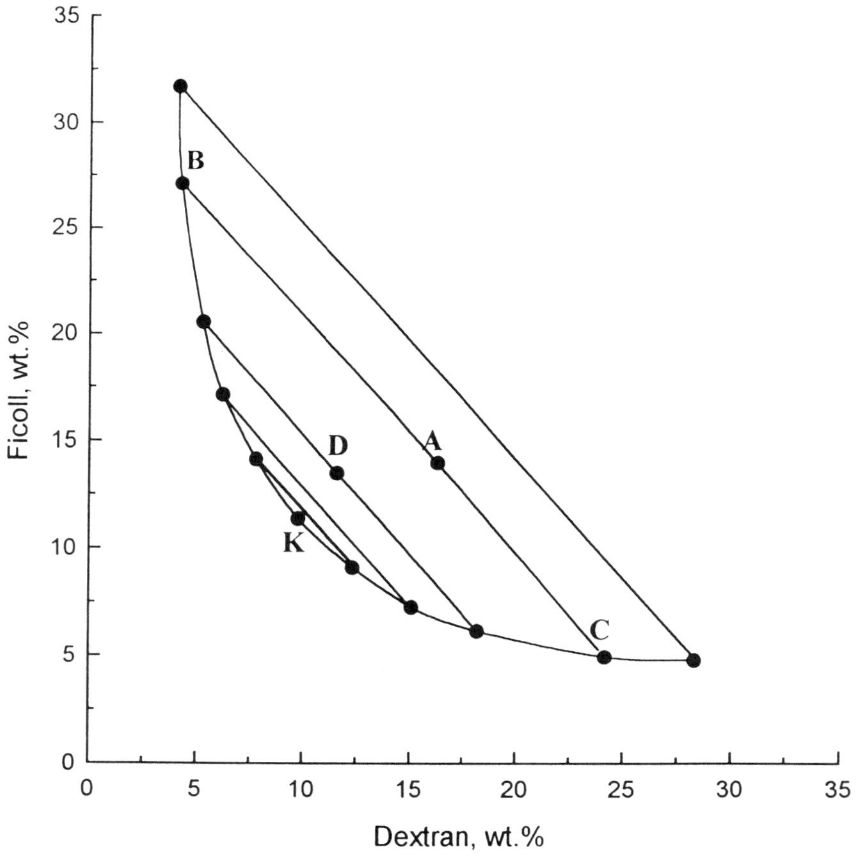

Figure 3.2. Rectangular phase diagram for Dex-70 — Ficoll-400 — water system at 23°C.

variation of the amount of each solvent is followed by a change in the relative volumes of the two phases but does not affect their compositions. Compositions of the phases can be changed only if one of the solvents is replaced by another solvent or if a third component, the so-called modifier, is added [11]. In the ternary (polymer-polymer-solvent or polymer-salt-solvent) aqueous polymer systems the same result, i.e., change in the compositions of the two phases, may be achieved by changing the total composition of a mixture in such a way that it is represented by the point, e.g., D, on any other tie line.

If we consider successive tie lines decreasing in length we encounter the point K at which two corresponding nodes coincide; this point is called the critical point. The critical point corresponds to the theoretical case in which the compositions and the volumes of the coexisting phases are equal. Position of the critical point (or the overall composition of a system corresponding to the critical point) is an important characteristic of the phase diagram of a polymer system [4]. An additional important characteristic of the phase diagram is supposed to be the so-called threshold point which is the point where the binodial line is tangential to the straight line cutting off equal segments on the axes of the phase diagram [4]. If the critical and the threshold points coincide the phase diagram is called symmetrical, the more different the positions of these points are, the more asymmetrical the phase diagram is.

In the literature on aqueous two-phase systems the rectangular form of a phase diagram is commonly used [1]. An illustration is given in Figure 3.2. The vertical axis is commonly used for the polymer which is enriched in the top phase. The difference between phase diagrams in Figures 3.1 and 3.2 is that in the latter the solvent concentration is omitted.

The best way to construct a phase diagram is to determine compositions of the coexisting phases of a series of systems in which the polymer concentrations are varied. Different methods for analysis of phase diagrams and the technical procedures employed are described in [12]. Much less precise method worth of notice is the cloud point determination. This method consists of adding an aqueous polymer (or salt) solution to the other polymer solution dropwise and finding the composition of the mixture at which turbidity appears. The first appearance of turbidity means that the system is about to enter the two-phase area. Its composition corresponds to the so-called cloud point.

The critical point is usually found by drawing a curve through the midpoints of a set of tie lines and extrapolating it to the point of intersection with the binodial.

When a material is partitioned in a given aqueous two-phase system under varied total composition of the system it is clearly necessary to use a single numerical measure of the compositions of the two phases between which it is partitioned. It was empirically found [13] that the tie line length (TLL) may be conveniently used as such a measure. The TLL value can be calculated from the polymer (salt) concentrations in the coexisting phases according to:

$$\text{TLL} = \{[C(P)_1 - C(P)_2]^2 + [C(Q)_1 - C(Q)_2]^2\}^{0.5} =$$
$$= \{[\Delta C(P)]^2 + [\Delta C(Q)]^2\}^{0.5} \quad (3.1)$$

where C(P) is the concentration of a polymer P in a given phase; C(Q) is the concentration of a polymer (or salt) Q in a given phase; subscripts 1 and 2 denote the coexisting phases; $\Delta C(P)$ and $\Delta C(Q)$ are the differences in the concentrations of the corresponding component P or Q between the two phases.

It has been recently found [14] that the slope of the tie line (STL) is also an important characteristic of a phase diagram (see below). The STL value is calculated as:

$$\text{STL} = \Delta C(P)/\Delta C(Q) \quad (3.2)$$

where $\Delta C(P)$ and $\Delta C(Q)$ are as defined above.

Analysis of the phase diagrams reported in the literature (see, for example, in [1]) implies that the STL value of an aqueous polymer system is usually constant, i.e. tie lines are parallel to each other (see typical data in Table 3.2).

It can be seen from the data in Table 3.2 that the STL value for a given phase diagram is constant within the experimental error limit.

It should be noted that in many nonaqueous polymer systems as well as in solvent two-phase systems the tie lines are generally nonparallel to each other in contrast to aqueous polymer systems. In my personal experience, averaging the STL values followed by checking out and repeating determination of the phase compositions in the cases away from the average STL value is probably the best way of obtaining reliable phase diagrams.

It should be emphasized, however, that the precise phase diagram data are not always necessary. It is usually the case when the factors influencing phase separation or partitioning of solutes (or particles) in a system under varied conditions are examined. For essentially all the other practical purposes, e.g., separation or isolation of biological materials, the exact knowledge of the polymer (salt) compositions of the phases is usually not required (see below).

It should be noted additionally that synthetic polymers are usually polydisperse and their molecular weight distributions may vary from lot to lot even when obtained from the same manufacturer. The positions of the binodials of the phase diagrams for the systems formed by different polymer lots differ accordingly. An example of the difference between the compositions of the phases of aqueous dextran (Dex) - PEG two-phase systems formed by different lots of the polymers is shown in Table 3.2 (compare the polymer concentrations in the two phases of the systems in which the concentration of PEG in the

Table 3.2.

Polymer Composition of the Coexisting Phases in Aqueous Dextran (M_w 40000) — PEG (M_w 20000) Systems at 20°C [a]

Bottom Phase		Top Phase		STL [b]	Ref.
PEG %wt.	Dex %wt.	PEG %wt.	Dex %wt.		
0.82	11.24	5.45	2.96	-0.559	[14]
0.51	14.46	7.35	1.58	-0.531	
- [c]	18.40	9.95	0.79	-0.565	
- [c]	30.84	18.06	0.10	-0.588	
			av.:	-0.558 ± 0.022	
2.10	9.57	7.52	1.79	-0.697	[16]
1.28	13.66	9.19	1.23	-0.636	
0.80	16.75	10.75	0.82	-0.625	
0.50	20.05	12.41	0.52	-0.610	
			av.:	-0.642 ± 0.038	
0.5	13.5	6.5	1.7	-0.508	[17]
0.4	16.9	8.8	0.9	-0.525	
0.4	19.2	10.4	0.5	-0.535	
0.3	22.8	13.0	0.3	-0.564	
			av.:	-0.533 ± 0.023	

[a] Experimental errors vary from 0.05 to 0.35 %wt.;
[b] STL is calculated according to: STL = ΔC(Peg)/ΔC(Dex);
[c] No concentration is given.

bottom phase is about 0.5 %wt.).

It is clear therefore that the position of the binodial for a given polymer system cannot be viewed as a physico-chemical constant of the system formed by a given mixture of polymers in a given solvent. This seems to be the major reason for that qualitative description of the basic features of phase separation phenomena provided by the theoretical models discussed below is usually supposed to be adequate for mixtures of polymers in a common solvent.

3.2. PHASE SEPARATION IN AQUEOUS SINGLE POLYMER SYSTEMS

Aqueous single polymer systems discussed here include solutions of a single polymer in pure water (binary systems) and solutions of a single polymer in water containing a low molecular weight additive (inorganic salt, glucose, etc.), i.e., ternary systems. The only difference of the latter from the two-polymer systems is that the third component is of low molecular weight.

It was mentioned above that at elevated or lowered temperatures aqueous solutions of certain polymers separate into two phases. One phase is relatively rich in the polymer, and the other phase is poor in the polymer. This temperature-dependent phase separation behavior of the 'closed loop' type is very similar to those observed in mixtures of two different solvents, such as water and 2-butanol [18], glycerol and m-toluidine (see, for example, in [19] and references cited therein), etc. Qualitatively this behavior is observed in the following manner.

A homogeneous single-phase liquid mixture (composed of, e.g., A and B molecules) at high temperature separates into two coexisting A-rich and B-rich phases as the temperature is lowered through the phase transition point, T_c. As the temperature is lowered further from inside the separated region another transition into the miscible single-phase system occurs at a lower critical point. Some of the mixtures, e.g., water and 2-butanol, exhibit not only upper and lower critical solution points, but also manifest another phase separation below the closed loop at even lower critical temperature. Systems of this type are beyond the scope of our discussion, however.

Walker and Vause [19] has attempted a quantitative treatment of phase separation in binary solvent mixtures on the basis of a generalized lattice model. In addition to the usual van der Waals interactions, the model includes energetically favorable highly directionally dependent short-ranged interactions (such as H-bonding) between unlike molecules. The high directionality of the bonding implies that the interaction is entropically disadvantageous as a large number of relative orientational states are rejected when a bond is formed [19]. Thus, the mixture exists as a single phase at high temperatures, since this state is entropically more favored (provided no bonding occurs). As temperature is lowered, the attractive van der Waals interactions between the like molecules are assumed [19] to dominate, and the system separates into two coexisting phases. At still lower temperatures the entropic disadvantage of the orientational bonding is supposed [19] to be overcome, and the system becomes homogeneous again so that bonding between unlike molecules can take place. This pattern suggested by Walker and Vause [19] is very similar to the one advanced by Kjellander and Florin [20] to explain temperature-dependent phase separation in aqueous solutions of PEG.

According to the model proposed by Kjellander and Florin [20] the entropically unfavorable structuring of water produced by PEG at low temperatures is overcome due to the large decrease in enthalpy. At higher temperatures, provided that the structure of water in the PEG hydration shell does not break down too rapidly with increasing temperature, the disadvantageous entropy contribution dominates and the system phase separates which decreases the extent of the enhanced structure. At even higher temperatures the structure essentially disappears and the system becomes homogeneous again, as this state is entropically more favored. Essentially the same model was suggested by Kjellander [21] to explain phase separation (observed as clouding) occurring in aqueous solutions of nonionic surfactants, such as poly(ethylene glycol) alkyl ethers. According to this model [21], phase separation in aqueous nonionic surfactants solutions at elevated temperature observed as a cloud point, is connected with an effect of the "solvation" force (a modified hydrophobic interaction) due to the overlap of the structured hydration shells of PEG chains belonging to different micelles.

It should be noted that when phase separation occurs at relatively high temperature the cloud point measurements are usually performed. The only phase diagram reported for a ternary aqueous single-polymer system at the relatively high temperature seems to be the one reported by Sjoberg et al. [22] for PEG-20,000 - glucose -water system at 90°C.

Introduction of an additive into an aqueous polymer solution may change the (clouding) temperature at which phase separation in a given polymer solution occurs at the fixed polymer concentration [23-28]. The effects of various inorganic salts on the cloud points of aqueous solutions of high molecular weight PEGs at the concentrations of 0.5 wt.% [23] are illustrated in Figure 3.3. The effectiveness of anions to depress the cloud point temperature clearly exceeds that of cations and follows the order: $I^- < Br^- < Cl^- < F^- < OH^- < SO_4^{2-} < CO_3^{2-} < PO_4^{3-}$. Among the alkali metal cations K^+ and Rb^+ appear to be the most effective while Li^+ is the least effective. The influence of the ions on the cloud point temperatures of aqueous solutions of poly(vinyl methyl ether) [29], non-ionic surfactants with ethylene glycol oligomers as polar groups [30], for the effects on the micelle formation of non-ionic surfactants in aqueous solutions [31], etc., increases in the similar order.

Data presented in Fig. 3.3 show that the water-structure-making salts (see above) decrease the cloud point temperature much more considerable than the water-structure-breaking salts (KI, KBr). The salt influence on the water

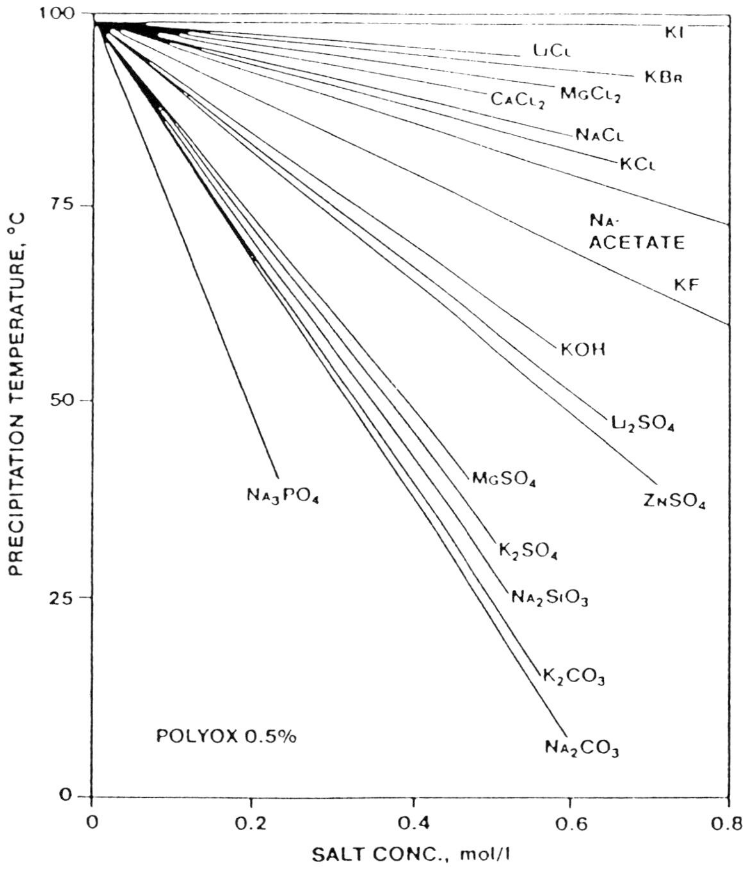

Figure 3.3. Effects of inorganic salts on the cloud point temperature for aqueous solution of 0.5 %wt. PEG of high molecular weight. (From K. P. Ananthapadmanabhan, E. D. Goddard, J.Colloid Interface Sci., 113, 294 (1986). Reprinted by permission of the American Chemical Society.)

structure may be quantitatively described in terms of the molal surface tension increment, σ_s, value (see Table 1.2). The initial effects of the salts on the cloud point temperature may be described as:

$$T_{c.p.} = T^{o}_{c.p.w} - \alpha_t \cdot C_s \tag{3.3}$$

where $T^{o}_{c.p.w}$ represents the cloud point temperature of 1.0 wt.% PEG in pure water (96.2°C) [28]; $T_{c.p.}$ is the cloud point temperature of aqueous solution of 1.0 wt.% PEG containing the total amount C_s (mole/kg) of a given salt; α_t is a constant.

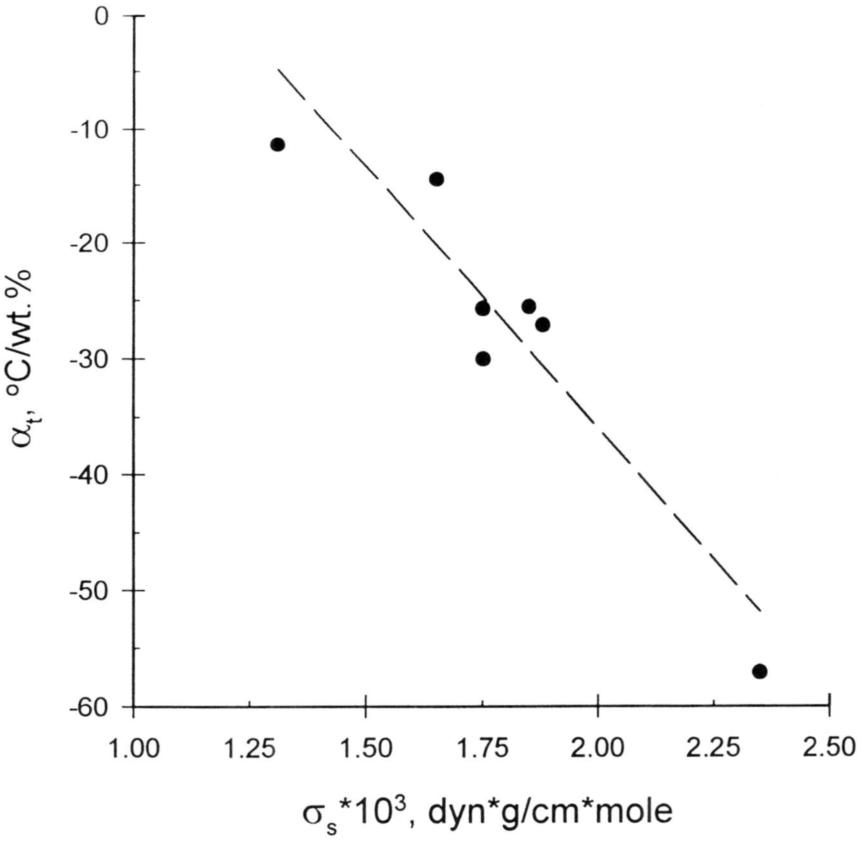

Figure 3.4. Coefficient α_t characterizing the salt effect on the cloud point of aqueous PEG solution according to Eqn. 3.3 versus the salt molal surface tension increment σ_s.

The data reported by Florin et al. [28] fit Equation 3.3 up to 0.6 mole/kg alkali metal chlorides, up to 1.0 mole/kg KF, and up to 1.5 mole/kg KBr. The data obtained for KI [28] do not fit Equation 3.3 probably due to specific interactions between PEG and KI [32]. The data given in [28] were treated according to Equation 3.3 and the α_t-values determined for different salts are plotted in Figure 3.4 versus the molal surface tension increments, σ_s, for these salts.

The observed relationship clearly implies that the salt influence on the cloud point temperature of the aqueous PEG solution [28] is mediated through the salt action on the water structure [33] (see Chapter 1).

Effects of non-ionic saccharides on the cloud point temperature of aqueous PEG 20,000 solutions at the polymer concentration of 10% wt. were reported by Sjoberg et al. [22]. The concentration effects studied in [22] are presented in Figure 3.5. The effectiveness of the saccharides to decrease the cloud point temperature follows the order: glucose < cellobiose, maltose < maltotriose < iso-maltose < iso-maltotriose, while ß-cyclodextrin increases the cloud point. It should be noted, first, that the concentration curves observed by Sjoberg et al. [22] are described by Equation 3.3 similar to those found for inorganic salts (Fig. 3.3).

As may be seen from Figure 3.5 the type of glycoside bond (α-1,4 for maltose, and ß-1,6 for iso-maltose) is more important for the saccharide effect than the saccharide size. The reason is assumed [22] to be the difference in the interactions of the saccharides with water. This assumption is clearly in line with the one above for the similar effects of inorganic salts.

The aforementioned theoretical model [20] has been extended by Florin et al. [28] to cover the influence of salts on the temperature-dependent phase separation in aqueous PEG solutions. The basic features of the model [28] are that the water structure is enhanced in the PEG hydration shell and that each macromolecule is surrounded by the water zone with the local salt concentration decreased in reference to the total salt concentration in the solution. The reason for the occurrence of such zone may be the so-called repulsive electrostatic image forces arising when charged species are brought into the vicinity of a dielectric discontinuity (see, e.g., in [27]). The assumption of the salt-poor zone in the polymer vicinity [28] is in agreement with the recent experimental NMR and neutron diffraction results [34] indicated that no direct ion-PEG interaction occur in aqueous solutions. According to the model [28] the overlap of the hydration shells of macromolecules leads to removal of some water from the PEG hydration shells. The differences observed for the salts examined (see Fig. 3.3b) were treated [28] in terms of varying degrees of salt penetration into the salt-poor water zone around the PEG macromolecules. It should be noted, however, that Florin et al. [28] did not take into consideration

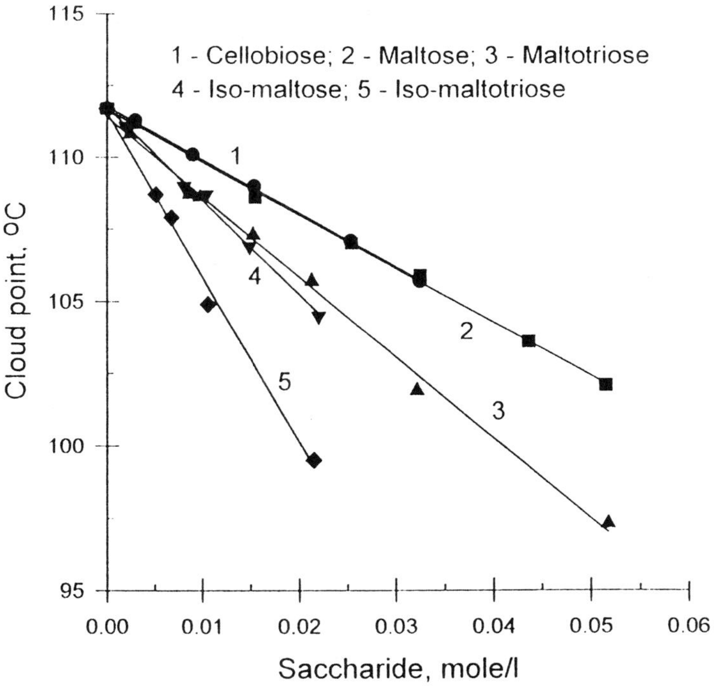

Figure 3.5. Effects of saccharides on the cloud point of aqueous solution of PEG-20000 at the concentration of 10% wt. Calculated from data in [22].

the salt effect on the overall water structure in the aqueous polymer solution. It has been noted above that the salt effect on the water structure may be viewed in terms of the change of the structure with an increase (or decrease) in temperature. Since the phase separation under discussion is temperature-dependent, analysis of the salts effects in these terms seems to be appropriate.

Considering that there is no direct PEG-ion interaction [27,28,34], and that the Flory-Huggins interaction parameter χ_{12} for aqueous solutions of PEG is essentially independent of the type of inorganic salt present [35,36] (see Table 3.3 below), it can be assumed that the water structure in the macromolecule hydration shell (i.e. the arrangement of water molecules "strongly affected" by PEG, see Chapter 2) is left unaltered by the ions present in the solution. The hydration shell of PEG may be viewed as completely free of ions, and the above repulsive PEG-ion interactions can be regarded as an exhibition of the incompatibility between the water structures in the hydration shells of ions and PEG macromolecules [37]. Under this assumption phase separation behavior of salt-containing aqueous PEG solutions can be described by the model suggested by Kjellander and Florin [20] with due regard for that instead of pure water the system contains a water-like solvent of modified features. Furthermore, the incompatibility between the hydration shells of a polymer and ions [37] may be the reason for isothermal phase separation in aqueous polymer-salt mixtures at ambient temperature [1,10,38]. It will be shown below that this assumption can also explain the influence of inorganic salt additives on phase separation in aqueous mixtures of two non-ionic polymers.

Phase separation is known to occur at room temperature in aqueous mixtures of water-structure-making salts with non-ionic polymers such as PEG,

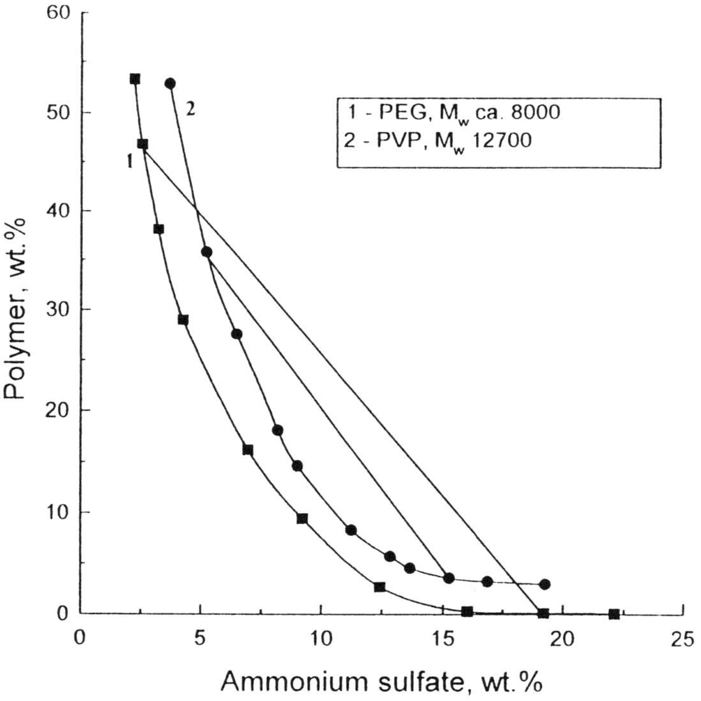

Figure 3.6. Phase diagrams for polymer—salt—water two-phase systems at 25°C.

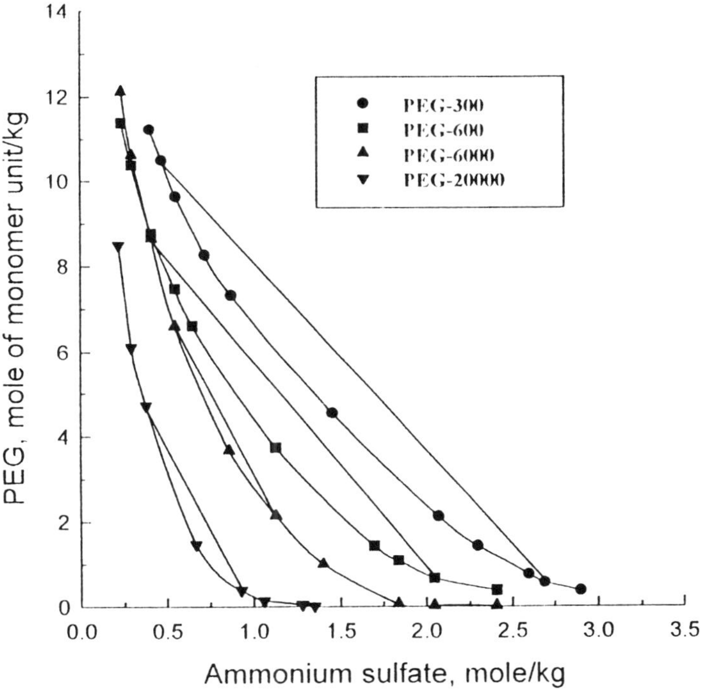

Figure 3.7. Phase diagrams for PEG—$(NH_4)_2SO_4$—water two-phase systems formed by PEGs of different molecular weights.

methoxypolyethylene glycol, polypropylene glycol [1], PVP, etc. Dextran seems to be completely miscible with inorganic salts in water up to high salt concentrations when gelation may occur. Furthermore, the results reported by Basedow et al.[39] indicate that aqueous salt solutions as well as solutions of many other non-ionic solutes, up to concentrations of 2.0 M, are even better solvents for Dex than pure water. Thus it may be concluded that the water structure in the hydration shell of dextran macromolecule is generally compatible with those in the hydration shells of different ions (for more detailed discussion see below).

Typical phase diagrams of PEG-ammonium sulfate-water [40] and PVP-ammonium sulfate-water systems [42] are shown in Figure 3.6.

Similar examples given in the book by Albertsson [1] clearly indicate that phase separation in an aqueous mixture of a polymer with a given salt depends on the polymer type more than on the polymer size.

Phase separation in the mixtures under discussion may be viewed as the result of limited mutual solubility of two water-like solvents. Solvent properties of an aqueous polymer solution depend on the polymer concentration (see above), and those of solution of a salt depend on the salt concentration. Hence, it is possible to assume that when dealing with an aqueous mixture of a polymer and a salt of varied total concentration we actually deal with a set of continuously changing pairs of two different water-like solvents.

Phase diagrams for different molecular weight PEGs- ammonium sulfate-water systems presented in Figure 3.7 show that the higher the polymer molecular weight, the lower the salt concentration required for phase separation in the system. The data presented in Fig.3.7 indicate that the amount of moles of monomer units of PEG needed for phase separation in the presence of a

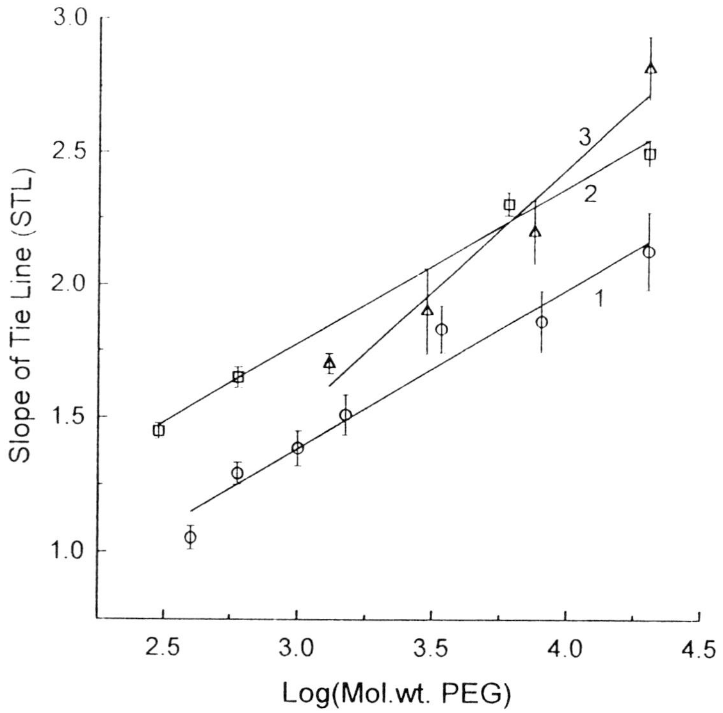

Figure 3.8. Slope of tie line (STL) for phase diagrams for PEG-salt-water two-phase systems as a function of the PEG molecular weight. Calculated from data reported in references indicated: 1: PEG— Potassium phosphate, pH 7 at 4°C [41]; 2: PEG—Ammonium sulfate at 22°C [40]; 3: PEG—Potassium phosphate at 20°C [1].

given amount of salt decreases with increasing size of the polymer macromolecule. This effect seems to agree with the aforementioned influence of the PEG molecular weight on the solvent polarity of its aqueous solutions (see Fig. 2.1b).

The effect of the PEG molecular weight on the slope of the tie line (STL) in the phase diagrams of the systems under discussion and in those of PEG-potassium phosphate-water systems described by Albertsson [1] and by Lei et al. [41] is shown in Figure 3.8.

The general trend observed for the curves in Fig. 3.8 is an increase in the STL value with increasing polymer molecular weight. An increase in the STL value indicates an increase in the difference between the polymer concentrations in the two phases at a given difference in the salt concentrations in the same phases. That implies a decrease in the mutual solubility of the aqueous polymer- and salt-containing media.

The polymer molecular weight effect may be due to an extension of the polymer hydration shell. The larger the size of these water regions of the particular water structure, the lower the total amount of the regions required for the overlap or fusion which seems to underlay the solvent properties of an aqueous polymer solution. Fusion of the macromolecules' hydration shells implies the compatibility of the water structures in the shells which seems to be common for water-soluble polymers. It should be mentioned that for low molecular weight solutes, e.g., polyhydroxy compounds, net repulsions are observed more often than attractions. For example, according to Franks [42, pp.12-13] net repulsions in an aqueous solution of sucrose molecules are estimated in the free energy terms by about 41.6 cal·mole^{-1}·(mole/kg)$^{-1}$, those of glycerol molecules by about 3.6 cal·mole^{-1}·(mole/kg)$^{-1}$, etc. The only exception mentioned in [42] seems to be inositol, its molecules in aqueous medium exhibit weak attraction estimated by about -62.1 cal·mole^{-1}· (mole/kg)$^{-1}$. In view of these examples, it may be assumed that the large size of a macromolecule is important for the compatibility between the water structures in the hydration shells of the macromolecules of the same type. It is also possible that not the large size but the cooperative action of hydrophilic and hydrophobic groups regularly positioned in the macromolecule is important for the compatibility of the water structures in the hydration shells. This question remains open at present.

Phase separation in polymer-salt-water systems at room temperature was observed mostly in the presence of the salts of water-structure-making type. Phase diagrams reported in [10] are shown in Figure 3.9 to illustrate the effects of the type of anion and cation on phase diagrams of aqueous PEG-salt mixtures. It can be seen from these phase diagrams that the tendency of salts to form aqueous two-phase systems in the mixtures with PEG is related to their position in the lyotropic series quantified, e.g., by the molal surface tension increment, σ_s, (see Table 1.2). This tendency is confirmed by the results reported

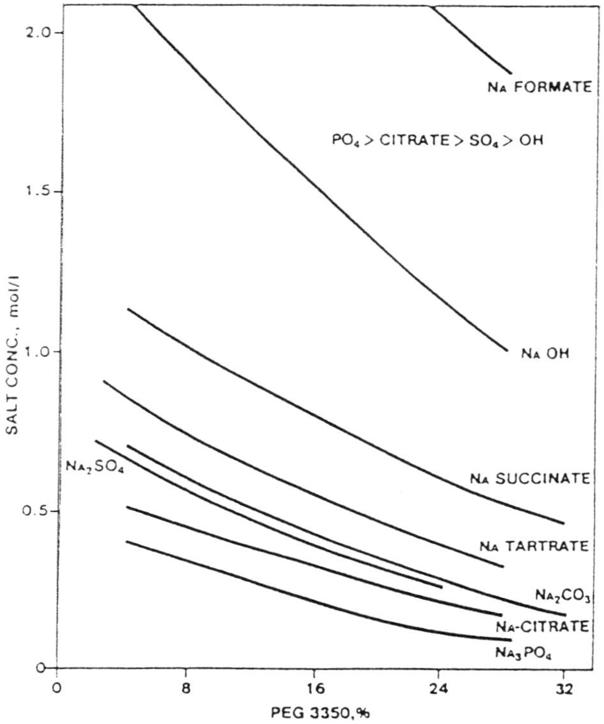

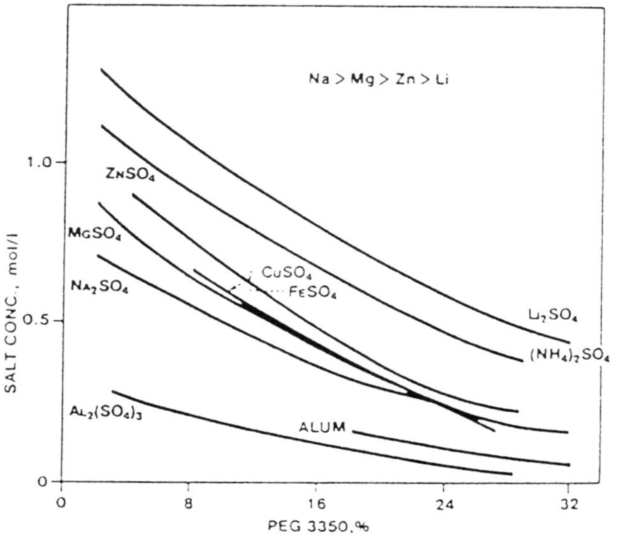

Figure 3.9. Binodial curves for (a) PEG—Na salt—water; and (b) PEG—Metallic sulfate—water two-phase systems. Phase separaqtion occurs above the curve. (From K. P. Ananthapadmanabhan, E. D. Goddard, Langmuir, 3, 25 (1987) by permission of the American Chemical Society.)

by Zvarova et al.[38]. According to Ananthapadmanabhan and Goddard [10] the binodials shown in Fig. 3.9 can be described by the Setschenov equation (see above) but the binodials of the phase diagrams determined in [40] as well as those given in the book by Albertsson [1] do not fit this equation.

An example of the temperature influence on the position of the binodial of PEG-salt-water system shown in Figure 3.10 is taken from [10].

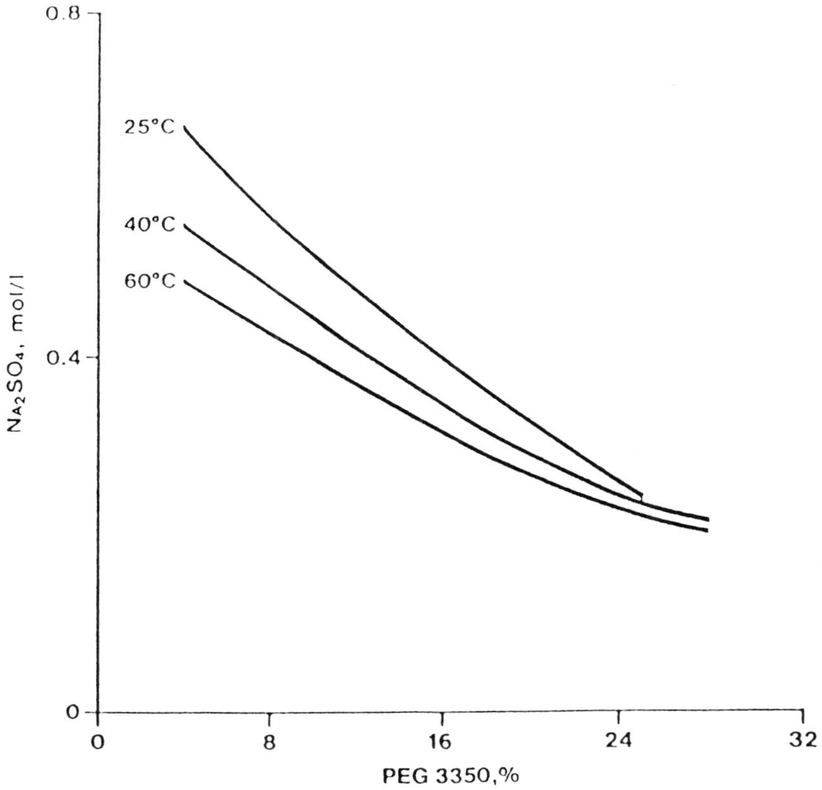

Figure 3.10. Effect of temperature on binodial curves for PEG— Na_2SO_4— water two-phase system. Phase separation occurs above the curve. (From K. P. Ananthapadmanabhan, E. D. Goddard, Langmuir, 3, 25 (1987) by permission of the American Chemical Society.)

The data given in the book by Albertsson [1] are in agreement with those in Fig. 3.10 [10]. The general trend seems to be the higher the temperature, the lower the concentration required for phase separation in aqueous polymer-salt-water systems. The temperature effect may be interpreted in terms of both aforementioned essentially similar models by Walker and Vause [19] and Kjellander and Florin [20], provided water molecules affiliated to the polymer- and salt-rich media are considered as the molecules with different characteristics.

The amounts of salt required for phase separation in aqueous solutions of PEG and commercially available random copolymer of ethylene oxide and propylene oxide (trade name UCON) at the fixed amount of a given polymer at different temperatures was shown by Ananthapadmanabhan and Goddard [43] to coincide with the cloud points. The authors [43] concluded that phase separation in polymer-salt-water systems and clouding observed in polymer solutions on heating are one and the same phenomenon. All the above experimental facts seem to be in favor of this conclusion.

All the experimental observations on phase separation in single polymer aqueous systems can be explained in terms of limited mutual solubility of two different polymer-rich and polymer-free aqueous solvents. This assumption may be regarded as a working hypothesis clearly demanding supporting evidence since only limited studies of the systems under consideration have been made up to the present. This hypothesis can be tested, however, by much more extensive body of experimental data on phase separation in aqueous two-polymer systems.

3.3. PHASE SEPARATION IN AQUEOUS TWO-POLYMER SYSTEMS: EXPERIMENTAL OBSERVATIONS

Phase separation in aqueous two-phase systems was studied mostly on those formed by two non-ionic polymers [1,12]. Therefore these systems for the most part will be considered below. The systems formed by ionic polymers will be mentioned very shortly.

The most important factor for phase separation in the systems under discussion obviously is the chemical nature of the two polymers. It seems that the type of chemical groups exposed to the solvent is the governing factor as implied, e.g., by that the water-soluble thermally denatured ovalbumin phase separates in aqueous mixture with the native ovalbumin [44,45].

Influence of molecular weight of phase polymers.

The effect of the molecular weight of a polymer 1 on the phase diagram of polymer 1 - polymer 2 - water system is generally as shown in Figure 3.11 for dextran-PEG-water system.

The higher the molecular weight of the polymer 1, the lower its

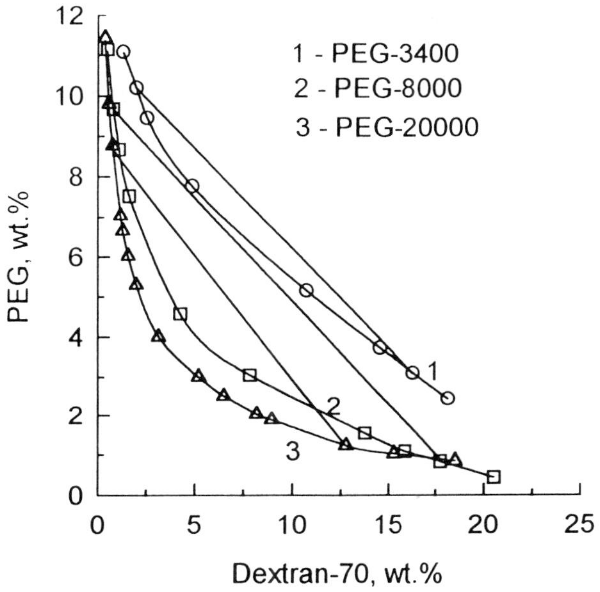

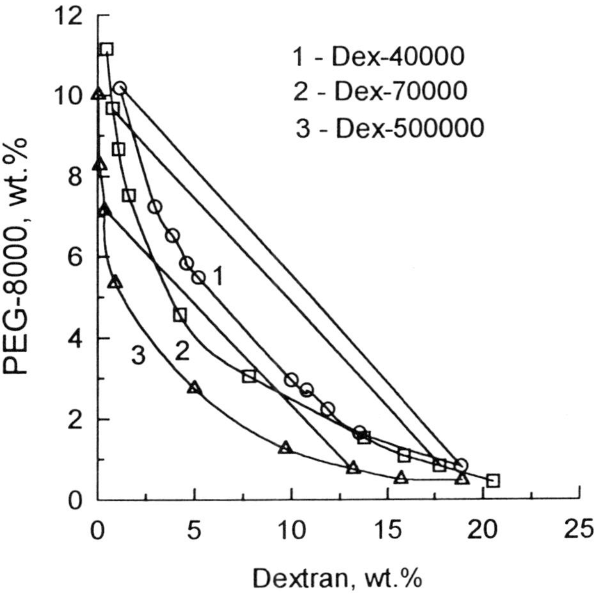

Figure 3.11. Phase diagrams for Dex—PEG—water systems formed by polymers of different molecular weights. (From A. D. Diamond, J. T. Hsu, Biotechnol. Techniques, 3, 119 (1989) by permission of Eaton Publishing Company.)

concentration required for phase separation [1,16,41,46,47]. The larger the difference in the molecular weights between the two polymers, the more asymmetrical is the binodial. It should also be mentioned that as shown by several

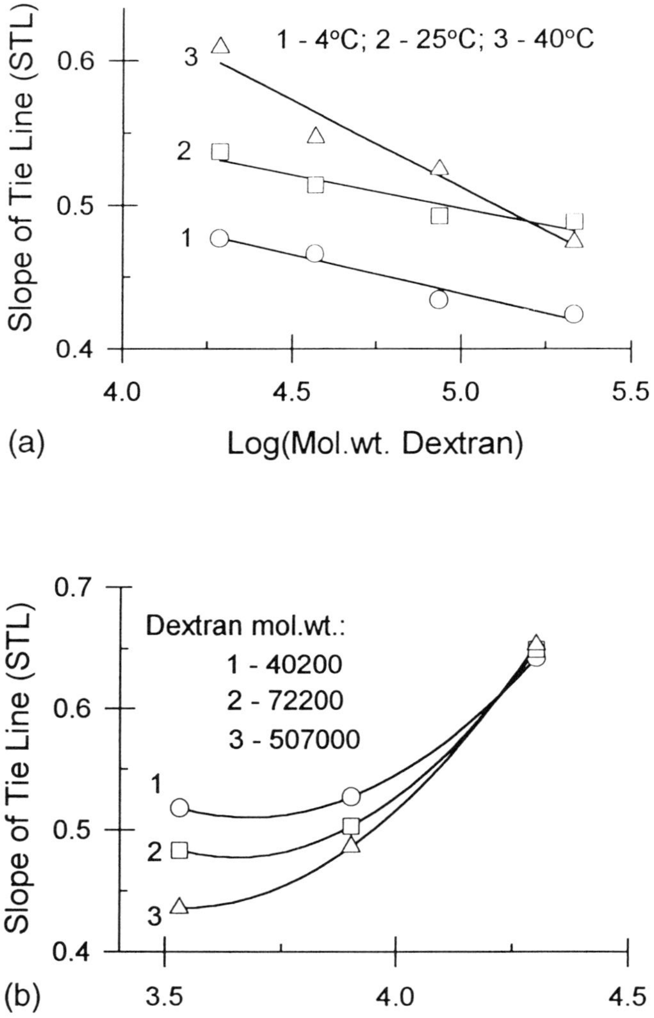

Figure 3.12. Slope of tie line (STL) for phase diagrams for Dex—PEG—water two-phase systems as a function of polymer molecular weight. (a) Calculated from data in [46]; (b) calculated from data in [16].

authors [1,48-50], the molecular weight distributions of phase polymers differ in the two phases. This difference seems to depend particularly on the polydispersity of a given phase polymer [47].

It should also be noted that the slope of the tie line (STL) in phase diagrams of aqueous two-polymer systems seems to be related to the molecular weight of the phase polymers. The relationship in question is illustrated in Figure 3.12 for several phase diagrams reported in the literature [16,46]. The STL value increases with increasing PEG molecular weight as observed in PEG-salt-water systems (see Fig. 3.8). The STL value in dextran-PEG-water systems, however, seems to decrease with increasing dextran molecular weight. Much more experimental data is needed for any conclusion in regard to general relationships between molecular weights of phase polymers and STL values.

Temperature effect.

The effect of temperature on phase separation in aqueous two-polymer systems differs from that observed in aqueous single polymer systems. In contrast to the data presented in Fig. 3.10, the concentrations of phase polymers required for phase separation in aqueous mixtures of two polymers usually increase with increasing temperature [1,14-17,43]. The examples of phase diagrams of three different aqueous two-polymer systems at different temperatures in the temperature range from 8°C up to 50°C [14] are given in Figure 3.13.

The binodials of the phase diagrams for the aqueous Dex-PEG (Fig. 3.13a) and Dex-polyvinylpyrrolidone (PVP) (Fig. 3.13b) systems at 38 and 50°C are so close to each other that only one line could be drawn in each case. The binodials for the aqueous Dex-polyvinyl alcohol (PVA) system at the temperatures from 23 to 50°C are also so close to each other that they could not be represented by separate lines (Fig. 3.13c). The difference in phase behavior of these systems at different temperatures can be seen, however, if the polymer compositions corresponding to the critical points of the phase diagrams and the slopes of the tie lines (STL) values are compared [14].

An increase in the temperature causes an increase in the amount of at least one of phase polymers needed for phase separation [1,14-17]. The STL value increases with increasing temperature [14] as shown in Figure 3.14.

It should be mentioned that Ananthapadmanabhan and Goddard [43] reported a linear dependence of the concentration of dextran required to form a two-phase system in the aqueous solution of PEG at the fixed concentration of 4.0 and 4.5 wt.% at temperatures in the ca.4°C to 21°C range. Similar treatment of the diagrams shown in Fig. 3.13 indicates that there is a linear relationship between the dextran concentration (at a given fixed concentration of a second polymer) and the temperature up to 38°C. At higher temperatures the relationship deviates from linearity significantly. The temperature value

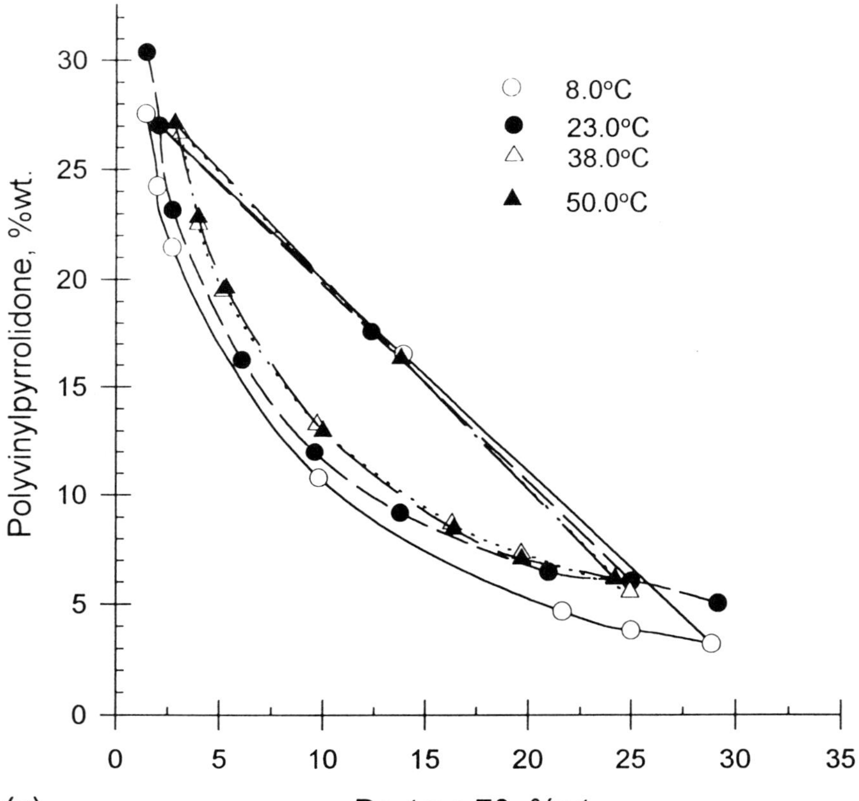

(a)

Figure 3.13. Phase diagrams of (a) Dex—PVP—water; (b) Dex—PVA—water; and (c) Dex—PEG—water at different temperatures.

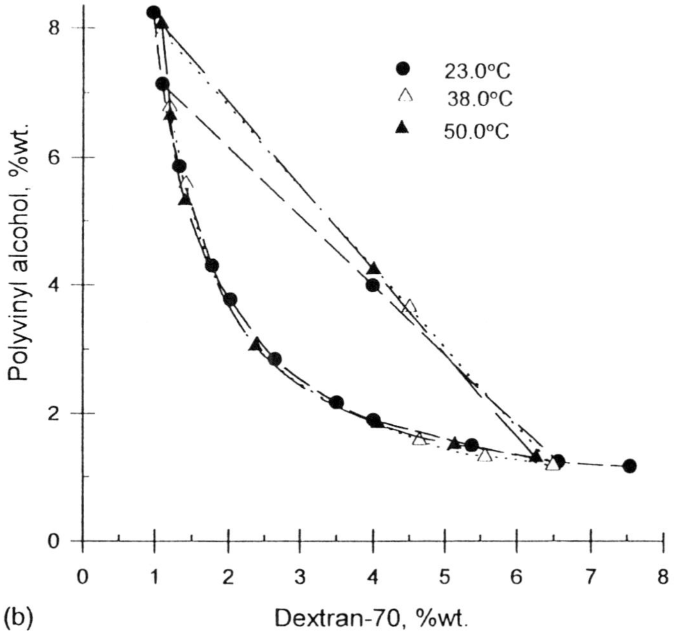
23.0°C
38.0°C
50.0°C
Polyvinyl alcohol, %wt.
0
2
4
6
8
0
1
2
3
4
5
6
7
8
Dextran-70, %wt.

(b)

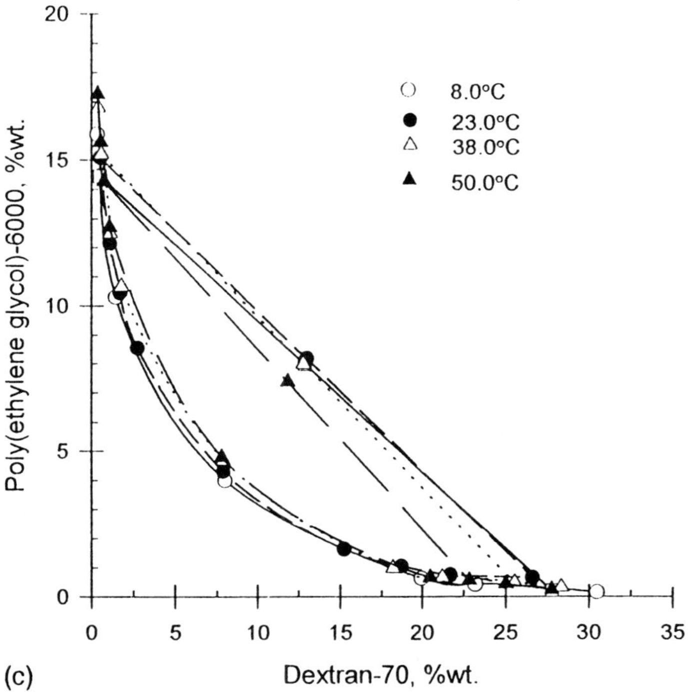
8.0°C
23.0°C
38.0°C
50.0°C
Poly(ethylene glycol)-6000, %wt.
0
5
10
15
20
0
5
10
15
20
25
30
35
Dextran-70, %wt.

(c)

obtained when the plot is extrapolated to zero dextran concentration depends on the second polymer concentration chosen and amounts not to -10°C reported in [43] for 4.0 and 4.5 wt.% PEG (mol.wt. 8000) in the presence of Dex (mol.wt. 472,000) but to -196°C at the same concentration of PEG of similar molecular weight in the presence of Dex of the molecular weight of 57,200. Similarly unrealistic temperature values, for example, -80°C for 6.0 wt.% PVP (mol.wt. 12,700) in the presence of the same Dex, are obtained for the aqueous Dex-PVP and Dex-PVA systems. Hence, it is difficult to accept the hypothesis [43] that aqueous PEG solutions in the absence of a second polymer should separate into two aqueous phases at the so-called hypothetical upper critical solution temperature.

The temperature dependence of the phase behavior of the systems shown above [14] as well as the data reported in the literature [1,15-17,41,46] appear to be in favor of the above hypothesis that phase separation in aqueous mixture of two polymers is a phenomenon similar to that occurring in a

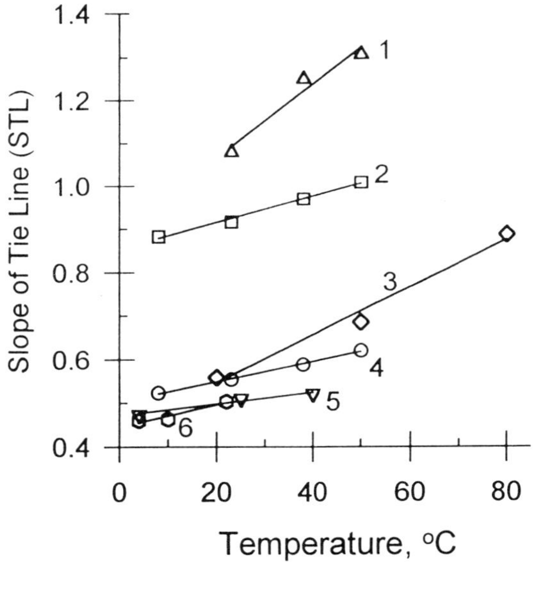

Figure 3.14. Slope of tie line (STL) for phase diagrams for Dex—polymer 2 —water two-phase systems as a function of temperature. Calculated from data reported in the references indicated. 1: Dex-70—PVA-50000 [14]; 2: Dex-70 —PVP-12000 [14]; 3: Dex-40—PEG-20000 [15]; 4: Dex-70—PEG-6000 [14]; 5: Dex-40—PEG-6000 [46]; 6: Dex70—PEG-8000 [16].

mixture of two different solvents of limited mutual solubility. In an aqueous two-polymer system the two different solvents are aqueous solutions of different polymers, e.g., those of Dex and PEG or PVP. Since an increase in the temperature disrupts the water structure, higher concentrations of phase polymers are required to achieve the threshold of the limited mutual solubility of their aqueous solutions. This conclusion is supported by the fact that addition of urea, the well known water-structure breaker, affects phase separation in aqueous two-polymer systems similarly to the temperature increase [14].

The examples of the effect of urea addition on the phase diagrams of the aqueous Dex-PEG, Dex-PVP, and Dex-PVA systems [14] are shown in Figure 3.15.

It can be seen that while an addition of urea produces the urea concentration-dependent shift of the binodial similar to that induced by the temperature increase, the STL values are changed in the presence of different amounts of urea in a more complex manner. That may be due to that urea not only disrupts the water structure but also replaces some of the water molecules in the macromolecule hydration shell [51]. An interpretation of the urea influence on the phase diagrams of the systems under consideration in any closer detail would clearly be premature.

Influence of inorganic salts.

According to the assumption accepted above as a working hypothesis, any water structure perturbing factor should affect phase separation in aqueous two-polymer systems. It has been, however, stated by Albertsson, the "father" of the aqueous two-phase partition technique, (see [1], pp.34-35) that the systems containing nonionic polymers only are hardly affected by the addition of, for example, 0.1 - 1 M sucrose or NaCl... It is only at higher salt concentrations that an effect on the phase system may be observed. This opinion has been generally accepted and was first seriously challenged only by the results obtained by Brooks et al.[52].

According to the data reported in [52], the addition of potassium sulfate to the aqueous Dex-PEG two-phase system changes the polymers' concentrations in the coexisting phases depending on the total concentration of the sulfate additive in the range from 0.03 to 0.4 mole/kg K_2SO_4. The authors [52] have studied the electrical properties of the phases (see below) and the salt distribution in the aqueous Dex-PEG system, but they have not pointed to the inconsistency between the results obtained and the generally accepted view [1].

Concentration effects of inorganic salts on phase separation in aqueous mixtures of Dex with PEG, PVP, and Fic at fixed polymers concentrations ratios [53] are shown in Figure 3.16 as the total polymer concentration required for phase separation versus salt concentration.

It can be seen from the data in Fig. 3.16 that the salt concentration

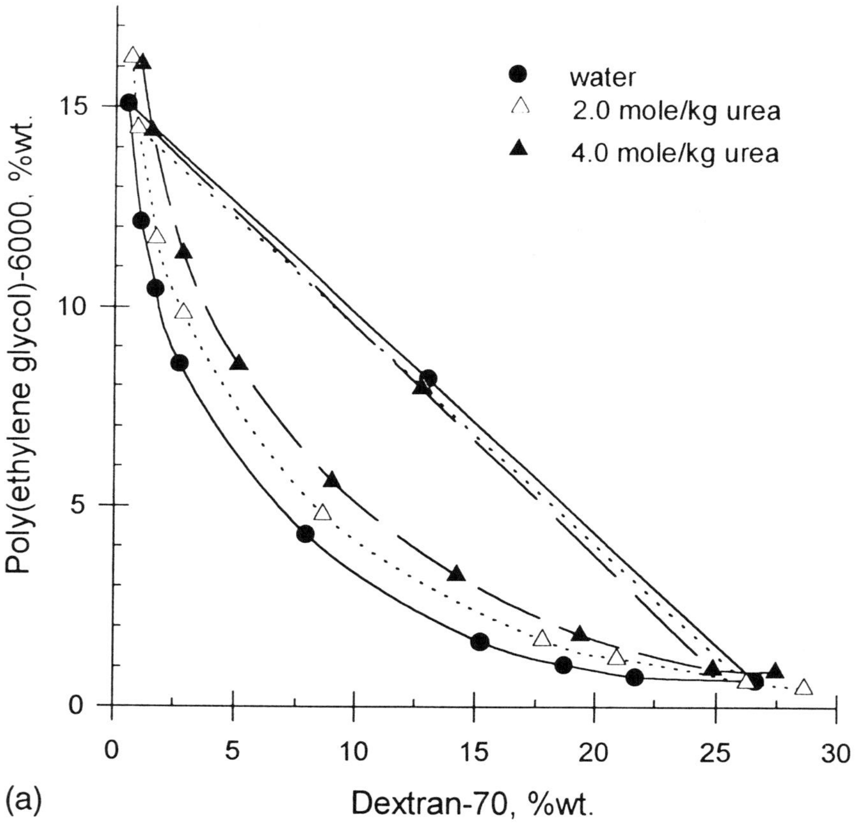

Figure 3.15. Effects of additives of urea on phase diagrams for (a) Dex—PEG—water; (b) Dex—PVP—water; and (c) Dex—PVA—water two-phase systems.

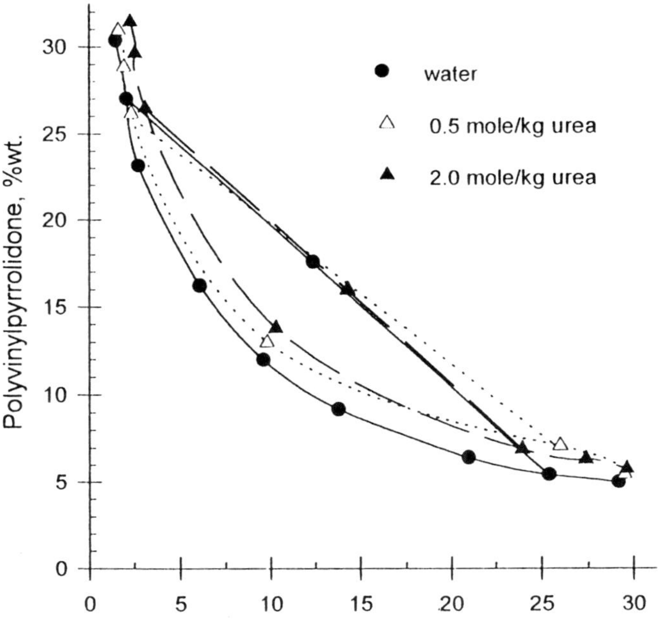
water
0.5 mole/kg urea
2.0 mole/kg urea
Polyvinylpyrrolidone, %wt.
Dextran-70, %wt.
0
5
10
15
20
25
30

(b)

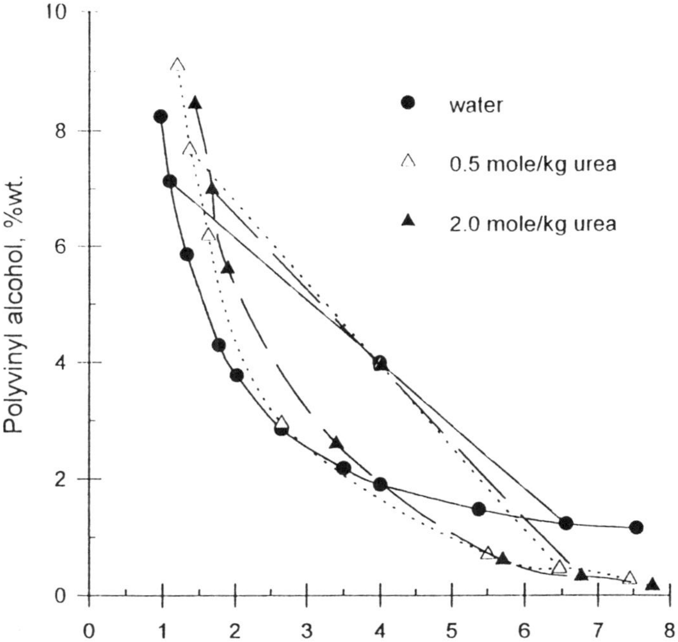
water
0.5 mole/kg urea
2.0 mole/kg urea
Polyvinyl alcohol, %wt.
Dextran-70, %wt.
0
2
4
6
8
10
1
3
5
7

(c)

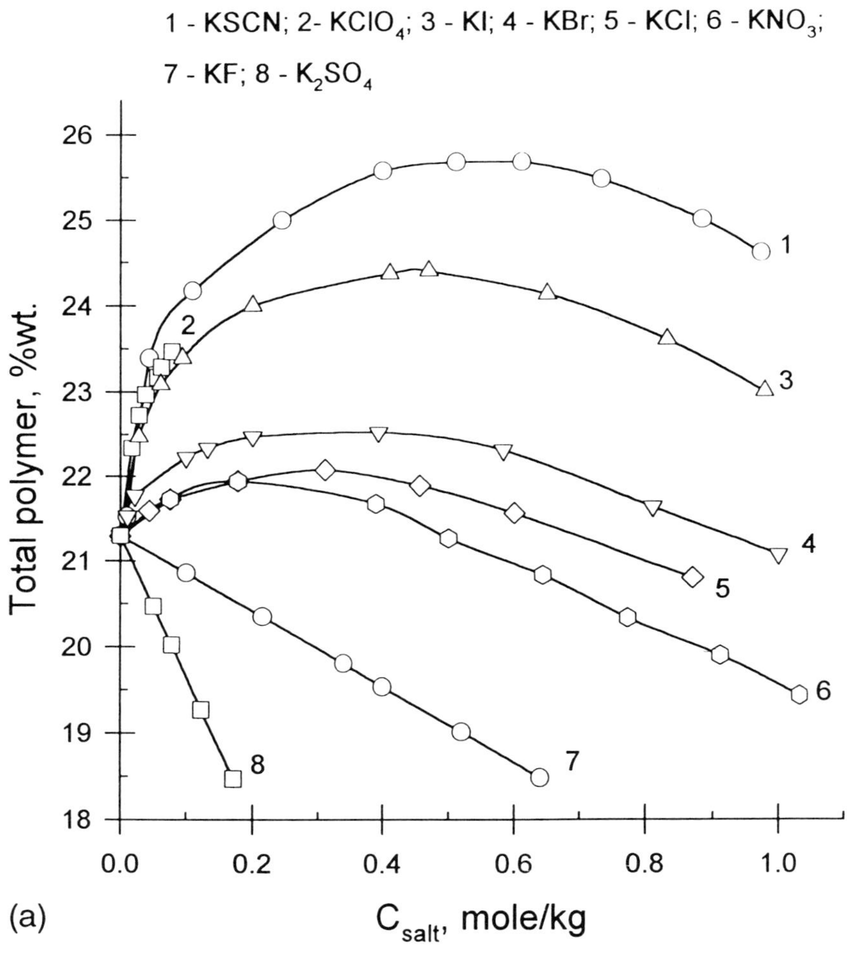

Figure 3.16. Concentration effects of inorganic salts on phase separation in aqueous mixtures of Dex with (a) Ficoll; (b) PVP; and (c) PEG at the fixed polymer/Dex concentrations ratios.

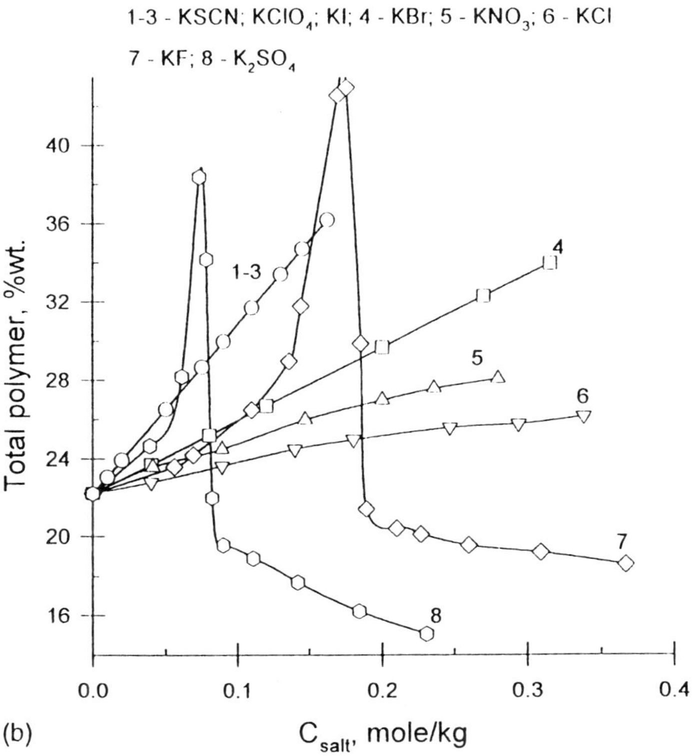
1-3 - KSCN; KClO4; KI; 4 - KBr; 5 - KNO3; 6 - KCl
7 - KF; 8 - K2SO4
Total polymer, %wt.
Csalt, mole/kg
1-3
4
5
6
7
8
16
20
24
28
32
36
40
0.0
0.1
0.2
0.3
0.4

(b)

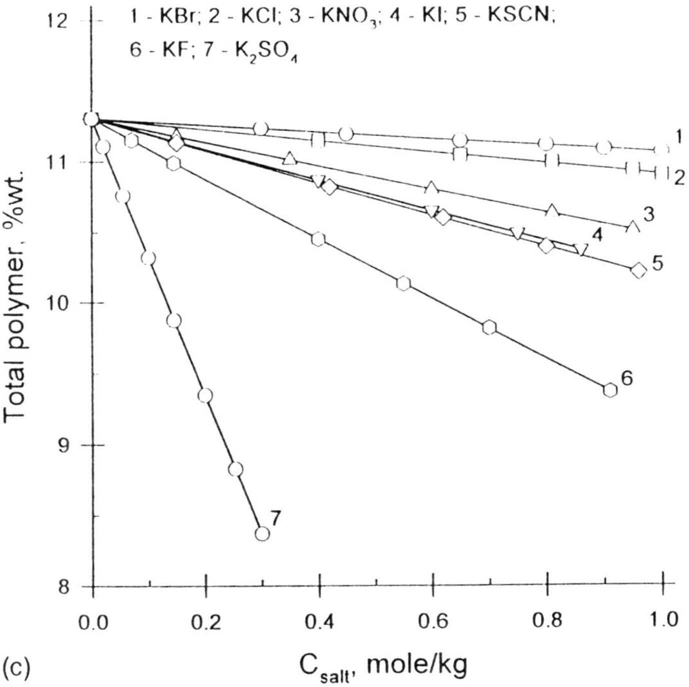
1 - KBr; 2 - KCl; 3 - KNO3; 4 - KI; 5 - KSCN;
6 - KF; 7 - K2SO4
Total polymer, %wt.
Csalt, mole/kg
1
2
3
4
5
6
7
8
9
10
11
12
0.0
0.2
0.4
0.6
0.8
1.0

(c)

effects on phase separation in the aqueous Dex-PEG mixture generally differs from those on phase separation in the mixtures of Dex with PVP or Ficoll. The effects of all the salts examined on phase separation in the aqueous mixture of Dex and PEG at the fixed PEG/Dex concentration ratio of 0.625 (Fig. 3.16a) are similar to those on the cloud point temperatures in the Dex-free aqueous solutions of PEG (see Fig. 3.3). The effectiveness of anions to reduce the total polymer concentration required for phase separation in the mixture follows the order similar to the one found for the anions' influence on the temperature-dependent phase separation in aqueous solutions of PEG [25,28,43]. Hence, it may be assumed as a first approximation that phase separation in the aqueous mixture of PEG and Dex occurs similarly to that in the Dex-free aqueous solution of PEG, i.e., due to the incompatibility between the water structure in the hydration shell of the PEG macromolecule and that in the salt-containing aqueous solution of dextran.

The concentration effects of the salts examined on phase separation in the aqueous mixtures of Dex with Ficoll or PVP (Fig. 3.16b,c) are clearly different with regard to the anions effects on the water structure. Addition of salts with the water-structure-breaking anions (ClO_4^-, SCN^-, I^-, Br^-, Cl^-, NO_3^-) initially increases the polymers compatibility. Above a certain salt concentration further increase in the salt amount decreases the total polymer concentration required for phase separation. This effect observed in the aqueous mixture of Dex with Ficoll (Fig. 3.16b) is not achieved in the Dex-PVP mixture (Fig. 3.16c). The possible reason may be the limited salt concentration range used due to the relatively large concentrations of the polymers in the Dex-PVP mixture.

The curves observed are typical for the salting-in/salting-out effects of salts on the aqueous solubility of amphiphilic solutes [33]. The water structure-making salts, K_2SO_4 and KF, decrease the polymers' compatibility in the aqueous polymer mixtures over the entire salt concentration range examined. The reasons for the unusual shape of the curves representing the concentration effects of these salts on the aqueous Dex-PVP mixture remain obscure as yet. The salt effects observed may be easily explained in terms of the influence of the salts on the mutual solubility of two different solvents. The different water structures in the hydration shells of PEG, PVP, and Ficoll is one possible reason for the difference in the salts effects on phase separation in the aqueous mixtures of Dex with these polymers.

More detailed study of the influence of inorganic salts additives on phase separation in aqueous Dex-PEG, Dex-PVP, Dex-Ficoll, and Dex-PVA systems [53-55] indicated that: a) the Dex-PEG system is the least sensitive to the salt presence and it is affected by the salt additives in the manner different from that typical for the other systems examined; and b) the effects of the salt additives on the phase diagrams of the aqueous Dex-PVP, Dex-Ficoll, and Dex-

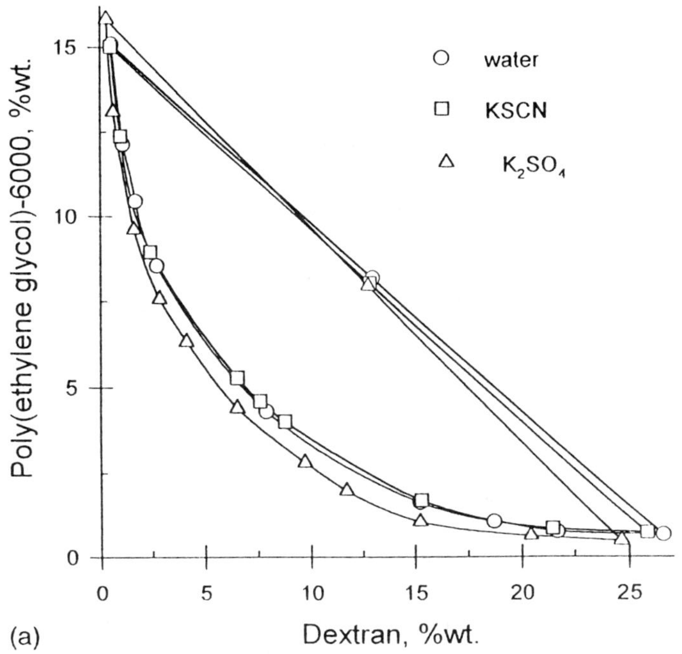

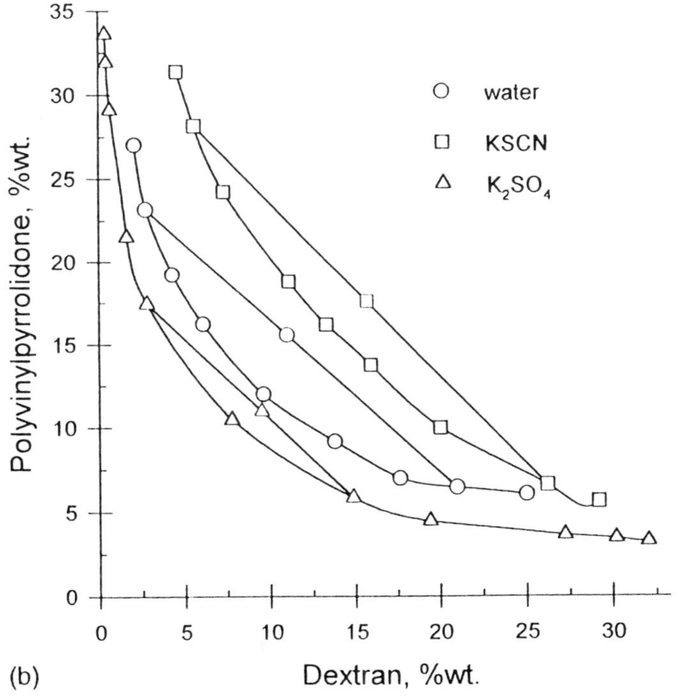

Figure 3.17. Phase diagrams for (a) Dex—PEG—water; and (b) Dex—PVP—water two-phase systems containing 0.1 mole/kg salt additive.

PVA systems are related to the salt action on the structure of water. To illustrate the difference in the salts effects on the phase diagrams of the Dex-PEG-water and Dex-PVP-water systems, the binodials for the systems containing 0.10 mole/kg KSCN, K_2SO_4 and for those without any salt added are shown in Fig. 3.17.

It can be seen that the presence of 0.10 mole/kg KSCN does not alter the binodial position for the Dex-PEG-water system (note the change in the STL value indicating that it does changes the polymer composition of the coexisting phases). The same salt, however, causes a considerable elevation of the binodial for the Dex-PVP-water system relatively to that for the salt-free system. The presence of 0.10 mole/kg K_2SO_4 causes a depression of the binodial relatively to that for the salt-free system in both systems.

The difference in the phase separation behavior is notable even more if the binodials for the aqueous Dex-PEG and Dex-PVA systems containing 0.50 mole/kg KSCN are compared (see Figure 3.18).

It can be seen that the salt slightly depresses the binodial for the Dex-PEG system while elevating that for the Dex-PVA system. The possible reasons for the different effects will be discussed below but it should be emphasized here that the above experimental evidence clearly dispels the aforementioned view [1] that inorganic salts additives affect phase separation in aqueous mixtures of two non-ionic polymers only when used in very large quantities.

The relationship between the effect of the salt additive on the phase diagram of a given aqueous two-polymer system and the action of the salt on the structure of water is clearly implied by the data obtained for the Dex-PVP-water system which seems to be the most sensitive to the type of a salt additive among the system examined [53-55].

To illustrate the relationship under discussion several typical examples of the phase diagrams of the Dex-PVP-water systems containing different salt additives [55] are shown in Figure 3.19.

The linear relationship between the polymer concentrations representing the critical points of the phase diagrams for the Dex-PVP-water systems containing 0.10 mole/kg of a given salt [55] and the molal tension increment of the salt, σ_s, used as a measure of the efficiency of the salt action on the water structure [33] is shown in Figure 3.20.

To describe the salt-dependent shift in the binodial of a given phase diagram the concentrations of the phase polymers required for phase separation with and without a given salt additive were determined [54] at the fixed polymers' concentrations ratio $C_i/C_{dextran} = 1.14$ (C_i is the concentration of PVP, Ficoll, or PVA; $C_{dextran}$ is the concentration of dextran). The difference between the C_i values corresponding to the systems with and without salt additive (ΔC_i) was used as a measure of the salt-induced shift of the binodial.

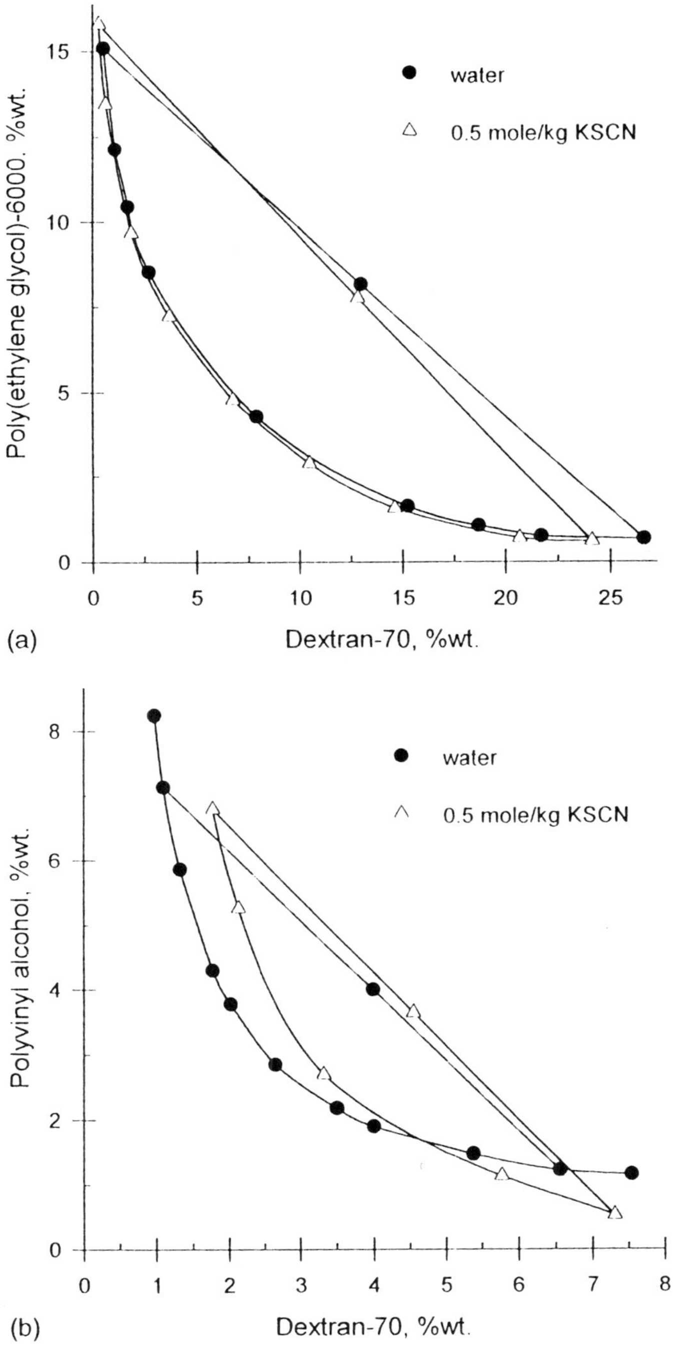

Figure 3.18. Effect of KSCN (0.5 mole/kg) on phase diagrams for (a) Dex—PEG—water; and (b) Dex—PVA—water two-phase systems.

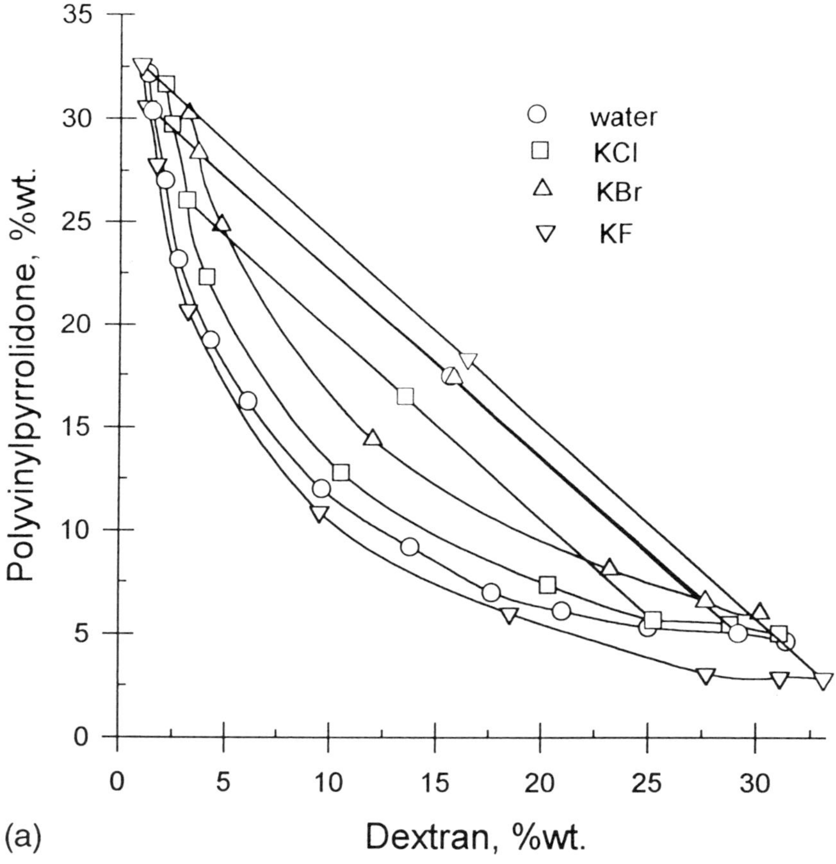

Figure 3.19. Phase diagrams for Dex—PVP—water two-phase systems containing 0.1 mole/kg salt additive.

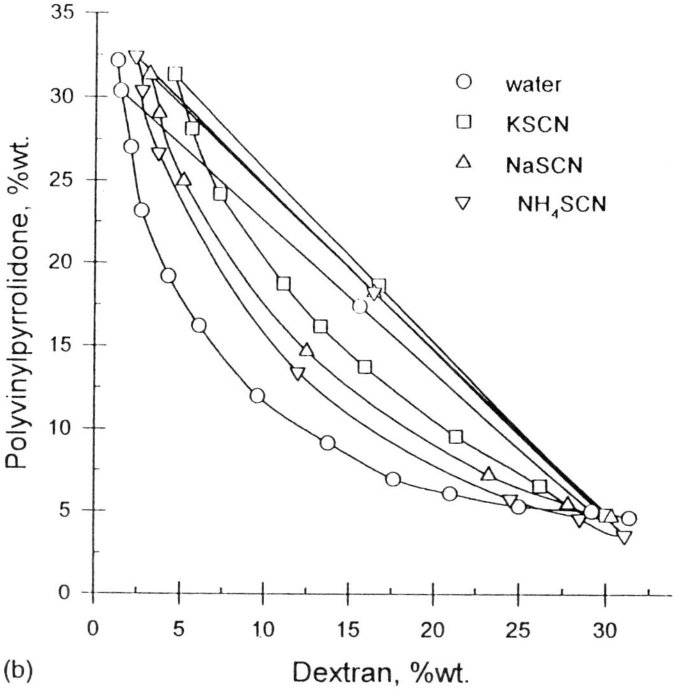

water
KSCN
NaSCN
NH_4SCN
Polyvinylpyrrolidone, %wt.
Dextran, %wt.

(b)

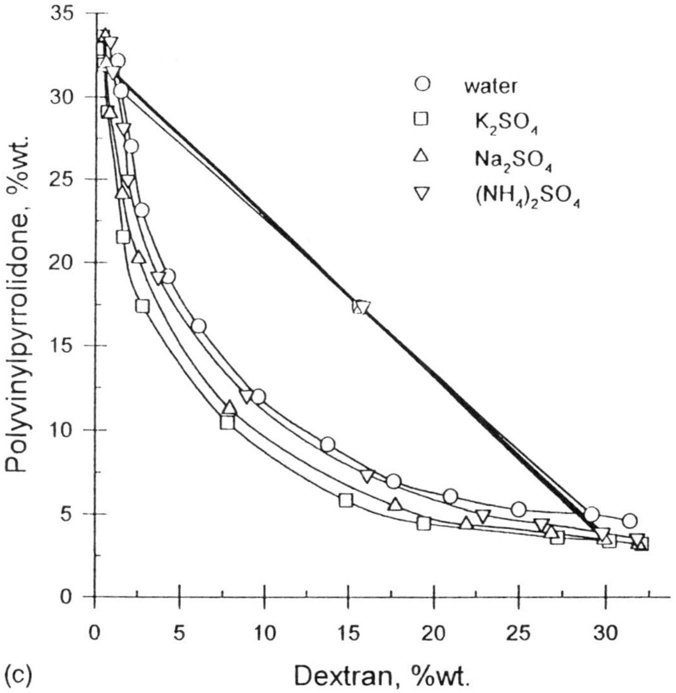

water
K_2SO_4
Na_2SO_4
$(NH_4)_2SO_4$
Polyvinylpyrrolidone, %wt.
Dextran, %wt.

(c)

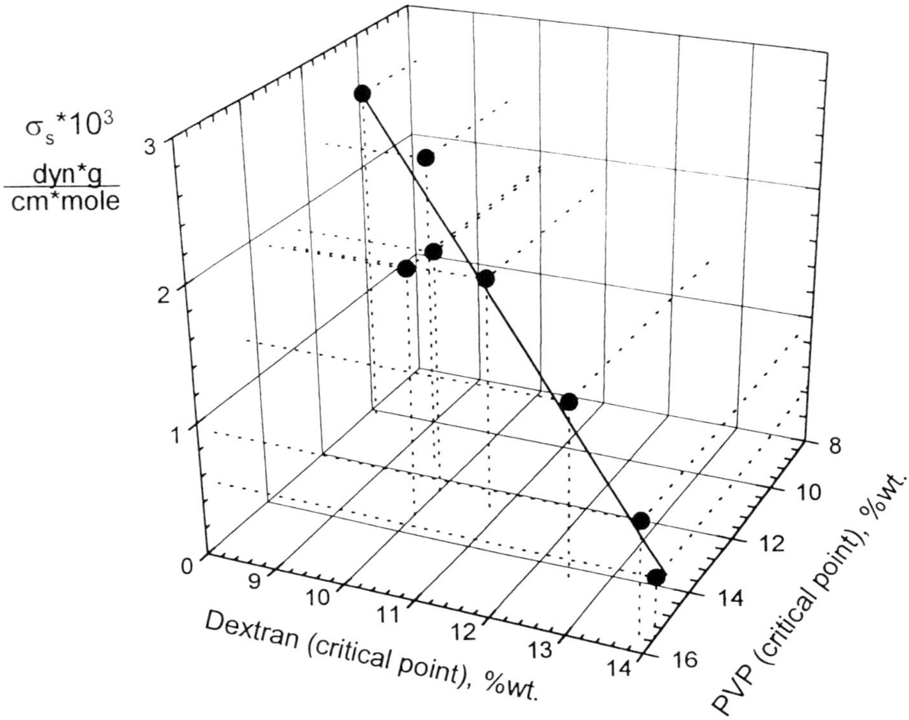

Figure 3.20. Relationship between the concentration of polymers corresponding to the critical points of Dex-PVP-water two-phase systems containing K_2SO_4, KSCN, KCl, KBr, KF, KCl, NaCl, CsCl at concentration of 0.1 mole/kg and the molal surface tension increment of the salt σ_s.

Figure 3.21 shows that the effect of a salt on the phase separation in the aqueous Dex-PVP, Dex-Ficoll, and Dex-PVA systems is linearly related to the lyotropy of the salt represented by the σ_s value. This relationship [54] fits both positive and negative ΔC_i values, i.e., salt-induced decrease and increase in the polymers' compatibility depending on the salt type. It can also be seen from the relationships plotted in Fig. 3.21 that the salt susceptibility of the systems examined in [54] decreases as follows: Dex-PVP (0.10 mole/kg salt) > Dex-Ficoll (0.10 mole/kg salt) > Dex-Ficoll (0.50 mole/kg salt) > Dex-PVA (0.50 mole/kg salt) > Dex-PEG (0.50 mole/kg salt). It must be noted, however, that this conclusion seems to hold for the case of the K-salts. The intersections of the lines representing the effects of 0.10 mole/kg salt and those of 0.50 mole/kg salt in Fig. 3.21 is hard to explain. Much more experimental study is clearly needed.

The water-structure-breaking salts generally elevate the binodial of an aqueous two-polymer two-phase system similarly to the temperature increase or the urea addition, while the water-structure-making salts depress the binodial of the system. In other words, when the water structure in the mixture of two

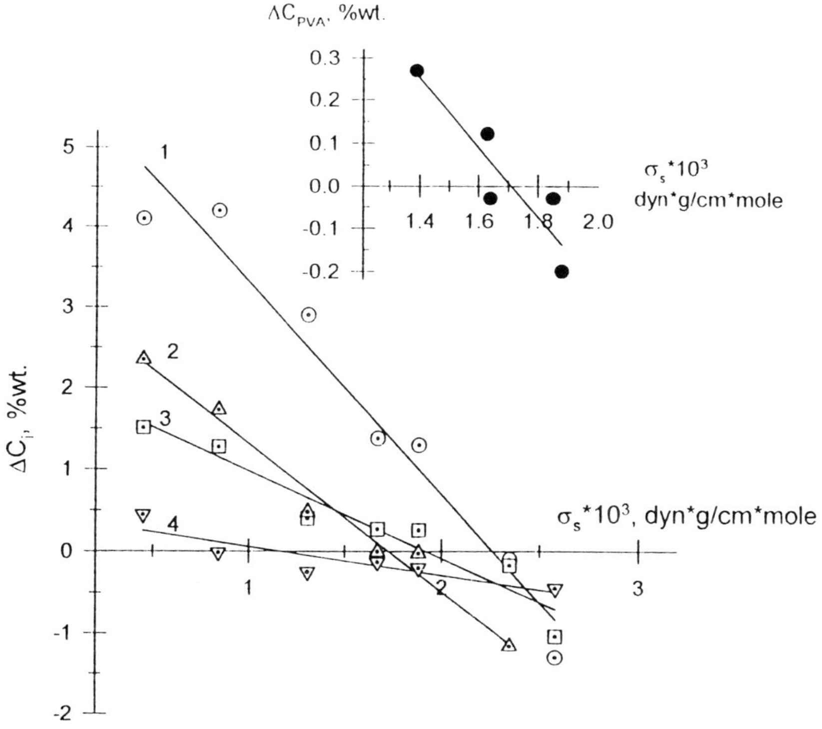

Figure 3.21. Shift in the compatibility (ΔC_i) of Dex with PVP (1), Ficoll (2,3), and PVA (4) at the constant polymer/Dex concentration ratio of 1.14 induced by the presence of 0.10 mole/kg (1,2) and 0.50 mole/kg (3,4) salt additives: KSCN, KI, KBr, KNO_3, KCl, KF, K_2SO_4 as a function of the salt lyotropy ($\Delta\sigma$). Inset: shift in the compatibility of Dex with PVA induced by the presence of 0.50 mole/kg KCl, CsCl, NaCl, LiCl, and NH_4Cl as a function of the salt lyotropy.

phase polymers is disrupted by a given factor (temperature increase, addition of urea or water-structure-breaking salt) the threshold amounts of the polymers required for phase separation in the mixture increase. If the factor (temperature decrease, addition of water-structure-making salt) enhances the water structure in the mixture the threshold amounts of the polymers are reduced. The data obtained for the Dex-PVP-water system and more limited data for aqueous Dex-Fic and Dex-PVA systems [53-55] are all in agreement with these trends. Dex-PEG-water system seems to be an exclusion in the sense that this system is much less susceptible to the salt effects.

Interrelationship between polymer and ionic composition of the two phases.

It has been found by Johansson [56,57] that in a salt-containing aqueous two-polymer system the salt concentrations in the coexisting phases are different. The generally accepted assumption was [1,56,57] that a salt additive distributes between the two phases creating the so-called interfacial electrostatic potential difference (see below) without essentially any effect on the polymer composition of the phases.

It has been established later [52,58] that the salt concentrations in the two phases change when the total polymer concentration is changed. In order to explore the contributions of the phase polymers in determining sodium phosphate, sodium chloride, and potassium sulfate distribution in the Dex-PEG-water two-phase system the equilibrium dialysis experiments with aqueous solutions of the individual polymers have been performed by Brooks et al.[52] and by Bamberger et al.[58]. The aqueous solutions of PEG and Dex were exhaustively dialyzed against various concentrations of salt and the salt concentrations inside and outside the dialysis bag were determined. It was found [58] that PEG rejected phosphate, sulfate and to a lesser extent chloride, in agreement to the data by Breen et al.[34], while the effect of Dex on the distribution of either salt was much smaller. The magnitude of the PEG effect on the salt distribution behavior was found to be essentially proportional to the polymer concentration [58]. The results reported in [52,58] particularly suggested an existence of a relationship between the polymer and salt concentrations in the two phases of the Dex-PEG-water-salt two-phase system.

Such a relationship was established by Zaslavsky et al.[55] analyzed the salt and polymer concentrations in the two phases of aqueous Dex-PEG, Dex-PVP, and Dex-Ficoll two-phase systems. The results reported in [55] indicated that the concentrations of a given salt additive in the phases are different. The salt concentrations in the two phases depend on the type and total concentration of phase polymers and on the type and total concentration of the salt additive [55].

In order to describe the relationship observed it was necessary to choose quantitative measures of the polymer and of the salt composition of the two phases. As we deal with a pair of two coexisting phases an adequate choice may be either the difference in the polymer (and salt) concentrations between the two phases or the ratio of these concentrations. In the attempts to describe partitioning of a solute [13] and certain physico-chemical features of the two phases, e.g., the interfacial tension [59,60] or the interfacial electrostatic potential difference [52] (see below), as a function of the total polymer concentration of the system the most commonly employed measure was the length of the tie line (TLL, see Equation 3.1).

It was shown by Johansson [13] particularly that TLL parameter is a measure of the dissimilarity of the compositions of the coexisting phases more appropriate than the distribution coefficient of a given phase polymer (defined as the ratio between the polymer's amounts in the two phases). Bamberger et al.[58,59] showed that in the Dex-PEG-water two-phase system the difference in the PEG (or Dex) concentrations in the two phases, ΔC(PEG) or ΔC(Dex), can be used instead of TLL parameter. It was shown above that the ΔC(polymer 1) and ΔC(polymer 2) are interchangeable as follows from Equation 3.2.

Recent results by Hsu et al.[61], Forciniti et al.[46], and Zaslavsky et al.[62] confirmed the possibility to use the difference in the concentrations of any phase polymer between the coexisting phases of an aqueous two-polymer two-phase system to describe quantitatively the polymer composition of the two phases. The advantage of this measure over the TLL parameter clearly follows from that the TLL value includes ΔC(polymer) and the additional constant STL value.

It was shown [55] that the salt composition of the two phases may be described by the ratio between the salt concentrations in the phases:

$$P(\text{salt}) = C(\text{salt})^i / C(\text{salt})^j \qquad (3.4)$$

where C(salt) is the concentration of a given salt in a given phase; superscripts "i" and "j" denote the phases enriched by the polymer i and by polymer j, respectively; P(salt) is the distribution ratio of the salt.

Polymer and ionic composition of the two phases of the aqueous Dex-PEG, Dex-PVP, and Dex-Ficoll systems containing various salt additives were examined in [55]. The results indicated that the salt composition of the phases is related to the ΔC(polymer i) value. These relationships illustrated in Figures 3.23 and 3.24 are described as:

$$\ln P(\text{salt}) = b(\text{salt}) \cdot \Delta C(\text{polymer i}) \qquad (3.5)$$

where b(salt) is a constant, with value depending on the type of phase polymers and on the type and total concentration of salt additive. The data reported by Brooks et al.[52] on the ionic composition of the phases of Dex-PEG-water system containing different amounts of potassium sulfate fit Equation 3.5.

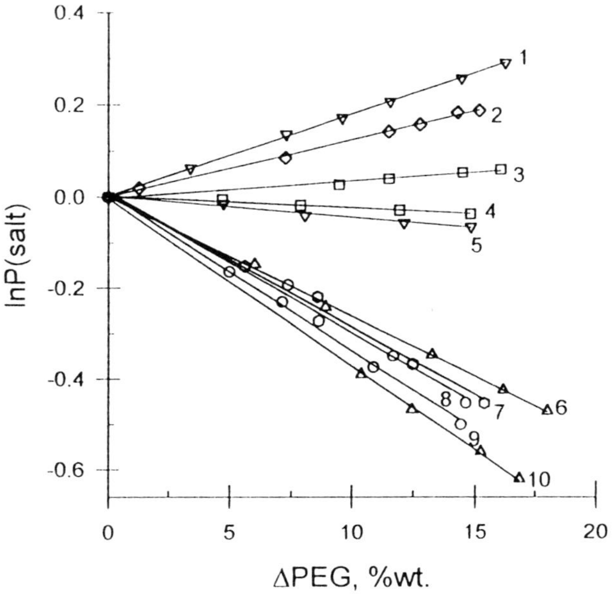

Figure 3.22. Distribution ratio of a salt additive in Dex—PEG—water two-phase system as a function of the difference in the PEG concentra-tions between the two phases. Salts (concentrations indicated in paren-thesis in mole/kg): 1 - NaSCN (0.1); 2 - KSCN (0.1-0.75); 3 - NH_4SCN (0.1); 4 - KCl (0.1); 5 - KCl (0.5); 6 - Na_2SO_4 (0.1); 7 - K_2SO_4 (0.05); 8 - K_2SO_4 (0.1); 9 - K_2SO_4 (0.25); 10 - $(NH_4)_2SO_4$ (0.1).

It can be seen from the data presented in Figs.3.22 and 3.23 that in both Dex-PEG-water and Dex-PVP-water systems the water-structure-promoting salts, e.g., KF, sulfates, concentrate in the lower Dex-rich phase, while the water-structure-breaking salts, for example, thiocyanates, $KClO_4$, concentrate in the upper PEG- or PVP-rich phase. The difference in the salt composition of the two phases increases with increasing difference in the polymer concentrations between the phases.

The effects of the phase polymers used in [55] on the water structure were discussed above (see in Chapter 2). These effects are likely to result in that water would be less structured in the Dex-rich phase relatively to PEG-, PVP, or Ficoll-rich phases. Experimental evidence for that is presented below. Hence, it follows from the data in [55] (Figs. 3.22 and 3.23) that the water-structure-making salts concentrate in the less structured aqueous medium of the Dex-rich phase, while the water-structure-breaking salts concentrate in the more

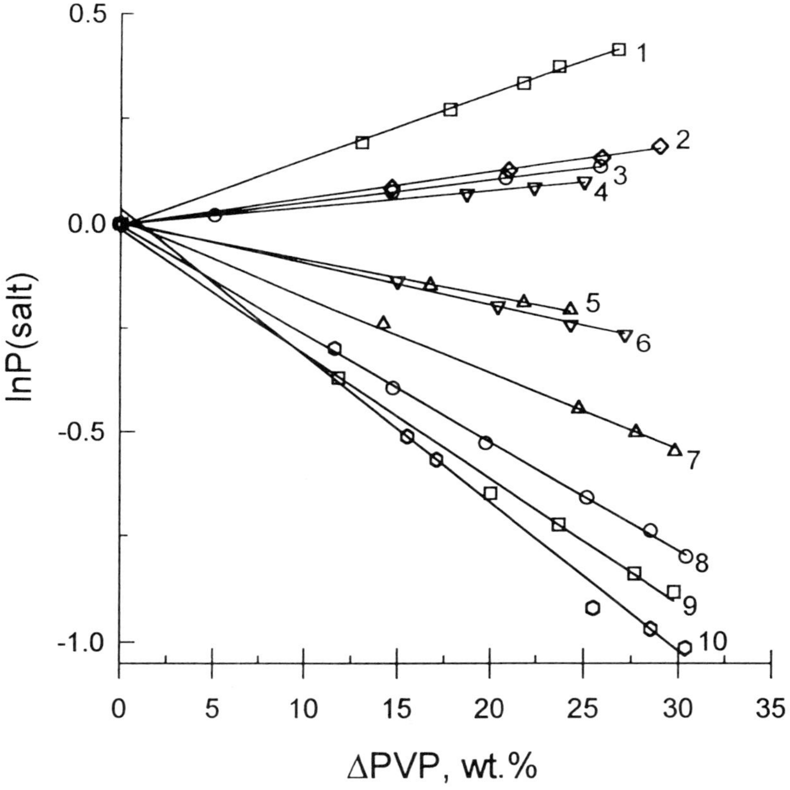

Figure 3.23. Distribution ratio of a salt additive in Dex—PVP—water two-phase system as a function of the difference in the PVP concentrations between the two phases. Salts (concentrations 0.1 mole/kg): 1 - NaSCN; 2 - NH_4SCN; 3 - KSCN; 4 - $KClO_4$ (0.05); 5 - KBr; 6 - KCl; 7 - KF; 8 - Na_2SO_4; 9 - $(NH_4)_2SO_4$; 10 - K_2SO_4.

structured medium in the PEG-, PVP-, or Ficoll-rich phase.

The possibility of direct salt-polymer interaction not taken into account in the above consideration seems to be highly unlikely as implied by an existence of a relationship between the b(salt) values for different aqueous two-phase systems examined in [55]. The relationships in question shown in Fig.3.24 are described as:

$$b(\text{salt i})_j = \alpha_j + \beta_j \cdot b(\text{salt i})_o \qquad (3.6)$$

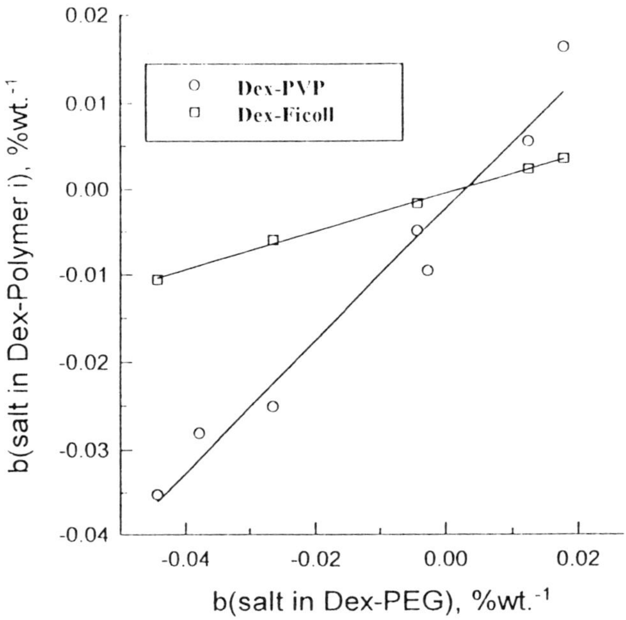

Figure 3.24. Relationship between the coefficients b(salt) for different salt additives in aqueous Dex—PEG and Dex—PVP and Dex—Ficoll two-phase systems (at the same salt concentration of 0.10 mole/kg).

where index "i" denotes salt additive; subscripts "j" and "o" denote Dex-polymer j (PVP, Ficoll)-water and Dex-PEG-water two-phase systems, respectively; α_j and β_j are constants.

An existence of the relationship described by Equation 3.6 implies a universal mechanism for salt behavior in the aqueous two-phase systems under discussion. This mechanism may be either the highly unlikely similar order in intensity of interactions of different salts with different phase polymers, or that salt behavior in different systems is governed by different influence of the polymers on the water structure in the phases.

The latter explanation seems to be more plausible as implied particularly by that the β_j values are directly proportional (see Fig. 3.24) to the relative immobilizing structuring influence of PEG, PVP, and Ficoll on water represented by the $B(\tau)$ value characterizing polymer effect on the dielectric orientational mobility of water (see Equation 2.22).

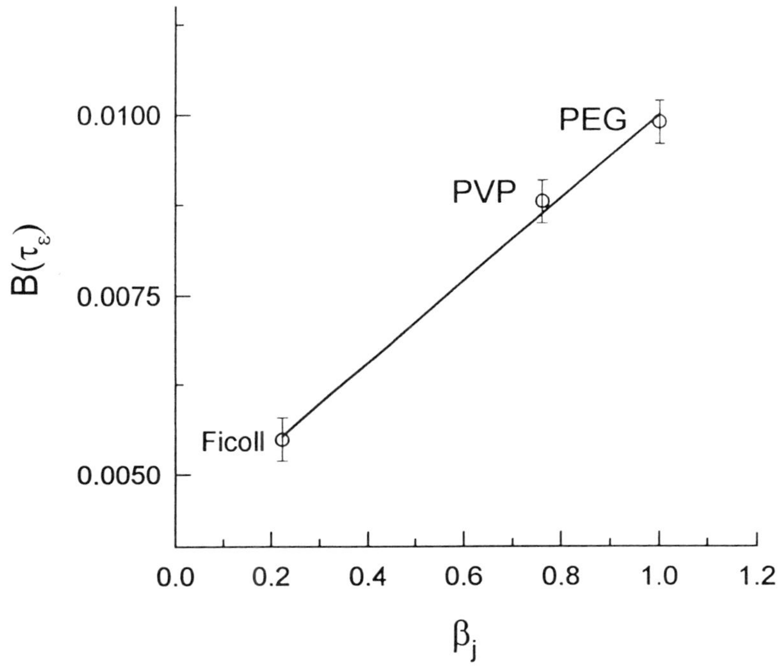

Figure 3.25. Relationship between coefficient β_j (Eqn. 3.6) and parameter $B(\tau_\varepsilon)$ (Eqn. 2.22). For explanation see text.

It should be stressed, however, that as shown above a salt additive does not merely distribute between the two phases of the fixed polymer composition. The additive itself seems to influence the water structure in the phases resulting in the redistribution of the polymers and the corresponding change in the polymer composition. Additionally to the data on the salt-induced shifts of phase diagrams given above, the results presented in Fig. 3.26 indicate that the tie line slope (STL) is directly related to the total salt concentration in an aqueous two-polymer two-phase system. The effect of a salt seems to be similar to that of temperature supporting the assumption that the effect is realized via salt influence on the water structure.

The role of a salt additive in an aqueous two-nonionic polymer two-phase system seems to be analogous to that of an organic modifier in water-organic solvent two-phase systems. The similarities in question warrant more comprehensive consideration.

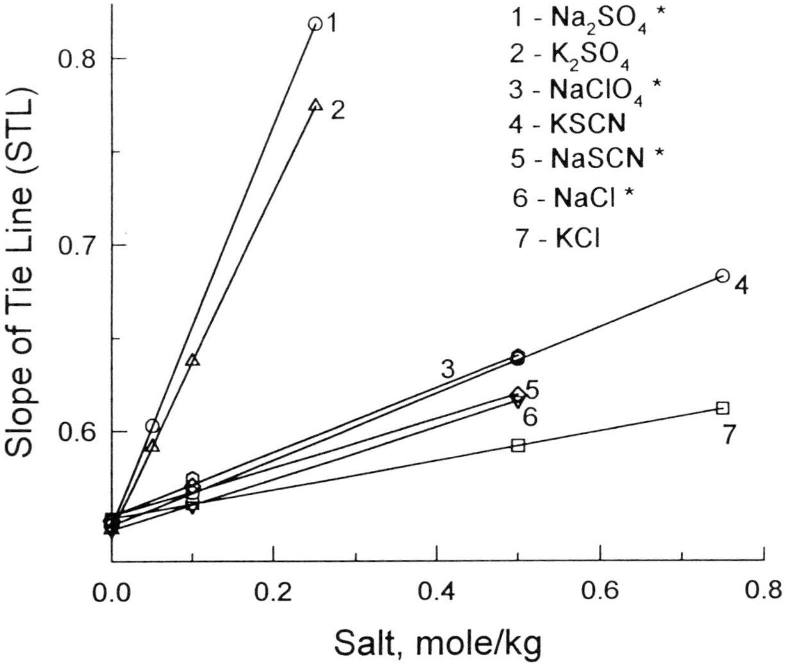

Figure 3.26. Slope of tie line (STL) for phase diagrams for Dex—PEG—water systems as a function of total concentration of salt additive. Calculated from data in [55] and from Zaslavsky et al., unpublished data. Systems denoted with * contains additionally 0.01 mole/kg universal buffer (composed of acetic acid, phosphoric acid, boric acid, and NaOH), pH 7.4.

Similarities between phase separation in aqueous polymer systems and in water-organic solvent systems.

Phase separation in solvent mixtures, such as water and chloroform, butanol, etc., is due to the limited mutual solubility of these solvents. These solvent two-phase systems are commonly used in liquid chromatography and extraction procedures. To vary the properties of the phases in a given water-organic solvent system the so-called organic modifier is usually added into the system [11]. There are generally two types of modifiers. The first type includes modifiers which are only slightly soluble in one of the two phases. An example is the system pentanol-heptane-water, where the composition of the aqueous phase is essentially independent of the amount of heptane additive in the

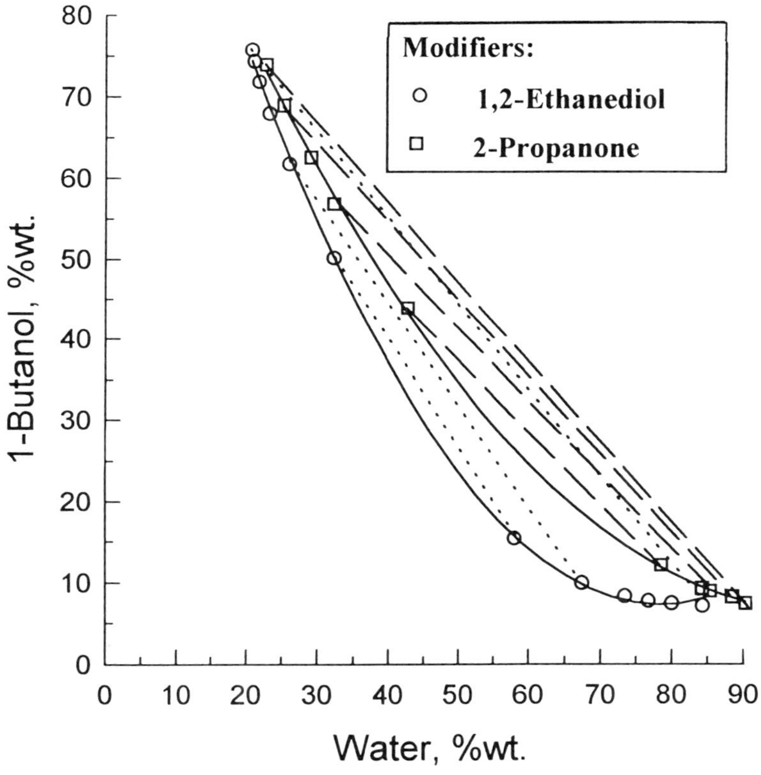

Figure 3.27. Rectangular phase diagrams for water—1-butanol two-phase systems at 27°C containing varied amounts of organic modifiers. Calculated from data in [63].

organic phase. The second type of modifiers covers additives miscible with both phases, e.g., methanol additive in water-chloroform two-phase system. Modifiers of this type appear to display effects similar to those of salt additives in aqueous two-nonionic polymer two-phase systems.

Ternary solvent systems under consideration are composed of water and two organic solvents. It is necessary to define which solvent is considered as a modifier and which as the "primary" organic solvent. The solvent defined as the primary is the one which has an excess over the other (defined as a modifier) in the nonaqueous phase. The compositions of the phases of ternary

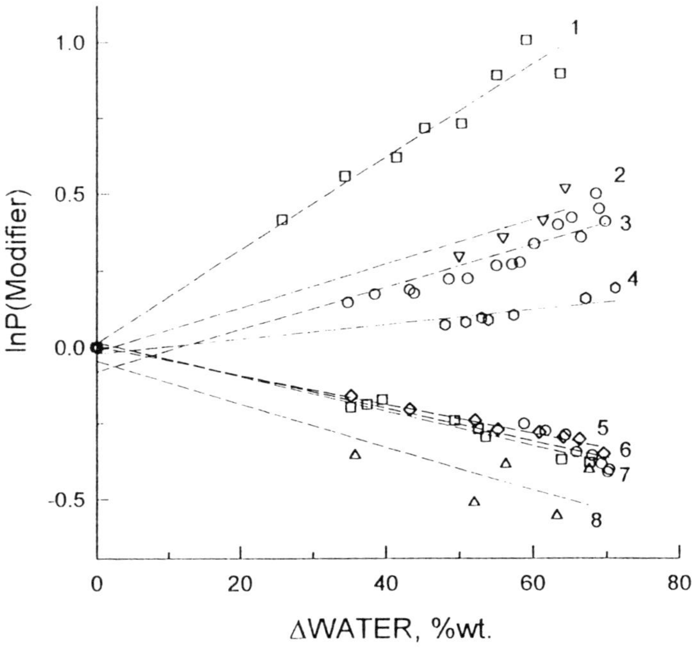

Figure 3.28. Distribution ratio of an organic modifier in 1-butanol—water two-phase system as a function of the difference in the water concentrations between the two phases. Calculated from data in [63].
Modifier: 1 - 1,2-ethanediol; 2 - 2,3-butanediol; 3 - methanol; 4 - 2-hydroxy-propanoic acid; 5 - acetic acid; 6 - succinic acid; 7 - ethanol; 8 - 2-propanone.

solvent systems listed in [63] covered by this definition were analyzed at random.

The results showed that, first, an organic modifier changes the solvent composition of the two phases like a salt additive does in aqueous two-nonionic polymer systems. An example is given in Figure 3.26. To emphasize the similarities under discussion phase diagrams for 1-butanol–water system containing different amounts of organic modifiers, 1,2-ethanediol and 2-propanone, are presented in rectangular form instead of commonly used [11,63] triangular form.

Analysis of these and other phase diagrams listed in [63] indicates that in very many though not all cases distribution ratio of an organic modifier is related to the difference in the concentration of water between the two phases

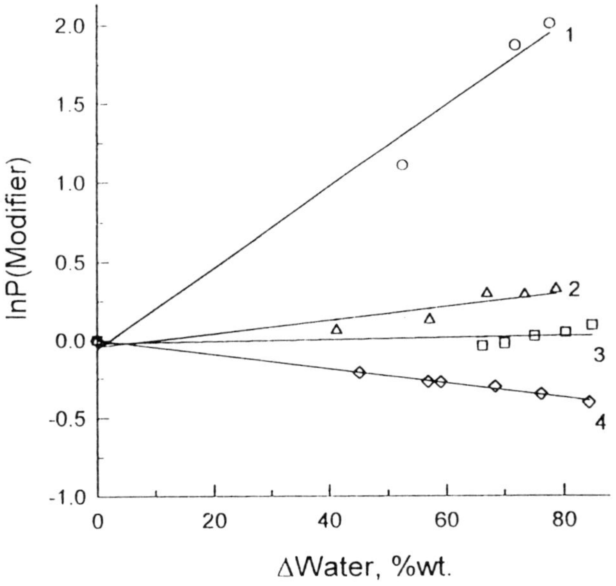

Figure 3.29. Distribution ratio of an organic modifier in ethyl ester of acetic acid—water two-phase system as a function of the difference in the water concentrations between the two phases. Calculated from data in [63]. Modifier: 1 - methanol; 2 - ethanol; 3 - acetic acid; 4 - 2-propanone.

exactly as shown above for salt additives in aqueous two-nonionic polymer systems [55]. These relationships are illustrated in Figures 3.27 and 3.28.

The relationships in question are described as:

$$\ln P(\text{modifier}) = b(\text{modifier}) \cdot \Delta C(\text{water}) \tag{3.7}$$

where P(modifier) is the distribution ratio of an organic modifier defined as the ratio between the concentrations (in wt.%) of the modifier in aqueous and organic solvent phases; ΔC(water) is the difference in the water concentration between the two phases; b(modifier) is a constant.

An existence of a relationship between b(modifier) values in different organic solvent-water two-phase systems was also explored. Unfortunately, the data on the same modifiers in different solvent systems [63] are limited but it was possible to establish that such the relationship does exists as shown in Fig. 3.30 for butanol-water and ethyl ester of acetic acid-water two-phase systems.

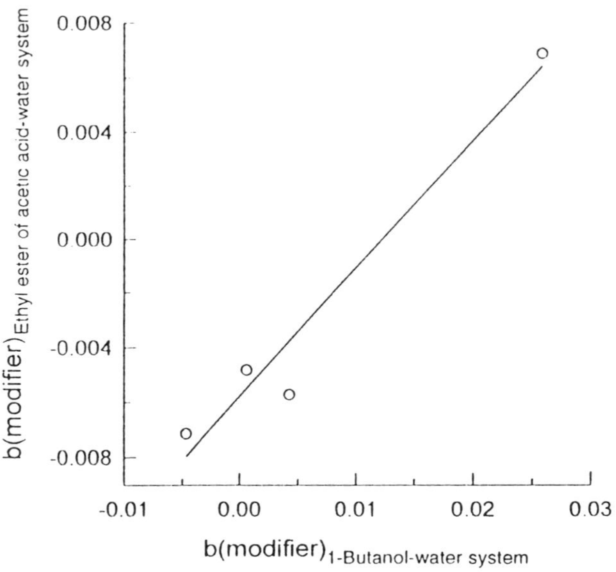

Figure 3.30. Relationship between the coefficients b(modifier) for different organic modifiers in two different organic solvent—water two-phase systems.

An additional similarity between effects of organic modifiers in the solvent two-phase systems and those of salt additives in aqueous two-nonionic polymer systems is observed when the effect of the total organic modifier concentration on the slope of tie line (STL) in a solvent system is analyzed. The STL value is determined as the ratio between the difference in water concentration between the two phases to that of an organic solvent. Results of analysis of the data reported by Conway [11] and by Gluck et al.[64] are given in Figure 3.31. The similarity between the data in Fig.3.31 and 3.26 is obvious.

Generality of the above trends for modifier-containing organic solvent-water two-phase systems should be explored more closely but it is beyond the scope of the present consideration.

General trends for phase separation in aqueous polymer systems outlined above and, particularly, the similarities between the trends observed for these systems and solvent two-phase systems imply a fundamental role of the solvent - water, in the phenomenon under consideration. These similarities, while hardly accidental do not provide any better insight into phase separation in aqueous polymer systems.

Theoretical treatments of phase separation in aqueous polymer systems suggested in the literature, however, usually do not take the role of water into consideration. These treatments are briefly discussed in the following section.

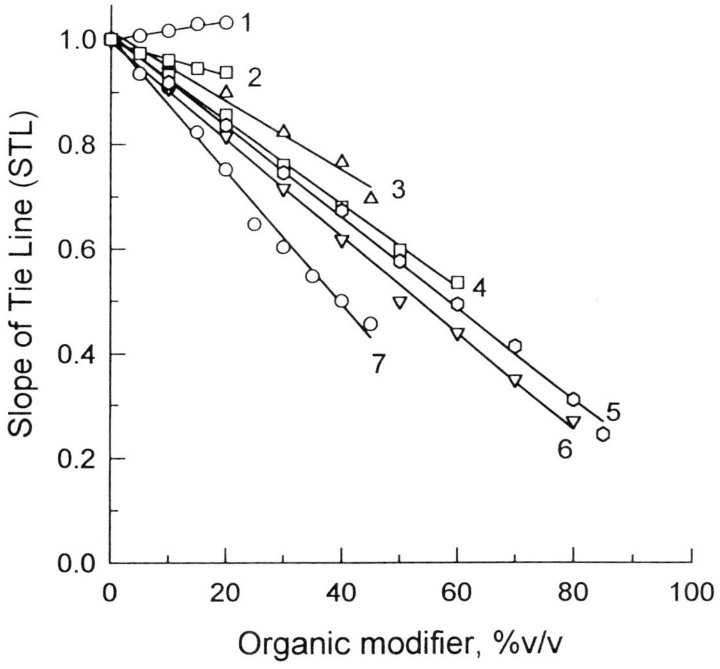

Figure 3.31. Slope of tie line (STL) for phase diagrams for organic solvent-water systems as a function of the concentration of organic modifier. Calculated from data in [11] and [64]. Solvent (modifier): 1 - Butanol (acetic acid); 2 - butanol (methanol); 3 - ethyl acetate (methanol); 4 - chloroform (methanol); 5 - heptane (2-propanol); 6 - ethylene dichloride (methanol); 7 - methyl tert.-butyl ether (acetonitrile).

3.4. THEORETICAL TREATMENTS OF PHASE SEPARATION

It should be pointed out, first, that no theory of phase separation in aqueous polymer systems capable of predicting phase behavior from the chemical structures of the components or from their individual physical features exists as yet. Theoretical treatments of phase separation in the systems published in the literature so far are mostly the attempts to fit experimental data into existing theoretical concepts. These attempts are fundamentally important for a better understanding of the interactions governing partitioning of solutes and particles in aqueous two-phase systems as these interactions are clearly related to those causing phase separation in the systems.

Theoretical treatments of phase separation reported so far may be roughly divided into two major categories according to the roles attributed to the components of the systems in the process under consideration. The majority of the treatments belongs to the first category. These treatments are based on the view that the macromolecular nature of phase polymers is of the primary importance for phase separation, while water as a solvent plays a secondary role. The second category includes the treatments based on the concept that the unique character of water as a medium plays the primary role in phase separation. The macromolecular nature of phase polymers is viewed in these treatments as an important factor providing the polymers with the strength of an influence on the water structure required for phase separation and usually lacking in low molecular weight compounds.

Only basic features of the concepts used for theoretical treatment of phase separation, their advantages and drawbacks are discussed below as well as how the results of these treatments agree with the experimental facts described above. For details of the theoretical models and treatments the reader is referred to the following references.

Treatments of phase separation based on the Flory-Huggins theory.

The original Flory-Huggins theory of polymer solutions [2] was first applied to polymer mixtures in a common solvent by Scott [65] and Tompa [4, 66]. Both authors used the expression for the free energy of mixing given by Flory [2] (see Equation 2.8) and derived the free energy of mixing of two polymers in a single common solvent as:

$$\Delta G^{mix} = (RT \cdot V/V_s) \cdot [\phi_s \cdot \ln\phi_s + \phi_1 \cdot \ln(\phi_1/v_1) + \phi_2 \cdot \ln(\phi_2/v_2) + + \chi_{12} \cdot \phi_1 \cdot \phi_2 + \chi_{1s} \cdot \phi_1 \cdot \phi_s + \chi_{2s} \cdot \phi_2 \cdot \phi_s] \quad (3.8a)$$

where V is the total volume of the mixture; V_s is the molar volume of the solvent; χ_{1s}, χ_{2s}, and χ_{12} are the Flory interaction parameters (components 1 and 2 are the two polymers; component s is the solvent); v_1 and v_2 are the ratios of the molar volumes of the polymer 1 and polymer 2 to the molar volume of the solvent.

Scott [65] examined the special "symmetrical" case (the interactions of both polymers with a solvent are similar, $\chi_{1s} = \chi_{2s}$, and the molecular weights of the polymers are identical, $v_1 = v_2$) for which the equations describing the chemical potentials of all the components of the system are symmetrical. In this case the phase diagram may be calculated without any special difficulty. The analysis performed by Scott [65] and Tompa [4,66] showed that the expressions derived for the polymer compositions of the coexisting phases and for the critical point of the phase diagram do not include parameters χ_{1s} and χ_{2s}. Therefore it was concluded that the polymer-solvent interactions are of no importance for phase separation in the ternary systems under consideration [2,4,65,66]. Phase separation was supposed to result solely from the repulsive polymer-polymer interactions ($\chi_{12} > 0$). Role of the solvent according to this theory [2,4] is merely to dilute the unfavorable polymer 1-polymer 2 contacts.

This conclusion was apparently supported by the experimental observations reported by Dobry et al.[67]. Of the 35 pairs of polymers examined by Dobry et al.[67], only 3 were found to be compatible up to moderate concentrations. Polymer pairs incompatible in one solvent were found [67] to be incompatible also in other solvents, in total agreement with the theoretically predicted secondary role of the solvent-polymer interactions. According to Nesterov et al.[3], the results reported by Dobry et al.[67] slowed down the advancement of studies of the mechanisms behind phase separation in ternary polymer systems for about 20 years.

An effect of a solvent on the compatibility of two polymers was observed first by Bank et al.[68] found that polystyrene and poly(vinyl methyl ether) are compatible in some solvents, for example, in toluene, benzene, or perchloroethylene, while incompatible in the others, e.g., chloroform, methylene chloride, etc.

These and numerous other experimental data are in line with the conclusions drawn by Zeman and Patterson [69] and Hsu and Prausnitz [70] from reexamination of the Scott [65] and Tompa [4,66] analysis. Relaxing the restriction of equal strength of different polymer-solvent interactions, the authors [69,70] showed that phase separation may be caused not only by the positive χ_{12} value but also by the difference in strengths of the polymer-solvent interactions, i.e., by
$|\chi_{1s} - \chi_{2s}| = |\Delta\chi|$. It was shown also [69,70] that if χ_{12} is negative (as for a compatible polymer pair) but $|\Delta\chi|$ is sufficiently large, the phase diagram will contain a closed region or loop within which phase separation occurs. For review of the literature on the so-called $\Delta\chi$-effect the reader is referred to [3,5].

It should be emphasized that the Flory-Huggins theory is based on the van der Waals model for intermolecular interactions. The theory has been developed by Flory [2] for non-polar polymer systems, and it was particularly stressed by Tompa [4, p.68] that the model should not be used to represent mixtures of polar components, where the interaction energies may depend on the mutual orientations of the molecules.

In spite of that, the theory has been numerously applied to aqueous polymer systems [25,26,35,36,50,71-80], though it has been shown [72,77,79-82] that aqueous polymer solutions exhibit significant deviations from the Flory-Huggins theory. These deviations may be listed as follows [80-82]:

i) Aqueous polymer solutions often phase separate on heating and sometimes they are characterized by phase diagrams of the "closed loop" type, i.e., the miscible single phase system is formed on heating from the two-phase system formed at lower temperature (solutions of PEG and polyvinyl alcohol (PVA) with some residual acetate groups offer the examples);

ii) Polymer molecules in aqueous medium are known to form aggregates or association complexes, and some of them show gelation upon heating or on cooling, or even only on standing (the typical examples are dextran, PVA, starch, etc.);

iii) Parameter χ which according to the current considerations [3] is concentration-dependent increases slightly with the volume fraction of the polymer ϕ_2 of a given polymer in some cases in agreement with the theory, but passes through a maximum and then decreases gradually at higher concentrations (the example is PVP [80]). In other cases the χ-parameter increases as predicted by the theory but decreases rapidly when $\phi_2 > 0.85$-0.9 (the examples given in [80] are PEG and polyvinyl acetal);

iv) The heat of dilution of an aqueous polymer solution Δ_H is generally negative, and changes up to a certain polymer concentration in agreement with the theory, but the Δ_H-concentration curve passes through a minimum and starts to increase [80] (the example is offered by PEG solutions).

The above deviations imply the shortcomings of the accepted polymer solution theory as applied to aqueous solutions. These shortcomings are supposed [80-82] to originate from the fundamental assumptions of the theory that: i) liquid lattice model provides an adequate representation of an aqueous solution; ii) the ideal entropy of mixing can be used; iii) deviations from ideal solution behavior can be simply accounted for in terms of the (enthalpic) polymer-solvent interaction parameter; and iv) the polymer chains are fully flexible.

It was found [14,25,26,50,73-76], however, that the Flory theory although unintended for use with strongly associated solvents such as water, may describe certain qualitative aspects of phase separation in polymer mixtures in aqueous media. That seems to be the reason for the opinion [74,83-85] that phase separation in an aqueous mixture of two polymers is due to the polymer-polymer repulsive interactions ($\chi_{12} > 0$) and that the solvent (water) is of no importance in the process. From this view [71,83-85] follows an erroneous (see below) conclusion that the partition behavior of a solute added at low concentration to an aqueous two-polymer two-phase system is governed by the interactions of the solute with the phase polymers.

The Flory-Huggins polymer-solvent interaction parameters reported in the literature for some water-soluble polymers are presented in Table 3.3. It should be noted that the χ_{1s} values listed in the Table are those determined for

Table 3.3.

Flory-Huggins Interaction Parameters χ_{1s} and Their Enthalpic Parts χ_H for Aqueous Polymer Solutions

Polymer [a]	χ_{1s}	χ_H	Ref.
Polyacrylic acid	0.498	-	[80,81]
Polymethacrylic acid	0.499	-	-"-
Polyacrylamide	0.44-0.49	0.075	-"-
Polymethacrylamide	0.490	-	-"-
Polyvinyl alcohol	0.8	-	-"-
Polyvinylpyrrolidone	0.48	-0.036	-"-
-"- (0.5 M NaCl)	-	-0.06	-"-
-"- (0.5 M KI)	-	-0.1	-"-
-"- (0.5 M $(NH_4)_2SO_4$)	-	-0.11	-"-
PEG-8000	0.45	-0.192	-"-
PEG-8000	0.436	-	[73]
PEG-8000 (5.9% wt.)	0.419	-	[77]
-"- (12.5% wt.)	0.441	-	-"-
-"- (22.5% wt.)	0.478	-	-"-
-"- (32.7% wt.)	0.510	-	-"-
PEG-3350 (5.0% wt.)	0.435	-	-"-
-"- (15.0% wt.)	0.451	-	-"-
-"- (24.7% wt.)	0.467	-	-"-
-"- (30.5% wt.)	0.499	-	-"-
PEG-20000	0.490	-	[35,36]
-"- (0.3 M K_2CO_3)	0.486	-0.147	-"-
-"- (0.3 M K_3PO_4)	0.491	-0.140	-"-
-"- (0.3 M $ZnSO_4$)	0.488	-0.138	-"-
-"- (0.39 M $MgSO_4$)	0.498	-0.040	-"-
-"- (0.3 M Na_2SO_4)	0.490	-0.116	-"-
-"- (0.3 M K_2SO_4)	0.490	-0.093	-"-
-"- (0.8 M KF)	0.490	-0.135	-"-
-"- (2.4 M NaCl)	0.484	-0.183	-"-
-"- (2.4 M KCl)	0.489	-0.152	-"-

[a] Concentration of the polymer in aqueous solution or the concentration of an inorganic salt in the solution is given in parenthesis.

single polymer solutions. The interaction parameters values reported for aqueous two-polymer systems will be discussed further on.

It can be seen from the χ_H values listed in Table 3.3 that only polyacrylamide exhibits endothermic mixing with water in agreement with the Flory-Huggins theory. In all the other polymer solutions (for which the χ_H values are given) $\chi_H < 0$, i.e., entropic effects seem to dominate.

The Flory-Huggins interaction parameter χ_{1s} values for PEG-20000 in aqueous solutions of different inorganic salts [35,36] are essentially identical, implying that the PEG-water interactions are independent of the type of salt present. Tthat clearly disagrees with the experimental observations of significantly different effects of various salts on the cloud point of aqueous PEG solutions (see above), and that aqueous PEG solutions in the presence of K_2SO_4 or Na_2SO_4 phase separate, while in the presence of other salts, for example, NaCl or KCl, no phase separation occurs. Hence the equal χ_{1s} values for aqueous PEG-20000 solutions containing 0.3 M K_3PO_4 and 2.4 M NaCl are misleading. It follows from the above that the Flory-Huggins interaction parameter values are of little or no use to predict phase separation in aqueous single polymer systems.

It should be noted also that all the χ_{1s} values listed in Table 3.3 for different water-soluble polymers cluster around $\chi_{1s} = 0.5$ (the only exception being PVA). For nonaqueous polymer solutions χ_{1s} values span a much wider range [3]. It appears that the χ_{1s} values for different water-soluble polymers are extremely close to each other even for the polymers mutually incompatible in water, e.g., PEG-PVP, PEG-polyacrylamide (PAAm), etc. [1,79,86].

When the Flory-Huggins theory is used to treat aqueous polymer systems it is usually noted that the interpretation of the interaction parameter χ has to be modified because of the aqueous medium (see, e.g., in [3,74]). The reason is the peculiar physical properties of water the highly ordered structure of which is not taken into account by the polymer solution theories. These theories, however, have been applied to describe phase separation in aqueous two-polymer systems [14,73,74,79].

Different approaches have been used in the literature for analysis of phase diagrams for aqueous two-polymer systems in terms of the Flory-Huggins theory. One approach is based on using the χ_{1s}-value for phase polymer 1, e.g., PEG, (determined in separate measurements on the aqueous solutions of this polymer) for calculations of the χ_{2s} and χ_{12} values (polymer 2 may be, e.g., dextran) from the phase diagram [73,76,83]. The other approach is based on using interaction parameters as adjustable variables to fit a given phase diagram under restrictions dependent on the particular theoretical model used [15,74,76]. Finally, the third approach [87] is based on calculations of the interaction parameters values from the phase diagram without any fixed or adjustable variables provided the polymer composition of the sufficient number of phases and that of the critical point are determined accurately enough.

The most obvious and important drawback of all these approaches is, first, that polymer-solvent interaction parameters χ_{1s} and χ_{2s} are assumed to be concentration-independent in direct conflict with the experimental observations (see, e.g., in [77]). That may be one of the reasons for the commonly observed discrepancies between the tie line slopes (STL) values for theoretically calculated phase diagrams and experimental data even when there is a reasonable fit between the binodial lines (see, for example, in [74]).

The other limitation specific for the first approach is an assumption that the polymer-solvent interactions are identical in a binary and ternary polymer systems. This assumption implies that the presence of the polymer 1 in a two-polymer system does not affect the interactions of the polymer 2 with the solvent. This assumption while possibly true for non-polar systems is hardly likely when applied to aqueous systems. Any deviation of thermodynamic properties of the corresponding binary polymer-solvent systems from additivity in a ternary (two-polymer) system is generally attributed to intermolecular interactions between the polymers. Hence parameter χ_{12} representing the deviation is viewed as a measure of the polymer-polymer interactions. In an aqueous polymer solution, however, the deviation in question may be caused not by the interactions between the polymers but by the mutually dependent effects of the polymers on the polymer-solvent interactions mediated through the solvent.

All the treatments under consideration are based on that the chemical potentials of each component in each phase must be equal. The chemical potential of a component is usually expressed using the Scott's approximation [65] of the Flory-Huggins theory. In contrast to the known experimental evidence it is usually assumed that the polymer-solvent interaction parameters are independent of the polymer concentrations over the entire concentration range used. The Flory-Huggins interaction parameters then are calculated from the polymer compositions of the coexisting phases and from the molecular weights of the polymers, e.g., as described in [87,88].

This procedure was employed to estimate the interaction parameters for the aqueous casein-Ficoll [87], Dex-PEG, Dex-Ficoll, Dex-PVP, Dex-PVA [14] and several other two-phase systems under additional assumption that a four-component system (two polymers, water, and inorganic salt or urea) may be treated as a pseudo-ternary system (salt-water viewed as a modified water) though salt concentrations in the two phases are different.

The interaction parameters reported for the aqueous Dex-Ficoll system [87] are 0.573 for casein-water interactions, 0.605 for Ficoll-water interactions, and 0.007 for casein-Ficoll interactions. When these values are compared with

Table 3.4

Flory-Huggins Interaction Parameters for Ternary Dex—Polymer i—Water Two-Phase Systems at 23°C (Dex M_w 57,200)

Polymer i	M_w	$\chi_{i\text{-water}}$	$\chi_{Dex\text{-water}}$	$\chi_{Dex\text{-}i}$
Ficoll	~400,000	0.447±0.018	0.493±0.020	-0.0002±0.001
PEG	~8,000	0.502±0.028	0.467±0.048	0.042±0.008
PVP	12,700	0.573±0.015	0.592±0.012	0.011±0.003
PVA	55,000	0.805±0.068	0.683±0.049	0.040±0.007

those given in Table 3.4 for the aqueous Dex-Ficoll system reported in [14], it can be seen that the Ficoll-water interaction parameter values are different in the systems compared. The reason may be that different lots of Ficoll with possibly different molecular weight distributions were used by Grinberg et al. [87] and by Zaslavsky et al.[14]. It was shown by Kang and Sandler [76], in particular, that dextran-water interaction parameter varies with dextran molecular weight. The interaction parameters values given in Table 3.4 for the aqueous Dex-PEG, Dex-PVP, Dex-Ficoll, and Dex-PVA systems were calculated from the phase diagrams determined for the systems using the same lot of Dex [14]. The data in Table 3.4 indicate that the Dex-water interaction parameter value clearly depends on the type of the second phase polymer. These data indicate also that the polymer-polymer interaction parameter value is by one-two orders of magnitude less than the polymer-water interaction parameters values. Kang and Sandler [76] used a different procedure for analysis of the aqueous Dex-PEG systems described by Albertsson [1] and reported the estimates of the interaction parameters consistent with those in Table 3.4.

To explain the difference in the polymer-polymer and polymer-solvent interaction parameters values the physical meaning of the polymer-polymer interaction parameter should be considered. This parameter is generally accepted to be a measure of polymer-polymer interactions. It should be taken into consideration, however, that polar polymers in an aqueous medium are strongly hydrated. The macromolecules are surrounded by hydration shells of different water structures depending on the nature, size and shape of the macromolecules. Therefore the presumable repulsive interactions between the two poly-

Table 3.5
Flory-Huggins Interaction Parameters for the Dex—PVP—Water Two-Phase Systems [a]

Additive (or Temperature) [b]	Concentration (mole/kg)	$\chi_{Dex\text{-}water}$	$\chi_{PVP\text{-}water}$	$\chi_{Dex\text{-}PVP}$
8.0°C		0.542	0.564	0.013
23.0°C		0.573	0.592	0.011
38.0°C		0.620	0.629	0.011
50.0°C		0.631	0.637	0.012
NH_4SCN	0.10	0.615	0.640	0.011
NaSCN	0.10	0.590	0.620	0.010
KSCN	0.10	0.612	0.640	0.009
$KClO_4$	0.05	0.562	0.581	0.010
KBr	0.10	0.506	0.537	0.010
KCl	0.10	0.613	0.628	0.011
KF	0.10	0.682	0.681	0.013
$(NH_4)_2SO_4$	0.10	0.651	0.661	0.013
Na_2SO_4	0.10	0.633	0.649	0.014
Cs_2SO_4	0.10	0.688	0.689	0.014
K_2SO_4	0.10	0.615	0.628	0.014
NaCl + PBS [c]		0.584	0.592	0.011
PBS [c]	0.11	0.638	0.637	0.015
Urea	0.50	0.638	0.638	0.010
Urea	2.00	0.631	0.646	0.010

[a] Estimated errors are less than 5% for all the χ-values given;
[b] Temperature is 23°C unless otherwise specified;
[c] 0.15 mole/kg NaCl in 0.01 mole/kg PBS; PBS - sodium phosphate buffer, pH 7.4

mers in an aqueous medium are likely to occur not between the macromolecules themselves but between the corresponding hydration shells. As these shells differ from each other due to the difference in the polymer-solvent interactions, the resulting effect may be regarded formally to be due to the $\Delta\chi$-effect (see above). As the polymer-polymer interaction parameter represents the repulsive interactions between the two differently arranged hydration shells, it is possible also to view it as a measure of the repulsion between two differently structured microphases of aqueous media in a given solution. Then the difference in the values between the polymer-polymer and polymer-solvent interaction parameters seems to be quite reasonable.

The data for the aqueous Dex-PVP systems furnish the most illustrative example of the general trends observed from the analysis of phase diagrams in terms of the Flory-Huggins theory [14]. The data in Table 3.5 indicate that an addition of a salt or urea, or an increase in the temperature producing significant changes of the phase diagram (see Figs. 3.14, 3.15, 3.19) does not affect the Dex-PVP interaction parameter $\chi_{Dex\text{-}PVP}$ value which amounts to 0.012 ± 0.002 for all the systems examined. On the one hand, this agrees with the above hypothesis that there are no direct interactions between the phase polymers and the ions or urea present in the systems. On the other hand, it indicates that the observed alterations of the polymers' compatibility in aqueous solution induced by a salt or urea or by the temperature increase are not due to any changes in the Dex-PVP interactions (if these interactions occur at all). Hence it indicates that the additives and the temperature alterations affect the interactions of the phase polymers with the solvent - water.

It can be seen from the data in Table 3.5 that the interaction parameters for the polymer-solvent interactions $\chi_{Dex\text{-}water}$ and $\chi_{PVP\text{-}water}$ depend on the type of additive and temperature of the system.

Analysis of the polymer-solvent interaction parameters values presented in Table 3.5 indicates that there is a relationship between these parameters described as [14]:

$$\chi_{Dex\text{-}water} = A + B \cdot \chi_{PVP\text{-}water} \tag{3.9}$$

where $A = -0.112 \pm 0.034$; $B = 1.157 \pm 0.054$ are constants; the correlation coefficient $r^2 = 0.964$ for $N = 19$ (number of the systems examined).

Using the familiar general expressions for the polymer-solvent interaction parameter given by the Flory-Huggins theory (see Equation 2.7) the Dex-water interaction parameter may be expressed as:

$$\chi_{Dex\text{-}water} = (z_{Dex} \cdot \Delta w_{Dex\text{-}water})/kT = (z_{Dex}/kt) \cdot [\tfrac{1}{2} \cdot (w_{Dex\text{-}Dex} + w_{water\text{-}water}) - w_{Dex\text{-}water}] \tag{3.10}$$

where z_{Dex} is the number of the Dex repeating units-water contacts; $\Delta w_{Dex\text{-}water}$ is the energy change associated with creating a new Dex repeating unit-water molecule contact; k is the Boltzmann constant; T the absolute temperature; $w_{Dex\text{-}Dex}$ is the free energy of the Dex-Dex repeating units interactions; $w_{water\text{-}water}$ is the free energy of water-water interactions; $w_{Dex\text{-}water}$ is the free energy of the Dex repeating unit-water interaction.

From the similar expression for the PVP-water interaction parameter it is derived:

$$\tfrac{1}{2}\cdot w_{water\text{-}water} = (kT/z_{PVP})\cdot\chi_{PVP\text{-}water} + w_{PVP\text{-}water} - \tfrac{1}{2}\cdot w_{PVP\text{-}PVP} \quad (3.11)$$

where the meaning of all the terms is the same as above but they are related to PVP.

Taking into account that phase separation in the aqueous Dex-PVP system occurs in a given aqueous medium Equations 3.10 and 3.11 can be combined and we obtain:

$$\chi_{Dex\text{-}water} = A + B\cdot\chi_{PVP\text{-}water} \quad (3.9a)$$

where

$$A = (z_{Dex}/kT)\cdot[\tfrac{1}{2}\cdot(w_{Dex\text{-}Dex} - w_{PVP\text{-}PVP}) - (w_{Dex\text{-}water} - w_{PVP\text{-}water})] \quad (3.12)$$

and

$$B = z_{Dex}/z_{PVP} \quad (3.13)$$

The constancy of the A and B coefficients in Equation 3.9 and the expressions 3.12 and 3.13 imply that the observed alterations of the phase diagrams for the Dex-PVP-water system induced by the temperature change or an addition of an inorganic salt or urea occur at the constant value of $(w_{Dex\text{-}water} - w_{PVP\text{-}water})$ term as the $(w_{Dex\text{-}Dex} - w_{PVP\text{-}PVP})$ term value is obviously independent of the solvent.

Taking into account Equation 3.10, it can be concluded that the observed alterations of the $\chi_{Dex\text{-}water}$ and $\chi_{PVP\text{-}water}$ values (see Table 3.5) are governed by the changes in the parameter $w_{water\text{-}water}$ value induced by the temperature alteration or addition of urea or inorganic salt affecting the structure of water in the system.

It should be noted that the Flory-Huggins interaction parameters determined in [14] for the aqueous Dex-PEG and Dex-PVA systems also fit the Equation 3.9 (with different values of the A and B constants).

Thus, the empirical relationship 3.9 between the polymer-solvent interaction parameters found for the systems [14] can be taken to imply that phase separation depends on the structure of water in the systems. This conclusion clearly agrees with the experimental observations discussed above, while it is in total disagreement with the basic assumptions of the theory. That means that the physical meaning of the interaction parameters as applied to aqueous polymer systems may be different from the generally accepted one. Consideration of that, however, is beyond the scope of the present discussion.

The indicated above drawbacks of the Flory-Huggins theory as applied to aqueous polymer systems are the common arguments for the advantages of the theoretical treatments based on the virial expansion model.

Theoretical treatments based on virial expansion model.

The virial expansion-type approach for treatment of phase separation in aqueous polymer systems has one definite advantage over those based on the Flory-Huggins theory. This approach does not use a lattice model to represent an aqueous polymer solution. It is based on description of the thermodynamic properties of a polymer solution by a power series in the polymer concentrations with empirically determined coefficients (see Equation 2.11).

Analysis of phase diagrams according to this approach is based on that the chemical potentials of each component in each phase must be equal. For example, the chemical potential of a solvent is described as:

$$\mu_1 - \mu^o_1 = -kT \cdot v_1 \cdot \rho_1 \cdot [m_2 + m_3 + ½ \cdot a_{22} \cdot (m_2)^2 + ½ \cdot a_{33} \cdot (m_3)^2 + a_{23} \cdot (m_2 \cdot m_3)] \quad (3.14)$$

where μ^o_1 is the chemical potential of pure solvent denoted by "1"; v_1 is the pure solvent molar volume; ρ_1 is the solvent density; m_2 and m_3 are the molalities of the two polymers denoted by "2" and "3"; and a_{ij} are the second virial coefficients of the ith and jth components.

In dilute polymer solutions, the interaction coefficients reflect the pairwise interactions, i.e., a_{22} reflects the strength of the interaction between two macromolecules of the polymer 2, a_{23} reflects the pairwise interaction between two macromolecules of polymers 2 and 3, etc. The higher-order terms, not shown in Equation 3.14, reflect the simultaneous interactions between three or more macromolecules in the solution.

According to Edmond and Ogston [73], the degree of incompatibility between two polymers in water is indicated by the sign and magnitude of coefficient a_{23}.

Kang and Sandler [50,76] extended the model suggested by Edmond and Ogston [73], and Prausnitz et al.[77,89-92] and Cabezas et al.[93-96] developed more rigorous calculation procedures including the effects of salts and polydispersity of phase polymers in their calculations of phase diagrams.

An example of the typical results of calculations of phase diagrams for aqueous polymer systems [77] on the basis of the approach under consideration is given in Figure 3.32.

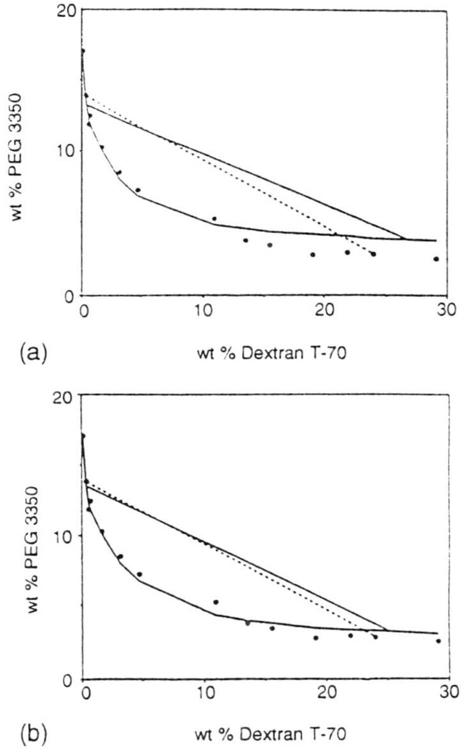

Figure 3.32. Comparison of experimental phase diagrams (broken line) with those predicted (solid line) using the osmotic virial expansion truncated (a) after the second virial coefficient and (b) after the third virial coefficient for the aqueous Dex-70—PEG-3350 two-phase system at 25°C. (From C. A. Haynes, R. A. Beynon, R. S. King, H. W. Blanch, J. M. Prausnitz, J. Phys. Chem., 93, 5612 (1989). Reprinted by permission of the American Chemical Society.)

The experimental data are compared in Fig. 3.32 with those calculated using virial expansions truncated at the second and third virial coefficients terms [77]. The calculated binodial and tie line in Fig. 3.32 fits the experimental data better when the higher-order terms are used [77]. In other words, the better fit is obtained when more adjustable parameters are used.

The virial expansion model represents an improvement over the treatments using the lattice model for aqueous polymer systems but it is still too far from reality. Essentially all the other drawbacks of the theoretical treatments based on the Flory-Huggins theory are shared by those based on the virial expansion model.

It was mentioned above that the application of the virial expansion truncated at the second-order terms is most appropriate for dilute polymer solutions. Aqueous polymer two-phase systems containing usually 5-25 wt.% of polymer in each phase, are certainly not in the dilute range, "and that may mean that the coefficients in the expansion are not true virial coefficients, but rather empirical parameters which may not have physical meaning" [97]. The other difficulty with the virial expansion approach is that similarly to the models discussed above the effects of two polymers on the thermodynamic activity of water are assumed to be independent of each other. The most fundamental problem, however, seems to be that the Edmond-Ogston expression (Equation 3.14) applies under the particular assumption that the solvent is non-interacting [98].

Recently a new approach to theoretical treatment of phase separation in aqueous polymer systems taking into consideration peculiar features of an aqueous medium has been suggested in the literature [99]. This approach based on the surface thermodynamics principles is outlined below.

Theoretical treatment based on the surface thermodynamics principles.

An advantage of the approach suggested by van Oss et al.[99, 100] is that it takes into account polar interactions. The basic features of the approach were outlined above (see in Chapter 2). It should be emphasized that the approach under consideration was not used to describe phase diagrams in aqueous polymer systems. Van Oss et al.[99, 100] addressed the most fundamental issue of the problem, namely, what is the driving force behind phase separation (in terms of surface thermodynamics principle).

The interfacial free energy change (per minimum effective surface area of contact between macromolecules of two polymers 1 and 2) associated with bringing together two macromolecules 1 and 2 initially present in solvent 3 with an effectively infinite layer of phase 3 separating surfaces of phase 1 and phase 2, ΔG_{132}, is used as a measure of the polymers compatibility in a given solvent.

Using the theoretical framework described above (see in Chapter 2) and experimental surface tension measurements van Oss et al.[99, 100] argue that "a van der Waals repulsion will not exist between Dex and PEG dissolved or suspended in water, because the van der Waals-Lifshitz surface tension, γ^{LW}, of water is only 21.8 mJ/m^2, while for both Dex and PEG γ^{LW} is of the order of 42 to 43 mJ/m^2, so that non-hydrated Dex and PEG would attract each other in water with an energy of about 7 mJ/m^2 ". (Note the contradiction between this statement and the basic concept of the Flory-Huggins theory.)

It was shown by van Oss et al.[99-103] that many proteins, polysaccharides, nucleic acids, etc., including Dex and PEG, are monopolar in the dried state, i.e., they are mainly Lewis bases (or electron donors). According to van Oss et al.[100] two monopolar solutes of the same type will repel each other, and phase separation results if the net repulsion energy exceeds the van der Waals energy of attraction.

From the surface tension measurements the hypothetical total energy of interaction between dry PEG and Dex macromolecules immersed in water was estimated [100] to be 46.7 mJ/m^2. The surface tension parameters for hydrated Dex and PEG could not be measured due to certain experimental difficulties but their values were assumed [100] to be close to those typical for hydrated biopolymers. For hy-drated biopolymers in water the total interaction energy ΔG_{132} according to van Oss et al.[99, 100], is of the order of -0.1 to -0.6 mJ/m^2. This value, compared to the ΔG_{132} for the dry macromolecules in water (46.7 mJ/m^2) is supposed [99, 100] to be almost zero.

It was assumed that the degree of dehydration of a polymer is linearly proportional to the total concentration of the polymer dissolved in water and that total dehydration is achieved at the polymer concentration of 100% wt. The total interaction energy at zero polymer concentration was taken as -0.6 mJ/m^2, and the total interaction energy at polymer concentration of 8 %(w/v) was estimated. This polymer concentration was chosen as it is close to the threshold polymer concentration for phase separation in aqueous Dex-PEG system [1]. The energy value found in [99, 100] amounts to +3 mJ/m^2, i.e., the mutual Dex-PEG repulsion energy amounts to about 1.5·kT as predicted by the model. Thus the reason for phase separation in an aqueous mixture of two monopolar polymers is suggested [99, 100] to be the mutual repulsion of the macromolecules provided the certain degree of the dehydration of the polymers is attained.

This model apparently makes an allowance for some role of water in phase separation in the systems under consideration. An assumption of a certain degree of dehydration of PEG at 8 %(w/v), however, clearly disagrees with the experimental facts (see in chapter 2). The model also cannot explain significant differences in phase diagrams for different polymer pairs in water, e.g., Dex-PEG and Dex-PVA or Dex-PVP, as well as the effects of water-structure-breaking salts (KSCN, NaSCN, etc.) or urea increasing the polymer concentrations required for phase separation and most of the other experimental observations outlined above. It is quite possible, however, that further development of the approach will overcome these problems.

Consideration of phase separation as the result of polymer effects on the water structure.

Zaslavsky et al. attempted to look at phase separation in aqueous polymer systems from the view point of effects of polymers on the water structure [14,40,53-55,62]. Once this view is adopted, everything seems to fall in place.

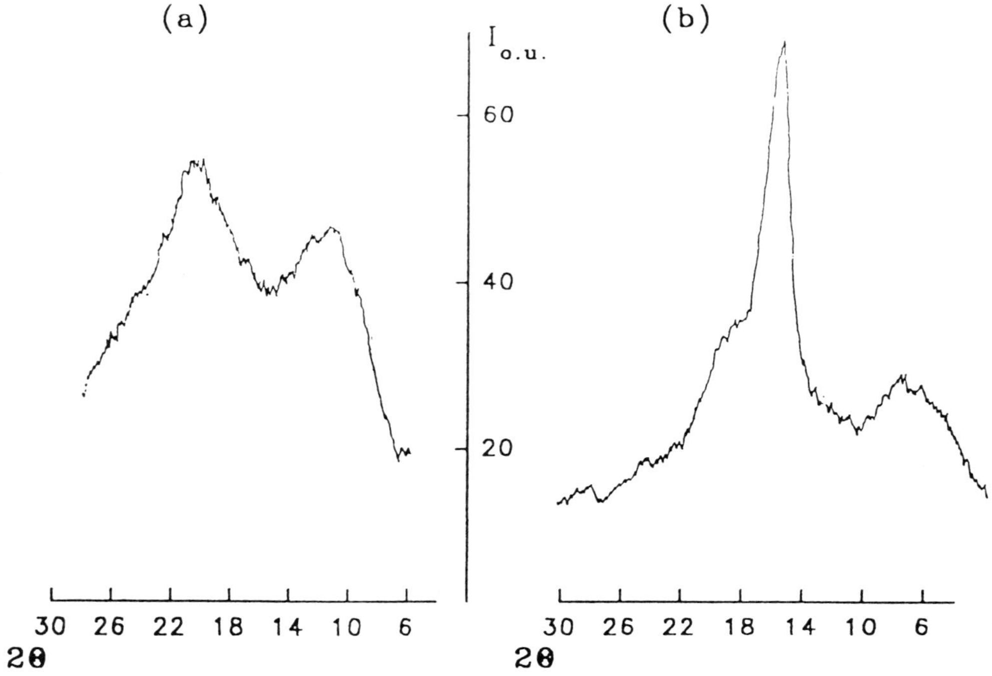

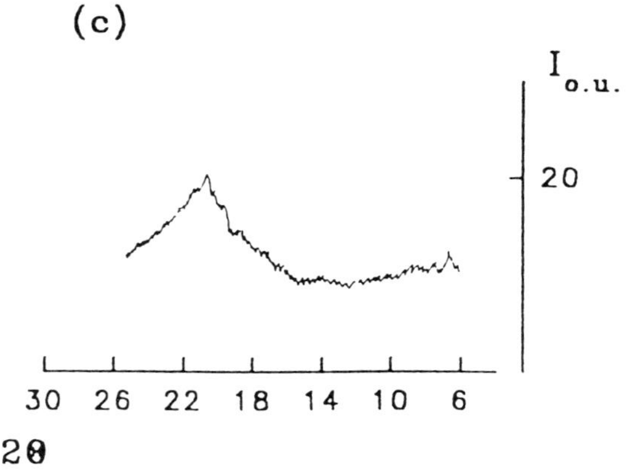

Figure 3.33. Diffractograms of dextran (a); polyvinyl alcohol (b); and 1:1 mixture of dextran and polyvinyl alcohol (c). Data from [104].

Examples of completely miscible polymer pairs in water seem hard to find in the literature probably due to that it is experimentally difficult to demonstrate compatibility of a two-polymers mixture in water conclusively. There seems to be no direct experimental evidence in favor or against the widely accepted opinion that water plays essentially no (or only secondary) role in phase separation in aqueous two-polymer systems.

In search of such experimental evidence the compatibility of several polymers' pairs in dry mixtures was studied [104] by means of the X-ray diffraction technique. The 1:1 mixtures of dextran with PEG, PVA, Ficoll, and PVP were examined and it was found that dextran is compatible with PVA in the absence of water. The X-ray diffraction data for these two polymers and for their dry 1:1 mixture are presented in Figure 3.33.

It can be seen from the data in Fig. 3.33 that the peaks specific for the individual polymers disappear in their dry 1:1 mixture. This is an unambiguous evidence that the polymers in question are totally compatible with each other in the absence of water. In aqueous medium the same polymers (Dex and PVA) phase separate at relatively low polymer concentration - composition of the critical point is 3.05% wt. Dex and 2.45% wt. PVA. (In terms of the Flory-Huggins theory that would mean that the polymers are highly incompatible.)

Hence it is unavoidable to conclude that phase separation at least in the case of the dextran-PVA pair is due not to the interactions of the polymers with each other but to their interactions with the solvent - water. It is possible that phase separation in aqueous mixtures of other polymer pairs is also due to the difference in the polymer-water interactions.

The differences in the interactions of different polymers with water as described above in Chapter 2 were established by various experimental methods. According to the solvatochromic, dielectric, and partition measurements as well as those of acid-base equilibria, the solvent features of aqueous medium are changed by the presence of polymers depending on the polymer type, molecular weight, and concentration.

The polymer influence on the solvent features of aqueous medium seems to decrease in the order: PEG > PVA > PVP > Ficoll > Dex, though positions of PVA and PVP appear to vary depending on the particular feature examined (see in Chapter 2). Polymer concentrations required for phase separation in water (without any additive) increase in the order Dex-PVA < Dex-PEG < Dex-PVP < Dex-Ficoll. Taking into account the difference in the polymers molecular weight, this order agrees with the above order of the polymer effect on the water features reasonably well.

An increase in the temperature as well as an addition of urea to the aqueous Dex-PEG, Dex-PVP, and Dex-PVA systems causes an increase in the amount of at least one of the polymers required for phase separation. Similar effect is produced by an addition of water-structure-breaking salts. An addition of water-structure-promoting salts affects the systems in the exactly opposite way. The general mechanism of influence of various water-structure-perturbing factors (temperature, urea and salt additives) on phase separation in aqueous mixtures of two non-ionic polymers is clearly implied by analysis [14] of the phase diagrams in terms of the Flory-Huggins theory. The only plausible explanation seems to be that the effects of all these factors on phase separation are realized via their influence on the water structure. The series of salts inducing phase separation in single-polymer aqueous systems (e.g., aqueous solutions of PEG) at room temperature includes the water-structure-promoting salts only (in line with the above considerations).

Behavior of salt additives in aqueous two-nonionic polymer systems is also readily explained from the viewpoint of the influence of the phase polymers and salt additives on the water structure. It is reasonable to assume that there is a condition of the minimal free energy of a two-phase system. To comply with this condition the difference between the water structures in the coexisting phases should be minimal. This may be achieved when the water-structure-making salts concentrate in the less water-structured (Dex-rich) phase and the water-structure-breaking salts concentrate in the other (PEG-, PVA-, PVP-, Ficoll-rich) phase with more structured water. The higher the polymer concentration in a given phase, the more structured water in the phase, and more water-structure-breaking salt amount in the phase is required to meet the above condition.

It may be assumed [14,105] that phase separation in aqueous polymer systems results from the influence of one or two polymers on the water structure. Phase separation may be considered in terms either of occurrence of two different water structures, or of limited mutual solubility of two different solvents of aqueous nature. In other words [14,105], phase separation in an aqueous mixture of two polymers results from the incompatibility of two different polymer-modified water structures. These structures are suggested [105] to be formed due to different arrangement of water molecules in the hydration shells of the unlike macromolecules with the possibility of association or overlapping between the shells of the similar macromolecules and mutual repulsion of those of the unlike macromolecules. As a result, two differently structured microphases of water ("filled" with the corresponding macromo-lecules) originate in the apparently homogeneous polymer mixture (below the binodial). This mixture may be viewed over the polymers concentrations range below the binodial as a thermodynamically stable microemulsion, i.e., as dispersion of liquid droplets of one phase in another phase with a given polymer as the core of the droplet. An increase in the amount of one or both poly-mers upsets the stability of this microemulsion and leads to "coagulation" of the droplets resulting in phase separation. Essentially similar model suggested by Treffry et al.[106] is illustrated in Figure 3.34.

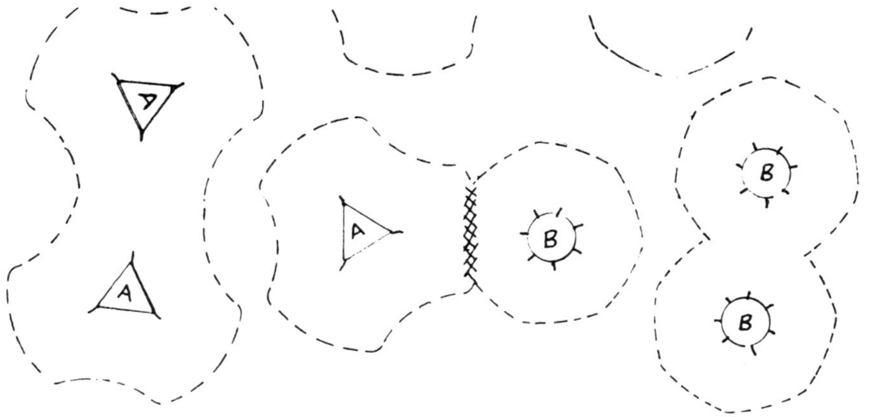

Figure 3.34. A and B represent polymer molecules surrounded by shells of solvent (water) molecules structured according to the steric arrangement of bonding sites on the polymer which may or may not match structure in the bulk solvent. Like shells may merge producing micro-aggregates (microphases). When unlike shells are forced together structural discontinuity leads to turbidity and phase separation. From T. E. Treffry, T. H. Lilley, P. J. Cheek, In: Separations Using Aqueous Phase Systems: Applications in Cell Biology and Biotechnology (D. Fisher and I. A. Sutherland, eds.), Plenum Press, New York, 1989, p.233. Reprinted by permission of Plenum Publishing Corp.

This model implies that the water structure and solvent properties of aqueous media in the coexisting phases of the systems must be different and the two phases may be viewed as the pair of two different solvents. The experimental evidence for this hypothesis and its biological implications are discussed further on.

Recently this concept was developed further using a new "statistical geometrical" theoretical treatment suggested by Guan, Treffry, and Lilley [107-109]. According to the authors [107], phase separation in aqueous two-phase systems must depend on the structure, i.e. the nature of the intermolecular forces, of the liquid phase.

The so-called binodal model [107-109] is based on the following assumptions. Molecules of the same species are assumed to be distributed at random in a homogeneous phase. The nature of this random molecular distribution is supposed not to be affected by molecular interactions. On the binodial curve the effects of the molecular interactions are suggested to be displayed as an adjustment of the average distances between the centers of the unlike molecules. The structure of the solution on the binodial curve is considered to be "geometrically saturated" [107]. That means that any point on the binodial curve of an aqueous two-polymer or single polymer-salt system is viewed as a saturated solution of polymer 1 in the solvent (aqueous solution of polymer 2 or salt) or as a saturated solution of a salt (or polymer 2) in the aqueous solution of polymer 1. The term "geometrically saturated" is used [107] to denote the space occupation and to indicate that the volume available for a given component is constrained to the geometrical limit.

The authors [109] emphasized that the above assumption means that "the solution structures before and after (or from) phase separations are different. Before phase separation (i.e., in the one-phase region), solute molecules are separated so that additional solute molecules can still be inserted into the free space which is present. At the point of phase separation, the solute molecules are now so closely packed that the solution is not able to accept any additional solute molecules and when the total solute concentration is increased, what happens is the formation of two geometrically saturated yet structurally quite different solutions. The further the system is from the critical point, the more different are the structures of these two saturated solutions." (The term "solute" is used by the authors [107-109] in reference to the phase-forming component (a polymer and/or salt.)

It should be noticed that the "geometrical saturation" model covers the aforementioned similarity between partially miscible water-organic solvent mixtures and aqueous mixtures of phaser-forming components of varied concentrations belonging to a given tie line. As indicated above both types of mixtures form two-phase systems with phases of variable weight (volume) ratio and invariable compositions.

The so-called effective excluded volume, <V>, was suggested as a factor representing the smallest spacing of species 1 which will accept an individual molecule 2 [107]. This parameter was shown [107-109] to be the only factor required to fit the experimental phase diagrams for aqueous Dex-PEG two-phase systems. The <V> value was found to be in the range of $(1\text{-}8)\cdot 10^5$ g/mole for these systems with the Dex molecular weight varying from 50,000 to 340,000 [107]. It was shown [109] also that calculated data are in a very good agreement with the experimental phase diagrams for the aqueous Dex-PEG systems with the ratio of molecular weights of the two phase-forming components greater than ca. 4. When the ratio is less than 4, significant deviations of the calculated data from experimental results sometimes were observed [109]. It was established [109] that the same parameter may be used to describe phase diagrams for aqueous PEG-salt two-phase systems as well.

The results reported by Guan, Treffry and Lilley [107-109] illustrate the first successful attempt to create a theoretical model covering phase separation in both two-polymer and single polymer-salt two-phase systems. It should be emphasized that this model is based on the assumption of different structures of the coexisting phases. That is likely to be the reason for this model to be the only one, to my knowledge, capable of description not only the binodial curves but also the slopes of tie lines in a good agreement with experimental data.

3.5. SUMMARY

I would consider my purpose achieved if from reading this part of the book a reader would recognize, first, that water and its peculiar solvent properties should be regarded as the component of aqueous polymer two-phase systems at least as important as phase polymers. All the experimental facts accumulated in the studies of these systems indicate that.

Secondly, according to F. Franks [110]: "...to most self-respecting polymer scientists water as a solvent was anathema. The reason is not hard to discern: water does not fit into the framework of the accepted and acceptable theories which govern the behavior of macromolecules in solution. On the other hand, alternative theories have not yet been developed. What is done instead is to <u>make</u> aqueous solutions fit into theory by using the technique of Procrustes' bed. Sooner or later, such fitting exercises tend to produce absurd values for some physical parameter or other." This should be kept in mind when considering the existent theoretical treatments of phase separation in aqueous polymer systems.

Finally, the similarity between aqueous two-phase systems and water-organic solvent systems should be emphasized. This similarity calls for comparative consideration of the numerous experimental facts accumulated in the studies of properties of solvent systems and in much more limited studies of those of aqueous polymer systems. That is done in the next part.

REFERENCES:

1. P. A. Albertsson, Partition of Cell Particles and Macromolecules, 3rd.ed., Wiley, New York, 1986.
2. P. J. Flory, Principles of Polymer Chemistry, Cornell University Press, Ithaca, New York, 1953.
3. A. E. Nesterov, Y. S. Lipatov, Thermodynamics of Polymer Solutions and Mixtures, Naukova dumka (Rus), Kiev, 1984.

4. H. Tompa, Polymer Solutions, Butterworths, London, 1956.
5. D. Patterson, Polymer Eng.Sci., 22, 64 (1982).
6. F. S. Bates, Science, 251, 898 (1991).
7. F. Tjerneld, S. Berner, A. Cajarville, G. Johansson, Enzyme Microb. Technol., 8, 417 (1986).
8. K. A. Giuliano, Anal. Biochem., 197, 333 (1991);
9. V. B. Tolstoguzov, Food Hydrocolloids, 2, 195; 339 (1988).
10. K. P. Ananthapadmanabhan, E. D. Goddard, Langmuir, 3, 25 (1987).
11. W.D. Conway, Countercurrent Chromatography: Apparatus, Theory, and Applications, VCH Publishers, New York, 1990.
12. S. Bamberger, D. E. Brooks, K. A. Sharp, J.M. Van Alstine, T. J. Webber, In: Partitioning in Aqueous Two-Phase Systems: Theory, Methods, Uses, and Applications to Biotechnology (H. Walter, D. E. Brooks, D. Fisher, eds.), Academic Press, Orlando, Fla, 1985, p.85.
13. G. Johansson, J.Chromatogr., 150, 63 (1978).
14. B. Y. Zaslavsky, T. O. Bagirov, A. A. Borovskaya, N. D. Gulaeva, L. M. Miheeva, A. U. Mahmudov, M. N. Rodnikova, Polymer, 30, 2104 (1989).
15. A. Sjoberg, G. Karlstrom, Macromolecules, 22, 1325 (1989).
16. A. D. Diamond, J. T. Hsu, Biotechnol. Techniques, 3, 119 (1989).
17. D. Forciniti, C. K. Hall, M. R. Kula, Biotech.Bioeng., 38, 986 (1991).
18. W. Dolgolenko, Zh.Phys.Khimia (Rus), 62, 339 (1980).
19. J. S. Walker, C. A. Vause, J.Chem.Phys., 79, 2660 (1983).
20. R. Kjellander, E. Florin, J.Chem.Soc., Faraday Trans.I, 77, 2053 (1981).
21. R. Kjellander, J.Chem.Soc., Faraday Trans.II, 78, 2025 (1982).
22. A. Sjoberg, G. Karlstrom, F. Tjerneld, Macromolecules, 22, 4512 (1989).
23. F. E. Bailey, R. W. Callard, J.Appl.Polym.Sci., 1, 56 (1959).
24. K. P. Ananthapadmanabhan, E. D. Goddard, J.Colloid Interface Sci., 113, 294 (1986).
25. S. Saeki, N. Kuwahara, M. Nakata, M. Kaneko, Polymer, 18, 1027 (1977).
26. S. Saeki, N. Kuwahara, M. Nakata, M. Kaneko, Polymer, 17, 685 (1976).
27. M. J. Garvey, I. D. Robb, J.Chem.Soc., Faraday Trans.I, 75, 993 (1979).
28. E. Florin, R. Kjellander, J. C. Eriksson, J. Chem.Soc., Faraday Trans. I, 80, 2889 (1984).
29. R. A. Horne, J. P. Almeida, A. F. Day, N. T. Yu, J.Colloid Interface Sci., 35, 77 (1971).

30. W. A. P. Luck, Topics Curr. Chemistry, 64, 113 (1976).
31. N. Nishikido, R. Matuura, Bull.Chem.Soc.Jpn., 50, 1690 (1977).
32. L. V. Alferova, E. B. Toporkova, K. K. Kalnin'sh, Vysokomolek. Sodinenia (Rus), 26A, 699 (1984).
33. W. Melander, C. Horvath, Arch.Biochem.Biophys., 183, 200 (1977).
34. J. Breen, L. Huis, J. de Bleijser, J. C. Leyte, Ber.Bunsenges. Phys. Chem., 92, 160 (1988).
35. E. A. Boucher, P. M. Hines, J.Polymer Sci., Polym.Phys.Ed., 16, 501 (1978).
36. M. Ataman, E. A. Boucher, J.Polymer Sci., Polym.Phys.Ed., 20, 1585 (1982).
37. I. D. Robb, "Polymer - small molecule interactions", In: Chemistry and Technology of Water-Soluble Polymers (ed. C. A. Finch), Plenum Press, New York, 1983, p.193.
38. T. I. Zvarova, V. M. Shkinev, G. A. Vorob'eva, B. Y. Spivakov, Y. A. Zolotov, Mikrochim. Acta, 1984, 449 (1984).
39. A. M. Basedow, K. H. Ebert, J.Polymer Sci.: Polymer Symp., 66, 101 (1979).
40. B. Y. Zaslavsky, N. D. Gulaeva, S. Djafarov, E. A. Masimov, L. M. Miheeva, J.Colloid Interface Sci., 137, 147 (1990).
41. X. Lei, A. D. Diamond, J.T. Hsu, J.Chem.Eng.Data, 35, 420 (1990).
42. F. Franks, Biophysics and Biochemistry at Low Temperatures, University Press, Cambridge, 1985.
43. K. P. Ananthapadmanabhan, E. D. Goddard, Colloids & Surfaces, 25, 393 (1987).
44. V. Polyakov, I. A. Popello, V. Y. Grinberg, V. B. Tolstoguzov, Nahrung, 30, 81 (1986)
45. V. Polyakov, I. A. Popello, V. Y. Grinberg, V. B. Tolstoguzov, Nahrung, 30, 365 (1986)
46. D. Forciniti, C. K. Hall, M. R. Kula, Fluid Phase Equilibria, 61, 243 (1991).
47. L. Piculell, S. Nilsson, L. Falck, F. Tjerneld, Polymer Comm. 32, 158 (1991).
48. D. Forciniti, C.K. Hall, M.R. Kula, J.Biotechnol., 16, 151 (1991).
49. A. K. Shishov, V. V. Krivobokov, Y. V. Chubarova, S.Y. Frenkel, Polymer Science U.S.S.R., 23, 1330 (1980).
50. C. H. Kang, S. I. Sandler, Macromolecules, 21, 3088 (1988).
51. B. P. Makogon, T. V. Stoopnikova, T. V. Vyschkina, I. L. Povh, Vysokomol. Soedinenia (Rus), 30B, 164 (1988).
52. D. E. Brooks, K. A. Sharp, S. Bamberger, C. H. Tamblyn, G. V. F. Seaman, H. Walter, J.Colloid Interface Sci., 102, 1 (1984).

53. B. Y. Zaslavsky, A. U. Mahmudov, T. O. Bagirov, A. A. Borovskaya, G. Z. Gasanova, N. D. Gulaeva, V. Y. Levin, N. M. Mestechkina, L. M. Miheeva, M. N. Rodnikova, Colloid & Polymer Sci., 265, 548 (1987).
54. B. Y. Zaslavsky, T. O. Bagirov, A. A. Borovskaya, G. Z. Gasanova, N. D. Gulaeva, V. Y. Levin, E. A. Masimov, A. U. Mahmudov, N. M. Mestechkina, L. M. Miheeva, N. N. Osipov, S. V. Rogozhin, Colloid & Polymer Sci., 264, 1066 (1986).
55. B. Y. Zaslavsky, L. M. Miheeva, Y. P. Aleschko-Ozhevskii, A. U. Mahmudov, T. O. Bagirov, E. S. Garaev, J.Chromatogr., 439, 267 (1988).
56. G. Johansson, Biochim.Biophys.Acta, 221, 387 (1970).
57. G. Johansson, Acta Chem.Scand., Ser.B, 28, 873 (1974).
58. S. Bamberger, G. V. F. Seaman, J. A. Brown, D. E. Brook, J.Colloid Interface Sci., 99, 187 (1984).
59. S. Bamberger, G. V. F. Seaman, K. A. Sharp, D. E. Brooks, J.Colloid Interface Sci., 99, 194 (1984).
60. S. Schurch, D. F. Gerson, D. J. L. McIver, Biochim.Biophys.Acta, 640, 557 (1981).
61. A. D. Diamond, J. T. Hsu, J.Chromatogr., 513, 137 (1990).
62. B. Y. Zaslavsky, L. M. Miheeva, G. Z. Gasanova, A. U. Mahmudov, J.Chromatogr., 403, 123 (1987).
63. J. M. Sorensen, W. Arlt, Liquid-Liquid Equilibrium Data Collection, Dechema, Frankfurt/Main, 1980, Vol.5, Parts 2 and 3.
64. S. J. Gluck, M. P. Wingeier, J.Chromatogr., 547, 69 (1991).
65. R. L. Scott, J.Chem.Phys., 17, 279 (1949).
66. H. Tompa, Trans.Faraday Soc., 45, 1142 (1949).
67. A. Dobry, F. Boyer-Kawenoki, J.Polym.Sci., 2, 90 (1947).
68. M. Bank, J. Leffingwell, C. Thies, Macromolecules, 4, 43 (1971).
69. L. Zeman, D. Patterson, Macromolecules, 5, 513 (1972).
70. C. C. Hsu, J. M.Prausnitz, Macromolecules, 7, 320 (1974).
71. D. E. Brooks, K. A. Sharp, D. Fisher, In: Partitioning in Aqueous Two-Phase Systems: Theory, Methods, Uses, and Applications to Biotechnology (H. Walter, D. E. Brooks, D. Fisher, eds.), Academic Press, Orlando, Fla, 1985, p.11.
72. G. J.Courval, D. G.Gray, Polymer, 24, 323 (1983).
73. E. Edmond, A. G. Ogston, Biochem.J., 109, 4, 569 (1968).
74. A. Gustafsson, H. Wennerstrom, F. Tjerneld, Polymer, 27, 1768 (1986).
75. E. I. Pozdnyakova, Z. A. Loogovaya, V. N. Tolmachev, Vysokomol. Soedinenia (Rus), 27B, 324 (1985).
76. C. H. Kang, S. I. Sandler, Fluid Phase Equilibria, 38, 245 (1987).

77. C. A. Haynes, R. A. Beynon, R. S. King, H. W. Blanch, J. M. Prausnitz, J.Phys.Chem., 93, 5612 (1989).
78. N. L. Abbott, D. Blankschtein, T. A. Hatton, Bioseparation, 1, 191 (1990).
79. R. J. Hefford, Polymer, 25, 979 (1984).
80. W. Burchard, In: Chemistry and Technology of Water-Soluble Polymers (C. A. Finch, ed.), Plenum Press, New York, 1983, p.125.
81. F. Franks, In: Chemistry and Technology of Water-Soluble Polymers (C. A. Finch, ed.), Plenum Press, New York, 1983, p.157.
82. D. H. Napper, Polymeric Stabilization of Colloidal Dispersions, Academic Press, London, 1983, Chapter 3.
83. J. N. Baskir, T. A. Hutton, U. W. Suter, Macromolecules, 20, 1300 (1987).
84. P. A. Albertsson, A. Cajarville, D. E. Brooks, F. Tjerneld, Biochim. Biophys.Acta, 926, 87 (1987).
85. W. Muller, Kontakte (Darmstadt), #3, 3 (1986).
86. P. Molyneux, Water-Soluble Synthetic Polymers: Properties and Behavior, CRC Press, Boca Raton, 1984, vol.2, pp.159.
87. V. Y. Grinberg, S. H. Dotdaev, Y. A. Borisov, V. B. Tolstoguzov, Vysokomol.Soed. (Rus.), 29B, 145 (1987).
88. G. Allon, G. Gee, J. P. Nicholson, Polymer, 1, 56 (1960).
89. R. S. King, H. W. Blanch, J. M. Prausnitz, AIChE J., 34, 1585 (1988).
90. C. A. Haynes, H. W. Blanch, J. M. Prausnitz, Fluid Phase Equilib., 53, 463 (1989).
91. C. A. Haynes, J. Carson, H. W. Blanch, J. M. Prausnitz, AIChE J., 37, 1401 (1991).
92. C. A. Haynes, F. J. Benitez, H. W. Blanch, J. M. Prausnitz, AIChE J., 39, 1539 (1993).
93. H. Cabezas, J. D. Evans, D. C. Szlag, Fluid Phase Equilib., 53, 453 (1989).
94. H. Cabezas, J. D. Evans, D. C. Szlag, In: Downstream Processing and Bioseparation: Recovery and Purification of Biological Products. ACS Symposium Series (J. Hamel, J. B. Hunter, S. K. Sikdar, eds.), Vol.419, p.38, American Chemical Society, Washington, D.C., 1990.
95. H. Cabezas, M. Kabiri-Badr, D. C. Szlag, Bioseparation, 1, 227 (1990).
96. H. Cabezas, M. Kabiri-Badr, In: Frontiers in Bioprocessing II (P. Todd, M. Bier, S. K. Sikdar, eds.), American Chemical Society, Washington, D.C., 1991/1992.
97. J. N. Baskir, T. A. Hatton, U. W. Suter, Biotechnol.Bioeng., 34, 541 (1989).

98. D. Forciniti, C. K. Hall, In: Downstream Processing and Bioseparation: Recovery and Purification of Biological Products. ACS Symposium Series (J. Hamel, J. B. Hunter, S. K. Sikdar, eds.), Vol.419, p.53, American Chemical Society, Washington, D.C., 1990.
99. C. J. van Oss, R. J. Good, M. K. Chaudhury, Sep.Sci. Technol., 22, 1515 (1987).
100. C. J. van Oss, M. K. Chaudhury, R.J. Good, Sep.Sci. Technol., 24, 15 (1989).
101. C. J. van Oss and R. J. Good, J.Macromol.Sci.-Chem., A26, 1183 (1989).
102. C. J. van Oss, R. J. Good, M. K. Chaudhury, J. Protein Chem., 5, 585 (1986).
103. C. J. van Oss, R. J. Good, M. K. Chaudhury, J.Chromatogr., 391, 53 (1987).
104. B. Y. Zaslavsky, V. Y. Levin, L. M. Miheeva, 1989, unpublished data.
105. B. Y. Zaslavsky, V. Y. Levin, A. U. Mahmudov, N. M. Mestechkina, L. M. Miheeva, S. V. Rogozhin, M. N. Rodnikova, Doklady Acad. Nauk USSR, 293, 649-652 (1987).
106. T. E. Treffry, T. H. Lilley, P. J. Cheek, In: Separations Using Aqueous Phase Systems: Applications in Cell Biology and Biotechnology (D. Fisher and I. A. Sutherland, eds.), Plenum Press, New York, 1989, p.233.
107. Y. Guan, T. H. Lilley, T. E. Treffry, J.Chem.Soc. Faraday Trans., 89, 4283 (1993).
108. Y. Guan, T. H. Lilley, T. E. Treffry, Macromolecules, 26, 3971 (1993).
109. Y. Guan, T. E. Treffry, T. H. Lilley, J.Chromatogr., 1994, in press.
110. F. Franks, In: Chemistry and Technology of Water-Soluble Polymers (C. A. Finch, ed.), Plenum Press, New York, 1983, p.vii.

PART 2.

PARTITIONING OF SOLUTES IN AQUEOUS TWO-PHASE SYSTEMS

General trends of partitioning of solutes in aqueous polymer two-phase systems and the physicochemical features of the phases will be discussed in this part of the book. Different models for the mechanism of partitioning of solutes suggested in the literature will be described, and the information provided by the partition technique will be considered.

The features important for handling the systems , e.g., viscosity, density, optical absorbance, etc., and their applications (mixing, speed of phase settling, etc.) are described in detail in [1,2] and will not be discussed here. Only features likely to govern partitioning of solutes in the systems will be considered below. To decide what particular features of the phases fall into this category an assumption about the possible mechanism of partitioning must be made.

There are two fundamentally different assumptions in the literature. The first one [3] is that partitioning of a solute in an aqueous polymer two-phase system is the result of the attractive and repulsive interactions between the solute and phase polymers. This mechanism is definitely realized in certain cases. The most illustrative example is offered by the so-called affinity partition technique (see below and, for example, in [1,4-6]). In the affinity partition mode an additive of a phase polymer-derivatized ligand is included into the system. This polymer-bound ligand additive usually concentrates in the parent polymer-rich phase together with the target solute associated with the ligand.

The other assumption [7,8] is that a nonuniform partitioning of a solute in an aqueous two-phase system results from the difference in the interactions of the solute not with the phase polymers but with the different aqueous media in the phases. This mechanism is realized in certain systems (see below) and is likely to be the part of the solute-polymer interaction partitioning mechanism in other systems.

The partitioning mechanism based on the difference in the solute-solvent interactions is realized in the most complete form in water-organic solvent two-phase systems. General trends of solute partitioning in such systems will be outlined below in comparison with those observed in aqueous polymer systems.

When aqueous two-phase systems are used for preparative purposes, e.g., for separation, isolation, or purification of a product, the knowledge of specific mechanism behind the product partitioning is important to increase the efficiency of the process or is of purely academic interest. If the partitioning technique is going to be used for analytical purposes, however, the knowledge of the partitioning mechanism is critically important for understanding the information obtained and its interpretation.

First, however, solvent features of the phases of aqueous polymer systems must be considered.

CHAPTER 4.

PHYSICOCHEMICAL PROPERTIES OF PHASES IN AQUEOUS POLYMER SYSTEMS

Solvent properties of various liquids may be characterized by numerous different methods (see, e.g., in [9]). Some of these methods were applied to analysis of the media in the phases of aqueous polymer systems. The results reported in the literature will be considered below in comparison with those obtained for water-organic solvent systems by the same or similar techniques.

It is also known that partitioning of structurally simple organic compounds and inorganic ions can be used to characterize the solvent properties of water-organic solvent systems (see, e.g., in [10-12]). This approach and the results of its application to analysis of solvent and aqueous polymer two-phase systems will be discussed. Comparison of partitioning of the same solutes in different two-phase systems will be considered as well as the electrochemical phenomena, and hydrophobic and ionic solvation in the phases.

4.1. FEATURES OF THE AQUEOUS MEDIA IN THE PHASES OF POLYMER TWO-PHASE SYSTEMS

Various physicochemical characteristics may be used to compare two different solvents. Among the most commonly used is the solvent polarity parameter, E_T, based on the transition energy for the solvatochromic absorption band of a dye used as a probe for a given solvent (see in Chapter 1). The solvent polarity of the coexisting phases of aqueous Dex-PEG and Dex-Ficoll systems as a function of the system polymer concentrations was examined by Zaslavsky et al.[13] using the water-soluble carboxylate-substituted betaine dye 6 [14] (Fig. 2.1). Positions of the phases of the aqueous two-phase systems on the normalized E_T^N solvent polarity scale (see in Chapter 2) calculated from the results obtained in [13] are shown in Figure 4.1 together with that of pure water.

The solvent polarity of the aqueous media in the two phases of the systems [13] is clearly different, the Ficoll-rich and PEG-rich phases being always less polar than the Dex-rich phase in agreement with the results obtained for aqueous solutions of these polymers (see Fig. 2.1a). The solvent polarity of the phases varies with their polymer composition, and the difference in the relative polarity between the two phases decreases with decreasing polymer concentrations.

It can be seen from the data [13] plotted in Fig. 4.2 (inset) that the solvent polarity difference, ΔE_T^N, is linearly related to the difference in the polymer concentration between the phases (ΔC(PEG) or ΔC(Ficoll), respectively).

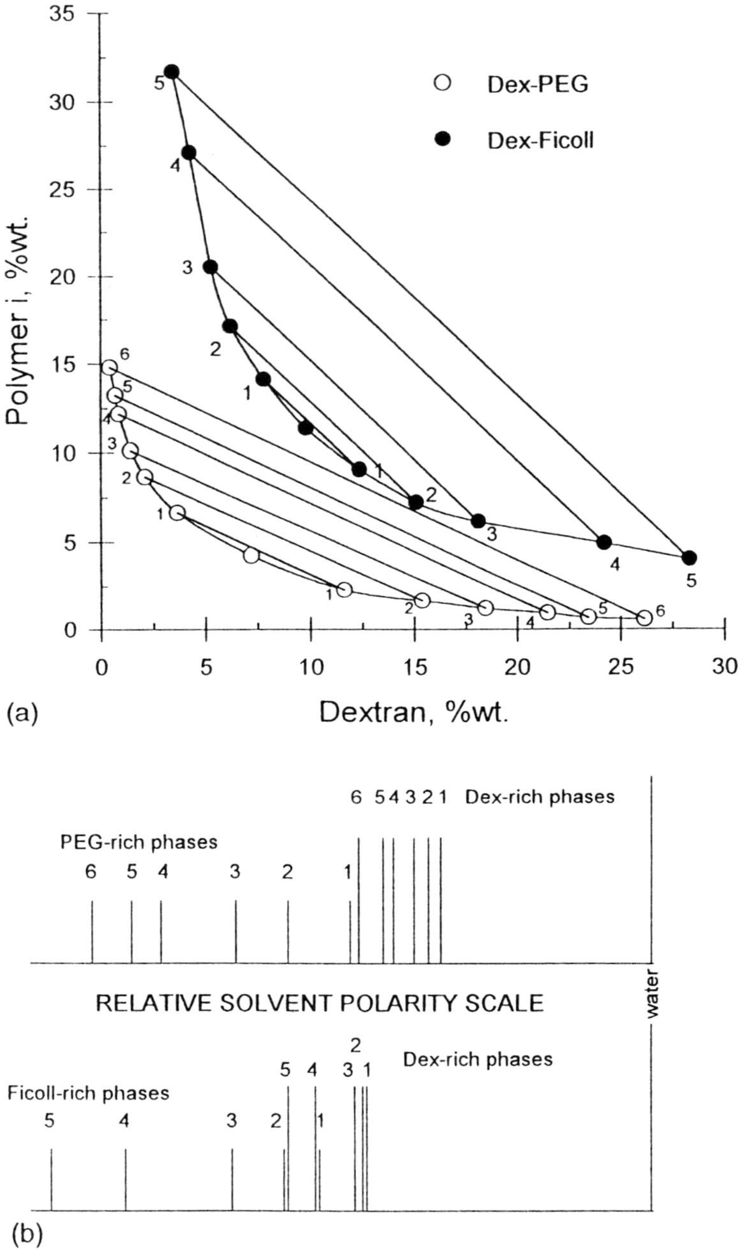

Figure 4.1. Phase diagrams and solvent polarity scale for the phases of aqueous Dex—PEG and Dex—Ficoll two-phase systems. Calculated from the data reported in [13]. Notations for the phases in phase diagram and solvent polarity scale are the same.

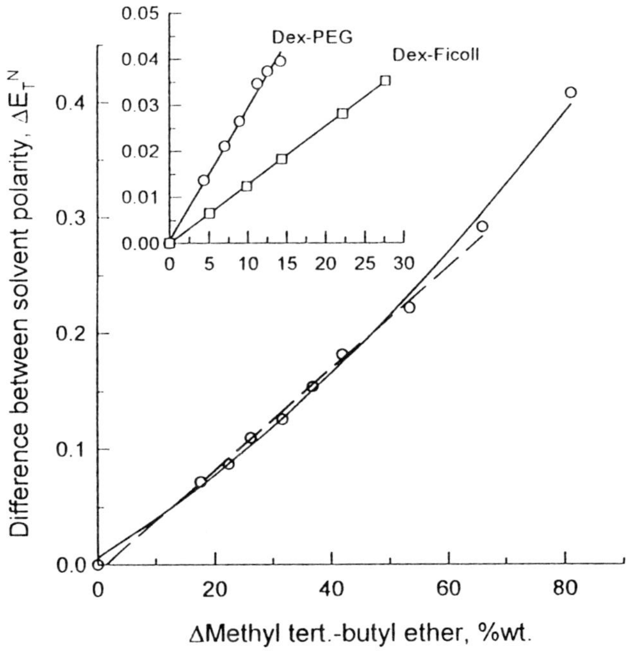

Figure 4.2. Difference in the solvent polarity between the phases of water—ethyl tert.-butyl ether (MtBE)—acetonitrile as a function of the difference in the concentration of MtBE between the phases. Inset: dfference in the solvent polarity between the phases of aqueous Dex—PEG and Dex—Ficoll systems as a function of the difference in the concentration of PEG or Ficoll, respectively. Calculated from the data reported in [13,15].

The solvent polarities of the two phases of the water - methyl tert-butyl ether (MtBE) two-phase system containing different amounts of acetonitrile (ACN) as an organic modifier were reported recently by Gluck and Wingeier [15]. The data calculated from the results in [15] are plotted in Fig. 4.2 as the solvent polarity difference vs. the difference in the MtBE concentration between the phases. The entire set of the data is described by a quadratic equation:

$$\Delta E_T^N = a + b\cdot\Delta[\text{MtBE}] + c\cdot(\Delta[\text{MtBE}])^2 \quad (4.1)$$

where ΔE_T^N is the difference in the normalized solvent polarity between the two phases; Δ[MtBE] is the difference in the MtBE concentration between the two phases; and a, b, and c are constants.

It can be seen, however, that the initial part of the data set (up to ca. 60 wt.% of Δ[MtBE] is described by a linear relationship, i.e., coefficient c = 0 in Equation 4.1. The results reported for the aqueous Dex-PEG and Dex-Ficoll two-phase systems [13] (see inset in Fig. 4.2) are also described by a linear relationship. It is possible that the observed linearity is due either to the limited set of the polymer compositions examined [13], or to the small difference in the solvent polarities between the phases specific for the solvents of the same nature. This question remains open as yet.

The similarly small difference in the features of the aqueous media between the two phases is observed in the measurements of complex permittivity of the phases of the aqueous Dex-PEG, Dex-PVP, and Dex-Ficoll two-phase systems [16]. The static dielectric constants, ε_s, and the dielectric relaxation time of water, τ, in the coexisting phases of these systems are different depending on the type and concentrations of phase polymers. The dielectric orientational mobility of water molecules is restricted in the PEG-, PVP-, and Ficoll-rich (less polar) phases to a larger extent than in the Dex-rich (more polar) phases [16]. The difference in the water mobility between the two phases amounts up to $0.7\cdot10^{-12}$ sec in the systems examined in [16]. Dielectric orientational mobility of pure water is $8.25\cdot10^{-12}$ sec which exceeds that in all the phases of the polymer systems investigated so far [16]. These results support the implication of those obtained in the solvatochromic measurements that the properties of aqueous media in the two phases of aqueous polymer systems are different from each other as well as from those of pure water.

Additional evidence is provided by the potentiometric measurements of pH in the coexisting phases of the aqueous Dex-PEG and Dex-Ficoll systems containing buffers of different initial pH [13]. The difference in the potentiometrically measured pH values between the two phases depends on the type and concentrations of phase polymers and on the initial pH of the buffer employed [13]. This difference indicates that the thermodynamic activity of protons in aqueous media of the two phases is different. It was found [13] particularly that the further away the initial buffer pH is from 7.0, the larger deviation from the initial pH value is observed in the phases. For example, pH-values in the phases of the aqueous Dex-Ficoll system containing 0.01 mol/kg universal buffer of the initial pH 8.31 is 7.84 in the Ficoll-rich phase, and 7.98 in the Dex-rich phase [13]. The pH difference between the two phases, e.g., in Dex-PEG system, reaches up to 0.14-0.16 pH unit. Such a pH difference might be theoretically achieved if the temperatures of the two phases would differ by ca. 5-10°C (see in Chapter 1).

Different properties of the aqueous media in the two phases of an aqueous polymer two-phase system are also observed in the partitioning studies.

Table 4.1

Some Structural and Spectral Features and Partition Coefficients of Sulphonephthalein Dyes* in Aqueous Dex—PEG and Dex—Ficoll Two-Phase Systems Containing 0.01 Mole/kg Universal Buffer, pH 7.15.

Dye	Constituents			lnK [a]	lnK [b]	pK_{a2} [c]	$\lambda_{max}(R^{2-})$	$E_T(R^{2-})$
	X_1	X_2	X_3				nm	kcal/mole
Phenol red	H	H	H	0.206	0.815	7.82	559	51.14
o-Cresol red	H	CH_3	H	0.263	1.041	8.19	572	49.98
Bromphenol blue	H	Br	Br	-	1.591	3.96	591	48.38
Bromcresol green	CH_3	Br	CH_3	0.530	1.866	4.67	614	46.56
Bromcresol purple	H	Br	CH_3	0.377	1.470	6.13	588	48.62
Bromthymol blue	CH_3	Br	i-C_3H_7	0.548	2.358	7.12	617	46.34
Thymol blue	CH_3	H	i-C_3H_7	0.428	1.882	9.13	597	47.89

* - general structure of a sulphonephthalein dye is

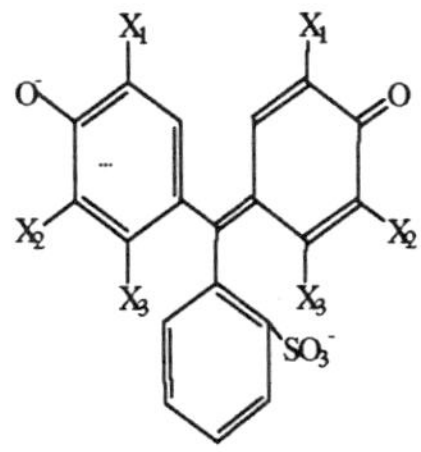

[a] - lnK for the dyes in the aqueous Dex-Ficoll two-phase system;
[b] - lnK for the dyes in the aqueous Dex-PEG two-phase system;
[c] - pK_{a2} values for the dyes in 0.01 M universal buffer are taken from [17].

Partitioning of a series of sulphonephthalein dyes in the aqueous Dex-PEG and Dex-Ficoll systems containing 0.01 mole/kg universal buffer of pH 7.15 was examined by Zaslavsky et al.[13]. The partition coefficient of a dye, K, was determined as the ratio of the dye concentration in the PEG- (or Ficoll-) rich phase to the dye concentration in the Dex-rich phase.

The ionic equilibria for all the sulphonephthalein dyes studied in [13] are represented by

$$RH^{-}(\text{yellow form}) <{=}{=}{=}{=}> R^{2-}(\text{alkaline form}) + H^{+} \qquad (4.2)$$

with the dissociation constant of the yellow form (RH^-) of the dye K_{a2}.

The partition coefficients of the dyes in the two-phase systems, the corresponding substituents X_1, X_2, and X_3 of the dyes, and the pK_{a2} values are presented in Table 4.1.

The R^{2-} ions of the dyes used in [13] are capable of electronic solvatochromism in aqueous polymer solutions [17,18]. The solvatochromic absorption band maximum, $\lambda_{max}(R^{2-})$, values are given in Table 4.1 together with the corresponding molar transition energies for these solvatochromic bands of the dyes, $E_T(R^{2-})$ (see in Chapter 1).

The data presented in Table 4.1 show that the changes in the structure affect the dyes' partition behavior as well as the other physicochemical properties of the dyes. The pK_{a2} values for the dyes vary significantly and equilibria of the dyes. There is a linear relationship, however, between the logarithm of the partition coefficient, lnK, of the dye and the molar electronic transition energy of the dye R^{2-} ion, $E_T(R^{2-})$. This relationship is shown in Figure 4.3a.

The chemical structure of the phenol red dye and the lowest wavelength value for the solvatochromic absorption band of its R^{2-} ion indicate that this dye experiences the most intensive interactions with pure water among the dyes examined. Among all these dyes phenol red distributes most equally between the phases in both systems. Hence phenol red was chosen [13] as a reference probe in the series used. If $E_T(R^{2-}{}_j)$ where "j" denotes a given dye is taken as a measure of the intensity of the dye interaction with water, the $E_T(R^{2-}{}_j)$ - $E_T(R^{2-}{}_o)$, where "o" denotes phenol red as the reference probe, will represent the relative intensity of solute-water interactions for the j-th dye.

The relationships in Fig. 4.3b indicate that the partitioning of a probe between the two phases is inversely correlated with the intensity of the probe-pure water interactions. The less intense probe-pure water interactions are, the larger lnK value for a given probe is. Thus, the partition coefficient of a solute may be considered as a measure of the intensity of the solute-water interactions (see below in more detail).

It has been established [19-21] that the distribution of nonionic solutes in different water-organic solvent two-phase systems is related to the solvato-

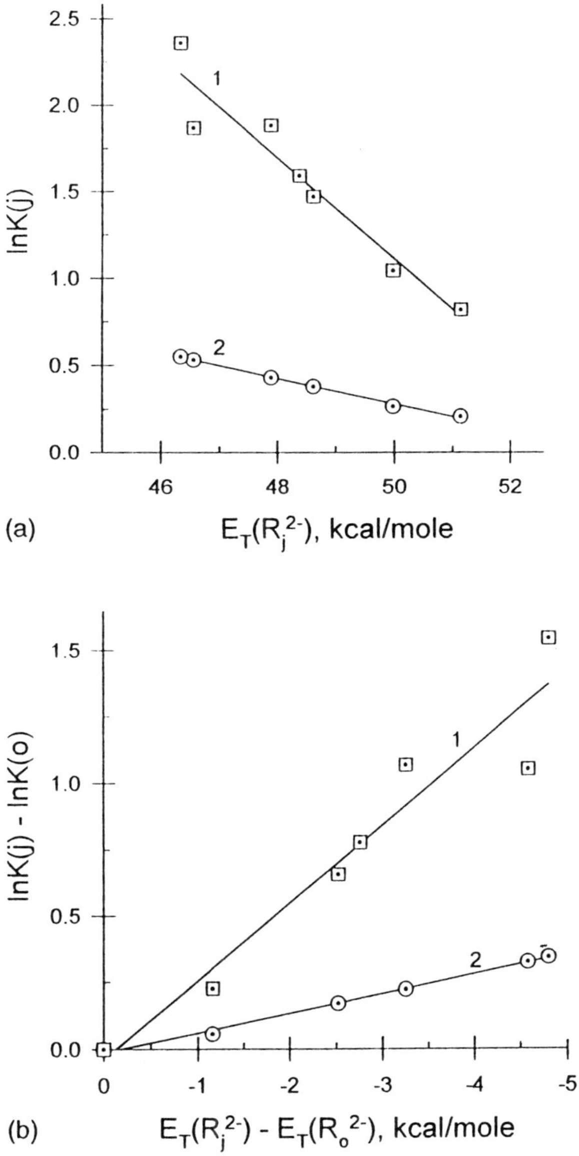

Figure 4.3. (a) Logarithms of the partition coefficients of the sulphonephthalein dyes in the aqueous Dex—Ficoll (1) and Dex—PEG (2) two-phase systems, pH 7.15 versus the molar transition energies for the solvatochromic bands of the R^{2-} ions of the dyes, $E_T(R^{2-})$; (b) The relative intensity of the dye-water interactions for a j-th dye, expressed as $E_T(R_j^{2-}) - E_T(R_o^{2-})$, versus the relative affinity of the j-th dye for the aqueous Dex-rich phase, expressed as $\ln K_j - \ln K_o$, where "o" denotes phenol red chosen as the reference probe.

chromic parameters of the solutes. The fact that the partition behavior of the dyes in aqueous Dex-PEG and Dex-Ficoll two-phase systems is related to these dyes' solvatochromic characteristic $E_T(R^{2-})$ is an additional indication of the similarity between solute partitioning in the solvent and aqueous polymer two-phase systems.

The indications provided by the results of partitioning of homologous series of structurally simple compounds are even more convincing.

4.2. PARTITIONING OF HOMOLOGOUS SERIES OF 'STRUCTURALLY SIMPLE' COMPOUNDS

When a solute distributes at constant temperature and pressure between two solvents i and j, which are immiscible or partially miscible, the chemical potential of the solute is the same in both phases at equilibrium:

$$\mu_i = \mu^o{}_i + RT\cdot\ln\mathbf{a}_i = \mu_j = \mu^o{}_j + RT\cdot\ln\mathbf{a}_j \qquad (4.3a)$$

where μ is the chemical potential of the solute in a given phase; μ^o is the chemical potential of the standard state; **a** is the solute activity. Equation 4.3a may be expressed as:

$$\mu^o{}_i - \mu^o{}_j = RT\cdot\ln(f_i/f_j) - RT\cdot\ln(C_j/C_j) \qquad (4.3b)$$

where f is the activity coefficient of the solute; C is the solute concentration. In the infinitely diluted system f = 1, and Equation 4.3b takes the form:

$$\Delta G_{tr(i\to j)} = -RT\cdot\ln(C_j/C_i) = -RT\cdot\ln K \qquad (4.3)$$

where $\Delta G_{tr(i\to j)}$ is the free energy of transfer of the solute from solvent i into solvent j; and K is the partition coefficient of the solute defined as the ratio between the concentrations of the solute in the solvents j and i.

When the partitioning of a solute in a two-phase system is to be considered in terms of thermodynamics, two conditions are postulated: a) both phases of the system are completely immiscible; and b) the behavior of the solute partitioned in the system is regarded as the ideal one (see, e.g., in [11]). Naturally, neither of the two conditions is fulfilled in reality.

As the partition coefficient is related to the difference between the chemical potentials of the solute in the two phases, it is clear that the partition coefficient value should be measured for the solute molecules being in the same form in both phases, i.e., for the nonprotonated or the ionized monomeric species. The basic methods of corrections for the degree of association of solute molecules in the phases of the common water-organic solvent partitioning systems are discussed in detail by Leo et al.[10]. Various methods of corrections for the effect of the ionization degree of a solute on its partition coefficient have been suggested by Alhaider et al.[22] and Martin [23]. No correction seems to be necessary in the case of the solute partitioning in an aqueous polymer two-phase system as the nature of the solvent is one and the same in both phases.

One of the primary goals of research in the field of solution thermodynamics has always been and still remains the prediction of the solution behavior from chemical structures and physical properties of solution components. One of the predictive methods which achieved a relative degree of success is the semi-empirical group contribution approach (see, e.g., in [11,23-26]).

This approach is based on the assumption that the free energy of the solution process is additively composed of independent contributions from the constituent functional groups. This assumption is widely and successfully used in studies of different processes in solution, e.g., adsorption, micelle formation, partitioning, dissolution, etc. [11,23-26].

Under 'structurally simple' compounds in this section are meant the organic compounds composed of the straight aliphatic alkyl chain and a single functional (polar) group, e.g., fatty acids, alkyl sulfates, aliphatic alcohols, amines, etc. Partitioning of several homologous series of monofunctional aliphatic compounds in water-organic solvent and aqueous polymer two-phase systems as shown in Figure 4.4 is described as:

$$\ln K = A + E \cdot N_C \qquad (4.4)$$

where K is the solute partition coefficient; N_C is the number of carbon atoms in the aliphatic alkyl chain of the partitioned solute molecule; A and E are constants.

Partition coefficient of a solute in an aqueous polymer two-phase system, K, (e.g., in the Dex-PEG system) will be defined throughout this text as a ratio of the solute concentration in the Dex-poor phase (e.g., PEG-rich phase) to the solute concentration in the Dex-rich phase. To differentiate between the aqueous polymer systems and water-organic solvent systems, the partition coefficient of a solute in water-organic solvent two-phase system will be denoted as P. The partition coefficient P is commonly defined as the ratio of the solute concentration in the organic solvent phase to the solute concentration in the aqueous phase. Equation 4.4 is true for both partition coefficients K and P in aqueous polymer and solvent two-phase systems.

It follows from the above basic assumption of the group contribution approach that the coefficient E in Equation 4.4 represents an average lnK increment per CH_2 group; and the coefficient A represents the total contribution of a part of the solute molecule which can not be expressed as a sum of CH_2 groups.

It follows from Equation 4.3 that coefficient E is related to the free energy of transfer of a CH_2 group from one to the other phase in a given two-

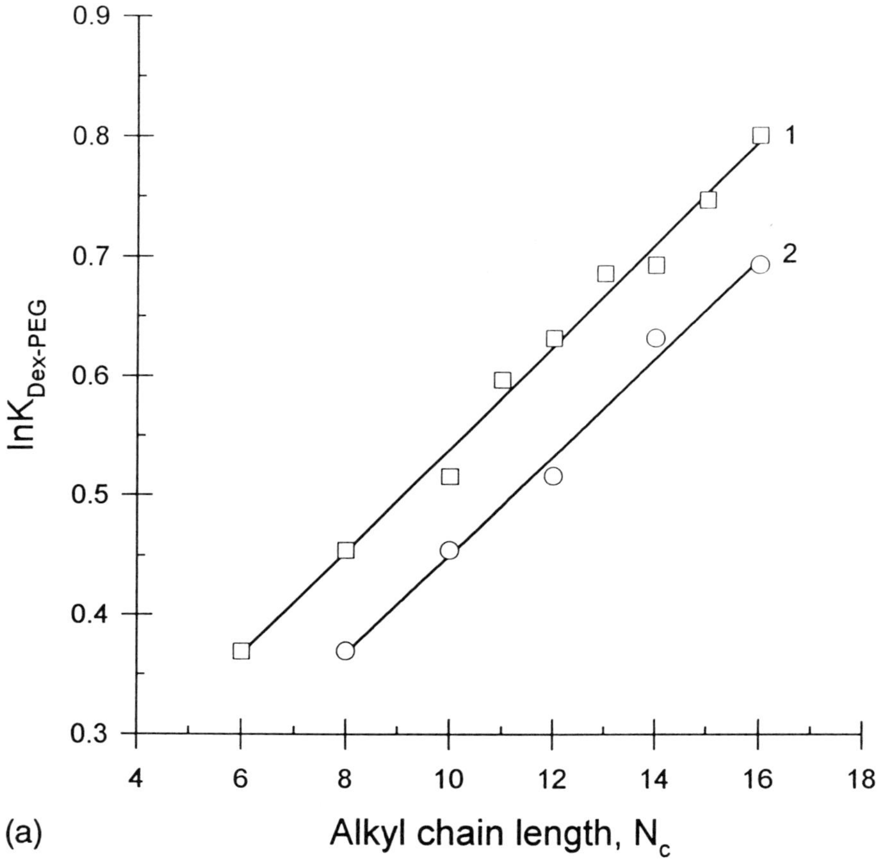

(a)

Figure 4.4. Partition coefficients of homologous series of solutes as a function of the solute alkyl chain length in: (a) Dex-500—PEG-6000—water two-phase system containing 1.0 mole/kg NaCl (Solutes: 1 - $C_nH_{2n+1}OSO_3Na$; 2 - $[C_nH_{2n+1}N^+(CH_3)_3]Br^-$). Calculated from data in [27]; (b) Octanol—water system (Solutes: 1 - $C_nH_{2n+1}COOH$; 2 - $C_nH_{2n+1}OH$). Calculated from the data reported in [10].

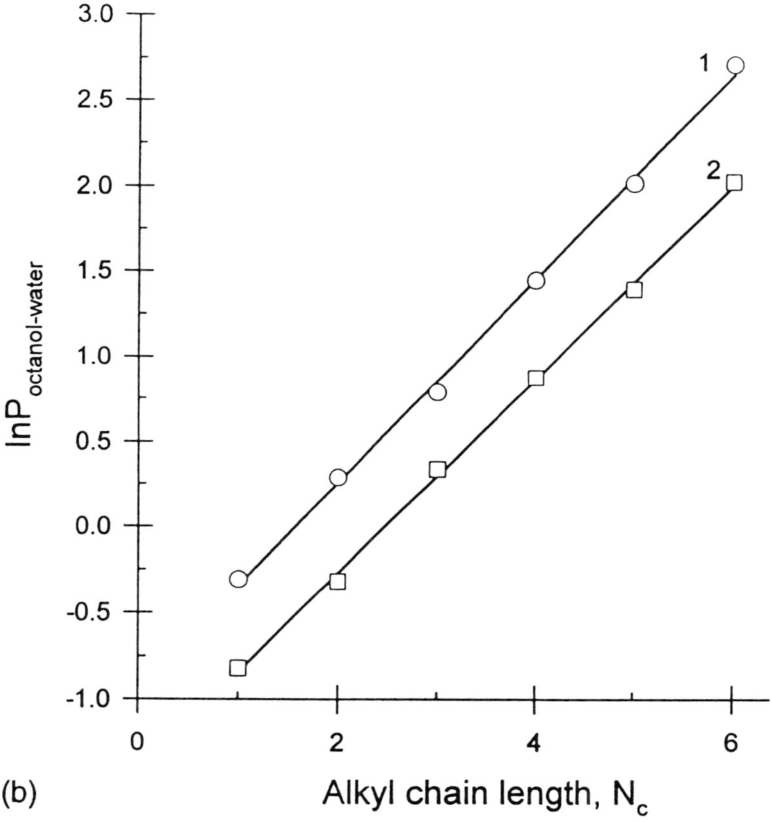

(b)

phase system, $\Delta G(CH_2)_{tr}$ as:

$$\Delta G(CH_2)_{tr} = -RT \cdot E \tag{4.5}$$

It should be particularly noted that the data [10,11,24-27] illustrated in Fig. 4.4 show that relationships described by Equation 4.4 for the different homologous series of aliphatic compounds in a given (aqueous polymer or solvent) two-phase system are essentially parallel. That clearly indicates that: (a) the methylene group increment, E, into lnK (or lnP) value is independent of the nature of the aliphatic solutes being partitioned, and hence (b) the methylene group increment E or $\Delta G(CH_2)_{tr}$ value may be used as a measure of the difference between the affinities of the two phases for a CH_2 group.

It can be seen from the data presented in Figure 4.5 that the E-value is

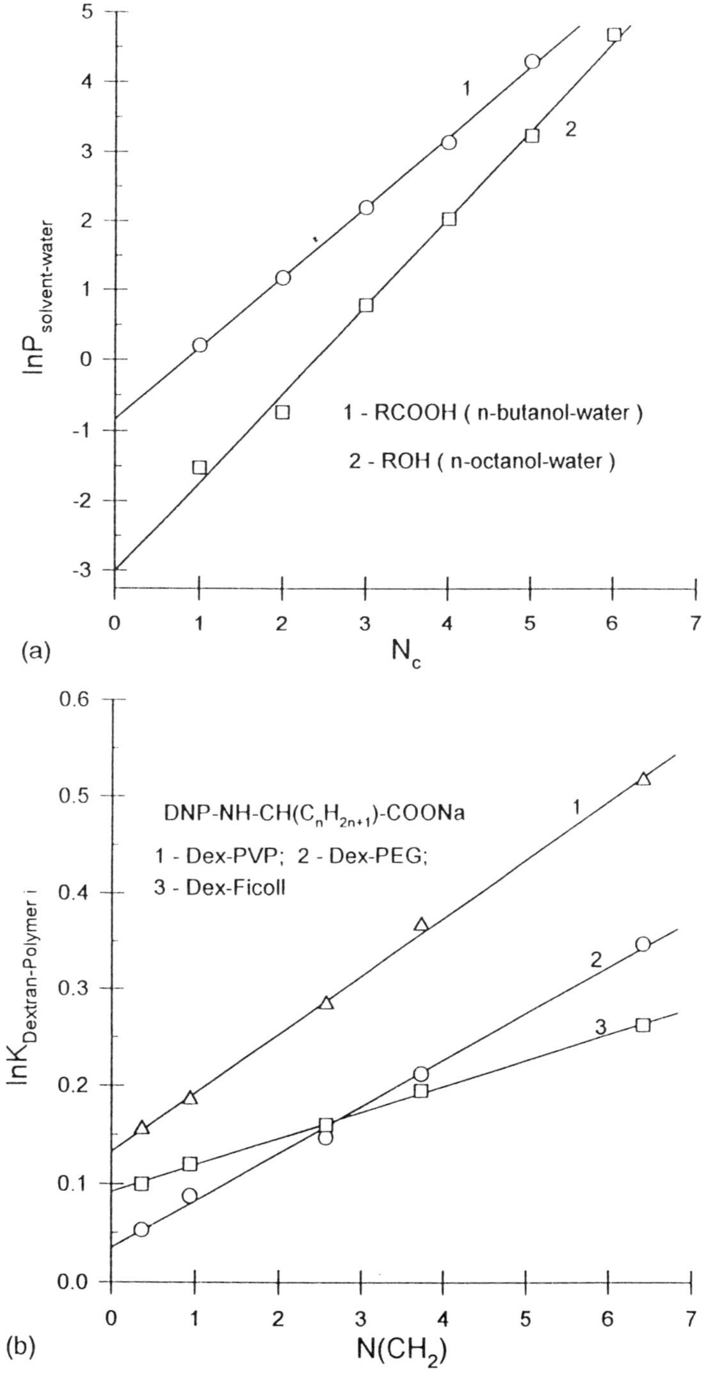

Figure 4.5. Partition coefficients of homologous series of solutes as functions of the alkyl chain length N_C or $N(CH_2)$ (difference in the terms is explained in the text) in (a) water-organic solvent and (b) aqueous polymer two-phase systems. Calculated from the data reported in [10,30].

different in different solvent systems as well as in different aqueous polymer two-phase systems. It may be noticed from the data in Fig. 4.5 that not the number of carbon atoms in the alkyl chain N_C but the averaged value, $N(CH_2)$, representing the equivalent number of methylene groups in the aliphatic side-chain of a sodium salt of homologous dinitrophenylated amino acid is used. Parameter $N(CH_2)$ for the alkyl chain of a given aliphatic compound may be equal to N_C or differ from it depending on the nature of the (polar) functional group of the compound, the length of the alkyl chain, and the composition of the solvent medium. Usually the shorter the chain, the larger the difference which commonly (but not always) disappears at the N_C value of 6-10 (see in more detail below).

The free energy of transfer of a methylene group, $\Delta G(CH_2)_{tr}$, (or parameter E) value may be used as a measure of the difference between the hydrophobic character of the two phases [11] and is one of the important characteristics in regard to the partitioning ability of a two-phase system (see below).

It should be mentioned that the positive E values implies the negative free energy of transfer of a methylene group from, e.g., Dex-rich phase to PEG-rich phase. In all the cases examined so far PEG-, PVP-, and Ficoll-rich phases displayed the affinity for a CH_2 group exceeding that displayed by the corresponding Dex-rich phases. In the aqueous PEG-salt systems the PEG-rich phases are more hydrophobic than the corresponding salt-rich phases though the number of the systems examined is too small to allow any generalization.

As can be seen from Figs. 4.4 and 4.5 the contribution of a polar group into lnK (or lnP) value represented in Equation 4.4 by coefficient A depends on both particular type and composition of a two-phase system and type of solute being partitioned. The physical meaning of coefficient A in the case of water-organic solvent two-phase systems is self-evident. It characterizes the difference between the interactions of the polar group with the different solvent media in the two phases. In aqueous two-phase systems the meaning of coefficient A is not as straightforward. If partitioning of a solute in the system is similar to that in water-organic solvent systems, the physical meaning of the coefficient A is clearly as indicated above. If, on the other hand, partitioning of a solute is driven by the direct solute-polymer interactions, the coefficient A represents the difference between the interactions of the corresponding polar group with the polymers in the two phases. This question will be considered in detail below.

Influence of Polymer Composition On the Difference Between the Hydrophobic Character of the Coexisting Phases.

It has been established by Zaslavsky et al.[28-30] and confirmed by Hsu et al.[31] that the free energy of transfer of a methylene group from one phase to the other, $\Delta G(CH_2)_{tr}$, or the corresponding coefficient E value (see

Equation 4.4) depends on the polymer composition of the phases.

Typical relationships observed [30] in aqueous polymer two-phase systems of different polymer composition between the logarithm of the partition coefficient K and the alkyl chain length of a solute are shown in Figure 4.6.

The results reported in [28-31] and illustrated in Figure 4.7 show that the difference between the hydrophobic character of the two phases (expressed as $\Delta G(CH_2)_{tr}$ or E value): (a) depends on the particular type of aqueous polymer two-phase system in use; and (b) is linearly dependent on the difference between the concentrations of a phase polymer between the two phases.

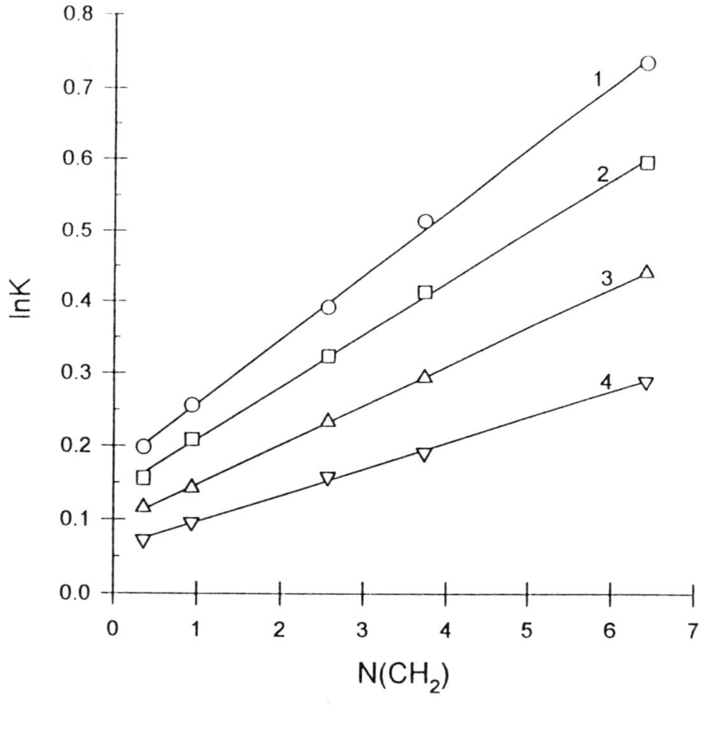

Figure 4.6. Partition coefficients of sodium salts of DNP-amino acids with aliphatic side chains as functions of the side chain length. System: Dex-70—PEG-6000—water—0.10 mole/kg K_2SO_4. Polymer concentrations (in % wt.): 1 - 7.96 % PEG; 12.77 % Dex; 2 - 6.70 % PEG; 10.78 % Dex; 3 - 5.28 % PEG; 8.50 % Dex; 4 - 4.67 % PEG; 7.45 % Dex.

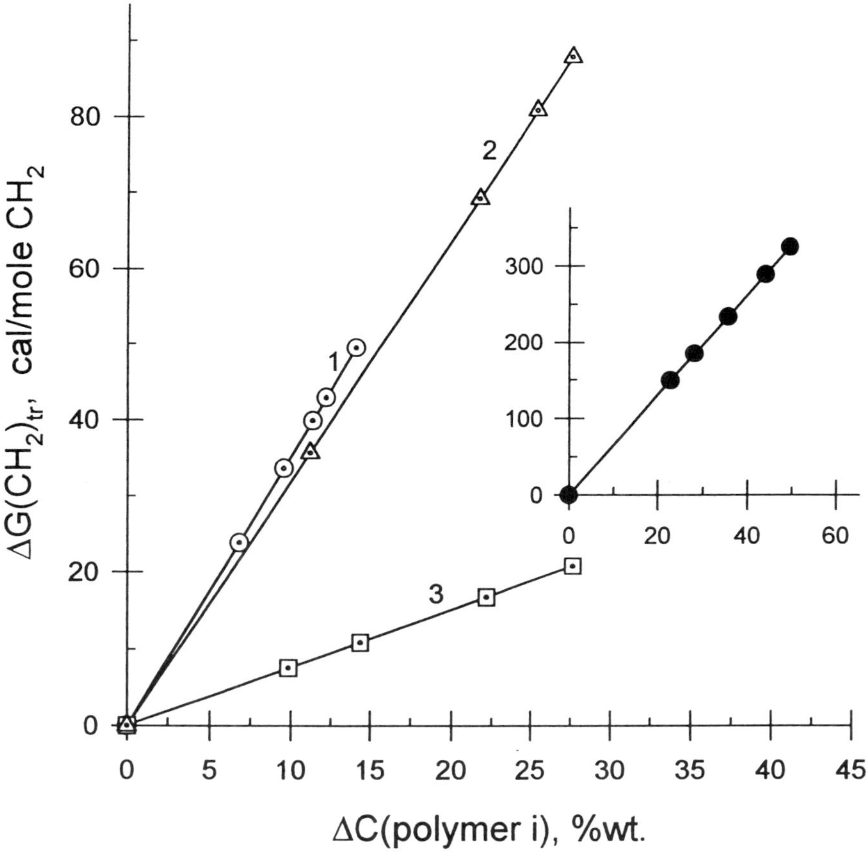

Figure 4.7. Difference between the hydrophobic character of the two phases of aqueous Dex-PEG (1), Dex-PVP (2), Dex-Ficoll (3), and PEG-600-$(NH_4)_2SO_4$ (inset) two-phase systems as a function of the difference between the concentrations of PEG, PVP, Ficoll, and PEG-6000 in the two phases, respectively [29, 30].

Only limited number of different aqueous polymer two-phase systems have been characterized by the parameter in question up to the present. It follows from the results reported in the literature [28-32] that among the systems examined the difference between the hydrophobic character of the two phases increases in the order: Dex-Ficoll < Dex-PVP < Dex-PEG < PEG-$(NH_4)_2SO_4$ <

PEG-potassium phosphate < PEG-$MgSO_4$.

The linear relationships shown in Fig. 4.7 may be described as:

$$E = e_E \cdot \Delta C(\text{polymer i}) \qquad (4.6a)$$

or

$$-\Delta G(CH_2) = e_E \cdot RT \cdot \Delta C(\text{polymer i}) = g_E \cdot \Delta C(\text{polymer i}) \qquad (4.6b)$$

where ΔC(polymer i) is the difference between the concentrations of phase polymer i in the two phases; coefficient $g_E = e_E \cdot RT$; and e_E (or g_E)is the constant characterizing the effect of polymer composition of the phases on the difference between the hydrophobic character of the phases.

It should be stressed that the differences between the hydrophobic character of the two phases in aqueous polymer two-phase systems are extremely small in comparison with that typical for water-organic solvent systems. To my knowledge, the smallest value of the free energy of transfer of a methylene group from an organic phase to an aqueous phase in water-organic solvent systems reported in the literature [11,33] amounts to 433 ± 6 cal/mole CH_2 for water-methyl ethyl ketone system. The highest value of the free energy of transfer of a CH_2 group from the polymer i-rich phase to the Dex-rich phase observed in aqueous two-polymer two-phase systems [30], e.g., Dex-PEG system, did not exceed 100 cal/mole CH_2 (theoretically possible higher values could not be achieved in the systems examined because of the experimental difficulties with handling highly concentrated polymer solutions).

The difference between the hydrophobic character of the two phases in aqueous PEG-salt two-phase systems [29,32] seems to exceed those typical for the systems formed by two nonionic polymers. The free energy of transfer of a methylene group from the PEG-rich phase to the salt-rich phase experimentally determined in the aqueous PEG-600-$(NH_4)_2SO_4$ two-phase system [29] amounts to 322 ± 11 cal/mole CH_2. According to Eiteman and Gainer [32], the free energy of transfer of a methylene group in the aqueous PEG-8000-$MgSO_4$ two-phase system may reach up to ca. 700 cal/mole CH_2. The results reported in [32] are presented in rather confusing graphical form which complicates their quantitative analysis. If the above estimate [32] is true it indicates that the difference between the hydrophobic character of the solvent media of the same aqueous nature may be as large as that observed in pairs of highly unlike solvents, such as water-octanol, water-nitrobenzene, etc. [11].

An influence of molecular weight of phase polymers on the difference between the hydrophobic character of the two phases has not been extensively studied as yet. The data reported in [29] indicate that variations of the PEG molecular weight from 300 to 20000 in the aqueous PEG-$(NH_4)_2SO_4$ two-phase system do not affect this characteristic (provided it is normalized in regard to

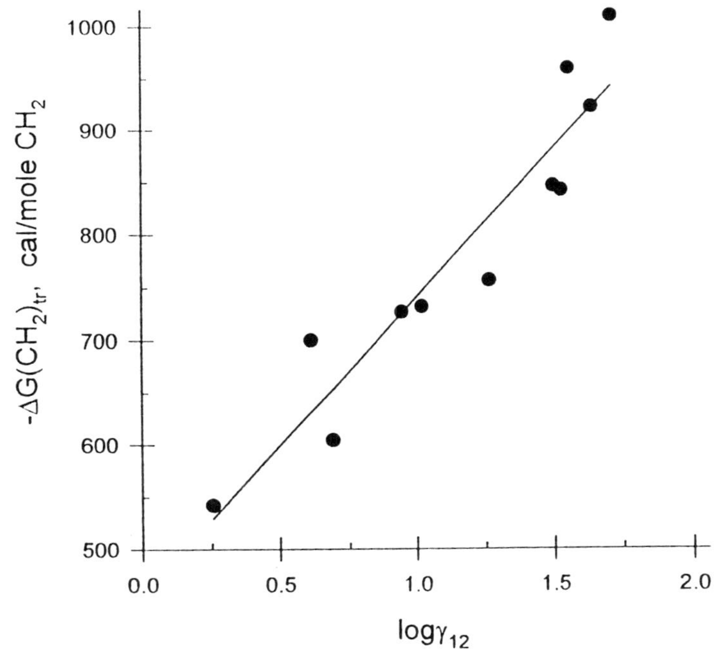

Figure 4.8. Relationship between the interfacial tension, γ_{12}, and the free energy of transfer of a CH_2 group from aqueous phase to nonaqueous phase for water-organic solvent two-phase systems.

the difference between the polymer concentrations in the two phases, see below). The $\Delta G(CH_2)_{tr}$ values reported for aqueous two-polymer two-phase systems [34] cannot be considered from this viewpoint, unfortunately, as the polymer composition of the phases was not examined. It is possible, however, to consider the influence in question indirectly. It was established [7] that the free energy of transfer of a CH_2 group between the two phases in water-polar organic solvent systems is related to the interfacial tension, γ_{12}, of the systems as shown in Figure 4.8.

If water-organic solvent systems are fundamentally similar to aqueous polymer two-phase systems the similar relationship should be expected. As the free energy of transfer of a CH_2 group between the two phases in a given aqueous polymer system is linearly related to the difference between the concentrations of a given phase polymer in the phases it is to be expected that the logarithm of the interfacial tension in the system would be also linearly related to the difference in the polymer concentrations.

The interfacial tensions in aqueous polymer systems are in the range of 0.5 to 500 $\mu N \cdot m^{-1}$, i.e., two-three orders of magnitude lower than those typical for water-organic solvent systems [1,35-37]. Recently Forciniti et al.[37] reported the first extensive set of data on the interfacial tension in different aqueous polymer systems of varied polymer concentrations formed by Dex and PEG of various molecular weights at pH 7.5 at three different temperatures 4, 25, and 40°C. The authors [37] analyzed their results in terms of logarithm of the interfacial tension as a function of logarithm of the tie line length, logarithm of the difference in Dex concentrations between the two phases, and logarithm of the difference in PEG concentrations between the two phases. The linear relationships were found in all the cases examined [37]. The effect of molecular weight of the phase polymers on the interfacial tension was not clearly established [37]. The trends observed were reported as "when systems of the same tie line length are compared, increasing the molecular weight of one of the polymers increases the interfacial tension" [37].

Limited data reported by Ryden et al. [1,35] implied the logarithm of the interfacial tension to be linearly related to the difference in a given phase polymer concentrations between the two phases. Forciniti et al.[37], however, fitted the experimental data to the double logarithm relationship [3] suggested by Bamberger et al.[36]. Analysis of the data given in [37] indicates that they do fit the expected linear relationship (an example of typical relationships is shown in Figure 4.9.)

It should be mentioned that in some cases the linear relationship fits the experimental data given in [37] only when one experimental point out of four is omitted but this point usually corresponds to the phases with polymer compositions clearly away from the averaged slope of the tie line (STL) value (see above).

Analysis of the experimental data reported in [37] according to the linear relation:

$$\log\gamma_{12} = \boldsymbol{a} + \boldsymbol{b}\cdot\Delta C(\text{PEG}) \tag{4.7}$$

shows that coefficients $\boldsymbol{a}$ and $\boldsymbol{b}$ which are constants for each individual two-phase system at a given temperature are linearly interrelated as $\boldsymbol{b} = k + z\cdot\boldsymbol{a}$, where k and z are constants independent of the temperature and molecular weights of the phase polymers. Coefficients $\boldsymbol{a}$, $\boldsymbol{b}$, k, and z were examined, and it was found that the interfacial tension in the aqueous Dex-PEG system formed by Dex of 40,000-500,000 molecular weight at 25°C depends on the polymer composition of the two phases and molecular weight of PEG as:

$$\log\gamma_{12} = A + B\cdot\Delta C(\text{PEG}) + D\cdot\log[M_w(\text{PEG})] + F\cdot\Delta C(\text{PEG})\cdot\log[M_w(\text{PEG})] \tag{4.8}$$

where M_w(PEG) is the molecular weight of PEG; A, B, D, and F are constants.

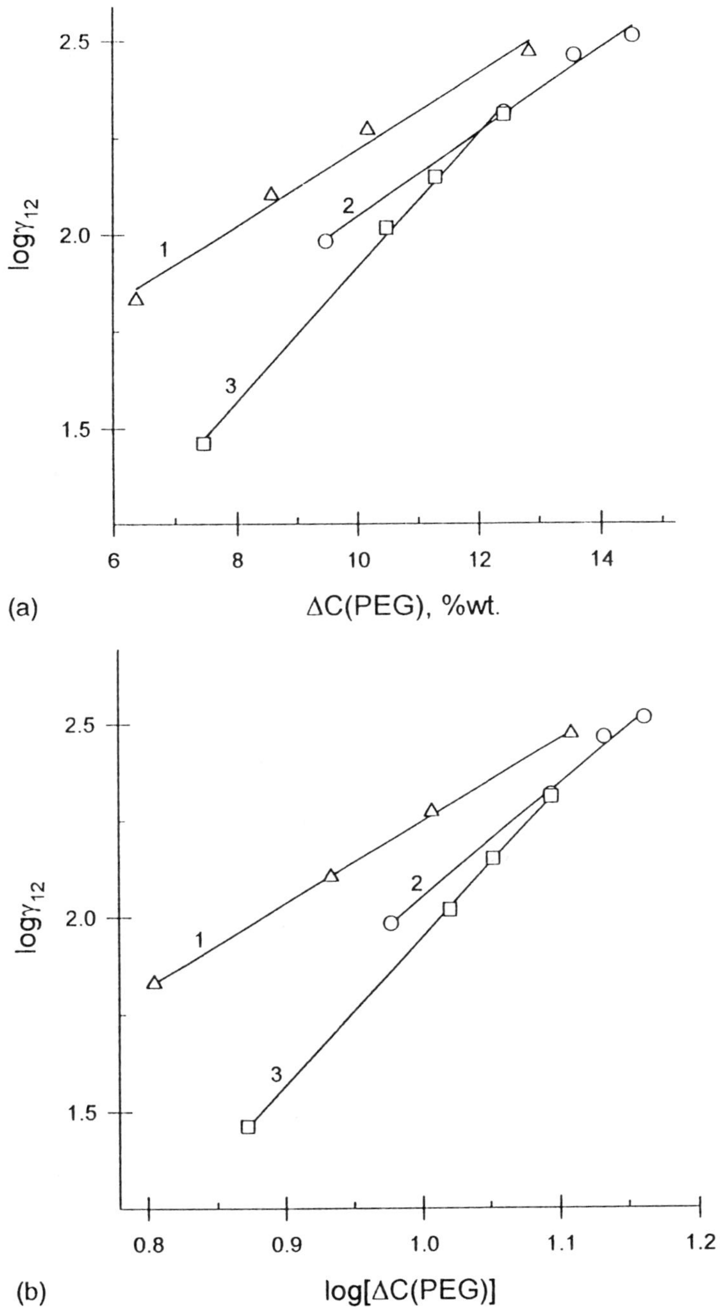

Figure 4.9. Logarithm of the interfacial tension as a function of (a) difference between the concentrations of PEG in the two phases, ΔC(PEG); (b) log[ΔC(PEG)]. Systems: (1) Dex-110—PEG-20000; (2) Dex-10—PEG-20000; (3) Dex-110—PEG-4000. Calculated from the data reported in [37].

An agreement between the interfacial values calculated with Equation 4.8 and the experimental values [37] is better than 25% in most of the cases, though the values of coefficients A, B, D, and F must be changed to fit the experimental data [37] for the Dex-PEG systems formed with Dex-10,000.

The fact that the interfacial tension increases with the difference in the polymer concentrations between the two phases according to Equation 4.7 implies that the relationship similar to that presented in Figure 4.8 [7] exists for aqueous polymer two-phase systems. That is an additional though indirect evidence for the similarity between the partition properties of aqueous polymer and water-organic solvent two-phase systems.

It also follows from Equation 4.8 that the molecular weight of PEG should affect the difference between the hydrophobic character of the two phases in aqueous Dex-PEG two-phase systems. This question, however, remains to be explored experimentally.

The difference between the hydrophobic character of the two phases, $\Delta G(CH_2)_{tr}$, is actually an important but merely the internal characteristic of a given two-phase system. It is quite possible that two different aqueous polymer two-phase systems with different solvent properties of the media in the phases would be characterized by the same $\Delta G(CH_2)_{tr}$ value (an example for water-organic solvent systems is, e.g., water-benzene and water-chloroform [33]). It is possible, however, to characterize the relative hydrophobic character of each phase separately using an organic solvent as an external reference medium [38].

The technique [38] is based on partitioning of a homologous series of aliphatic monofunctional compounds in two-phase systems formed by octanol (or any other organic solvent taken as reference) and a separate (upper or bottom) phase from a given aqueous polymer two-phase system exactly as described above (see in Chapter 2) for aqueous solutions of individual polymers. As the principle of the technique was described in detail above, only the results reported in [38] are outlined below. These results are shown in Figure 4.10 in terms of the corresponding $\Delta G(CH_2)_{tr}$ values.

It may be seen from the results [38] shown in Fig. 4.10 that: (i) it is possible to characterize the hydrophobic character of the phases of aqueous polymer systems relative to an external reference medium; (ii) the difference between the free energies of transfer of a methylene group from the same organic solvent to separate aqueous phases is in perfect agreement with the experimentally determined value of the free energy of transfer of the group between these two aqueous phases; and (iii) as mentioned above, the same value of $\Delta G(CH_2)_{tr}$ for two different aqueous two-phase systems does not mean that the solvent properties of the phases are the same.

The data presented in Fig. 4.10 were averaged over those reported in

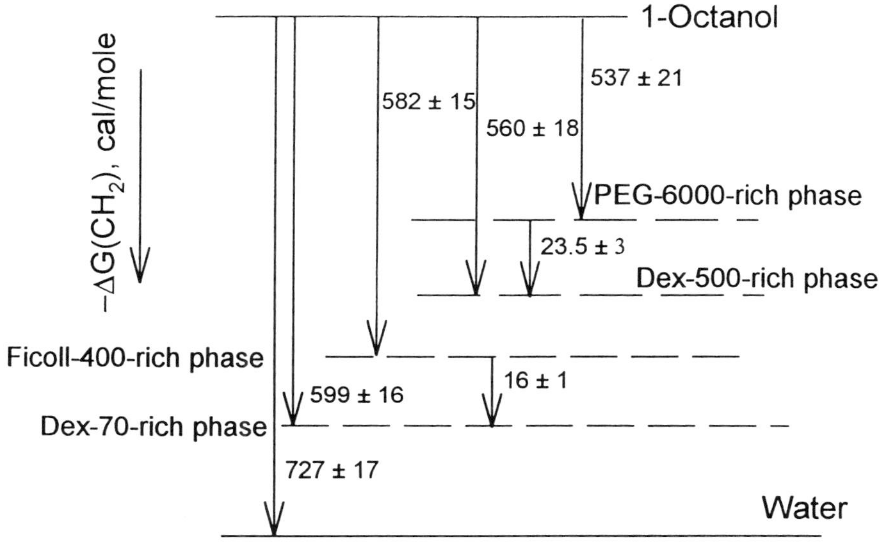

Figure 4.10. Free energy of transfer of a CH_2 group between two phases of aqueous Dex-Ficoll and Dex-PEG two-phase systems and between octanol and the separate phases of these systems. Total composition of the systems: (a) 10.8 %wt. Dex-70, 12.5 %wt. Ficoll-400; (b) 7.0 %wt. Dex-500, 4.4 %wt. PEG-6000; both systems contain either 0.15 mole/kg NaCl in 0.01 mole/kg sodium phosphate buffer, pH 7.4 or 0.11 mole/kg sodium phopshate buffer, pH 7.4. Values are given in cal/mole CH_2.

[38] for the two-phase systems of two different salt compositions. That does not imply, however, that the presence of a salt additive does not affect the difference between the hydrophobic character of the two phases.

<u>Influence of Salt Composition On the Difference Between the Hydrophobic Character of the Coexisting Phases.</u>

As shown above, the presence of inorganic salt additives change the polymer composition of the phases of aqueous two-phase systems formed by non-ionic polymers. Since the polymer and salt composition of the two phases of the systems are interrelated (see above), it should be expected that the properties of aqueous media in the phases would depend on both polymer and salt composition of the phases.

The effects of the polymer and salt composition of the two phases on the difference between the hydrophobic character of the phases were studied by Zaslavsky et al.[28,30,39]. All the relationships obtained so far fit Equation 4.6b. The g_E coefficients reported in [28,30,39] for the aqueous Dex-PEG and Dex-PVP systems are listed in Table 4.2. These coefficients characterize an increase in the free energy of interfacial transfer of a CH_2 group with increasing polymer concentration in the system. It can be seen that all salts affect the difference between the hydrophobic character of the two phases of the systems. The g_E values [30] for the Dex-PEG, Dex-PVP, and Dex-Ficoll systems of the same total salt composition support the above conclusion that the effect of the total polymer concentration on the difference between the hydrophobic character of the two phases decreases in the order: Dex-PEG > Dex-PVP > Dex-Ficoll, apparently independent of the type of salt present in the system (provided the same salt is used in each system). Hence, it was concluded [30] that the hydrophobic character of the aqueous media in the phases of an aqueous polymer two-phase system is governed mainly by the polymer composition of the phases though it is clearly affected by the salt additive as well.

The data in Table 4.2 indicate that an addition of a salt seems to decrease the g_E value in reference to that for the salt-free system in line with the above assumption that a salt additive (up to 0.10 mole/kg total concentration) distributes so that it decreases the difference between the properties of the two phases. An increase of the total salt concentration in the aqueous Dex-PEG systems is followed by an increase in the g_E value. Only in the systems with relatively high total salt concentration the coefficient g_E values exceed that for the salt-free system.

It may also be concluded that any study of the salt effects on the aqueous two-phase system partitioning ability must include analysis of the phase polymer composition. In the case the effects are explored in a system of a single fixed polymer composition, an erroneous conclusion may be easily made. An example is offered by the results reported by Zaslavsky et al.[27] on the effect of the replacement of NaCl for KCl (at the same total concentration of 0.10 mole/kg) in the aqueous Dex-500 (7.0% wt.) - PEG-6000 (4.4% wt.) two-phase system on the difference between the hydrophobic character of the two phases. It appears from the data [27] shown in Figure 4.11 that the $\Delta G(CH_2)_{tr}$ increases in the order H_2O < NaCl (0.10 mole/kg) < KCl (0.10 mole/kg). This order is directly opposite to the one established [30] when the difference between the hydrophobic character of the phases was examined as a function of the polymer composition of the phases. The conclusion that the replacement of NaCl for KCl increases the difference between the hydrophobic character of the two phases [27] was erroneous because the change in the polymer composition of the two phases in the presence of a salt additive has not been taken

Table 4.2
Coefficients g_E for Aqueous Polymer Systems. [28,30,39]

Salt	(mole/kg)	g_E, cal·(mole CH_2)$^{-1}$·(% wt.)$^{-1}$	
		Dex-PVP	Dex-PEG
-	-	3.19 ± 0.01	3.52 ± 0.01
univ. buffer [a]	0.01	-	3.04 ± 0.12
KSCN	0.10	1.53 ± 0.05	2.93 ± 0.03
KSCN	0.50	-	3.21 ± 0.07
KSCN	0.75	-	4.09 ± 0.11
NH_4SCN	0.10	2.58 ± 0.01	-
NaSCN	0.10	1.82 ± 0.01	2.56 ± 0.01
NaSCN	0.10	-	3.04 ± 0.17
NaSCN	0.50	-	3.45 ± 0.12
NaCl [b]	0.10	-	3.04 ± 0.17
NaCl [b]	0.50	-	3.04 ± 0.17
NaCl [c]	0.15	2.36 ± 0.09	2.75 ± 0.01
KCl	0.10	2.36 ± 0.01	2.49 ± 0.01
KCl	0.50	-	3.21 ± 0.01
KCl	0.75	-	3.90 ± 0.03
$NaClO_4$ [b]	0.10	-	3.04 ± 0.12

Table 4.2 (continued)

Salt	(mole/kg)	g_E, cal·(mole CH_2)$^{-1}$·(% wt.)$^{-1}$	
		Dex-PVP	Dex-PEG
KF	0.10	2.00 ± 0.01	-
Na_2SO_4 [b]	0.05	-	3.04 ± 0.17
Na_2SO_4	0.10	2.19 ± 0.01	2.91 ± 0.03
Na_2SO_4 [c]	0.25	-	4.39 ± 0.34
K_2SO_4	0.05	-	3.17 ± 0.01
K_2SO_4	0.10	2.31 ± 0.01	3.49 ± 0.01
K_2SO_4	0.25	-	5.11 ± 0.01
Phosph. buffer [d]	0.11	2.30 ± 0.11	3.93 ± 0.04

[a] 0.01 mole/kg universal buffer, pH 7.5; [b] salt at a given concentration in 0.01 mole/kg universal buffer, pH 7.5; [c] 0.15 mole/kg NaCl in 0.01 mol/kg sodium phosphate buffer, pH 7.4; [d] 0.11 mole/kg sodium phosphate buffer, pH 7.4.

into account.

Since the polymer and salt compositions of the two phases are interrelated (see above), it is difficult to estimate the separate effects of the phase polymers and salt additives on the difference between the hydrophobic character of the two phases.

Up till now only coefficient E in Equation 4.4 and its dependence on the polymer and salt composition of the phases was discussed. This characteristic of a system while clearly important does not describe all the properties of the aqueous media in the phases governing partitioning of solutes in aqueous two-phase systems. Coefficient A in Equation 4.4 is at least equally important. As mentioned above the coefficient A represents the contribution of a (polar and/or ionized) part of the solute molecule into the logarithm of the solute partition coefficient. As outlined in Chapter 1, the interactions of polar and ionized groups with water differ from those of non-polar groups. Hence the

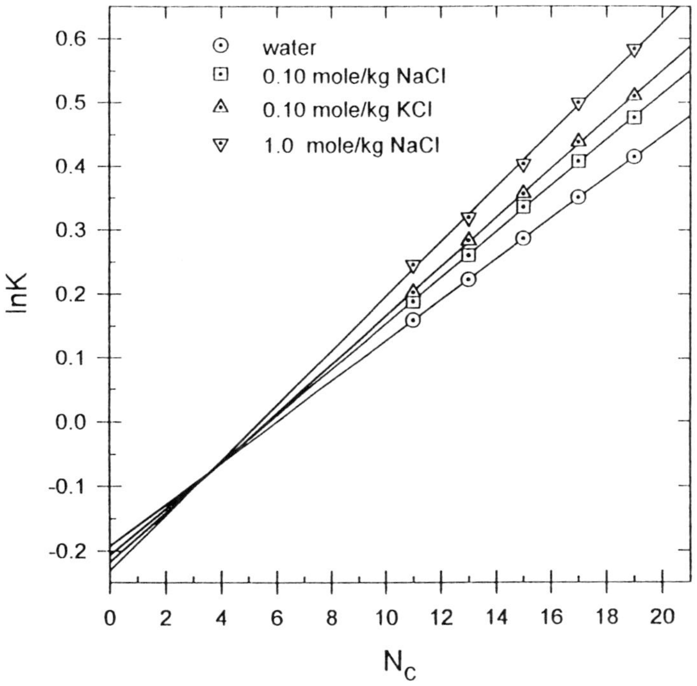

Figure 4.11. Logarithm of the partition coefficient, lnK, for alkyltrimethylammonium bromides as a function of the total number of carbon atoms, N_C, in the solute molecule. Aqueous Dex-500 (7 %wt.) — PEG-6000 (4.4 %wt.) two-phase system.

physical meaning of coefficient A and the influence of polymer and salt composition of aqueous two-phase systems on the coefficient A value are discussed in separate section.

4.3. INFLUENCE OF POLAR GROUPS OF A SOLUTE ON THE SOLUTE PARTITIONING IN AQUEOUS TWO-PHASE SYSTEMS

The nature of (van der Waals) interactions of any non-polar group of a solute molecule with an aqueous medium is essentially the same. Polar groups, however, may participate in highly specific interactions with water, such as dipole-dipole, dipole-induced dipole, H-bonding, etc. If the group is ionized the additional electrostatic ion-ion and ion-dipole interactions occur. As mentioned above (Chapter 1), the type and intensity of these polar interactions depend on the specific features of the particular group (as well as on those of the solvent medium).

Coefficient A(j) in Equation 4.4 represents the contribution of the j-th polar group of the solute molecule into the logarithm of the solute partition coefficient. Hence it is related to the free energy of transfer of this group from one phase to the other phase in a given two-phase system, $\Delta G(\text{polar group j})_{tr}$, as:

$$\Delta G(\text{polar group j})_{tr} = -RT \cdot A(j) \tag{4.9}$$

The coefficient A(j) values for different polar groups will be discussed below in terms of the $\Delta G(\text{polar group j})_{tr}$ values.

The results reported in [27] indicated that the free energies of transfer of different polar groups between the two phases of a given system are different as expected. In the aqueous Dex-500 (7.0%wt.) - PEG-6000 (4.4%wt.) two-phase system at pH > 8.0 the free energy of transfer of the polar $->N^+OH^-$ group $\Delta G(->N^+OH^-)_{tr}$, was found [27] to be +572.2 ± 3.9 cal/mole while that of a CH_2 group, $\Delta G(CH_2)_{tr}$, is -18.4 ± 0.7 cal/mole. It was found [27] also that an addition of 0.1 mole/kg NaCl or KCl or 1.0 mole/kg NaCl to the system did not affect the value of $\Delta G(->N^+X^-)_{tr}$ (X^-: OH^-, Cl^-, Br^-) significantly, while induced noticeable changes in the difference between the hydrophobic character of the two phases (as indicated above). The reasons for this apparent independence of the $\Delta G(->N^+X^-)_{tr}$ value of the salt composition of the system will be considered below.

The difference between the signs of the $\Delta G(CH_2)_{tr}$ and $\Delta G(->N^+X^-)_{tr}$ values seems to be in agreement with an intuitive feeling that as the PEG-rich phase displays the affinity for a non-polar CH_2 group exceeding that of the Dex-rich phase, the opposite should be expected for a polar group. This feeling is misleading, however, as indicated by the estimate of the free energy of transfer of the polar ionic $-OSO_3Na$ group, $\Delta G(-OSO_3Na)_{tr}$ reported [27] to be -76.4 ± 2.0 cal/mole in the same system containing 1.0 mole/kg NaCl. That means that the affinities of both polar $-OSO_3Na$ group and non-polar $-CH_2$-group for the PEG-rich phase under the indicated conditions exceed those for the Dex-rich phase. It should be mentioned here that in the same salt-free system the lnK - N_C relationship for sodium alkyl sulfates did not fit Equation 4.4 and the $\Delta G(-OSO_3Na)_{tr}$ value could not be estimated [27]. The possible explanation will be discussed below.

Analysis of partitioning of homologous series of aliphatic monofunctional solutes, such as sodium alkyl sulfates, fatty acids, alkyltrimethylammonium bromides, etc., is complicated by that the solute concentrations cannot be measured directly in the phases. The generally employed methods require an extraction procedure and the results obtained are not as accurate as necessary.

An alternative choice is the chromophore-containing compounds, e.g., dinitrophenylated (DNP-) amino acids with an aliphatic alkyl side-chain. The concentrations of these compounds in the phases may be determined by direct optical absorbance measurements increasing significantly the accuracy of determination of the partition coefficient value.

The homologous series of sodium salts of 2.4-dinitrophenyl (DNP) - amino acids with aliphatic alkyl side-chain: DNP-glycine (DNP-Gly), DNP-alanine (DNP-Ala), DNP-norvaline (DNP-NVal), DNP-norleucine (DNP-NLeu), and DNP-2-amino-n-octanoic acid (DNP-NAO) was almost exclusively used in the studies of the effects of polymer and salt composition and buffer pH on the coefficients A and E values for characterization of aqueous polymer two-phase systems [7,13,28-30,33,34,38,39]. That certainly limits our discussion about an influence of the polar group-solvent media interactions in the two phases on the solute partitioning in aqueous polymer systems. The trends reported in the literature are outlined below but the reader should keep in mind that the generality of some of these trends at least remains to be verified experimentally.

It has been established independently by Diamond and Hsu [31] (with certain reservations, see below) and Zaslavsky et al.[28,30,39] that the logarithm of the solute partition coefficient for different solutes is linearly dependent on the difference between the concentrations of the ith phase polymer in the two phases:

$$\ln K(j) = k_{ji} \cdot \Delta C(\text{polymer i}) \qquad (4.10)$$

where $\Delta C(\text{polymer i})$ is the difference between the concentrations of phase polymer i in the two phases; k_{ji} is the constant characterizing the effect of the polymer composition of the system on the partition coefficient K(j) of the solute j.

It follows from Equations 4.4, 4.6, and 4.10 that the coefficient A (or the free energy of transfer of a given polar group) is described by the similar relationship:

$$A(j) = a_A(j) \cdot \Delta C(\text{polymer i}) \qquad (4.11a)$$

or

$$-\Delta G(\text{polar group j})_{tr} = g_A(j) \cdot \Delta C(\text{polymer i}) \qquad (4.11b)$$

where $a_A(j)$ or $g_A(j)$ is the constant characterizing the effect of the polymer composition of the phases on the difference between the interactions of the polar group j with the phases.

Since both Equations 4.6 and 4.11 for a given two-phase system contain the same variable, $\Delta C(\text{polymer i})$, it is obvious that coefficients A and E are interrelated.

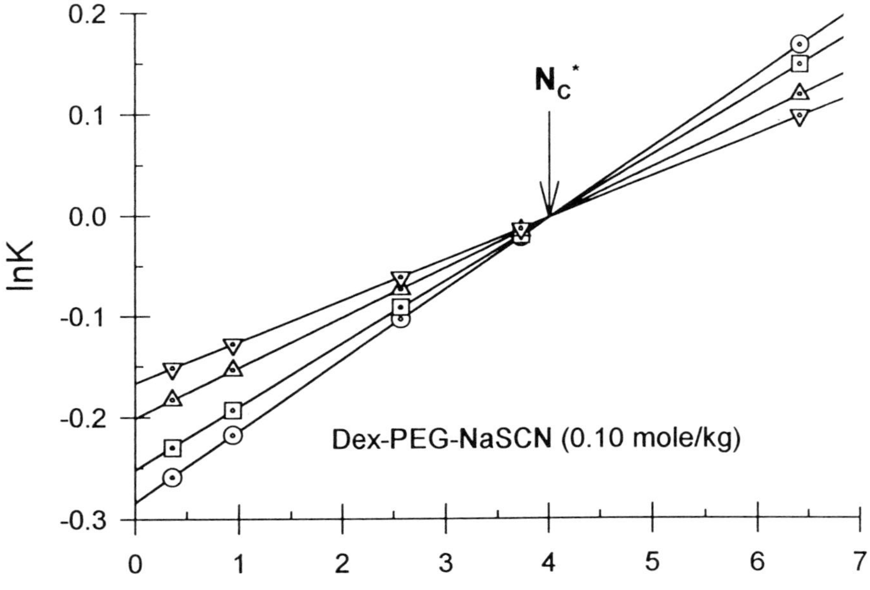

(a)

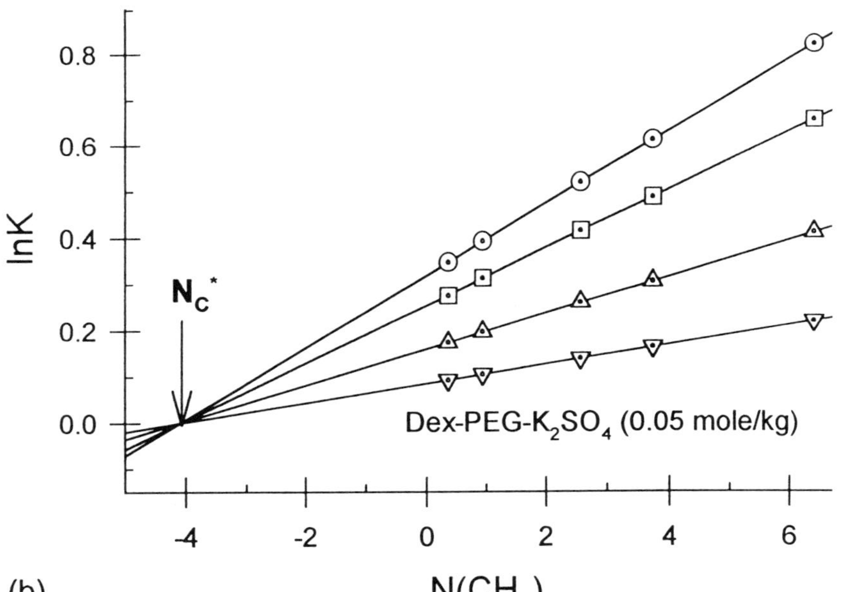

(b)

Figure 4.12. Logarithm of partition coefficient of sodium salt of DNP-amino acid in the aqueous two-phase systems of different polymer compositions as a function of the alkyl side chain length. Calculated from data reported in [39].

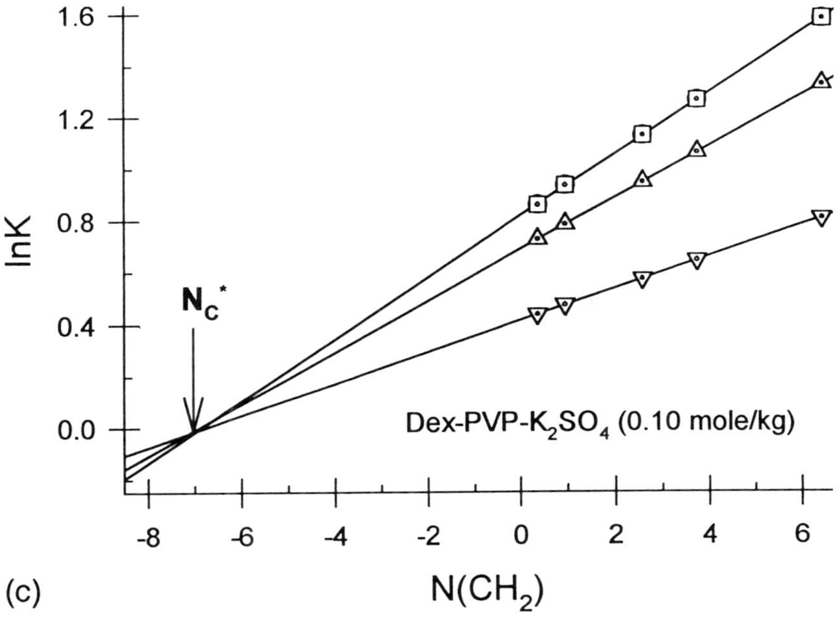

(c)

This interrelationship may be described as:

$$A(j) = n^*(j) \cdot E \tag{4.12}$$

or

$$\Delta G(\text{polar group } j)_{tr} = n^*(j) \cdot \Delta G(CH_2)_{tr} \tag{4.12a}$$

where coefficient $n^*(j)$ is the ratio

$$n^*(j) = a_A(j)/e_E = g_A(j)/g_E \tag{4.13}$$

Combining Equations 4.4 and 4.12 we obtain

$$\ln K(j) = A(j) + E \cdot N_C = (n^*(j) + N_C) \cdot E \tag{4.14}$$

Equation 4.14 implies that in a given j-th homologous series of solutes there is a hypothetical or real member of the series with the alkyl chain length $N_C{}^* = -n^*(j)$ characterized by the uniform partitioning ($K = 1$) independent of the polymer composition of the phases in a given aqueous two-phase system. This conclusion is supported by the experimental evidence as shown in Figure 4.12.

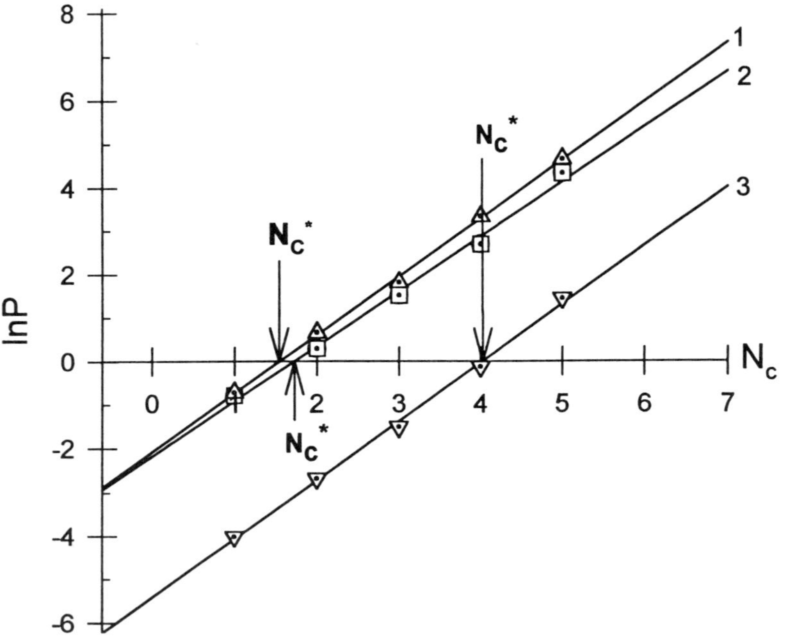

Figure 4.13. Logarithm of partition coefficient of aliphatic carbon acid $C_nH_{2n+1}COOH$ in a water-organic solvent system as a function of the alkyl chain length, N_C. 1 - water-octanol; 2 - water - diethyl ether; 3 - water-benzene. Calculated from the data reported in [10].

As shown in Figure 4.13, Equation 4.14 holds also for solutes being partitioned in water-organic solvent systems. The physical meaning of the ratio $n^*(j)$ equal to A(j)/E in water-organic solvent two-phase systems is clear. It is interpreted [10] as that for the solute with the j-th polar group and the alkyl chain of the length $N_C^* = -n^*(j)$ the forces of interactions of the polar group with water and organic solvent cancel those of the alkyl chain. The balance between these interactions leads to the uniform partitioning of the solute.

This interpretation is supported by that the $N_C^*(j)$ values in different water-organic solvent two-phase systems are related to the water content in the organic phase [33] as shown in Figure 4.14. The relationships are described as:

$$N_C^*(j)_m = [A(j)/E]_m = a + b \cdot \log(S_{water}^{org})_m \tag{4.15}$$

where S_{water}^{org} is the concentration of water (solubility) in an organic phase of a given water-organic solvent two-phase system; subscript "m" denotes the two-phase system; a and b are constants depending on the j-th polar group in

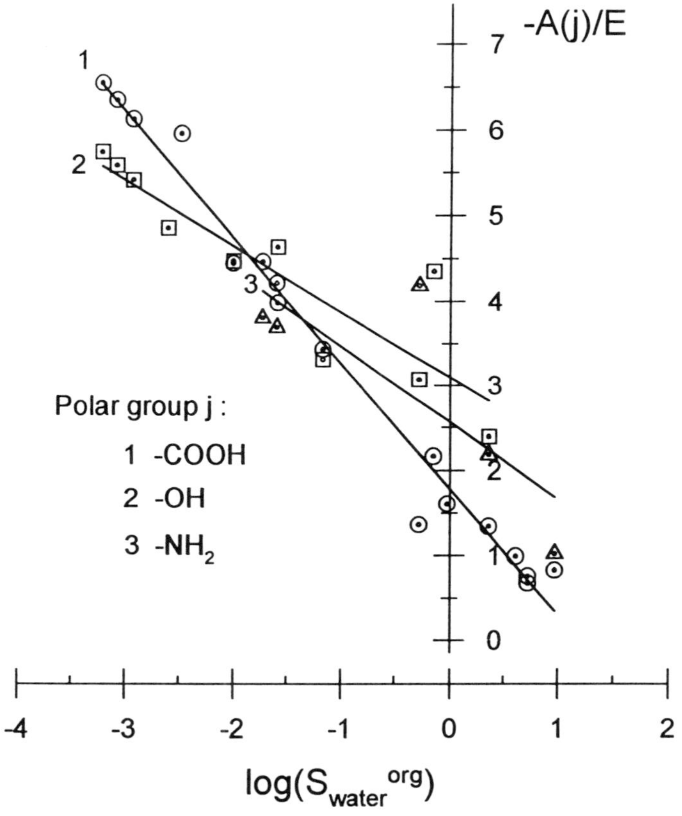

Figure 4.14. Relationships between the ratio -A(j)/E and the water content of nonaqueous phases in different water-organic solvent two-phase systems.

the molecules being partitioned.

We will return to the data presented in Fig. 4.14 further to discuss several important implications of the relationships described by Equation 4.15. The only conclusion from these data to be considered at the moment is that the A/E ratio may be used as a relative measure of the interactions of polar groups with the solvent media in the phases of <u>different</u> solvent two-phase systems.

As illustrated in Fig. 4.12, the effect of the polymer composition of the phases on the contribution of a given polar group into the solute partition coefficient represented by parameter $g_A(j)$ depends on the type of the salt additive. This effect depends also on the total concentration of the additive and on the type of phase polymers employed as may be seen from the parameter $g_A(j)$

values [28,30,39] listed in Table 4.3 for the aqueous Dex-PEG and Dex-PVP two-phase systems.

The g_A values for the DNP-NH-CH-COONa moiety given in Table4.3 indicate that the salt effect on the $g_A(j)$ parameter is much more significant than on the g_E parameter. Under all the conditions used the g_E values in both aqueous Dex-PEG and Dex-PVP two-phase systems are varied over relatively narrow range from $1.5 \cdot 10^{-3}$ to $3.2 \cdot 10^{-3}$ cal·(mole $CH_2)^{-1} \cdot (wt.\%)^{-1}$ in the Dex-PVP and from $2.5 \cdot 10^{-3}$ to $5.1 \cdot 10^{-3}$ cal·(mole $CH_2)^{-1} \cdot (wt.\%)^{-1}$ in the Dex-PEG two phase system. The g_A values under the same conditions vary from -11 to +19.0 cal·(mole polar group)$^{-1} \cdot (wt.\%)^{-1}$ in the aqueous Dex-PEG and from -2.3 to +18 cal·(mole polar group)$^{-1} \cdot (wt.\%)^{-1}$ in the aqueous Dex-PVP two-phase system.

The data reported in [29] on the aqueous PEG-$(NH_4)_2SO_4$ two-phase systems with PEGs of different molecular weights from 300 to 20,000 support the above conclusion. These data [29] indicate that the difference in the polar interactions between the phases (probed by the same polar DNP-NH-CH--COONa group) is governed mainly by the salt composition of the phases. The g_A coefficient values for the aqueous PEG-$(NH_4)_2SO_4$ two-phase systems examined [29] vary from 40.3 to 61.7 cal·(mole polar group)$^{-1} \cdot (wt.\%)^{-1}$ depending on the PEG molecular weight.

The data [39] given in Table 4.3 indicate an increase or decrease in the free energy of transfer of the DNP-NH-CH-COONa group from Dex-rich phase to PEG- or PVP-rich phase with increasing polymer concentration depending on the type and total concentration of salt additive. These data imply that the order of the salt effect on the g_A parameter follows the order of the water-structure-affecting properties of the salts for both aqueous Dex-PVP and Dex-PEG two-phase systems [39]. That is, the more strongly water-structure-making salts (sulfates, phosphates, KF) give positive g_A values, whereas the more strongly water-structure-breaking salts (rhodanides) give negative g_A values.

The other trend for the DNP-NH-CH-COONa group implied by the data in Table 4.3 follows from comparison of the g_A values with those of the coefficient b(salt). As indicated in Chapter 3, all the water-structure-making salt additives examined so far tend to accumulate in the less structured aqueous medium in the Dex-rich phase. That is represented by the corresponding nega-tive values of the coefficient b(salt) in Equation 3.5. The water-structure-breaking salts inclined to accumulate in the aqueous medium with more pro-nounced water structure in the PEG- (or PVP-, Ficoll-, etc.) -rich phase are characterized by the positive values of the coefficient b(salt). Comparison of the signs of g_A and b(salt) coefficients indicate that the polar group under conside-ration displays an increased affinity for the relatively salt-poor phase. The salt type seems to be more important than the salt concentration (or ionic strength) as follows from

Table 4.3

Coefficients g_A for Aqueous Polymer Systems.

Salt	mole/kg	g_A, cal.mole^{-1}·(% wt.)$^{-1}$		b(salt) ·10^3	b(salt) ·10^3
		Dex-PVP	Dex-PEG	Dex-PVP	Dex-PEG
-	-	7.03 ± 0.05	1.97± 0.01	-	-
Univ.buffer [a]	-	-	7.33 ± 0.74	-	-16.9 ± 0.2
KSCN	0.10	-1.91 ± 0.01	-11.05 ± 0.12	5.5 ± 1.4	12.4 ± 0.7
KSCN	0.50	-	-5.31 ± 0.28	-	12.4 ± 0.7
KSCN	0.75	-	-1.38 ± 0.04	-	12.4 ± 0.7
NH_4SCN	0.10	-2.34 ± 0.04	-	6.5 ± 1.2	-
NaSCN	0.10	-8.48 ± 0.06	-10.32 ± 0.12	16.3 ± 2.1	17.7 ± 2.1
NaSCN [b]	0.10	-	-9.14 ± 0.40	-	5.3 ± 0.3
NaSCN [b]	0.50	-	-5.91 ± 0.06	-	10.2 ± 0.8
NaCl [b]	0.10	-	-2.40 ± 0.12	-	-6.8 ± 0.4
NaCl [b]	0.50	-	-4.26 ± 0.12	-	-1.7 ± 0.7
NaCl [c]	0.15	2.88 ± 0.12	-4.87 ± 0.09	-4.9 ± 1.2	2.5 ± 0.9
KCl	0.10	1.33 ± 0.02	-5.5 ± 0.01	-9.6 ± 0.8	-2.7 ± 0.4
KCl	0.50	-	-4.53 ± 0.01	-	-4.6 ± 0.5
KCl	0.75	-	-3.24 ± 0.07	-	-7.9 ± 0.6
$NaClO_4$ [b]	0.10	-	-7.83 ± 0.30	-	7.5 ± 1.1

Table 4.3 (continued)

Salt	mole/kg	g_A, cal.mole^{-1}·(% wt.)$^{-1}$		b(salt) ·10^3	b(salt) ·10^3
		Dex-PVP	Dex-PEG	Dex-PVP	Dex-PEG
$NaClO_4$ [b]	0.50	-	-7.96 ± 0.36	-	16.2 ± 0.9
KF	0.10	7.63 ± 0.01	-	-18.3 ± 1.1	-
Na_2SO_4 [b]	0.05	-	8.06 ± 0.36	-	-29.9 ± 2.5
Na_2SO_4	0.10	16.33 ± 0.02	8.65 ± 0.17	-25.1 ± 1.8	-26.5 ± 1.3
Na_2SO_4 [b]	0.25	-	16.78 ± 0.42	-	-39.7 ± 2.6
K_2SO_4	0.05	-	12.97 ± 0.03	-	-35.7 ± 0.4
K_2SO_4	0.10	16.11 ± 0.06	17.26 ± 0.01	-35.2 ± 1.5	-44.4 ± 0.1
K_2SO_4	0.25	-	19.23 ± 0.01	-	-45.6 ± 0.1
Phos.buf. [d]	0.11	18.06 ± 0.14	15.57 ± 0.09	-28.1 ± 3.2	-37.9 ± 1.5

[a] 0.01 mole/kg universal buffer, pH 7.5; [b] Salt at a given concentration in 0.01 mole/kg universal buffer, pH 7.5; [c] 0.15 mole/kg NaCl in 0.01 mole/kg sodium phosphate buffer, pH 7.4; [d] 0.11 mole/kg sodium phosphate buffer, pH 7.4.

the g_A and b(salt) values for the Dex-PEG systems containing 0.10 mole/kg K_2SO_4 and 0.25 mole/kg Na_2SO_4.

Similar trend was observed in the aqueous Dex-500-PEG-6000 two-phase system [27] for the ->N^+X^- (X^-: OH^-, Cl^-, Br^-) moiety, and the opposite effect seems to be true for the -OSO_3Na group. Much more data are clearly necessary to establish general trends for different polar groups.

Coefficient b(salt) according to Equation 3.5 characterizes the relationship between the salt and polymer concentrations in the phases of the aqueous two-phase system formed by a given polymer pair and salt additive at the fixed total concentration. This coefficient is independent of the (variable) concentrations of phase polymers. All the solvent features of the aqueous media in the phases are clearly governed by their polymer and salt concentrations (at the

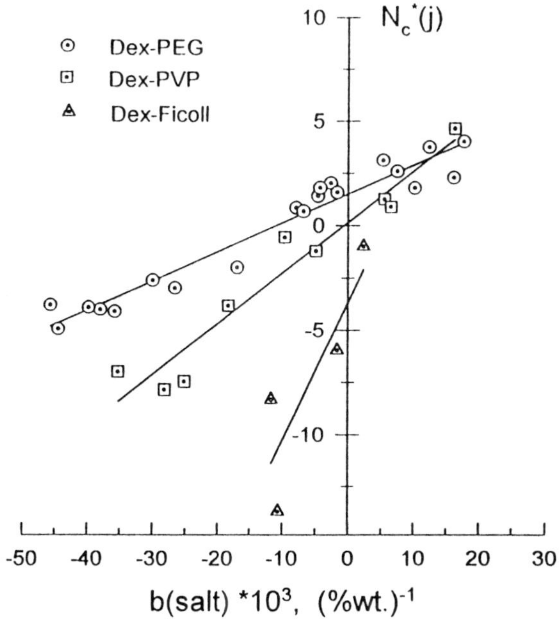

Figure 4.15. Relationship between parameter $N_C^*(j)$ for sodium salts of DNP-amino acids with aliphatic side chain and coefficient b(salt) in aqueous polymer two-phase systems of different polymer and salt composition. Calculated from the data reported in [28,30,39].

fixed temperature and pressure). Hence it seems possible to use the b(salt) coefficient as a relative measure of the variable polymer and salt compositions of the two phases in a given aqueous two-polymer system. This measure is in a way analogous to the water content of an organic phase, S_{water}^{org}, used for comparison of different water-organic solvent systems.

The $N_C^* = -g_A(j)/g_E$ values for different aqueous polymer systems and the corresponding b(salt) values are interrelated as shown in Figure 4.15. The relationships for the aqueous Dex-PEG, Dex-PVP, and Dex-Ficoll two-phase systems containing different salt additives are described as:

$$N_C^*(j) = -g_A(j)/g_E = F + W \cdot b(salt) \tag{4.16}$$

where *F* and *W* are constants depending on the phase polymers used.

Examination of coefficients *F* and *W* indicates that they are (a) linearly interrelated; and (b) related to parameter $B(\tau)$ representing the concentration effect of a given polymer i (PEG, PVP, Ficoll) on the dielectric orientational mobility of water molecules (see Equation 2.22). Analysis of these relationships for the examined aqueous Dex-polymer i-salt two-phase systems leads to:

$$N_C^*(j) = k_o + k_1 \cdot b(salt) + k_2 \cdot B_i(\tau) \cdot k_3 \cdot B_i(\tau) \cdot b(salt) \qquad (4.17)$$

where k_o, k_1, k_2, and k_3 are constants; $B_i(\tau)$ is the coefficient in Equation 2.22 characterizing the concentration influence of polymer i in its salt-free aqueous solution on the dielectric relaxation time of water; $N_C^*(j)$ and b(salt) are as defined above.

The $N_C^*(j)$ parameter represents the ratio of the difference in the interactions of the j-th polar group with the aqueous media in the two phases to that of a CH_2 group. It is possible to consider this parameter as a measure of the relative solvation (hydration) power of the media in the two phases in regard to both j-th polar group and non-polar methylene group. This relative hydration power of the media in the two phases governs the solute partitioning in a given two-phase system. Hence Equation 4.17 implies that the partitioning capability of an aqueous Dex—polymer i—salt two-phase system depends on the relative structuring influence of the polymer i on water and on the polymer and salt composition of the phases (represented by the b(salt) coefficient).

One of the additional important implications of Equation 4.17 is that the aqueous two-phase systems formed by the same pair of polymers and the same salt additive should be viewed as different systems when the total concentrations of the salt additive are different. The same is likely to be true for ternary water—organic solvent—organic modifier two-phase systems with different concentrations of the modifier. An experimental evidence for this conclusion will be discussed below.

This conclusion implies also that aqueous polymer two-phase systems at different pHs should be viewed as different systems since different pH values in the system are provided by varying the relative total amounts of buffer salts.

The results of the recent study [40] of aqueous Dex-70—PEG-6000 two-phase systems containing 0.01 mole/kg "universal" buffer of varied pH and different salt additives support the above conclusion. The "universal" buffer used [40] includes acetic acid, boric acid, o-phosphoric acid, and sodium hydroxide. By varying the relative amounts of the components the pH values of 4.8, 7.5, and 9.4 were obtained. The polymer composition of the phases was found [40] to be independent of the buffer pH value within the experimental error limits. Partitioning of sodium salts of DNP-amino acids with aliphatic side-chains was studied in the systems of varied polymer concentrations containing 0.01 mole/ kg universal buffer at different pHs without any additional salt and with 0.05 mole/kg Na_2SO_4, 0.1 mole/kg NaCl, NaSCN, $NaClO_4$, 0.25

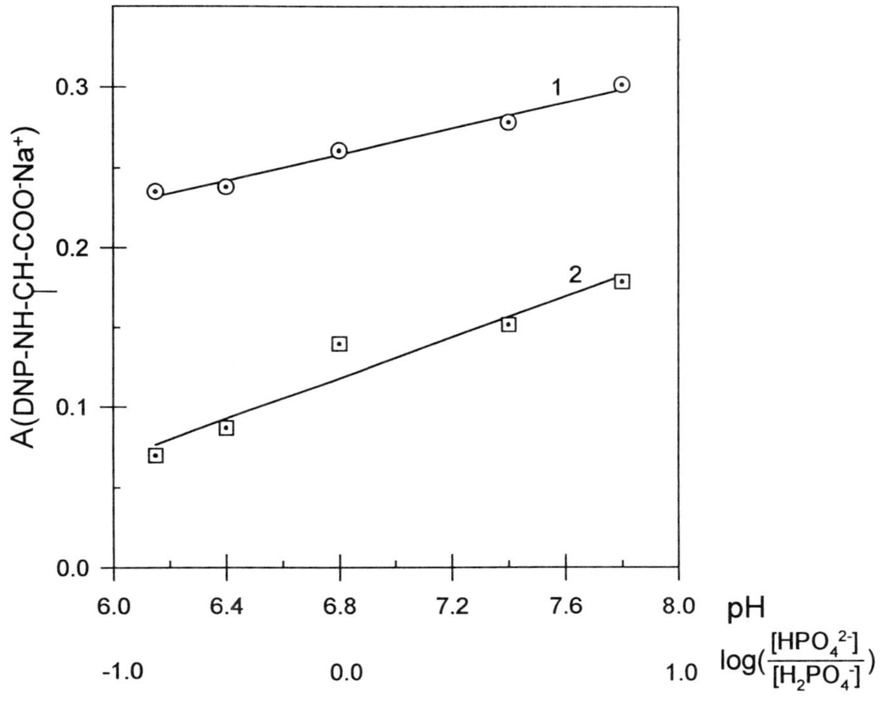

Figure 4.16. Coefficient A as a function of pH and/or ratio between the amounts of phosphate-ions in the aqueous (1) Dex-70-Ficoll-400 and (2) Dex-40-Ficoll-400 two-phase systems containing 0.11 mole/kg sodium phosphate buffer.

mole/kg Na_2SO_4, and 0.5 mole/kg NaCl, NaSCN, and $NaClO_4$. The results obtained [40] are presented in Figure 4.17 as the corresponding pH-functions of g_A coefficients. It should be mentioned that the effects of the salts on the polymer compositions of the phases were examined [40] and taken into account in determinations of the g_A and g_E coefficients.

The results [40] indicate that the pH change in the aqueous Dex-PEG two-phase system containing 0.01 mole/kg universal buffer without any additional salt affects the polar interactions in the phases (as represented by coefficient g_A) noticeably - ten-fold change in the g_A value follows the pH change from 4.8 to 9.3 (see Figure 4.17). As the properties of the DNP-NH-CH-COONa group are not changed significantly over this pH range, the effect seems to support the above conclusion that the systems with different total salt compositions (and pH) are to be viewed as the different systems with different partitioning capability. The difference between the hydrophobic character of the two phases is independent of the pH variations [40]. That agrees with the above conclusion that the polymer composition of the phases is the major factor affecting interactions of non-polar groups with the aqueous media in the phases.

The similar trend was observed [41] in the aqueous Dex-500—PEG-6000 and Dex-70—Ficoll-400 two-phase systems containing 0.11 mole/kg sodium phosphate buffer with pH varied over the 6.15 to 7.8 range. Changes in the polar interactions (represented by the A(DNP-NH-CH-COONa group)-value) [41] are illustrated in Figure 4.16.

The difference between the hydrophobic character of the two phases represented by the $\Delta G(CH_2)_{tr}$ value was found [42] to be independent of the $[HPO_4^{2-}]/[H_2PO_4^-]$ concentration ratio (or pH value) in the aqueous Dex-500—PEG-6000 system containing 0.11 mole/kg sodium phosphate buffer. In the aqueous Dex-70—Ficoll-400 two-phase system, on the other hand, the $\Delta G(CH_2)_{tr}$ was noticeably dependent upon the composition (or pH) of the same buffer [42]. Apparently different effects of the buffer composition (and pH) on the $\Delta G(CH_2)_{tr}$ in these two systems may be due to different influence of the buffer composition on the polymer concentrations in the phases (see below). This influence was not explored and taken into account in [41,42]. It follows from Equation 4.6b, however, that if a change in the buffer composition changes the polymer concentrations in the phases, i.e., ΔC(polymer i), an alteration in the $\Delta G(CH_2)_{tr}$ may occur under invariable g_E value. That possibly explains the apparent influence of the $[HPO_4^{2-}]/[H_2PO_4^-]$ concentration ratio on the hydrophobic properties of the phases in the aqueous Dex-500—PEG-6000 two-phase system containing 0.15 mole/kg NaCl in 0.01 mol/kg sodium phosphate buffer [41].

An addition of small amounts (up to 0.12 mole/kg) of NaCl to the aforementioned Dex-500—PEG-6000 two-phase system containing 0.01 mole/kg sodium phosphate buffer, pH 6.8 was observed [42] to reduce the difference between the hydrophobic character of the phases. An increase in the NaCl concentration to 0.15 mole/kg increases the $\Delta G(CH_2)_{tr}$ which appears to be constant in the range of 0.15 to 0.50 mole/kg NaCl. These data [42] are in partial agreement with those obtained in the more recent study [40]. It was found [40] that the aqueous Dex-70—PEG-6000 two-phase systems containing 0.01 mole/kg universal buffer without any additional salt or containing additional 0.1 or 0.5 mole/kg NaCl, 0.1 mole/kg NaSCN or $NaClO_4$, or 0.05 mole/kg Na_2SO_4 are characterized by the same g_E coefficient value of 3.04 ± 0.17 cal·(mole CH_2)$^{-1}$·(wt.%)$^{-1}$ over the examined 4.8 to 9.3 pH range.

Salt effects on the polymer composition of the phases should be mentioned here. An example of the differences in the phase compositions induced by the salts in the aqueous Dex-PEG two-phase system of the fixed total poly-

Table 4.4

Partitioning Properties of the Phases of Aqueous Dex-70—PEG-6000 Two-Phase Systems of Fixed Total Polymer Composition (8.5% wt. Dex; 5.3% wt. PEG) Containing 0.01 Mole/kg Universal Buffer, pH 4.8-9.3 and Different Salt Additives [40]. [a]

Salt (mole/kg)	$\Delta G(CH_2)_{tr}$ cal/mole CH_2	ΔG(polar group)$_{tr}$ cal/mole	ΔC(PEG) % wt.	-STL [b]	g_E cal/mole·(%wt.)$^{-1}$	g_A cal/mole·(%wt.)$^{-1}$
Univ. buf. (0.01)	-21.4 ± 0.5	-52.6 ± 3.0	7.17	0.552	2.98 ± 0.06	7.33 ± 0.8
NaCl (0.10)	-21.1 ± 0.6	17.0 ± 2.5	7.10	0.554	2.97 ± 0.08	-2.40 ± 0.8
Na_2SO_4 (0.05)	-23.9 ± 0.8	-61.3 ± 3.0	7.60	0.600	3.14 ± 0.11	8.06 ± 0.6
NaSCN (0.10)	-20.8 ± 1.5	66.4 ± 1.8	7.30	0.579	2.85 ± 0.20	-9.1 ± 0.8
$NaClO_4$ (0.10)	-22.7 ± 0.5	58.7 ± 2.0	7.50	0.583	3.03 ± 0.05	-7.83 ± 0.6
NaCl (0.50)	-22.3 ± 0.6	32.0 ± 2.0	7.50	0.617	2.97 ± 0.07	-4.26 ± 0.8
Na_2SO_4 (0.25)	-52.2 ± 3.2	-187 ± 17	11.15	0.818	4.68 ±0.30	16.78 ±1.2
NaSCN (0.50)	-27.1 ± 0.8	46.4 ± 3.2	7.85	0.617	3.45 ± 0.10	-5.91 ± 0.8
$NaClO_4$ (0.50)	-31.2 ± 0.9	68.8 ± 3.0	8.65	0.638	3.61 ± 0.10	-7.96 ± 1.0

[a] for pH-dependent parameters the listed values are given at pH 7.5; [b] slope of tie line (STL) is defined according to Equation 3.2: STL = ΔC(PEG)/ΔC(Dex)

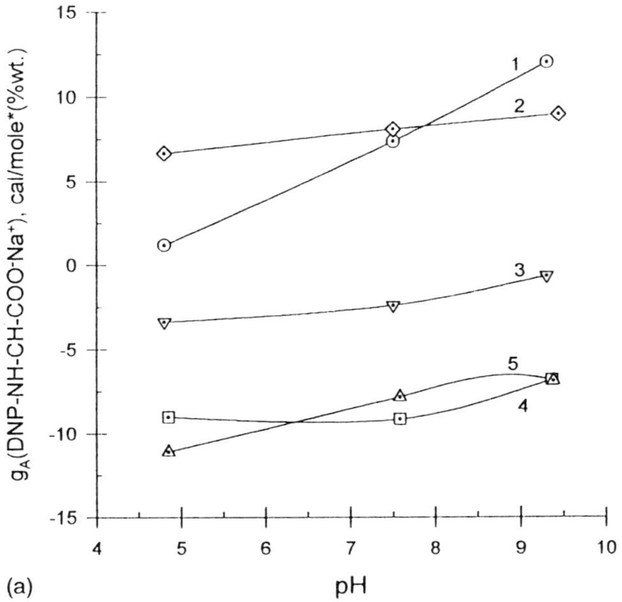

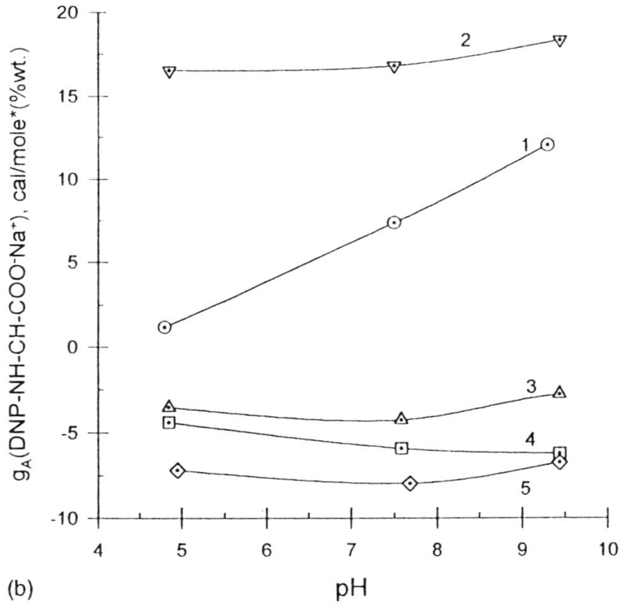

Figure 4.17. pH-Dependence of the coefficient g_A(polar group) in the aqueous Dex-70—PEG-6000 two-phase systems of varied polymer concentra-tions and different salt compositions. Systems contain: (a) 0.01 mole/kg universal buffer without any additional salt (1); buffer and: (2) 0.05 mole/kg Na_2SO_4; (3) 0.10 mole/kg $NaClO_4$; (4) 0.10 mole/kg NaCl; (5) 0.10 mole/kg NaSCN;
(b) 0.01 mole/kg universal buffer without any additional salt (1); buffer and: (2) 0.25 mole/kg Na_2SO_4; (3) 0.50 mole/kg NaCl; (4) 0.50 mole/kg NaSCN; (5) 0.50 mole/kg $NaClO_4$.

mer concentrations is given in Table 4.4. The data presented in Table 4.4 indicate, particularly, the order of the salt effect on the hydrophobic properties of the phases. Changes in the universal buffer composition over the pH range from 4.8 to 9.3 seem not to affect the hydrophobic properties of the phases.

The difference between the interactions of the polar DNP-NH-CH--COONa group with the aqueous media in the two phases, on the other hand, is clearly affected by changes in the buffer composition though in the presence of additional salt not as dramatically as in the system containing solely 0.01 mole/kg buffer. The g_A coefficients are plotted as functions of pH in the aqueous Dex-70—PEG-6000 two-phase systems containing up to 0.1 mole/kg additional salt in Figure 4.17a and in the systems containing up to 0.5 mole/kg additional salt in Figure 4.17b.

The data given in Figure 4.17 indicate that the presence of an additional salt in the system reduces the buffer composition (expressed as pH) effect on the polar interactions in the phases. The higher the salt concentration, the less pronounced effect is observed (the slopes of the curves in Fig. 4.17a exceed those in Fig. 4.17b).

An additional salt, however, not merely reduces the buffer composition (pH) effect. It also increases the difference between the polar interactions in the two phases at any given buffer composition (or pH) relatively to that for the system containing buffer without any additional salt. This effect is characterized by the vertical displacement of a curve from the $g_A = 0$ line. The further away from this line the curve is, the larger the difference between the polar interactions in the two phases.

Both type of phase polymers and type of salt additives govern the properties of the aqueous media in the two phases. These properties determine in particular the relative affinity of a given polar group for the two phases. In the aqueous Dex-PEG two-phase systems containing the water-structure-making salts, e.g., Na_2SO_4, Na_2HPO_4, etc., the affinity of the polar DNP-NH-CH--COONa group for the PEG-rich phase seems generally to exceed that for the Dex-rich phase. That means that both non-polar CH_2 and the above polar group favor the same (PEG-rich) phase. Hence the $N_C{}^*(j)$ value is negative implying that the solutes of the general structure DNP-NH-CH(C_nH_{n+1})-COONa cannot be uniformly distributed in the system. Theoretically the uniform distribution ($K = 1$) may occur for the solute of this structure provided its polarity exceeds that of the above polar group, i.e., for the solute with different polar group.

In the aqueous Dex-PEG two-phase systems containing the water-structure-breaking salts, e.g., NaSCN or $NaClO_4$, the same polar group and non-polar CH_2 group have different affinities for the different phases. Hence parameter $N_C{}^*(j)$ is positive, meaning that a certain number of non-polar CH_2

groups in necessary for the non-polar interactions to cancel out the polar interactions for the solute of the above general structure to be uniformly distributed (K = 1) in the system.

There are several important implications to be noticed from the above data. Contribution of a polar group into partition coefficient of a solute in an aqueous polymer two-phase system is interrelated with that of a nonpolar group and hence they cannot be manipulated separately. The contribution of a non-polar group, however, is strongly dependent upon the type of phase polymers and to a smaller degree upon the salt composition of the system. The contribution of a polar group, on the other hand, while being dependent on the type of phase polymers as well, is affected by the salt composition of the system to a much larger degree. The contribution in question may even be changed from the negative to the positive one by the appropriate change in the salt composition of the system. More experimental study of this issue is needed to gain better understanding of the general trends for different polar groups.

One important conclusion may be reached even on the basis of the limited experimental data discussed above. It is that aqueous polymer two-phase systems of different salt compositions including those with the same buffer at different pHs are to be viewed as different systems with likely different partitioning capabilities.

An important additional issue related to the contribution of a polar group into partition coefficient of a solute in an aqueous two-phase system is the specific role of charge and the sign of the charge in the case of the ionized polar group. To discuss these issues it is necessary, first, to consider electrochemical phenomena in aqueous two-phase systems.

4.4. ELECTROCHEMICAL PHENOMENA IN AQUEOUS TWO-PHASE SYSTEMS

Most of the biological solutes (and particles) being partitioned in aqueous polymer two-phase systems under commonly used conditions are charged. Additionally to all the intermolecular interactions experienced by nonionic species, ionic solutes experience electrostatic ion-ion and ion-dipole solute-solvent interactions. Inorganic salt additives impose local elctrostatic fields affecting the partitioned ionic solute-solvent electrostatic interactions in the two phases. Therefore electrochemical phenomena in the aqueous two-phase systems are important for the solute partitioning.

An asymmetry of dipole orientation at the gas (air, vapor, etc.)/polar liquid interface results in a finite surface potential difference, $\Delta\chi$, having a value between zero and ca. ± 1 V (see, e.g., in [43-45]). If the polar liquid contains an electrolyte the unequal adsorption of ions at the gas/liquid interface results in an electrical double layer contributing to the surface potential difference, the contribution being dependent on the type and concentration of the electrolyte [46]. The surface potential at the air (vapor)/aqueous solution interface depends on the arrangement of the water molecules and electrolyte ions near the interface which may be only partially representative of the electrochemical properties of the bulk solution [43,46,47].

At the interface of two immiscible polar solvents both containing a dissolved salt, two electrical double-layers arise [43,48], due to the differential adsorption of cations and anions in each phase, near their common interface (similar to the single double-layer at the vapor/liquid interface). As the result, an interfacial potential difference or so-called distribution potential, $\Delta\psi$, arises at the interface between the two phases.

Consideration of the process of partitioning of a solute between two solvent phases in thermodynamic terms usually involves comparison of the energies of transfer of the solute from the hypothetical gas phase into each of the solvent phases (see, e.g., in Chapter 1). That is, formally, the energy of

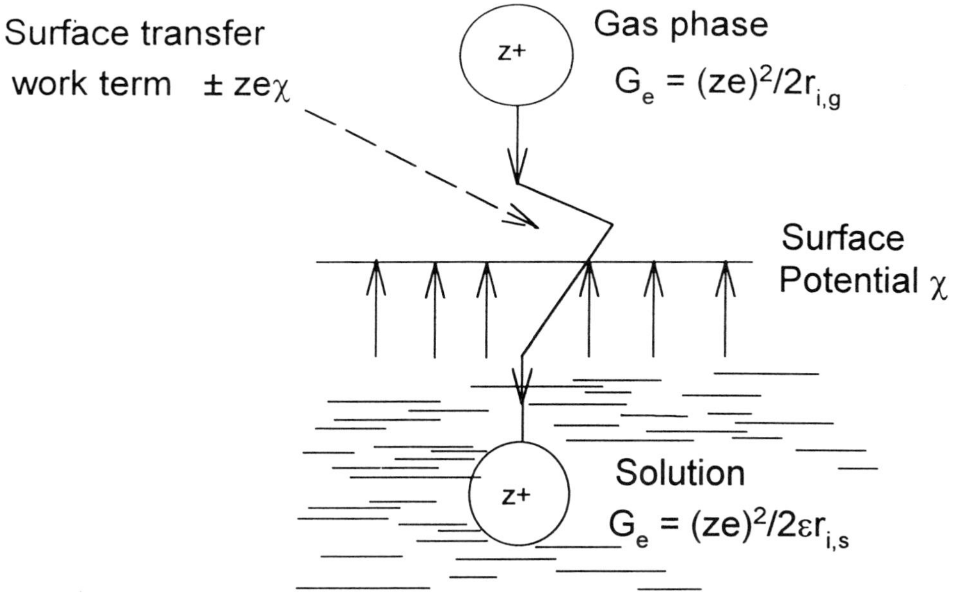

Figure 4.18. Schematic illustration of transfer of an ion of charge $z^+(e)$ across a gas/solution interface at which there is a surface potential χ arising from oriented dipoles (arrows, schematic - normally incomplete orientation occurs) and possibly unequal adsorption of cations and anions. G_e - energy of charging the ion in a given phase. (After B. E. Conway, Ionic Hydration in Chemistry and Biophysics, Elsevier, Amsterdam, 1981 by permission of Elsevier Scientific Publishing Company.)

transfer is defined as the difference between the energy of the solute in the solvent medium and in the gas phase in some appropriate standard states (normally unit pressure or fugacity of 1 atm in the gas phase and unit activity in the liquid phase at 298°K).

If the solute is charged, this definition implies that the solute ion is initially formed in the gas phase and involves an hypothetical transfer of the ion across the gas/liquid interface.

Because of the presence of the surface potential at the gas/solvent interface, the energy of transfer of a mole of ions of charge Z across the interface from the gas phase involves the energy $Z \cdot F \cdot \chi$ as well as the energy change due to interactions of the ions with the solvent medium (see Figure 4.18).

The free energy of transfer of a mole of ions, $\Delta G(ion)^{gas->solvent}{}_{tr}$, is described as

$$\Delta G(ion)^{gas->solvent}{}_{tr} = \Delta\mu \pm Z \cdot F \cdot \chi \qquad (4.18)$$

where $\Delta\mu$ is the difference between chemical potentials of the ion in the gas phase and the solvent phase; $Z \cdot F \cdot \chi$ is the electrical energy of transfer of an ion with charge Z across the interface with the potential difference χ; and F is Faraday constant.

Because of the differences between the surface potentials (to their vapors) of two different solvents, the free energy of transfer of a mole of ions of charge Z between the two solvent phases, $\Delta G(ion)^{1->2}{}_{tr}$ is as derived from Equation 4.18:

$$\Delta G(ion)^{1->2}{}_{tr} = \mu_2 - \mu_1 \pm Z \cdot F \cdot \Delta\chi \qquad (4.19)$$

where μ_1 and μ_2 are the chemical potentials of the ion in the solvent phases 1 and 2, respectively; $\Delta\chi$ is the difference in the surface potentials of the two phases.

The surface potential difference between the two phases, $\Delta\chi$, is usually small in comparison with the so-called distribution potential, $\Delta\psi$, arising from an unequal distribution of a salt additive between the two phases. Hence the free energy of transfer of a mole of ions between the two polar solvent phases takes the form:

$$\Delta G(ion)^{1->2}{}_{tr} = -RT \cdot \ln K(ion) = \mu_2 - \mu_1 \pm Z \cdot F \cdot \Delta\psi \qquad (4.19a)$$

where K(ion) is the partition coefficient of a given ion; all other terms are as defined above.

It should be noticed that the partition coefficient K(ion) is purely theoretical term which cannot be measured. The reason is as follows. It must be emphasized that for the two phases at equilibrium there is the requirement for the phases to be electrically neutral. To meet this requirement, transfer of an electrolyte $M_{m+}N_{n-}$ dissociating into m_+ ions of positive charge Z_+ and n_- ions of negative charge Z_- may occur only as transfer of electrically equivalent amounts of both ions, so that $m_+ \cdot Z_+ \cdot F \cdot \Delta\psi = n_- \cdot Z_- \cdot F \cdot \Delta\psi$ and the electrical interfacial work terms cancel out.

The partition coefficient of an ionic solute characterizes partitioning not of a single ion but of the solute an an electrically neutral combination of the corresponding ions. The total free energy of transfer of an electrically neutral combination of ions (or macro-ions) does not involve any electrical interfacial work terms because these terms cancel.

Widely used incorrect and misleading terms, such as, e.g., "charged phases", "charge-sensitive" phases, "charge-dependent" and "charge-independent partitioning", etc., create a misconception often encountered in the aqueous two-phase partition literature. The misconception is that one phase in the system with a measurable interfacial electrostatic potential is considered as being charged in relation to the other phase. Hence partitioning of an ionic solute (or particle) in such a "charged" system is often viewed as that occurring under influence of an external electrical field. In other words, it is believed sometimes that partitioning of an ionic solute between the two phases is directly affected by the sign of the solute charge and the signs of the charges of the phases.

It should be clear from the condition of electroneutrality that the phases are not charged in relation to each other. The intensity of the electrostatic ion-ion and ion-dipole interactions in the two phases may be different due to different dielectric properties of the aqueous media in the phases and different concentrations of the supporting electrolyte in these phases. The two phases of an aqueous polymer system are similar in this regard to, e.g., water and octanol, or two aqueous solutions of NaCl at different concentrations.

It clearly follows from the above considerations that any ionic solute, e.g., protein, nucleic acid, etc., being partitioned between the two phases is distributed not as a single macro-ion but as an electrically neutral combination of the macro-ion and the electrically equivalent amount of the corresponding counter-ions. The result is that (i) the condition of electroneutrality of the phases is maintained; (ii) the electrical energy contributions in the partition coefficient are canceled out; and (iii) the partition coefficient value for an ionic solute characterizes partition behavior not of a single ion (poly-ion) but of the corresponding salt.

That being the case, the question is what is the physical meaning of an electrostatic interfacial potential difference measurable between the two phases and what purpose does it serve to study the potential difference in aqueous two-phase systems.

An interfacial electrostatic potential difference arises in two-phase systems from the free charge at the interface due to an unequal distribution of cations and anions across the interface [45]. The resulting potential difference, $\Delta\psi$, is called the distribution potential. The Nernst's theory of distribution potentials as applied to two-phase systems was outlined, e.g., by Brooks et al. [3]. It was particularly pointed out in [3, p.32] that the potential difference between the two phases is determined primarily by the difference in the intrinsic interactions of the potential-creating anion and cation with the phases.

The theory of the interfacial distribution potential was developed for water-polar organic solvent two-phase systems [43-45,49]. It has been shown [44,49] that the distribution potential is described as:

$$\Delta\psi = \tfrac{1}{2}\cdot F^{-1}\cdot[(\Delta G^{o}_{M+} - \Delta G^{o}_{X-})^{w} - (\Delta G^{o}_{M+} - \Delta G^{o}_{X-})^{org}] \tag{4.20}$$

where ΔG^{o}_{M+} and ΔG^{o}_{X-} are the free energies of solvation of cation M^+ and anion X^- of the potential-determining salt MX; superscripts "w" and "org" denote the aqueous and organic phases of the two-phase system; F is Faraday constant.

Equation 4.20 clearly defines the distribution potential as parameter representing the difference between the solvation energies of the ions taking part in the distribution equilibrium [44]. The distribution potential value is known [43-49] to characterize particularly the features of electrical double layers formed at the interface of a given two-phase system.

It is generally accepted that all the physical and physico-chemical features of an interface, e.g., interfacial tension, interfacial electrostatic potential difference, etc., reflect the corresponding properties of the bulk phases being in a thermodynamic equilibrium [50]. In respect to the electrical double layer(s) at the interface this principle means that the potential difference in a two-phase system reflects the difference in the electrostatic properties of the coexisting phases. This difference in an aqueous two-phase system may be particularly viewed as the difference between the capabilities of the aqueous media in the two phases to participate in the ion-dipole hydration interactions with an ionic solute being partitioned. The question is, if the interfacial potential determinable in a given system may serve as an adequate measure of this difference.

Technical procedures used in experimental measurements of an interfacial electrostatic potential difference in aqueous two-phase systems were described at length by Bamberger et al.[2]. Two basic approaches to the potential difference determination discussed in [2] are based on: (i) analysis of partitioning of a solute of known and variable charge, and (ii) direct measurements with reversible, nonpolarizable electrode.

The first approach was used by Johansson [51-53]. According to this approach the electrostatic potential difference is determined from the slope of a relationship between the logarithm of the ionic solute partition coefficient and the net charge of the ion. Proteins have been used as the "probes" of the interfacial potential [51,52], and the net charges of the protein macro-ions have been determined separately by titration at different pHs. The drawbacks of this approach have been discussed in detail by Bamberger et al.[2]. The major problem with this approach not commented upon in [2] is the implication that an ionic solute partitioning occurs in the form of a single ion or poly-ion ignoring the contradiction between this implication and the condition of the electroneutrality of the two phases.

It is impossible to measure separate ionic thermodynamic characteristics, e.g., ionic solvation free energies, enthalpies, partition coefficients, etc., because one cannot study a solution of, e.g., Na^+ which does not at the same time contain an equivalent amount of Cl^- or some other negative ion. (Overall electrical neutrality of the solution must be preserved.) Hence the partition coefficient value of an ionic solute, e.g., sodium alkyl sulfate, protein, nucleic acid, peptide, etc., characterizes partitioning of both an ion and counter-ion(s).

The second method is based on direct measurements using silver/silver chloride or calomel electrode connected to the system under study by salt bridges. The latter consist of microcapillaries filled with 1-3 M KCl or capillaries filled with a KCl-saturated gel, e.g., agar. The electrodes are connected to a high-impedance voltmeter. The tips of the capillaries are immersed in one phase of a system, then one of the electrodes is moved into the other phase and the difference in voltmeter readings is taken as the potential under measurement. The precision of the method is about 0.05 mV. The details of the technique may be found in [2] and references cited therein.

The results obtained in the potential $\Delta\psi$ studies were reviewed by Bamberger et al.[2] and the main conclusion seems to be that the currently available data are too limited for any generalization. That is certainly true but it must be added that to interpret the information provided by the potential measurements in a given two-phase system it should be accompanied by the experimental data on the polymer and ionic composition and the dielectric properties of the aqueous media in the phases. An additional data on the interfacial tension and partitioning of some "simple" solutes, e.g., monofunctional aliphatic compounds or inorganic salts (see below), would be highly desirable as well. This information must be obtained for a few aqueous polymer systems at least in order to develop an approach allowing one to relate an interfacial potential difference measurable in a given system with certain features of the solute partition behavior in the system. No information of this type has been reported, to my knowledge, as yet.

As explained above, the interfacial potential difference value reflects (in a rather complicated way) the differences between the ion-solvent interactions for the potential-determining ions. Theoretical analysis by Neogi [54] indicating that an interfacial potential difference cannot affect partitioning of proteins (and other ionic solutes) should be mentioned here. The most important question in this regard is if the potential difference may be used as a measure of the capability of the aqueous media in the phases to participate in the ion-dipole interactions experienced by an ionic solute being partitioned. Partial indirect answer to this question may be found from the experimental data reported by Miheeva et al.[55] and Zvarova et al.[56].

Miheeva et al.[55] studied partitioning of alkali halides MeX in the aqueous Dex-70 (14.0 %wt.) — Ficoll-400 (16.3%wt.) two-phase system containing 0.15 mole/kg NaCl in 0.01 mole/kg sodium phosphate buffer, pH 7.4. The total concentration of each alkali halide being partitioned in the system was varied over the concentration range from ca. $1 \cdot 10^{-4}$ to $3 \cdot 10^{-3}$ mole/kg. The concentrations of an alkali halide in the two phases were determined by atomic absorbance measurements of the corresponding metal concentrations.

The authors [55] suggested that the possible induction and dispersion interactions of the ions being partitioned with the phase polymers, e.g., with axial OH-groups, should be blocked by the supporting electrolyte the amount of which is 50-500 times as much as that of the salt being partitioned. This assumption agrees with the observed independence of the partition coefficient of a salt on the total salt concentration in the system. The constancy of the partition coefficient value implies additionally that the degree of dissociation for a given salt is one and the same in the two coexisting phases of the system.

The aqueous medium in the Ficoll-rich phase is structured to a larger degree than water in the Dex-rich phase (see in Chapters 2 and 3). The data [55] presented in Table 4.5 indicate that NaCl and lithium halides concentrate in the Ficoll-rich phase, while all the potassium, rubidium, and cesium halides concentrate in the Dex-rich phase. Using the terminology suggested by Samoilov (57) the positively hydrated ions have the affinity for the Ficoll-rich phase, while the negatively hydrated ions favor the Dex-rich phase. In the cases of KF and CsF it seems that the contribution of the cation into the free energy of transfer of the salt between the phases dominates over that of the anion. The salts examined in [55] seem to distribute between the coexisting phases in accordance to the principle of the least disturbance of the water structure in the phases.

It is rather surprising that partitioning of potassium, rubidium, and cesium halides in the presence of a large excess of NaCl depends on the type of the halide anion (see Figure 4.19). Note that the partition coefficients were determined by measurements of the metal concentrations. This experimental finding seems to imply that halides of the aforementioned alkali metals exist in the aqueous phases of the polymer system employed [55] in the form of ion-pairs.

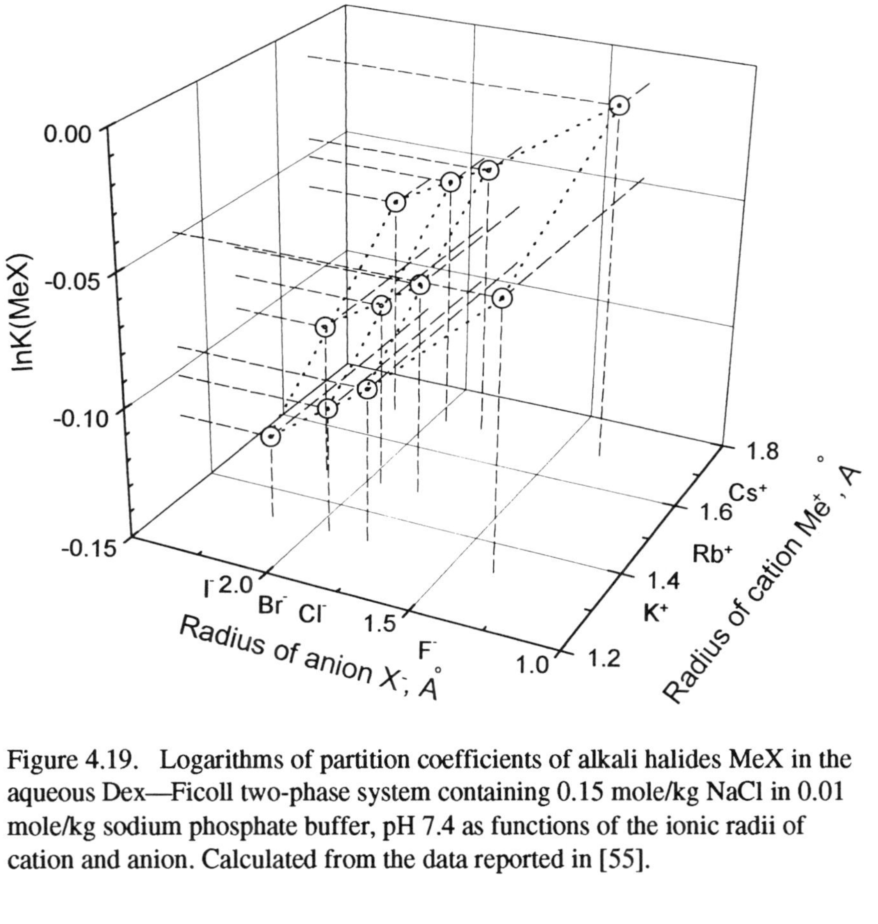

Figure 4.19. Logarithms of partition coefficients of alkali halides MeX in the aqueous Dex—Ficoll two-phase system containing 0.15 mole/kg NaCl in 0.01 mole/kg sodium phosphate buffer, pH 7.4 as functions of the ionic radii of cation and anion. Calculated from the data reported in [55].

The data [55] presented in Fig. 4.19 indicate that for potassium, rubidium, and cesium halides there are similar relationships between the partition coefficient of a given MeX salt and the ionic radius of its anion R_X. All the relationships observed are described as:

$$\ln K_{MeX} = A + B \cdot \ln R_X \tag{4.21}$$

where A and B are constants the values of which depend on the cation type.

Analysis of the above A and B values [55] indicates that both parameters are related to the cations radii values as:

$$A = 0.970 + 0.1 \cdot \ln R_{Me} \tag{4.22}$$

and

$$B = -0.22\cdot(R_{Me})^{-1.3} \tag{4.23}$$

Combining the above Equations it is readily derived:

$$K_{MeX} = 0.971\cdot(R_{Me})^{0.1} / (R_X)^{0.22\cdot(R_{Me})^{-13}} \tag{4.24}$$

The experimental partition coefficients values for all alkali halides reported in [55] are in good agreement with those calculated according to Equation 4.24.

The physical meaning of Equation 4.24 remains obscure at present as essentially all physical and physicochemical properties of inorganic salts or ions in aqueous solutions are related to the ionic radii. According to the authors [55], Equation 4.24 may be viewed as an indication that partitioning of inorganic salts under the conditions used is governed by the steric factors as well as by the ion-water interactions (negative or positive hydration) both related to the difference in the water structure in the two coexisting phases.

The aforementioned difference in the hydration properties of the two phases of the system used in [55] has been characterized separately [30] by partitioning of a homologous series of sodium salts of dinitrophenylated amino acids with aliphatic side-chains. The free energy of transfer of a CH_2 group from the dextran-rich phase to the Ficoll-rich phase amounts to -20 ± 4 cal/mole CH_2, and that of transfer of the polar ionic group DNP-NH-CH-COONa to -96 ± 4 cal/mol [30]. Analysis of Equation 4.24 indicates that the free energy of transfer of the hypothetical alkali halide with unit radii of both ions amounts to -546 cal/mol.

The discrepancy between the two estimates of the difference in the ionic hydration ability of the phases is likely to originate from [55]: a) both sodium cation and anionic carboxylic group in the polar ionic DNP-NH-CH--COONa moiety are known to be positively hydrated in aqueous solutions while the latter estimate was determined from the data for salts of potassium, rubidium, and cesium, i.e. for those of negatively hydrated cations; and b) the size of the above ionic moiety is much larger than that of the hypothetical alkali halide with unit radii of both ions. It should be repeated that partitioning of halides of positively hydrated lithium and sodium does not fit Equation 4.24 implying discontinuity of the relationship between the ionic radii and partition coefficients of differently hydrated cations.

It is generally believed (see ,e.g., in [43]) that there is the spasmodic rearrangement of the water structure in the close vicinity of an ion when passing from negatively hydrated ions to the positively hydrated ones. This rearrangement is usually attributed to the requirements of the electrostatic field of an ion in regard to activation energies of neighboring water molecules and is assumed to be due to the spasmodic change of the coordination number of ions

Table 4.5

Partition Coefficients of Inorganic Salts in Aqueous Dex—Ficoll and PEG—$(NH_4)_2SO_4$ Two-Phase Systems.

Salt	Partition coefficient K Dex-Ficoll [a]	Salt	Partition coefficient K PEG-$(NH_4)_2SO_4$ [b]
LiF	1.030 ± 0.006	Na_2SO_4	0.126
LiCl	1.034 ± 0.007	K_2SO_4	0.158
LiBr	1.033 ± 0.006	$ZnSO_4$	0.63
LiI	1.030 ± 0.006	$CoSO_4$	0.63
NaCl	1.054 ± 0.005	$CdSO_4$	0.5
KF	0.953 ± 0.007	$CuSO_4$	0.63
KCl	0.913 ± 0.006	$Ce_2(SO_4)_3$	0.25
KBr	0.903 ± 0.005	$Lu_2(SO_4)_3$	0.4
KI	0.889 ± 0.008	$Cr_2(SO_4)_3$	0.06
RbCl	0.934 ± 0.006	$Eu_2(SO_4)_3$	0.25
RbBr	0.924 ± 0.007	$In_2(SO_4)_3$	0.71
RbI	0.912 ± 0.006		
CsF	0.988 ± 0.003	NH_4Br	3.16
CsCl	0.957 ± 0.006	NH_4I	10.0
CsBr	0.950 ± 0.004	NH_4SCN	5.0
CsI	0.939 ± 0.005	$(NH_4)_3PO_4$	0.25

[a] Dex-Ficoll two-phase system containing 0.15 mole/kg NaCl in 0.01 mole/kg sodium phosphate buffer, pH 7.4. Data from [55]; [b] PEG-$(NH_4)_2SO_4$ system containing unspecified amount of H_2SO_4, pH 4.0. Data from [56] are presented as K-values for the sulfates or ammonium salts, while in original publication [56] same values are given for the cations or anions (see explanation in text).

or to the transitions from contact ion-pairs to separated ion-pairs and to independently hydrated ions.

It was concluded by Miheeva et al.[55] that it is impossible to chose certain ionic moiety or solute for reference to characterize adequately all types of ion-dipole interactions in the phases of an aqueous polymer two-phase system. Different types of these interactions should be characterized by different reference ionic groups or solutes.

This conclusion seems to be supported by the data reported by Zvarova et al.[56] examined partitioning of inorganic salts in the aqueous PEG-2,000 (15.0 %wt.) — $(NH_4)_2SO_4$ (14.4 %wt.) two-phase system containing unspecified amounts of H_2SO_4. Partition coefficients of different salts with radioactive anion or cation were determined with radiometric measurements. The total concentration of each salt being partitioned in the system was 10^{-5} M. Possibly presuming that all salts being partitioned in the presence of a large excess of $(NH_4)_2SO_4$ should be viewed as ammonium salts or sulfates, the authors [56] did not even indicate what actual salts have been used in the experiments.

Zvarova et al.[56] presented the results as the partition coefficients of the (radioactive) cations and anions likely due to the aforementioned misconception that the partition coefficient of an ionic solute characterizes the partition behavior of the ion being monitored. In view of the above theoretical considerations and the experimental data by Miheeva et al.[55] this presentation is clearly misleading. Let us assume, however, that the data [56] in question characterize partitioning of the corresponding ammonium salts and sulfates. The data reported by Zvarova et al.[56] are partially presented in Table 4.5. (Instead of presenting the partition coefficients for cations or anions as reported in the original publication [56], the corresponding sulfates and ammonium salts are indicated in Table 4.5.)

All the sulfates and ammonium phosphate clearly favor the $(NH_4)_2SO_4$-rich phase, while ammonium bromide, iodide, and thiocyanate prefer the PEG-rich phase. This partition behavior of the salts in the aqueous PEG-$(NH_4)_2SO_4$ two-phase system [56] is in line with the trend observed in the distribution of salt additives in aqueous Dex-PEG and Dex-PVP systems [58, 59]. The trend is that all sulfates and phosphate (salts of strong water-structure-making anions) appear to avoid the PEG-rich phase, while the salts of the water-structure-breaking anions (SCN^-, Br^-, and I^-) tend to favor this phase. The principle of the least disturbance of the water structure in the phases observed by Miheeva et al.[55] in the aqueous Dex-Ficoll salt-containing two-phase system seems to operate in the PEG-$(NH_4)_2SO_4$ system as well. The solvent properties of the aqueous media in the phases of the latter system are much more different than those in the aqueous two-polymer two-phase systems (see above). That possibly explains why the partitioning of salts in the PEG-salt system is much more one-sided than in the Dex-PEG, Dex-PVP, or Dex-Ficoll systems. It should be mentioned that a given aqueous Dex-Ficoll or PEG-$(NH_4)_2SO_4$ two-phase system is characterized by an interfacial potential difference of the particular value. Different inorganic salts partition in a given system quite differently [55,56,60]. Hence the interfacial potential difference may not be an adequate general measure of the capability of the aqueous media in the phases to participate in the ion-dipole interactions.

The issue of an electrostatic interfacial potential difference in aqueous two-phase systems and its role in the ionic solute partitioning is one of the two most controversial issues in the field of aqueous polymer two-phase systems (the other is the problem of direct solute-phase polymer interactions discussed in the next chapter). Therefore it seems necessary to finish this consideration with a brief summary to emphasize the conclusions.

1. Partitioning of an ionic solute in an aqueous polymer two-phase system differs from partitioning of nonionic solute in that it is affected by ion-ion and ion-dipole interactions absent in the case of the nonionic species. Both polymer and salt compositions of the phases influence these interactions. Polymers are likely to act through their effects on the dielectric properties of the media, and salts via the local electrostatic fields and their own ion-ion and ion-dipole interactions imposed on the media.

2. An electrostatic interfacial potential difference in an aqueous two-phase system characterizes the difference between the hydration of the anions and cations of the potential-determining salt additive.

3. Both phases are electrically neutral and <u>neither phase is electrically charged in relation to the other</u>. Misleading terms, such as "charge-dependent partitioning", "charged phases", etc., should be excluded from the literature on aqueous two-phase systems. <u>An interfacial potential difference does not affect partitioning of ionic solutes</u> because the electrical interfacial work terms cancel out as only electrically equivalent amounts of ions of opposite charges may be transferred across the interfacial boundary to preserve the electroneutrality of the phases.

4. <u>The partition coefficient of an ionic solute characterizes partition behavior of the electrically neutral combination of both positively and negatively charged ions (poly-ions) independent of the particular ion being monitored.</u> The question of the type of the counter-ions (in regard to the monitored ion) remains open and may not be readily resolved for the salt-containing two-phase

systems.

5. Specificity of ion-water interactions does not allow one currently to make a rational choice of a salt or ionic moiety as a reference for the relative scaling of the solvent ability of the aqueous media in the phases in regard to ion-dipole interactions. The only possibly promising approach seems to be for those involved in the aqueous two-phase systems research to reach an agreement to use any set of easily monitored salts or ionic moieties as a set of references to accumulate additional information to resolve the issue in the future.

4.5. HYDROPHOBIC AND POLAR HYDRATION IN AQUEOUS TWO-PHASE SYSTEMS

As mentioned above, separate consideration of contributions of nonpolar and polar (nonionic and ionic) groups into the solute partition coefficient is due to the different nature of interactions of these groups with an aqueous medium (see in Chapter 1).

The possibility to describe the partitioning results obtained in aqueous two-phase systems using the concepts developed for water-organic solvent systems will be considered below. This possibility, once established, serves as a strong argument for the above hypothesis about the fundamental similarity between aqueous polymer and water-organic solvent two-phase systems.

Theoretical considerations of a distribution of a solute between two phases were mostly developed to describe the solute distribution either between thermodinamically equilibrated aqueous and organic phases (see, e.g., in [25]] or between a mobile phase and a stationary phase during the chromatographic separation process (see, e.g., in [61,62]). In each case the partition behavior of a solute clearly results from the forces that operate between solute molecules and molecules of each phase. The following types of intermolecular interactions are generally considered: directional dipole-dipole interactions, induction, dispersion, H-bonding, electron pair donor-acceptor interactions, and Coulomb ion-dipole and ion-ion as well as solvophobic interactions.

Intermolecular forces considered as nonspecific are often called van der Waals forces, since van der Waals recognized these forces as the reason for the nonideal behavior of real gases. These forces comprise the directional, induction, and dispersion forces which are supposed to be "more physical" in nature as compared to "more chemical" H-bonding and electron pair donor-acceptor forces [61].

Hydrophobic (or, more generally, solvophobic [63]) interactions differ from all the other forces. The difference is described by Kaliszan [61,62] as follows. Commonly, a force between two particles is a function of the intrinsic properties of the particles themselves. For example, Coulomb forces arise from the charges situated in the particles. In the case of hydrophobic (solvophobic) interactions the forces are mainly dependent on the properties of the solvent and not on the solute. To differentiate between the solute-solvent and solute-solute interactions often misleadingly called by the same hydrophobic interaction term, Ben-Naim [63] suggested to use the term hydrophobic hydration to describe the peculiar interactions between water and non-polar molecule or molecular fragment.

Hydrophobic hydration may be described as the process of reorganization of water molecules in the vicinity of a non-polar solute (or fragment) resulting in the more ordered state of this local water region. Because of that inert non-polar solutes are usually viewed as "water-structure-makers". The terminology of structure promotion and breaking is currently recognized, however, as "probably too limited to provide a good description of the underlying molecular reality. Instead, what appears to happen is a reorganization of (solvent) molecules or hydrogen bonds that can be perceived either as structure making or breaking depending on the experimental probe used to study it" [64]. It should be added that the interactions between water and a non-polar solute (group) include not solely hydrophobic hydration interaction, but the dispersion forces as well. The dispersion forces, however, being part of van der Waals interactions are approximately ten times weaker than the H-bonding water-water interactions governing the hydrophobic hydration forces.

The basic theory of distribution of a solute between two phases has been developed by Scott [65] and extended by Scott and co-workers [66] later to include reversed-phase chromatography mode. The distribution coefficient of a solute between the mobile and stationary phases in a chromatography system was defined by Scott [65] as a ratio between the magnitudes of total forces acting on solute in the two phases. These forces were divided into ionic, polar, and dispersive intermolecular interactions. It was assumed and experimentally verified [65-67] to be possible to affect these different interactions separately (to a certain limited degree) by varying the solvent composition of the mobile phase. The solute distribution behavior is changed by affecting polar interactions through varied concentration of a polar solvent (e.g., isopropanol) in a dispersion medium (e.g., heptane). The ionic interactions can be affected predominantly by varying buffer salt concentration in the mobile phase [65,68]. The dispersive interactions can be influenced by varying the density of the dispersion solvent at constant polar component concentration, i.e., by using several mobile phases each composed of a fixed amount of polar solvent (e.g., ethyl acetate) in binary mixtures with different hydrocarbons [67].

The other theoretical approach to chromatography process suggested by Karger et al.[69] includes essentially the same division of intermolecular interactions into polar interactions (including orientation, induction, and H-bonding interactions together) and dispersion interactions. Theoretical consideration of chromatographic separation in the reversed-phase chromatography mode suggested by Horvath et al.[70] is based on the aforementioned Sinanoglu theory of solvophobic interactions (see in Chapter 1). The approach developed by Horvath et al.[70] was extended later [71] to describe the effects of inorganic salts on the aqueous solubility of proteins. This model offers a simple theoretical framework for consideration of effects of polymer and salt composition of the phases upon contributions of polar and non-polar groups into the solute partition coefficient in an aqueous two-phase system.

According to the model by Melander and Horvath [71] the free energy of transfer of a non-polar solute from a hypothetical gas phase into an aqueous saline solution, ΔG^o, is determined by the free energy change for formation of a cavity in the solvent to accommodate the solute molecule, ΔG_{cav}, the free energy change due to van der Waals interactions between the solute and solvent, ΔG_{VdW}, and the additional $RT \cdot \ln(RT/PV_s)$ term accounting for the "free volume" of the solute in the solvent of the molecular volume V_s. The free energy change for cavity formation can be written in simplified form as [71]:

$$\Delta G_{cav} = k \cdot (\gamma^o + \sigma_s \cdot m_s) \cdot V_1^{2/3} \quad (4.25)$$

where γ^o is the surface tension of pure water; m_s is the molality of a given salt; σ_s is the molal surface tension increment of the salt (see in Chapter 1); V_1 is the solute molar volume; k is the constant accounting for correction of the macroscopic surface tension of the solvent to molecular dimensions.

Thus, the hydration of a non-polar solute in an aqueous saline solution according to the model [71] is described as:

$$\Delta G^o = \Delta G_{cav} + \Delta G_{VdW} = k \cdot (\gamma^o + \sigma_s \cdot m_s) \cdot V_1^{2/3} + \Delta G_{VdW} \quad (4.26)$$

where all the terms are defined above.

When Equation 4.26 is applied to the hydration of a non-polar molecule or moiety in a given phase of an aqueous polymer two-phase system, one should take into account that the phase polymers present in the phase alter the surface tension of the medium, i.e., the γ^o-value, as well as the ΔG_{VdW} value. Therefore the free energy of transfer of a non-polar solute (or moiety), e.g., CH_2 group, between the two phases, $\Delta G(CH_2)_{tr}$, expressed in the above terms may be written as:

$$\Delta G(CH_2)_{tr} = k \cdot \Delta\gamma^o + k \cdot \sigma_s \cdot \Delta m_s + \Delta G_{VdW1} - \Delta G_{VdW2} \quad (4.27)$$

where $\Delta\gamma^o$ is the difference between the surface tensions of the two hypothetically salt-free phases; Δm_s is the difference between the molalities of a given salt in the two phases; indexes "1" and "2" denote the coexisting phases; all the other terms are as defined above.

According to Equation 4.27 the free energy of interfacial transfer of a non-polar CH_2 group (or any other non-polar group or molecule) is determined by the difference between the free energies for cavity formation and van der Waals interactions in the two phases. Both polymer and salt compositions of the phases are likely to affect the difference between the energy of van der Waals interactions in the two phases as indicated by their influence on the solvent polarity and dielectric constant of the aqueous medium (see above). The energy of van der Waals interactions is, however, about ten times weaker than that of H-bonding interactions predominately governing the cavity formation in an aqueous medium.

The energy of cavity formation is also affected by both polymer and salt compositions as represented by the separate $\Delta\gamma^o$-containing term (polymer effect) and $\sigma_s \cdot \Delta m_s$-containing term (salt effect). The difference between the polymer composition of the two phases exceeds that between the salt composition significantly. Therefore the polymer composition effect on the difference between the hydrophobic character of the two phases should dominate relatively to the salt composition effect in total agreement with the experimental data.

The salt composition effect should be expected to increase with increasing σ_s and Δm_s values. That is observed experimentally when the effects of different salt additives, e.g., 0.1 mole/kg K_2SO_4 and KSCN are compared. The b(salt) coefficient value for K_2SO_4 in the aqueous Dex-PEG two-phase system exceeds that for KSCN by a factor of ca. 3.6 indicating that the Δm_s-values for K_2SO_4 would exceed those for KSCN at all possible polymer compositions of the phases. The σ_s value for K_2SO_4 also exceeds that for KSCN (see Table 1.2). Hence according to the above assumptions, the addition of 0.1 mole/kg K_2SO_4 should affect the difference between the hydrophobic character of the phases in the aqueous Dex-PEG two-phase system to a larger extent than the addition of the same amount of KSCN. Comparison of the parameter g_E values for the systems containing these salts (see Table 4.2) indicates that this pediction is in total agreement with the experimental data.

If a solute being partitioned possesses an ionic group j the term accounting for the difference in the energies of electrostatic ion-dipole interactions between the solute and solvent in the two phases is to be included into Equation 4.27:

$$\Delta G(\text{ionic group j})_{tr} = k \cdot \Delta\gamma^o + k \cdot \sigma_s \cdot \Delta m_s + (\Delta G_{VdW1} - \Delta G_{VdW2}) + \\ + (\Delta G(\text{e.s.})_1 - \Delta G(\text{e.s.})_2 \quad (4.28)$$

where ΔG(e.s.) accounts for the free energy change due to electrostatic ion-dipole interactions between the solvent and solute in a given phase; all the other terms are as defined above.

The energy of electrostatic ion-dipole interactions is clearly governed by the type and concentration of a salt present in a given phase (see above).

Therefore it is to be expected that the contribution of the ionic group j into the solute partition, i.e., the $\Delta G(\text{ionic group j})_{tr}$ value, should be dependent primarily upon the salt composition. It should also be affected to a lesser degree by the polymer composition of the phases due to the polymer influence on the dielectric constant of the aqueous medium. Thus, the experimental data discussed above are in total agreement with the theoretical considerations.

The contribution of a polar but non-ionic group into the solute partition coefficient may be expected to depend on both salt and polymer composition, the salt effect being less dramatic than in the case of ionic groups. There is no experimental evidence to support this point of view as yet. The question of contribution of any particular polar non-ionic or ionic group into the solute partition coefficient is complicated by that the hydration interactions of any particular polar group is likely to be highly specific and any generalization may hardly be expected.

It seems possible to use the free energy of an interfacial transfer of a non-polar CH_2 group, $\Delta G(CH_2)_{tr}$, as a relative measure of the difference between the total ability of water to form a cavity and participate in van der Waals interactions in the two phases. It is much more speculative but also possible to view the free energy of an interfacial transfer of a given ionic group as a relative measure of the difference between the ability of water to participate in electrostatic ion-dipole interactions in the two phases. What ionic group should be chosen as a reference one is the question hard to answer at present. Even in the case of 1:1 monoatomic inorganic salts the choice of a reference salt (positively and/or negatively hydrated anion and cation?) is questionable. In the case of an ionic group, e.g., -COOH or $-NH_2$, the issue is even more complicated. There currently seems to be no strong arguments in favor of any particular ionic group to be used as a reference (see above). The majority of the experimental results in regard to the issue under discussion was obtained for compounds possessing a DNP-NH-CH-COONa group. The disadvantages of using this particular group (voluminous and containing substituted aromatic ring) as a probe of solely electrostatic ion-dipole interactions are obvious. Only to the first approximation the free energy of transfer of this group between the two phases of an aqueous polymer two-phase system may be viewed as a measure of the ability of water to participate in any one particular kind of intermolecular interactions. According to the above considerations and the experimental g_A parameter values, however, this approximation is not as far from reality as might be expected.

If the $\Delta G(\text{DNP-NH-CH-COONa})_{tr}$ is taken as a measure of the difference between the relative ability of water to participate in ion-dipole interactions, the interrelationship established between this parameter and $\Delta G(CH_2)_{tr}$, or between g_A and g_E coefficients, (see Equations 4.12, 4.13, 4.16, and 4.17) is to be expected. The reason for this interrelationship seems to be that the ability of water molecules to participate in any kind of intermolecular interactions in an aqueous medium of a given composition is determined by the thermodynamic state and/or structure of water in the medium, i.e., by the arrangement of the water molecules and H-bonds in the medium. The relationship between the solvatochromic E_T parameter values for different probes and their partition coefficients in aqueous two-phase systems (see Fig. 4.3) as well as the aforementioned results on the partitioning of inorganic salts support the above assumption. All kinds of solute-solvent interactions (for a given type of solute) originate from the nature of the solvents used in a given two-phase system. That is likely to be the reason for the constant $n^*(j)$ and $N_C{}^*(j)$ values (see Equation 4.12) for any particular series of solutes in a given water-organic solvent or aqueous polymer two-phase system.

Biological solutes commonly separated or studied by the technique of partitioning in aqueous two-phase systems are usually neither purely non-polar, nor merely ionic. These solutes generally possess ionic and non-ionic polar groups together with clearly non-polar molecular fragments.

The simplest example of solutes capable of different hydration interactions are compounds with clearly distinct polar (possibly ionic) and non-polar regions in the molecule. These compounds are covered by the general term amphiphiles. The most simple of these are monofunctional compounds with aliphatic alkyl chain.

Even these "simple" compounds, e.g., sodium alkyl sulfates, alkyltrimethylammonium chlorides, etc., under certain conditions display partition behavior that is not readily understood. That is typical most of all for compounds with relatively short alkyl chains. Figure 4.20 presents the logarithm of the solute partition coefficient as a function of the alkyl chain length for aliphatic carboxylic acids in water—i-butanol and for sodium salts of DNP-amino acids with aliphatic side chains in the aqueous Dex—PEG—0.1mole/kg K_2SO_4 two-phase systems as an illustration of this behavior.

The likely explanation for the nonlinear relationships shown in Fig. 4.20 is that the interactions of the solvent with the polar group of a solute affect those with the non-polar part of the solute molecule. This effect decays exponentially in agreement with the current hydration force models (see, e.g., in [72]) and clearly depends on the solvent and the nature of the solute polar group. The effect in question may be estimated quantitatively as follows. Partition behavior of the same homologous series of solutes with relatively long alkyl chains (with $N_C \geq 8\text{-}10$) usually fits the linear relationship described by Equation 4.4. Once the A and E parameters are determined, the partition coefficients of the solutes with "short" alkyl chains are fitted to the linear relationship using the N_C as an adjustable parameter. As the result, for example, the

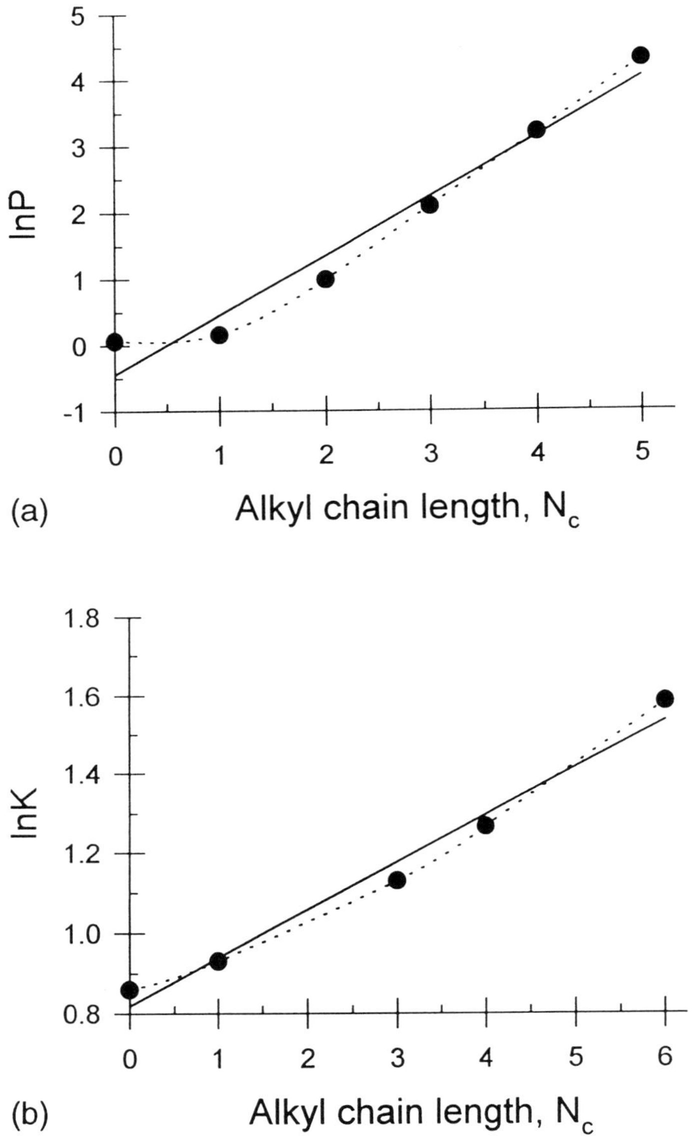

Figure 4.20. Logarithm of the solute partition coefficient as a function of the alkyl chain length: (a) aliphatic carbon acids in water—butanol two-phase system (calculated from the data reported in [10]); (b) sodium salts of DNP-amino acids in aqueous Dex-70—PEG-6000 two-phase system containing 0.10 mole/kg K_2SO_4 (calculated from the data reported in [30]).

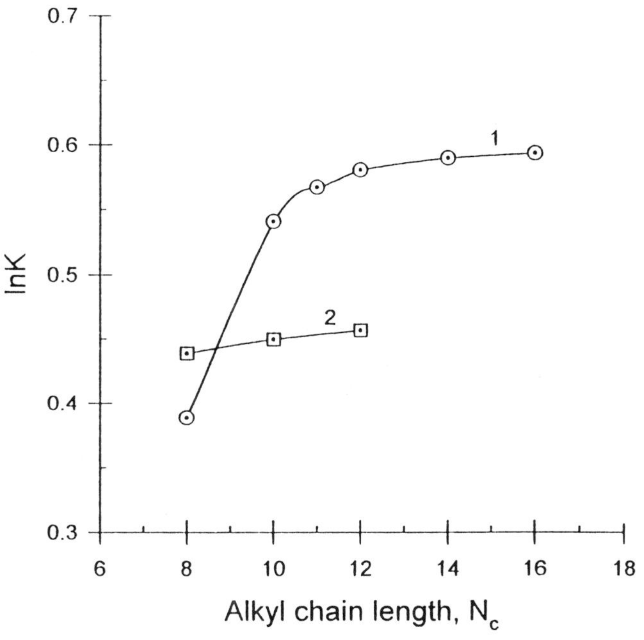

Figure 4.21. Logarithm of the partition coefficient of sodium alkyl sulfate in aqueous Dex-500—PEG-6000 two-phase system (1) without any salt additive, and (2) in the presence of 0.10 mole/kg NaCl as a function of the alkyl chain length.

aliphatic side-chain of dinitrophenylated sodium salt of norvaline (-CH_2-CH_2-CH_3) is characterized by the N_C value not of 3 but 2.43. The difference between these "theoretical" and "empirical" values may be viewed as a measure of the influence of the polar group-water interactions on the non-polar fragment-water interactions. An additional problem hard to resolve in this case is the possible difference between the intensity of hydrophobic hydration interactions for methylene (-CH_2-) and methyl (-CH_3) groups. The hydrophobic properties of these two groups are known to be different (see, e.g., in [11]) but the difference in question is hard to estimate quantitatively. This difference may be one of the reasons for that the aliphatic side-chain of dinitrophenylated sodium salt of 2-amino-n-octanoic acid is characterized by the N_C value of 6.43 instead of the expected value of 6.

An influence of the polar group-water interactions on the adjacent non-polar fragment-water interactions are possibly explained as follows. The polar group-water H-bonds or dipole-dipole interactions may distort the arrangement of water molecules nearby the non-polar fragment possibly reducing the intensity of the water-water interactions in this local area and consequently decreasing the intensity of the hydrophobic hydration of a given nonpolar fragment. This effect should be expected to spatially decay and disappear at a certain distance from the polar group. That is in line with the observed disappearance of the deviation of the lnK-N_C relationship from linear curve with increasing alkyl chain length N_C.

The effects under discussion were explored by studying the partitioning of amphiphilic solutes in water-organic solvent systems, solubility of the solutes in water and organic solvents, surfactants micelle formation, etc. (see, e.g., in [11,24-26]). The free energies of transfer of a methylene group between aqueous and nonaqueous phases are relatively large in the range of 600 to 1000 cal/mol CH_2. The experimental error is usually also rather large in the range of 100 to 150 cal/mol CH_2. Therefore the effects under discussion may be masked by the experimental error leading to the generally accepted conclusion [24] that the effect of a polar group on the adjacent nonpolar group-water interactions do not extend more than over 1-2 methylene groups from the polar moiety.

The more sensitive technique of partitioning in aqueous two-phase systems indicates the above conclusion to be incorrect. That is shown particularly by the data [27] on partitioning of sodium alkyl sulfates in the aqueous Dex—PEG two-phase system without any salt additive and in the presence of 0.10 mole/kg NaCl presented in Figure 4.21.

The nonlinear character of the lnK-N_C relationship observed [27] may be explained by that the effect in question influences the interactions of water with the alkyl chain as long as C_{16}. The data obtained in [27] are too limited to use them as a basis for a quantitative model of the effect under discussion. It should be indicated particularly that the effect seems to reduce significantly or disappear completely in the presence of 1.0 mole/kg NaCl [27] (see Fig. 4.4). This may be viewed as indication of the dependence of the effect in question upon the ionic composition of the medium. If this assumption is correct it may have important biological implications. For example, the ionic composition of an aqueous extracellular medium may influence the state of the lipid matrix of biological membranes due to effect on the polar groups of lipids affecting the relative affinity of their alkyl chains for the aqueous medium and nonpolar environment in the membrane bilayer. This question remains to be explored experimentally.

4.6. SUMMARY

It was argued in the previous chapter that phase separation in aqueous polymer systems results from different effects of two polymers (or a single polymer and a salt) on the water structure. The implication of this hypothesis is that the solvent features of aqueous media in the coexisting phases are to be different. The final question raised in the previous chapter was if this implication is true.

The experimental data discussed above indicate that the solvent features of aqueous media in the two phases are different. The difference in question is established by dielectric, solvatochromic, and potentiometric measurements as well as by studies of partitioning of homologous series of monofunctional aliphatic compounds.

Results of partitioning of structurally simple solutes indicate that the basic rules of solute partitioning in aqueous two-phase systems are similar to those in water-organic solvent systems. These rules are consistent with the current concepts on solute-solvent interactions in aqueous systems, e.g., on hydrophobic hydration and ionic hydration phenomena.

Characteristic peculiarity of aqueous polymer two-phase systems is that the differences between the properties of the two phases relatively to those typically observed in water-organic solvent systems are very small as should be expected for a pair of solvents of the same (aqueous) nature. The small difference between the solvent features of the two phases in aqueous two-phase systems provides certain advantage from the viewpoint of enlarged sensitivity of solute partitioning toward modifications in the solute structure. This issue will be considered in detail below.

REFERENCES:

1. P. Å. Albertsson, Partition of Cell Particles and Macromolecules, 3rd.ed., Wiley, New York, 1986.
2. S. Bamberger, D. E. Brooks, K. A. Sharp, J. M. Van Alstine, T. J. Webber, In: Partitioning in Aqueous Two-Phase Systems: Theory, Methods, Uses, and Applications to Biotechnology (H. Walter, D. E. Brooks, D. Fisher, eds.), Academic Press, Orlando, Fla, 1985, pp.85-130.
3. D. E. Brooks, K. A. Sharp, D. Fisher, In: Partitioning in Aqueous Two-Phase Systems: Theory, Methods, Uses, and Applications to Biotechnology (H. Walter, D. E. Brooks, D. Fisher, eds.), Academic Press, Orlando, Fla, 1985, pp.11-84.
4. H. Walter, G. Johansson, D. E. Brooks, Anal.Biochem., 197, 1 (1991).

5. A. Cordes, J. Flossdorf, M. R. Kula, Biotechnol.Bioeng., 30, 514 (1987).
6. S. S. Suh, F. H. Arnold, Biotechnol.Bioeng., 35, 682 (1990).
7. B. Y. Zaslavsky, N. M. Mestechkina, L. M. Miheeva, S. V. Rogozhin, J.Chromatogr., 256, 49 (1983).
8. B. Y. Zaslavsky, E. A. Masimov, Topics Curr.Chem., 146, 171 (1988).
9. C. Reichardt, Solvents and Solvent Effects in Organic Chemistry, 2nd ed., Verlag Chemie, New York, 1986.
10. A. Leo, C. Hansch, D. Elkins, Chem.Rev., 71, 525 (1971).
11. S. S. Davis, T. Higuchi, J. H. Rytting, Adv.Pharm.Sci. (H. S. Bean, ed.) Acad.Press, 73 (1974).
12. Y. Marcus, J.Phys.Chem., 91, 4422 (1987).
13. B. Y. Zaslavsky, A. A. Borovskaya, N. D. Gulaeva, L. M. Miheeva, Biotechnol.Bioeng., 40, 1 (1992).
14. C. Reichardt, E. Harbusch-Gornert, G. Schafer, Liebigs Ann. Chem., 839 (1988).
15. S. J. Gluck, M. P. Wingeier, J.Chromatogr., 547, 69 (1991).
16. B. Y. Zaslavsky, L. M. Miheeva, M. N. Rodnikova, G. V. Spivak, V. S. Harkin, A. U. Mahmudov, J.Chem.Soc., Faraday Trans.I, 85, 2857 (1989).
17. B. Y. Zaslavsky, L. M. Miheeva, N. D. Gulaeva, A. A. Borovskaya, M. I. Rubtsov, L. L. Lukatskaya, N. O. Mchedlov-Petrossyan, J.Chem.Soc. Faraday Trans., 87, 931 (1991).
18. B. Y. Zaslavsky, L. M. Miheeva, E. A. Masimov, S. Djafarov, C. Reichardt, J.Chem.Soc. Faraday Trans., 86, 519 (1990).
19. M. J. Kamlet, R. M. Doherty, M. H. Abraham, Y. Marcus, R. W. Taft, J.Phys.Chem., 92, 5244 (1988).
20. Y. Migron, Y. Marcus, J.Chem.Soc. Faraday Trans., 87, 1339 (1991).
21. Y. Marcus, J.Phys.Chem., 95, 8886 (1991).
22. A. A. Alhaider, C. D. Selassie, S. O. Chua, E. J. Lien, J. Pharm. Sci., 71, 89 (1982).
23. Y. C. Martin, In: Drug Design (E. J. Ariens, ed.), Vol.8, New York, Academic Press, 1979, pp.1-72.
24. C. Tanford, The Hydrophobic Effect: Formation of Micelles and Biological Membranes, New York, Wiley, 1973.
25. R. F. Rekker, The Hydrophobic Fragmental Constant: Its Derivation and Application. A Means of Characterizing Membrane Systems, Amsterdam, Elsevier, 1977.
26. C. Hansch, A. Leo, Substituent Constants for Correlation Analysis in Chemistry and Biology, New York, Academic Press, 1979.
27. B. Y. Zaslavsky, L. M. Miheeva, S. V. Rogozhin, Biochim. Biophys. Acta, 510, 160 (1978).

28. B. Y. Zaslavsky, L. M. Miheeva, G. Z. Gasanova, A. U. Mahmudov, J.Chromatogr., 403, 123 (1987).
29. B. Y. Zaslavsky, N. D. Gulaeva, S. F. Djafarov, E. A. Masimov, Miheeva L. M., J.Colloid Interface Sci., 137, 147 (1990).
30. B. Y. Zaslavsky, A. A. Borovskaya, N. D. Gulaeva, L. M. Miheeva, J.Chem.Soc., Faraday Trans.I, 87, 141 (1991).
31. A. D. Diamond, J. T. Hsu, Biotechnol.Bioeng., 34, 1000 (1989).
32. M. A. Eiteman, J. L. Gainer, Biotechnol.Prog., 6, 479 (1990).
33. B. Y. Zaslavsky, L. M. Miheeva, S. V. Rogozhin, J.Chromatogr. 216, 103 (1981).
34. B. Y. Zaslavsky, L. M. Miheeva, N. M. Mestechkina, L. G. Shchyukina, M. A. Chlenov, L. I. Kudrjashov, S. V. Rogozhin, J.Chromatogr., 202, 63 (1980).
35. J. Ryden, P. Å. Albertsson, J.Colloid Interface Sci., 37, 219 (1971).
36. S. Bamberger, G. V. F. Seaman, K. A. Sharp, D. E. Brooks J.Colloid Interface Sci., 99, 194 (1984).
37. D. Forciniti, C. K. Hall, M. R. Kula, J.Biotechnol., 16, 279 (1990).
38. B. Y. Zaslavsky, L. M. Miheeva, S. V. Rogozhin, J.Chromatogr., 212, 13 (1981).
39. B. Y. Zaslavsky, A. A. Borovskaya, N. D. Gulaeva, L. M. Miheeva, J.Chem.Soc., Faraday Trans.I, 87, 137 (1991).
40. B. Y. Zaslavsky, A. A. Borovskaya, N. D. Gulaeva, L. M. Miheeva, 1990, unpublished results.
41. B. Y. Zaslavsky, L. M. Miheeva, N. M. Mestechkina, S. V. Rogozhin, Biochim. Biophys. Acta, 253, 149 (1982).
42. B. Y. Zaslavsky, L. M. Miheeva, N. M. Mestechkina, S. V. Rogozhin, Biochim. Biophys. Acta, 253, 139 (1982).
43. B. E. Conway, Ionic Hydration in Chemistry and Biophysics, Elsevier, Amsterdam, 1981.
44. L. I. Boguslavsky, Electrochemical Phenomena and Interface, Nauka, Moscow, 1978, pp.95-106.
45. J. T. Davies, E. Rideal, Cand.J.Chem., 33, 947 (1955).
46. N. L. Jarvis, M. A. Scheiman, J.Phys.Chem., 72, 74 (1968).
47. B. E. Conway, Adv. Colloid and Interface Science, 8, 91 (1977).
48. R. Aveyard, B. Vincent, Progress in Surface Science, 8, 59 (1977).
49. Z. Koczorowski, S. Minc, Electrochem.Acta, 8, 645 (1963).
50. M. J. Jaycock and G. D. Parfitt, Chemistry of Interfaces, Ellis Horwood, Chichester, 1981.
51. G. Johansson, Mol.Cell.Biochem., 4, 169 (1974).
52. G. Johansson, J.Chromatogr., 150, 63 (1978).
53. G. Johansson, J.Chromatogr., 322, 425 (1985).
54. P. Neogi, J.Colloid Interface Sci., 159, 261 (1993).

55. L. M. Miheeva, E. D. Maximova, Y. P. Aleschko-Ozhevskii, B. Y. Zaslavsky, J. Solution Chem., 20, 607 (1991).
56. T. I. Zvarova, V. M. Shkinev, G. A. Vorob'eva, B. Y. Spivakov, Y. A. Zolotov, Microchim.Acta, 3, 449 (1984).
57. O. Y. Samoilov, Structure of Aqueous Solutions of Electrolytes and Hydration of Ions, Acad.Sci.USSR, Moscow, 1957.
58. D. E. Brooks, K. A. Sharp, S. Bamberger, C. H. Tamblyn, G. V. F. Seaman, H. Walter, J.Colloid Interface Sci., 102, 1 (1984).
59. B. Y. Zaslavsky, L. M. Miheeva, Y. P. Aleschko-Ozhevskii, A. U. Mahmudov, T. O. Bagirov, E. S. Garaev, J.Chromatogr., 439, 267 (1988).
60. R. D. Rogers, A. H. Bond, C. B. Bauer, Sep.Sci.Technol., 28, 1091 (1993).
61. R. Kaliszan, Quantitative Structture-Chromatographic Retention Relationships, Wiley, New York, 1987, pp.25-48.
62. R. Kaliszan, Anal.Chem., 64, 619A (1992).
63. A. Ben-Naim, Hydrophobic Interactions, Plenum Press, New York, 1980.
64. N. Muller, Acc. Chem. Res., 23, 23 (1990).
65. R. P. Scott, J.Chromatogr., 122, 35 (1976).
66. E. D. Katz, K. Ogan, R. P. W. Scott, J.Chromatogr., 352, 67 (1986).
67. R. P. Scott, P. Kucera, J.Chromatogr., 112, 425 (1975).
68. J. P. Knox, A. Pryde, J.Chromatogr., 112, 71 (1975).
69. B. L. Karger, L. R. Snyder, C. Econ, J.Chromatogr., 125, 71 (1976).
70. C. Horvath, W. Melander, J. Molnar, J.Chromatogr., 125, 129 (1976).
71. W. Melander, C. Horvath, Arch.Biochem.Biophys., 183, 200 (1977).
72. G. Cevs, J.Chem.Soc.Faraday Trans., 87, 2733 (1991).

CHAPTER 5.

GENERAL TRENDS IN SOLUTE PARTITION BEHAVIOR

General trends reported in the literature for the solute partition behavior in aqueous polymer two-phase systems are outlined below. These trends are discussed as compared to those observed in water-organic solvent systems to emphasize the fundamental similarity between partitioning of solutes in these apparently different systems. Effects of phase composition of the systems and those of the structure of solutes being partitioned on their partition behavior are considered and different explanations of these effects suggested in the literature are discussed. Finally, the most important question - namely, what information about a given solute is provided by the solute partition coefficient in an aqueous two-phase system, is addressed.

Due to space limitations it is practically impossible to cover all the experimental information accumulated in the literature on solute partitioning in aqueous two-phase systems and especially on that in water-organic solvent systems. Hence the very difficult selection of the data to be discussed below had to be made. The following criteria for the selection have been used. First, the data reported in the literature in the numerical rather than graphical form are easier to consider. Secondly, and even more important, the detailed information about the phase composition of the particular system used is usually needed to analyze the factors affecting partition behavior of a given solute. Additionally, an illustrative character of the data in regard to being consistent with or contradictory to the general trends was considered.

Many variable factors affect partitioning of a solute in aqueous two-phase systems. These factors include the type, molecular weight and concentration of phase polymers, type and concentration of additives, pH, temperature, etc. Hence there are many different ways to manipulate the solute partitioning. It is, however, difficult to chose the partition conditions appropriate for a particular mixture of proteins, nucleic acids, etc., as most of these variables are mutually dependent and their influence on the solute partitioning is not well understood. For example, it is impossible to change the temperature without changing both polymer and salt composition of the phases. To simplify the problem it is necessary, however, to consider the effects of these variable factors on the solute partitioning separately. It seems logical to start with the effects of the polymer composition of the phases.

5.1. EFFECT OF POLYMER COMPOSITION OF THE PHASES ON SOLUTE PARTITIONING

The polymer composition of a given phase term is generally used in the literature in a very broad sense. This term in the case of an aqueous two-polymer two-phase system actually includes six different variable parameters. These parameters are: (i) type or chemical structure of the phase polymer 1 predominantly present in the phase; (ii) molecular weight of the polymer 1; (iii) concentration of the polymer 1 in the phase; (iv), (v), and (vi) identical to (i), (ii), and (iii), respectively, but characterizing the phase polymer 2 present in the phase under consideration in lesser amount. In the case of an aqueous polymer-salt two-phase system the number of factors is reduced by one as the salt molecular weight is not included. It should be noticed that while the polymer type and molecular weight can be chosen at will, the concentrations of the two polymers in a given phase are mutually dependent and hence may be manipulated only to a very limited extent.

There is not a great deal known about the effect of the type of phase polymers on the solute partitioning as yet. Most of the experimental data relevant to the issue under discussion have been obtained using the aqueous Dex-PEG two-phase systems [1-3]. The aqueous two-phase systems formed by other polymer pairs [4-6] became the object of the experimental studies only recently. The data obtained in these systems will be discussed below.

The effects of the polymer composition of the phases on the solute partitioning have been explored almost exclusively in the aqueous Dex-PEG two-phase systems [4,7-11]. There are several general rules for the influence of the polymer composition of the phases on the solute partitioning in the aqueous Dex-PEG two-phase systems agreed upon in the literature [1,2]. These rules are: (1) At the composition of a system close to the critical point a solute distributes almost equally between the two phases, i.e. the solute partition coefficient K is close to 1; (2) when the total polymer concentration is increased the solute partitioning becomes more one-sided, i.e. K-value increases above 1 or decreases below 1; and (3) an increase in the molecular weight of one of the phase polymers decreases the tendency of the solute to partition to the phase enriched in that polymer.

The first rule is readily understandable. The critical point of the phase diagram corresponds to the theoretical system with the two phases of identical composition. At the system composition close to the critical point the compositions of the two phases in the aqueous Dex-PEG system are rather similar and hence the solute partitioning must be relatively equal, i.e., K-value should be close to 1.

The second rule is also clear. When the polymer concentration in a system is increased the system composition is removed from the critical point and the difference between the compositions of the two phases increases. That means that the difference between the solvent properties of the aqueous media in the phases increases, providing an increasing driving force for the unequal partitioning of a solute. To understand the effect of the polymer composition of the phases on the solute partition behavior the quantitative analysis of a relationship between the solute partition coefficient and the difference between the polymer composition of the phases is necessary.

To discuss the relationship in question it is necessary, first, to describe the experimental procedures used to determine the solute partition coefficient.

Determination of the Solute Partition Coefficient - How It May Be Done and How It Should Be Done.

The solute partition coefficient K in an aqueous Dex-PEG two-phase system is generally defined as the ratio of the solute concentration in the PEG-rich phase to the solute concentration in the Dex-rich phase. The commonly used procedure of the so-called "single-step partition" (to differentiate from the widely used countercurrent chromatography mode) consists of the following simple technical operations (for details see in [1,2]).

Concentrated stock solutions of Dex and PEG in water are prepared as well as those of salts or other additives to be used. The solute to be partitioned is dissolved in the appropriate medium (water or buffer solution). The stock solutions of phase polymers, salts, and the solute to be partitioned are mixed in the amounts appropriate to bring the system to the desired composition. Then the system may be centrifuged for 10-20 min at ca. 4400g to speed phase settling or allowed to settle for 12 to 24 hours (depending on the polymers' molecular weights) at the chosen temperature. Aliquots of the settled phases are pipetted from both top and bottom phases and analyzed for the concentrations of the solute being partitioned.

Different assay methods may be used depending on the nature of the partitioned solute. The most common assay methods are direct spectrophotometry (e.g., with colored solutes, hemoproteins, or biopolymers with a high specific absorption coefficients) and direct spectrofluorometry (e.g., for peptides or proteins with a high content of tryptophan residues). Other common peptide or protein detection techniques include dye binding reactions with Coomassie G-250 [12] or fluorescamine [13].

Whatever analytical assay method is used, it is usually advantageous to use a separate blank assay to avoid or correct the potential error caused by the contribution of the phase polymers, e.g., to the total optical absorbance measured. The same is true when enzyme concentrations in the phases are measured by specific enzyme activity assay techniques, because phase polymers may alter the kinetics or interfere with the enzyme reaction in some other way. Commonly used phase polymers can interfere with the biuret reaction protein assays in the phases. The analytical methods compatible with the partition technique were reviewed in detail in [2].

There are essentially two different ways to determine the solute partition coefficient K using the above procedures.

The one most commonly employed is to perform two-three or more partition experiments with a fixed amount of the solute introduced into a system. (This amount must be sufficiently low in order for the solute not to perturb phase equilibrium in the two-phase system. The solute amount not exceeding 0.1 % of total weight of the system is usually appropriate.) Aliquots of the settled phases are analyzed for the solute concentrations, the ratio between those is calculated, and thus the partition coefficient K value is determined. The advantage of this method is its time- and labor-efficiency but these come with a price. The price is the decreased accuracy even if the precision may be quite good, e.g., the K values are usually reported to be reproducible in the 3 to 5 % range. An additional consideration is that the method under discussion lacks an internal check on the possibility of the K-value being affected by the solute self-association or dissociation, direct solute-polymer interactions, etc., indicating that this value should not be treated as a thermodynamic entity (see below).

The other method is more time- and labor-consuming but it is much more accurate. This method is based on the same procedures but four to five different amounts of a solute varied over about 1-1.5 order of magnitude range are introduced into separately prepared systems of the same composition. To assure the reproducibility of the measurements the experiments are repeated

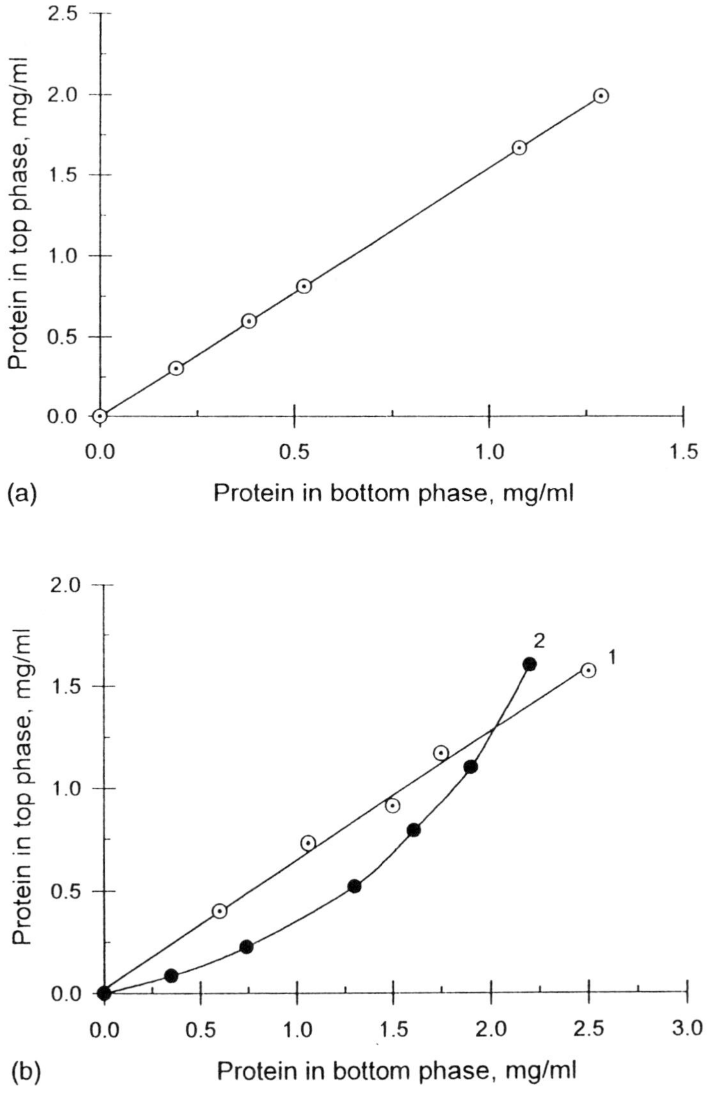

Figure 5.1. (a) Concentrations of human γ-globulin in the two phases of aqueous Dex—Ficoll two-phase system containing 0.11 mole/kg sodium phosphate buffer, pH 7.4 under varied total concentration of the protein in the system; (b) concentrations of egg white lysozyme in in the two phases of aqueous Dex—Ficoll two-phase system containing (1) 0.11 mole/kg sodium phosphate buffer, pH 7.4 and (2) 0.01 mole/kg universal buffer, pH 4.4 under varied total concentration of the protein in the system.

two to four times with each amount of the solute. To increase the reliability of the assay measurements they are performed using a series of four-five different dilution of at least two aliquots separately withdrawn from each phase.

The partition coefficient value for a given solute is determined by the standard linear regression analysis as the slope of the linear function described in the case of the aqueous Dex-PEG two-phase system as:

$$C(\text{solute})_{PEG} = a + K \cdot C(\text{solute})_{Dex} \tag{5.1}$$

where C(solute) is the solute concentration in a given phase; subscripts "PEG" and "Dex" denote the phases enriched in the corresponding polymers; intercept "a" is usually close to zero.

The procedure in question was described in detail in [14, 15]. Typical data obtained according to this method in the partition experiments with human γ-globulin and egg white lysozyme in the aqueous Dex-Ficoll two-phase system containing 0.11 mole/kg sodium phosphate buffer, pH 7.40 are presented in Figure 5.1. Dimerization of lysozyme known to occur at alkaline pH is the likely reason for the observed deviation of the curve from linearity (Fig. 5.1b). Partitioning of lysozyme in the same system containing 0.01 mole/kg universal buffer, pH 4.4, i.e. under conditions not inducing the protein association, is described by a linear plot similar to that observed in Fig. 5.1a for human γ-globulin. Similar example of the partition coefficient determination is given by Albertsson in the monograph [1, pp.82-83] with the comment that the K-value as a function of the protein concentration has not been studied for a large number of proteins. This is true, unfortunately, as the relationship such as the one shown in Fig. 5.1b indicates immediately an existence of some kind of interactions affecting the solute partition behavior that has to be taken into account if the partition coefficient is supposed to be treated thermodynamically. It should be emphasized that it seems not to be necessary to use the latter method if the solute partition coefficient is to be used, for example, for choosing the appropriate partition conditions for a given mixture to be separated or a target protein to be isolated, etc. This method, however, should be strongly recommended if the purpose of the partition experiment is to study thermodynamic relationships, the mechanism of the partition process, or the factors affecting the process. It is to be mentioned that this method was always used in the experimental studies performed by my co-workers and me. The difference in the methodology used may be at root of some disagreements between the results reported in the literature. These disagreements are considered below.

The quantitative relationship between the solute partition coefficient K and the difference between the concentrations of a phase polymer i in the two phases was derived by Diamond and Hsu [8] from the Flory-Huggins theory of polymer thermodynamics (see below) and confirmed experimentally for peptides and certain proteins. This relationship is similar to the one discussed above for "structurally simple" monofunctional aliphatic compounds:

$$\ln K(j) = k_{ji} \cdot \Delta C(\text{polymer i}) \qquad (4.10)$$

where K(j) is the partition coefficient of the jth protein or peptide; ΔC(polymer i) is the difference between the concentrations of the i-th phase polymer between the two phases (in aqueous Dex-PEG systems PEG is usually chosen as polymer i); and k_{ji} is a constant the value of which depends on the solute being partitioned and on the particular aqueous two-phase system employed.

According to Diamond and Hsu [8], partitioning of different dipeptides and proteins, such as trypsin, ribonuclease a from bovine pancreas, horse heart cytochrome C, and chicken egg lysozyme (all proteins with molecular weights under 25,000) fits Equation 4.10 fairly well. Partitioning of proteins with molecular weights exceeding 35,000 (for example, rhodanese from bovine liver, human transferrin, bovine serum albumin, turkey egg ovalbumin, etc.) in the aqueous Dex-PEG and PEG-potassium phosphate two-phase systems [8,9] did not fit Equation 4.10 but might be described by an empirical relationship:

$$\ln K = A \cdot \Delta PEG + b \cdot (\Delta PEG)^2 \qquad (5.2)$$

where A and b are the coefficients depending on the type of the protein being partitioned and the particular two-phase system employed.

It should be emphasized that Equation 4.10 has a theoretical basis [8]. It is also consistent with the experimental results described above for "structurally simple" compounds and with the linear dependence of all the studied solvent properties of the aqueous media in the phases and single polymer solutions upon the polymer concentration (see in Chapters 2-4).

Equation 5.2 is purely empirical and seems not to have any physical meaning. If Equation 5.2 is correct the conclusion by Walter et al.[16] that "the interactions (of proteins) with the physical properties of the phases is complex and, hence, the properties reflected by partitioning can neither be simply defined nor unequivocally stated" seems to be unavoidable. This conclusion though favored by some biologists is unacceptable to physical chemists as thermodynamics of any process has nothing to do with the simplicity or complexity of the participants in the process.

According to the experimental results discussed below [Zaslavsky et al., 1990, unpublished data], partitioning of biopolymers of relatively high molecular weights from human serum albumin to nucleic acids in the aqueous Dex-PEG and Dex-Ficoll two-phase systems fit Equation 4.10 perfectly. The only difference between the data reported by Diamond and Hsu [8,9] and those obtained by Zaslavsky et al. seems to be the different methodologies used to determine the partition coefficients of solutes. The immediately obvious question,

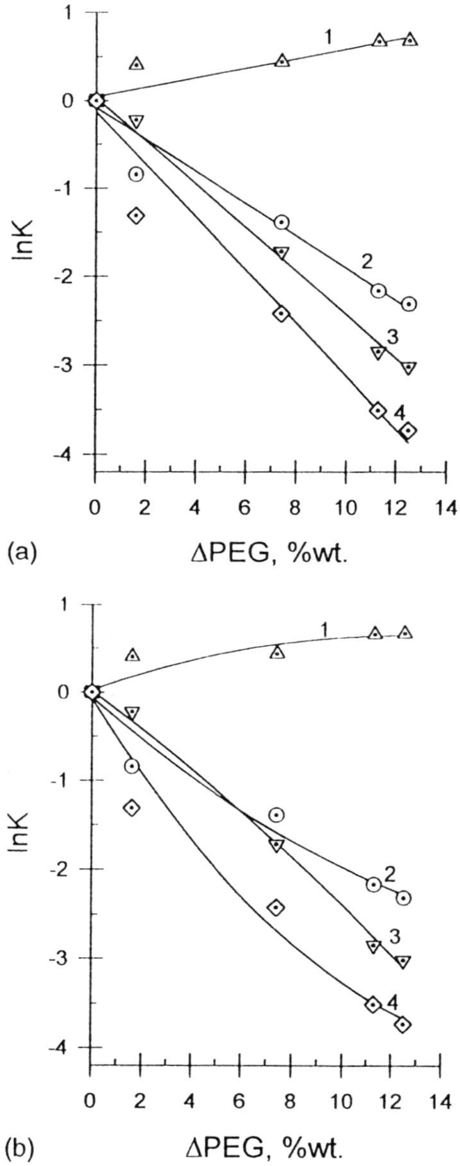

Figure 5.2. Logarithms of partition coefficients of proteins in aqueous Dex-40—PEG-6000 two-phase systems at pH close to the isoelectric point of a given protein: 1 - chymotrypsinogen A; 2 - bovine serum albumin; 3 - human transferrin; 4 - catalase; (a) according to Equation 4.10; (b) according to Equation 5.2. Calculated from the data reported in [10].

why the methodological difference affects the results for relatively large proteins and does not affect those for small proteins remains open as yet.

When there is a contradiction between the experimental results obtained by two different research groups the data reported by an other independent group are of principal interest. Recently Forciniti et al.[10,17-19] reported an extensive set of experimental data on partitioning of proteins in aqueous Dex-PEG two-phase systems formed by polymers of different molecular weights as well as on the phase compositions in the systems.

Analysis of the experimental data set reported by Forciniti et al.[10,19] in terms of lnK - ΔPEG relationships for five different proteins in 16 different aqueous Dex-PEG two-phase systems indicates several trends worthy of particular notice.

First, the lnK - ΔPEG curves for all the protein examined [10,19] are of convex, concave, or linear shape with no particular trend in regard to the phase polymers' molecular weights or particular protein being partitioned. Second, the observed deviation of a curve from linearity is often determined by merely one experimental point on the curve. Usually this point corresponds to the same ΔPEG value for different proteins. Comparison of the data sets obtained for the same protein at different pHs [19] indicates that this point is skewed of the linear curve in the lnK - ΔPEG relationships for all reported experimental data sets at four different pH. Analysis of the polymer composition of the two phases for a given system indicates that the "suspected" point usually belongs to the ΔPEG and ΔDex values corresponding to the slope of the tie line (STL = ΔPEG/ΔDex) away from the averaged STL value for a given phase diagram (see in Chapter 10). An example is shown in Figure 5.2.

The problem with the lnK - ΔPEG relationship seems to be that both lnK and ΔPEG values are determined experimentally and the propagation of experimental error may be very significant. Two ways to improve the situation may be recommended: (a) to measure the solute partition coefficient as a function of the total solute concentration in a given two-phase system; and (b) to use the average tie-line slope (STL) value as an internal reference for the correctness of the determinations of the polymer composition in the coexisting phases. These two procedures insures one at least partially from propagation of experimental errors out of control.

Even though the above two procedures have not been employed, the experimental data reported by Forciniti et al.[10,19] may be fitted to Equation 4.10 with the correlation coefficient r^2 exceeding 0.99 in the most of the cases. The values of the coefficient k_{ji} in Equation 4.10 were determined and analyzed in regard to the molecular weights of the phase polymers Dex and PEG but no clear relationship could be found.

It was shown above that variations in the phase polymer molecular

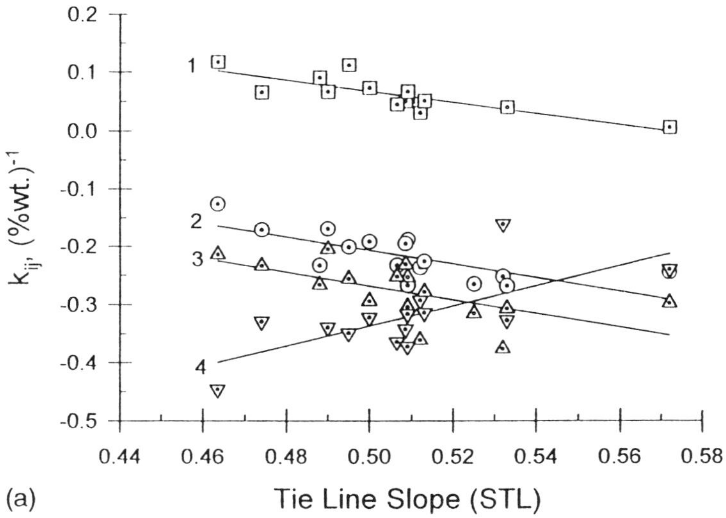

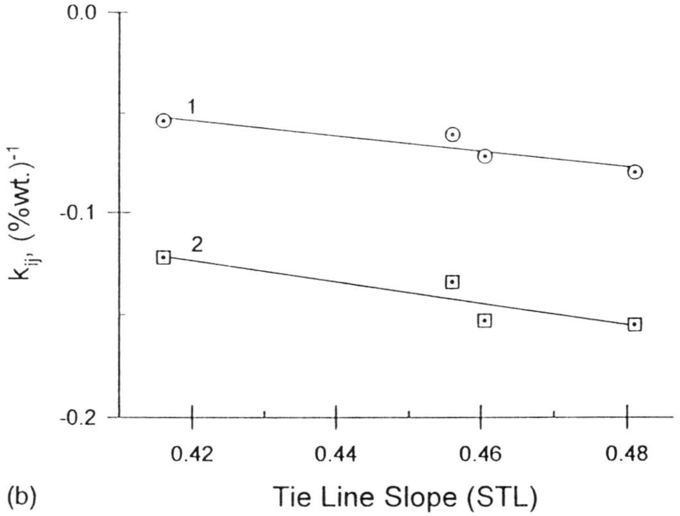

Figure 5.3. Coefficient k_{ji} for several proteins versus tie line slope for aqueous Dex—PEG two-phase systems formed by polymers of different molecular weights: (a) 1 - chymotrypsinogen A, pH 5.6; 2 - bovine serum albumin, pH 4.6; 3 - human transferrin, pH 5.6; 4 - catalase, pH 5.2 (all proteins were partitioned at pH close to the corresponding isoelectric point). Calculated from the data reported in [10]; (b) 1 - trypsin; 2 - chicken egg lysozyme. Calculated from the data reported in [8].

weight affect the tie-line slope value for the phase diagram. The effect of Dex molecular weight on the STL was shown to differ from that of the PEG molecular weight (see Fig. 3.12). Hence an existence of a relationship between the k_{ji} values for a given j-th protein and the STL_i values calculated from the data reported in [10,19] was explored. The results obtained are given in Figure 5.3a.

The relationships in question are described as

$$k_{ji} = A + B \cdot STL_i \tag{5.3}$$

where A and B are constants.

The experimental data reported by Diamond and Hsu [8] were also treated according to Equation 5.3, and the results obtained are presented in Figure 5.3b. The fit of the latter data to Equation 5.3 is clearly much better than that of the former ones which may be attributed to that the accuracy of the phase diagram determinations by Diamond and Hsu [8] clearly exceeds that by Forciniti et al.[10,19] as judged by the errors in the average STL values.

The physical meaning of the relationships described by Equation 5.3 remains obscure at present. An existence of these relationships, however, strongly indicates the fundamental importance of the STL parameter for the phase diagram description.

The above results indicate clearly that the effect of the polymer molecular weight on the solute partitioning may not be separated from that of the polymer concentration [19]. Hence attempts to study the effect of the phase polymer molecular weight on protein partitioning, such as the one reported in [7] when the systems formed by polymers of different molecular weights at a single fixed polymer concentration are used may not be considered as providing useful information.

Influence of the chemical structure of phase polymers on the solute partitioning has not been explored systematically as yet. The limited data on partitioning of different solutes in aqueous two-phase systems formed by various pairs of polymers will be briefly commented on further in this Chapter.

The following conclusions in regard to the influence of the concentration and molecular weight of phase polymers on the solute partitioning in aqueous two-phase systems may be drawn. It should be emphasized here that none of the following conclusions are suggested to be true for the affinity partitioning mode (see below) or partitioning of a solute in aqueous two-phase systems formed by polyelectrolytes or in the systems containing additives affecting the solute partitioning through specific additive-solute interactions.

1. The solute partitioning in an aqueous two-phase system (represented by the lnk value) is linearly dependent upon the difference between the concentrations of a phase polymer in the two phases. That allows one to predict partition behavior of a solute from the partition experiments in a system of a single fixed polymer concentrations provided the phase diagram is accurately

determined.

2. The effect of molecular weight of phase polymers on the solute partitioning is realized through the influence on phase diagram and is completely taken into account once the phase diagram is determined. It was shown above that the STL parameter is needed in addition to ΔPEG parameter to describe the effect of the polymer molecular weight on the solute partitioning. That implies that the presence of minor amounts of the "second" phase polymer in the phase enriched in the "first" polymer is important for the solute partitioning and cannot be ignored (see below).

3. Partitioning of solutes in an aqueous two-phase system of varied polymer concentrations is in line with the corresponding variations in the solvent features of the aqueous media in the phases.

5.2. EFFECTS OF LOW MOLECULAR WEIGHT ELECTROLYTE AND NONELECTROLYTE ADDITIVES

Addition of low molecular weight additives, ionic, e.g., inorganic salts, and nonionic, e.g., urea, dimethylformamide, etc., may strongly affect partitioning of solutes in aqueous two-phase systems. There are essentially two fundamentally different mechanisms for the effects of these additives. The first one is based on the effect of the additive on the composition and properties of the phases of a given system. The other mechanism may be realized through the effect of the additive on the properties of a solute. The latter mechanism is realized, for example, when a protein being partitioned changes its size depending on the ionic strength of the medium [20], undergoes conformational changes in the presence of a particular additive, or binds the additive. In all these cases the nature and/or relative amount of particular groups of the protein macromolecule exposed to the solvent is changed. That means that the solute-solvent interactions of this "changed" solute differ from those of the same solute in the absence of a given additive. These effects are highly specific and beyond the scope of the present discussion though sometimes it is difficult to differentiate these effects from the effects of additives on the properties of the phases. The latter effects and their influence on the solute partitioning are of our primary concern here.

The data reported on the effects of different inorganic salts on partitioning of two non-ionic solutes, 4-nitrophenyl-α-D-mannopyranoside (Man) and 4-nitrophenyl-N-acetyl-ß-D-glucos-aminide (N-Ac-Glu) in the aqueous Dex-PEG and Dex-PVP two-phase systems containing different inorganic salts [21,22] provide an example of the study of this kind. Partitioning of both glycosides was found [22] to be linearly dependent on the difference between the polymer concentrations between the two phases in agreement with Equation 4.10. The slopes of the linear relationships characterized by the coefficient k_i

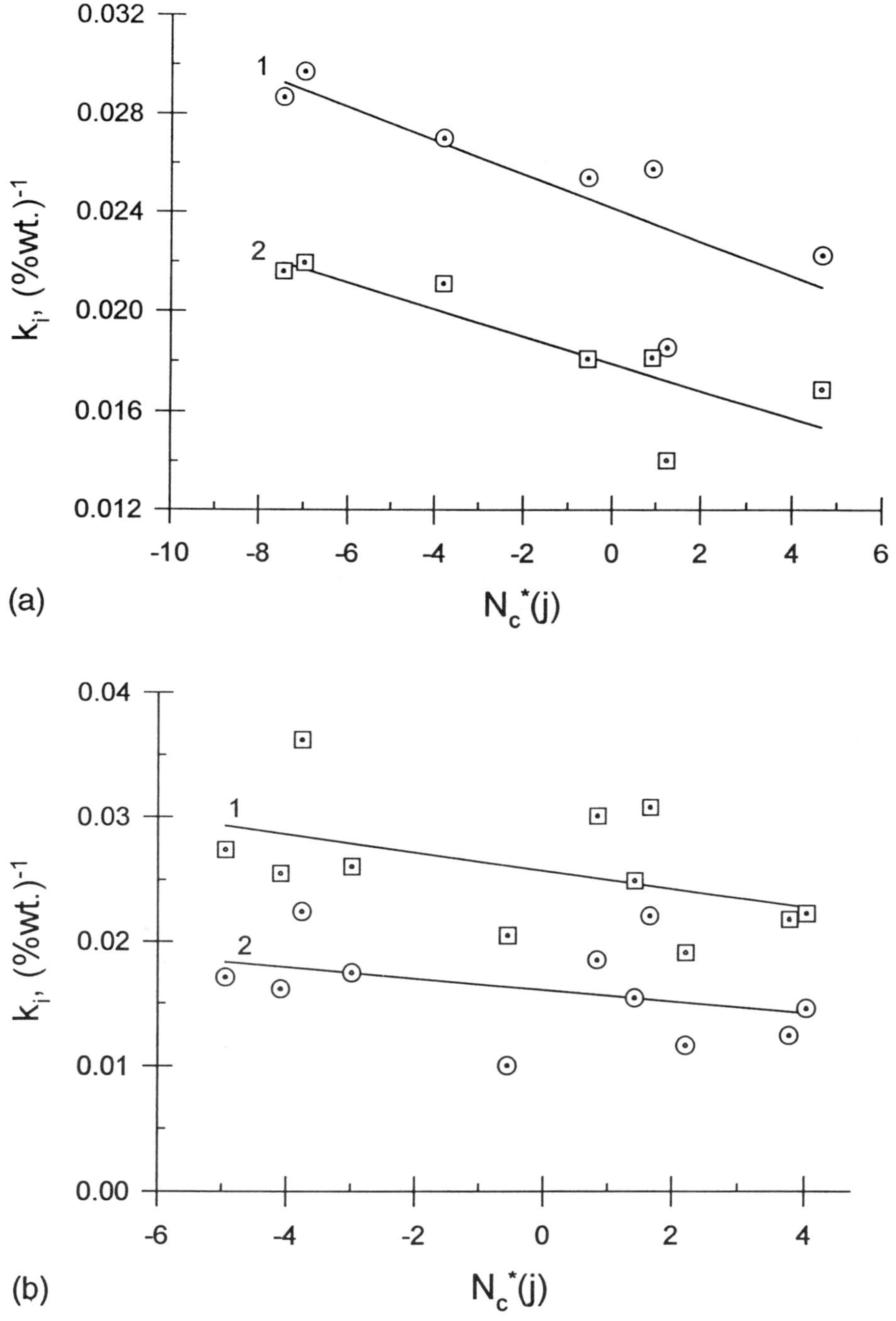

Figure 5.4. Coefficients k_i for 4-nitrophenyl-N-acetyl-ß-D-glucosaminide (1) and 4-nitrophenyl-α-D-mannopyranoside (2) versus parameter $N_C{}^*(j)$ values for the aqueous Dex—PVP (a) and Dex—PEG (b) two-phase systems containing various inorganic salt additives.

values for a given glycoside depend on both type of phase polymers (Dex-PEG or Dex-PVP) and salt additives used [22]. No relationship is found between the k_i values reported in [22] and the slope of tie line (STL) values for the systems employed. That seems to indicate that the effect of a salt additive is realized primarily not through the salt influence on the polymer composition of the phases.

It was suggested above that partitioning of a solute in an aqueous two-phase system is governed by the difference between the solute-solvent interactions in the coexisting phases exactly as in a water-organic solvent system. Partitioning of a given solute is affected by a change in the composition of the phases due to alteration of the ability of the solvent media to participate in the interactions induced by this change.

The interactions in question were discussed above and it was shown that they may be divided into polar (including ionic) and hydrophobic (solvophobic) interactions. According to the experimental data presented in the previous chapter, the difference between the total ability of aqueous media in the two phases to participate in both polar and hydrophobic interactions with a solute being partitioned may be represented by the parameter $N_C^*(j) = -n^*(j)$ (see Equations 4.13, 4.15, 4.17).

Therefore a relationship between the coefficient k_i characterizing the glycoside partition behavior and parameter $N_C^*(j)_i$ should be expected. The data from [22] are plotted against the $N_C^*(j)$ values (calculated from the data given in Tables 4.2 and 4.3 according to Equation 4.13) in Figure 5.4. The linear relationships observed in Fig. 5.4 support the above view of the solute partitioning in aqueous two-phase systems as similar to that in water-organic solvent systems.

Parameter $N_C^*(j)$ is determined from partitioning of a given j-th homologous series of solutes (used as reference solutes) in a given two-phase system under varied polymer composition of the phases. That requires to perform partitioning studies additionally to experimental analysis of the system's phase diagram. Fortunately, phase diagram provides one with the parameter describing the interrelationship between the polymer and salt composition of the phases, i.e. the b(salt) coefficient (see Equation 3.5). It was shown above that parameter $N_C^*(j)$ is linearly related to the b(salt) coefficient in aqueous two-phase systems containing different salt additives (see Equation 4.16 and Fig. 4.15). The practical advantage of this linear relationship is that it allows one to use coefficient b(salt) instead of parameter $N_C^*(j)$ as a representative of the measure of the system partitioning ability without additional partitioning experiments. It also allows one to analyze the experimental data reported in the literature for the systems lacking the information needed to determine the $N_C^*(j)$ values.

It must be verified, first, however, that the replacement of parameter $N_C^*(j)$ with coefficient b(salt) does not lead to distortion of a character of the k_i

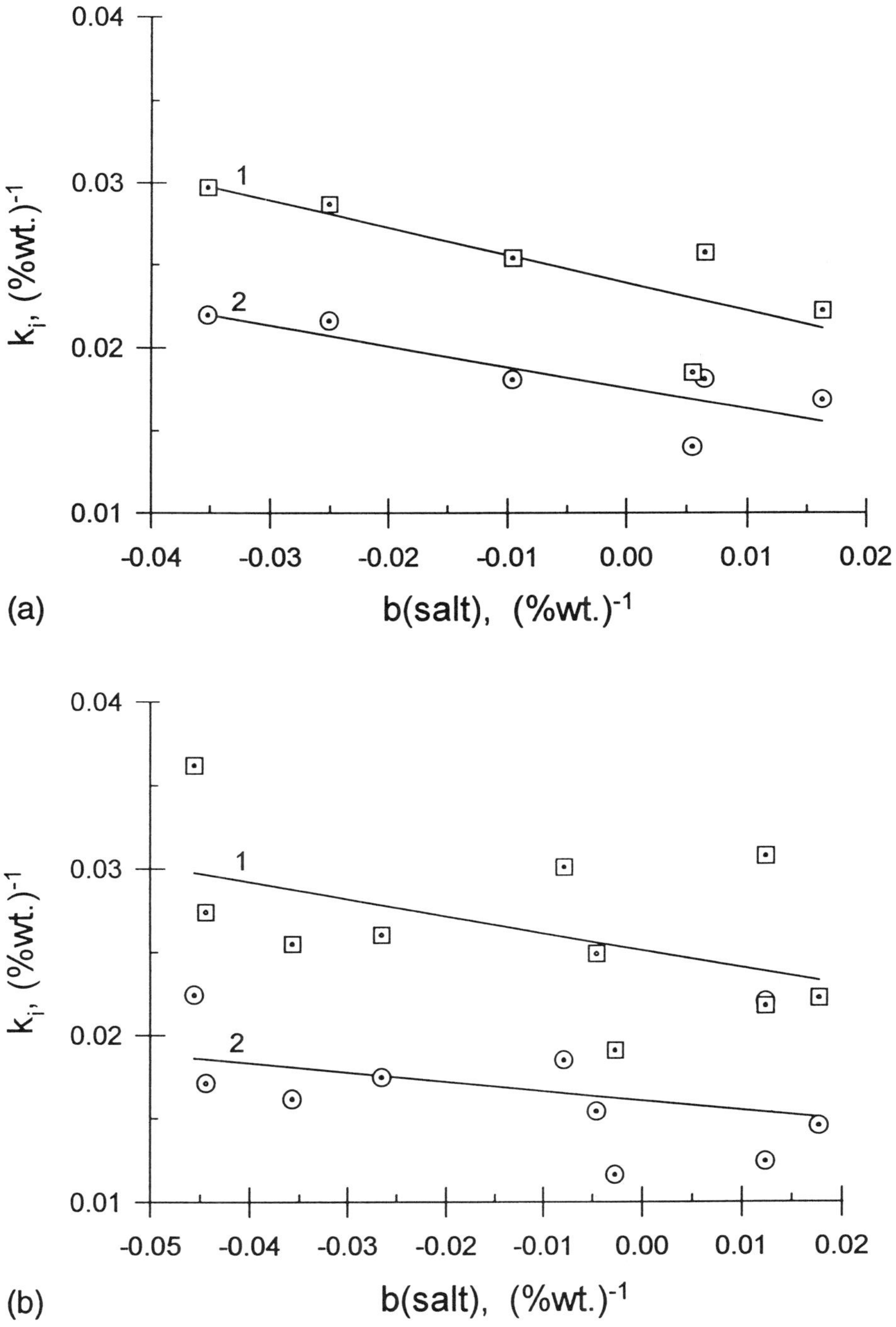

Figure 5.5. Coefficients k_i for 4-nitrophenyl-N-acetyl-ß-D-glucosaminide (1) and 4-nitrophenyl-α-D-mannopyranoside (2) versus coefficient b(salt) values for the aqueous Dex—PVP (a) and Dex—PEG (b) two-phase systems containing various inorganic salt additives.

- $N_C^*(j)$ relationship. The data presented in Figure 5.5 indicate that while the positions of the curves in reference to those in Fig. 5.4 are changed (as should be expected according to Equation 4.16) the linear shape of the relationships is maintained. Thus, the relationships presented in Figure 5.5 support the assumption that the b(salt) coefficient may be used as a representative of the above measure of the relative partitioning ability of a given aqueous two-phase system.

If this assumption is correct, several questions must be answered before the physical meaning of this parameter may be discussed in more detail. Among the most obvious questions are:

1. Is it possible to use the coefficient b(additive) similar to b(salt) for aqueous two-phase systems containing non-electrolyte additives?
2. Is the similar measure applicable to water-organic solvent systems containing the so-called organic modifier additives? and
3. Is it possible to use the same measure, i.e. b(additive), to estimate the relative partitioning ability of aqueous two-phase systems toward biological macromolecules.

Only preliminary answers to these questions may be found from the current literature, unfortunately, since the data needed are very limited.

To answer the first question, the data reported by Johansson and Joelsson [23] on the partitioning of benzoic acid in the aqueous Dex-500—PEG-8000 two-phase system containing different amounts of N,N-dimethylformamide (DMF) were analyzed. Using the analogy with, e.g., aqueous Dex-PEG two-phase systems containing different total amounts of KCl or K_2SO_4 (see above), the systems containing different total amounts of DMF were treated separately as different systems. The coefficients k_j for benzoic acid in these systems were calculated according to Equation 4.10. The coefficient b(DMF) values were calculated in accordance to Equation 3.5, i.e. b(DMF) = $(\ln[DMF]^{PEG} - \ln[DMF]^{Dex})/\Delta PEG$, where [DMF] is the concentration of DMF in a given phase; superscripts "PEG" and "Dex" denote the phase rich in the corresponding polymer.

The results obtained are plotted in Figure 5.6a. The relationship is linear for the experimental points corresponding to the aqueous Dex-PEG two-phase systems containing from zero to 35% wt. DMF. The point corresponding to the system containing 45% wt. DMF is clearly away from the linear curve. This system, however, may hardly be viewed as the aqueous polymer system containing an organic additive, since the total amount of DMF (45%wt.) in the system exceeds that of water (44.1%wt.). The system in question may probably be considered as the two-phase system formed by Dex and PEG in dimethylformamide containing water as an additive. For the aqueous Dex-PEG systems the partition behavior of benzoic acid is linearly related to the b(DMF) coeffi

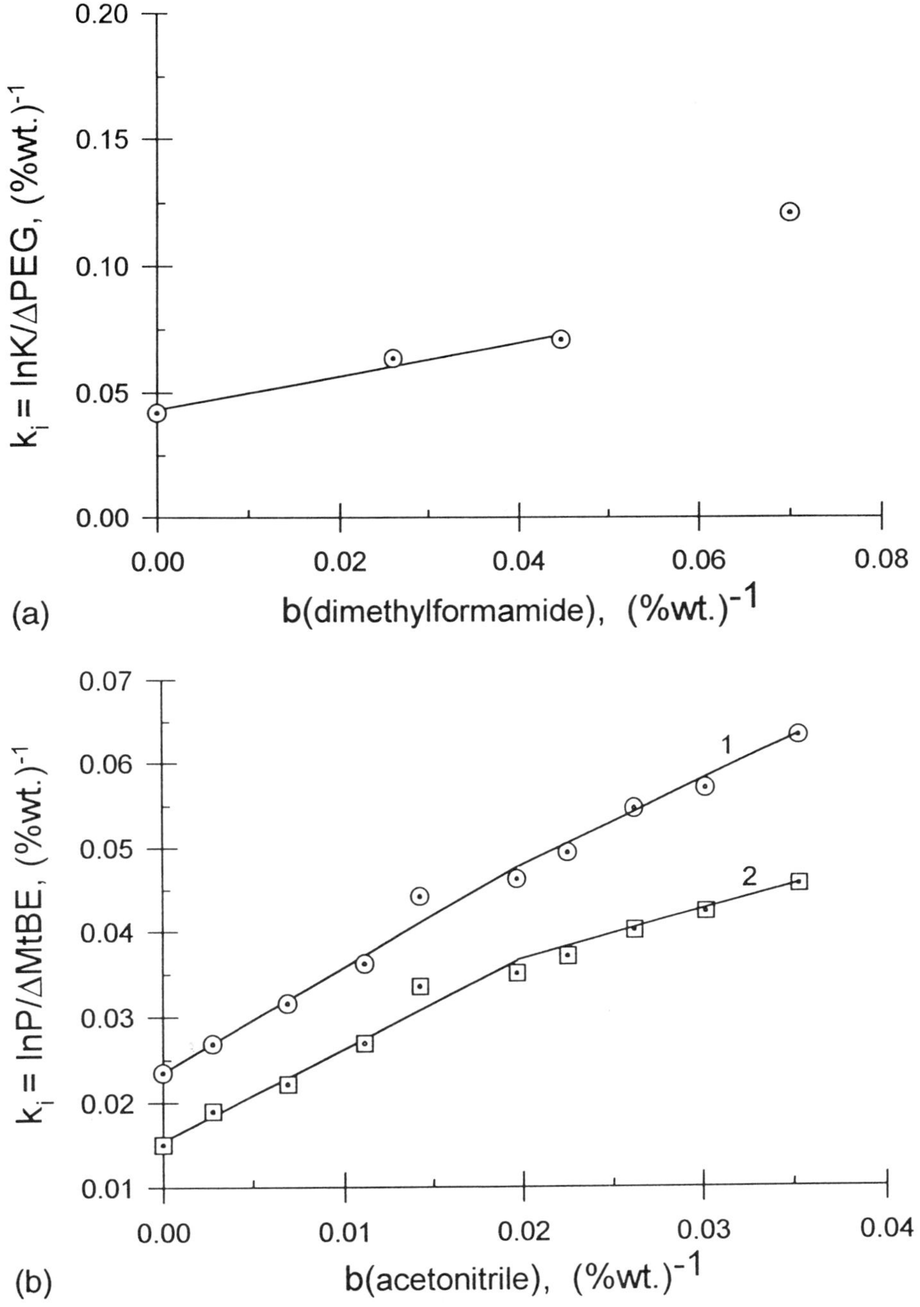

Figure 5.6. Coefficients k_i for solutes in different two-phase systems as functions of the coefficient b(additive): (a) aqueous Dex—PEG systems containing different amounts of dimethylformamide; solute - benzoic acid. Calculated from data in [23]; (b) water—methyl tert.-butyl ether (MtBE) systems containing different amounts of acetonitrile; solutes: ethyl acetate (1); butanol (2). Calculated from data in [24].

cient in the manner similar to that observed for glycosides in different salt-containing aqueous two-phase systems.

Thus, the preliminary answer to the first question is positive. The coefficient b(additive) appears to be an equally adequate measure of the partitioning ability of aqueous two-phase systems containing electrolyte as well as non-electrolyte additives. Much more experimental data are clearly necessary in order to confirm or refute this conclusion.

To answer the second question, the data available in the literature on the solute partitioning in ternary water—organic solvent—organic modifier two-phase systems [24,25] had to be analyzed. Surprisingly, the amount of experimental data reported in numerical form was found to be almost as limited as in the case of aqueous polymer two-phase systems.

The data reported by Gluck and Wingeier [24] offer, to my knowledge, the only example of numerical data on solute partitioning and phase composition in a ternary solvent system, namely, water — methyl tert.-butyl ether (MtBE) — acetonitrile (ACN) two-phase system. The data from [24] were treated as described above using the equations: $k_i = \ln P/\Delta MtBE$, and $b(ACN) = (\ln[ACN]^{MtBE} - \ln[ACN]^{aq.})/\Delta MtBE$, where [ACN] is the concentration of acetonitrile in a given phase; superscripts "MtBE" and "aq." denote the MtBE-rich phase and aqueous phase, respectively; ΔMtBE is the difference between the concentrations of MtBE in the two phases.

The linear relationships presented in Figure 5.6b are clearly similar to those in Figures 5.5 and 5.6a supporting the preliminary conclusion that the b(modifier) coefficient may serve as a representative of the measure of the system partitioning ability for ternary water—organic solvent—modifier systems similar to the coefficient b(additive) in aqueous two-phase systems.

It should be noticed that each curve in Fig. 5.6b consists of two linear fragments with different slopes. The initial fragment corresponds to the systems with MtBE as a dominant component in the nonaqueous phase, and the other one corresponds to those with the amount of ACN in the nonaqueous phase exceeding that of MtBE. Analysis of the data on solute partitioning in various ternary solvent systems presented in the book by Conway [25] in the graphic form leads to linear lnK/ΔSolvent — b(modifier) relationships similar to those shown in Figure 5.6.

It should be noticed that the linear relationship established for solute partitioning in ternary solvent systems is fundamentally important for liquid-liquid chromatography as it allows one to predict the solute partition behavior, once the phase diagram and partition coefficients for a given solute in the systems of two-three different compositions are known. It should be particularly stressed that the relationship in question was established for ternary solvent systems from the analogy between these systems and aqueous polymer systems.

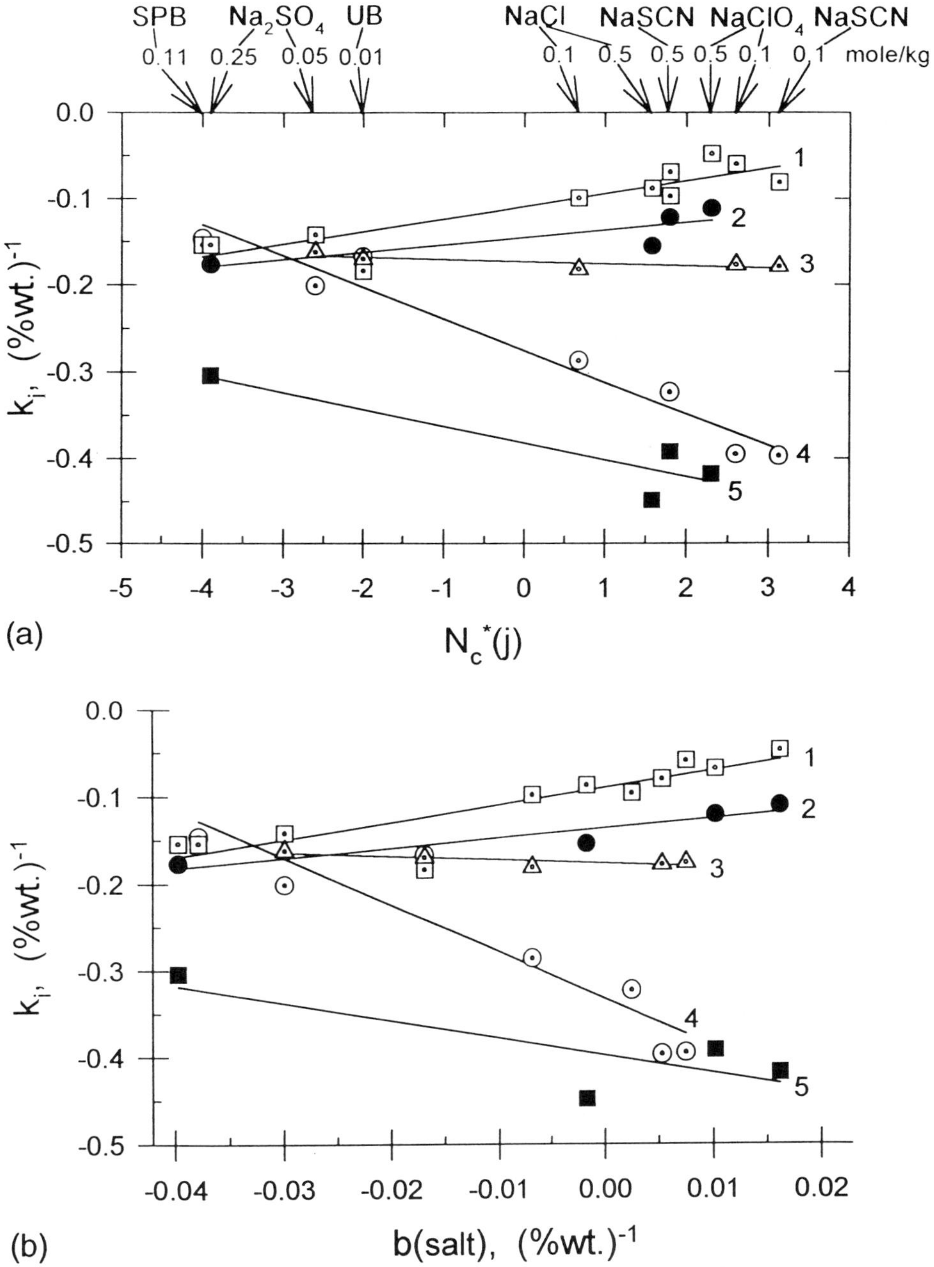

Figure 5.7. Coefficients k_i for proteins in aqueous Dex—PEG two-phase systems containing 0.01 mole/kg universal buffer (UB), pH 7.45 and different salt additives versus (a) parameter $N_C{}^*(j)$; and (b) coefficient b(salt). Proteins: 1 - cytochrome C (0.1 and 0.5 mole/kg salt); 2 - oxyhemoglobin (0.5 mole/kg salt); 3 - oxyhemoglobin (0.1 mole/kg salt); 4 - bovine serum albumin (0.1 mole/kg salt); 5 - bovine serum albumin (0.5 mole/kg salt). SPB - sodium phosphate buffer, pH 7.4.

Thus, the answer to the second question seems to be clear. Yes, the b(additive) parameter may be used as a representative of the system partitioning ability for water—organic solvent—organic modifier two-phase systems.

The last question to be answered is, if the b(salt) coefficient may be used to describe the partitioning ability of aqueous polymer two-phase systems toward biological macromolecules.

Partitioning of several proteins in aqueous Dex-70—PEG-6000 two-phase systems of varied polymer concentrations and containing different salt additives was studied by Zaslavsky et al.[26]. Partitioning of bovine serum albumin, horse cytochrome C and human oxyhemoglobin in the systems of varied polymer concentrations fits Equation 4.10. The coefficients k_i for the proteins in the systems with different salt additives are plotted versus the b(salt) values in Figure 5.7.

It can be seen from the data in Figure 5.7 that partitioning of albumin and oxyhemoglobin depends on the salt additive concentration in the system, while that of cytochrome C fits the linear relationship independent of the salt concentration. Partitioning of albumin and oxyhemoglobin in the presence of 0.5 mole/kg NaCl do not fit the corresponding relationships possibly due to some specific salt-induced changes in the conformations of these proteins. The data obtained [26] for DNA from *Escherichia coli* and polynucleotides poly A, poly C and poly U indicate that partition behavior of nucleic acids follows the general trend similar to those shown in Fig. 5.7 for proteins.

Thus, the answer to the last question is positive. The b(salt) coefficient may be used to describe the partitioning ability of aqueous polymer two-phase systems toward biological macromolecules. Actually, all the experimental data discussed so far indicate that partitioning of both biological macromolecules and small solutes follows the same general trends.

Practical importance of the fact that the coefficient b(additive) may be used as a measure of the partitioning ability of an aqueous two-phase system is clear. It may be suggested that to decide, for example, upon the particular salt composition to be used for separation of a given mixture of solutes in a given type of an aqueous two-polymer two-phase system it is sufficient to estimate partitioning of the mixture components in two systems of different salt composition. After that the reasonable decision can be made based on the knowledge of the b(salt) coefficients for the systems of any salt composition.

The fact that the same general trend fits partition behavior of biological macromolecules and small organic solutes indicates that the mechanism of the partition process is fundamentally the same for all these solutes. Similar trends in partitioning of solutes in additive-containing aqueous polymer and solvent two-phase systems confirms the hypothesis about the fundamental similarity between these systems.

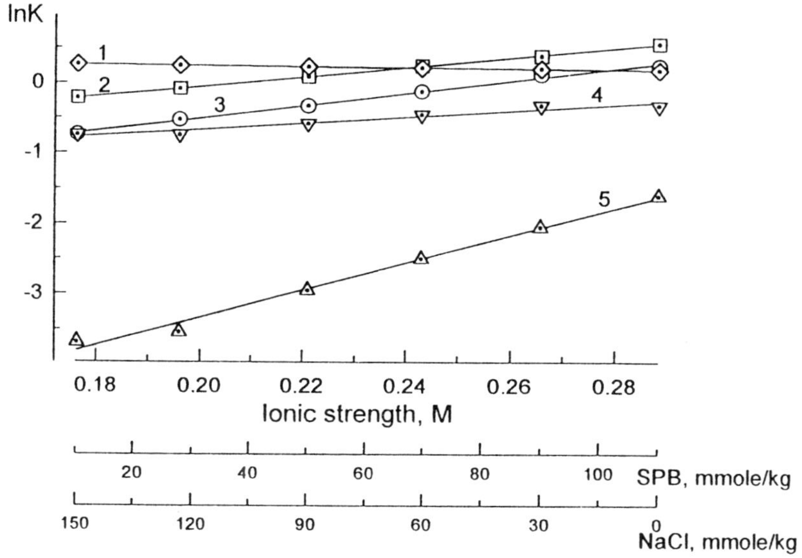

Figure 5.8. Logarithm of the solute partition coefficient, K, as a function of ionic composition in the aqueous Dex—Ficoll two-phase system containing NaCl and sodium phosphate buffer (SPB), pH 7.4. Solutes: 1 - peptide Tyr-D-Ala-Gly-Phe-N_2H_2-Leu; 2 - human γ-globulin; 3 - human serum albumin; 4 - poly C; 5 - poly U. Calculated from the data reported in [29,31,33].

An additional issue worthy of particular notice is the effect of the total additive concentration on the solute partitioning. First, it must be pointed out once more that the systems formed by same polymers and same additive but with different total concentrations of this additive should be considered as totally different systems. That is the case for the aqueous polymer systems with salt or non-electrolyte additives as well as for water-organic solvent systems containing different amounts of an organic modifier.

For the salt-containing aqueous polymer systems the ionic strength seems to be a measure preferable to the total concentration of the additive. That follows, for example, from the fact that the linear relationships for albumin and oxyhemoglobin in Fig. 5.7 fit the systems containing 0.5 mole/kg 1:1 salts (NaSCN, $NaClO_4$, etc.) and 0.25 mole/kg Na_2SO_4 or the systems containing 0.1 mole/kg 1:1 salts and 0.05 mole/kg Na_2SO_4, i.e. the systems with the same ionic strength. The data reported in the studies of partitioning of different solutes as a function of the salt composition in the aqueous Dex-Ficoll and Dex-PEG systems containing varied amounts of sodium phosphate buffer, pH 7.4

and NaCl [27-30] provide an additional evidence. Typical data reported in [29,31,33] are given in Figure 5.8.

Partitioning of proteins, polynucleotides, and low molecular weight ionic solutes, e.g., peptides, mononucleotides, amino acids, etc., was studied [27-33] in aqueous Dex-Ficoll and Dex-PEG two-phase systems containing different amounts of sodium phosphate buffer, pH 7.4 and NaCl varied in the way to keep the overall salt concentration isotonic. The condition of isotonicity while clearly important for partitioning of cells may seem totally unwarranted for that of solutes. That is perfectly true when the partition technique is used for preparative purposes, e.g., separation or isolation of a given biopolymer.

If the partition technique is used as an analytical method, however, it may be important to simulate biological conditions. Isotonicity of an aqueous medium is one of the factors complying with this requirement that may be easily provided. To keep the overall salt concentration isotonic under varied salt composition in the above systems the sodium phosphate buffer concentration was reduced with concomitant increase in the NaCl concentration [27-33]. The two most different isotonic salt compositions in this case were 0.11 mole/kg sodium phosphate buffer, pH 7.4 (ionic strength 0.288 M) and 0.15 mole/kg NaCl in 0.01 mole/kg sodium phosphate buffer, pH 7.4 (ionic strength 0.176 M). These two salt compositions are commonly used in biomedical practice as essentially interchangeable, and the issue of the difference between the ionic strengths of these media was discussed in the biochemical literature only as relevant to measurements of the electrophoretic mobility of cells. The difference in question (ca. 0.11 M) is relatively small, and hence the clearly different partition behavior of solutes in the systems of the above salt compositions is worthy of particular notice.

It was shown [28] that the logarithm of the partition coefficient of an ionic solute in the aqueous Dex-Ficoll and Dex-PEG two-phase systems of the above salt composition is linearly related to the ionic strength of the system according to:

$$\ln K = C + B \cdot I \tag{5.4}$$

where K is the solute partition coefficient; I is the ionic strength of the system; C and B are constants.

It is possible that the effect in question is due to the influence of the salt composition (represented by the ionic strength value) on the polymer composition of the phases. This issue has not been examined as it was believed at the time that the salt additives at concentrations up to ca. 0.5 M do not affect phase diagram for an aqueous two-phase system formed by two nonionic polymers [1]. As shown above, this assumption was found later to be incorrect but the issue under discussion has not been explored as yet. Whatever the reason, the linear relationship (Equation 5.4) between the ionic strength of the systems

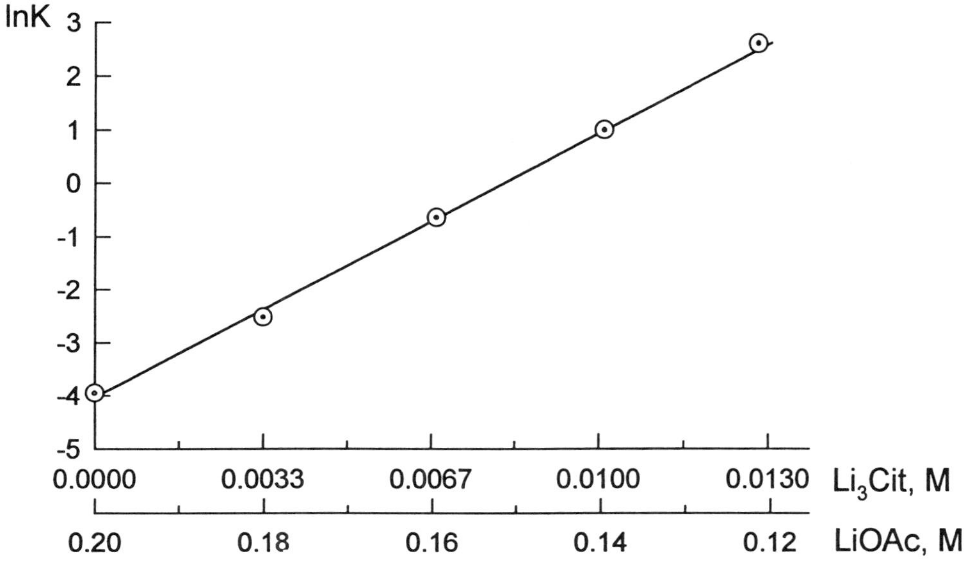

Figure 5.9. Partition coefficient of linearized plasmid DNA (PDS1, 3829 bp) in the aqueous Dex-PEG two-phase system containing cacodylate buffer, pH 6.0 and different amounts of lithium citrate (Li_3Cit) and lithium acetate (LiOAc) at 37°C. (From W. Muller, Bioseparation, 1, 265 (1990). Reprinted by permission of Kluwer Academic Publishers.)

containing the same type of salts, NaCl and sodium phosphate buffer of the fixed pH value and the logarithm of the ionic solute partition coefficient has been established in aqueous Dex-Ficoll and Dex-PEG two-phase systems beyond any doubt. The ionic strength of the system is actually purely formal parameter, as the salt concentrations in the two phases are different meaning the ionic strengths of the media in the phases are different. The relationship in question is practically convenient, however, as it allows one to compare partition behavior of different ionic solutes under varied ionic composition of the media (see below).

Walter and Anderson [16] argued that the use of the ionic strength value as a formal representative of the system salt composition is misleading as it is applicable only to the systems with the same type of ions (for example, partitioning of a solute in the systems of the same ionic strength but containing KCl instead of NaCl does not fit the relationship). Equation 5.4 is definitely limited to the systems containing the same ions and probably even only to those with the ionic strength varied over the indicated narrow range. More general relationship if in existence would be much more convenient. For the time being, however, only this limited relationship has been established [27] and its usefulness will be demonstrated further on.

The salt effect cannot be reduced to the ionic strength influence as clearly indicated, for example, by the dependence of DNA partition behavior in the aqueous Dex-PEG two-phase systems with additives of lithium acetate and lithium citrate varied in the way to keep the constant ionic strength of 0.2 M [34]. The data reported in [34] are shown in Figure 5.9.

The data in Fig. 5.9 illustrate the effect of a partial replacement of lithium acetate with lithium citrate on the DNA partition coefficient. It should be particularly noticed that the effect in question is induced by the partial replacement of supporting anions on partitioning of polyanionic DNA. That implies that it is not the effect of counter-ion (cation is the same) and it may be assumed to be the result of the salt effect on the properties of the phases. It should also be pointed out that the replacement described in [34] is actually similar to that performed with the buffer salts when the system pH is changed (see below).

The general problem with the analysis of the solute partitioning, especially, that of biological macromolecules, in aqueous two-phase systems of different salt compositions is the difficulty in separation of the effects of salts on the properties of the phases from those on the properties of a solute. Our current knowledge of the influence of different salts on the interactions of biopolymers with aqueous medium is very limited. Much more experimental information including effects of salt additives on phase diagrams and various physicochemical properties of the aqueous media in the phases is needed to overcome this difficulty. That is also true in regard to pH-effects on the solute partitioning in aqueous two-phase systems considered below.

5.3. pH-EFFECTS ON THE SOLUTE PARTITIONING

Manipulations of the pH values in aqueous two-phase systems are commonly used to steer partitioning of biopolymers, proteins, in particular [1-3]. Most of the studies of the pH-effect on the protein partitioning were performed in two-phase systems of a fixed polymer composition with or without various salt additives. These studies have been extensively reviewed by Albertsson [1], and Johansson [35] and will be only briefly commented on below.

It has been suggested recently by Eiteman and Gainer [36-41] that a pH difference between the phases of an aqueous polymer two-phase system may be used as a measure of the physico-chemical properties of the phases governing partition behavior of a charged solute relative to that of the uncharged solute of the same structure. The suggestion seems to be reasonable. Actually the measurable pH difference represents the difference between the thermodynamic activity of H^+ ion. It is similar to the interfacial potential difference in the sense that it may be viewed as a characteristic of the difference in the ionic hydration ability of the phases (see in Chapter 4).

There is a difficulty with the model suggested by Eiteman and Gainer [36-41], however. The model developed [36-41] is intended for predicting partition behavior of ionic solutes as a function of pH in the aqueous two-phase system with previously calibrated properties of the phases. Calibration of the phases' properties [36-38] involves in essence determination of the A and E values in Equation 4.4. Eiteman and Gainer [39-41] suggested this calibration to be performed for a given system at a single alternatively chosen pH value assuming that pH variations do not affect the phases' properties. This assumption is incorrect as follows from the data presented in Table 4.4. Hence while the pH difference may be a convenient measure of the ionic hydration properties of the phases, the model suggested [36-41] is hardly acceptable in regard to predicting partition behavior of solutes.

It should be emphasized that the pH value of a buffered solution is determined by the ratio between the buffer salts. This ratio must be changed to change the solution pH. The change in question is equivalent to a change in the salt composition of the medium. That is why the pH-effects on the solute partitioning in aqueous two-phase systems are hard to differentiate from those of salt additives.

On the other hand, molecules of biopolymers, e.g., proteins, peptides, etc., contain a large variety of acidic and basic groups with different pK_a values, differently charged at different pH values. The pH changes not only alter the solute net charge, however, they may also induce conformational changes in the structure of a given biopolymer, as well as association or dissociation into subunits, etc.

Thus, there are two major physico-chemical factors governing the pH-dependence of the solute partitioning in a given aqueous two-phase system. One is the changes in the properties of the aqueous media in the phases induced by a given pH change and/or the corresponding change in the buffer salt composition. The other factor is the pH-induced changes in the solute resulting in the change of the interactions of the solute, e.g., protein, with an aqueous medium.

The generally accepted view of the pH-effects on the solute partitioning in aqueous two-phase systems [1,35] is based on the aforementioned misconception in regard to the role of the interfacial electrostatic potential difference in partitioning of ionic solutes. The pH dependence of the protein partition coefficient is usually treated in terms of an apparent electrical potential difference between the two coexisting phases under influence of which charged proteins partition [42]. The issue of the electrostatic potential difference was considered above in Chapter 4. Two experimental observations commonly used in the literature (see, for example, in [1,35,42]) as the argument in favor of the direct influence of the potential difference on the ionic solute partition behavior should be discussed here as both originate from the pH-dependence of the protein partitioning.

The argument is based on the empirical relationship observed between the net charge Z and partition coefficient K of a protein [35]. This relationship is presented in the recent publications [35,43] as:

$$\ln K = \ln K(o) + \gamma Z \tag{5.5}$$

where K(o) is the partition coefficient of the solute at the pH value corresponding to the solute isoelectric point pI; and γ is defined as a factor that depends on the polymer composition of the system, the salt additives used, and the temperature.

Equation 5.5 seems to be of rather limited value, unfortunately. The linear character of the lnK - Z relationship predicted by the equation is generally not realized in practice, as can be seen from the complex curves obtained experimentally (see, e.g., in Fig. 5.10). An existence of a relationship between the solute charge and lnK, linear or otherwise, may be readily interpreted in terms of solute-solvent interactions.

The net charge on a solute macromolecule depends on the amount of its ionic groups and hence is related to the contribution of the ion-dipole interactions experienced by the solute in the two phases into the total energy of the solute-solvent interactions. The factor γ supposedly related to the interfacial electrostatic potential difference [35,42,44] may be viewed as a representative of the difference between the capabilities of the aqueous media in the two phases to participate in the ion-dipole hydration interactions with an ionic solute being partitioned (see in Chapter 4). Therefore an existence of the lnK - Z relationship, linear or not, cannot be considered as an evidence for the direct influence of the potential difference on the ionic solute partitioning.

The major limitation of Equation 5.5 which makes it hard to accept, however, is not merely the physical meaning of the terms included but the implication that the partition coefficient of the solute with zero net charge is independent of the system salt composition. Numerous experimental data discussed above indicate this implication to be generally untrue.

The results reported on the partition coefficients of proteins as functions of pH in the aqueous Dex-PEG two-phase systems containing different salt additives (typically but not always NaCl and Na_2SO_4) [1,35,45-47] indicate that the lnK - pH plots usually cross at the point (cross-point) corresponding to pH value equal or very close to the protein isoelectric point pI (the technique known as cross-partition). The cross-partition results are explained in the literature [1,35] as due to that the protein with zero net charge at pI distributes between the two phases with the partition coefficient K(o) independent of the salt

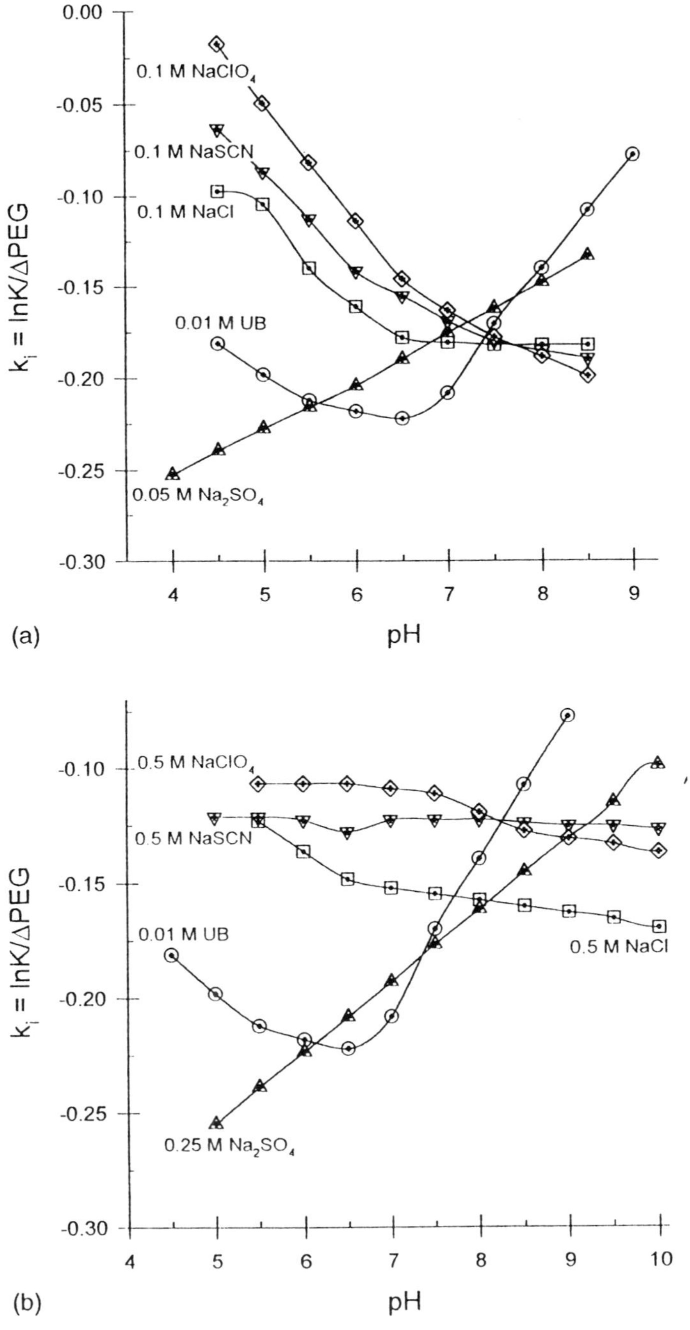

Figure 5.10. Coefficients k_i for proteins as functions of pH in aqueous Dex—PEG two-phase systems containing 0.01 mole/kg universal buffer (UB) without any salt additive, and with salt additives at the concentrations (in mole/kg) indicated: (a, b) - human hemoglobin; (c, d) - bovine serum albumin; (e, f) - horse cytochrome C.

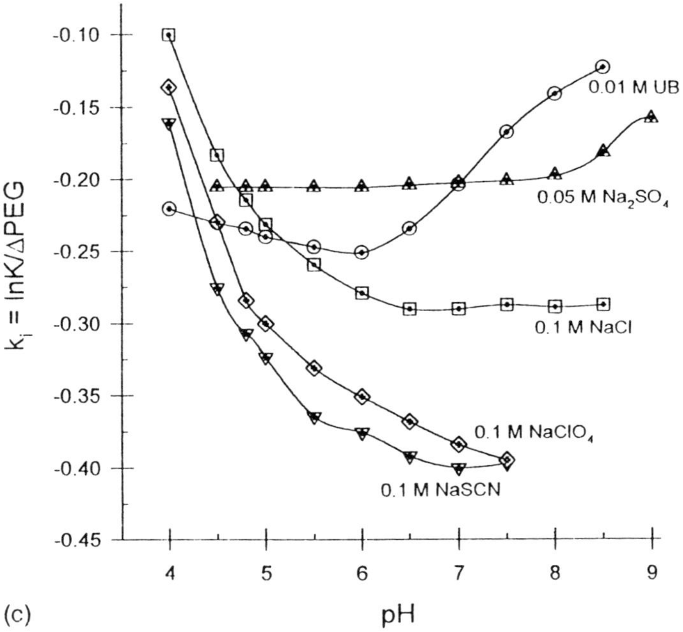
-0.10
-0.15
-0.20
-0.25
-0.30
-0.35
-0.40
-0.45
k_i = lnK/ΔPEG
4
5
6
7
8
9
pH
0.01 M UB
0.05 M Na2SO4
0.1 M NaCl
0.1 M NaClO4
0.1 M NaSCN

(c)

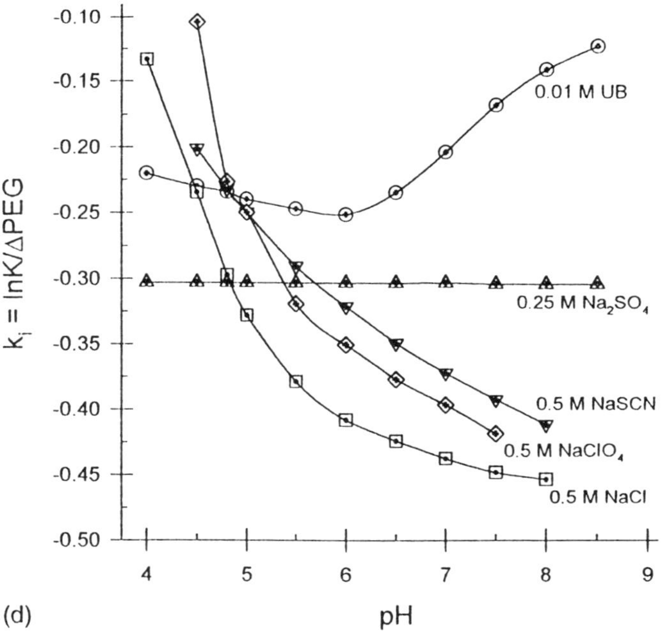
-0.10
-0.15
-0.20
-0.25
-0.30
-0.35
-0.40
-0.45
-0.50
k_i = lnK/ΔPEG
4
5
6
7
8
9
pH
0.01 M UB
0.25 M Na2SO4
0.5 M NaSCN
0.5 M NaClO4
0.5 M NaCl

(d)

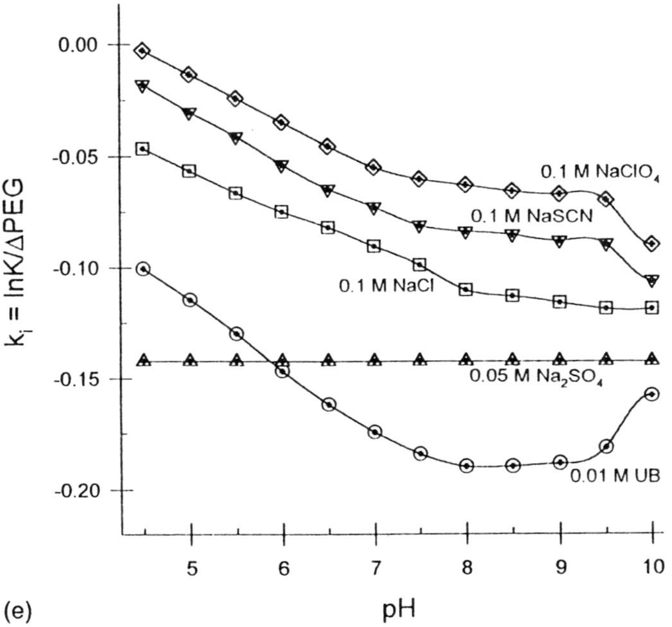
0.00
-0.05
-0.10
-0.15
-0.20
$k_i = \ln K/\Delta PEG$
0.1 M NaClO$_4$
0.1 M NaSCN
0.1 M NaCl
0.05 M Na$_2$SO$_4$
0.01 M UB
5
6
7
8
9
10
pH

(e)

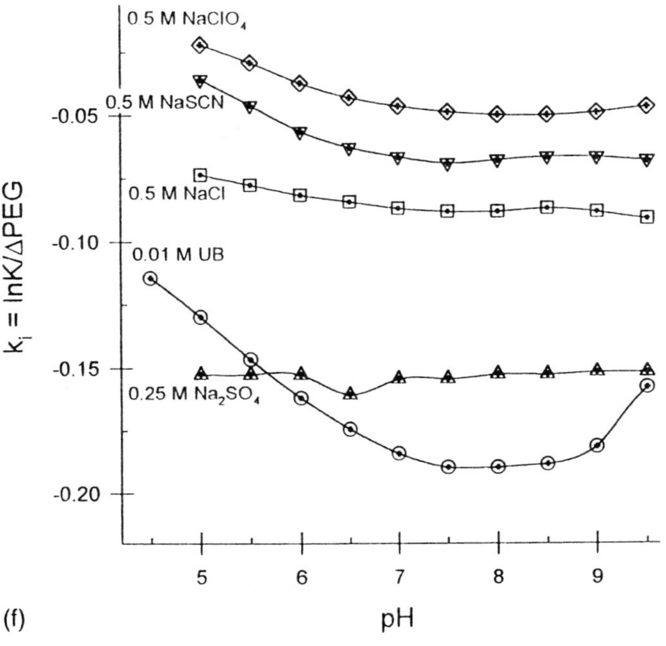
0.5 M NaClO$_4$
0.5 M NaSCN
0.5 M NaCl
0.01 M UB
0.25 M Na$_2$SO$_4$
-0.05
-0.10
-0.15
-0.20
$k_i = \ln K/\Delta PEG$
5
6
7
8
9
pH

(f)

composition of the system (see Equation 5.5).

The cross-partition results reported in the literature [45-47] were obtained in the aqueous Dex-500-PEG-6000 two-phase systems. The aqueous Dex-70—PEG-6000 systems of varied polymer concentrations containing 0.01 mole/kg universal buffer and different salt additives with the solvent features of the phases examined by the "simple solutes" partitioning described above (see Fig. 4.17) were used in the cross-partition experiments performed by Zaslavsky and co-workers [48]. The results obtained are presented in Figure 5.10 as pH-functions of the k_i coefficients (see Equation 4.10) for the proteins examined.

Partitioning of human hemoglobin (Fig. 5.10a) and bovine serum albumin (Fig. 5.10c) in the systems containing 0.01 mole/kg universal buffer without any salt additive and with 0.1 mole/kg salt (NaCl, NaSCN, $NaClO_4$) or 0.05 mole/kg Na_2SO_4 as a function of pH appears to be in agreement with the cross-partition concept. Not all five but at least several curves intersect with each other over the relatively narrow pH ranges reasonably close to the isoelectric points of the proteins (pI = 6.95 for hemoglobin, and pI = 4.6 for bovine serum albumin) [49]. It should be noticed that the pH values presented in Fig. 5.10 are as measured in the initial buffer solution and somewhat different from those in the phases (see in Chapter 4). In the case of cytochrome C, however, there is no cross-point nearby the isoelectric point of the protein (pI = 9.28) [49] at both salt concentrations used (Fig. 5.10 e, f).

According to the data presented in Fig. 5.10, an increase in the salt concentration from 0.1 to 0.5 mole/kg (NaCl, NaSCN, $NaClO_4$) or from 0.05 to 0.25 mole/kg Na_2SO_4 changes the partition behavior of the proteins and the positions of cross-points become widely scattered over pH depending on the particular salt-specific curves compared.

According to the aforementioned hypothesis, partitioning of a solute in an aqueous two-phase system is governed by the solute-solvent interactions in the two phases. These interactions for a given solute are determined by the polymer and salt composition of each phase. It follows from all the data discussed so far that a common point of intersection (cross-point) of the pH - partition coefficient curves for a given protein in the presence of different salt additives may be observed only under specific condition of essentially identical solute-solvent interactions in the phases of aqueous two-phase systems containing different salt additives, for example, NaCl and Na_2SO_4. The characteristics of the solvent features of the media in the phases of the systems employed [48] at different pH, i.e., g_A and g_E values (see Chapter 4), does not allow one, however, to explain the identical partition coefficients of a protein at a given pH value in the systems of different salt compositions.

The data presented in Fig. 5.10 were obtained in the aqueous two-phase systems formed by Dex-70 and PEG-6000, while those reported in the

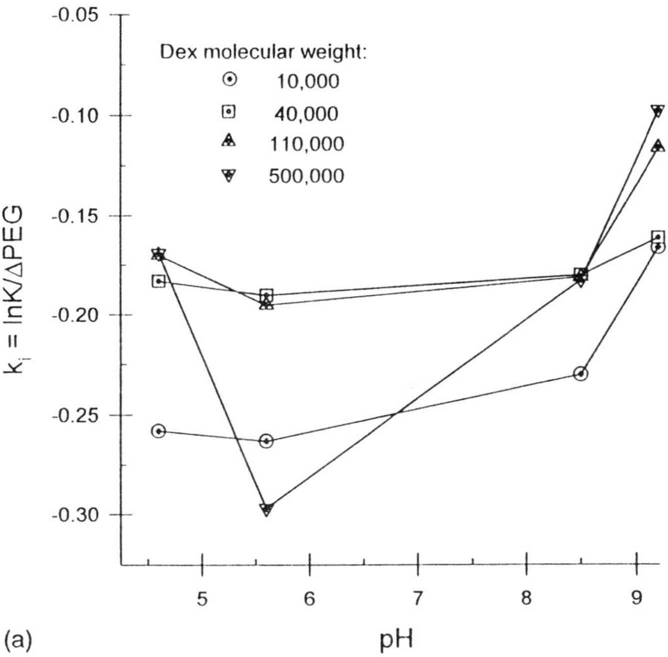

(a)

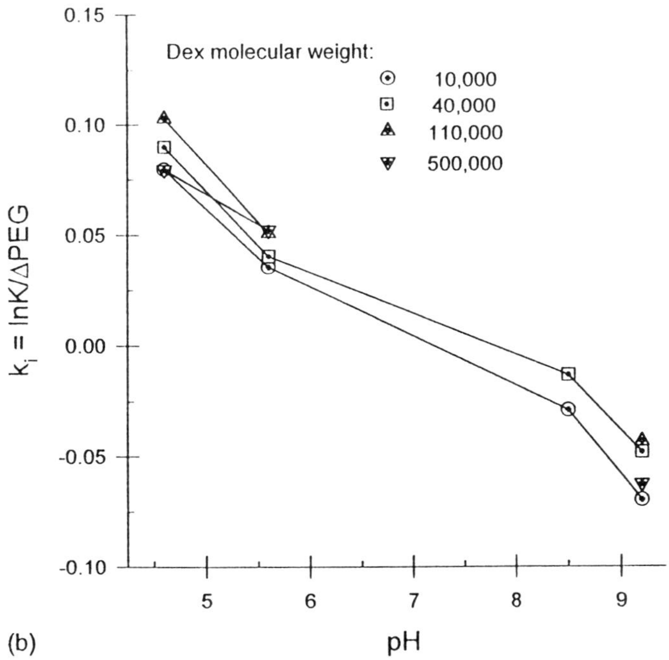

(b)

Figure 5.11. Coefficients k_i for (a) bovine serum albumin and (b) lysozyme as functions of pH in aqueous Dex—PEG two-phase systems formed by PEG-6000 and Dex of different molecular weights. Calculated from the data reported in [19].

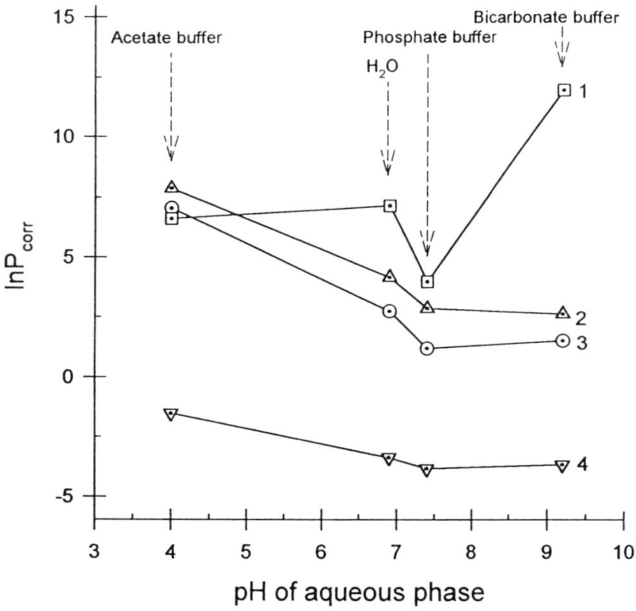

Figure 5.12. Partition coefficients of drugs in octanol—water and octanol—buffer two-phase systems as functions of pH in aqueous phase. Drugs: 1 - salicylic acid; 2 - atropine sulfate; 3 - procainamide; 4 - phthalimide. P_{corr} - partition coefficient corrected for the degree of ionization of a given drug at a given pH. Calculated from the data reported in [50].

literature on cross-partitioning were obtained using Dex-500—PEG-6000 systems. The effect of the dextran molecular weight as a reason for the observed discrepancy between the data discussed above [48] and those reported by other authors [35,45-47] seems to be highly unlikely, however. The data obtained by Forciniti et al.[19] indicate that the dextran molecular weight affects the pH-dependent partitioning of proteins just slightly if at all (see in Figure 5.11).

The coefficients k_i for bovine serum albumin and lysozyme calculated from the experimental data reported in [19] are plotted as functions of pH in the aqueous Dex—PEG two-phase systems formed by PEG-6000 and dextrans of molecular weights from 40,000 to 500,000. There is clearly no dextran molecular weight effect on the partitioning of lysozyme, and the effect on that of albumin seems to be related to the experimental errors in the partition coefficient determinations (see above).

The likely reason for the discrepancy in question may be the difference in the types of buffers used in [48] (universal buffer) and in [45-47] (usually phosphate buffer).

The effect of the buffer type on the partitioning of solutes in buffer-octanol two-phase systems was explored by Wang and Lien [50].

The partition coefficient of a solute in a two-phase system is constant only if a single molecular species is distributed between two immiscible phases. Therefore thermodynamic partition coefficient of a solute in water—octanol two-phase system, P_{corr}, called true or corrected partition coefficient is defined as the one characterizing the transfer of only undissociated molecular species. The true partition coefficient of the undissociated solute form is determined as $P_{corr} = P_{app}/(1 - \alpha)$ where α is the degree of ionization, and P_{app} is the apparent partition coefficient measured as the ratio between the total solute concentrations in the two phases. The α value can be readily calculated from $\alpha = 1/[[1 + + \text{antilog}(pK_a - pH)]$ for acidic solutes and $\alpha = 1/[1 + \text{antilog}(pH - pK_a)]$ for basic solutes.

It was found [50], particularly, that the solutes partition coefficients P_{corr} (corrected for the solute ionization degree at a given pH) are clearly dependent on the type of the buffer in use. For example, the $\log P_{corr}$ values reported in [50] for atenolol are 0.43 ± 0.01 in phosphate buffer, pH 7.4 - octanol system and 0.27 ± 0.03 in tris(hydroxymethyl)aminomethane buffer, pH 7.4 - octanol system, for ephedrine the $\log P_{corr}$ values are 0.75 ± 0.02 and 0.87 ± 0.02, respectively, while for sulfadiazine sodium the $\log P_{corr}$ values are the same -0.03 ± 0.02. If the partition coefficients of solutes depend on the type of buffer in water-octanol system, the dependence in aqueous two-phase systems is likely to be even more significant.

Wang and Lien [50] explored the effects of different buffers at different pH values on the partitioning of solutes in the buffer—octanol two-phase systems. Several examples of the results reported in [50] are presented in Figure 5.12.

The observed pH dependence of the corrected partition coefficients P_{corr} for undissociated solutes may be attributed to specific buffer salts ions - solute interactions and/or to different effects of various buffers salts on the solvent features of the aqueous phase. The latter assumption complies with the data considered in Chapter 1 as well as with those on the effects of different salt additives on the partition ability of aqueous polymer two-phase systems. If this assumption is correct, it implies that the effects of the different buffers salts on the solvent properties of the phases in aqueous two-phase systems are likely to be much more significant than in water-octanol system.

Additionally, most of the solutes being partitioned in water-octanol system are conformationally stable in contrast to proteins and other biological solutes commonly studied by partitioning in aqueous two-phase systems. An interpretation of a pH-dependent partition behavior of a given biopolymer is complicated by the possible effects of the pH-induced conformational changes. Depending on the type and amount of groups additionally exposed to or eliminated from the direct contact with the solvent, a conformational change may or may not affect biopolymer-solvent interactions and hence the biopolymer partitioning in an aqueous two-phase system. Practically any experimental observation may be attributed to the possible conformational change in the biopolymer being partitioned.

It is theoretically possible to use the partition technique to study an influence of the biopolymer conformational changes (induced by pH or other factors) on the biopolymer interactions with an aqueous medium (see below). For this possibility to be realized in practice, however, an information on contributions of different chemical moieties into the solute partition behavior is required.

Currently available experimental data on the effects of different chemical groups and molecular structure of a solute on the solute partitioning in aqueous two-phase systems are considered below.

5.4. EFFECT OF THE STRUCTURE AND MOLECULAR WEIGHT OF A SOLUTE

Effect of the Solute Size.

Following the assumption by Albertsson [1], it is commonly believed that small solutes such as amino acids or glycosides generally distribute evenly in aqueous two-phase systems. The assumption is clearly erroneous as indicated by many experimental data considered above and to be discussed below. It is undoubtedly true, however, that larger proteins and nucleic acids generally tend to partition more one-sidedly than small ones. The trend in question is usually illustrated by the plot reproduced from [47] in Figure 5.13.

The correlation presented in Fig. 5.13, however, is not a general one as follows from the data by the same authors [47] on the partition behavior of 16 different hemoproteins independent of the protein molecular weight.

The data reported by Albertsson et al.[7] apparently supporting the hypothesis about the importance of the solute molecular weight for its partition behavior actually contradict it as indicated by the following. Partitioning of large ß-galactosidase (molecular weight, mol.wt. 540,000) in the aqueous Dex-70-PEG-8000 two-phase system containing 0.01 M sodium phosphate buffer, pH 6.8, for example, is less one-sided (K = 0.38) than that of relatively small cytochrome C (mol.wt. 12,384) or bovine serum albumin (mol.wt. 69,000), both characterized by K = 0.18 [7]. Partition coefficient of catalase (mol.wt. 250,000), K = 0.79, is similar to K = 0.78 for ovalbumin (mol.wt. 45,000) in

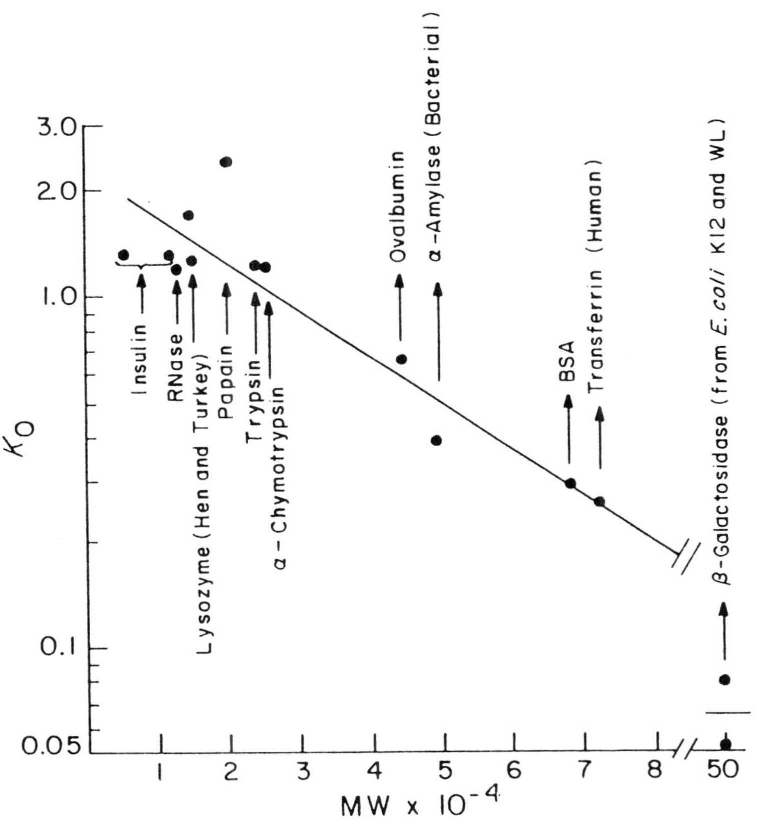

Figure 5.13. Relationship between the protein molecular weight (MW) and partition coefficient, K_o, at the isoelectric point in the aqueous Dex-500 (7% wt.) — PEG-6000 (4.4%wt.) system containing 0.1 M NaCl or 0.05 M Na_2SO_4 and 0.01 M phosphate or glycine buffer at 20°C. (From S. Sasakawa and H. Walter, Biochemistry, 11, 2760 (1972). Reprinted by permission of the American Chemical Society.)

the aqueous Dex-500-PEG-8000 system of the same salt composition [7]. Partition of human γ-globulin (mol.wt. ca. 150,000) in the aqueous Dex-70-Ficoll-400 system containing 0.15 mole/kg NaCl in 0.01 mole/kg sodium phosphate buffer, pH 7.4 is much less one-sided (K = 0.806) than that of human serum albumin (mol.wt. 69,000), K = 0.532 [31].

Thus, the effect of the molecular weight of a solute on its partition behavior while clearly observed for some solutes seems to be counterbalanced by some other factors or nonexistent for the other solutes.

Additional evidence for the existence of the effect in question was offered by Diamond and Hsu [8,9,51]. According to the aforementioned data reported in [8,9,51], partition coefficients of proteins with molecular weights above ca. 25,000 measured under varied concentrations of phase polymers in aqueous Dex-PEG two-phase systems apparently deviate from linear dependence described by Equation 4.10. The likely methodological reasons for these results [8,9,51] were discussed above. Analysis of the experimental data reported by Forciniti et al.[10,19] indicates that partitioning of, for example, transferrin, catalase, etc., considered as "large" proteins by Diamond and Hsu [8,9,51] fits Equation 4.10 fairly well (see Fig. 5.2).

There are two important implications of the presumable difference between the partition behavior of large and small solutes in aqueous two-phase systems. One is that partitioning of large molecules may be treated in terms of the Flory-Huggins polymer thermodynamics theory as a phase polymer 1—phase polymer 2—biopolymer 3 (being partitioned) system [7,44]. The other even more important implication is that partition behavior of biological macromolecules cannot be understood by studying that of small fragments of their structures. The experimental facts considered so far contradict this implication as follows from that partitioning of both small and large solutes fits the same relationships, for example, between the partition coefficient and parameter $N_C^*(j)$ or coefficient b(salt) (see Figs. 5.5 - 5.7) as well as Equation 4.10 (see Fig. 5.2).

It is hard to distinguish the possible effects of the solute size (or molecular weight) and chemical structure if, for example, a protein and a peptide or different DNA restriction fragments (see, e.g., in [52]) are compared. The only type of solutes allowing one to explore the effect of a solute size (or molecular weight) seems to be synthetic homopolymers as an increase in the polymer size is not accompanied by a change in the chemical structure.

Partitioning of various molecular weight fractions of PEG, polyacrylamide (PAAm), polyvinylpyrrolidone (PVP), and poly(vinyl alcohol) (PVA) of different degrees of acetylation in the aqueous Dex-70-Ficoll-400 two-phase system containing 0.15 mole/kg NaCl in 0.01 mole/kg sodium phosphate buffer, pH 7.4 was examined by Zaslavsky et al.[53,54]. Partition coefficients of the polymers examined are plotted in Figure 5.14 as functions of the polymers molecular weights.

The data [53] in Fig. 5.14 show that the affinity of polyacrylamides (PAAm) of different molecular weights for the upper Ficoll-rich phase increases with increasing molecular weight in agreement with the aforemen-

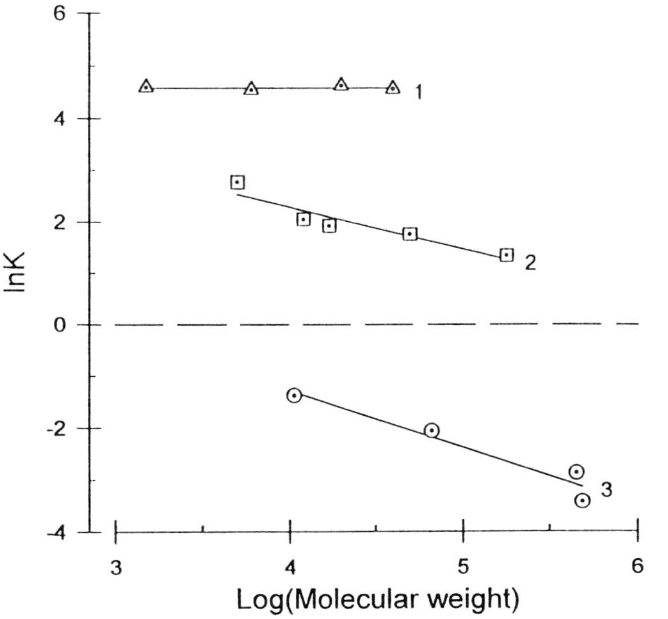

Figure 5.14. Partition coefficients of different molecular weight fractions of PEG (1), polyvinylpyrrolidone (PVP) (2), and polyacrylamide (3) in aqueous Dex-70—Ficoll-400 two-phase system containing 0.15 mole/kg NaCl in 0.01 mole/kg sodium phosphate buffer, pH 7.4 as functions of the molecular weight of the polymer fraction being partitioned.

tioned hypothesis that the higher the molecular weight of a solute, the more one-sided is the solute partition. Partition behavior of PVPs of different molecular weights, however, contradicts the hypothesis. The data in Fig. 5.14 show that the partition coefficient of the PVP fraction with the molecular weight increasing from 5,000 to 180,000 decreases from ca. 16 to 3.8, i.e. becomes less one-sided. Partition behavior of both PEG and PVA was found [53,54] to be independent of the polymer molecular weight over the range examined (from 1,500 to 40,000 for PEG and from 20,000 to 100,000 for PVA). The possible reasons for the different partition behavior of all these polymers will be discussed further on.

It should be pointed out that the acetylation degree of PVA in contrast to the polymer molecular weight does affects its partition behavior [54]. Partition coefficient of PVA increases with the acetylation degree, i.e. partial replacement of hydroxyl groups with acetate groups, as shown in Figure 5.15a. Partition behavior of different ß-1,4-glucomannanes from different plant

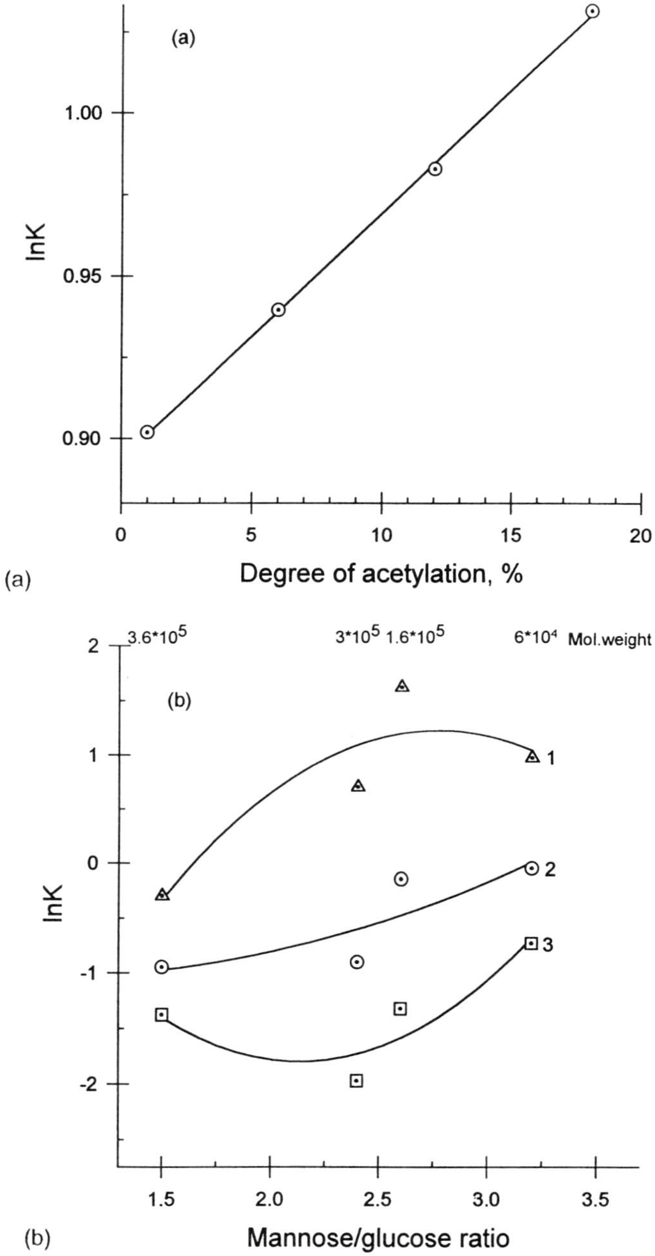

Figure 5.15. Partition coefficients of (a) fractions of polyvinyl alcohol (PVA) with different molecular weight from $2 \cdot 10^4$ to $1 \cdot 10^6$ as functions of the PVA acetylation degree; (b) β-1,4-glucomannanes as functions of the mannose/glucose ratio in the polysaccharide structure in aqueous Dex—Ficoll two-phase systems containing (1) 0.11 mole/kg sodium phosphate buffer, pH 7.4; (2) 0.09 mole/kg NaCl in 0.05 mole/kg sodium phosphate buffer, pH 7.4; and (3) 0.15 mole/kg NaCl in 0.01 mole/kg sodium phosphate buffer, pH 7.4.

sources in the aqueous Dex-70-Ficoll-400 two-phase system containing varied amounts of NaCl (from zero to 0.15 mole/kg) and sodium phosphate buffer (from 0.11 to 0.01 mole/kg) [55] confirms that the structure of a solute affects the solute partitioning more than the solute molecular weight. An example of the polysaccharides partition behavior is given in Fig. 5.15b.

The data in Fig. 5.15b show that partitioning of the glucomannanes in the system containing 0.09 mole/kg NaCl in 0.05 mole/kg sodium phosphate buffer, pH 7.4 becomes more one-sided with decreasing molecular weight and increasing the relative content of mannose residues, while in the systems containing 0.11 mole/kg sodium phosphate buffer, pH 7.4 or 0.15 mole/kg NaCl in 0.01mole/kg buffer partition behavior of the same polysaccharides is more complicated. In neither case partitioning of polysaccharides follows the pattern consistent with the hypothesis about the direct effect of the solute molecular weight on its partition behavior.

The experimental data discussed above indicate that the size of a solute is not a primary factor affecting the solute partition behavior. It does not mean, however, that the size does not play any role in the solute partitioning. It is well established that the solute size represented by the molecular volume or any other descriptor such as a solvent accessible surface area, molecular weight, etc., is a factor important for partitioning of a solute in water-organic solvent systems, for solubility, toxicity and other physicochemical and biological properties [56].

The role of the solute size in partitioning clearly follows from the requirement to create a cavity in the solvent to accommodate the solute being transferred (see above). It was shown in different water-organic solvent two-phase systems that the solute partition coefficient is related not only to the solute size descriptor but also to the differences between the sizes of the solute and the solvents molecules in the two phases:

$$\Delta G_{tr} \approx -RT\cdot \ln K_V - RT\cdot [V_r(aq) - V_r(org)] \quad (5.6)$$

where ΔG_{tr} is the free energy of transfer of a solute between the two phases; K_V is the solute partition coefficient expressed in volume fraction units; and $V_r(aq)$ and $V_r(org)$ are the ratios of the molar volume of solute to the molar volume of solvent in the aqueous and organic phases, respectively.

Equation 5.6 was used by Sharp et al.[57] to analyze particularly the free energies of transfer of n-alkanes from hydrocarbon solvents to water. Equation 5.6 allows one to account for part of the discrepancy between the "microscopic" hydrophobic effect estimates based on the solubility measurements and "macroscopic" hydrophobic effect estimates based on the macroscopic surface tension measurements [57].

The effect of a solute size on the solute partitioning in aqueous two-phase systems may hardly be accounted for by Equation 5.6, however. Both terms $V_r(1)$ and $V_r(2)$, where 1 and 2 denote the two phases, should cancel each other in Equation 5.6 due to the same nature of the solvent in both phases. Additionally, the difference between the free energies of creating the cavity of a given size in the two phases is likely to be small in comparison with that in water-organic solvent systems. Hence the effect of a solute size on the solute partitioning in aqueous two-phase systems should not be large and may be viewed as insignificant.

This conclusion is confirmed by the theoretical treatment of biopolymer hydration [58]. According to Ben-Naim et al.[58], the hydrogen bonding between the solvent and the groups located at the surface of the proteins is the most significant part of the total free energy of solvation of globular proteins. The authors [58] divided the solvation free energy into the hard-core interactions, depending on the entire volume of the solute and equivalent to creation of a cavity in the solvent, van der Waals interactions, and hydrogen bonding. The solution lacking charged ions was considered and no charge-charge or charge-dipole interactions were included into the treatment [58]. From estimates of the contributions of these three terms into the total free energy of the protein-solvent interactions for nine different proteins it was shown [58] that the larger the protein, the more dominant is the hydrogen bonding term. The effect of this term seems to be highly sensitive to the orientations and the extent of the exposure of the functional groups at the protein surface to the solvent [59].

Thus, in view of the above theoretical considerations and experimental data it may be concluded that the chemical structure of a solute and not the solute size is the primary factor governing the solute partition behavior in a given aqueous two-phase system.

Effect of the Solute Chemical Structure.

Before the experimental data on the solute structure effect upon the solute partitioning in aqueous two-phase systems will be considered, it is necessary to briefly outline the rationales generally used in the studies of these effects on the solute partitioning in water-organic solvent systems.

As mentioned above (Chapter 4), the group contribution approach is widely applied to analysis of partition behavior of solutes in two-phase systems. The approach is based on the assumption that the free energy of the solution process is additively composed of independent contributions from the constituent functional groups in the solute structure. The approach was used above to estimate the contributions of a CH_2 group and DNP-NH-CH-COONa group into the solute partition coefficient in aqueous two-phase systems.

The partition coefficient of a solute in a water-organic solvent system is generally accepted to be an additive-constitutive property of the solute molecule. The additive term means that multiple substituents exert an influence equal to the sum of the individual substituents. The constitutive term indicates that the effect of a substituent may differ depending on the molecule to which it is attached or its environment (see, e.g., in [60-63]). It was shown above that each successive addition of a CH_2 group into a molecule changes the logarithm of the solute partition coefficient by a constant increment depending on the two-phase system in use but usually not on the solute type (see Figs. 4.4 - 4.6). This increment specific for a given chemical moiety is commonly used as a measure of the effect of the moiety on the solute partitioning in a given two-phase system.

The so-called substituent constant, π_X, was defined [64] as the change in the logarithm of the partition coefficient of a solute YH brought about by substitution of a hydrogen atom with a substituent X:

$$\pi_X = \log P_{YX} - \log P_{YH} \tag{5.7}$$

where P is the octanol/water partition coefficient of a solute; subscripts "YX" and "YH" denote the derivative YX and parent YH compounds. Equation 5.7 was defined and verified experimentally using water-octanol system mostly but it was shown to be applicable to solute partitioning in any water-organic solvent system.

Nys and Rekker [65,66] suggested a different fragmental constant, f_X, taking into account that the substituent constant for hydrogen atom, π_H, has a finite value. The two constants are related as:

$$\pi_X = f_X - 0.20 \tag{5.8}$$

A compilation of π_X values for a large variety of chemical groups may be found in the book by Hansch and Leo [61], and that of f_X values in the monograph by Rekker [67].

As mentioned above, the contribution of a substituent into the solute partition coefficient depends on the structure to which it is attached. The flexibility and conformation of a molecule, branching, presence of unsaturated bonds, intramolecular bonding, proximity interactions, etc., all affect the solute partition behavior. To account for these effects, correction factors are used in calculations of the solute partition coefficient (see, for example, in [60-63,67]).

Analysis of partition coefficients for the solutes with relatively rigid structure in aqueous two-phase systems indicates that the solute partition behavior in these systems may be treated similarly to that in water—organic solvent systems.

Results of partitioning of DNP-amino acids with nonionic side-chains in the aqueous Dex-Ficoll two-phase system containing 0.15 mole/kg NaCl in 0.01 mole/kg sodium phosphate buffer, pH 7.4 [32] are plotted in Figure 5.16a

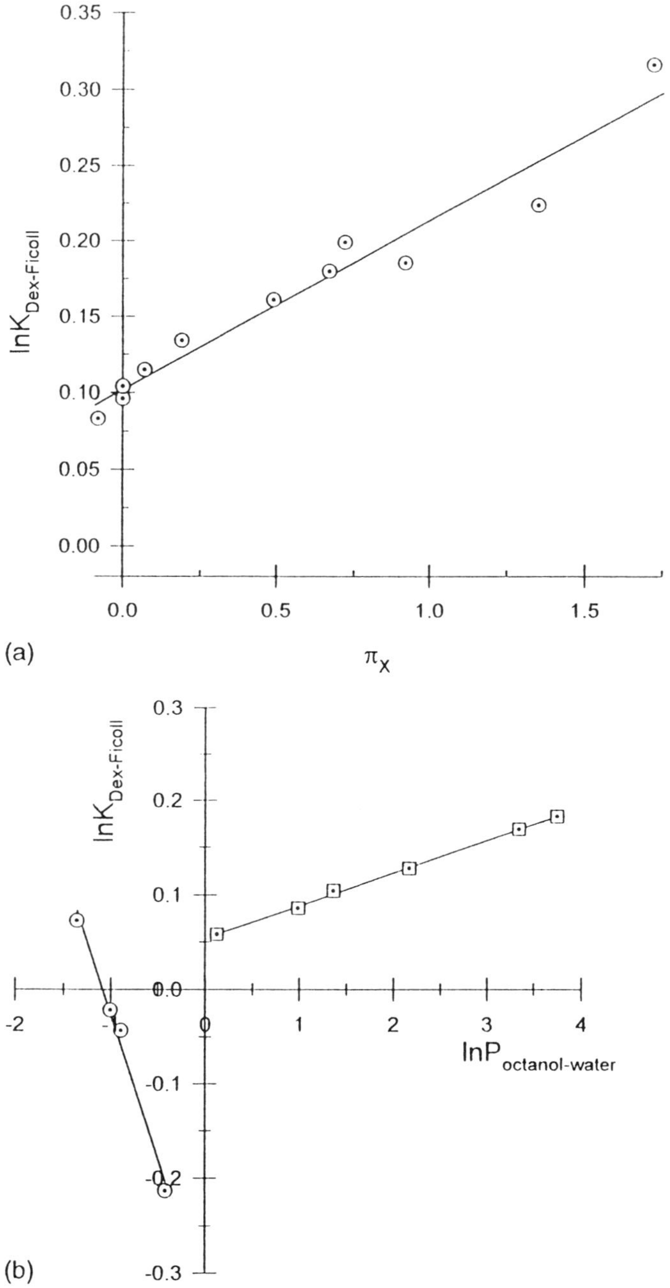

Figure 5.16. Relationships between partition coefficients of different sets of solutes in aqueous Dex—Ficoll two-phase system containing 0.15 mole/kg NaCl in 0.01 mole/kg sodium phosphate buffer, pH 7.4 and (a) π_X for amino acids with nonionic side-chains; (b) partition coefficients in octanol—water system for p-nitrophenyl glycosides (1), and morphine and its derivatives (2). Calculated from the data reported in [68-71].

against the substituent constant π_X values for the corresponding side-chains reported by Akamatsu and Fujita [68].

The data in Fig. 5.16a indicate that partition behavior of amino acids with nonionic side-chains in the aqueous two-phase system is linearly correlated with that in water-octanol system as well as in other water-organic solvent systems (see below).

Similar relationships may be found for other types of solutes, e.g., p-nitrophenyl glycosides [69,70], morphine-like drugs [71] (see in Fig. 5.16b), ribonucleosides [72,73], and dinucleosidephosphates and their (2'-5')-isomers [33,74,75]. These relationships will be discussed in more detail below. The variety of solutes examined in aqueous two-phase systems is too limited for far reaching general conclusion. The above relationships, however, strongly suggest that the solute partitioning in the aqueous two-phase systems used and water-organic solvent systems is fundamentally similar.

The data reported in the literature on the partitioning of relatively short peptides [71,76-78] indicates that the principle of additivity of contributions from the constituent functional groups in the solute structure is fulfilled but in rather limited sense. Partition behavior of a large variety of dipeptides in the aqueous PEG—K_2HPO_4/KH_2PO_4 two-phase system was studied by Diamond et al.[77]. According to the results reported in [77] partitioning of the dipeptides of the general structure X-Y where Y is the amino acid residue with the side-chain bulkier than that of the residue X differs from partitioning of the peptides of the reversed structure Y-X. The data reported by Diamond et al.[77] is plotted in Figure 5.17.

It can be seen from the data in Fig. 5.17a that the partition coefficients of the peptides examined in [77] are correlated as:

$$\ln K_{Y\text{-}X} = 0.09_{(\pm 0.04)} + 0.93_{(\pm 0.04)} \cdot \ln K_{X\text{-}Y} \tag{5.9}$$
$$N = 23;\ r^2 = 0.9794$$

where K is the dipeptide partition coefficient; subscripts denote the structures of the dipeptides; X and Y the amino acids residues, the latter with the side-chain more voluminous than the former; N is the number of pairs of peptides examined; and r^2 correlation coefficient.

It follows from Equation 5.9 that the partition coefficient of a dipeptide with the bulkier side-chain at the amino terminal generally exceeds that of the dipeptide of the reversed structure. Similar effects are observed in the water—octanol two-phase system where the contribution of the steric effect of the side-chain substituent at the N-terminal was reported [68] to be about 2.2 times greater than that at the C-terminal of the molecule.

Analysis of the data reported by Diamond et al.[77] reveals that the substituent constant for a given side-chain varies depending on the position of

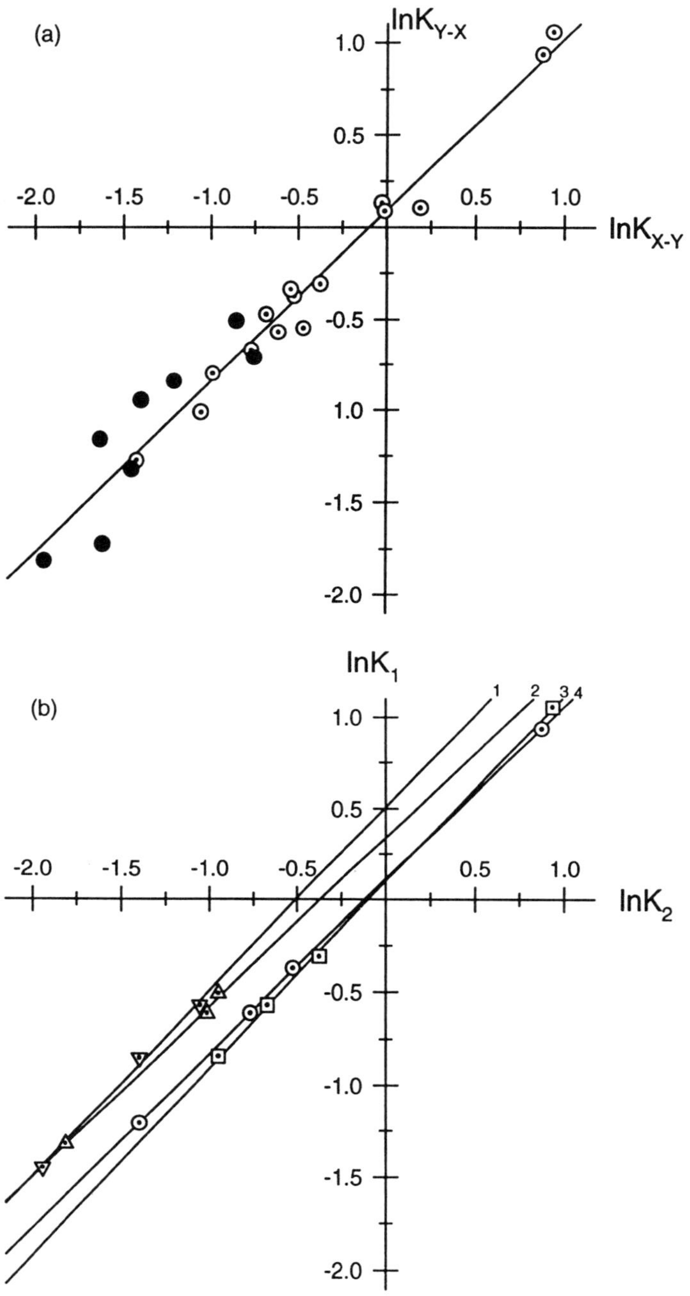

Figure 5.17. Relationships between the partition coefficients of dipeptides of reversed structure in the aqueous PEG-3400—potassium phosphate two-phase system: (a) Y-X versus X-Y; filled symbols denote dipeptides containing amino acid residues with ionic side-chains; (b) (1) Gly-X versus Val-X; (2) X-Gly versus X-Val; (3) X-Gly versus X-Ala; 4 - Gly-X versus X-Val. (X and Y - amino acids residues.) Calculated from the data reported in [77].

the side-chain in the solute structure in regard to the C- or N-terminal. Typical results [77] presented in Figure 5.17b indicate that, for example, the contribution of a CH_3 group (Ala- vs. Gly-containing dipeptides) into the logarithm of the peptide partition coefficient in the system used amounts to 0.113 ± 0.004 in the dipeptides of the Ala-X structure as compared to 0.100 ± 0.013 in the dipeptides of the X-Ala structure. The contribution of the valine side-chain, $-CH(CH_3)_2$, is the same within the limits of the experimental error amounting to 0.340 ± 0.063 for X-Val dipeptides and to 0.509 ± 0.110 for Val-X dipeptides, being ca. 3.76 times that of the contribution of the Ala side-chain. Comparison of the contributions of the same side-chains in the partitioning of amino acids in aqueous polymer or water-organic solvent systems (see below) indicates that the difference in these contributions exceeds that in amino acids (the contribution of the valine side-chain is ca. 3.06 times that of the alanine side-chain in the case of amino acids or their derivatives). The likely reason for the variable character of the contribution of an amino acid residue side-chain is the mutually dependent influence of the neighboring groups in the peptide structure on the total solute-solvent interactions.

The effects of the neighboring groups on the solvation of different substituents have been considered in the literature in detail (see, e.g., in [62,63,67, 68,79-81]). These effects were estimated for a number of peptides by partitioning in the aqueous PEG-$MgSO_4$ two-phase systems of varied phase compositions [78]. The effect in question was measured as the difference between the experimentally determined free energy of transfer of a peptide between the two phases and that calculated according to the additivity principle. In some cases the effects reported in [78] are very significant. For example, the effect in question amounts to +910 cal/mol for tetrapeptide Tyr-Gly-Gly-Phe (with total experimentally observed free energy of transfer of 4170 cal/mol), or -470 cal/mole for dipeptide Trp-Ala (with experimental free energy of transfer 2050 cal/mol), i.e. amounts up to 22-23% of the partition coefficient value [78].

The deviation of the solute partition coefficient from the value calculated using the additivity principle generally increases with the peptide chain-length likely due to the peptide folding. Partition behavior of α- and γ-endorphins in the aqueous Dex-Ficoll two-phase system containing 0.15 mole/kg NaCl in 0.01 mole/kg sodium phosphate buffer, pH 7.4 is characterized by the partition coefficients 1.062 and 1.066, respectively, similar to that of approximately 3 times smaller Leu-enkephalin ($K = 1.026$) [71]. These results comply with that the solute-solvent interactions of conformationally flexible compounds are governed not by the total solute structure but solely by the groups accessible to the solvent. Because of that partition behavior of proteins, glycoproteins, etc., cannot be described on the basis of the additivity principle.

Partition behavior of proteins as well as the other types of biopolymers is highly specific [1,35]. Our current knowledge of the protein tertiary structure in solution is too limited to allow one any detailed interpretation of the protein partition coefficients. The only promising approach currently seems to be the (bio)chemical modification of a protein structure.

Partitioning of amino acids and peptides in aqueous PEG—salt two-phase systems [77,78] indicates that tryptophan has the highest affinity for the PEG-rich phase and the tryptophan-containing peptides are characterized by the partition coefficients generally exceeding those of the tryptophan-free peptides of the similar structures. Tryptophan residues exposed on the surface of ß-galactosidase from *E.coli* were assumed to be the reason for the partitioning of this protein strongly in favor of the PEG-rich phase in the aqueous PEG-potassium phosphate two-phase system [82]. The assumption was confirmed by that the genetically engineered proteins containing additional poly-tryptophan sequences partition into the PEG-rich phase much more strongly than the original protein [83]. Partition behavior of genetically engineered proteins containing additional poly-aspartic acid and poly-arginine structures is currently under study [84].

Effects of covalent modifications of α-chymotrypsin by different reagents on the protein partitioning in the aqueous Dex-Ficoll two-phase system containing 0.15 mole/kg NaCl in 0.01 mole/kg sodium phosphate buffer, pH 7.4 were studied by Mozhaev et al.[85]. Introduction of 12 phenyl groups into the protein molecule by treatment of the protein with benzoyl chloride changes the protein partition coefficient ca. two-fold from 1.104 to 2.197 [85]. Replacement of all accessible amino groups in the protein with different numbers (from 1 to 5) of carboxylic groups by acylation with o-phthalic anhydride, pyromellitic and mellitic anhydrides gradually alters the partition coefficient of α-chymotrypsin from 1.104 to 0.106 [85]. Chemical and/or biochemical modification of proteins seems to be the most promising approach to better insight into partition behavior of a protein as a function of the protein structure.

Partition behavior of nucleosides, mono- and poly-nucleotides, and nucleic acids follows the general trends similar to those observed in partitioning of peptides and proteins. For example, analysis of the partition coefficients of 5'-ribonucleotides and 5'-deoxyribonucleotides in the aqueous Dex-Ficoll two-phase systems containing varied amounts of NaCl and sodium phosphate buffer, pH 7.4 indicates that the contribution of the ribose C-2 hydroxyl group into the logarithm of the nucleoside or nucleotide partition coefficient is constant and amounts to -0.039 ± 0.007 [72]. The same value was reported for the hydroxyl group contribution from the results of partitioning of amino acids [32] which complies with the above additivity principle.

On the other hand, partition coefficients of cyclic AMP and AMP (0.981 and 0.746, respectively [72]) differs much more significantly than the additivity principle would suggest. Partition coefficients of AMP and ATP are essentially identical [72] indicating insignificant contribution of the phosphate groups. At the same time the change of an ionic composition in the Dex-Ficoll system from 0.15 mole/kg NaCl in 0.01 mole/kg sodium phosphate buffer, pH 7.4 to 0.11 mole/kg sodium phosphate buffer, pH 7.4 increases the logarithms of the partition coefficients of GMP, CMP, TMP, and GTP by exactly the same value of 0.157, and the same change in the ionic composition increases the partition coefficients of AMP and ATP by 0.228 [72]. These results suggest that the interactions of the phosphate groups with the aqueous medium are affected by adenine in the way different from that characteristic for the other nucleic bases.

Partition of dinucleosidephosphates and their structural analogues with the ribose ring replaced with acyclic hydroxyalkyl substituents in the above aqueous Dex-Ficoll two-phase system [33] also showed that different bases affect the phosphate group-solvent interactions differently. Partition coefficients of ApA and UpU in the system containing 0.11 mole/kg sodium phosphate buffer, pH 7.4 are 1.418 and 1.362, while that of ApU amounts to just 1.028 [33]. Partition of (2'-5')-isomers of UpU and ApA under the same conditions differs from that of the corresponding naturally occurring dinucleosidephosphates, their partition coefficients are 1.008 and 0.940, respectively [33]. The effects of the structure and conformation of these compounds on the solute-solvent interactions will be considered in more detail below.

It follows from the data [33,72] on partitioning of nucleotides and their derivatives in the aqueous Dex-Ficoll two-phase systems with varied amounts of NaCl and sodium phosphate buffer, pH 7.4 that partition of these compounds is generally more sensitive to the ionic composition of the system than that of amino acids derivatives and peptides. The same conclusion has been drawn by Albertsson [1] from comparison of partition behavior of nucleic acids and proteins in aqueous Dex-PEG two-phase systems. Partitioning of nucleic acids and polynucleotides in aqueous Dex-Ficoll and Dex-PEG two-phase systems with varied polymer concentrations was found to fit Equation 4.10 [86]. There was no data published, to my knowledge, on the partition behavior of nucleic acids in aqueous two-phase systems with experimentally determined compositions of the two phases. That complicates the interpretation of the data reported in the literature in quantitative terms.

General trends in the partition behavior of nucleic acids worth of particular notice are [1,34,52,87]: (i) polypyrimidine nucleotides distribute in aqueous Dex-PEG two-phase systems with the partition coefficients usually exceeding those of polypurine nucleotides and decreasing in the order: poly(U) > poly(C) > poly(A) > poly(G); (ii) the effect of the secondary structure of nucleic acids exceeds that of the primary structure (i.e., only the groups exposed to the solvent govern the partition behavior); and (iii) changes in the partition coefficients of nucleic acids induced by ligand binding are generally much more significant than those observed for proteins. The extremely high sensitivity of the partition behavior of nucleic acids to the salt composition of aqueous two-phase systems as compared to that of proteins may be explained by the presence of the much larger amount of the solvent accessible ionic groups in the structure of nucleic acids.

The solute partition behavior in aqueous two-phase systems was discussed above in terms of the solute-solvent interactions. It is still a much debated question in the literature, however, if the solute partitioning is governed predominantly by its interactions with the different aqueous media in the two phases or by the direct interactions of the solute with phase polymers. This issue is fundamentally important for analytical applications of the partition technique. To resolve this issue it is necessary to compare the solute partition behavior in aqueous two-phase systems formed by different polymers.

5.5. COMPARISON OF SOLUTE PARTITION BEHAVIOR IN DIFFERENT AQUEOUS TWO-PHASE SYSTEMS

It is well known among those dealing with aqueous two-phase systems that replacement of a given phase polymer with the same polymer from a different manufacturer or even from a different batch produced by the same manufacturer may cause all sorts of troubles. These may be, e.g., changes in the relative volumes of the phases (indicating a change in the polymer compositions of the phases) at the same total polymer concentrations, changes in the solute (or particles) partition behavior without any noticeable changes in the visual appearance of the system, etc. According to one of the prominent experts in cell partitioning, H. Walter [88], there are examples when different laboratories could not even obtain two-phase systems in the aqueous mixtures of Dex and PEG and Dex and Ficoll using the concentrations reported in the literature presumably because of the differences between the polymers produced by different manufacturers. To overcome these difficulties the usual practice is for a laboratory to purchase large amounts of the polymers samples and use only these samples for all the partition studies. Once in a while the laboratory runs out of the polymers samples, and the troubles begin.

The important implication of the issue under discussion is also that it is virtually impossible to compare the partition data reported by different authors using different polymer samples in quantitative terms. It would obviously make no sense to try to use the partition technique as an analytical method before this issue is resolved. The similarity between the solute partitioning in aqueous polymer and water-organic solvent two-phase systems, fortunately, comes to the rescue.

About forty years ago it has been shown by Collander [89] that the partition coefficient values for different solutes in two different water-organic solvent systems are interrelated as:

$$\log P_i = a_i \cdot \log P_o + b_i \tag{5.10}$$

where a and b are constants; subscripts "i" and "o" denote the systems under comparison. Equation 5.10 is universally known as the Collander equation or solvent regression equation. The water-octanol system was chosen as the reference system [90], i.e., as the system denoted by "o" in Equation 5.10, due to several practical reasons outlined below. The constants in Equation 5.10 may vary not only from one solvent system to another but also with the type of solutes being partitioned.

Leo and Hansch [90,91] tried to rationalize the problem of variability of the constants in Equation 5.10 by considering the hydrogen-bonding ability of the solutes and solvents. They found that if the organic solvent in the system "i" had properties similar to those of 1-octanol, one equation could be derived to fit all different solutes. Various alcohols, e.g., oleyl alcohol, primary butanols, primary, secondary, and tertiary pentanols, cyclohexanol, etc., as well as methyl isobutyl ketone, and ethyl acetate are the examples. If organic solvents with the properties different from those of octanol, e.g., benzene, cyclohexane, chloroform, ether, etc., were employed various a_i and b_i values in Equation 5.10 had to be used for different solutes. The solutes were grouped into electron acceptors and electron donors, and different a_i and b_i were allocated to each group [90,91]. The difference between the $\log P_o$ and $\log P_i$ in water—cyclohexane system for a series of phenols was reported [92] to be quantitatively explained by hydrogen bonding.

The partition coefficient of a solute in a given water—organic solvent system represents the difference between the energies of the solute-solvent interactions in the two phases. The differences between the solvent features of aqueous phases at equilibrium with different organic solvents are generally assumed to be negligible as compared to those between the solvents. Equation 5.10 is viewed therefore [90] as characterizing the difference between the interactions of a given solute with two different organic solvents (saturated with water) in reference to the solute-water interactions. Hence the physical meaning of the coefficients a_i and b_i are of theoretical as well as practical importance.

The coefficient a_i in Equation 5.10 was suggested [90,91] to be viewed as a measure of the solvent system's sensitivity to changes in lipophilicity (hydrophobicity) of solutes. The intercept (coefficient b_i), however, has gained much more attention than the slope due to that its value was found to be varied depending on the solutes examined much more significantly than the coefficient a_i value [90]. It is clear that the intercept value in Equation 5.10 for a given solvent system "i" is the $\log P_i$ for a solute which distributes equally between the phases of water-octanol system. It was argued [90] that a negative intercept b_i value indicates that a given solvent is more hydrophobic than octanol, and a positive value indicates that it is less hydrophobic. It was particularly shown [90] that there is a good correlation between the b_i coefficient values for different solvent systems (in reference to water—octanol system) and the water solubility in the corresponding organic solvents:

$$\log(S_{water}{}^{org})_i = 1.077 \cdot b_i + 0.249 \quad (5.11)$$

$$N = 17;\ r^2 = 0.979;\ s = 0.217$$

where $(S_{water}{}^{org})_i$ is the water content at saturation of the organic phase in the i-th water—organic solvent system; N is the number of the solvent systems examined; r^2 correlation coefficient; s is standard deviation from regression.

It should be noticed that Equations 5.11 and 4.15 (see above) indicate the linear relationship between coefficient b_i and parameter $N_C{}^*(j)_i$ characterizing the partition ability of a two-phase system (see in Chapter 4).

Analysis of the literature data on partition coefficients measured in various solvent systems for homologous series of fatty acids, aliphatic alcohols and amines [93] revealed the physicochemical meaning of coefficients a_i and b_i in the Collander equation described in natural logarithm terms:

$$\ln P_i = a_i \cdot \ln P_o + b_i \quad (5.10a)$$

where all the terms are as defined above.

It was particularly found [93] that the slope (coefficient a_i in Equation 5.10a) is related to the free energies of transfer of a methylene group between the two phases of the systems compared:

$$a_i = \Delta G(CH_2)_{tr,i} / \Delta G(CH_2)_{tr,o} \quad (5.12)$$

where $\Delta G(CH_2)_{tr}$ is the free energy of transfer of a CH_2 group from the non-aqueous to the aqueous phase of the system; superscripts "i" and "o" denote the systems under comparison, "o" referring to the system (water—octanol) used as the reference.

The physical meaning of the intercept of the Collander equation (coefficient b_i in Equation 5.10a) follows from analysis of partition coefficients for a given homologous series of solutes described by Equation 4.4 as:

$$\ln K = A + E \cdot N_C \quad (4.4)$$

where K is the solute partition coefficient; N_C is the number of carbon atoms in the aliphatic alkyl chain of the solute molecule; A is the contribution of a polar group into the logarithm of the solute partition coefficient, and E represents that of a CH_2 group.

Combination of Equation 5.10a and Equation 4.4 leads to the simple conclusion [93]:

$$b_i = E\cdot(A_i/E_i - A_o/E_o) \tag{5.13}$$

or in a different form:

$$b_i = A_i - a_i\cdot A_o \tag{5.14}.$$

The relationship between a_i and b_i described by Equation 5.14 clarifies why both are considered [67,90,91] to be "slightly different" measures of the relative hydrophobic character of the organic solvent in a given solvent system.

Unlike the constant a_i, coefficient b_i in the Collander equation 5.10a depends not only on the solvent systems being compared but also on the nature of the solutes. According to Rekker [67], the essential role of the b_i term in the equation is to account for the difference between one functional group of a solute and another when transferred from the aqueous phase to the nonaqueous phase of the solvent system, and *vice versa*. For one functional group this transfer in a given system is much easier than for another, depending on local dehydration-solvation possibilities. The relationship described by Equation 5.11 is likely to be due to that the water content of a given solvent may regulate the affinity of a functional group for this solvent phase. The relationships between A/E for different functional groups and water solubility in various organic solvents discussed above (see Fig. 4.14 and Equation 4.15) are clearly consistent with this interpretation.

Partition coefficients for a set of different solutes were reported in eight aqueous Dex—Ficoll two-phase systems prepared using polymers produced by different manufacturers and different lots of polymers from the same manufacturer [94]. An example of typical relationships observed between the partition coefficients of various solutes in different aqueous Dex-Ficoll systems is given in Figure 5.18.

The data in Fig. 5.18 indicate that the relationships obtained are described by the Collander equation. It was shown particularly that Equation 5.10a fits partition coefficients of DNP-amino acids, sodium alkyl sulfates, different peptides and proteins, for example, cytochrome C, myoglobin, albumin [94], total human plasma proteins, ferritin, γ-globulins, [30], glycosides, nucleotides, polynucleotides, etc. [86]. It should be pointed out that while the partition coefficients of non-ionic solutes, e.g., glycosides, nucleosides, purine and pyrimidine bases, etc., fit Equation 5.10a independent of the salt composition of the aqueous Dex–Ficoll two-phase systems under comparison that is not always the case with ionic solutes. The solute partitioning seems not to fit Equation 5.10a if the salt composition affects the solute-solvent interactions specifically, for example, via changes in the solute conformation, through direct solute-salt interactions, etc., allowing one to view the solute as two essentially different

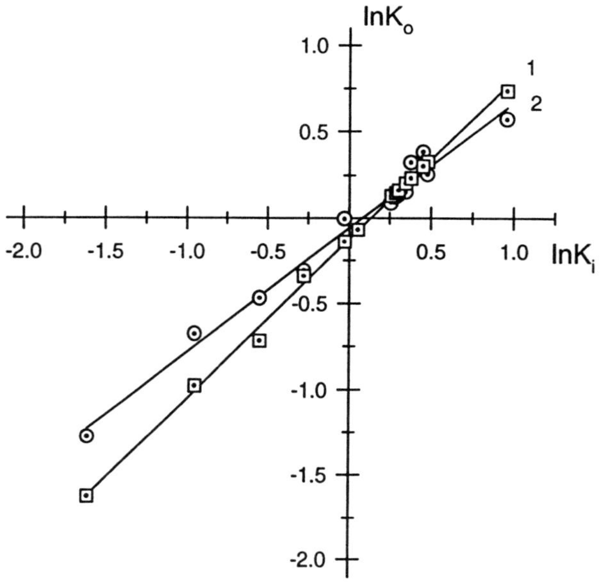

Figure 5.18. Relationships between the partition coefficients of different solutes in aqueous Dex—Ficoll two-phase systems formed by different samples of polymers. All systems contain 0.11 mole/kg sodium phosphate buffer, pH 7.4. K_o - partition coefficients of solutes in the system formed by 10.8% wt. Dex-70 (Minmedprom, Moscow, former USSR, Lot 580870) and 12.5% wt. Ficoll-400 (Pharmacia, Uppsala, Sweden, Lot 11069); K_i - partition coefficients of solutes in the system formed by (1) 12.0% wt. Dex-70 (Minmedprom, Moscow, Lot 390476) and 14.0% wt. Ficoll-400 (Pharmacia, Uppsala, Lot 6594); and (2) 11.5% wt. Dex-40 (Loba, Austria) and 13.0% wt. Ficoll-400 (Pharmacia, Uppsala, Sweden, Lot 15215).

compounds in an aqueous medium at different salt compositions.

Similar observations were reported by Wang and Lien [50] in buffer—octanol two-phase systems. In contrast to nonionic compounds, partitioning of ionic solutes was affected by the type of buffer (and pH) used in the system and required using different values for coefficients in the Collander equation. Wang and Lien [50] explained their observations by the various ability of the different buffer-originated counter-ions to affect the mechanism of intermolecular or intramolecular bonding forces in the partitioning behavior.

The partition behavior of a given ionic solute in some cases may fit the Collander equation with the coefficients determined for a set of different non-ionic (and ionic) solutes in the aqueous Dex-Ficoll systems of different salt compositions. That may be used as an indication of the lack of specific solute-ion interactions and ion-induced conformational changes of the ionic solute in question.

Thus, partition behavior of different solutes in aqueous two-phase systems formed by Dex and Ficoll from different manufacturers may be compared quantitatively using Equation 5.10, provided the total salt composition is the same in the systems under comparison.

Similar results were obtained in the study of partitioning of various solutes in aqueous Dex-PEG two-phase systems [86]. The Collander equation fits the partition coefficients of solutes examined in the Dex—PEG two-phase systems prepared with the polymers from different manufacturers if the salt composition in the systems under comparison is the same. An example of typical relationships observed between the solute partition coefficients in aqueous Dex—PEG two-phase systems formed by different polymers based on the data reported by Forciniti et al.[19] is given in Figure 5.19.

Much more important, however, is the finding [86] that partition coefficients of the solutes of different chemical nature (from amino acids, glycosides, nucleotides to proteins, nucleic acids, etc.) in the aqueous Dex—PEG system may be compared quantitatively with those in the aqueous Dex—Ficoll two-phase system, again provided that both systems have the same total salt composition. An example of this relationship is given in Figure 5.20.

It was verified experimentally [86] that the slope of the relationship under consideration, i.e. coefficient a_i, may be calculated from the free energies of transfer of a CH_2 group between the two phases of the systems being compared in accordance with Equation 5.12. Coefficient b_i may be determined from Equation 5.10a and its value was found to be independent of the solutes examined so far [86].

The constancy of the coefficient b_i value is particularly important due to the following reasons. The coefficient b_i value reflects the difference between specific intermolecular interactions in two water-organic solvent systems, e.g., hydrogen-bonding effects, electronic effects of substituents on the solvation of a given solute, etc. [95]. That agrees with Equation 5.14 implying that coefficient b_i represents the difference between the polar interactions of functional groups of a solute with the solvents in the two systems under comparison. Various groups are known to interact differently with different solvents [95-97] providing various b_i values depending on the chemical nature of the group(s) in question (and the solvents, indeed).

Practical importance of the constancy of coefficients b_i and a_i in Equation 5.10a for aqueous Dex—PEG and Dex—Ficoll two-phase systems is that it allows one to replace the polymers used with minimal difficulties. It is neces-

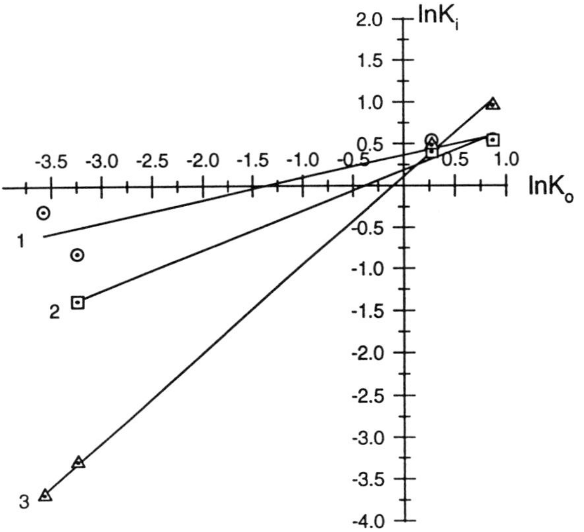

Figure 5.19. Relationships between the partition coefficients of different solutes in aqueous Dex—PEG two-phase systems formed by polymers of various molecular weights at pH 4.6. Solutes: lysozyme, chymotrypsinogen A, bovine serum albumin, catalase. K_o - partition coefficients in the system formed by 12.19% wt. Dex-10 and 8.39% wt. PEG-20000; K_i - partition coefficients in the systems formed by: (1) 8.64% wt. Dex-500 and 5.00% wt. PEG-6000; (2) 9.96% wt. Dex-40 and 5.61% wt. PEG-6000; (3) 10.29% wt. Dex-40 and 7.64% wt. PEG-20000. Calculated from the data reported in [19].

sary just to examine partitioning of a set of 10-12 randomly chosen solutes in the "old" and "new" systems to predict the partition coefficient of a solute in the "new" system (provided it was determined in the "old" systems). That allows one to employ the partition technique (using aqueous Dex-PEG and Dex-Ficoll two-phase systems) for analytical purposes (see below). Theoretical importance of the issue under discussion is equally critical.

Constancy of the coefficient b_i value in Equation 5.10a independent of the chemical nature of solutes partitioned in the aqueous Dex—PEG and Dex—Ficoll two-phase systems (of the same total salt composition) [86] implies either that the solute-phase polymer interactions are identical for all the different solutes examined, or that the solute-polymer interactions do not occur and/or do not affect the solute partitioning in these systems. Since the first assumption seems to be highly unlikely, the second one must be taken into consideration. It is always much harder to prove the lack of interactions than their occurrence,

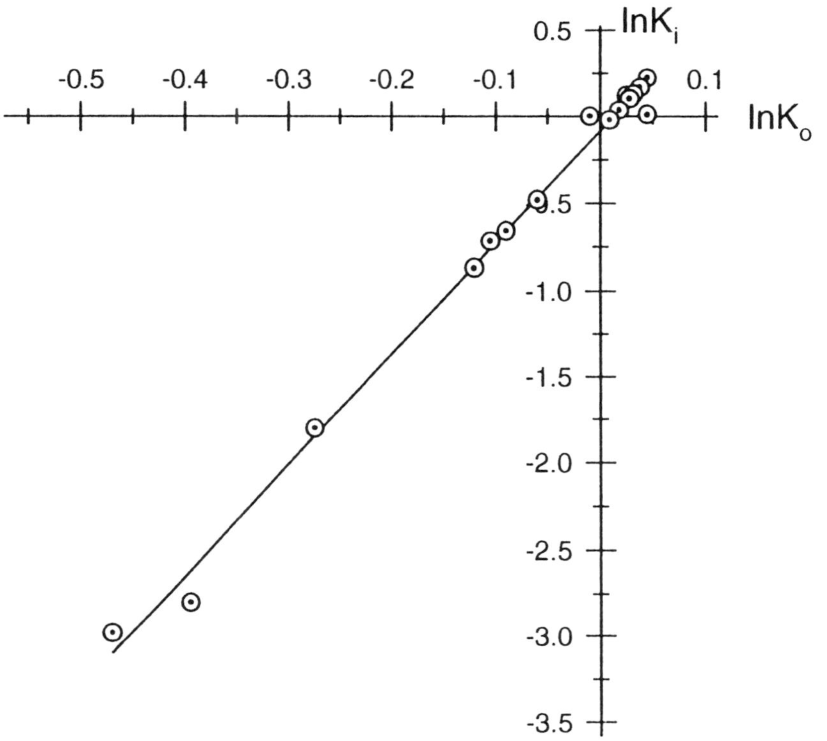

Figure 5.20. Relationship between the partition coefficients of various solutes in aqueous Dex-70 (10% wt.) — Ficoll-400 (11.7% wt.) two-phase system (K_o) and aqueous Dex-70 (9.6% wt.) — PEG-6000 (5.9% wt.) two-phase system (K_i). Both systems contain 0.15 mole/kg NaCl in 0.01 mole/kg sodium phosphate buffer, pH 7.4.

as the absence of the response from a probe or a method in use may be attributed to the ineptitude or low sensitivity of the probe or the method in question. The results presented here, however, together with those discussed above are self-consistent and comply with all the numerous experimental observations accumulated in the field of the solute partitioning in water-organic solvent systems. Therefore these results may be viewed as an unambiguous (though non-direct) evidence of the lack of the interactions in question.

It must be stressed that the conclusion as well as the evidence under discussion is applicable at present only to aqueous Dex—PEG and Dex—Ficoll two-phase systems. There clearly should be some cases when a particular solute will participate in direct interactions with one or both phase polymers in these systems. The above evidence obviously does not cover all solutes and all aqueous two-phase systems, for example, systems including additives of polymer-bound affinity ligands, systems formed by polyelectrolytes, etc. It was found particularly that partitioning of certain proteins in the aqueous Dex—PVP two-phase system does not fit Equation 5.10a (with Dex—Ficoll or Dex—PEG system as the reference) probably due to the direct protein-PVP interactions implied additionally by the slight but noticeable deviation of the protein partition coefficient from Equation 4.10 at relatively high PVP concentrations in the system.

If the above evidence and conclusion about the absence of direct solute-polymer interactions in the aqueous Dex—PEG and Dex—Ficoll two-phase systems for the majority of the solutes are accepted, the information provided by the solute partition behavior in these systems seems rather straightforward. Before discussing this issue, however, we must consider different theoretical considerations on the solute partitioning process in aqueous two-phase systems present in the current literature.

5.6. THEORETICAL TREATMENTS OF THE SOLUTE PARTITIONING

The practical purpose of the theoretical consideration of any process is to predict the process course from the chemical structures and/or physical properties of the components of the process as well as the effects of various variable factors. The ultimate goal is clearly the quantitative prediction but even qualitatively correct one usually is an indication of the reasonably high level of the understanding of the process in question. The theoretical apprehension of the solute partitioning in aqueous two-phase systems in these terms may be viewed as existing in a rudimentary stage if existing at all. The reason is the enormous complexity of the process under consideration.

Theoretical views on pure water and aqueous solutions of small solutes or inorganic salts are in an early stage as yet, let alone those on aqueous solutions of polymers at moderate concentrations or on behavior of biological macromolecules in an aqueous medium. Hence the current theoretical treatments of the solute partitioning in aqueous two-phase systems are actually the attempts to apply various existent theoretical frame-works to the process in question. These attempts are briefly outlined here in comparison with the semi-empirical approaches used in the studies of water-organic solvent systems. Possible promising lines of further research in this area are also discussed.

All theoretical treatments of the solute partitioning in aqueous two-phase systems may be divided into two groups according to the theoretical frame-works they are based on. The first group covers the treatments based on the Flory-Huggins theory, and the second one includes those based on the virial expansion model. The basics of these theoretical approaches were outlined above (see Chapter 2).

Treatments of Solute Partitioning Based On the Flory-Huggins Theory.

Brooks et al.[44] developed the first lattice model for solute partitioning applying the Flory theory of polymer-solvent mixing to aqueous two-phase systems, treating the solute being partitioned as a third polymer component. The expression for the solute partition coefficient was derived from the equality of the solute chemical potentials in the two phases using only first order terms in the polymer concentration differences between the phases [44]. According to this expression the solute partition behavior is governed by the molecular volumes of the components of the system and the Flory interaction parameters χ describing the solute interactions with the solvent and each phase polymer (for details see [44] or reviews [42,98,99]). Brooks et al. [7, 38] showed that the treatment used predicts that the protein partitioning should be more one-sided with increasing protein molecular weight or increasing total polymer concentrations in the system and that the protein partitioning should increase to the phase rich in the polymer with decreased molecular weight. These apparent agreements with the qualitative features of protein partitioning according to Abbott et al. [42] may be fortuitous, however, as the origin of the predicted trends arises from the Flory-Huggins form of entropy of mixing.

A simplified version of the Flory-Huggins expression derived by Brooks et al.[44] was inferred by Diamond and Hsu [8] based on the constancy of the tie-line slope (STL, see above). The linear Equation 4.10 was derived by Diamond and Hsu [8] theoretically with the slope k_i as the parameter depending on the molecular weights of the phase polymers and the solute and on the interactions of the solute with the polymers and the solvent.

One of the approximations inherent in the treatments based on the Flory-Huggins theory is the representation of dense compact macromolecules of globular proteins as diffuse random coiling species. This view lacks a physical basis [44] and according to Baskir et al.[99] and Abbott et al.[42] may lead to unrealistic conclusions.

To avoid this hindrance the lattice model was modified by Baskir et al.[100-102] for a spherical protein interacting with the polymer phase. The protein is represented in the model [100-102] as a rigid, impenetrable and spherical body of known size with homogeneous surface. Additionally to the conventional polymer-solvent interaction parameter, a polymer segment-protein surface interaction parameter was defined as characterizing the free energy change for the displacement of solvent segments from the protein surface by a statistical polymer segment. More detailed review of the model [100-102] and some criticism may be found in [42,98,99].

An approach not based on the lattice model but fitting in this group of theoretical treatments of the solute partitioning considers the protein-polymer interactions in terms of the de Gennes scaling theory of polymer solutions [103]. The approach suggested by Abbott et al.[104-108] is focused on the development of a molecular level understanding of the interactions between flexible phase-forming polymers and globular proteins and the extent to which these interactions influence the (protein) partitioning behavior. The protein is considered by Abbott et al.[104] as a compact colloid, and the phases are divided in those formed by low and high molecular weight polymers. Protein partition behavior in aqueous Dex-PEG two-phase systems is suggested [105] to result from the balance of repulsive steric and weak attractive protein-polymer interactions. The protein shape (conformation) is argued [105] to be a significant factor in its partition behavior.

Two major shortcomings shared by all the theoretical treatments outlined above does not allow one to regard them as helpful for better insight into the process of solute partitioning. First, the solute molecular weight and not its chemical structure is considered as a primary factor. Second, the experimentally established fact that the solvent features of the aqueous media in the two phases are different is ignored. Even in the case when direct phase polymer-solute interactions do take place in the system, the two above factors must be taken into account. Ignoring them obviously simplifies the theoretical consideration but this oversimplification may not be adequate even to a first approximation.

Specific features of aqueous medium in the two phases are also ignored in the theoretical treatments based on the virial expansion approach.

<u>Treatments of Solute Partitioning Based On the Virial Expansion Approach.</u>

The essence of the virial expansion approach was described above. Application of this approach to the solute partitioning is usually based on a generalization of the treatment suggested by Edmond and Ogston [109] to describe phase separation in aqueous polymer mixtures. According to this treatment, the chemical potentials of phase polymers, solvent, and the solute being partitioned are expressed as functions of the components concentrations and second (and higher-order) virial coefficients (see Equation 3.14). If the expansion is cut off at the second-order terms, the coefficients of the terms can be related to measurable quantities such as osmotic pressure [109]. In <u>dilute</u> polymer solutions, the virial coefficients a_{ij} (see Equation 3.14) are viewed as reflecting the pairwise interactions of molecules i and j. The higher-order terms are supposed to represent the simultaneous interactions between three or more macromolecules in the solution. The limitations of this approach were considered in Chapter 3.

This approach is, however, clearly advantageous over the other theoretical treatments as indicated by a satisfactory agreement between the calculated and experimental partition coefficients for several proteins in the aqueous Dex-PEG two-phase systems of varied polymer and salt composition reported by King et al.[110]. Protein partition coefficients were calculated from the experimentally determined interaction parameters and an electrostatic interfacial potential difference measured between the two phases and included into the chemical potentials expressions. The interactions parameters were measured using low-angle laser light scattering in buffered aqueous protein PEG-containing, Dex-containing, and protein free Dex- and PEG-containing solutions at a fixed salt concentration and pH used in the two-phase system [110]. The approach used by King et al.[110] may be viewed as semi-empirical rather than theoretical [98] but the good agreement between the partition coefficients for several proteins determined experimentally and calculated from the separate independent measurements shows it to be fruitful.

An attempt to calculate the partition coefficients of bovine serum albumin in aqueous Dex-PEG-salt two-phase systems from the osmotic second virial coefficients was reported by Haynes et al.[111]. The authors [111] developed a molecular-thermodynamic model based on the constant volume McMillan-Mayer solution theory [112] (the osmotic virial expansion) and Guggenheim's extension [113] of the Debye-Huckel theory to account for the long-range electrostatic ion-ion interactions. Calculated partition coefficients for albumin showed the qualitatively correct trends but disagreed with the experimental values [111]. Haynes et al.[111] concluded that the virial expansion expression might be improved by including higher-order virial coefficients. The model was improved in more recent publications [114,115], and it has been shown [115] that the interfacial potential difference may be calculated from the ratio of the salt additive concentrations in the phases. Restriction of viewing aqueous medium as a dielectric continuum inherent in the model [115] seems to limit its usefulness, however.

Forciniti and Hall [116] reported a theoretical treatment of solute partitioning using the virial expansion derived from the constant pressure Hill solution theory [117]. The expansion reduces to the Edmond-Ogston expression used by King et al.[110] under assumptions of a non-interacting solvent and an incompressible system. Proteins being partitioned were represented in the treatment [116] as impenetrable spheres, and phase polymers (Dex and PEG) as impenetrable spheres, cylinders (PEG), and as flexible coils. The osmotic virial coefficients inferred from these different excluded-volume models were included into the virial expansion expressions for protein partition coefficient and checked against experimentally observed trends for the protein partition coefficients at the isoelectric point as functions of the molecular weights of the protein and phase polymers [116]. It was concluded [116] that attractive protein-polymer interactions must be included into the model additionally to the repulsive excluded volume forces. It should be particularly noted that as pointed out by Forciniti et al.[116] the process of protein partitioning in an aqueous polymer two-phase system may not be treated in terms used for treatment of the polymer (PEG)-induced protein precipitation.

The extensive set of experimental data obtained by Forciniti et al. [10, 17-19] on the partition coefficients of several proteins under different conditions in 16 different aqueous Dex—PEG two-phase systems of varied polymer concentrations with characterized phase compositions was intended [116] to improve theoretical treatment of the solute partitioning using the virial expansion approach. (The experimental data [10,17-19] were discussed above to illustrate certain important general trends in the solute partitioning.) These data, however, did not lead to significant development of the theoretical treatment. The reason seems to be similar to those mentioned above in regard to the treatments based on the Flory-Huggins theory. The different solvent features of aqueous media in the two phases cannot be ignored.

An attempt to include the concept of the solvent structure into consideration of the interactions occurring in the polymer- and salt-containing aqueous medium was undertook by Forciniti and Hall [118]. A strong correlation between non-electrostatic and electrostatic contributions to the free energy of a mixture was found [118] in complete agreement with the experimental observations discussed in Chapter 4.

An attempt to take into account specific features of water as a solvent in aqueous two-phase polymer systems was made by van Oss et al.[119] using the surface thermodynamics principles. The approach used by van Oss et al. [119] was considered above (see Chapter 3), and the model suggested is too underdeveloped as yet to discuss it here. It is possible that further developments may lead to better understanding of the solute partitioning. The presentation of all different kinds of biological solutes as "monopolar Lewis bases (electron donors)" or "monopolar Lewis acids (electron acceptors)" used in the current model [119] seems to be too indiscriminating to provide accurate picture of different partition behavior of various biological solutes.

It seems possible to conclude that current theoretical treatments fail to provide self-consistent physical model of solute partitioning in aqueous two-phase systems. These treatments do not allow one not only to predict partition behavior of a given solute but even to get a better insight into the properties of the phases and phase polymers governing the solute partitioning. The limitations of the current theoretical treatments of aqueous solutions seem to be "amplified" when applied to aqueous polymer two-phase systems even though these treatments may be successful when applied to such complex problems as polymer-induced protein precipitation, membrane fusion, etc.

It should be pointed out that no successful theory of much less complicated process of solute partitioning in water-organic solvent systems currently exists, to my knowledge. The efforts in exploring solute partitioning in water-organic solvent systems tremendously exceed those in aqueous two-phase systems in terms of time, money, and pure numbers of research groups dealing with the problem, solutes and systems being investigated, etc. The structure of organic solutes being examined in the solvent systems is usually much simpler than that of biological solutes analyzed in aqueous two-phase systems. The information accumulated in the solvent systems exceeds that in the aqueous polymer systems many times. In spite of that no substantial quantitative theory of solute partitioning in different solvent systems exists, and the most successful approach currently in use may be considered as merely semi-empirical.

The semi-empirical approach used for analysis of solute partition in different solvent systems is briefly outlined below. This approach may indicate the promising directions of further investigations of solute partition in aqueous two-phase systems.

Semi-Empirical Approach Used for Treatments of Solute Partitioning in Water-Organic Solvent Systems.

It was shown by Kamlet et al.(see in [120] and references cited therein) that partition of solutes in different solvent systems as well as their solubilities and other properties related to the solute-solvent interactions are well correlated by equations (similar to Equations 1.9a and 1.9b, see in Chapter 1) including linear combinations of dependencies on up to five solute parameters. These equations include a cavity term, depending on the solute volume, dipolarity/polarizability term accounting for the dipole-dipole and dipole-induced dipole solute-solvent interactions, and hydrogen bonding terms, pertaining to the hydrogen bond donation and acceptance properties of the solutes and solvents. It was shown [120] particularly that partition coefficients of a large set of different nonionic solutes (including non-hydrogen bonding, hydrogen bond acceptors, and weak and strong hydrogen bond donors) in the water—octanol system are described as:

$$\log P = (0.32 \pm 0.04) + (5.35 \pm 0.05)\cdot V_1/100 - (1.04 \pm 0.04)\cdot\pi^* + (0.35 \pm 0.03)\cdot\delta - (3.84 \pm 0.05)\cdot\beta_m + (0.10 \pm 0.04)\cdot\alpha_m \quad (5.15)$$

$$N = 245;\ r^2 = 0.9959;\ s = 0.131$$

where P is the solute partition coefficient in the water-octanol two-phase system; V_1 is the intrinsic volume of the solute; π^* is the solute solvatochromic dipolarity parameter (see in chapter 1); δ is a polarizability correction; β_m characterizes the solute hydrogen bond accepting ability; and α_m the solute hydrogen bond donation ability; N is the number of solutes in correlation; r^2 is the correlation coefficient; s the standard deviation of the fit.

Marcus [96] applied the expression similar to Equation 5.15 to partitioning of different monofunctional aliphatic and mono- and bifunctional aromatic solutes in 25 different water-"dry" organic solvent systems. The term "dry" was used [96] to cover the water-saturated solvents that have a sufficiently low water content (mole fraction of water less than 0.13) and properties essentially the same as those of neat solvents. It was established [96] that the solute partition coefficient in any of the 25 water-"dry" solvent systems examined is described as:

$$\log P = A_V \cdot V_2 \cdot (\Delta\delta_{HI})^2 + A(\alpha) \cdot \alpha_{m2} \cdot \Delta\beta_1 + A(\beta) \cdot \beta_{m2} \cdot \Delta\alpha_1 \qquad (5.16)$$

where V_2 is the solute intrinsic volume; α_{m2} and β_{m2} are the characteristics of the solute hydrogen bond donation and accepting abilities, respectively; $(\delta_{HI})^2$ is the solvent cohesive energy density (square of the Hildebrand solubility parameter); β_1 and α_1 characterize the solvent hydrogen bond accepting and donation properties, respectively; Δ signifies the difference between the property of the solvent and that of water; A_V, $A(\alpha)$, and $A(\beta)$ are the coefficients.

Essentially the same coefficients A_V, $A(\alpha)$, and $A(\beta)$ values were found [96] for all the solvent systems examined, including aliphatic and aromatic hydrocarbons, halogenated hydrocarbons, ethers, and esters. Equation 5.16 was rationalized in terms of the solute-solvent interactions in the two phases, rewritten as [96]:

$$\Delta G^o_{tr} = -RT \cdot \ln P = \text{cavity term} - 43.8 \cdot \alpha_{m2} \cdot \Delta\beta_1 - 26.4 \cdot \beta_{m2} \cdot \Delta\alpha_1 \qquad (5.17)$$

and used to analyze the process of transfer of ionic solutes from water to organic solvents (mostly water-miscible but "dry" in the above sense) [121]. The analysis [96] provided the "pseudosolvatochromic" characteristics of the cations' acidity α_+ (the ability of cations to attract electron pair donation from basic solvents) and anions' basicity β_- (the ability of anions to donate an electron pair to an acidic solvent). These characteristics of ions allow one to scale the ionic solute-solvent interactions together with those of nonionic solutes [96].

Solvatochromic characteristics are presently established for a vast variety of organic compounds including water-soluble ones. Hence it should be possible to study partitioning of the solutes with known dipolarity/polarizability and hydrogen bond donation/accepting properties in aqueous polymer two-phase systems. Using the approach outlined above it may be possible to obtain the characteristics (used to quantify the properties of organic solvents) for the aqueous media in the phases of the polymer two-phase systems. The first step might be the application of the approach to aqueous polymer solution-octanol (or other organic solvent) two-phase systems with varied polymer type and concentration. That may lead to the better insight into the polymer-induced changes of the solvent features of aqueous medium. The second step would be applying the approach to aqueous polymer two-phase systems. It is hard to predict how successful the outcome of the research along these lines may be, but it seems highly likely that plenty of new information important for understanding of the peculiarities of solute-solvent interactions in aqueous medium would be obtained using this approach. This is merely one of the many possible approaches to study the mechanisms behind the solute partitioning in aqueous two-phase systems originating from the similarities between aqueous polymer and water-organic solvent two-phase systems. Other more successful approaches may be developed. In any case, I believe that pursuing different new experimental and semi-empirical approaches may presently be much more fruitful than trying to fit aqueous two-phase systems into theoretical frameworks developed for systems of totally different types.

The indicated above similarity between the fundamental features of aqueous polymer and water-organic solvent systems and the solute partitioning in these systems allows one to consider the information provided by the solute partition behavior in a given aqueous polymer two-phase system.

5.7. WHAT INFORMATION IS PROVIDED BY THE SOLUTE PARTITION BEHAVIOR IN AN AQUEOUS TWO-PHASE SYSTEM ?

Summing up the above experimental observations, the following features of aqueous two-phase systems and solute partition behavior in these systems should be noticed:

1. Certain aqueous polymer two-phase systems, for example, Dex—PEG and Dex—Ficoll systems, can be viewed as systems formed by two mutually immiscible, though water-like, solvents. Partitioning of a solute between the two phases of such a system can be regarded as transfer of the solute from the aqueous medium with one set of properties into the aqueous medium with a different set a properties. All the following aspects are related to the systems of this category.

2. The two phases of a given aqueous polymer system comprise two aqueous media of different chemical composition.

3. The different chemical composition of the two phases causes the difference in the solvent features of the aqueous media in the phases.

4. The difference between the solvent features of the media of the two phases consists in the different structure and/or thermodynamic state of water in the phases. That creates the difference between the free energies of formation of a cavity to accommodate a solute in the phases and the difference between the capabilities of water in the phases to participate in van der Waals interac-

tions, hydrogen bonding, electrostatic ion-dipole interactions, etc., with the solute being partitioned.

5. Effects of phase polymers (type, molecular weight, concentration), low molecular weight additives (type and concentration), pH, temperature, etc., on the solute partition behavior are realized through the influence of these variables on the solute-solvent interactions either via their effects on the solvent features of the aqueous media in the two phases, or on the solute capability to participate in some of these interactions, for example, by inducing changes in the solute conformation, ionization degree, association, etc.

6. The primary factor governing the solute partition behavior in an aqueous polymer two-phase systems is the difference between the solute-solvent interactions in the two phases. This difference in the system of fixed composition depends on the type, spacial arrangement, and number of the solvent-accessible groups in the solute structure. The solute molecular size may influence the solute partition behavior due to increased amount of the solvent-accessible groups in comparison to the solute of smaller size.

It should be emphasized that the above considerations do not cover the affinity partition technique (see below) or certain cases in which a given solute may interact directly with one or both phase polymers. Even in these systems, however, the above considerations should be taken into account, as the difference between the solvent features of the media in the two phases is likely to influence the solute-polymer interactions.

The small difference between the properties of the phases in aqueous polymer two-phase systems seems to explain the ability of the systems to separate closely related biological molecules. This effect is being used to advantage when the systems are employed for separating biological materials (see, for example, in [1, 35,98]). The same high sensitivity of the method of partitioning in aqueous polymer two-phase systems can be exploited for the analysis of biological materials.

The analytical information provided by the solute partition behavior in an aqueous Dex-PEG or Dex-Ficoll two-phase system is clearly related to the difference between the solute-solvent interactions in the two phases, i.e., solute interactions with aqueous media of different chemical composition. The solute-solvent interactions are known to be highly specific for a given solute structure.

The partition coefficient for a given solute in a water—organic solvent system, for example, water—octanol system, is a constant feature of the solute [61,62,67] similar to other physico-chemical constants such as the specific absorption coefficient, critical micelle concentration, chromatographic retention index, etc. The partition coefficient for a given solute in an aqueous two-phase system of fixed composition was numerously shown (see, for example, in [69]) to be a highly sensitive characteristic of the solute. Typically, albumin samples from different manufacturers that contain traces of different contaminants but are indistinguishable by standard analytical methods (for example, HPLC and electrophoresis) are easily differentiated by using the partition technique [30]. Many other examples including the difference between the partition coefficients values for reversed dipeptides [78], isomers of dinucleosidephosphates [33], etc., were presented above and more examples will be given below.

The main conclusion (to be substantiated in more detail below) is [69] that the partition coefficient K for an individual biological solute can be used as a simple, highly sensitive, and cost-effective relative measure of solute identity and/or purity if the K value for a standard reference solute is known. The K value is similar to the melting point, which is widely used as a simple test of the purity of a synthetic product.

Finally, the solute partition coefficient provides unique quantitative information about a given biological solute in regard to its interactions with aqueous medium. The solute partition coefficient in an aqueous Dex—PEG or Dex—Ficoll two-phase system may by used as a measure of the solute's relative hydrophobicity. Theoretical considerations and experimental data substantiating this assertion are discussed in the next part of the book together with the role of the solute hydrophobicity in different biological processes.

REFERENCES:

1. P.Å. Albertsson, Partition of Cell Particles and Macromolecules, 3rd. ed., Wiley, New York, 1986.
2. S. Bamberger, D.E. Brooks, K.A. Sharp, J.M. Van Alstine, T.J. Webber, In: Partitioning in Aqueous Two-Phase Systems: Theory, Methods, Uses, and Applications to Biotechnology (H. Walter, D.E. Brooks, D. Fisher, eds.), Academic Press, Orlando, Fla, 1985, pp.85-130.
3. Separations Using Aqueous Phase Systems: Applications in Cell Biology and Biotechnology (Eds., D. Fisher and I. A. Sutherland), Plenum Press, New York, 1989, pp.233-238.
4. B.Y. Zaslavsky, A.A. Borovskaya, N.D.Gulaeva, L.M. Miheeva, J. Chem. Soc. Faraday Trans., 87, 141 (1991).
5. F. Tjerneld, S. Berner, A. Cojarville, G. Johansson, Enzyme Microb. Technol., 8, 417 (1986).
6. F. Tjerneld, G. Johansson, Bioseparation, 1, 255 (1990).
7. P. Å. Albertsson, A. Cajarville, D. E. Brooks, F. Tjerneld, Biochim. Biophys. Acta, 926, 87 (1987).

8. A. D. Diamond, J. T. Hsu, Biotechnol.Bioeng., 34, 1000 (1989).
9. A. D. Diamond, J. T. Hsu, J.Chromatogr., 513, 137 (1990).
10. D. Forciniti, C. K. Hall, M. R. Kula, Biotechnol.Bioeng., 38, 986 (1991).
11. B. Y. Zaslavsky, L. M. Miheeva, G. Z. Gasanova, A. U. Mahmudov, J.Chromatogr., 403, 123 (1987).
12. M. M. Bradford, Anal. Biochem., 72, 248 (1976).
13. S. Udenfriend, S. Stein, P. Bohlen, W. Dairman, Science, 178, 871 (1972).
14. L. M. Miheeva, B. Y. Zaslavsky, S. V. Rogozhin, Bioorg.Khim. (Rus), 2, 1245 (1976).
15. L. M. Miheeva, E. D. Maximova, Y. P. Aleschko-Ozhevskii, B. Y. Zaslavsky, J.Solution Chem., 20, 607 (1991).
16. H. Walter, J. L. Anderson, FEBS Lett., 131, 73 (1981).
17. D. Forciniti, C. K. Hall, M. R. Kula, J.Biotechnol., 16, 279 (1990).
18. D. Forciniti, C. K. Hall, M. R. Kula, Fluid Phase Equilibria, 61, 243 (1991).
19. D. Forciniti, C. K. Hall, M. R. Kula, Chem.Eng.Science, 47, 165 (1992).
20. M. Potschka, J.Chromatogr., 441, 239 (1988).
21. B. Y. Zaslavsky, L. M. Miheeva, G. Z. Gasanova, A. U. Mahmudov, J. Chromatogr., 392, 95 (1987).
22. B. Y. Zaslavsky, A. A. Borovskaya, N. D. Gulaeva, L. M. Miheeva, J.Chem. Soc.Faraday Trans., 87, 137 (1991).
23. G. Johansson, M. Joelsson, J.Chromatogr., 464, 49 (1989).
24. S. J. Gluck, M. P. Wingeier, J.Chromatogr., 547, 69 (1991).
25. W. D. Conway, Countercurrent Chromatography: Apparatus, Theory, and Applications, VCH Publishers, New York, 1990.
26. B. Y. Zaslavsky, L. M. Miheeva, 1990, unpublished data.
27. B. Y. Zaslavsky, L. M. Miheeva, N. M. Mestechkina, V. Pogorelov, S. V. Rogozhin, FEBS Letts., 94, 77 (1978).
28. B. Y. Zaslavsky, L. M. Miheeva, N. M. Mestechkina, S. V. Rogozhin, J.Chromatogr., 253, 139 (1982).
29. B. Y. Zaslavsky, N. M. Mestechkina, L. M. Miheeva, S. V. Rogozhin, J.Chromatogr., 256, 49 (1983).
30. B. Y. Zaslavsky, N. M. Mestechkina, S. V. Rogozhin, J.Chromatogr., 260, 329 (1983).
31. B. Y. Zaslavsky, N. M. Mestechkina, S. V. Rogozhin, Biochim. Biophys. Acta, 579, 463 (1979).
32. B. Y. Zaslavsky, N. M. Mestechkina, S. V. Rogozhin, J.Chromatogr., 240, 21 (1982).

33. L. G. Shchyukina, B. Y. Zaslavsky, S. V. Rogozhin, V. L. Florentiev, Mol.Biologia (Rus.), 18, 1128 (1984).
34. W. Muller, Bioseparation, 1, 265 (1990).
35. G. Johansson, In: Partitioning in Aqueous Two-Phase System: Theory, Methods, Uses, and Applications to Biotechnology (H. Walter, D. E. Brooks, D. Fisher, eds.), Academic Press, Orlando, Fla, 1985, pp.161-226.
36. M. A. Eiteman, J. L. Gainer, Bioseparation, 2, 31 (1991).
37. M. A. Eiteman, J. L. Gainer, Chem. Eng. Commun., 105, 17 (1991).
38. M. A. Eiteman, J. L. Gainer, J.Chromatogr., 586, 341 (1991)
39. M. A. Eiteman, Sep. Sci. Technol., 29, 685 (1994).
40. M. A. Eiteman, J. Chromatogr., 668, 21 (1994).
41. L. M. Cohen, M. A. Eiteman, J. L. Gainer, Sep. Sci. Technol., 1994, in press.
42. N. L. Abbott, D. Blankschtein, T. A. Hatton, Bioseparation, 1, 191 (1990).
43. G. Johansson, J.Chromatogr., 322, 425 (1985).
44. D. E. Brooks, K. A. Sharp, D. Fisher, In: Partitioning in Aqueous Two-Phase Systems: Theory, Methods, Uses, and Applications to Biotechnology (H. Walter, D. E. Brooks, D. Fisher, eds.), Academic Press, Orlando, Fla, 1985, pp.11-84.
45. P. Å. Albertsson, S. Sasakawa, H. Walter, Nature, 228, 1329 (1970).
46. H. Walter, P. Å. Albertsson, S. Sasakawa, Biochemistry, 11, 3880 (1972).
47. S. Sasakawa, H. Walter, Biochemistry, 11, 2760 (1972).
48. B. Y. Zaslavsky, L. M. Miheeva, N. M. Mestechkina, 1989, unpublished data.
49. P. G. Righetti, T. Caravaggio, J.Chromatogr., 127, 1 (1976).
50. P. H. Wang, E. J. Lien, J.Pharm.Sci., 69, 662 (1980).
51. A. D. Diamond, J. T. Hsu, AIChE J., 36, 1017 (1990).
52. W. Muller, In: Partitioning in Aqueous Two-Phase Systems: Theory, Methods, Uses, and Applications to Biotechnology (H. Walter, D. E. Brooks, D. Fisher, eds.), Academic Press, Orlando, Fla, 1985, pp.227-266.
53. B. Y. Zaslavsky, A. V. Baevskii, S. V. Rogozhin, A. V. Gedrovich, A. V. Shishkov, A. A. Gasanov, A. A. Masimov, J.Chromatogr., 285, 63 (1984).
54. B. Y. Zaslavsky, L. M. Miheeva, S. V. Rogozhin, Y. A. Davidovich, A. V. Gedrovich, A. V. Shishkov, A. A. Gasanov, A. A. Masimov, J.Chromatogr., 291, 203 (1984).

55. N. K. Shcherbukhina, B. Y. Zaslavsky, V. D. Shcherbukhin, D. P. Hasaev, A. A. Masimov, Appl.Biochem.Microbiol. (Rus.), 18, 548 (1982).
56. J. C. McGowan, A. Mellors, Molecular Volumes in Chemistry and Biology: Applications including Partitioning and Toxicity, Ellis Horwood, Chichester, 1986.
57. K. A. Sharp, A. Nicholls, R. F. Fine, B. Honig, Science, 252, 106 (1991).
58. A. Ben-Naim, K.-L. Ting, R. L. Jernigan, Biopolymers, 28, 1309 (1989).
59. A. Ben-Naim, K.-L. Ting, R. L. Jernigan, Biopolymers, 28, 1327 (1989).
60. S. S. Davis, T. Higuchi, J. H. Rytting, Adv.Pharm.Sci. (H. S. Bean, ed.) Acad.Press, 73 (1974).
61. C. Hansch, A. Leo, Substituent Constants for Correlation Analysis in Chemistry and Biology, New York, Academic Press, 1979.
62. Y. C. Martin, Quantitative Drug Design: A Critical Introduction, Marcel Dekker, New York, 1978.
63. J. T. Chou, P. C. Jurs, In: Physical Chemical Properties of Drugs, Eds. S. H. Yalkowsky, A. A. Sinkula, S. C. Valvani, Marcel Dekker, New York, 1987, pp.163-199.
64. T. Fujita, I. Iwasa, C. Hansch, J.Am.Chem.Soc., 86, 5175 (1964).
65. G. G. Nys, R. F. Rekker, Chimica Therapeutic., 9, 521 (1973).
66. G. G. Nys, R. F. Rekker, Chimica Therapeutic., 10, 361 (1974).
67. R. F. Rekker, The Hydrophobic Fragmental Constant: Its Derivation and Application. A Means of Characterizing Membrane Systems, Amsterdam, Elsevier, 1977.
68. M. Akamatsu, T. Fujita, J.Pharm.Sci., 81, 164 (1992).
69. B. Y. Zaslavsky, Anal.Chem., 64, 765A (1992).
70. R. D. Poretz, I. J. Goldstein, Arch.Biochem.Biophys., 125, 103 (1968).
71. B. Y. Zaslavsky, N. M. Mestechkina, L. M. Miheeva, S. V. Rogozhin, G. Y. Bakalkin, G. G. Rjazhsky, E. V. Chetverina, A. A. Asmuko, J. D. Bespalova, N. V. Korobov, O. N. Chichenkov, Biochem. Pharmacol., 31, 3757 (1982).
72. B. Y. Zaslavsky, L. G. Shchyukina, S. V. Rogozhin, Mol. Biologia (Rus), 15, 1315 (1981).
73. J. F. Tinker, G. B. Brown, J.Biol.Chem., 173, 585 (1948).
74. Z. E. Rassi, C. Horvath, In: Practical Aspects of Modern HPLC (I. Molnar, ed.) Walter de Gruyter, New York, 1983, pp.1-14.
75. J. Jacobson, Z. E. Rassi, C. Horvath, In: Practical Aspects of Modern HPLC (I. Molnar, ed.) Walter de Gruyter, New York, 1983, pp.15-23.
76. S. Sasakawa, H. Walter, Biochemistry, 13, 29 (1972).

77. A. D. Diamond, X. Lei, J. T. Hsu, Biotechnol.Techniques, 3, 271 (1989).
78. M. A. Eiteman, J. L. Gainer, Bioechnol.Prog., 6, 479 (1990).
79. J. Draffehn, B. Schonecker, K. Ponsold, J.Chromatogr., 216, (1981).
80. Y. C. Martin, J.Med.Chem., 24, 229 (1981).
81. C. T. Chiou, J. H. Block, M. Manes, J.Pharm.Sci., 71, 1307 (1982).
82. S. O. Enfors, K. Kohler, A. Veide, Bioseparation, 1, 305 (1990).
83. K. Kohler, C. Ljungquist, A. Kondo, B. Nilsson, A. Veide, Proceeding of 7th Internatiuonal Conference on Partitioning in Aqueous Two-Phase Systems, New Orleans, 1991, Abst. #2.
84. J. Luther, C. E. Glatz, Proceeding of 7th Internatiuonal Conference on Partitioning in Aqueous Two-Phase Systems, New Orleans, 1991, Abst. #15.
85. V. V. Mozhaev, V. A. Siksnis, N. S. Melik-Nubarov, N. Z. Galkantaite,G. J. Denis, E. P. Butkus, B. Y. Zaslavsky, N. M. Mestechkina, K. Martinek, Eur.J.Biochem., 173, 147 (1988).
86. T. Bagirov, Ph.D. thesis, Azerbaidzhan State University, Baku, Azerbaidzhan, 1990.
87. W. Muller, Liquid-Liquid Partition Chromatography of Biopolymers, GIT Verlag, Darmstadt, 1988.
88. H. Walter, 1980, personal communication.
89. R. Collander, Acta Chem.Scand., 5, 774 (1951).
90. A. Leo, C. Hansch, D. Elkins, Chem.Rev., 71, 525 (1971).
91. A. Leo, C. Hansch, J.Org.Chem., 36, 1539 (1971).
92. C. Hansch, W. J. Dunn III, J.Pharm.Sci., 61, 1 (1972).
93. B. Y. Zaslavsky, L. M. Miheeva, S. V. Rogozhin, J.Chromatogr., 216, 103 (1981).
94. B. Y. Zaslavsky, L. M. Miheeva, N. M. Mestechkina, L. G. Shchyukina, M. A. Chlenov, L. I. Kudrjashov, S. V. Rogozhin, J.Chromatogr., 202, 63 (1980).
95. C. Yamagami, N. Takao, T. Fujita, J.Pharm.Sci., 82, 155 (1993).
96. Y. Marcus, J.Phys.Chem., 95, 8886 (1991).
97. N. E. Tayar, R. S. Tsai, B. Testa, P. A. Carrupt, A. Leo,J. Pharm. Sci., 80, 590 (1991).
98. H. Walter, G. Johansson, D. E. Brooks, Anal.Biochem., 197, 1 (1991).
99. J. N. Baskir, T. A. Hatton, U. W. Suter, Biotechnol.Bioeng., 34, 541 (1989).
100. J. N. Baskir, T. A. Hatton, U. W. Suter, Macromolecules, 20, 1300 (1987).
101. J. N. Baskir, T. A. Hatton, U. W. Suter, J.Phys.Chem., 93, 969 (1989).
102. J. N. Baskir, T. A. Hatton, U. W. Suter, J.Phys.Chem., 93, 2111 (1989).

103. P. G. de Gennes, Scaling Concepts in Polymer Physics, Cornell University Press, Ithaca, New York, 1979.
104. N. L. Abbott, D. Blankschtein, T. A. Hatton, Macromolecules, 24, 4334 (1991).
105. N. L. Abbott, D. Blankschtein, T. A. Hatton, Macromolecules, 25, 3917 (1992).
106. N. L. Abbott, D. Blankschtein, T. A. Hatton, Macromolecules, 25, 3932 (1992).
107. N. L. Abbott, D. Blankschtein, T. A. Hatton, Macromolecules, 25, 5192 (1992).
108. N. L. Abbott, D. Blankschtein, T. A. Hatton, In: Colloid-Polymer Interactions (P. L. Dubin, P. Tong, eds.), ACS Symposium Series, American Chemical Society, Washington, D.C., 1992.
109. E. Edmond, A. G. Ogston, Biochem.J., 109, 4, 569 (1968).
110. R. S. King, H. W. Blanch, J. M. Prausnitz, AIChE J., 34, 1585 (1988).
111. C. A. Haynes, H. W. Blanch, J. M. Prausnitz, Fluid Phase Equilibria, 53, 463 (1989).
112. W. G. McMillan, Jr., J. E. Mayer, J.Chem.Phys., 13, 276 (1945).
113. E. A. Guggenheim, Trans. Faraday Soc., 62, 3446 (1966).
114. C. A. Haynes, J. Carson, H. W. Blanch, J. M. Prausnitz, AIChE J., 37, 1401 (1991).
115. C. A. Haynes, F. J. Benitez, H. W. Blanch, J. M. Prausnitz, AIChE J., 39, 1539 (1993).
116. D. Forciniti, C. K. Hall, ACS Symp. Ser., 419, 53 (1990).
117. T. L. Hill, J.Chem.Phys., 30, 93 (1959).
118. D. Forciniti, C. K. Hall, J.Chem.Phys., 1994, in press.
119. C. J. van Oss, M. K. Chaudhury, R. J. Good, Sep.Sci.Technol. 22, 1515 (1987).
120. M. J. Kamlet, R. M. Doherty, M. H. Abraham, Y. Marcus, R. W. Taft, J.Phys.Chem., 92, 5244 (1988).
121. Y. Marcus, M. J. Kamlet, R. W. Taft, J.Phys.Chem., 92, 3613 (1988).

PART 3.

ANALYTICAL APPLICATIONS OF THE PARTITION TECHNIQUE

There are a number of fundamental similarities between solute partitioning in water-organic solvent and in aqueous polymer two-phase systems. The most important one is that solute partition behavior in these systems is governed by the difference between the solute-solvent interactions in the two phases suggesting analytical applications of the technique of partitioning in aqueous two-phase systems.

It was mentioned above that the water-organic solvent partitioning serves as a basis for extraction and counter-current chromatography procedures. It is also widely used for estimating solute hydrophobicity. This characteristic of chemical compounds is critically important for analysis of quantitative structure-activity relationships (QSARs) in drug design, pharmacology, toxicology, biochemistry, etc. Specific features of biological solutes, especially macromolecules, do not permit the use of water-organic solvent systems for assessing the hydrophobicity of biomolecules. Since aqueous two-phase systems are compatible with biological molecules, these systems may be useful for the analysis of the hydrophobicity of biological solutes.

To discuss this application of the aqueous two-phase partition technique and its advantages and limitations, it is necessary, first, to consider the physical meaning of the characteristic in question, what it is needed for, and how it is commonly measured. These issues are discussed in Chapter 6 together with the experimental data supporting the suggestion to use the technique for hydrophobicity measurements. Particular examples of these measurements for a variety of biological and synthetic solutes are discussed in Chapter 7. Additional applications for analysis of individual biopolymers and their multicomponent mixtures, based on the unique quantitative information provided by the technique, are considered and illustrated in Chapter 8. Finally, the use of the concepts developed here to applications of the partition technique for separation of biological and inorganic materials are considered in Chapter 9.

CHAPTER 6.

HYDROPHOBICITY OF BIOLOGICAL SOLUTES: HOW TO MEASURE IT AND ITS APPLICATIONS

When solubility, extraction, chromatographic behavior, and other properties of a solute related to the solute-solvent interactions are considered, the term hydrophobicity is commonly used. The same term is usually encountered in the discussions of pharmacological, toxicological, and other biological effects of chemical compounds. Various properties and biological functions of biopolymers, e.g., stability against denaturating agents, interactions with ligands, receptors, biological membranes, etc., are also often discussed in terms of the hydrophobic properties of biopolymers [1-3]. The definition of the term, however, is often neglected and that may lead to misinterpretation of experimental facts and incorrect conclusions. Therefore it is necessary, first, to define the terms and discuss the basic concepts and methods used to estimate the hydrophobicity of solutes.

Advantages and limitations of the methods commonly used for estimating the hydrophobicity of solutes are discussed below from both theoretical and practical viewpoints. That means that the principles of quantitative structure-activity relationship (QSAR) studies as applied to drug design, toxicology, etc., are briefly outlined, and the unresolved questions relevant to the methods of the hydrophobicity measurements are commented on.

Finally, the above assertion that the solute partition coefficient in an aqueous Dex-PEG or Dex-Ficoll two-phase system may be used as a measure of the solute's relative hydrophobicity is discussed in detail, the experimental evidence is presented, and advantages and limitations of the technique are considered.

6.1. MAIN CONCEPTS AND DEFINITIONS

The energy of a solute present in a solvent environment may be described as a sum of two distinct terms:

$$E = E_g + E_s \tag{6.1}$$

where E_g is the energy of the inherent molecular motions of an isolated solute molecule which may be calculated by certain quantum chemistry methods; E_s is the energy of solvation which in turn may be presented as a sum of a number of contributions:

$$E_s = E_e + E_r + E_p + E_{VdW} + E_{cav} \tag{6.2}$$

where E_e is the energy of electrostatic interactions between the solvent and solute; E_r the energy of repulsion; E_p the polarization energy; E_{VdW} accounts for the energy of van der Walls interactions between the solute and solvent; E_{cav} expresses the energy of formation of a cavity in the solvent to accommodate the solute molecule.

It should be mentioned that the above resolution of the E_s parameter according to Equation 6.2 is rather arbitrary; the other formulations can also be used [4]. For example, E_r and E_p in some cases are included into the E_{VdW} term, in other cases E_p is combined with E_e, and so forth. The most generally used form of Equation 6.2 seems to be the following simplified one:

$$E_s + E_e + E_{VdW} + E_{cav} \tag{6.3}$$

According to the current conceptions the E_s value is a measure of the lyophilic or lyophobic character of the solute, i.e., an index of the intensity of the solute-solvent interactions. Hydrophobicity and hydrophilicity of a solute are reflections of the lyophobic and lyophilic character of the solute.

The hydrophilicity and hydrophobicity are the measures of the intensity of molecular interactions of a solute or the surface of a solid phase with water in dispersed systems, the dispersing medium in which is water [5].

Hydrophilicity (as the lyophilicity in general) is specified by the value of the free energy of hydration (solvation) of a given compound or a solid phase surface. Hydrophobicity should be regarded as a small extent of hydrophilicity, since all substances possess the latter property to a ceratin degree [5]. Even the most hydrophobic pure hydrocarbon surface of paraffin absorbs water, i.e., it is hydrophobic only in the sense of being of the very slightly hydrophilic character. The concept of hydrophilicity and hydrophobicity is applicable not only to the solid phases, for which it is the property of a surface, but also to single molecules, their fragments, atoms, and ions. Electrostatically charged and polar groups having a dipole moment are usually hydrophilic. These groups increase the aqueous solubility of the molecules possessing such groups, whereas the nonpolar hydrophobic fragments incorporated in the molecules decrease their solubility in water. Thus, solubility of a compound in water and nonpolar organic solvents is an overall result of the interactions of hydrophobic and hydrophilic groups of the compound molecule with a given solvent environment.

It should be emphasized that the value of the E_s term in Equation 6.1 may be calculated only within the framework of models based on the approximations of the classic or quantum mechanics; the E_s value cannot be determined experimentally.

Therefore, in order to estimate the hydration energy of a given solute experimentally, the free energy change for transferring the solute molecule from the pure solute phase to water, from the gas phase into water, or from one solvent to another one (see, e.g., in [6]) is examined.

As the result of analysis of thermodynamic characteristics of the above types of transfer, some simplistic definitions of hydrophobicity and hydrophilicity have appeared in the literature. For example, the term hydrophobic is often used for the compounds which are readily soluble in many nonpolar organic solvents and only sparingly soluble in water [7]. Martin [2, p.12] defined hydrophobicity as the physical property of the molecule which governs its partitioning into a nonaqueous solvent. According to Tanford [7], the hydrophobicity of a solute is represented by the free energy of transfer of the solute between water and nonpolar organic solvent. The sign of the corresponding value of the free energy change of transfer is indicative of the hydrophobic or hydrophilic character of the solute under study.

There are two problems with regard to the above simplistic definitions. The most obvious one is the implication of the total neutrality of an organic solvent towards the solute being examined. This implication is clearly incorrect but may be used as a first approximation (see below). The other implication is that the hydrophobicity is an intrinsic property of a solute similar to, for example, molecular weight, melting point, etc. This implication is totally incorrect.

It follows from Equation 6.2 that the energy of solvation (hydration) E_s should depend, on the one hand, upon the properties of the solute and, on the other hand, upon the properties of the solvent (water). This obvious fact is for no apparent reason often ignored in the literature on the hydrophobic-hydrophilic properties of chemical compounds. In some cases it leads to an inadequate interpretation of experimental facts.

It was shown above that the water structure-perturbing influence of additives may affect the energy of formation of a cavity in the solvent to accommodate a solute molecule, i.e., the value of the term E_{cav} in Equation 6.2. Melander and Horvath [8] examined particularly the relationship between the effect of inorganic salts on the surface tension of the aqueous solution and the free energy change for formation of a cavity of a given size in the solution. Experimental results obtained by Zaslavsky and Masimov et al.[9-11] on the effect of macromolecules on the relative affinity of their aqueous solutions for a CH_2 group (see in Chapter 2) imply that the energy of formation of a cavity and the ability of water to participate in van der Waals interactions with a solute, i.e., the values of terms E_{cav} and E_{VdW} in Equation 6.2, depend upon the chemical composition of an aqueous solution. The contribution of the energy of electrostatic solute-solvent interactions E_e in Equation 6.2 to the overall hydration energy E_s depends upon the presence of electrolyte additives [8] and upon the thermodynamic state of water dipoles [12] which may be affected by the solution components modifying the structure and/or state of water in the solution.

It follows, therefore, that the hydration energy E_s (or the hydrophobicity or hydrophilicity, which is one and the same) of a solute depends not only on the chemical nature and structure of the solute molecule, but also on the structure and/or thermodynamic state of water in a given aqueous medium, the latter being governed at a given temperature and pressure by the chemical composition of the medium.

6.2. METHODS OF ANALYSIS OF THE RELATIVE HYDROPHOBICITY OF CHEMICAL COMPOUNDS

As mentioned above, based on the simplistic definitions of hydrophobicity (or hydrophilicity) of chemical compounds [7], the hydrophobic character of a solute may be measured by the free energy of transfer of the solute from water to a nonpolar environment of an organic solvent. To quantify the free energy value, one of the three following methods is usually employed: analysis of solubility of compounds in water and organic solvent; partitioning of compounds in water-organic solvent systems; and partition chromatography.

Analysis of Solubility of a Solute in Water and in Organic Solvents.

Measurements of solubility of a substance in water and organic solvent to estimate the relative hydrophobicity of the substance are based on the general thermodynamic condition of the equality of the chemical potentials of a given solute in the phase of the pure solute and in the saturated solution of the solute. It follows from this condition that the free energy change for transferring the solute from water into an organic solvent is:

$$\Delta G_{tr}^{w\to s} = \mu^o{}_s - \mu^o{}_w = RT\cdot\ln(f_w/f_s) - RT\cdot\ln(C_s/C_w) \qquad (6.4)$$

where μ^o is the standard chemical potential of a solute; C the solubility of the solute, i.e., the molar concentration of the solute in the saturated solution; f the activity coefficient of the solute; and subscripts "w" and "s" denote water and organic solvent, respectively.

When the solubility of a compound in water and in the organic solvent is sufficiently low, the activity coefficients f_w and f_s are close to unity, and Equation 6.4 becomes:

$$\Delta G_{tr}^{w\to s} = -RT\cdot\ln(C_s/C_w) \qquad (6.5)$$

As mentioned above, the dissolution of a solute in water may be theoretically divided into three hypothetical steps [13-16]. These steps include: removal of the solute molecule from its initial environment; formation of a cavity in water to accommodate the solute molecule; introduction of the solute molecule into the cavity.

The two latter steps depend on the size and the effective surface area (or volume) of the solute molecule, and on the solute-water and water-water interaction energies. The water structure-perturbing effects of various additives were discussed above (Chapters 1 and 2). It is known that the solubility of a substance in water and in an aqueous solution of a salt or some other additive may differ. This must be taken into account when studying the water solubility of readily soluble compounds [17] as the saturated aqueous solution of such a compound should be regarded as the aqueous medium, the structure of water in which has been modified by the dissolved compound (even assuming the lack of the solute-solute interactions).

The first step of the solution process, i.e., removal of a solute molecule from its original environment, depends upon the intensity of intermolecular interactions in the pure phase of the solute [18-20]. Amidon et al.[18] examined the aqueous solubilities of various aliphatic hydrocarbons, olefins, and monofunctional aliphatic compounds in context of their molecular surface areas. Their results [18] indicate that the functional group contributions to the free energy of solution in water are nearly equivalent if the pure solute standard state is used, but differ significantly when the gas phase (1 mm Hg) standard state is chosen. If the difference between the solubilities of a solute in water and an organic solvent is analyzed, the contributions of the solute-solute interactions in the pure solute phase to the free energies of transfer cancel out. For this and other aforementioned reasons, the difference in the solubility of solutes in water and an organic solvent and not just the aqueous solubilities are used to estimate the relative hydrophobicity of solutes.

The estimates of the relative hydrophobicity of solutes obtained by measurements of the comparative solubility in water and organic solvents are usually in agreement with those obtained by the partition technique (see below).

Numerous efforts (see, e.g., [21,22]) were undertaken to find out how aqueous solubility and the partition coefficients of different solutes in water-organic solvent systems are related to the size of the solute molecule. Since the solute packing into the solvent clearly depends on the solute surface, a relationship between surface area and solution thermodynamics would be expected.

The relationship between the aqueous solubility of a homologous set of solutes and their molecular surface area is generally described as:

$$-\ln C_w = \beta_o \cdot F_m + w_o \qquad (6.6)$$

where F_m is the solute molecular surface area accessible for the solvent; C_w the aqueous solubility of the solute; β_o and w_o are constants.

Hermann [14] estimated the coefficient β_o values for a series of alkanes and cycloalkanes and for a number of alkyl-benzenes. The β_o values are 33 $Å^{-2}$ for alkanes and cycloalkanes and 30 $Å^{-2}$ for aromatic systems [14]. From the data reported by Amidon et al.[18] the β_o values for different monofunctional aliphatic compounds examined appear to be constant and amount to 22.6 $Å^{-2}$. A similar β_o value of 22 $Å^{-2}$ was reported by Chothia [23] for the nonpolar side-chains of amino acids - those of alanine, valine, leucine, and phenylalanine. For side-chains of serine, threonine, histidine, methionine, and (for no apparent reason) tryptophan, the coefficient $ß_o$ value is ca.13-15 $Å^{-2}$ [23].

Two implications of the above results should be noted. First, the aqueous solubility of solutes of similar chemical nature is linearly related to the solute molecular surface area. Secondly, the value of the $ß_o$ coefficient appears to decrease with increasing hydrophilicity of the solute. These results suggest that the hydration interactions accompanying transfer of an amphiphilic molecule into water [24] are likely to oppose to some degree the stabilizing effect of the nonpolar fragment of the molecule on the local water structure. It remains unclear, however, whether differences in the $ß_o$ values are due to the hydration effect or to the difference in intensity of the intermolecular interactions in the pure solute phase.

An additional complication inherent in the method under discussion is related to the appropriate choice of an organic solvent to serve as a nonaqueous medium. Data reported by Nozaki and Tanford [17] and Fendler et al.[25] on the solubility of various amino acids in water, aqueous dioxane and ethanol solutions, and in n-hexane provide a typical example of the difficulties accompanying this choice.

Nozaki and Tanford [17] measured solubilities of different amino acids in water and in progressively increasing concentrations of ethanol and dioxane in water. Solubilities of the amino acids were extrapolated to pure organic solvents and the free energy of transfer for the amino acid from pure solvent to water was calculated. Using glycine as a reference, and subtracting its free energy of transfer from that of all the other amino acids, the relative hydrophobicities of the side-chains of the amino acids were estimated [17]. The same approach was used by Fendler et al.[25] with n-hexane as the organic solvent. The values of the free energy of transfer from ethanol and from dioxane to water were reported for five amino acids [17]. For three of these amino acids (tryptophan, tyrosine, and histidine) the free energy of transfer values for the side-chains appear to be independent of the organic solvent used [17]. For the phenylalanine side-chain, the observed values differ for ethanol and dioxane within the experimental error range (100 cal/mole), whereas for the leucine side-chain the values differ more than by 800 cal/mole [17]. When the hydrophobicity estimates for the amino acid side-chains reported by Nozaki and Tanford [17] are compared with those for the same side-chains reported by Fendler et al.[25], it appears that the values in question agree within the experimental error range only for the side-chains of three amino acids (valine, histidine, and phenylalanine) and differ considerably for those of five other amino acids - alanine, leucine, isoleucine, serine, and threonine. Hence the agreement between the estimates obtained for a given solute using different organic solvents may be accidental.

The data on the solubility of adenine and thymine in water and in ethanol and n-propanol reported by Herskovits et al.[26] support this conclusion. The difference between the hydrophobicity values of adenine and thymine measured by the free energies of transfer of the solutes from water into an organic solvent is 105 cal/mole whith ethanol, but 250 cal/mole with n-propanol used as a nonaqueous medium [26]. Scruggs et al.[27] observed particularly that the solubility of adenine in chloroform is markedly affected by the presence of water in the solvent.

Thus, the main limitations of the method of analysis of the hydrophobicity of solutes based on measurements of the solutes solubility in water and in an organic solvent are:

1. The method can be used only for the study of the compounds of moderate solubility both in water and organic solvents.
2. The choice of a given organic solvent to be used as nonpolar medium is usually open to objection.
3. The method cannot be used to study labile biological solutes (proteins, nucleic acids, etc.), the intact features of which are affected by organic solvents.
4. Relative hydrophobicity estimates obtained by the method may be compared only for the solutes of similar chemical nature.

Since the first of these limitations can be bypassed with an approach based on the study of partitioning of solutes in water—organic solvent two-phase systems, most of the hydrophobicity estimates for chemical compounds have been obtained by this method.

<u>Partitioning of Solutes in Water-Organic Solvent Systems.</u>

When a solute distributes at a constant temperature between two solvents, which are immiscible or partially miscible, the equality of the chemical potentials of the solute in the two phases may be described in the form (see above):

$$\Delta G_{tr}^{w \to s} = -RT \cdot \ln(c_s/c_w) = -RT \cdot \ln P \qquad (6.7)$$

where P is the solute partition coefficient in a given water-organic solvent system; c_s and c_w represent the equilibrium concentrations of the solute in the organic solvent phase and in the aqueous phase of the system, respectively.

To estimate the hydrophobicity of a solute it is necessary to determine the difference between the chemical potentials of the solute in the two phases. Hence, the partition coefficient value must be measured for the solute molecules being in the same form in both phases, i.e., for the nonprotonated or the ionized tion of solute molecules in the phases of the partitioning systems are discussed in detail in [28]. Various methods of corrections for the effect of the ionization degree of a solute on its partition coefficient were suggested by Alhaider et al. [29] and Martin [2,30].

When the relative hydrophobicity of solutes is estimated by partitioning in the solvent systems, one is faced with the aforementioned problem arising when measuring the solubility of substances in water and an organic solvent, i.e., which solvent should be used to simulate a nonpolar medium.

In terms of the above discussion, ideally a hydrocarbon solvent such as n-hexane should be used to measure the relative hydrophobicity of a solute [8]. One important advantage of the use of an alkane solvent is the absence of specific solute-solvent interactions which are highly sensitive to molecular structure [31]. The disadvantage of such solvents, however, is that most polar substances are essentially insoluble, with the result that partition coefficients cannot be measured with sufficient accuracy to be useful. Moreover, when a polar molecule does dissolve in such solvents, it brings with it water molecules. Dissolved molecules in hydrocarbon solvents also tend to associate with each other rather than with the solvent. The net result is that when hydrocarbon solvents are used for hydrophobicity measurements, the organic phase will contain several different species of solute. Hence, the measured partition coefficient may not be easily interpreted in terms of fundamental molecular interactions [2,32]. One should also bear in mind that the features of nonaqueous phases in biological systems differ from those of hydrocarbon solvents. These phases as a rule contain considerable amounts of water linked with the polar and ionized group of biological molecules present in the phases [32].

A number of more polar organic solvents were used as a model nonaqueous phase: diethyl ether, chloroform, olive oil, oleyl alcohol, n-octanol, n-butanol, etc. [1,2,19,28,31-33]. When choosing an organic solvent to simulate a nonpolar medium in a partitioning system, one must take into account [28,32,33]: a) the mutual solubility of water and the solvent; b) the solvation capacity of a solvent in relation to the solute being partitioned; and c) the hydrogen bond-donating and accepting properties of a solvent.

The most generally-used solvent at present is 1-octanol [1-3,28,32,33]. Both because of its hydroxyl group and the relatively high concentration of water (2.3 M at saturation), octanol appears to be a good solvent for most organic compounds. Water-saturated octanol is sufficiently polar so that dissolved molecules tend to associate with the solvent rather than with each other. It has a regular structure which is supposed not to be changed by the addition of a solute [32]. Additionally, octanol is chemically stable, commercially available, non-volatile, and it does not absorb ultraviolet light. All these characteristics are of practical importance. The use of n-octanol is preferable as compared to that of other alcohols since historically, the water—n-octanol two-phase system was used to determine partition coefficients of a vast number of chemical compounds in order to study their relative hydrophobicity [1-3,28,33].

Parameters considered when choosing an organic solvent for a partitioning system [1-3,28,32, 33] are essential for postulating a physical model for the partitioning process. Such a model [33] differs from that of the solution process in that instead of the pure solute phase the solution of the solute in an organic solvent is considered. According to Rekker [33], transfer of a solute molecule from a nonaqueous phase into an aqueous one may be simulated by a cavity-to-cavity (or "hole-to-hole") transfer, dependent upon the difference in the free energies required to form an appropriate cavity in both phases of the system. Rekker [33] suggested that an adequate description of the solute partitioning in a solvent two-phase system should take into account not only the size of the solute molecule but primarily the structural features of the media in both phases of a given system.

The "hole-to-hole" model advanced by Rekker [33] seems to be the most adequate one to account for the relationship known to exist between the aqueous solubility of solutes and their partition behavior in the solvent systems [22,34] as well as for the differences observed between the solute partition behavior in various solvent systems [19,28,32,33]. This concept appears also to be supported by the data [35] on the relationship between the partition coefficients of non-polar solutes in the water-octanol system and the molecular surface area of the solutes. The model [33] also seems to agree with that the value of the free energy of transfer of polar solutes from water into an organic solvent depends upon the specific solute-solvent interactions to a much greater degree than it does upon the surface area of the solute molecule [16].

For a more complete description of the partitioning systems Rekker [33] suggested to use the so-called "discriminating power" of the systems. This parameter [33] denotes the spread any given solvent system imparts to the partition coefficient values for a set of molecular structures subjected to partitioning in that system. Davis et al.[19] suggested to measure the relative hydrophobic character of a solvent system by the fee energy of the hypothetical transfer of a CH_2 group between the phases of the system, the value of which varies from 450 cal/mole to ca. 1000 cal/mole CH_2 depending on the type of the solvent system (see above).

According to the aforementioned additivity principle, the contributions of the polar groups into the logarithm of the solute partition coefficient in a given solvent system are usually viewed as estimates of the relative hydrophobicity of these polar groups. Zaslavsky et al.[36] showed, however, that the values of the ratio between the contributions of a polar group and CH_2 group, A/E, (see Equation 4.15) and not just the separate values of these contributions should be considered to analyze the estimates of the relative hydrophobicity of polar groups of solutes obtained in different solvent systems. More specifically, it was found [36] that the relationships between the A/E ratio value and the solubility of water in a given organic solvent for aliphatic carboxyl acids, alcohols, and amines are represented by the mutually intersected linear curves (see Fig. 4.14). Since the curves intersect [36], the apparent hydrophobicity of one polar group in regard to the other group may be reversed depending on the particular solvent system used. These results [36] support the assertion that the relative hydrophobicity only of the solutes of the same chemical nature may be examined by the method of partitioning in water-organic solvent two-phase systems.

Moreover, while the hydrophobicity estimates obtained from the solubility measurements generally correlate fairly well with those derived from partition experiments on more hydrophobic solutes [22], the correlation seems to be rather poor or nonexistent for hydrophilic solutes such as, e.g., amino acids [37]. According to Yunger and Cramer [37], the additive-constitutive rules may be quantitatively different for highly hydrophilic solutes in reference to those for hydrophobic solutes.

Thus, an employment of the technique of partitioning in water-organic solvent system in studies of the hydrophobic character of solutes is limited by that:

1. The estimates of the hydrophobic character of solutes depend upon the choice of a particular solvent partitioning system;
2. The estimates obtained can be used for comparison of solutes of the same chemical nature only; and
3. The method cannot be used to study labile biological solutes which are liable to denaturation or conformation changes induced by organic solvents.

Partition Chromatography.

The shaking-flask method has been and still is the most generally used method for determining the water-octanol partition coefficient of a solute as the solute hydrophobicity index. This method, however, is very tedious, requires relatively large amounts of pure solutes to be examined, and is limited to $\log P_{oct}$ values between -2 and +4. These practical disadvantages of the method have led researchers to investigate alternative methods for determination of $\log P_{oct}$ values.

Different chromatographic methods such as thin-layer chromatography [38,39], centrifugal partition chromatography [40], and reversed-phase high performance liquid chromatography (RP-HPLC) [41-44] have been found successful alternatives to the shaking-flask technique. The use of the different versions of partition chromatography in the hydrophobicity studies is based on that the chromatographic behavior of solutes (characterized by retention index, R_F value, etc.) correlate well with the logarithms of the solutes partition coefficients in the water-octanol two-phase system [38-44]. The main advantages and limitations of the use of the chromatography in the study of the hydrophobic character of chemical compounds have been discussed at length in a number of papers and reviews (see, e.g., in [38-44] and references cited therein); therefore only the most important of these are very briefly mentioned here.

Practical advantages of the partition chromatography methods over the methods discussed above are as follows: they are relatively fast and labor saving, and they allow one to work with rather impure compounds when only small amounts of sample are available. The major drawback is the need to use an organic solvent or a mixture of solvents which restricts the applicability of the method for the study of many biological solutes and, as indicated above, leads to an ambiguity of the estimates of the relative hydrophobicity for polar organic compounds.

It is generally assumed that transfer of a solute from water into a non-polar solvent simulates roughly such biological processes as transfer of proteins from blood plasma to cellular membranes, penetration of drugs through skin, binding of ligands and drugs to non-polar sites in protein macromolecules, etc. Hence analysis of thermodynamic quantities of transfer of solutes, particularly of biological origin, from water into an organic solvent are believed to be of both theoretical and practical importance.

It should be taken into account, however, that the approximation used is extremely rough. Firstly, the medium of a non-polar organic solvent appears to be rather inadequate model of the nonaqueous compartments or phases in biological systems. Secondly, organic solvents employed in the commonly used partitioning systems are far from being inert toward the solutes being partitioned. Usually the effect of the solute-solvent interactions on the partition coefficient of the solute cannot be quantified. Hence the estimates of the hydrophobic character of solutes may be used only for a relative rating of the solutes of the same or very similar chemical nature. Furthermore, these experimental methods cannot, as was noted before, be used in the studies of biological solutes, since properties of the latter may be altered by organic solvents.

Transfer of a Solute from Water to a Gaseous Phase.

As most biochemical processes in living systems occur in aqueous media, attempts have been made to develop a method to characterize "the absolute tendencies of solutes to leave water and enter a featureless cavity of unit dielectric constant that neither attracts nor repels the solutes" [45]. These attempts are aimed at creating a method of direct estimation of the hydration energy E_S in Equation 6.1 (see above).

The method developed by Wolfenden [45] is based on measuring the dimensionless equilibrium constant for transfer of a substance from the dilute vapor phase, in which each molecule exists in virtual isolation, to an aqueous solution so dilute that each solute molecule is completely surrounded by water, and solute-solute interactions can be neglected. This may be accomplished by measuring solubilities of a gas under known pressure or, for less volatile compounds, by determining concentrations of solute in the gas space over solutions of known concentrations. In the cases of highly hydrophilic solutes, measured volumes of an inert carrier gas can be bubbled through an aqueous solution of known concentration, and then through an efficient trap that recovers the solute quantitatively from the vapor phase [45]. Specific methodical details of this technique may be found in the papers by Wolfenden et al. [45-48].

From the viewpoint of studying the hydrophobic character of biological solutes, the important advantage of the approach is that it excludes the need for an organic solvent. The possibility to use the method for the study of the relative hydrophobicity of highly hydrophilic biological macromolecules is, however, open to question. Though the correlation observed between the "vapor-to-water" partition coefficients and the partition coefficients for the same solutes in water-organic solvent systems is concluded by Wolfenden to be "not bad" [45], it does not seem to be sufficiently conclusive. For example, Wolfenden and co-workers [45,48] have found tryptophan and tyrosine to be highly hydrophilic, while the same amino acids were reported [17] to be highly hydrophobic from solubility measurements. It should be mentioned, however, that the authors [45-48] did not suggest the method to be applicable for studying the relative hydrophobicity of biological molecules. They advanced the term "hydration potential" of a solute [48] which seems to be related to the hydrophobic character of the solute. The outlook for the applicability of the approach to study the relative intensity of the hydration interactions of biological molecules seems to be rather uncertain at present.

Different hydrophobicity estimates for various chemical moieties comprising biological solutes structures obtained using the above techniques are commonly employed for estimating the hydrophobic properties of biological macromolecules (see below). Significant differences between the estimates obtained by different methods were discussed by Rose et al.[49]. These differences (considered in more detail below) prompted development of a number of special methods for analysis of the hydrophobic properties of biological macromolecules. These methods will be discussed below.

In order to explain why the hydrophobic properties of biological solutes are important, it is necessary, first, to outline the basic principles of quantitative structure-activity relationships (QSAR) approach used in drug design, toxicology and other biomedical studies.

6.3. INFLUENCE OF THE RELATIVE HYDROPHOBICITY OF CHEMICAL COMPOUNDS ON THEIR BIOLOGICAL ACTIVITY

A large body of literature exists on quantitative relationships between the structure and biological activity of chemical compounds. Only the basic principles of these relationships and some issues related to the hydrophobic properties of solutes will be outlined here. For a more comprehensive treatment of the subject the reader should refer to [1-3,22,30, 33,50-53].

Studies of quantitative structure-activity relationships (QSAR) are based on the two following assumptions. A first assumption is that the response of a biological system (*in vivo* or *in vitro*) to a drug or other xenobiotic introduced into the system results from the interaction of the drug with a certain component of the system (receptor). A second assumption is that the response may be described in terms of physicochemical characteristics of the interaction.

Attractive and simple as this hypothesis appears, there is a difficulty with its experimental justification, since not only the mechanism of the interaction in question but even the chemical nature of the particular component of the biological system involved is often unknown. To bypass this difficulty, two additional assumptions have been adopted (see, e.g., in [30]. One is that the chemical structure of a drug may be described quantitatively in terms of its physicochemical properties. The other assumption is that the physico-chemical characteristics of the drug may be used as the indexes of its interaction with the unknown component in the biological system. In other words, it is assumed that the biological potency of a drug is quantitatively related to its physicochemical characteristics.

Numerous successful studies of quantitative relationships between biological potencies of drugs and their different physicochemical characteristics [1-3,28,50-53] justify the above assumptions. The most important physicochemical characteristics were found to be those related to electronic, steric, and hydrophobic properties of the drug molecule. The generally accepted explanation is that essentially all types of interactions (covalent, electrostatic, van der Waals, etc.) possibly experienced by the drug in a biological system are determined to some degree by the above properties of the molecules.

The reciprocal of the concentration or dose of a given drug producing a given response from the biological system, for example, decrease in blood pressure, hemolysis of erythrocytes, decrease in the rate of equilibrium constant for a given enzymic reaction, etc., is commonly used as a measure of the drug biological potency.

It should be mentioned that the study of QSAR does not allow one to predict the biological effect of the xenobiotic with a given chemical structure. If the effect of a drug is known, however, the QSAR analysis helps to rationalize modifications of the drug structure required to increase or reduce the effect in question. Additionally, the QSAR analysis provides one with some understanding and helps to create a working hypothesis of the pertinent chemistry of the drug in the biological system (see, e.g., in [2]).

The substituent effects on the biological potency-relevant physicochemical properties of a molecule are usually described in terms of various substituent constants. The Hammett σ_X value of a substituent represents the effect of the substituent X on the electronic properties of the molecule. It is determined as the logarithm of the effect of the substituent on the acid dissociation constant of benzoic acid (see, e.g., in [54]). The Taft E_S value represents the effect of a substituent on the steric properties of the molecule. E_S is experimentally determined from the relative rates of hydrolysis of the acyl-substituted methyl ester compared to that of methyl acetate (for spherically symmetrical substituents E_S is proportional to the radius of the substituent). The Hansch π_X value or the Rekker f_X value (see above) represents the substituent effect on the hydrophobic properties of the molecule. In some cases, instead of the substituent constants, the characteristics of the whole molecule are used, e.g., molar refractivity, derived from the refractive index; parachor; logarithm of the compound partition coefficient in a water-organic solvent system; dipole moment of the molecule, etc. Additionally, the so-called indicator ("dummy") variables are used to account for a discontinuity in the structural features of the molecules which are not represented by the usual physico-chemical properties. These variables are arbitrarily assigned one value to indicate the presence of a particular feature, e.g., 1, and another to indicate its absence, e.g., 0. Examples of the use of these variables may be found in [2].

Numerous quantitative structure-activity relationships reported in the literature include binding of drugs to proteins, their interactions with enzymes, cells, and tissues, action as antimicrobial agents, anesthetics, antitumor agents, etc., the drugs' toxicity, mutagenicity, and carcinogenicity, etc. It should be mentioned that among all these relationships only a few do not include the characteristic of hydrophobic properties of the molecule (or substituents). The reason seems to be clear. The response of a given biological system to a drug is supposed to depend, among other factors, on the ability of the drug to reach a given receptor, i.e., the specific component of the system capable of interaction with the drug resulting in the response under study.

Distribution of a drug in an organism is controlled by physiological and drug-related factors [55]. Drug-related factors are hydrophobicity, ionization constant, and the presence and location of certain functional groups governing its electronic and steric properties. All these factors determine the binding of a drug to intravascular and extravascular binding sites [56]. The hydrophobicity of drugs is generally viewed as the parameter in control of the transport behavior of drugs from their site of administration to the site of action through a number of different compartments and biomembranes, as well as the drugs binding with proteins and hydrophobic receptor sites.

All the models of the biological potency - hydrophobicity relationship may be divided into linear and non-linear ones.

The linear relationship is typically described as:

$$\log(1/C) = k_1 + k_2 \cdot \log P \tag{6.8}$$

where C is the concentration or dose of a drug producing the effect being monitored in a given biological system; logP is the logarithm of the drug partition coefficient in the water-octanol system (instead of logP other hydrophobicity parameters, e.g., π_X, f_X, etc., may be used); k_1 and k_2 are constants.

The linear relationship described by Equation 6.8 is generally assumed to be fulfilled only within the limited range of the hydrophobic properties of the drugs being examined. It is clear that the biological potency in a given series of drugs with increasing drug hydrophobicity cannot increase indefinitely. After achieving the maximal potency possible for the drugs of a given type the further increase in the drug hydrophobicity will be accompanied by decrease in the potency. To explain this type of relationship several non-linear models have been suggested in the literature. These models are discussed at length by Martin [30].

The nonlinear relationships are commonly described as:

$$\log(1/C) = -k_1 \cdot (\log P)^2 + k_2 \cdot \log P + k_3 \quad \text{(parabolic)} \tag{6.9}$$

$$\log(1/C) = k_1 \cdot \log P - k_2 \cdot \log(\beta \cdot P + 1) + k_3 \quad \text{(bilinear)} \tag{6.10}$$

where k_1, k_2, k_3, ß are constants.

There are other types of mathematical expressions describing the non-linear potency-hydrophobicity relationships as well (see, for example, in [30,51, 52]. It follows from any of these expressions that among a given series of drugs there is the drug with the structure corresponding to the "optimal" hydrophobicity ($\log P_o$) providing the maximal biological effect possible for the drugs of the series.

The nonlinear models are typically based on the treatment of the transfer of a drug from the site of administration to the site of action in terms of the "random-walk" process, kinetics of transfer or equilibrium distribution between the various "phases" and compartments of the biological system. A typical example is the equilibrium model by Higuchi and Davis [57] where the biological system is viewed as a multiphase equilibrium system. A drug introduced into such a system distributes between the phases according to the Gibbs-Duhem equation, i.e., under the condition of equality of the chemical potentials of the drug in all the phases. The model [57] explains the nonlinear shape of the potency-hydrophobicity relationship as due to that the "optimal" hydrophobicity of a given drug results in the drug distribution with the maximal concentration in the 'receptor compartment' (i.e., maximal occupancy of the receptor sites) at the minimal total amount of the drug introduced into the system.

A typical example of parabolic relationship between the rate of metabolism of primary aliphatic alcohols by uridine diphosphate glucutonosyltransferase (glucuronidation rate) and the logarithm of the water-octanol partition coefficient for the series of phenol derivatives reported by Kim [58] is presented in Figure 6.1.

The nonlinear shape of the relationship (parabolic, bilinear, etc.) indicates that among a series of drugs examined there is the drug with "optimal" hydrophobicity ($logP_o$) providing the maximal biological effect possible for the drugs of the series. The QSAR analysis, thus, provides clues for molecular modifications leading to the structure with maximal possible biological potency in a given series of drugs.

Additionally, the identical values of $logP_o$ for different drugs series in the same biological system may be viewed as an indication of an identical mechanism of the biological action of these series. The difference in the $logP_o$ for two series of drugs may indicate the different mechanisms of action. For example, Jeppson [59] reported toxicity of three series of aliphatic hydrocarbons, ethers, and ketones given to mice intravenously. Toxicity of the substances expressed as a median lethal dose, LD_{50}, i.e. the dose producing death in 50% of the animals in a group, was analyzed by Kubinyi [60] as a function of the substance hydrophobicity (measured as logP). The "optimal" hydrophobicity values $logP_o$ were found to be 2.62 for ethers, 2.72 for ketones, and 4.85 for alkanes [60]. Is it possible to conclude that the mechanisms of the toxic action of ethers and ketons is similar and different from that of alkanes? The question remains open and, in spite of its obvious importance, is infrequently discussed in the QSAR literature. The obvious reason seems to be the aforementioned difficulty of comparison of the hydrophobicity estimates for chemically different compounds obtained by the water-octanol partition technique.

The method providing the possibility to compare hydrophobicity of compounds of various structures and chemical nature would allow one to resolve the above questions. That is clearly important for better understanding of mechanisms of biological action of drugs and other xenobiotics. The technique of partitioning in aqueous two-phase systems provides this possibility as will be shown below. An additional advantage of the technique is that it may be applied to proteins, glycoproteins, and other biological solutes.

The most efficient and selective chemical regulators of physiological processes are clearly those designed by nature itself, such as hormones, proteins, glycoproteins, etc. The availability of large quantities of natural or modified substances, for example, immunosuppressants, biological response

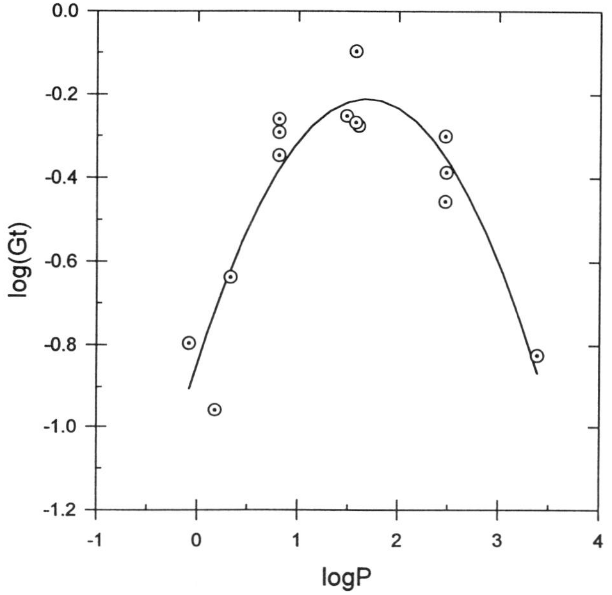

Figure 6.1. Relationship between the glucuronidation rate (expressed as logarithm of the rate constant for glucuronide formation, Gt) and hydrophobicity for a series of phenol derivatives (expressed as logarithm of the octanol-water partition coefficient, $P_{octanol\text{-}water}$). Calculated from the data reported in [58].

modifiers, hormones, etc., due to recombinant DNA-derived technology recently created a new field of biopharmaceuticals. The substances being used as "biological drugs" are much more complex than the common pharmaceutically active chemicals.

Mammalian-cell-expressed recombinant glycoproteins that are approved or under development as pharmaceutical agents include tissue plasminogen factor, erythropoietin, ß-interferon, etc. Recombinant DNA-derived human growth hormone is widely used to treat growth hormone-deficient children. Various so-called biological response modifiers, for example, interleukin-2, α-interferon, tumor necrosis factor, transforming growth factor, etc., are currently under different stages of laboratory and clinical trials as promising anticancer pharmaceutical agents. Covalent conjugation of polyethylene glycol (PEG) to proteins has recently become a method of dramatically altering a protein's pharmacology and immunogenicity.

Analysis of structure-activity relationships for these agents may lead to new highly efficient drugs of low toxicity. For this analysis to be productive, however, it must be quantitative as in the case of common chemical drugs. Application of the standard QSAR methods to biopharmaceutical agents is hindered by the overwhelming complexity of their structures. It seems currently impossible to describe the structure of a peptide, let alone glycoprotein, in terms of electronic, steric, and hydrophobic substituent constants. The obvious reason is the complexity of the molecule structure, and the cooperative effects of all different fragments of a conformationally flexible molecule on its physicochemical properties.

Physicochemical descriptors for such molecules are currently under active investigation. Hydrophobicity of a biological solute may be one of the most important descriptors [61] due to several obvious reasons. First, as indicated above, hydrophobic properties is one of the leading factors in QSAR analysis for common pharmaceutical agents. Secondly, hydrophobicity of a molecule represents the intensity of the molecule interactions with an aqueous medium. That means that any change in the molecule conformation affecting the groups accessible to the solvent would affect the molecule hydrophobicity. Therefore, hydrophobicity of a molecule may be viewed as a factor representing its functionally active conformation. Third, as suggested by Tanford [62], the distribution of a biological solute throughout the living body liquids and tissues is likely to be governed by the difference between the solute-solvent interactions in various phases or places in the body. There is a clear similarity between this hypothesis [62] and the aforementioned model suggested by Higuchi and Davis [58] for pharmaceutical agents.

Thus, it is clear that the hydrophobicity may be used as a physicochemical descriptor of a biological solute for QSAR analysis. The reliability of the biological solute hydrophobicity estimate obviously depends on the method used. Experimental methods for estimating hydrophobic properties of biological macromolecules are discussed below.

6.4. METHODS FOR STUDYING HYDROPHOBIC PROPERTIES OF BIOLOGICAL MACROMOLECULES

The importance of the hydrophobic properties of biological macromolecules with regard to their function and structural organization has long been recognized. It is generally believed that the genetic code in its most primitive form could only differentiate between two classes of amino acids, i.e., hydrophilic and hydrophobic, and the grouping of codons and amino acids by similar hydrophobicity criteria has been advocated [63, 64]. It is also well known that interactions with water govern tertiary structure of proteins (see, for example, in [65,66]), conformation of nucleic acids [67,68], protein-protein, protein-biomembrane, and protein- and nucleic acid-ligand interactions (see, e.g., in [68-71]), etc.

Numerous efforts have been made to use different hydrophobicity or polarity scales to analyze packing in globular proteins [72,73], predict secondary structure [74-77] and transmembrane segments [73,78-84], to evaluate amphiphilicity (i.e., hydrophobic-hydrophilic balance) of α-helices [76,85,86], etc. It has recently become clear that some crucial intermolecular recognition processes are guided mostly by general structural features and not by an exquisite discrimination among closely related structures [87] as was generally assumed. Signal peptides offer an illustrative example, as their information content is virtually independent of a precise amino acid sequence [87-89]. The proper targeting functions of signal peptides can be retained after dramatic changes in sequence, so long as the overall non-polar nature of the constituent residues is maintained [87,90, 91]. The structures of membrane-spanning fragments of membrane proteins are also believed to be related to the overall hydrophobicity of the fragment [82-84].

The methods most generally used in the analysis of the hydrophobic character of biological molecules have been extensively reviewed in the literature [6,92-94]. Only the basic features, advantages and limitations of these methods are considered below.

All the methods currently used for estimating the hydrophobicity of biological solutes, proteins and peptides, in particular, may be divided into two groups: semi-empirical and empirical.

Semi-Empirical Methods.

All semi-empirical methods for estimating the hydrophobicity of biological solutes, primarily of peptides and proteins, are based on the hydrophobicity-hydrophilicity classification of amino acids.

The earliest attempts to classify amino acids according to their hydrophobicity were based upon considerations of their chemical nature and steric structure of their side chains. According to the first classification, all the amino acids have been grouped into two classes: hydrophobic and hydrophilic ones. Capaldi et al.[95] suggested to divide amino acids not into two but into three groups - hydrophilic, hydrophobic, and intermediate ones. For many amino acids, attribution to one or the other group varied depending on the particular authors' opinion [95, 96]. Various qualitative classifications have been used to estimate the differences in the relative hydrophobicity of membrane-bound proteins and lipoproteins as compared to globular proteins [95,97]. As a quantitative measure of the relative hydrophobicity of a given protein, the sum of the residue mole percentages of hydrophilic amino acids in the protein macromolecule (the so-called polarity index) has been proposed [95]. The attempts to employ this criterion to estimate the relative hydrophilic (or hydrophobic) character of proteins [95,97] have failed mainly due to two reasons. First reason is a qualitative and often incorrect division of amino acids into hydrophilic and hydrophobic. The second one is that the authors [94,96] did not take into account the tertiary structure of the molecule and the fact that only the solvent-accessible amino acid residues determine the interaction of a protein macromolecule with the aqueous medium, i.e., the relative hydrophobicity of the protein.

To quantify the hydrophobicity of amino acids, many different scales have been proposed [17,25,37,45,48,98-107]. Most of the scales are based on the estimates for the free energies of transfer of the amino acid side chains from water to an organic solvent obtained from partitioning of amino acids, their derivatives, or analogues in water (or buffer)-organic solvent two-phase systems [25,37,97, 98,101-105] or solubility measurements [17, 99]. Several scales have been constructed from solvation energies calculated from vapor pressures of side-chain analogues [45,48], the surface tension measurements in aqueous solutions of amino acids [106], studies of distribution of amino acids and their derivatives into lipid bilayers [107], etc. All these scales are usually only partially correlated.

The best agreement is generally observed between the hydrophobicity estimates obtained by different techniques for amino acids with non-polar side-chains. The estimates for these amino acids vary depending on the organic solvent used as the nonaqueous medium. For example, the free energy of transfer of an amino acid side chain from water (or aqueous buffer solution) to an organic solvent varies for phenylalanine from -2300 cal/mole (to octanol [37]) to -2770 cal/mole (to methanol [100]), for tryptophan from -35 cal/mole (to octanol [104]) to -3400 cal/mole (to ethanol [17]), for valine from -1160 cal/mole (to octanol [37]) to -2360 cal/mole (to CCl_4 [98]), and for leucine from -1800 cal/mole (to ethanol [17]) to -3070 cal/mole (to CCl_4 [98]).

The values reported for alanine vary more noticeably from +420 cal/mole (to hexane [25]) to -770 cal/mole (to methanol [100]). The reason for that may be the compactness of the alanine molecule. The side-chain composed of a single CH_3 group is situated very close to the polar carboxyl and amino groups. The mutual effects of these groups on their interactions with a solvent may increase the sensitivity of the hydrophobicity estimate for the alanine side-chain toward the particular organic solvent used.

The same high sensitivity is observed for the amino acids with polar side-chains. The free energy of transfer of the side- chain from water (buffer) to an organic solvent varies for threonine from +40 cal/mole (to hexane [25]) to -400 cal/mole (to ethanol [17]), and for serine from +770 cal/mole (to hexane [25]) to -55 cal/mole (to octanol [104]). The ability of the side-chain hydroxyl group in these amino acids to participate in hydrogen bonding with the solvent is likely to explain the above variations in the hydrophobicity estimates. The estimates for the proline side-chain vary significantly (as might be expected) from -780 cal/mole (to octanol [37]) to -250 cal/mole (to hexane [25]). Very few estimates for the amino acids with amide side-chains and charged side-chains have been reported in the literature [37] possibly due to understanding that the interactions of these side-chains with a solvent are too specific for the estimates to be reliable.

These estimates have been used to analyze hydrophobicity of segments of a polypeptide chain by summing the contributions of the constituent side-chains, i.e., using the principle of additivity of the substituent effects and the invariability of the backbone atoms. The so-called hydrophobicity profile approach is a simple way to quantify the concentration of hydrophobic residues along the linear polypeptide chain [108]. The approach is based on the assumption that residues with hydrophobic side-chains tend to bury themselves within macromolecules, away from solvent water. It was used with some success in predictions of trans-membrane segments [82-84,108], antigenic determinants [109], packing in globular proteins (see, e.g., in [93]), segments responsible for the protein-induced fusion of biological membranes [110], etc. The hydrophobicity profile approach based on the analysis of the protein amino acid sequence does not allow one, however, to estimate the actual overall hydrophobicity of a protein macromolecule which is governed solely by the residues exposed to an aqueous environment.

It has been established by Chothia [23,111] that the hydrophobicity estimates of amino acids are poorly correlated with the extent to which the residues are buried in the protein macromolecule. As mentioned above, the hydrophobicity of a non-polar solute is linearly related to the solute surface area. This observation together with the assumption that the more hydrophobic the amino acid residue, the more completely buried in the protein interior it will be, served as the basis for the so-called solvent accessibility scale (see, e.g., in [93]). By means of the solvent accessibility approach using a computer-based analysis of known protein structures, it has been shown that as much as 40 to 60 percent of the surface area of many globular proteins are taken up by non-polar amino acids residues [111,112]. The non-polar amino acids residues located at the macromolecule surface are assumed to account for the biospecific conformation of the protein as well as for its ability to complex or aggregate with other types of biological molecules. This line of investigation of function-related hydrophobic properties of proteins seems to be of a great interest.

The above approaches are seriously limited from the viewpoint of their applications to estimating the overall protein or peptide hydrophobicity as a function-related descriptor. The most obvious limitation relates to the inadequate hydrophobicity estimates for the amino acids with charged and highly polar side chains usually located at the molecule surface and participating in the interactions with an aqueous medium. Secondly, the estimates for amino acid residues do not take into account the highly likely influence of neighboring residues. The influence in question is observed in many different cases for reversed dipeptides [113] tripeptides [114,115], steroids [116], etc. Finally, while different hydrophobicity estimates for the same amino acids have been reported by authors using the same partition technique but different aqueous buffer-octanol systems [37,104] (see below), it seems impossible to analyze the effect of the composition of an aqueous medium on the hydrophobicity of proteins and peptides with the above semi-empirical approaches. That is why various empirical methods for estimating hydrophobicity of biological solutes have been developed.

Empirical Methods.

Among the biological solutes subjected to the QSAR analysis the leading role belongs to peptides. The reason is, first, that a large number of natural and synthetic peptides has the clinical potential as the promising pharmaceutical agents, and, second, the structures of peptides are not as complex as those of proteins and may be, in principle, treated in terms of physicochemical properties [117-119]. In contrast to relatively small rigid structures of common drug molecules, however, the larger flexible peptides are believed to exist in solution as the population of different conformers. The functionally active conformation is determined both by the peptide amino acid sequence and its environment (solvent, closeness of a biological membrane or protein macromolecule, etc.) [118,119]. Analysis of the receptor binding of various native peptides and their conformationally constrained synthetic analogues led Taylor et al. [118] to the conclusion that the affinities and selectivities of peptides determined by all the structural and conformational features may be represented by a limited number of the structure descriptors. It is currently unclear what descriptor may be used to represent the functionally active conformation of a peptide but the hydrophobicity index seems to be among the most promising ones.

The extensive set of the partition coefficients of 124 di- to pentapeptides with nonionic side-chains in the octanol — 0.1 M phosphate buffer system was reported recently by Akamatsu and Fujita [114]. An empirical equation correlating the variations in the peptide hydrophobicity (measured as the logP value) with physicochemical descriptors for the side chain substituents and substructures was obtained [114]. It should be mentioned that the conformational potential index derived from the Chou-Fasman ß-turn potential parameter (see, e.g., in [120] had to be included in the equation [114] indicating that the peptide partition behavior is representative of the peptide conformation.

Only one attempt to use the hydrophobicity of amino acid residues as measured by the water-octanol partitioning to analyze the potency-structure relationship for opioid peptides in QSAR terms [117] has been reported in the literature, to my knowledge. Fauchere [117] found that to describe the ability of a series of enkephalin derivatives of the general structure H-Tyr-D-Ala-Gly-X-Y-NH_2 (X and Y are variable amino acid residues) to depress the contractions of electrically stimulated guinea pig ileum and mouse vas deferens preparations, a number of structural descriptors was required. The descriptors necessary for QSAR were [117] the total hydrophobicity ($\pi_X + \pi_Y$) of the side-chains of the residues X and Y, the electronic factor for X, and the sum of the steric factors for X and Y. Analysis of the established correlation indicated [117] that the descriptors used are important for an overall potency rather than a selective opiate activity of the peptides examined. Therefore this attempt [117] at QSAR analysis was judged by Schwyzer [119] as not totally successful.

It is well known that when placed in nonaqueous solvents proteins usually denature, and enzymes may exhibit a new substrate specificity [121] due to conformational changes originating from alterations in the biopolymer-solvent interactions. Since similar changes in peptides are likely (see, for example, in [122]), the use of the water-octanol partitioning for estimating the function-related properties of conformationally flexible biological solutes is highly questionable. Additionally, partitioning of peptides with ionic side-chains in a water-organic solvent system presents a difficult and as yet unresolved problem.

Finally, the important question remaining open is the possibility for comparing the partition coefficients of peptides of different structure. As mentioned above, the hydrophobicity estimates obtained by the water-octanol partition technique may be compared only for a series of solutes with similar structure. An illustrative example is offered, e.g., by the data reported in [123]. Tayar et al.[123] showed that the human skin permeability coefficients for steroid hormones are well correlated with their hydrophobicities as measured by the logP values, and that completely separated correlation as good as the one for steroids exists for aliphatic alcohols. No global correlation was observed, however, indicating that "logP cannot be viewed as a unifying variable" [123].

According to the X-ray diffraction analysis of dry and water-saturated octanol [124] "octanol molecules are arranged in roughly spherical aggregates with a polar center (containing octanol hydroxyl groups and water, when present) and an apolar outer region (consisting of octanol hydrocarbon chains) with a polar-apolar interface between them. Thus, hydrated octanol possesses a range of localized environments (rather than a single 'averaged' environment), mimicking the properties of biological binding sites as diverse as lipid bilayers, nucleic acids, and apolar binding pockets on proteins. Even disregarding likely solute-induced changes in local structure, it is clear that these different microenvironments could accommodate a wide variety of different types of solutes. For example, polar solutes will tend to enter the polar centers of the aggregates, apolar solutes will partition into the large apolar regions, and amphipathic molecules will be in both of these regions and straddle the interface between them". In other words, Franks et al.[124] suggested that the water—octanol partitioning of solutes of different polarity occurs into different "regions" of the nonaqueous phase, implying that the partition coefficients for solutes of different hydrophobicity, for example, peptides containing relatively non-polar amino acid residues and those containing charged and polar residues, may not be compared. Before this issue is resolved completely, any use of the water—octanol partition technique for estimating the hydrophobicity of conformationally flexible biological solutes remains open to criticism.

All of the above limitations are shared by the different chromatographic techniques such as TLC, HPLC, etc., used to estimate the hydrophobicity of peptides and other biological solutes (see, for example, in [42,pp.243,265] and [125] and references cited therein) by analogy with that of drugs and other chemicals.

The incorrect use of the term hydrophobicity have caused certain confusion in the literature (see, e.g., in [126]) worthy of particular notice. For example, the results of the hydrophobic interaction chromatography of proteins have been viewed in some cases as providing an information about the proteins' hydrophobicity. Typically, chromatographic behavior of human skin fibroblast proteoglycans and related oligosaccharides eluted with gradient of Triton X-100 on Octyl-Sepharose in 4 M guanidinum hydrochloride was interpreted [127] in terms of the relative hydrophobicity of the solutes. It should be emphasized that the hydrophobic interaction chromatography technique is based on the interactions of non-polar sites or 'pockets' on the surface of a macromolecule with a given hydrophobic ligand coupled to an inert insoluble matrix. The method of hydrophobic chromatography provides an important information about the macromolecule ability to participate in hydrophobic interactions with a given ligand but does not allow one to estimate the hydrophobicity of the macromolecule. This is also true for the methods based on the analysis of interactions of proteins with free hydrophobic or amphiphilic probes, e.g., by fluorometry [128] or by "hydrophobic partition", i.e. study of the protein partition behavior in an aqueous Dex-PEG two-phase system with part of PEG replaced by, e.g., palmitoyl-PEG (see, for example, in [129, pp.88-91].

It should be repeated once again that whatever experimental technique is used, the data on the binding ability of a macromolecule in regard to a given nonpolar or amphiphilic probe do not provide a means for estimating the hydrophobic character of the macromolecule which is a measure of the intensity of its interactions with an aqueous medium.

Effects of Inorganic Salts on the Aqueous Solubility of Proteins.

The conception that the hydrophobicity of a protein is the property of its surface was used as the basis of the technique suggested by Melander and Horvath [8]. This technique consists of an analysis of the effects of inorganic salts on the aqueous solubility of proteins. According to the model considered by Melander and Horvath [8], the free energy of solvation of a protein macromolecule in aqueous solution is described by Equation 6.3. The presence of a salt alters the protein solubility due to the concentration-dependent effect of the salt on the free energy of formation of a cavity in water, E_{cav}, and the free energy of electrostatic interactions of protein with the solvent, E_e [8]. The water-structure-perturbing effect of inorganic salt was suggested [8] to be quantified by a molal surface tension increment of a given salt.

It was shown [8] that analysis of relationship between the salting-out constants for a given protein and the salts' molal surface tension increments allows one to estimate "the relative surface hydrophobicity" calculated as the ratio of the non-polar surface area to the molecular weight of the protein. According to Melander and Horvath [8], the hydrophobic character of a protein macromolecule is supposed to be constant at both high and 'physiological' concentrations of different inorganic salts, which seems to be untrue in most cases (see below).

The approach [8] under discussion, although not commonly used, deserves special attention as it seems to be the only one taking into account the fact that hydration interactions of a biological macromolecule depend upon the concentration of the component of an aqueous solution affecting the structure and/or thermodynamic state of water in the solution. At the same time the authors of the model [8] assumed the constancy of the protein hydrophobicity which is a measure of the above interactions varying with the composition of an aqueous medium. This example seems to be typical of that even in the case of clear contradiction between experimental results and interpretation of the results from the conventional point of view, the conventional ideas take the upper hand. In this particular case [8] that means that the hydrophobicity of a solute is considered as an intrinsic property of the solute and not as a measure of the intensity of the interaction between the solute and the solvent which depends on properties of both solute and solvent.

Finally, the treatment of solute-solvent interactions based on the surface thermodynamics principles described above (Chapter 2) may also be used for estimates of the solutes hydrophobicity and hydrophilicity [130-132].

Measurements of Interfacial Interactions.

According to C.J. van Oss [130-133] "... hydrophobicity and hydrophilicity is the competition between the interfacial free energy of cohesion of the solid (particles or molecules), immersed in the liquid, and the free energy of cohesion of the liquid (water in this case)". On the one hand, van Oss stated [131] that "the most rigorous way of expressing hydrophilicity (hydrophobicity) of any material 'i' is in terms of its free energy of adhesion to water, ΔG_{iw}, i.e., its free energy of hydration". On the other hand, the interfacial interaction energy, ΔG_{iwi}, between, for example, proteins of the same species, immersed in water was suggested [131] to be used as a hydrophilicity (hydrophobicity) measure.

Interfacial interaction energy for a given biopolymer "i", ΔG_{iwi}, may be determined as [131]:

$$\Delta G_{iwi} = -2(\sqrt{\gamma_i^{LW}} - \sqrt{\gamma_w^{LW}})^2 - 4(\sqrt{\gamma_i^+} \cdot \sqrt{\gamma_i^-} + \sqrt{\gamma_w^+} \cdot \sqrt{\gamma_w^-} - \sqrt{\gamma_i^+} \cdot \sqrt{\gamma_w^-} - \sqrt{\gamma_i^-} \cdot \sqrt{\gamma_w^+}) \quad (6.11)$$

where all terms are as defined above in Chapter 2. All the surface tension components and parameters may be determined from the measurements of contact angles of a liquid on a surface of the biopolymer using a series of apolar and polar liquids (see in Chapter 2).

It clearly follows from the definition that the positive value of $\Delta G_{iwi} > 0$ means that the solute is hydrophilic, as its molecules repel each other, i.e., their affinity for water exceeds that for each other. The negative value of $\Delta G_{iwi} < 0$ indicates the solute to be hydrophobic. The ΔG_{iwi} value may be used as a quantitative measure of the relative hydrophobicity (hydrophilicity) of a solute.

For the measurements of the contact angles of different liquids on surfaces of dry proteins [131,134], a protein solution in distilled water (usually) is spread over a glass slide and allowed to dry (and kept for a few days in a vacuum desiccator, in the presence of a dehydrating material). For the contact angles measurements on surfaces of hydrated proteins, a concentrated solution must be further concentrated in an ultrafilter. Measurements of contact angles are then performed on the hydrated protein layer on top of the ultrafilter membrane [134].

Results of these measurements for human serum albumin (HSA) reported by van Oss et al.[131,135] offer an illustrative example. The interfacial free energy ΔG_{iwi} values are -52.5 mJ/m^2 for dry HSA at pH 4.9; -22.9 mJ/m^2 for dry HSA at pH 7.0; +62.0 mJ/m^2 for hydrated HSA (with 1 layer of hydration water) at pH 7.0; and +20.9 mJ/m^2 for hydrated HSA (with 2 layers of hydration water) at pH 7.0. The free energy of hydration ΔG_{iw} derived from these ΔG_{iwi} values were reported [131] as: -91.6 mJ/m^2 for dry HSA at pH 4.9; -105.3 mJ/m^2 for dry HSA at pH 7.0; -143.9 mJ/m^2 for hydrated HSA (with one layer of hydration water) at pH 7.0; and -145.3 mJ/m^2 for hydrated HSA (with two layers of hydration water) at pH 7.0.

It was concluded by van Oss [131] that all proteins with the free energy of hydration $\Delta G_{iw} > |\approx 115|$ mJ/m^2 are hydrophilic, and proteins with $\Delta G_{iw} < |\approx 115|$ mJ/m^2 are hydrophobic. No protein was found to be more hydrophilic than water (ΔG_{ww} = -145.6 mJ/m^2), though HSA hydrated with two layers of water is rather close.

An important advantage of the approach developed by van Oss [130-135] to estimating the relative hydrophobicity of biological (and synthetic) materials is that it is applicable to cells and other particles as well as to soluble materials such as polysaccharides, proteins, etc. The limitation of the technique is, however, that the conformation of a biological macromolecule in the surface layer (dried or hydrated) may differ from those in solution. Additionally, the technique may hardly be used to study the effects of the composition of an aqueous medium on the relative hydrophobicity of biopolymers (see below).

Using the advantages and limitations of the methods discussed above as the guidelines, it is possible to define the requirements for an 'ideal' method for estimating the hydrophobicity of biological solutes. These requirements are:

1. The method should not employ an organic solvent as there seems to be no solvent capable of dissolving polar compounds and inert toward different functional groups in a solute under examination.
2. Biological solutes being analyzed should maintain their functions under the conditions employed by the method.
3. The hydrophobicity estimates provided by the method for solutes of different structure and chemical nature should be comparable, i.e., should be possible to be viewed on the unified scale.
4. The hydrophobicity estimates provided by the method for "inert" solutes should be correlated with those obtained by the water-octanol partition technique and other related techniques. The term "inert" here covers the solutes unlikely to participate in specific interactions with octanol and other organic solvents.
5. The method should allow one to explore an influence of the chemical composition of an aqueous medium on the hydrophobicity of different solutes.

The technique of solute partitioning in aqueous polymer two-phase systems seems to meet almost all of these requirements.

6.5. PARTITIONING IN AQUEOUS TWO-PHASE SYSTEMS AS A METHOD FOR ESTIMATING THE RELATIVE HYDROPHOBICITY OF SOLUTES

In order to discuss if the method of partitioning in aqueous two-phase systems meets the above requirements for an "ideal" technique for estimating the solute hydrophobicity, it is necessary, first, to summarize the features of the method relevant to the application under discussion. These features are:

(i) Both phases of a given aqueous two-phase system comprise two aqueous polymeric media of different chemical composition. No organic solvent is present. According to the above experimental evidence (see Figure 5.20), phase polymers in the aqueous Dex—PEG and Dex—Ficoll two-phase systems do not participate in direct interactions with the solutes being partitioned. It is possible, hence, to view the process of a solute partitioning in these systems as transfer of the solute from the aqueous medium with one set of properties into the aqueous medium with a different set of properties.

(ii) It is well known [129] that biological solutes being subjected to partitioning in aqueous polymer two-phase systems usually maintain their function and biological activity when the partition is carried out at a suitable pH and salt concentration. Phase polymers in many cases seem even stabilize proteins against inactivation (see, for example, in [129, pp.99-102]).

(iii) The media in both phases is of the same aqueous nature. Hence the only difference between the partition behavior of solutes of various chemical nature and structure may arise from the difference between their interactions with the same aqueous media, i.e., from the difference between the hydrophobicity of the solutes. That means that the hydrophobicity estimates for different solutes may be regarded on the unified scale.

(iv) As shown below, the relative hydrophobicity estimates provided by the method of partitioning in aqueous two-phase systems for non-polar solutes are correlated with those obtained by the water-organic solvent partition technique.

(v) The method of partitioning in aqueous two-phase systems allows one to study an influence of the chemical composition of an aqueous medium on the relative hydrophobicity of different solutes (see below).

The shortcomings of the method considered below do not allow one to view it as an "ideal" method for estimating the hydrophobicity of biological solutes. It seems, however, to be a much better approximation of the "ideal" technique than any of the other methods discussed above.

Partitioning of amino acids and their dinitrophenylated (DNP-) derivatives in the aqueous Dex-Ficoll two-phase systems with salt composition varied from 0.11 mole/kg sodium phosphate buffer, pH 7.4 to 0.15 mole/kg NaCl in 0.01 mole/kg sodium phosphate buffer, pH 7.4 was studied by Zaslavsky et al.[136]. Using DNP-glycine as a reference, and subtracting its free energy of transfer between the two phases from that of all the other DNP-amino acids, the free energies of transfer of the amino acids' side-chains were determined [136]. The values obtained [136] are presented in Table 6.1 together with those for the same amino acids' side-chains determined in different water-organic solvent systems by various authors.

The data given in Table 6.1 show that the values determined in the aqueous Dex-Ficoll two-phase system [136] differ from those reported in the literature by one to two orders of magnitude.

The only reasonable explanation of the disagreement observed seems to be related to the different properties of the phases in the systems used. The differences in question are indicated by the different values of the free energy of transfer of a methylene group, $\Delta G(CH_2)$, given in Table 6.1. It was shown above that to compare the interactions of a j-th polar group with the solvent media in the phases of different solvent two-phase systems, the A_m/E_m or $\Delta G(\text{polar group j})_m/\Delta G(CH_2)_m$ ratio (subscript "m" denotes the particular solvent system used) should be used as a measure of the interactions (see Equation 4.15 and Fig. 4.14). The reason seems to be that the use of the free energy of transfer of a CH_2 group as a denominator allows for the difference in the nonspecific properties of the phases. The nonspecific properties mean here the properties originated from the packing of the solvent and its ability to participate in van der Waals interactions. The hydrogen bonding and electrostatic properties of a solvent are displayed depending on the particular solute being partitioned and contribute to the total free energy of the solute transfer between the phases, $\Delta G(\text{solute})_{tr}$.

The ratio expressed as

$$\Delta G(\text{solute})_{tr}/\Delta G(CH_2) = n(CH_2) \qquad (6.12)$$

has been defined as the equivalent quantity of methylene groups and suggested by Zaslavsky et al.[136] to be used as a measure of the relative hydrophobicity of a solute (or a moiety).

A positive value of $n(CH_2)$ means that a given solute (or a moiety) is hydrophobic and its relative hydrophobicity is equal to that of n amount of methylene groups. A negative value of $n(CH_2)$ means that the solute (or a moiety) is hydrophilic and its relative hydrophobicity is the reverse of that of n amount of CH_2 groups.

The relative hydrophobicity estimates for the side-chains of different amino acids obtained in [136] and those calculated from the data reported in the literature according to:

$$n(CH_2)^{\text{side-chain i}} = [\Delta G(\text{amino acid i})_{tr} - \Delta G(\text{glycine})_{tr}]/\Delta G(CH_2) \qquad (6.13)$$

are listed in Table 6.2.

The data given in Table 6.2 indicate that the relative hydrophobicity estimates for the amino acids with aliphatic side-chains are in a good agreement. The $n(CH_2)$ values determined in [136] for the side-chains of DNP-tryptophan and DNP-phenylalanine appear to be too high likely due to the effect of the dinitrophenyl moiety on interactions of these side-chains with an

Table 6.1 Free Energies of Transfer of Amino Acids' Side Chains (ΔG_{tr}, cal/mole)*.

Amino acid	Ficoll → Dex	EtOH	Hexane	Air/H_2O	i-AmOH	$CHCl_3$	CCl_4	Hexanol	MetOH	Octanol	Octanol	Octanol*	Octanol*
	[136][a]	[17]	[25]	[106]	[98]	98]	[98]	[98]	[100]	[98,105]	[101]	[37] [b]	[114] [c]
Trp	275.7	3400	-	-	-	-	-	-	3390	2940	2936	2730	2635
Phe	112.1	2500	2490	2330	-	-	-	-	2770	2540	2308	2293	2144
norLeu	61.1	2600	-	-	2230	2550	3170	2380	-	2890	-	-	-
Leu	52.2	1800	2700	2460	2070	2460	3070	2170	2490	3070	1570	1802	1748
Ileu	61.7	-	2650	2260	-	-	-	-	-	2080	-	1938	1598
Met	49.2	1300	-	-	-	-	-	-	-	1830	-	1693	1270
norVal	49.2	-	-	-	1590	2040	2340	1710	-	2180	-	-	-
Val	38.5	1500	1680	1560	1430	2000	2360	1560	1690	1590	1338	1160	1120
Ala	22.5	500	-420	200	440	740	830	480	770	670	341	505	-328
Tre	11.9	400	-40	-	-	-	-	-	-	460	-	-	341
Pro	5.3	-	250	980	-	-	-	-	-	-	-	778	1174

Gly[d]	0	0	0	0	0	0	0	0	0	0	0	0	0
Ser	-7.7	-300	-770	-390	-	-	-	-	-	-68	-	54	55
Gln	5.3	-	-	200	-	-	-	-	-	140	-	-	-
Aspn	-7.7	-	-	-	-	-	-	-	-	-218	-	-	-
Glu	-19.45[e]	-	-	300	-	-	-	-	-	-1280	-1775	-396	-
Asp	-56.5[e]	-	-	-	-	-	-	-	-	-	-	-	-
Arg	-1.1[e]	-	-	120	-	-	-	-	-	-1290	-1679	-5489	-
Lys	-59.9[e]	-	-	350	-	-	-	-	-	-81	-1188	492	-
$\Delta G(CH_2)$	16.1	589	692	640	604	846	922	587	708	727	727	568	473

* free energy of transfer of a given side chain from the solvent indicated into water;

a ionic composition of the system - 0.11 mole/kg sodium phosphate buffer, pH 7.4 [136];

b 0.01 M phosphate buffer, pH 7.0 was used in [37] as the aqueous phase in the octanol-water system;

c 0.10 M phosphate buffer, pH 7.0 was used in [114] as the aqueous phase in the octanol-water system;

d free energy of transfer of the side chain of glycine (-H) is zero by definition;

e at the ionic composition of the system of 0.15 mole/kg NaCl in 0.01 mole/kg sodium phosphate buffer, pH 7.4 the free energies of transfer of the side chains of the denoted amino acids are (in cal/mole): Glu -62.8; Asp -72.5; Arg +22.6; Lys -7.1.

aqueous medium. The estimates for the same side-chains obtained from the partition experiments with free amino acids are given in parenthesis and seem to be in a better agreement with those calculated from the literature data.

The relative hydrophobicity of the phenyl moiety may be calculated as the difference between the estimates given in Table 6.2 for the side-chains of phenylalanine and alanine. It amounts to +3.4 equivalent CH_2 groups as calculated from the data reported in the aqueous Dex-Ficoll two-phase system [136] as compared to the value of +3.3 ± 1.0 equivalent CH_2 groups calculated as an average over all the other data for the same side-chains listed in Table 6.2

Using the additivity principle, it is possible also to estimate the relative hydrophobicity of an aliphatic hydroxyl group from comparison of the $n(CH_2)$ values for the side-chains of serine and threonine with that for alanine. The value in question corresponds to -1.7 ± 0.1 equivalent CH_2 groups when measured in the aqueous two-phase system. This estimate is in a fair agreement with those calculated from the data obtained in water-alcohol systems, e.g., -1.2 equivalent CH_2 groups (using ethanol [17]) and -1.15 equivalent CH_2 groups in water-octanol system [37,98,105], and as might be expected it differs significantly from -0.5 equivalent CH_2 group calculated from the data obtained in the water-hexane system [25].

For the other polar groups, the agreement between the relative hydrophobicity estimates may also be viewed as fair. The relative hydrophobicity of ε-amino group in lysine, for example, may be estimated from comparison of the data in Table 6.2 for the side-chains of lysine and norleucine derivatives. The estimate amounts to -7.5 equivalent CH_2 groups in the aqueous medium containing 0.11 mole/kg sodium phosphate buffer, pH 7.4 and -4.2 equivalent CH_2 groups in the aqueous medium containing 0.15 mole/kg NaCl in 0.01 mole/kg sodium phosphate buffer, pH 7.4, i.e., it clearly depends on the salt composition of the medium [136]. The octanol-water partition technique provides the relative hydrophobicity estimate for the same ε-amino group in lysine corresponding to -4.6 ± 0.5 equivalent CH_2 groups as calculated from the data reported in [98,99,101,105].

The relative hydrophobicity of the side-chain amide group corresponds to -1.8 ± 0.3 equivalent CH_2 groups independent of the salt composition of the aqueous Dex-Ficoll two-phase system [136]. The estimate for the same group calculated from the data obtained by the octanol-water partition technique [98, 99,105] amounts to -1.6 ± 0.4 equivalent CH_2 groups.

The relative hydrophobicity estimates for the side-chain carboxyl group vary with the salt composition of the medium in the aqueous Dex-Ficoll two-phase system [136]. These estimates amount to -5.9 ± 0.1 equivalent CH_2 groups in the presence of 0.15 mole/kg NaCl in 0.01 mole/kg sodium phosphate buffer, pH 7.4, and to -4.0 ± 0.8 equivalent CH_2 groups in the presence of 0.11

mole/kg sodium phosphate buffer, pH 7.4. The estimate for the carboxyl group in the side-chain of glutamic acid calculated from the octanol-water partitioning data [98,99,101,105] amounts to -3.8 ± 0.1 equivalent CH_2 groups. The agreement between the two latter values is likely to be purely coincidental, however.

Thus, the results obtained by partitioning of amino acids derivatives in the aqueous Dex-Ficoll two-phase systems [136] indicate that the relative hydrophobicity estimates for hydrocarbon fragments as well as for certain polar moieties such as hydroxyl and amide groups are essentially identical with those obtained by the water-organic solvent partition technique or the solubility measurements, provided the estimates are presented in terms of equivalent methylene groups. The data [136] given in Table 6.2 show also that the method of partitioning in aqueous two-phase systems allows one to study the influence of the composition of an aqueous medium on the relative hydrophobicity of solutes (moieties).

The data on partitioning of adenine and uridine in various Dex-PEG and Dex-Ficoll two-phase systems [137] listed in Table 6.3 emphasize that the partition technique under discussion provides the estimates of the relative hydrophobicity and not the absolute hydrophobicity of a solute.

The data presented in Table 6.3 show that the partition behavior of both solutes represented by the free energies of interfacial transfer varies depending on the particular system, salt composition, and pH employed. The estimates of the solutes' hydrophobicities given as the equivalent quantities of CH_2 groups vary accordingly. The difference between the estimates, however, is constant and amounts to 4.7 ± 0.1 equivalent CH_2 groups [137]. That means that the relative hydrophobicity of adenine exceeds that of uridine by 4.7 ± 0.1 equivalent CH_2 groups. The solute-water interactions are likely to vary with changes in the polymer and salt composition of the two phases. That is the likely reason for the variations in the hydrophobicity estimates observed for both solutes.

According to Equations 4.6b and 4.10 considered above, both free energy of transfer of a given solute and free energy of transfer of a CH_2 group are directly proportional to the difference between the concentrations of a phase polymer in the two phases. Hence the relative hydrophobicity of the solute represented by the equivalent quantity of methylene groups should be independent of the concentrations of phase polymers in the system employed. That was verified experimentally [138].

Partitioning of two nonionic glycosides, 4-nitrophenyl-N-acetyl-ß-D-glucosaminide and 4-nitrophenyl-α-D-mannopyranoside, was studied in aqueous Dex-PEG and Dex-PVP two-phase systems of varied polymer concentrations and various salt compositions [138]. The results obtained [138] are

Table 6.2 Relative Hydrophobicity of Amino Acids' Side Chains (expressed in equivalent quantities of methylene groups [a]).

Amino acid	Ficoll → Dex [136][b]	EtOH [17]	Hexane [25]	Air/H_2O [106]	i-AmOH [98]	$CHCl_3$ [98]	CCl_4 [98]	Hexanol [98]	MetOH [100]	Octanol [98,105]	Octanol [101]	Octanol* [37][c]	Octanol* [114][d]
Trp	17.1 (8.1)[e]	5.8	-	-	-	-	-	-	4.8	4.0	4.0	4.8	5.6
Phe	7.0 (4.7)[e]	4.2	3.6	3.6	-	-	-	-	3.9	3.5	3.2	4.0	4.5
norLeu	3.8	4.4	-	-	3.7	3.0	3.4	4.0	-	4.0	-	-	-
Leu	3.2	3.0	3.9	3.8	3.4	2.9	3.3	3.7	3.5	4.2	2.2	3.2	3.7
Ileu	3.8	-	3.8	3.5	-	-	-	-	-	2.9	-	3.4	3.4
Met	3.1	2.2	-	-	-	-	-	-	-	2.5	-	3.0	2.7
norVal	3.1	-	-	-	2.6	2.4	2.5	2.9	-	3.0	-	-	-
Val	2.4	2.5	2.4	2.4	2.4	2.4	2.6	2.6	2.4	2.2	1.8	2.0	2.4
Ala	1.4 (1.3)[e]	0.8	-0.6	0.3	0.7	0.9	0.9	0.8	1.1	0.9	0.5	0.9	-0.7

Tre	0.7	0.7	-0.1	-	-	-	-	-	-	0.6	-	-	0.7
Pro	0.3	-	0.4	1.5	-	-	-	-	-	-	-	1.4	2.5
Ser	-0.5	-0.5	-1.1	-0.6	-	-	-	-	-	-0.1	-	0.1	0.1
Gln	0.3	-	-	0.3	-	-	-	-	-	0.2	-	-	-
Aspn	-0.5	-	-	-	-	-	-	-	-	-0.3	-	-	-
Glu	-1.2[f]	-	-	0.5	-	-	-	-	-	-1.8	-2.4	-0.7	-
Asp	-3.5[f]	-	-	-	-	-	-	-	-	-	-	-	-
Arg	-0.07[f]	-	-	0.2	-	-	-	-	-	-1.8	-2.3	-9.7	-
Lys	-3.7[f]	-	-	0.5	-	-	-	-	-	-0.1	-1.6	0.9	-

[a] equivalent quantity of methylene groups, $n(CH_2)$, is determined as $n(CH_2) = \Delta G(\text{solute})_{tr}/\Delta G(CH_2)$; for definition, see text;
[b] ionic composition of the system - 0.11 mole/kg sodium phosphate buffer, pH 7.4 [136];
[c] 0.01 M phosphate buffer, pH 7.0 was used in [37] as the aqueous phase in the octanol-water system;
[d] 0.10 M phosphate buffer, pH 7.0 was used in [114] as the aqueous phase in the octanol-water system;
[e] values in parentheses are the relative hydrophobicity estimates obtained from analysis of partitioning of free amino acids;
[f] at the ionic composition of the system of 0.15 mole/kg NaCl in 0.01 mole/kg sodium phosphate buffer, pH 7.4 the relative hydrophobicities of the denoted side chains are (in equivalent quantity of CH_2 groups): Glu -3.9; Asp -4.5; Arg +1.4; Lys -0.4.

Table 6.3 Relative Hydrophobicity of Adenine and Uridine as Measured in Several Different Aqueous Two-Phase Systems.

System [a]	pH	Concentration of SPB mole/kg	Concentration of NaCl mole/kg	$\Delta G(CH_2)$ cal/mole	$\Delta G(\text{solute})_{tr}$ cal/mole		$n(CH_2)$ [b]		$\Delta n(CH_2)$ [c]
					Adenine	Uridine	Adenine	Uridine	
I	6.8	0.11	0	22.18	716.30	614.35	32.29	27.70	4.59
I	6.8	0.085	0	20.04	705.80	610.79	35.18	30.48	4.70
I	6.8	0.01	0.15	17.20	692.62	609.01	40.27	35.41	4.86
I	6.8	0.01	0.20	17.32	689.07	607.82	39.78	35.09	4.69
II	6.15	0.11	0	8.42	662.97	623.84	78.74	74.09	4.65
II	7.40	0.11	0	16.07	683.14	603.67	42.51	37.56	4.95
II	7.80	0.11	0	17.79	690.25	607.82	38.80	34.17	4.63

[a] system I: 7.0% wt. Dex-500, 4.4% wt. PEG-6000; system II: 10.8% wt. Dex-70; 12.5% wt. Ficoll-400; [b] equivalent quantity of methylene groups determined according to Equation /6-11/; [c] difference between the relative hydrophobicity of adenine and uridine determined as: $\Delta n(CH_2) = n(CH_2)^{adenine} - n(CH_2)^{uridine}$.

Table 6.4

Relative Hydrophobicity of 4-nitrophenyl-N-acetyl-β-D-glucosaminide (NAcGl) and 4-nitrophenyl-α-D-mannopyranoside (Mann)[a].

Salt	Conc.	Dex-PEG			Dex-PVP		
	mole/kg	Mann	NAcGl	$\Delta n(CH_2)^b$	Mann	NAcGl	$\Delta n(CH_2)^b$
-	-	1.68	3.45	1.77	3.14	4.84	1.70
KSCN	0.10	2.49	4.37	1.88	5.44	7.19	1.75
KSCN	0.50	4.07	5.68	1.61	-	-	-
NH_4SCN	0.10	-	-	-	4.16	5.91	1.75
NaSCN	0.10	-	-	-	5.51	7.26	1.75
KCl	0.10	2.77	4.55	1.78	4.55	6.39	1.84
KCl	0.50	2.85	4.60	1.75	-	-	-
KCl	0.75	2.82	4.58	1.76	-	-	-
KF	0.10	-	-	-	6.27	8.02	1.75
Na_2SO_4	0.10	3.56	5.30	1.75	5.86	7.78	1.92
K_2SO_4	0.10	3.02	4.78	1.75	-	-	-
K_2SO_4	0.10	2.91	4.66	1.75	5.63	7.62	1.99
K_2SO_4	0.10	2.60	4.20	1.60	-	-	-

[a] expressed in equivalent quantities of CH_2 groups determined according to Equation 6.11 as measured by partitioning in aqueous Dex—PEG—salt and Dex—PVP—salt two-phase systems;

[b] difference between the relative hydrophobicity of the two glycosides is determined as $n(CH_2)^{NAcGl} - n(CH_2)^{Mann}$.

(Calculated from the data reported in [138].)

presented in Table 6.4 as the relative hydrophobicities of the glycosides expressed in equivalent quantities of methylene groups.

The relative hydrophobicity of both glycosides varies depending on the salt composition of the particular system employed [138]. The difference between the relative hydrophobicities of the solutes is constant, however. It follows from these data [138] that the relative hydrophobicity of 4-nitrophenyl-N-acetyl-ß-D-glucosaminide exceeds that of 4-nitrophenyl-α-D-mannopyranoside by 1.77 ± 0.09 equivalent CH_2 groups under all the different conditions explored.

The data on the partitioning of proteins, human serum albumin, horse myoglobin, and equine heart cytochrome C, in different aqueous Dex-Ficoll two-phase systems formed by the polymers of various molecular weights and concentrations [139] support the above assertion that the relative hydrophobicity of a solute is independent of the concentrations of phase polymers in the aqueous two-phase system used. The differences between the relative hydrophobicities of the proteins in the aqueous medium of a given salt composition were found to be constant [139].

It follows from the above experimental data that the technique of partitioning in aqueous two-phase systems meets practically all the aforementioned requirements for an "ideal" method for estimating the hydrophobicity of biological (and synthetic) solutes. The technique does not use an organic solvent and provides the relative hydrophobicity estimates for solutes of different nature on the unified scale. The estimates provided by the technique for solutes of low and "mild" polarity are well correlated with those provided by the water-organic solvent partitioning and the related methods. Biological solutes being analyzed by the technique maintain their functions. An influence of the chemical composition of an aqueous medium on the relative hydrophobicity of solutes may also be explored by the technique.

The major drawback of the technique is that it does not allow one to measure the hydrophobicity of a solute as defined above by the term E_s in Equation 6.1. What it does allow one to measure instead, is the difference between the E_s values for a given solute in two aqueous media of different composition. In other words, the technique provides an information about the sensitivity of the intensity of the solute-water interactions for a given solute toward changes in the composition of an aqueous medium.

It should be noticed, that the water-organic solvent partition and the related methods give an information about the difference between the intensity of the solute-solvent interactions in a given organic solvent and water. That means that these methods do not permit the measurement of solute hydrophobicity, E_s, as well. The similarity of the solvent in both phases of an aqueous two-phase system offers an obvious advantage over a water-organic solvent system.

What is measured by the technique under discussion is the hydrophobicity of a solute in reference to the solute (or moiety) chosen for standard as a function of the composition of an aqueous medium. This measure may be defined as the **relative** hydrophobicity of the solute.

It should be stressed particularly that the difference in the relative hydrophobicity of two solutes is constant and independent of the specific aqueous two-phase system employed provided the composition of an aqueous medium is the same. The medium composition in this case covers all the additives (salts, low- and high-molecular-weight compounds) present in the system but the phase polymers used.

An influence of inorganic salts on the solute-solvent interactions in aqueous two-phase systems, i.e. on the relative hydrophobicity of a given solute, especially in the case of ionic solutes (see above), must always be taken into account. From this viewpoint the results reported by Hsu and his colleagues [113] warrant more detailed consideration.

Partition behavior of amino acids and several dipeptides was studied by Diamond et al.[113] in the aqueous PEG-3400-potassium phosphate systems. Unfortunately, the systems of varied polymer and salt concentrations used in [113] have not been characterized in terms of the free energy of interfacial transfer of a CH_2 group. Using the average relative hydrophobicity values for the side-chains of leucine (+3.4), isoleucine (+3.5), and valine (+2.3) from Table 6.2, however, it is possible to estimate the $\Delta G(CH_2)$ value for the aqueous PEG-salt system used by Diamond et al. [113]. The estimate amounts to 101.18 ± 7.67 cal/mole CH_2 (for transfer from the PEG-rich phase into the salt-rich phase). Using this value, the relative hydrophobicity estimates for the side-chains of amino acids examined in [113] may be calculated according to Equation 6.13. It is also possible to estimate the relative hydrophobicities of the side-chains of amino acid residues at the N- and C-termini of dipeptides from the partition coefficients values for different dipeptides reported in [113].

The relative hydrophobicity of a given side-chain of an amino acid residue may be estimated according to:

$$n(CH_2)_C^{\text{side-chain } i} = [\Delta G(\text{Gly-X}_i)_{tr} - \Delta G(\text{Gly-Gly})_{tr}]/\Delta G(CH_2) \qquad (6.14)$$

and

$$n(CH_2)_N^{\text{side-chain } i} = [\Delta G(\text{X}_i\text{-Gly})_{tr} - \Delta G(\text{Gly-Gly})_{tr}]/\Delta G(CH_2) \qquad (6.15)$$

where Gly-Gly, X_i-Gly, and Gly-X_i denote dipeptides of the indicated structures; subscripts C and N denote the C-terminal or N-terminal position of the X_i amino acid residue in the dipeptide structure.

Estimates of the relative hydrophobicities of the amino acid residues side-chains calculated according to Equations 6.14 and 6.15 and those calcu-

Table 6.5

Relative Hydrophobicity of Side Chains of Amino Acid Residues [a].

Amino acid	$n(CH_2)$ [b]	$n(CH_2)^{C}$ [c]	$n(CH_2)^{N}$ [d]	$n(CH_2)^{average}$ [e]
Trp	8.30	8.86	9.04	8.73 (0.39)
Phe	5.0	4.85	5.50	5.12 (0.34)
Leu	3.69 [f]	2.78	3.45	3.31 (0.47)
Ileu	3.40 [f]	2.78	3.62	3.27 (0.44)
Met	2.30	2.21	2.86	2.46 (0.35)
Val	2.09 [f]	1.88	2.31	2.09 (0.22)
Ala	0.64	0.46	0.72	0.61 (0.13)
Pro	1.41	0.69	1.56	1.22 (0.46)
Tyr	4.69	5.00	5.61	5.10 (0.47)
Ser	-0.60	-0.50	-0.03	-0.38 (0.30)
Gln	0.88	-	-	-
Aspn	0.23	-	-	-
Glu	1.00	-	-	-
Asp	0.23	-0.85	0.93	0.10 (0.90)
His	-0.53	-1.48	0.30	-0.57 (0.89)
Lys	-2.40	-2.60	-2.13	-2.38 (0.24)

[a] expressed in equivalent quantities of methylene groups, $n(CH_2)$ determined according to Equation 6.12;

[b] calculated from the partition coefficients for free amino acids according to Equation 6.12;

[c] amino acid residue at the C-terminal position in a dipeptide structure; calculated from the partition coefficients for dipeptides of the general structure Gly-X and Gly-Gly according to Equation 6.13;

[d] amino acid residue at the N-terminal position in a dipeptide structure; calculated from the partition coefficients for dipeptides of the general structure X-Gly and Gly-Gly according to Equation 6.14;

[e] average for all the three estimates of the relative hydrophobicity for a side chain of a given amino acid residue; deviation is given in parenthesis;

[f] average value for the relative hydrophobicity of the side chain of a given amino acid from Table 6.2 was used to calculate $\Delta(CH_2)$ for the aqueous PEG-salt two-phase system [113] under consideration.

(Calculated from the data reported in [113].)

lated according to Equation 6.13 from the data for dipeptides and free amino acids reported in [113] are presented in Table 6.5.

It follows from the data in Table 6.5 (compare with Table 6.2) that: (i) there a fair agreement between the estimates obtained in different two-phase systems for non-polar and certain polar side-chains, e.g., those of threonine and serine; (ii) the estimates for polar and ionic side-chains calculated from the data obtained in the aqueous PEG-salt two-phase system [113] are closer to those obtained from the data in water-organic solvent systems that to those calculated from the data in the aqueous Dex-Ficoll two-phase system [136]; and (iii) the effect of an amino acid residue position in the dipeptide structure on the relative hydrophobicity of the residue side-chain is within the experimental error limit for the most of the residues examined in [113]. The effect in question seems to be significant for only two amino acid residues among those examined by Diamond et al.[113], namely for aspartic acid and histidine residues. In both cases the affinity of an ionic side-chain for an aqueous medium appears to increase when the residue is at the C-terminal position in the dipeptide structure, and the side-chain-water interactions seem to be hindered when the residue amino group is free, the more so when the α-carboxyl group of the residue is engaged in the peptide bond. Rather surprisingly, no positional effect seems to exist for the side-chain of the lysine residue. The effects in question seem to be marginal. Further study is needed before the position effect may be considered as an established experimental fact.

The aforementioned fact that some of the estimates of the relative hydrophobicity of the amino acid side-chains given in Table 6.5 are closer to those obtained from the solute partitioning in water-organic solvent systems (see Table 6.2) than to those determined by the technique of partitioning in aqueous two-phase systems requires an explanation. The likely reason is that the relatively large salt concentration in the salt-rich phase of the aqueous PEG-salt system used [113] affects the solute-solvent interactions in this phase. For some solutes, this effect may lead to results similar to those observed in water-organic solvent systems where the nature of counter-ions may affect the partition behavior of ionic solutes [104,140] significantly (see above). Additionally, aqueous PEG-salt two-phase systems may not allow one to examine the effect of the chemical composition of an aqueous medium on the relative hydrophobicity of solutes. That is why the attempts to analyze the relative hydrophobicity of peptides using aqueous PEG-salt two-phase systems [113,141-144] should be considered with caution.

Thus, the technique of the solute partitioning in aqueous two-polymer two-phase systems may be concluded to be the most suitable for studying the relative hydrophobicity of biological (and synthetic) solutes, the aqueous Dex-PEG and Dex-Ficoll two-phase systems currently being the systems to be used.

6.6. SUMMARY

Summing up the above considerations, it should be repeated that the quantitative structure-activity relationships (QSAR) analysis is among the most successful approaches to rational drug design and better understanding of possible mechanism of an action of a substance on a biological system. That is well established for common drugs and chemicals and it is theoretically true for naturally occurring substances (peptides, nucleotides, proteins, etc.) and their derivatives.

Various physicochemical descriptors are used in the QSAR analysis to represent the structure of a given substance. One of the most fundamental descriptors is known to be the hydrophobicity of the substance presumably due to its relation to the substance transport from the site of administration to the site of action (receptor compartment) as well as to the substance-receptor interactions.

According to the definition, the hydrophobicity of a substance is a measure of the overall intensity of the total interactions of the substance with an aqueous medium (including hydrogen bonding, van der Waals, electrostatic interactions, etc.). The common methods for estimating the hydrophobicity of chemical compounds are based on measurements of the free energy of transfer of a compound from an organic solvent into water using the water-organic solvent partitioning technique, solubility measurements, etc. Using these methods for hydrophobicity measurements implies the assumption of total inertness of an organic solvent toward the solute under examination. This assumption is clearly incorrect for the most of the polar solutes being studied. Therefore these methods actually provide the possibility to measure the difference between the intensity of the solute-water interactions and that of the solute-organic solvent interactions. That is the likely reason why the hydrophobicity estimates obtained by these techniques may be compared only for the solutes of the same or close chemical nature. (For these solutes the intensity of the solute-organic solvent interactions may be viewed as similar.)

An additional limitation of the common methods for estimating the solute hydrophobicity is that due to the employment of an organic solvent these methods are poorly suited for analysis of highly polar and ionic compounds, and especially conformationally flexible biological solutes.

The method of partitioning in aqueous two-phase systems is suited for studying biological and highly polar and ionic solutes much better than the water-organic solvent partitioning and the related techniques. The advantages of the method in question are: (i) no organic solvent is used; (ii) the solvent media in the two phases is of the same aqueous nature; and (iii) the biological function of a solute being subjected to analysis is usually maintained. Additionally, the partition coefficients of solutes of low and mild polarity in an aqueous polymer two-phase system are well correlated with those in, e.g., water-octanol system.

The limitation of the method of partitioning in aqueous two-phase systems is that it allows one to measure actually not the hydrophobicity of a solute (as defined above) but the difference between the intensities of the solute-aqueous medium interactions in the two aqueous media of different chemical composition. In other words, the method allows one to measure the sensitivity of the solute hydrophobicity toward the chemical composition of an aqueous medium. As the estimates provided by the method for solutes or moieties of low and mild polarity are identical with those provided by the standard commonly used techniques, however, the estimates in question may be viewed to a first approximation as those of the relative hydrophobicities of the solutes. It is particularly important that the difference between the estimates for different solutes is independent of the particular aqueous polymer, e.g., Dex-PEG or Dex-Ficoll, two-phase system employed (i.e., independent of the nature and concentration of phase polymers, and for nonionic solutes independent of the nature and concentration of salt additives present as well).

In order to compare the relative hydrophobicity estimates obtained in different two-phase systems it is convenient to use the equivalent quantity of methylene groups, $n(CH_2)$ defined as the ratio between the free energy of transfer of a given solute from one phase into the other phase and the free energy of the similar transfer of a CH_2 group.

A positive value of $n(CH_2)$ means that a given solute (or a moiety) is hydrophobic and its relative hydrophobicity is equal to that of n amount of methylene groups. A negative value of $n(CH_2)$ means that the solute (or a moiety) is hydrophilic and its relative hydrophobicity is the reverse of that of n amount of CH_2 groups.

The relative hydrophobicity of a solute represented by a given quantity of methylene groups is independent of the particular two-phase system used provided there are no specific interactions of the solute with an organic solvent or (in aqueous two-phase systems) with a phase polymer or additives (inorganic salt, polymer additive, etc.) present in the system.

Using the method under consideration and the equivalent quantity of methylene groups as a measure of the solute relative hydrophobicity, it is possible to construct a unified scale of the relative hydrophobicity of various biological and synthetic compounds. While the correctness of the absolute position of a given solute on the scale is uncertain, it is definitely correct in reference to those for all the other solutes. It must be noticed additionally that a given scale is valid only for the specified composition of an aqueous medium. Due to different effects of the chemical composition of an aqueous medium on the relative hydrophobicity of various solutes, different scales exist for different compositions of the medium, and a change in the composition may lead to a different scale with the significant changes in the relative positions of different solutes, ionic ones, in particular. This issue will be addressed in more detail below.

It should also be noted that aqueous Dex-PEG and Dex-Ficoll two-phase systems seem currently to be the most suitable systems for studying the relative hydrophobicity of solutes by the technique under consideration.

Finally, the last but not least important aspect of the possible analytical applications of the method relates to the solute partition coefficient value which, in a given aqueous two-phase system of a fixed polymer and salt composition, is the constant feature of a given solute related to the solute biological potency. This result implies that the partition coefficient of an individual biological solute may be used as a simple, highly sensitive, and cost-effective relative measure of solute identity and/or purity if the partition coefficient value for a standard reference solute is known. The technique of partitioning in aqueous two-phase systems may be particularly valuable for assessing lot-to-lot consistency of production of recombinant proteins, glycoproteins, etc. These and some other applications together with measurements of the relative hydrophobicity of biological and synthetic compounds are discussed in the next two chapters.

REFERENCES:

1. R. Franke, Theoretical Drug Design Methods, Elsevier, Amsterdam, 1984.
2. Y. C. Martin, Quantitative Drug Design: A Critical Introduction, Marcel Dekker, New York.
3. C. Hansch, A. Leo, Substituent Constants for Correlation Analysis in Chemistry and Biology, New York, Academic Press, 1979.
4. L. G. Gorb, I. A. Abronin, N. V. Harchevnikova, G. M. Zhidomirov, Rus.J.Phys.Chem., 58, 9 (1984).
5. P. A. Rebinder: Hydrophilicity and Hydrophobicity, In: Kratkaya Chim. Encyclopedia, Vol.1, Sovetskaya Encyclopedia, Moscow, 1961, p.937.
6. B. Y. Zaslavsky, E. A. Masimov, Topics Curr.Chem., 146, 171 (1988).
7. Ch. Tanford, The Hydrophobic Effect: Formation of Micelles and Biological Membranes, New York, Wiley, 1973.
8. W. Melander, C. Horvath, Arch.Biochem.Biophys., 183, 200 (1977).
9. B. Y. Zaslavsky, E. A. Masimov, L. M. Miheeva, S. V. Rogozhin, D. P. Hasaev, Doklady Acad.Nauk USSR (Rus), 261, 669 (1981).
10. E. A.Masimov, B. Y. Zaslavsky, A. A. Gasanov, S. V. Rogozhin, J.Chromatogr., 284, 337 (1984).
11. E. A. Masimov, B. Y. Zaslavsky, A. A. Gasanov, Y. A. Davidovich, S. V. Rogozhin, J.Chromatogr., 284, 349 (1984).

12. H. Planche, H. Renon, J.Phys.Chem., 85, 3924 (1981).
13. R. B. Hermann, J.Phys.Chem., 75, 363 (1971).
14. R. B. Hermann, J.Phys.Chem., 76, 2754 (1972).
15. R. B. Hermann, J.Phys.Chem., 79, 163 (1975).
16. S. M. J. Harris, T. Higuchi, J. H. Rytting, J.Phys.Chem., 77, 2694 (1973).
17. Y. Nozaki, Ch. Tanford, J.Biol.Chem., 246, 2211 (1971).
18. G. L. Amidon, S. H. Yalkowsky, S. T. Anik, S. C. Valvani, J.Phys.Chem., 79, 2239 (1975).
19. S. S. Davis, T. Higuchi, J. H. Rytting, Adv.Pharm.Sci., 4, 73 (1974).
20. B. D. Anderson, R. A. Conradi, J.Pharm.Sci., 69, 424 (1980).
21. J. C. McGowan, A. Mellors, Molecular Volumes in Chemistry and Biology: Applications including Partitioning and Toxicity, Ellis Horwood Ltd., Chichester, 1986.
22. K. C. James, Solubility and Related Properties, Marcel Dekker, New York, 1986.
23. C. Chothia, Nature, 248, 338 (1974).
24. B. D. Anderson, J. H. Rytting, T. Higuchi, J.Pharm.Sci., 69, 676 (1980).
25. J. H. Fendler, F. Nome, J. Nagyvary, J.Mol.Evol., 6, 215 (1975).
26. T. T. Herskovits, J. P. Harrington, Biochemistry, 11, 4800 (1972).
27. R. L. Scruggs, E. K. Achter, P. D. Ross, Biopolymers, 11, 1961 (1972).
28. A. Leo, C. Hansch, D. Elkins, Chem.Rev., 71, 525 (1971).
29. A. A. Alhaider, C. D. Selassie, S. O. Chua, E. J. Lien, J.Pharm.Sci., 71, 89 (1982).
30. Y. C. Martin, In: Drug Design (Ed. E. J. Ariens), Vol.8, Acedemic Press, New York, pp.1-72 (1979).
31. B. D. Anderson, In: Physical Chemical Properties of Drugs (S. H. Yalkowsky, A. A. Sinkula, S. C. Valvani, eds.), Marcel Dekker, New York, 1979, pp. 231-266.
32. R. N. Smith, C. Hansch, M. Ames, J.Pharm.Sci., 64, 599 (1975).
33. R. F. Rekker, The Hydrophobic Fragmental Constant: Its Derivation and Application. A Means of Characterizing Membrane Systems, Amsterdam, Elsevier, 1977.
34. S. H. Yalkowsky, S. C. Valvani, T. J. Roseman, J.Pharm.Sci., 72, 866 (1983).
35. A. Leo, C. Hansch, P. Y. C. Jow. J.Med.Chem., 19, 611 (1976).
36. B. Y. Zaslavsky, L. M. Miheeva, S. V. Rogozhin, J.Chromatogr., 216, 103 (1981).
37. L. M. Yunger, R. D. Cramer, Mol.Pharmacol., 20, 602 (1981).
38. E. Tomlinson, J.Chromatogr., 113, 1 (1975).

39. T. L. Hafkensheid, E. Tomlinson, Adv.Chromatogr., 25, 1 (1986).
40. H. Terada, Y. Kosuge, W. Murayama, N. Nakaya, Y. Nunogaki, K. I. Nunogaki, J.Chromatogr., 400, 343 (1987).
41. R. F. Rekker, J.Chromatogr., 300, 109 (1984).
42. R. Kaliszan, Quantitative Structure-Chromatographic Retention Relationships, Wiley, New York, 1987, pp.25-48.
43. A. Bechalany, A. Tsantili-Kakoulidou, N. E. Tayar, B. Testa, J.Chromatogr., 541, 221 (1991).
44. E. Sanchez-Moyano, C. Seco, A. Santolaria, S. Fabra-Campos, M. Herraez, A. Martin-Villodre, J.Pharm.Sci., 81, 720 (1992).
45. R. Wolfenden, Science, 222, 1087 (1983).
46. R. Wolfenden, Biochemistry, 17, 201 (1978).
47. P. M. Cullis, R. Wolfenden, Biochemistry, 20, 3024 (1981).
48. R. Wolfenden, L. Andersson, P. M. Cullis, C. C. B. Southgate, Biochemistry, 20, 849 (1981).
49. G. D. Rose, L. M. Gierasch, J. A. Smith, Adv.Prot.Chem., 37, 1 (1985).
50. A. Burger, A Guide to the Chemical Basis of Drug Design, New York, Wiley, 1983.
51. E. J. Lien, SAR: Side Effects and Drug Design, Marcel Dekker, New York, 1987.
52. Physical Chemical Properties of Drugs, Eds. S. H. Yalkowsky, A. A. Sinkula, S. C. Valvani), Marcel Dekker, New York, 1979.
53. Quantitative Structure-Activity Relationshps of Drugs (J. G. Topliss, ed.) Academic Press, New York, 1983.
54. C. Hansch, A. Leo, R. W. Taft, Chem.Rev., 91, 165 (1991).
55. A. J. Atkinson, Jr., T. I. Ruo, M. C. Frederiksen. Trends Pharm. Sci., 12, 96 (1991).
56. J. Clausen, M. H. Bickel, J.Pharm.Sci., 82, 345 (1993).
57. T. Higuchi, S. S. Davis, J.Pharm.Sci., 59, 1376 (1970).
58. K. H. Kim, J.Pharm.Sci., 80, 966 (1991).
59. R. Jeppson, Acta Pharmacol.Toxicol., 37, 56 (1975).
60. H. Kubinyi, J.Med.Chem., 20, 625 (1977).
61. C. Hansch, J. P. Bjorkroth, A. Leo, J.Pharm.Sci., 76, 663 (1987).
62. C. Tanford, Science, 200, 1012 (1978).
63. C. R. Woese, D. H. Dugre, M. Kondo, W. C. Saxinger, Cold Spring Harbor Symp. Quant. Biol., 31, 723 (1966).
64. M. Bina, R. J. Feldmann, R. G. Deeley, Proc.Natl.Acad.Sci. USA, 77, 1278 (1980).
65. M. M. Teeter, Annu.Rev.Biophys.Biophys.Chem., 20, 577 (1991).
66. J. A. Rupley, G. Careri, Adv.Prot.Chem., 41, 37 (1991).
67. W. Saenger, Annu.Rev.Biophys.Biophys.Chem., 16, 93 (1987).

68. E. Westhof, Annu.Rev.Biophys.Biophys.Chem., 17, 125 (1988).
69. T. Stegmann, R. W. Doms, A. Helenius, Annu.Rev.Biophys.Biophys. Chem., 18, 187 (1989).
70. D. C. Rau, V. A. Parsegian, Science, 249, 1278 (1990).
71. M. F. Colombo, D. C. Rau, V. A. Parsegian, Science, 256, 655 (1992).
72. G. D. Rose, S. Roy, Proc.Natl.Acad.Sci. USA, 77, 4643 (1980).
73. J. Kyte, R. F. Doolittle, J.Mol.Biol., 157, 105 (1982).
74. H. Cid, E. Bunster, E. Arriagada, M. Campos, FEBS Lett., 150, 247 (1982).
75. D. Eisenberg, R. M. Weiss, T. C. Terwilliger, Proc.Natl.Acad.Sci. USA, 81, 140 (1984).
76. J. Finer-Moore, R. M. Stroud, Proc.Natl.Acad.Sci. USA, 81, 155 (1984).
77. H. R. Guy, Biophys.J., 47, 61 (1985).
78. D. M. Engelman, T. A. Steitz, Cell, 23, 411 (1981).
79. J. K. Argos, J.K. M. Rao, P. A. Hargrave, Eur.J.Biochem., 128, 565 (1982).
80. G. Von Heijne, Eur.J.Biochem., 116, 419 (1981).
81. G. Von Heijne, Eur.J.Biochem., 120, 275 (1981).
82. G. D. Fasman, W. A. Gilbert, Trends Biochem.Sci., 15, 89 (1990).
83. F. Jahnig, Trends Biochem.Sci., 15, 93 (1990).
84. M. Crinmi, M. D. Esposti, Trends Biochem.Sci., 16, 119 (1991).
85. D. Eisenberg, R. M. Weiss, T. C. Terwilliger, Nature, 299, 371 (1982).
86. H. R. Guy, Biophys.J., 45, 249 (1984).
87. S. H. Gellman, Biochemistry, 30, 6633 (1991).
88. L. M. Gierasch, Biochemistry, 28, 923 (1989).
89. L. L. Randall, S. J. S. Hardy, Science, 243, 1156 (1989).
90. C. A. Kaiser, D. Preuss, P. Grisafi, D. Botstein, Science, 235, 312 (1987).
91. D. A. Kendall, S. K.Doud, E. T. Kaiser, Biopolymers, 29, 139 (1990).
92. G. D. Rose, L. M. Gierasch, J. A. Smith, Adv.Prot.Chem., 37, 1 (1985).
93. Predictions of Protein Structure and the Principles of Protein Conformation, Ed., G. Fasman, Plenum Press, New York, 1989.
94. J. L. Cornette, K. B. Cease, J. H. Margalit, J. L. Spouge, J. A. Berzofsky, C. DeLisi, J.Mol.Biol., 195, 659 (1987).
95. R. A. Capaldi, G. Vanderkooi, Proc.Natl.Acad.Sci. USA, 69, 930 (1972).
96. V. G. Daschevskii, Mol.Biologia (Rus), 14, 105 (1980).
97. M. A. Davis, H. Hauser, R. B. Leslie, M. C. Phillips, Biochim. Biophys.Acta, 317, 214 (1973).

98. K. Nandi, Int.J.Peptide Protein Res., 8, 253 (1976).
99. V. Pliska, M. Schmidt, J.L. Fauchere, J.Chromatogr., 216, 79 (1981).
100. K. Gekko, J.Biochem., 90, 1633 (1981).
101. J. L. Fauchere, K. Q. Do, P. Y. C. Jow, C. Hansch, Experentia, 36, 1203 (1980).
102. D. J. Abragham, A. J. Leo, Proteins, 2, 130 (1987).
103. J. L. Fauchere, V. Pliska, Eur.J.Med.Chem.Chim.Ther., 18, 369 (1987).
104. M. Akamatsu, Y. Yoshida, H. Nakamura, M.Asao, H. Iwamura, T. Fujita, Quant.Struct.-Act.Relat., 8, 195 (1989).
105. R. Klein, M. Moore, M. Smith, Biochim.Biophys.Acta, 233, 420 (1971).
106. H. B. Bull, K. Breese, Arch.Biochem.Biophys., 161, 665 (1974).
107. R. E. Jacobs, S. H. White, Biochemistry, 28, 3421 (1989).
108. G. D. Rose, J. E. Dworkin, In: Predictions of Protein Structure and the Principles of Protein Conformation (G. Fasman, ed.), Plenum Press, New York, 1989, pp.625-633.
109. T. P. Hopp, K. R. Woods, Proc.Natl.Acad.Sci. USA, 78, 3824 (1981).
110. T. Korte, K. Ludwig, A. Herrmann, Biosci.Rep., 12, 397 (1992).
111. C. Chothia, J.Mol.Biol., 105, 1 (1976).
112. C. Chothia, C. Janin, Nature, 256, 705 (1975).
113. A. Diamond, K. Yu, J. T. Hsu, In: Protein Purification: From Molecular Mechanisms to Large-Scale Processes (M. R. Ladisch, R. C. Willson, C. C. Painton, S. E. Builder, eds.), ACS Simposium Series, Vol. 427, Amercian Chemical Society, Washington, D.C., 1990, pp. 52-65.
114. M. Akamatsu, T. Fujita, J.Pharm.Sci., 81, 164 (1992).
115. J. F. Reading, G. R. Hedwig, J.Chem.Soc. Faraday Trans., 86, 3117 (1990).
116. J. Draffehn, B. Schonecker, K. Ponsold, J. Chromatogr., 216, 63 (1981).
117. J. L. Fauchere, J.Med.Chem., 25, 1428 (1982).
118. J. W. Taylor, G. Osapay, Acc.Chem.Res., 23, 338 (1990).
119. R. Schwyzer, Biopolymers, 31, 785 (1991).
120. P. Prevelige, Jr., G. D. Fasman, In: Predictions of Protein Structure and the Principles of Protein Conformation (G. Fasman, ed.), Plenum Press, New York, 1989, pp.391-416.
121. A. Zaks, A. M. Klibanov, J.Am.Chem.Soc., 108, 2767 (1986).
122. S. E. Blondelle, K. Buttner, R. A. Houghten, J.Chromatogr., 625, 199 (1992).
123. N. E. Tayar, R. S. Tsai, B. Testa, P. A. Carrupt, C. Hansch, A. Leo, J.Pharm.Sci., 80, 744 (1991).

124. N. P. Franks, M. H. Abraham, W. R. Lieb, J.Pharm.Sci., 82, 466 (1993).
125. M. C. Wilce, M. I. Aguilar, M. T. W. Hearn, J.Chromatogr., 536, 165 (1991).
126. A. Ben-Naim, Hydrophobic Interactions, Plenum Press, New York, 1980.
127. A. Schmidtchen, L. A. Fransson, Biomed.Chromatogr., 7, 48 (1993).
128. M. L. Marin, D. Selgas, C. Pin, C. Casas, J.Sci.Food Agric., 60, 505 (1992).
129. P. Å. Albertsson, Partition of Cell Particles and Macromolecules, 3rd. ed., Wiley, New York, 1986.
130. C. J. van Oss, Colloids and Surfaces A, 78, 1 (1993).
131. C. J. van Oss, In: Protein Interactions (Ed., H. Visser), VCH Verlag, Weinheim, 1992, pp.25-55.
132. C. J. van Oss, Interfacial Forces in Aqueous Media, Marcel Dekker, 1994.
133. C. J. van Oss, R. F. Giese, Clays and Clay Miunerals, 1994, in press.
134. C. J. van Oss, R. J. Good, J. Protein Chem., 7, 179 (1988).
135. C. J. van Oss, M. K. Chaudhury, R. J. Good, Chem.Rev., 88, 927 (1984).
136. B. Y. Zaslavsky, N. M. Mestechkina, L. M. Miheeva, S. V. Rogozhin, J.Chromatogr., 240, 21 (1982).
137. B. Y. Zaslavsky, L. M. Miheeva, N. M. Mestechkina, S. V. Rogozhin, J.Chromatogr., 253, 139 (1982).
138. B. Y. Zaslavsky, A. A. Borovskaya, N. D. Gulaeva, L. M. Miheeva, J.Chem.Soc. Faraday Trans., 87, 137 (1991).
139. B. Y. Zaslavsky, L. M. Miheeva, N. M. Mestechkina, L. G. Shchyukina, M. A. Chlenov, L. I. Kudrjashov, S. V. Rogozhin, J.Chromatogr., 202, 63 (1980).
140. P. H. Wang, E. J. Lien, J.Pharm.Sci., 69, 662 (1980).
141. A. D. Diamond, X. Lei, J. T. Hsu, Biotechnol.Techniques, 3, 271 (1989).
142. M. A. Eiteman, J. L. Gainer, Bioechnol.Prog., 6, 479 (1990).
143. R. Kuboi, K. Yano, H. Tanaka, I. Komasawa, J.Chem.Eng.Japan, 26, 286 (1993).
144. W. H. Wang, R. Kuboi, I. Komasawa, Chem.Eng.Sci., 47, 113 (1992).

CHAPTER 7.

MEASUREMENTS OF THE RELATIVE HYDROPHOBICITY OF BIOLOGICAL SOLUTES BY THE AQUEOUS TWO-PHASE PARTITION TECHNIQUE

The method of partitioning in aqueous two-phase systems (particularly, Dex-PEG and Dex-Ficoll systems) provides an estimation of the relative hydrophobicity of biological and synthetic compounds. To discuss the advantages and limitations of the method it is necessary to consider the results of its applications to particular solutes. First, results obtained with the technique in the studies of solutes with rather simple structure are discussed. These results are considered in regard to the information about the solute-water interactions provided for the same solutes by other physicochemical methods. Next, the estimates of the relative hydrophobicity of conformationally flexible peptides are discussed in terms of QSAR analysis. The relative hydrophobicity of synthetic polymers, proteins, their complexes with ligands, etc., is then discussed in regard to the possible role of aqueous medium in regulation of the biopolymers' functions *in vivo*.

7.1. SOLUTES OF RELATIVELY SIMPLE STRUCTURE

Some of the data discussed below were discussed earlier from a different perspective but must be analyzed again in terms of relative hydrophobicity.

An example of compounds with relatively simple molecular structure is offered by sulphonephthalein dyes of the general structure shown in Table 7.1. There are also shows estimates of relative hydrophobicity of the dyes obtained by partitioning in the aqueous Dex-PEG and Dex-Ficoll two-phase systems, both containing 0.01 mol/kg universal buffer at pH 7.15, calculated from the data reported in [1].

Comparison of the relative hydrophobicity estimates for o-cresol red and phenol red and those for bromcresol green and bromphenol blue allows one to calculate the contribution of a CH_3 group. This contribution, calculated as an average of the differences between the relative hydrophobicities of the above dyes determined in the two aqueous two-phase systems, amounts to 1.53 ± 0.12 equivalent methylene groups for the CH_3 group in the X_2 position, and 1.81 ± 0.07 equivalent methylene groups for the CH_3 group in the X_1 position.

The contribution of the Br moiety may be calculated from the estimates given in Table 7.1 for bromthymol blue and thymol blue. The average value of the contribution of the Br moiety in the X_2 position may be estimated as 3.23 ± 0.45 equivalent CH_2 groups. The contribution of the Br moiety into the logarithm of the octanol-water partition coefficient of an aromatic compound is known [2] to depend on the moiety position. This contribution when measured by the water-octanol partition technique may be estimated as varying from 1.68 to 2.38 equivalent CH_2 groups.

The fulfillment of the additivity concept as well as the role of the constituent position cannot be judged here because of the limited number of molecular structures examined [1].

The aforementioned relationship between the partition coefficients of the dyes in aqueous two-phase systems [1] and the solvatochromic effects of the dyes (see Fig. 4.3) should aslo be mentioned. This relationship supports the assertion that the solute partition coefficient in the aqueous Dex-PEG (Dex-Ficoll) two-phase system reflects the relative intensity of the solute-water interactions, i.e., the relative hydrophobicity of the solute.

The effect of the constituent position upon the relative hydrophobicity of a solute, was explored by studying partitioning of a series of p-nitrophenyl-glycosides [3] in the aqueous Dex-Ficoll two-phase systems containing different amounts of sodium phosphate buffer, pH 7.4 and NaCl.

It is known that hydration of carbohydrates is governed by their stereochemistry, the most important factors being the relative position of the next-nearest-neighbor hydroxyl groups within the carbohydrate molecule (see in [4, 5] and references cited therein). The extent of hydration is supposed [4]

Table 7.1

Structural Features and Relative Hydrophobicity of Sulphonephthalein Dyes* as Measured by Partitioning in Aqueous Dex-PEG and Dex-Ficoll Two-Phase Systems (0.01 mole/kg universal buffer, pH 7.15).

Dye	Constituents			$n(CH_2)^a$	$n(CH_2)^b$
Phenol red	H	H	H	11.44 ± 1.24	10.72 ± 0.75
o-Cresol red	H	CH_3	H	14.61 ± 1.90	13.70 ± 0.95
Bromphenol blue	H	Br	Br	-	20.93 ± 1.20
Bromcresol green	CH_3	Br	CH_3	29.44 ± 3.82	24.55 ± 1.70
Bromcresol purple	H	Br	CH_3	20.94 ± 2.72	19.34 ± 1.30
Bromthymol blue	CH_3	Br	i-C_3H_7	30.44 ± 3.95	31.03 ± 2.11
Thymol blue	CH_3	H	i-C_3H_7	23.78 ± 3.09	24.76 ± 1.68

* - general structure of a sulphonephthalein dye is

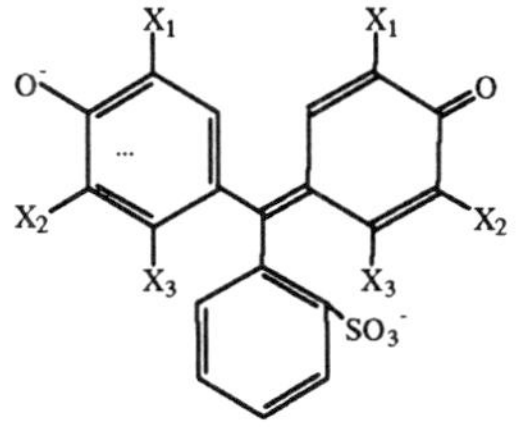

[a] - $n(CH_2)$ determined in the aqueous Dex-Ficoll two-phase system;
[b] - $n(CH_2)$ determined in the aqueous Dex-PEG two-phase system.

to be governed mostly by the position of the OH(4) in conjunction with the relative position of the OH(2). The estimates of the relative hydrophobicity of the p-nitrophenyl-glycosides given in Table 7.2 essentially agree with the above view.

It should be mentioned that variations in the salt composition of the aqueous medium within the range used (see Table 7.2) do not affect the relative hydrophobicity of the glycosides.

The most hydrophilic of the glycosides examined appears to be p-nitrophenyl-α-D-galactopyranoside [3]; this agrees with the conclusion [4] that galactose perturbs the three-dimensional hydrogen-bonded structure of water

Table 7.2
Relative Hydrophobicity* of 4-nitrophenyl-glycosides as Measured by Partitioning in Aqueous Dex-Ficoll Two-Phase System [a] and Octanol-Water System.

p-nitrophenyl-	$n(CH_2)$ [b]	$n(CH_2)$ [c]
N-acetyl-β–D-glucopyranoside	5.7 ± 0.6	-
α-D-xylopyranoside	2.9 ± 0.3	-
β-D-galactopyranoside	2.7 ± 0.4	-1.18
β-L-fucopyranoside	2.0 ± 0.3	-
β-D-glucopyranoside	-0.8 ± 0.3	-0.87
α-D-glucopyranoside [d]	-1.6 ± 0.4	-0.77
α-D-mannopyranoside[d]	-7.9 ± 0.6	-0.37
α-D-galactopyranoside	-8.2 ± 0.5	-

* relative hydrophobicity expressed in equivalent quantities of methylene groups;
[a] salt composition: sodium phosphate buffer, pH 7.4, varied from 0.01 mole/kg to 0.11 mole/kg; and NaCl varied from 0.15 mole/kg to zero;
[b] $n(CH_2)$ determined in the aqueous Dex-Ficoll two-phase system [3];
[c] $n(CH_2)$ calculated from the data obtained in octanol-water system [7];
[d] denoted are the glycosides examined at only one salt composition of the aqueous Dex-Ficoll two-phase system, namely, 0.15 mole/kg NaCl in 0.01 mole/kg sodium phosphate buffer, pH 7.4;

more than other carbohydrates, such as glucose, mannose, or xylose.

Galema and Hoiland [4] reported similar hydration of methyl-α-galactopyranoside and methyl-ß-galactopyranoside. The data presented in Table 7.2 show, however, that the relative hydrophobicity of the p-nitrophenyl derivative of ß-D-galactopyranoside exceeds that of α-D-galactopyranoside by 10.9 equivalent CH_2 groups. There may be two reasons for this substantial discrepancy. The most likely reason is that different properties are detected by the methods used in [3] and [4].

Galema and Hoiland [4] measured the density and compressibility of carbohydrate solutions in water, permitting calculation of partial molar volumes, isentropic partial molar compressibility, and the hydration number values. Comparison of the partial molar compressibility for a given saccharide with a molar compressibility of pure water was used to judge the disturbance of the hydration layer of the saccharide in reference to pure water [5]. Hence, not

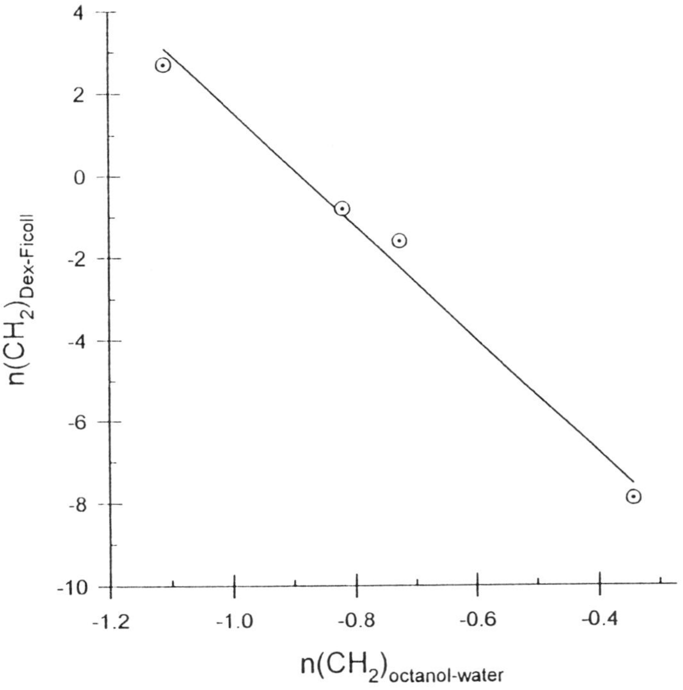

Figure 7.1. Correlationship between the relative hydrophobicity estimates for p-nitrophenyl-glycosides obtained by partitioning in aqueous Dex-Ficoll and octanol-water two-phase systems. Calculated from the data reported in [3,7].

the relative intensity of the solute-solvent interactions determined by the partition technique, but the disturbance of the solvent resulting from these interactions was examined in [5]. The other likely reason for the disagreement between the data in Table 7.2 and those reported by Galema and Hoiland [4] is that p-nitrophenyl-glycosides were studied in [3] instead of methyl-glycosides or free carbohydrates examined in [4].

Significant difference observed between the relative hydrophobicities of the relative intensity of the solute-solvent interactions determined by the partition technique, but the disturbance of the solvent resulting from these α- and ß-anomers of p-nitrophenyl-glucopyranoside and galactopyranoside, in particular, seems to comply with the different hydroxyl-proton chemical shifts in the NMR spectra of the anomers of glucose and galactose in aqueous solutions at low temperature (-6°C) [6].

It should be mentioned also that the p-nitrophenyl moiety may affect the relative hydrophobicity of the glycosides and the effect may depend on the stereochemistry of the saccharide. The presence of the moiety may also explain the discrepancy observed between the relative hydrophobicity estimates for the glycosides determined by the aqueous two-phase partition technique [3] and those calculated from the octanol-water partitioning data [7].

The estimates in question are correlated (Fig. 7.1) but the relationship is diametrically opposed to what might be expected. Data presented in Table 7.2 and plotted in Fig. 7.1 show that while the relative hydrophobicity estimates determined in the two systems under comparison agree for p-nitrophenyl-ß-D-glucopyranoside, they differ for the other glycosides. The more hydrophobic a given glycoside appears in the water-octanol system [7], the less hydrophobic it is as measured by the aqueous two-phase partition technique [3]. The situation seems to be similar to that observed for a given polar group in different water-organic solvent systems (see Fig. 4.14).

It should be repeated here that when the ratio $\Delta G(\text{polar group j})_{tr}/\Delta G(CH_2)$ values for two different polar groups in various water-organic solvent systems are compared, the ratio value (relative hydrophobicity) for a given polar group may exceed that for the other group in one solvent system, and vice versa in the other solvent system. This was attributed [8] to different polar group-solvent interactions in various solvent systems.

It seems likely that the interactions of the p-nitrophenyl moiety with octanol may affect the p-nitrophenyl-glycoside partitioning in the water-octanol system. These interactions may depend on the stereochemistry of the carbohydrate fragment of the molecule, and that may be the reason for the discrepancy between the relative hydrophobicity estimates obtained in the water-octanol and aqueous Dex-Ficoll two-phase systems. The agreement between the estimates for p-nitrophenyl-ß-D-glucopyranoside is likely to be coincidental. It seems reasonable to suggest that the estimates obtained by the aqueous two-phase partition technique are more adequate since the technique in question does not use an organic solvent capable of unknown effects on the solute partition behavior.

The results obtained in the study of anthracycline antibiotics [9] support the above conclusion. Partitioning of adriamycin, rubinomycin, and carminomycin was examined in two different systems. Octanol-buffer and aqueous Dex-Ficoll two-phase systems of the same salt composition (see in Table 7.3) were used, and the total antibiotic concentration in a given two-phase system was varied form $1 \cdot 10^{-6}$ to $1 \cdot 10^{-5}$ M range to avoid dimerization [9]. Different fluorescence spectra of the antibiotics in the aqueous and octanol phases [9] imply different solute-solvent interactions in the two media. The relative hydrophobicity estimates measured in the two systems [9] are listed in Table 7.3. The data given in Table 7.3 indicate that while the relative hydrophobicity estimates determined in the two systems are directly related (in contrast to those for glycosides, Fig. 7.1), the additivity principle appears to be fulfilled only par-

Table 7.3

Relative Hydrophobicity* of Anthracycline Antibiotics as Measured by Partitioning in Aqueous Dex-Ficoll and Octanol-Buffer Two-Phase Systems.

Antibiotic	$n(CH_2)$ [a]	$n(CH_2)$ [b]
Rubinomycin	20.1 ± 1.7	1.58 ± 0.05
Adriamycin	16.1 ± 1.5	-0.25 ± 0.05
Carminomycin	29.3 ± 1.5	3.09 ± 0.05

* relative hydrophobicity expressed in equivalent quantities of methylene groups;
[a] $n(CH_2)$ determined in the aqueous Dex-Ficoll two-phase system containing 0.15 mol/kg NaCl in 0.01 mol/kg sodium phosphate buffer, pH 7.4;
[b] $n(CH_2)$ determined in the octanol-buffer (0.15 M NaCl in 0.01 M phosphate buffer, pH 7.4) two-phase system.

tially. First, the relative hydrophobicity of rubomycin exceeds that of adriamycin by 4 ± 3.2 equivalent CH_2 groups as measured by the aqueous two-phase partition technique and by 1.83 ± 0.10 equivalent CH_2 groups as determined in the octanol-buffer system instead of 2.65 equivalent CH_2 groups as might be expected from the π_{OH} values [2,10]. Second and more important, replacement of the CH_3O- group in the rubomycin molecule for hydroxyl group (in carminomycin) according to the additivity concept [2, 10] should decrease the relative hydrophobicity of the molecule. The data in Table 7.3 indicate, however, that the replacement in question actually increases the relative hydrophobicity of carminomycin in reference to that of rubomycin. This unexpected observation may derive from the conformational change induced by the chemical modification of the antibiotic structure [9]. The quantum-chemical calculations of preferred conformations of carminomycin and rubomycin used to estimate the free energies of hydration of aglycon and carbohydrate fragments of the antibiotics molecules failed to predict the observed effect [9].

An important role of stereochemistry and conformation of a molecule in its interactions with an aqueous medium is well illustrated by the data [11, 12] on the relative hydrophobicity of nucleotides and their synthetic analogues. Partitioning of nucleotides, nucleosides, etc. was examined [11,12] in the aqueous Dex-Ficoll two-phase systems containing different amounts of sodium phosphate buffer, pH 7.4 and NaCl. Typical data reported in [11,12] are presented in Figure 7.2 as the relative hydrophobicity of a solute versus the salt composition of the aqueous medium.

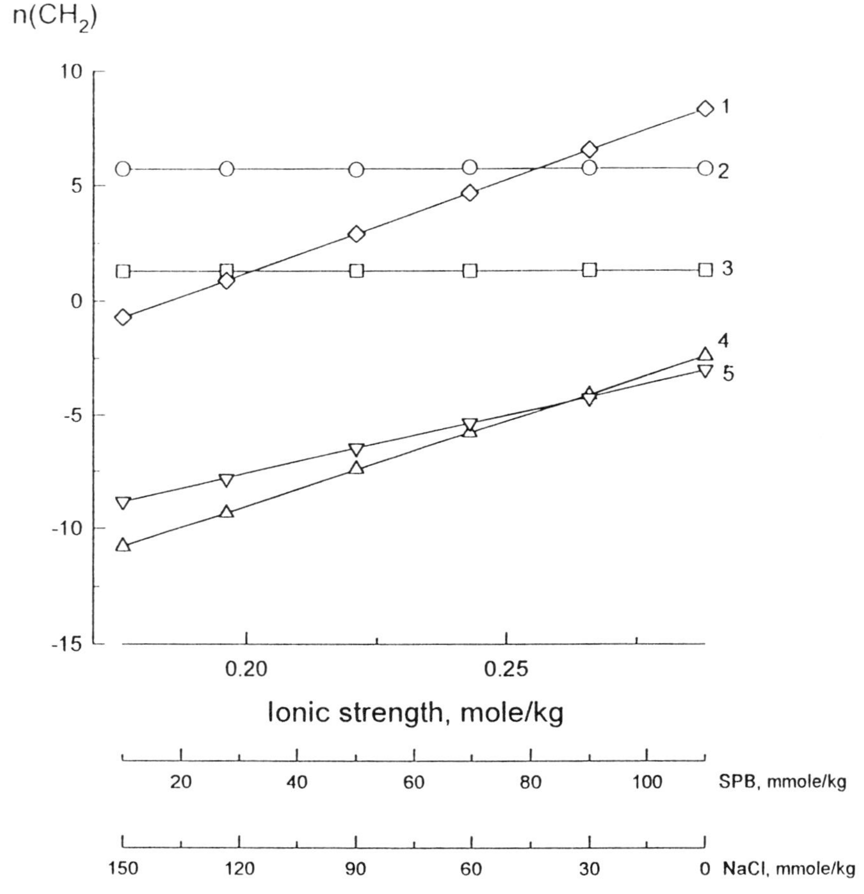

Figure 7.2. Relative hydrophobicity of solutes as a function of the salt composition of the aqueous medium containing NaCl and sodium phosphate buffer (SPB), pH 7.4. Solutes: (1) cyclic AMP; (2) adenine; (3) deoxycytidine; (4) adenosine 5'-phosphate (AMP); (5) - guanosine 5'-phosphate (GMP).

The data plotted in Fig. 7.2 show that the relative hydrophobicity of a solute may be described as a function of the salt composition of the aqueous medium as:

$$n(CH_2)_i = \alpha_i + \beta_i \cdot I \tag{7.1}$$

where $n(CH_2)_i$ is the relative hydrophobicity of the ith solute (expressed in terms of equivalent quantity of CH_2 groups); I is the ionic strength of the medium; α_i and β_i are constants specific for the solute being examined.

The ionic strength I value is used in Equation 7.1 solely as a quantitative index of the salt composition only for the particular composition employed, i.e., NaCl and sodium phosphate buffer, pH 7.4, in the case under consideration.

The suggested use of the ionic strength value as a quantitative index of salt composition in an aqueous two-polymer two-phase system [13] has been criticized [14]. Walter and Anderson [14] showed that the partition coefficients of the same solute may differ in the aqueous two-phase systems with different salt composition, for example, with NaCl replaced for KCl, both being at the same ionic strength value. It is certainly true that the ionic strength value is inadequate measure of the salt composition. This measure, however, seems to be the only one available so far, and it may be used as a quantitative index to compare the aqueous media containing the same salts at different concentrations. Even then, it is possible that the concentration range over which the ionic strength value may be used as a quantitative index of the salt composition is limited.

It would be highly desirable to use the coefficient α_i value to a first approximation as a measure of the relative hydrophobicity of the i-th solute in the salt-free aqueous medium. This approximation, however, implies an assumption of the validity of Equation 7.1 over an entire ionic strength range down to the zero ionic strength value. This assumption is very doubtful. It also needs to explore experimentally if the α value for a given solute is the same when derived from several relative hydrophobicity - ionic strength relationships observed in the media containing different salts. No experimental study of this issue has been undertaken so far, to my knowledge, and hence this question remains open.

Keeping the foregoing in mind, Equation 7.1 may be used to treat the experimental data [11,12] under consideration.

The α_i and β_i values calculated from the data reported in [11,12] are listed in Table 7.4 together with the relative hydrophobicity estimates for the solutes at two different salt compositions of the aqueous medium - 0.15 mole/kg NaCl in 0.01 mole/kg sodium phosphate buffer, pH 7.4, and 0.11 mole/kg sodium phosphate buffer, pH 7.4.

The data presented in Table 7.4 indicate, first, that the relative hydrophobicity of the two purine bases (adenine and guanine) exceeds that of thymine and cytosine in agreement with their molecular structures and the literature data [15].

The data in Table 7.4 show that the relative hydrophobicity of the deoxyribonucleosides examined [11] is independent of the salt composition of the aqueous medium under the conditions employed. Comparison of the relative hydrophobicity estimates for deoxyribonucleosides and the corresponding bases indicates that the additivity principle is not fulfilled. Comparison of the estimates for AMP and dAMP, CMP and dCMP, however, shows the principle to be met. We can therefore calculate the contribution of the hydroxyl group at C(2) of the ribose residue into the relative hydrophobicity of ribonucleotide molecule as -1.4 ± 0.3 equivalent CH_2 groups in a perfect agreement with the estimate derived above from the relative hydrophobicities of amino acids.

The relative hydrophobicities of ribonucleosides may be calculated from the estimates for deoxyribonucleosides listed in Table 7.4 using the additivity concept. The difference between the calculated estimates for adenosine (+4.5 equivalent CH_2 groups) and cytidine (-0.1 equivalent CH_2 group) agrees well with the data by Tinker and Brown [15] obtained by the solute partitioning in the octanol - 1.0 M phosphate buffer, pH 6.5 system. The data by Tinker and Brown [15] expressed in equivalent quantities of CH_2 groups indicate that the relative hydrophobicity of adenosine exceeds that of cytidine by 3-4 equivalent CH_2 groups. According to the same data [15] the relative hydrophobicity of guanosine exceeds that of cytidine by 0.5 equivalent CH_2 group, while this difference is estimated by 0.4 equivalent CH_2 group as measured by the aqueous two-phase partition technique [11].

The relative hydrophobicity of mononucleotides depends upon the salt composition of the aqueous medium. The effect of the salt composition (represented by the ionic strength value) on the relative hydrophobicity of a solute is characterized by the corresponding β_i value. It may be noticed that the β_i values are essentially the same for guanosine, cytidine, and thymidine 5'-mono- and 5'-triphosphates. These values differ significantly, however, from those for deoxyadenosine and adenosine 5'-mono- and 5'-triphosphates. The relative hydrophobicity of nucleosides is independent of the salt composition ($\beta_i = 0$) under the conditions employed [11]. Hence it seems reasonable to assume that the observed dependence of the relative hydrophobicity of the mononucleotides upon the salt composition is due to specific features of the phosphate group. Provided the assumption is correct, the data in Table 7.4 imply that the nonionic fragment, e.g., adenine or cytosine, may affect the ionic group interactions with an aqueous medium much more than incorporation of two additional ionic groups (β_i values are the same for mono- and tri-phosphates of the same nucleoside). Formation of intramolecular bond may also affect the relative hydrophobicity of a molecule more than incorporation of additional ionic groups as follows from the relative hydrophobicity estimates given in Table 7.4 for

Table 7.4

Relative Hydrophobicity of Nucleosides, Bases, Nucleotides, Dinucleosidephosphates and Their Synthetic Analogues as Measured by Partitioning in Aqueous Dex-Ficoll Two-Phase System.

Compound	α_i^a	β_i^a	$n(CH_2)^b$	$n(CH_2)^c$
Adenine	5.7	0	5.7	5.7
Guanine	5.4	0	5.4	5.4
Cytosine	0.8	0	0.8	0.8
Thymine	2.1	0	2.1	2.1
Deoxyadenosine	5.6	0	5.6	5.6
Deoxyguanosine	1.7	0	1.7	1.7
Deoxycytidine	1.3	0	1.3	1.3
Deoxythymidine	1.7	0	1.7	1.7
AMP	-24.0	75.0	-10.8	-2.4
dAMP	-22.3	75.0	-9.1	-0.7
cAMP	-14.9	80.4	-0.7	8.3
ATP	-23.9	75.0	-10.7	-2.3
GMP	-17.9	51.8	-8.8	-3.0
GTP	-19.7	51.8	-10.6	-4.8
CMP	-18.4	50.9	-9.4	-3.7
dCMP	-17.2	50.9	-8.2	-2.5
CTP	-20.3	51.8	-11.2	-5.4
TMP	-15.5	51.8	-6.4	-0.6
TTP	-16.9	51.8	-7.8	-2.0
ApA	-14.2	93.8	2.4	12.9
(2'-5')-ApA	-2.3	0	-2.3	-2.3
AEpA[d]	-14.8	84.8	0.1	9.6
APpA[d]	-39.4	177.7	-8.1	11.8
ABpA[d]	-28.0	139.3	-3.5	12.1
UpU	-21.5	114.3	-1.4	11.4
(2'-5')-UpU	-17.2	60.7	-6.5	0.3
UEpU[d]	-6.5	16.1	-3.7	-1.9
UPpU[d]	-8.6	45.5	-0.6	4.5
UBpU[d]	-18.4	75.0	-5.2	3.2
ApU	1.0	0	1.0	1.0
AEpU[d]	-16.8	81.2	-2.5	6.6
APpU[d]	-27.4	126.8	-5.1	9.1
ABpU[d]	-3.7	72.3	9.0	17.1
UEpA[d]	-17.1	77.7	-3.4	5.3

Table 7.4 Continued.

UPpA[d]	-18.5	94.6	-1.8	8.8
UBpA[d]	-6.6	49.1	2.1	7.6

[a] α_i and β_i are coefficients in Equation 7.1; experimental errors of the values given do not exceed 5%;

[b] relative hydrophobicity of a solute in the aqueous medium containing 0.15 mole/kg NaCl in 0.01 mole/kg sodium phosphate buffer, pH 7.40;

[c] relative hydrophobicity of a solute in the aqueous medium containing 0.11 mole/kg sodium phosphate buffer, pH 7.40;

[d] synthetic analogues of dinucleosidephosphates of the following structures:

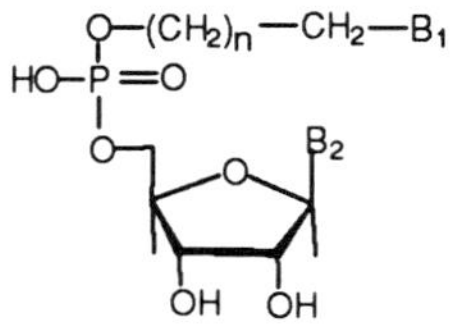

AEpA:	$B_1 = B_2$ = adenine, n = 1;		
APpA:	$B_1 = B_2$ = adenine, n = 2;		
ABpA:	$B_1 = B_2$ = adenine, n = 3;		
UEpU:	$B_1 = B_2$ = uracil, n = 1		
UPpU:	$B_1 = B_2$ = uracil, n = 2		
UBpU:	$B_1 = B_2$ = uracil, n = 3		
AEpU:	B_1 = adenine,	B_2 = uracil;	n = 1;
APpU:	B_1 = adenine,	B_2 = uracil;	n = 2;
ABpU:	B_1 = adenine,	B_2 = uracil;	n = 3;
UEpA:	B_1 = uracil,	B_2 = adenine;	n = 1;
UPpA:	B_1 = uracil,	B_2 = adenine;	n = 2;
UBpA:	B_1 = uracil,	B_2 = adenine;	n = 3.

AMP, ATP, and cAMP. Both effects are hardly predictable on the basis of the additivity concept.

Violation of the additivity principle is best illustrated by the data in Table 7.4 for (3'-5')- and (2'-5')-isomers of dinucleosidephosphates, and a series of their synthetic analogues with the ribose ring replaced with acyclic hydroxyalkyl substituents [12]. Significant differences between the relative hydrophobicity estimates and β_i values for (3'-5')- and (2'-5')-isomers of ApA and UpU should be mentioned. These differences indicate that the stereochemistry of a compound may greatly affect its relative hydrophobicity. It should be noticed particularly that in contrast to (3'-5')-isomer of ApA, the relative hydrophobicity of the (2'-5')-isomer is independent of the salt composition of the aqueous medium under the conditions employed [12]. That complies with the experimentally observed [16] conformational stability of (2'-5')-isomer of ApA noticeably exceeding that of the (3'-5')-isomer. Similar though less pronounced trend is observed for isomers of UpU.

Comparison of the β_i values for dinucleosidephosphates ApA, UpU, and ApU indicates that interactions of an ionic phosphate group with an aqueous medium depend significantly on the bases' structures [12].

Replacement of the ribose residue with 2-oxyethyl fragment in the structure of ApU increases the relative hydrophobicity of a molecule as might be expected. The same replacement in both ApA and UpU molecules reduces the relative hydrophobicity of the compounds. An elongation of the oxyalkyl fragment in the analogues of the dinucleosidephosphates resulted in the decreased hydrophobicity of a molecule in almost every case [12] directly contrary to the additivity concept. These results [12] support the conclusion that the relative hydrophobicity of a solute depends on its conformation.

It should also be noted that the relative hydrophobicity and β_i values for the AEpU and UEpA are reasonably similar and seem to be essentially independent of the type of base — adenine or uracil, positioned in the Np- or -pN part of the molecule. When the length of the oxyalkyl fragment is increased, i.e., for $N_1PpRibN_2$ and $N_1BpRibN_2$, an exchange of the N_1 and N_2 bases is followed by noticeable changes in the β_i values and relative hydrophobicity of compounds [12]. These results are clearly at variance with the additivity concept and support the above conclusion in regard to the effect of the solute conformation on its relative hydrophobicity.

It should finally be repeated that the relative hydrophobicity of naturally occurring dinucleosidephosphates ApA, UpU, and ApU given in Table 7.4 (in the aqueous medium containing 0.15 mole/kg NaCl in 0.01 mole/ kg sodium phosphate buffer, pH 7.4) are well correlated with the retention indexes for the compounds in the reversed-phase HPLC mode [17,18]. Data reported by Jacobson et al.[18] showed that the chromatographic behavior of the dinucleosidephosphates depends on the position of a particular base in the Np- or -pN part of the molecule. This finding is in agreement with the dependence, noted above, of the relative hydrophobicity of the compounds on the location of the bases in the molecular structure.

Thus, studies on the relative hydrophobicity of low molecular weight biological solutes and their synthetic analogues by the aqueous two-phase partition technique indicate that the relative hydrophobicity of a solute depends upon: (a) the stereochemistry and conformation of the solute molecule; and (b) the salt composition of an aqueous medium.

Both these factors must be taken into consideration in QSAR analysis as shown below for enkephalin-like peptides and their analogues.

7.2. PEPTIDES AND QSAR ANALYSIS

Among biological solutes, peptides are an obvious choice for QSAR analysis. First, as mentioned above, peptides and their analogues have been attracting attention recently as potential therapeutical drugs and some of them are currently applied as such. Peptides therefore are among biological solutes with defined function and measurable potency.

Natural enkephalins and enkephalin-like peptides offer an example of endogenous biological solutes and their analogues that interact with opiate receptors in much the same way as morphine and morphine-like drugs. Therefore, QSAR analysis may be undertaken for compounds of different chemical nature using the relative hydrophobicity of a compound as a physicochemical descriptor of its structure (see below).

The structure descriptors commonly used for chemical drugs (see above) usually fail to represent the molecular structure of peptides due to their conformational flexibility. The possibility to use the relative hydrophobicity of peptides as their potency-related structure descriptor is important for QSAR analysis. Once established, this possibility would also imply strongly that the relative hydrophobicity may be used as a function-related descriptor of proteins.

The relative hydrophobicity of peptides of different lengths reported by Zaslavsky et al.[19,20] was measured by partitioning in the aqueous Dex-Ficoll two-phase systems containing different amounts of sodium phosphate buffer, pH 7.4 and NaCl. The typical data [19,20] are presented in Figure 7.3 as the relative hydrophobicity of a peptide versus the salt composition (represented by the ionic strength value) of the aqueous medium.

The data from [19,20] treated according to Equation 7.1 are shown in Table 7.5 as the relative hydrophobicities of the peptides at two different salt compositions of the aqueous medium, 0.15 mole/kg NaCl in 0.01 mole/kg sodium phosphate buffer, pH 7.4, and 0.11 mole/kg sodium phosphate buffer, pH 7.4, and the corresponding α_i and β_i values.

Values shown in Table 7.5 indicate, first, that the relative hydrophobicity of di- and tri-peptides composed of the amino acid residues with nonionic side-chains is independent of the salt composition of the aqueous medium under the conditions employed [19,20]. Similar results were obtained under the same conditions for free amino acids [21]. The effects of the salt composition on the carboxyl and amino groups seem to cancel each other. The data for Leu- and Met-enkephalins (Tyr-Gly-Gly-Phe-Leu-OH and Tyr-Gly-Gly-Phe-Met-OH, respectively) imply that the structure of the C-terminal amino acid residue affects interactions of the carboxyl group with an aqueous medium as characterized by the β_i value. This conclusion is supported by the data for tetrapeptides Tyr-D-Ala-Gly-Phe-OH and Tyr-D-Ala-Gly-Phe(NO_2)-OH (β_i values are 72.5 M^{-1} and 37.1 M^{-1}, respectively).

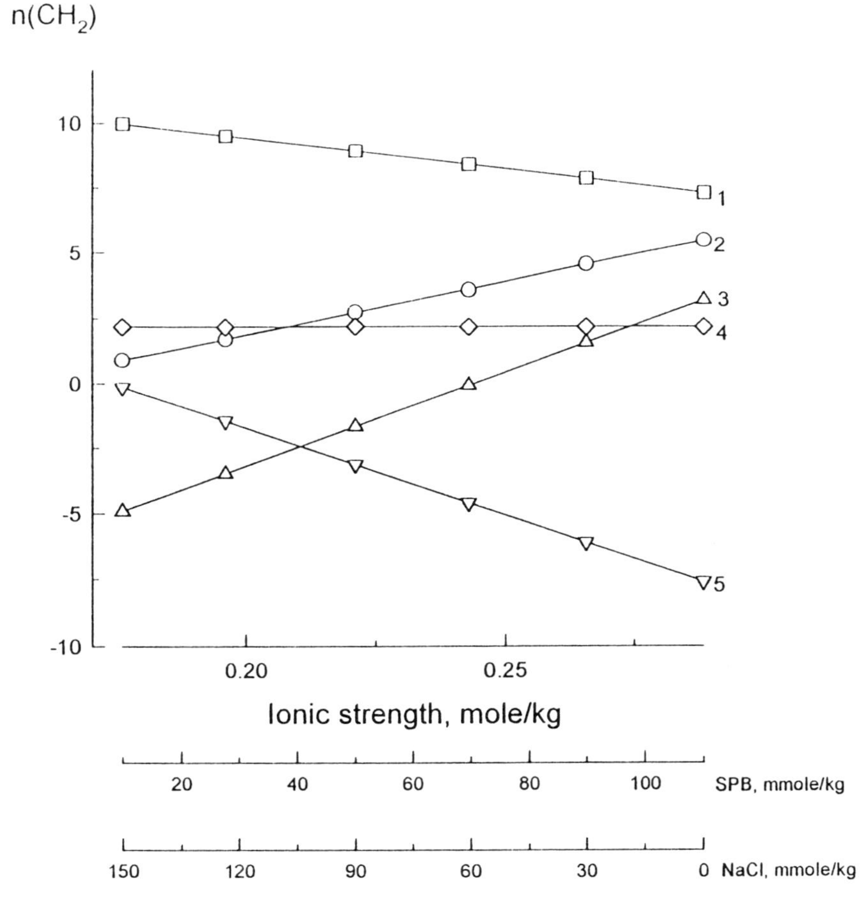

Figure 7.3. Relative hydrophobicity of peptides as a function of the salt composition of the aqueous medium containing NaCl and sodium phosphate buffer (SPB), pH 7.4. Peptides: 1 - Tyr-D-Ala-Gly-Phe-N_2H_2-Leu; 2 - Tyr-Gly-Gly-Phe-Leu-OH; 3 - Tyr-D-Ala-Gly-Phe-OH; 4 - α-endorphin; 5 - Tyr-Gly-Gly-Phe-Leu-Arg-Lys-Arg-OH.

Since the $ß_i$ values are zero for all the tripeptides examined, the dependence of the relative hydrophobicity of peptides on the salt composition is likely determined not only by the presence of ionic groups (a necessary but not sufficient condition) but also by the three-dimensional structure of the peptide molecule in solution. This assumption agrees with the concept that the dependence in question is displayed by tetra- and larger peptides capable of intramolecular bonding, possibly leading to differences in the ability of ionic groups to interact with the solvent.

It follows from the data in Table 7.5 that the additivity principle is fulfilled only partially and only for peptides of limited size. The relative hydrophobicity of the leucine side-chain calculated from the estimates given in Table 7.5 for the dipeptides Gly-Leu-OH and Gly-Gly-OH amounts to 2.5 equivalent CH_2 groups in a fair agreement with the estimate of 3.2 equivalent CH_2 groups derived from the data for DNP-Leu and DNP-Gly (see in Table 6.2). The estimate for the same side-chain derived from the data in Table 7.5 for the dipeptides Leu-Gly-OH and Gly-Gly-OH, however, is much lower. It amounts to 1.9 equivalent CH_2 groups, in agreement with the data [19,20] on the contributions of an amino acid residue into the total hydrophobicity of a peptide being dependent on the C- or N-terminal position of the residue.

The relative hydrophobicity of the D-alanine side-chain calculated from the data given in Table 7.5 for pentapeptides Tyr-D-Ala-Gly-Phe(NO_2)-Leu-OH and Tyr-Gly-Gly-Phe(NO_2)-Leu-OH amounts to ca. 1 equivalent CH_2 group. Somewhat lower estimate of 0.5 equivalent CH_2 group is obtained for the same side-chain from the data for the hexapeptides Tyr-D-Ala-Gly-Phe-Leu-D-Arg-OH and Tyr-Gly-Gly-Phe-Leu-D-Arg-OH. Both estimates are in a reasonable agreement with those calculated above from the data obtained by various techniques for amino acids and their derivatives (see in Table 6.2). The estimate for the same D-alanine side-chain derived from the data in Table 7.5 for hexapeptides Tyr-D-Ala-Gly-Phe-D-Leu-Arg-OH and Tyr-Gly-Gly-Phe-D-Leu-Arg-OH amounts to -0.5 equivalent CH_2 group, and that derived from the data for octapeptides Tyr-D-Ala-Gly-Phe-Leu-Arg-Lys-Arg-OH and Tyr-D-Ala-Gly-Phe-Leu-Arg-Lys-Arg-OH varies from 1.98 to 3.67 equivalent methylene groups, depending on the salt composition of the aqueous medium. It may be concluded that the additivity principle is applicable to compounds of limited conformational flexibility only and may not be applied to those of highly flexible structure where not all fragments are equally accessible to the solvent. Replacement of the L-leucine residue with its enantiomer (D-Leu residue) in the hexapeptide Tyr-D-Ala-Gly-Phe-Leu-Arg-OH increases the relative hydrophobicity of the peptide by noticeable 2.8 equivalent CH_2 groups, while the similar replacement of the L-arginine residue (with D-Arg) reduces the relative hydrophobicity of the peptide by 2.3 equivalent CH_2 groups. Both changes are likely to be due to the changes in the peptide conformations.

The contribution of a nitro group into the relative hydrophobicity of solutes as measured by the water-organic solvent partition technique [2] is equivalent to that of -0.28 CH_2 group. Contribution of the same group into the relative hydrophobicity of a peptide measured by the aqueous two-phase partition technique depends on the peptide structure. It follows from the data for tetrapeptides Tyr-D-Ala-Gly-Phe-NH_2 and Tyr-D-Ala-Gly-Phe(NO_2)-NH_2 and those for pentapeptides Tyr-Gly-Gly-Phe-Leu-OH and Tyr-Gly-Gly-Phe(NO_2)-Leu-OH that the contribution in question amounts to 1.5 equivalent methylene groups. Data shown in Table 7.5 for the tetrapeptides Tyr-D-Ala-Gly-Phe-OH and Tyr-D-Ala-Gly-Phe(NO_2)-OH indicate that the contribution of a nitro group amounts to 8.61 to 4.65 equivalent CH_2 groups depending on the salt composition of the aqueous medium under the conditions employed. In view of these results, the variations observed may be attributed to different conformations of the peptides dependent on the particular peptide structure and salt composition of the aqueous medium. These results [19,20] support the conclusion that the additivity concept is not applicable to conformationally flexible peptides.

This conclusion is also confirmed by the data obtained for the relatively large peptides, e.g., α- and γ-endorphins composed of 16 and 17 amino acid residues, respectively. The primary structure of α-endorphin differs from that of γ-endorphin in that it lacks C-terminal leucine residue, and still the relative hydrophobicity of the latter peptide exceeds that of the former one by merely 0.2 equivalent CH_2 group. It should be recalled that the relative hydrophobicity of the leucine side-chain amounts to 2.4 equivalent CH_2 groups (see Table 6.2). Similarly, elimination of the N-terminal tyrosine residue from the γ-endorphin molecule barely reduces the relative hydrophobicity of the original peptide by merely 0.1 equivalent CH_2 group. Both changes in the relative hydrophobicity of the peptides are unpredictable in terms of the additivity concept. The likely reason seems to be the difference between the solvent accessibility of various fragments of the peptides due to different conformations.

Moreover, it follows from the foregoing that calculations of the relative hydrophobicity of peptides or peptide fragments, based on the hydrophobicities of amino acids side-chains and the additivity principle, may not be adequate. The HPLC technique seems more suitable for the purpose but also not totally adequate. According to the data reported by Sasagawa et al.[24], for example, the retention time of β-endorphin exceeds significantly those (identical) of β-endorphin and Leu-enkephalin. According to the generally accepted view [24-27], that means that the relative hydrophobicity of β-endorphin exceeds those of α-endorphin and Leu-enkephalin. These results may represent the reality taking into account that the mixture of the aqueous solution of trifluoroacetic acid with acetonitrile was used as an eluent [24].

Table 7.5

Relative Hydrophobicity of Peptides as Measured by Partitioning in Aqueous Dex—Ficoll Two-Phase System.

Peptide	α_i [a]	β_i [a]	$n(CH_2)$ [b]	$n(CH_2)$ [c]
Gly-Gly-OH	-5.8	0	-5.8	-5.8
Gly-Leu-OH	-3.3	0	-3.3	-3.3
Leu-Gly-OH	-3.9	0	-3.9	-3.9
Tyr-Arg-OH	-4.4	0	-4.4	-4.4
Gly-Trp-OH	6.4	0	6.4	6.4
Gly-Gly-Gly-OH	-5.6	0	-5.6	-5.6
Phe-Leu-Arg-OH	1.3	0	1.3	1.3
Phe-Leu-Gly-OH	1.5	0	1.5	1.5
Ileu-His-Pro-Phe-OH	0	0	0	0
Tyr-D-Ala-Gly-Phe-OH	-17.68	72.5	-4.91	3.21
*Tyr-D-Ala-Gly-Phe(NO_2)-OH	-2.8	37.14	3.7	7.86
*Tyr-D-Ala-Gly-Phe-NH_2	3.8	0	3.8	3.8
Tyr-D-Ala-D-Phe-NH_2	-2.3	0	-2.3	-2.3
*Tyr-D-Ala-Gly-Phe(NO_2)-NH_2	5.3	0	5.3	5.3
*Tyr-D-Ala-Gly-Phe-N_2H_3	3.9	0	3.9	3.9
*Tyr-D-Ala-Gly-Phe(NO_2)-N_2H_3	7.5	0	7.5	7.5
Phe(NO_2)-D-Ala-Gly-Phe-NH_2	4.9	0	4.9	4.9
Tyr-D-Ala-Gly-Phe-Arg-OH	8.3	-23.93	4.08	1.4
*Tyr-Gly-Gly-Phe(NO_2)-Leu-OH	-4.71	40.18	2.37	6.87
*Tyr-D-Ala-Gly-Phe(NO_2)-Leu-OH	-3.53	39.55	3.44	7.87
*Tyr-Gly-Gly-Phe-Leu-OH	-6.28	40.8	0.91	5.48
*Tyr-Gly-Gly-Phe-Met-OH	-5.24	37.14	1.3	5.46
*Tyr-D-Ala-Gly-Phe(NO_2)-N_2H_2-Met	0.3	0	0.3	0.3

*Tyr-D-Ala-Gly-Phe(NO_2)-N_2H_2-Gly	5.2	0	5.2	5.2
*Tyr-D-Ala-Gly-Phe-N_2H_2-His	12.56	-22.77	8.55	6.0
*Tyr-D-Ala-Gly-Phe-N_2H_2-Leu	14.2	-23.93	9.98	7.3
*Tyr-D-Ala-Gly-Phe(Me)-Met-OH	4.6	0	4.6	4.6
*Pro-Tyr-D-Ala-Gly-Phe-NH_2	1.4	0	1.4	1.4
*Lys-Tyr-D-Ala-Gly-Phe-NH_2	7.3	-18.04	4.14	2.12
*Arg-Tyr-D-Ala-Gly-Phe-NH_2	7.75	-25.18	3.31	0.49
Tyr-Pro-Phe-Pro-Gly-OH	3.01	0	3.01	3.01
*Tyr-D-Ala-Gly-Phe-Leu-Arg-OH	1.4	0	1.4	1.4
*Tyr-D-Ala-Gly-Phe-Leu-D-Arg-OH	3.8	0	3.8	3.8
Tyr-Gly-Gly-Phe-Leu-D-Arg-OH	3.3	0	3.3	3.3
Tyr-Gly-Gly-Phe-D-Leu-Arg-OH	4.7	0	4.7	4.7
*Tyr-D-Ala-Gly-Phe-D-Leu-Arg-OH	4.2	0	4.2	4.2
*Tyr-D-Ala-Gly-Phe-D-Leu-D-Arg-OH	1.9	0	1.9	1.9
*Tyr-Gly-Gly-Phe-Leu-Tre-OH	-7.37	42.5	0.12	4.88
*Tyr-D-Ala-Gly-Phe-Leu-Arg-Lys-Arg-OH	10.93	-51.52	1.85	-3.92
*Tyr-Gly-Gly-Phe-Leu-Arg-Lys-Arg-OH	11.61	-66.61	-0.13	-7.59
*α-endorphin	2.2	0	2.2	2.2
*β-endorphin	7.05	-57.59	-3.1	-9.55
*γ-endorphin	2.4	0	2.4	2.4
*des-Tyr-γ-endorphin	2.3	0	2.3	2.3

[a] α_i and β_i are coefficients in Equation 7.1; experimental errors of the values given do not exceed 5%;

[b] relative hydrophobicity of a solute in the aqueous medium containing 0.15 mole/kg NaCl in 0.01 mole/kg sodium phosphate buffer, pH 7.40;

[c] relative hydrophobicity of a solute in the aqueous medium containing 0.11 mole/kg sodium phosphate buffer, pH 7.40;

* peptides denoted with asterisk display opioid activity (see below).

Quantitative Structure-Activity (QSAR) Analysis.

Peptides in Table 7.5 displaying opiate activity are denoted with asterisk. It is known [28] that the opioid peptides may interact with multiple heterogeneous receptors. Therefore three different bioassay systems: depression of the electrically-induced contractions of (a) mouse vas deferens and (b) guinea pig ileum preparations, and (c) inhibition of the binding of [^{3}H]naloxone to rat brain homogenates were used in the QSAR study [20]. Additional biological activity measurements performed later [29, 30] included analgesic effect of the peptides under intravenous and intracisternal administration in rats. The opioid activities displayed by the peptides in three bioassay systems [20,29,30] are shown in Table 7.6.

The data [20,29,30] partially shown in Table 7.6 indicate that there is no direct correlation between the potencies of the peptides in different bioassays. The differential effects of enkephalin analogues in the guinea pig ileum and mouse vas deferens are usually explained by the diversity of opiate receptors [31,32]. It has been demonstrated [31] that the enkephalins' and endorphins' action in mouse vas deferens occurs on receptors different from those on which morphine and its classical surrogates act. These receptors, called by Lord et al.[31] σ-receptors, are not responsible for the action of enkephalins in the quinea pig ileum. In this preparation the peptides are supposed to interact mainly with the μ-receptors which mediate the action of the classical morphine-like compounds. According to the physicochemical theory of the mode of drug-receptor interaction [33], the hydrophilic-hydrophobic balance of the opiate receptor and its complex with a drug is of fundamental importance for the pharmacological action of the opiates. One of the key factors essential for the expression of the opiate action taken into consideration when designing the theory [33] was the relative hydrophobicity of the drug.

As may be seen from Fig. 7.3, the relative hydrophobicity of many of the peptides examined [20] depends on the salt composition (ionic strength) of the medium. The influence of the salt composition on the relative hydrophobicity of some of the peptides appears to be almost diametrically opposite depending on the structure of the peptide in question. Nevertheless, the quantitative hydrophobicity-activity relationships were established [20] for these peptides. The correlations between the relative hydrophobicity of the peptides and their potency in different bioassay systems are presented in Figure 7.4.

There is a strong correlation between the affinity of the opioid peptides for [^{3}H]-naloxone binding sites in rat brain homogenate and the relative hydrophobicity of the peptides at the fixed salt composition corresponding to the ionic strength value of 0.170 M [20]. This correlation shown in Fig. 7.4 (curve 2) is described as:

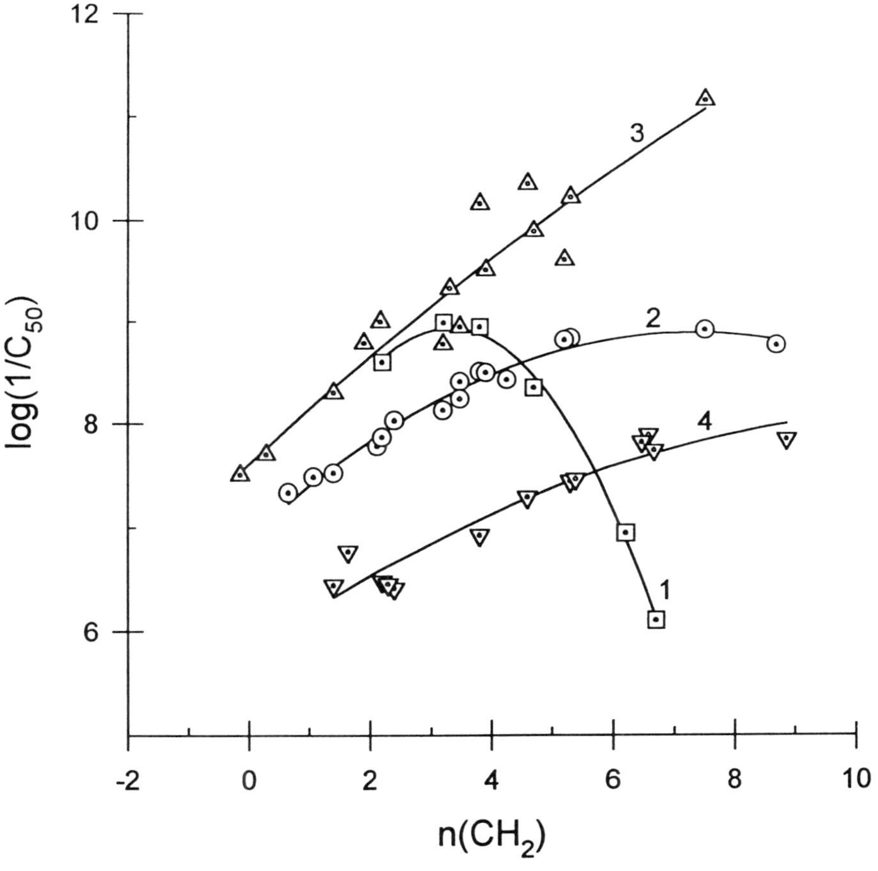

Figure 7.4. Relationships between the activity of morphine-like drugs (1) and opioid peptides (2-4) in different bioassays (expressed as $\log(1/C_{50})$) and the relative hydrophobicity of compounds (expressed in terms of $n(CH_2)$) at the corresponding salt composition/ionic strength of the aqueous medium. (1) and (2) - rat brain receptor binding assay [20]; (3) - analgesic effect under intracisternal administration in rats [30]; (4) - mouse vas deferens assay system [20].

Table 7.6

Activity of Opioid Peptides* in the Rat Brain Receptor Binding (RBR)[a] and Mouse Vas Deferens (MVD)[b] Assays, and as Analgesic Agents Under Intracisternal Administration in Rats (IC)[c].

PEPTIDE	RBR[a] $\cdot 10^9$, M	MVD[b] $\cdot 10^8$, M	IC[c] $\cdot 10^{10}$, M
Tyr-D-Ala-Gly-Phe(NO_2)-OH	3.89	1.4	11.2
Tyr-D-Ala-Gly-Phe-NH_2	3.24	11.7	0.692
Tyr-D-Ala-Gly-Phe(NO_2)-NH_2	1.45	3.63	0.603
Tyr-D-Ala-Gly-Phe-N_2H_3	3.16	-	3.16
Tyr-D-Ala-Gly-Phe(NO_2)-N_2H_3	1.20	-	0.069
Tyr-Gly-Gly-Phe(NO_2)-Leu-OH	1.66	-	-
Tyr-D-Ala-Gly-Phe(NO_2)-Leu-OH	7.41	0.0078**	16.6
Tyr-Gly-Gly-Phe-Leu-OH	4.68	1.3	-
Tyr-Gly-Gly-Phe-Met-OH	3.31	1.5	-
Tyr-D-Ala-Gly-Phe(NO_2)-N_2H_2-Gly	1.51	-	2.51
Tyr-D-Ala-Gly-Phe-N_2H_2-His	1.70	3.5	-
Tyr-D-Ala-Gly-Phe-N_2H_2-Leu	-	1.8	-
Tyr-D-Ala-Gly-Phe(Me)-Met-OH	-	5.1	0.45
Pro-Tyr-D-Ala-Gly-Phe-NH_2	0.302	36.0	-
Lys-Tyr-D-Ala-Gly-Phe-NH_2	3.72	36.0	-
Arg-Tyr-D-Ala-Gly-Phe-NH_2	5.75	-	-
Tyr-D-Ala-Gly-Phe-Leu-Arg-OH	-	-	50.1
Tyr-D-Ala-Gly-Phe-Leu-D-Arg-OH	-	-	4.8
Tyr-D-Ala-Gly-Phe-D-Leu-Arg-OH	-	-	1.3
Tyr-D-Ala-Gly-Phe-D-Leu-D-Arg-OH	-	-	16.2
Tyr-Gly-Gly-Phe-Leu-Tre-OH	-	-	320.0

Tyr-D-Ala-Gly-Phe-Leu-Arg-Lys-Arg-OH	-	-	10.0
Tyr-Gly-Gly-Phe-Leu-Arg-Lys-Arg-OH	-	-	200.0
α-endorphin	1.35	0.33	-
γ-endorphin	9.33	0.39	-
des-Tyr-γ-endorphin		0.355	-

* all activities are presented as $\log(1/C_{50})$, where C_{50} is the peptide concentration (in mole/l) producing a given biological affect; ** the potency denoted did not fit Equation 7.5 and has not been included in the relationship.

[a] inhibition of the binding of $[^3H]$-naloxone to rat brain homogenates; [b] depression of the electrically-induced contractions of the mouse vas deferens preparations; [c] analgesic effect of the peptide administered intracisternally in rats (peptide concentration producing analgesic effect in 50% of animals in a group.

$$\log(1/C_{50})^{RBR} = 6.88 + 0.55 \cdot n(CH_2)^* - 0.038 \cdot [n(CH_2)^*]^2 \quad (7.2)$$

$$N = 16;\ r^2 = 0.987;\ s = 0.044$$

where $n(CH_2)^*$ is the relative hydrophobicity of a peptide at the salt composition of an aqueous medium corresponding to the ionic strength of 0.170 M value under the conditions employed [20]; C_{50} is the peptide concentration producing 50% inhibition of the binding of $[^3H]$-naloxone to rat brain homogenate.

The activities of morphine, nalorphine, d-methadone, levorphanol, codeine, and naloxone in the rat brain receptors binding assay as reported by Terenius [34] are plotted in Fig. 7.4 (curve 1) versus the relative hydrophobicity of the drugs measured by the aqueous two-phase partitioning technique [20]. Even though the number of drugs is very limited, the relationship presented in Fig. 7.4 (curve 1) seems to be statistically significant. It is described as:

$$\log(1/C_{50})^{RBR} = 6.23 + 1.63 \cdot n(CH_2) - 0.25 \cdot [n(CH_2)]^2 \quad (7.3)$$

$$N = 6;\ r^2 = 0.998;\ s = 0.044$$

where $n(CH_2)$ is the relative hydrophobicity of a morphine-like drug (independent of the salt composition of an aqueous medium under the conditions used [20]); C_{50} is the drug binding to rat brain homogenate [34].

Equation 7.2 indicates that the "optimal" relative hydrophobicity, $n_o(CH_2)^*$, of the enkephalin-like peptide (at the salt composition corresponding to the ionic strength value of 0.170 M under the conditions employed [20]) displaying the maximal potency in the rat brain homogenate assay is equivalent to that of 7.3 ± 0.3 CH_2 groups. The "optimal" value for opiates appear to be much lower, it amounts to 3.3 ± 0.1 equivalent CH_2 groups. The difference between the relationships described by Equations 7.2 and 7.3 and the corresponding "optimal" $n_o(CH_2)$ values for the peptides and morphine-like drugs is in line with the hypothesis [31] that opioid peptides and opiates interact with different receptors.

The relationship observed between the peptide relative hydrophobicity and analgesic potency, when administered intracisternally, is presented in Fig. 7.4 as curve 3. This may be described as:

$$\log(1/C_{50})^{IC} = 7.61 + 0.55 \cdot n(CH_2)^* - 0.01 \cdot [n(CH_2)^*]^2 \quad (7.4)$$

$$N = 15;\ r^2 = 0.948;\ s = 0.044$$

where $n(CH_2)^*$ is the relative hydrophobicity of a peptide at the salt composition of an aqueous medium corresponding to the ionic strength of 0.170 M value under the conditions employed [20]; C_{50} is the peptide concentration producing analgesic effect in 50% of rats in a group when the peptide is administered intracisternally [30].

It seems likely that due to the limited number of the peptides examined, Equation 7.4 (and curve 3 in Fig. 7.4) describes merely a part of the left branch of a relationship (of parabolic-like shape) existent between the analgesic potency and relative hydrophobicity of the peptides. The "optimal" relative hydrophobicity value in this case, unfortunately, cannot be determined. It should be noticed, however, that, first, the relative hydrophobicities of the peptides fitting Equation 7.4 are the same as those fitting Equation 7.2 established for the peptides potency in a different bioassay. Both these assays have in common that the biological effect being monitored results from the interactions of the peptides with rat brain receptors. It may be assumed that the salt composition characterized by the ionic strength value of 0.170 M is similar or close to that in rat brain tissue or, more specifically, in the rat brain opioid receptors compartment. Secondly, the slope of the left branch of the curves 2 and 3 in Fig. 7.4 is the same within the experimental error limits. That may be due to the same nature of the peptide-receptor interactions in both bioassays.

No correlation could be found between the analgesic effect of the peptides administered in rats intravenously [32] and their relative hydrophobicity. The lack of the relationship in question may be due to different stabilities and the blood-brain barrier permeation abilities of the peptides in addition to their different affinities for the rat brain receptors. A more detailed study of much larger number of enkephalin-like peptides would be necessary to resolve this issue.

A fairly good correlation was found [20] between the potency of the peptides in the mouse vas deferens assay and the relative hydrophobicity of the peptides at the salt composition corresponding to the ionic strength value of

0.315 M under the conditions employed. The relationship shown in Fig. 7.4 as curve 4 may be described as:

$$\log(1/C_{50})^{MVD} = 5.80 + 0.40 \cdot n(CH_2)^{**} - 0.02 \cdot [n(CH_2)^{**}]^2 \qquad (7.5)$$

$$N = 15;\ r^2 = 0.948;\ s = 0.044$$

where $n(CH_2)^{**}$ is the relative hydrophobicity of a peptide at the salt composition of an aqueous medium corresponding to the ionic strength of 0.315 M value under the conditions employed [20]; C_{50} is the peptide concentration producing the 50% depression of the electrically-induced contractions of the mouse vas deferens preparations. (It should be mentioned that Tyr-D-Ala-Gly-Phe(NO_2)-Leu-OH with the activity in the mouse vas deference assay exceeding those of all the other peptides examined [20] by three to four orders of magnitude does not fit Equation 7.5 and was not included in the relationship.)

It should be noticed that the relative hydrophobicity of the peptides fits Equation 7.5 on condition that the salt composition of the aqueous medium under the conditions employed [20] corresponds to the ionic strength value of 0.315 M, quite different from that required for Equations 7.2 and 7.4. That agrees with the σ-receptors in mouse vas deferens interacting with opioid peptides under conditions unlike those that exist in the rat brain homogenate or tissue ("conditions" are the local membrane environment, for example, ionic composition, and probably the state of the ionizable groups of the receptor).

No correlation could be found [20] between the potency of the peptides on the guinea pig ileum and their relative hydrophobicity.

The data [20,29,30] considered offer the first and, to my knowledge, only example so far of successful QSAR analysis of peptides using the experimental estimates of the relative hydrophobicity as the single physicochemical descriptor of the peptide structure.

These data imply, first, that the relative hydrophobicity of a peptide determined by the aqueous two-phase partition technique may be used as a potency-related descriptor of the peptide structure. The relative hydrophobicity of conformationally flexible peptides clearly depends on the peptide conformation. The above results of QSAR analysis indicate that the effect of the composition of an aqueous medium on the relative hydrophobicity of biological solutes must be taken into consideration. The effects of the medium composition and the solute conformation on the solute relative hydrophobicity are likely to be interdependent, but this issue remains to be examined. The aqueous medium composition, e.g., ionic, macromolecular composition, etc., is known to be important for the solute biological activity and, likely, function due to the influence of the "molecular surroundings" on the preferred conformation of the solute [35] and on the solute relative hydrophobicity [36,37].

The effects in question may be expected to become more pronounced with increasing molecular weight and size of a solute. Results of the studies of the relative hydrophobicity of biological and synthetic macromolecules by the aqueous two-phase partition technique are discussed below.

7.3. SYNTHETIC MACROMOLECULES

Synthesis and study of macromolecular drugs and polymers that may be used as carriers for common pharmaceutical agents are currently under active investigation. From the standpoint of medicinal chemistry, the relative hydrophobicity of a given polymer is important as a factor known to influence the distribution of the polymer throughout the body tissues. Studying synthetic macromolecules as compared to biopolymers is easier in terms of interpretation. It is possible, first, to increase the molecular size without any change in its chemical structure. Linear macromolecules may be studied and compared to those forming random coils, and certain physicochemical methods not applicable to biopolymers may be used to examine synthetic polymers.

The relative hydrophobicity of synthetic homo- and heteropolymers was studied by Zaslavsky et al.[38-40]. The results reported [38-40] are briefly outlined below in regard to their biomedical implications.

Different nonionic polymers with varied molecular weights were examined by partitioning in the aqueous Dex-Ficoll two-phase systems containing different amounts of sodium phosphate buffer, pH 7.4 and NaCl. The data reported [38,39] are presented in Table 7.7 as the relative hydrophobicities for the polymer fractions of different molecular weights.

The data given in Table 7.7 indicate that the relative hydrophobicity of PEG is independent of the polymer molecular weight in the range from 1500 to $4 \cdot 10^4$ and amounts to 169.5 ± 2.6 equivalent CH_2 groups. This independence [38] seems to agree with the fact that the hydrophobic substituent constants, π_X, are 0.5 for a CH_2 group and -0.98 for a -O- moiety [2], resulting in a net value of approx. zero for the PEG monomeric unit ($-CH_2-CH_2-O-$). Hence it appears reasonable that the effect of the molecular weight on the relative hydrophobicity of PEG is negligible or absent. It should be mentioned that the aforementioned study of the effect of PEG on the solvent features of aqueous medium [41] indicated that the effect in question is also independent of the PEG molecular weight.

The data of Table 7.7 show that the relative hydrophobicity of polyacrylamide (PAAm) and polyvinylpyrrolidone (PVP) in contrast to that of PEG and poly(vinyl alcohol) depends on the molecular weight of the polymers according to the general relation:

$$n(CH_2) = A \cdot (M_w)^B \tag{7.6}$$

Table 7.7

Relative Hydrophobicity of Synthetic Nonionic Homopolymers as Measured by Partitioning in Aqueous Dex-Ficoll Two-Phase Systems.

POLYMER	MOLECULAR WEIGHT	$n(CH_2)$
Poly(ethylene glycol)	$1.5 \cdot 10^3$	170 ± 2.0
	$6.0 \cdot 10^3$	168 ± 3.5
	$2.0 \cdot 10^4$	171 ± 2.4
	$4.0 \cdot 10^4$	169 ± 2.0
Polyacrylamide	$1.06 \cdot 10^4$	-51.0 ± 1.2
	$6.65 \cdot 10^4$	-76.4 ± 3.1
	$4.54 \cdot 10^5$	-106.3 ± 1.2
	$4.90 \cdot 10^5$	-126.8 ± 2.2
Polyvinylpyrrolidone	$5.0 \cdot 10^3$	102.5 ± 7.1
	$1.2 \cdot 10^4$	75.8 ± 2.4
	$1.7 \cdot 10^4$	71.0 ± 1.5
	$5.0 \cdot 10^4$	64.8 ± 1.6
	$1.8 \cdot 10^5$	49.3 ± 3.6
Poly(vinyl alcohol)*:	$2 \cdot 10^4$ - $1 \cdot 10^5$	
1% Ac		33.4 ± 1.2
6% Ac		34.8 ± 1.0
12% Ac		36.4 ± 0.7
18% Ac		38.2 ± 1.0

* Ac = acetate groups

where $n(CH_2)$ - relative hydrophobicity of a given polymer expressed in terms of equivalent quantity of methylene groups; M_w is the polymer molecular weight; A and B are constants specific for a given polymer [38]. The reader interested in polymer-solvent interactions as derived from the results shown in Table 7.7 should be referred to [42].

In addition to synthetic homopolymers, the relative hydrophobicity of naturally occurring nonionic heteropolymers of the carbohydrate nature was examined [42]. Partitioning of several plant ß-1,4-glucomannanes was studied in the aqueous Dex-Ficoll two-phase systems containing different amounts of sodium phosphate buffer, pH 7.4 and NaCl [42]. Results of the treatment of the data reported in [42] according to Equation 7.1 are given in Table 7.8 as the

Table 7.8

Relative Hydrophobicity of β-1,4-glucomannanes as Measured by Partitioning in Aqueous Dex-Ficoll Two-Phase Systems.

Source	Mol.wt.	Ac % [a]	Gl:Man [b]	$n(CH_2)$ [c]	$n(CH_2)$ [d]	α_i [e]	β_i [e]
E. comosus	$6 \cdot 10^4$	2.7	1 : 3.2	-51 5	-11 5	-113.9	357.1
E. fuscus	$1.58 \cdot 10^5$	2.6	1 : 2.6	-73 12	26 12	-228.8	883.9
Tuber-salep	$3 \cdot 10^5$	2.0	1 : 2.4	-49 13	60 13	-220.5	973.2
E. hissaricus	$3.6 \cdot 10^5$	1.8	1 : 1.5	-27 10	36 10	-126.1	562.5

[a] Ac = acetate groups; [b] Gl:Man is the ratio of the glucose and mannose residues in a given polysaccharide; [c] relative hydrophobicity of a polysaccharide in the aqueous medium containing 0.15 mole/kg NaCl in 0.01 mole/kg sodium phosphate buffer, pH 7.40; [d] relative hydrophobicity of a polysaccharide in the aqueous medium containing 0.11 mole/kg sodium phosphate buffer, pH 7.40; [e] α_i and β_i are coefficients in Equation 7.1.

relative hydrophobicities of the polysaccharides at two different salt compositions of the aqueous medium and the corresponding α_i and β_i values.

It is important to note that the relative hydrophobicity of the nonionic polysaccharides depends on the salt composition of an aqueous medium under the conditions employed [42]. This result indicates that the effect of the salt composition on the relative hydrophobicity of a solute is not related solely to the processes of ionic hydration. The effect in question may possibly be attributed to the influence of the salt composition of an aqueous medium on the solute-water hydrogen bonding.

As discussed above, the relative hydrophobicity of synthetic polymers was found to be interrelated with the polymer effects on the dielectric properties [43], overall solvent polarity of the aqueous medium [44], etc. (see Chapter 2). The correlation between the relative hydrophobicity and surface activity of synthetic acrylic acid - 2-methyl-5-vinylpyridine copolymers reported in [40] should also be mentioned. It was found [45] also that the relative hydrophobicity of the polymers is quantitatively related to the polymer influence on the relative hydrophobic character of an aqueous medium (see Chapter 2). It should be repeated that the relative hydrophobic character of solvent term covers the

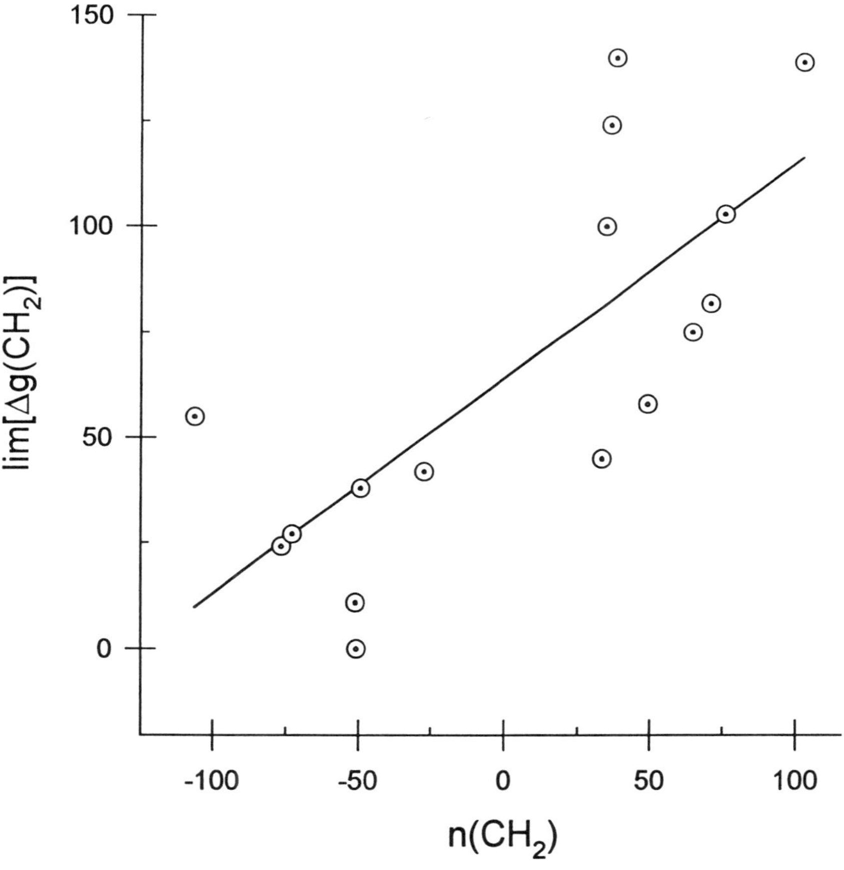

Figure 7.5. Relationship between the relative hydrophobicity of polymers, $n(CH_2)$, and the maximum value of the relative hydrophobic character of the polymers' aqueous solutions, $\lim[\Delta g(CH_2)]$, at the salt composition of 0.15 mole/kg NaCl in 0.01 mole/kg sodium phosphate buffer, pH 7.4.

thermodynamic affinity of the solvent medium for non-polar groups and solutes, namely for a CH_2 group, in reference to that of a solvent (water, octanol-saturated water, etc.) chosen as a reference medium. As shown above, an increase of the polymer concentration is usually followed by a change in the relative hydrophobic character of the solution up to a certain limit, lim[$\Delta g(CH_2)$], specific for a given polymer. This limiting value may be viewed [45] as a measure of the effect of the polymer on the hydrophobic character of the solution.

The relationship in question [45] is shown in Figure 7.5 as a plot of lim[$\Delta g(CH_2)$] values for different polymer solutions versus the relative hydrophobicities of the corresponding polymers. The relationship plotted in Fig. 7.5 is described as:

$$\lim[\Delta g(CH_2)] = -64.12(\pm 7.68) - 0.51(\pm 0.12) \cdot n(CH_2) \tag{7.7}$$

$$N = 16;\ r^2 = 0.746;\ s = 30.63$$

where lim[$\Delta g(CH_2)$] is the maximum value (in cal/mole) of the relative hydrophobic character attainable for the aqueous solution of a given polymer in the presence of 0.15 mole/kg NaCl in 0.01 mole/kg sodium phosphate buffer, pH 7.4; $n(CH_2)$ is the relative hydrophobicity of the polymer in the aqueous medium of the same salt composition; N is the total number of different polymers examined; r^2 is the correlation coefficient; s is the standard deviation from the regression.

The correlation between the affinity of a solute for an aqueous medium and the effect of the solute on this medium [45] appears to be theoretically sound. The relative hydrophobicity of a solute by definition (see above) is a measure of the total free energy of the solute-aqueous medium interactions. It seems likely that these interactions perturb the intermolecular hydrogen bonds in the aqueous medium, which due to a cooperative effect results in an alteration of the thermodynamic state of the bulk medium. The most spectacular example of the relationship between the affinity of a solute for water and its effect on the water structure appears to be the relationship between the position of an ion in the Hofmeister lyotropic series and the relative hydrophilicity (hydrophobicity) of the ion [46].

The most important feature of the relationship [45] seems to be that Equation 7.7 may be used to estimate the relative hydrophobic character of aqueous solutions of biological macromolecules which cannot be determined experimentally. For example, the hydrophobic character of aqueous protein solutions cannot be estimated by the method used (see in Chapter 2) as it is impossible to form a two-phase system using octanol and an aqueous- protein solution. The relative hydrophobicity of proteins can be estimated, however, by the aqueous two-phase partitioning technique and the corresponding lim[$\Delta g(CH_2)$] values can be calculated using Equation 7.7. This approach, as shown below, may be used for estimating the relative hydrophobic character of biological tissues.

7.4. RELATIVE HYDROPHOBICITY OF PROTEINS

It has been frequently shown in the literature (see, for example, in [47-49]) that the technique of partitioning in aqueous two-phase systems is highly sensitive to the individual features of a protein being partitioned. The features important for the protein partition behavior are known to be related to the type, amount, and topography of the chemical groups located at the surface of a macromolecule or, in other words, to the amino acid residues exposed to the solvent [47-49]. One example among many, indicating the sensitivity of the partition technique to the protein modifications, is offered by Desbuquois and Aurbach [50]. For example, the partition coefficient of insulin in the aqueous Dex-500—PEG-6000 two-phase system containing 0.02 M citrate-phosphate buffer, pH 5.0 decreases by ca. 2.5-fold from 1.00 to 0.38 as the result of 1:1 iodination [50]. The partition coefficient of intact glucagon in the same two-phase system amounts to 1.15, while the 1:1 and 1:2.1 iodinated derivatives are characterized by the partition coefficient values of 0.45 and 0.24, respectively [50].

Unfortunately, the most of the data reported on protein partitioning in aqueous two-phase systems cannot be considered in terms of relative hydrophobicity of the proteins because of the lack of the physicochemical characteristics of the particular systems used by different authors. The limited set of data on the relative hydrophobicity of proteins, lectins, etc., generated by Zaslavsky et al.[51-56], using the aqueous Dex-Ficoll two-phase systems containing different amounts of sodium phosphate buffer, pH 7.4 and NaCl, is discussed below together with the results obtained by Kuboi et al.[57-64] in the aqueous Dex-PEG systems.

The approach suggested by Kuboi et al. [57-64] for estimating the relative hydrophobicity of proteins is essentially identical to the one described above. Terminology used in the publications [57-64] is, however, slightly different from those used throughout this text and must be explained.

To characterize the difference between the relative hydrophobic character of the two phases of an aqueous polymer system, partitioning of a series of free amino acids in the system of a given fixed composition is studied [57, 58]. The logarithm of the partition coefficient of a given amino acid is plotted versus the index of the amino acid relative hydrophobicity RH. The index RH used by Kuboi et al.[57,58] is the free energy of transfer of the side-chain of the amino acid from ethanol to water as determined by Nozaki and Tanford [65]. The linear plot is obtained and described as [57,58]:

$$\ln K_i = \ln K_{Gly} + HF \cdot RH_i \tag{7.8}$$

where K_i is the partition coefficient of a given i-th amino acid in an aqueous two-phase system of a fixed polymer and salt composition. Relative hydrophobicity RH_i of the i-th amino acid is defined as $RH_i = \Delta G(\text{side-chain i})_{EtOH\rightarrow water}$ where $\Delta G(\text{side-chain i})_{EtOH\rightarrow water}$ is the free energy of transfer of the i-th amino acid side-chain from ethanol to water as reported by Nozaki and Tanford [65]; K_{Gly} is the partition coefficient of glycine, and $\Delta G(\text{Glycine side-chain})_{EtOH\rightarrow water}$ is zero by definition [65]. The slope HF is the so-called hydrophobicity factor, i.e., a constant of the value specific for a given aqueous two-phase system [57-64].

The similarity between Equations 7.8 and 4.4 is pretty obvious as the RH_i or $\Delta G(\text{side-chain i})_{EtOH\rightarrow water}$ parameter may be expressed as (see Chapter 6):

$$RH \equiv \Delta G(\text{side-chain i})_{EtOH\rightarrow water} = \Delta G(CH_2)_{EtOH\rightarrow water} \cdot n(CH_2)_i \qquad (7\text{-}9)$$

where $n(CH_2)_i$ is the equivalent quantity of methylene groups for the side-chain of the i-th amino acid, and $\Delta G(CH_2)_{EtOH\rightarrow water}$ is the free energy of transfer of a CH_2 group from ethanol to water (particular values see in Tables 6.1 and 6.2).

Replacing $\Delta G(\text{side-chain i})_{EtOH\rightarrow water}$ in Eqn. 7.8 with the product $\Delta G(CH_2)_{EtOH\rightarrow water} \cdot n(CH_2)_i$ leads to

$$\ln K_i = \ln K_{Gly} + HF \cdot \Delta G(CH_2)_{EtOH\rightarrow water} \cdot n(CH_2)_i \qquad (7.10)$$

where all terms are as defined above.

Combining Equations 4.4 and 7.10 the hydrophobicity factor HF_j for a j-th two-phase system may be expressed as:

$$HF_j = E_j / \Delta G(CH_2)_{EtOH\rightarrow water} \qquad (7.11)$$

or (since $\Delta G(CH_2)_{EtOH\rightarrow water} = 589$ cal/mole CH_2, see in Table 6.1)

$$\Delta G(CH_2)_i \text{ [cal/mole } CH_2] = -RT \cdot \Delta G(CH_2)_{EtOH\rightarrow water} \cdot HF_j \approx$$

$$\approx 1483.7 \cdot HF_j \text{ [mole/kJ]}.$$

The HF_j values reported by Kuboi et al.[60] for different aqueous two-phase systems being expressed in $\Delta G(CH_2)_i$ values are in a good agreement with the $\Delta G(CH_2)_{tr}$ values presented above (see data in Table 4.2). These values [60] amount to 10.4 cal/mol CH_2 for the aqueous two-phase system composed of 9% wt. Dex-100 and 10.8% wt. PEG-1540; 13.4 cal/mole CH_2 for the system composed of 9% wt. Dex-70 and 13% wt. PEG-1540; 18.7 cal/mole CH_2 for the system composed of 9% wt. Dex-100 and 9% wt. PEG-4000; 29.7 cal/mole CH_2 for the system composed of 9% wt. Dex-100 and 9% wt. PEG-6000. Addition of salts such as NaCl or Na_2SO_4 to aqueous Dex-PEG two-phase systems was observed [63] to increase the difference between the relative hydrophobic character of the phases. Addition of 2.0 M NaCl, for example, increases this difference up to about 150 cal/mole CH_2 [63], the observation clearly in agreement with the results discussed in Chapter 4. Unfortunately, the authors [63] did not always specify not only the molecular weights of the phase polymers but even the polymers' concentrations in the systems examined.

It should be pointed out also that the difference between the relative hydrophobic character of the two phases in aqueous PEG-potassium phosphate systems commonly exceeds those typical for Dex-PEG systems [63]. The difference in question amounts to 148.8 cal/mole CH_2 in the system composed of 12% wt. PEG-1540 and 12% wt. potassium phosphate, pH 7.5 at 4°C [64] and increases up to about 370-400 cal/mole CH_2 with increasing concentration and molecular weight of PEG and pH (i.e., the concentrations and ratio of mono- and dibasic potassium phosphates) [63].

Even larger differences between the relative hydrophobic character of the two phases are achieved in the aqueous PEG-$MgSO_4$ systems [63]. The aqueous two-phase system formed in the mixture of 20% wt. PEG-6000 and 6.8% wt. $MgSO_4$ is characterized [61] by about 400 cal/mole CH_2 in agreement with the estimate reported by Eiteman and Gainer [66]. Addition of 1.0 M Na_2SO_4 to an aqueous PEG-$MgSO_4$ system increases the difference between the relative hydrophobic character of the two phases from 148.4 cal/mole CH_2 to about 520 cal/mole CH_2.

The hydrophobicity factor HF is used by Kuboi et al. [58-64] for estimating the so-called surface hydrophobicity HFS of various proteins. The surface hydrophobicity of a given i-th protein is determined as [58-64]:

$$HSF_i = \ln K_{ij}/HF_j \qquad (7.12)$$

where K_{ij} is the partition coefficient of the i-th protein in the j-th aqueous two-phase system; other terms are as defined above.

Equation 7.12 is clearly identical to Eqn. 6.11 though expressed in slightly different terms, and the surface hydrophobicity of a protein HSF may be expressed in the equivalent quantity of CH_2 groups.

Kuboi et al.[58-64], however, do not actually use Equation 7.12. Presumably in order to increase the reliability of the hydrophobicity estimate, the authors [58-64] plot the values of the logarithms of the protein partition coefficients determined in several different aqueous two-phase systems versus the HF values for these systems. The observed plot is treated in linear regression terms and the slope obtained is considered as the true hydrophobicity value HSF. This apparently reasonable procedure instead of enhancing the method actually worsens it. The reason is the aforementioned relationship between the partition coefficients of a solute in different (aqueous) two-phase systems described by Equation 5.10a. The intercept b_i in equation 5.10a is likely to be different for different pairs of the systems compared and seems to be at least one of the reasons for the scatter of the data observed in the linear plots reported by Kuboi et al.[58-64].

Fortunately, however, the value of the coefficient b_i is usually small (see in Chapter 5) and it should not affect the protein surface hydrophobicity values HSF reported by Kuboi et al.[57-64] to a very significant degree. The HSF values reported for various proteins may be readily expressed in terms of equivalent CH_2 groups. The relative hydrophobicity (net surface hydrophobicity in terms used by the authors [57-64]) of the most of the proteins was examined at pH corresponding to their respective pI values.

According to the data reported [57], the relative hydrophobicity of papain (at pH = pI = 8.8) is equivalent to that of +17.5 CH_2 groups, that of ovalbumin (at pH = pI = 4.85) and cytochrome C (at pH = pI = 10.05) may be expressed as -39.3 equivalent CH_2 groups. Bovine serum albumin is characterized by the relative hydrophobicity equivalent to that of -92.3 CH_2 groups, bovine hemoglobin (at pH = pI = 7.0) by -83.1 equivalent CH_2 groups, and ribonuclease a by that of -24.8 equivalent CH_2 groups. The authors [57-64] did not always clearly specify the ionic composition of the systems used for the protein partitioning and that hinders interpretation of the values reported. These values appear to be similar to those determined by partitioning in aqueous Dex-Ficoll two-phase systems (see in chapter 6 and below), at least by the order of magnitude.

The data by Kuboi et al.[57-64] deserving particular attention are, in my view, those indicating strong dependence of the protein hydrophobicity upon salt composition of the aqueous medium. For example, the relative hydrophobicity of ß-galactosidase from *E.coli* increases from -181 equivalent CH_2 groups in the salt-free medium to -2.4 equivalent CH_2 groups in the presence of 1.0 M NaCl [59].

It has been particularly shown [60] that conformational changes in protein macromolecule induced, for example, by addition of small amounts of HCl are accompanied by changes in the protein relative hydrophobicity readily detected by the partition technique. Strong linear correlationships between the relative hydrophobicity and ellipticity for cytochrome C, papain, lysozyme, and apomyoglobin were observed [60] in the presence of varied amounts of HCl. The difference between the relative hydrophobicity of a protein in its native state and unfolded state (defined as that corresponding to the protein ellipticity of zero value) was found to be about 146 equivalent CH_2 groups for cytochrome C, 196 equivalent CH_2 groups for horse apomyoglobin, and 101 equivalent CH_2 groups for papain (the relative hydrophobicity of a protein in completely unfolded state being always more hydrophobic than that of the protein in the native state) [60].

Relative hydrophobicity of bovine carbonic anhydrase was shown [62] to be changed in the presence of different amounts of guanidine hydrochloride [62]. Denaturation of the enzyme in the presence of 2 M guanidine hydrochloride is accompanied by the drastic increase in the enzyme relative

Table 7.9

Relative Hydrophobicity of Human Serum Proteins as Measured by Partitioning in Aqueous Dex—Ficoll Two-Phase Systems.

Protein	α_i [a]	β_i [a]	$n(CH_2)$ [b]	$n(CH_2)$ [c]
Albumin-1 [d]	-84.6	326.8	-27.0	9.6
Albumin-2 [d]	-76.3	303.6	-22.8	11.2
γ-Globulin	16.2	0	16.2	16.2
α-Globulin	-107.9	461.6	-26.6	25.1
α_1 and α_2-Globulins	-65.1	245.5	-21.8	5.7
Oxyhemoglobin	11.4	32.1	17.1	20.7
Oxyhemoglobin-1 (mod.) [e]	-5.3	0	-5.3	-5.3
Oxyhemoglobin-2 (mod.) [e]	-65.7	419.6	8.2	55.2

[a] α_i and β_i are coefficients in Equation 7.1; [b] relative hydrophobicity of a protein in the aqueous medium containing 0.15 mole/kg NaCl in 0.01 mole/kg sodium phosphate buffer, pH 7.40; [c] relative hydrophobicity of a protein in the aqueous medium containing 0.11 mole/kg sodium phosphate buffer, pH 7.40; [d] albumin-1 and albumin-2 are samples from different manufacturers, ICN Pharmaceuticals, USA, and Central Institute of Hematology and Blood Transfusion, Moscow, Russia, respectively; [e] oxyhemoglobin-1 produced by treatment of intact oxyhemoglobin with glutaraldehyde in the presence of pyridoxal 5-phosphate, $M_w = 2.5 \cdot 10^4$; $M_w/M_n = 2.2$; oxyhemoglobin-2 is fraction of oxyhemoglobin-1 with $M_w = 4 \cdot 10^5$; $M_w/M_n = 1.2$.

hydrophobicity (from about -33.6 equivalent CH_2 groups to about +112 equivalent CH_2 groups). Strong correlation between the relative hydrophobicity and molecular ellipticity of the protein in the presence of various amounts of guanidine hydrochloride was also observed [62]. The important conclusion is that the relative hydrophobicity of a biological macromolecule is highly sensitive to conformational changes.

Essentially all the data reported by Zaslavsky et al.[51-56] have been treated according to Equation 7.1 and they are presented further on as the relative hydrophobicities of the proteins at two different salt compositions of the aqueous medium, 0.15 mole/kg NaCl in 0.01 mole/kg sodium phosphate buffer, pH 7.4, and 0.11 mole/kg sodium phosphate buffer, pH 7.4, and the corresponding α_i and β_i values.

The relative hydrophobicities of several human serum proteins under the above conditions [51] are presented in Table 7.9.

It should be noted, first, that the two samples of albumin partition differently. The difference between the samples could not be detected by standard electrophoretic and chromatographic procedures and amino acid analysis [51]. It was reasoned [51,52] that the difference between the protein samples detected by the partition technique was due to the different type and amount of lipids present in the samples. In this context it should be mentioned that the immunologic properties of human serum albumin are unaltered by removal of the fatty acids, as are the circular dichroism spectra and other features of the tertiary structure of the protein [67-69]. The data in Table 7.9 (as well as the other data given below) support the aforementioned assertion that the molecular weight of a protein is not of significant importance for protein-solvent interactions. The nature of the residues exposed to the solvent is of much greater significance. This conclusion is confirmed by the results presented below for other proteins.

It was suggested above that the relative hydrophobicity of proteins may be related to their functional activity. The data of Table 7.9 indicate in particular that the relative hydrophobicity of human serum proteins varies with the ionic composition of an aqueous medium. That agrees, for example, with the idea that protein solubility in water is dependent on the presence of inorganic salts. It is rather surprising, however, that the relative hydrophobicity of, e.g., α-globulin varies as drastically as from -26.6 to +25.1 equivalent CH_2 groups under rather limited change in the ionic composition from 0.15 mole/kg NaCl in 0.01 mole/kg sodium phosphate buffer, pH 7.4 to 0.11 mole/kg sodium phosphate buffer, pH 7.4.

It was suggested [51,52] that the observed effect of an ionic composition of an aqueous medium on the relative hydrophobicity of a biopolymer (peptide, protein, etc.) may be regulate the biopolymer function. The ionic composition of different tissues and biological liquids is known [70] to be different and it is also known to vary (over a limited range), e.g., in the blood stream depending on the particular organ or tissue localization of a blood vessel. That means that the ionic microenvironment of a given protein in the blood stream may vary producing noticeable changes in the relative hydrophobicity of the protein. These changes may lead to adhesion (or desorption) of the protein to endothelial cells lining the blood vessel wall or to membranes of the blood cells, binding or release of a certain ligand by the protein in different parts of the blood stream, etc., affecting the protein participation in different

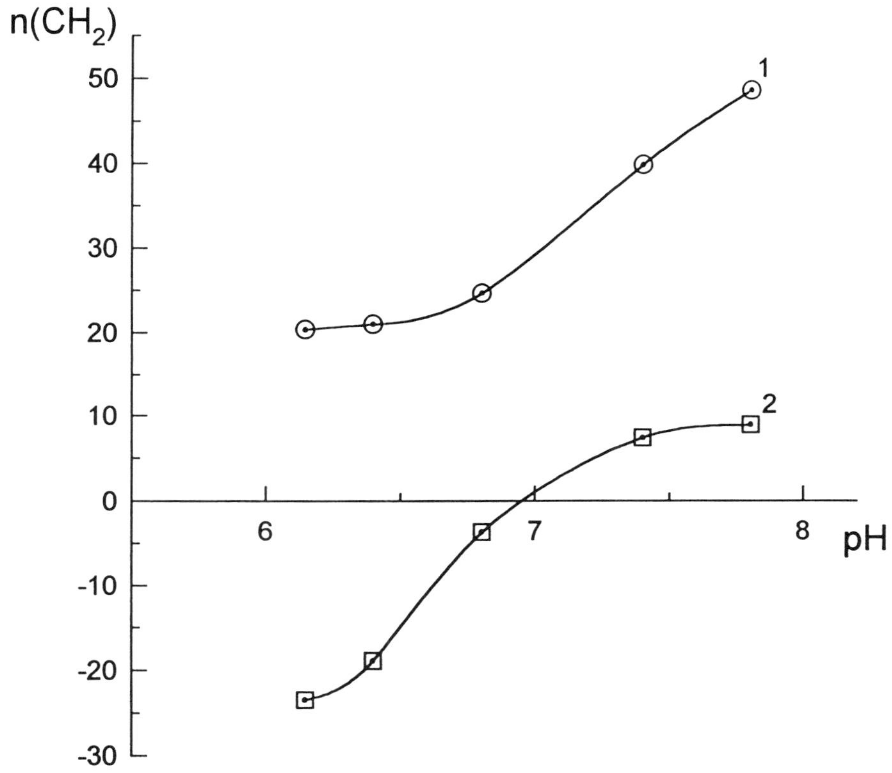

Figure 7.6. Relative hydrophobicity of oxyhemoglobin (1) and human serum albumin (2) as a function of pH in the aqueous medium containing sodium phosphate buffer of constant ionic strength value of 0.165 M.

physiologically important processes. This assumption clearly calls for an experimental verification.

Data presented in Table 7.9 for chemically modified oxyhemoglobins appear to agree with the assumption in question. Treatment of oxyhemoglobin with glutaraldehyde in the presence of pyridoxal 5-phosphate produces a preparation which being administered in rats intravenously is rapidly removed from

Table 7.10

Influence of pH and Ionic Composition on the Relative Hydrophobicity of Human Serum Albumin and Oxyhemoglobin as Measured by Partitioning in Aqueous Dex—Ficoll Two-Phase Systems.

pH	α_i [a]	β_i [a]	$n(CH_2)$ [b]	$n(CH_2)$ [c]
Albumin				
6.15	-101.6	262.5	-55.4	-26.0
6.40	-88.8	237.5	-47.0	-20.4
6.80	-76.3	236.6	-34.6	-8.1
7.40	-84.6	326.8	-27.0	9.6
7.80	-92.7	357.1	-29.8	10.2
Oxyhemoglobin				
6.15	80.3	-227.7	40.2	14.7
6.40	57.5	-139.3	33.0	17.4
6.80	30.9	-33.0	25.1	21.4
7.40	11.4	32.1	17.1	20.7
7.80	6.2	49.1	14.9	20.4

[a] α_i and β_i are coefficients in Equation 7.1;
[b] relative hydrophobicity of a protein in the aqueous medium containing 0.15 mole/kg NaCl in 0.01 mole/kg sodium phosphate buffer of indicated pH value;
[c] relative hydrophobicity of a protein in the aqueous medium containing 0.11 mole/kg sodium phosphate buffer of indicated pH value.

the body [71]. The relative hydrophobicity of this preparation (oxyhemoglobin-1) as shown in Table 7.9 is independent of the salt composition ($ß_i = 0$) under the conditions employed. The high molecular weight fraction of this preparation (oxyhemoglobin-2) is characterized by a prolonged time of the blood circulation [74] and its relative hydrophobicity is dependent upon by the salt composition of an aqueous medium significantly ($ß_i = 419.6$). It should not be assumed that the salt composition dependence of the relative hydrophobicity of the protein preparation is the decisive factor, however, as the relative hydrophobicity of α-globulin, for example, is independent of the salt composition variations under the same conditions [51,52].

The relative hydrophobicity of a protein is usually affected by pH, as is predicted from data on the pH and salt influence on the protein partitioning in aqueous two-phase systems discussed in Chapter 5. This data (Figs. 5.10 - 5.12) was obtained, however, mostly at the salt compositions rather far from those usually encountered by proteins in biological systems. The relative hydrophobicity of human serum albumin and oxyhemoglobin was examined [19] as a function of the more "physiological" salt composition in the aqueous medium containing NaCl and sodium phosphate buffer at different pH values over the 6.15 - 7.8 pH range. The results [19] treated according to Equation 7.1 are given in Table 7.10. It is known (for human oxyhemoglobin, in particular [72]) that the properties of a protein in solution depend upon both pH and ionic strength. Hence the relative hydrophobicity of the two proteins was additionally studied [19] as a function of pH in the aqueous medium containing sodium phosphate buffer with pH varied from 6.15 to 7.8 at the constant ionic strength value of 0.165 M. The data reported in [19] are plotted in Figure 7.6 as the protein relative hydrophobicity versus the buffer pH value.

The data given in Table 7.10 indicate that the effect of the ionic composition of the aqueous medium on the relative hydrophobicity of both proteins under the conditions employed [19] increases with increasing pH. The relative hydrophobicity of human serum albumin appears to be more affected by the ionic composition of an aqueous medium than that of oxyhemoglobin. This result is in agreement with the hypothesis described above with regard to the possible functional role of the effect of ionic composition on the solute hydrophobicity: albumin exists in the blood stream as a carrier of different ligands, while hemoglobin is likely found in a much more conservative microenvironment inside the red blood cell. The pH-dependence of the proteins relative hydrophobicity shown in Fig. 7.6 indicates that the relative hydrophobicity of serum albumin varies with pH from -23.5 equivalent CH_2 groups at pH 6.15 to +9.0 equivalent CH_2 groups at pH 7.8, i.e. from being highly hydrophilic to rather hydrophobic, more significantly than that of oxyhemoglobin (from +20.4 to +48.6 equivalent CH_2 groups over the same pH range), again in line with the

Table 7.11

Relative Hydrophobicity of Albumins from Different Sources as Measured by Partitioning in Aqueous Dex—Ficoll Two-Phase Systems.

Source	α_i [a]	β_i [a]	$n(CH_2)$ [b]	$n(CH_2)$ [c]
Sheep	-50.9	160.7	-22.6	-4.6
Bovine-1 [d]	-65.7	221.4	-26.7	-1.9
Bovine-2 [d]	-57.3	202.7	-21.6	1.1
Rabbit	-73.5	256.2	-28.4	0.3
Rat	-61.0	231.2	-20.3	5.6
Horse	-64.1	225.0	-24.5	0.7
Porcine	-66.3	237.5	-24.5	2.1
Dog	-73.2	243.8	-30.2	-2.9
Human-1 [e]	-84.6	326.8	-27.0	9.6
Human-2 [e]	-76.3	303.6	-22.8	11.2
Chicken	-106.3	392.8	-37.1	6.9
Egg	-35.6	142.8	-10.4	5.6

[a] α_i and β_i are coefficients in Equation 7.1; [b] relative hydrophobicity of a protein in the aqueous medium containing 0.15 mole/kg NaCl in 0.01 mole/kg sodium phosphate buffer, pH 7.40; [c] relative hydrophobicity of a protein in the aqueous medium containing 0.11 mole/kg sodium phosphate buffer, pH 7.40; [d] bovine albumin-2 in contrast to bovine albumin-1 contains less than 0.05% of fatty acids (the data is given for two different samples of bovine albumin-1 manufactured by Sigma Chemical Co., USA, and Calbiochem, USA); [e] human albumin-1 and human albumin-2 are from different manufacturers, ICN Pharmaceuticals, USA, and Central Institute of Hematology and Blood Transfusion, Moscow, Russia, respectively.

above hypothesis.

The above hypothesis also implies that the relative hydrophobicity of proteins with a similar function but different cellular or tissue localization should be different. The results of the study [55] of the relative hydrophobicity of inorganic pyrophosphatases from different sources confirm this assumption. The relative hydrophobicities of the enzymes have been measured in the aqueous medium containing 15 mM NaCl, 1 mM $MgSO_4$, 1 mM dithioerythritol, and 10 μM EDTA in 0.01 M Tris HCl buffer, pH 7.5 using the partitioning in aqueous Dex-Ficoll two-phase system technique [55]. As found by Unguryte et al.[55], the relative hydrophobicity of rat liver mitochondrial inorganic pyrophosphatase under the above conditions amounts to -6.9 ± 2.0 equivalent CH_2 groups and greatly exceeds those of the cytosolic pyrophosphatases from rat liver (-42 ± 0.5 equivalent CH_2 groups), from baker's yeast (-36 ± 1 equivalent CH_2 groups), and from *Escherichia coli* (-32 ± 1 equivalent CH_2 groups). These results appear to be in line with the membrane localization of the (more hydrophobic) mitochondrial enzyme.

The relative hydrophobicities of serum albumins from different sources reported by Zaslavsky et al.[52] are shown in Table 7.11. The differences in the relative hydrophobicity between different albumins are quite noticeable. This is of interest since it is generally supposed [69] that the folding of these proteins is very similar. This conclusion, however, was made on the basis of detergent binding studies [69]. The data in Table 7.11 indicate that the solvent-protein interactions may differ when the ligand-binding sites of the macromolecules are similar. That supports the aforementioned view that the relative hydrophobicity of a protein is essentially a surface property of the macromolecule which is only marginally (if at all) affected by the residues buried in its interior or by those forming a binding pocket.

The aforementioned hypothesis that the effect of the ionic composition on the relative hydrophobicity of biological solutes may regulate the solute functions seems to be supported by the results of comparison of the effects of salt composition on the hydrophobic properties of serum albumins and red blood cells from different mammalian species [52,73]. Partitioning of erythrocytes and albumins in aqueous Dex-Ficoll two-phase systems containing different amounts of sodium phosphate buffer, pH 7.4 and NaCl was found [52,73] to fit Equation 5.4:

$$\ln K = C + B \cdot I \tag{5.4}$$

where K is the partition coefficient of a protein (or particle); I is the ionic strength value used as a quantitative index of the system salt composition (see above); and C and B are constants. It should be mentioned particularly that the cell partition coefficient is defined differently from that of a soluble substance

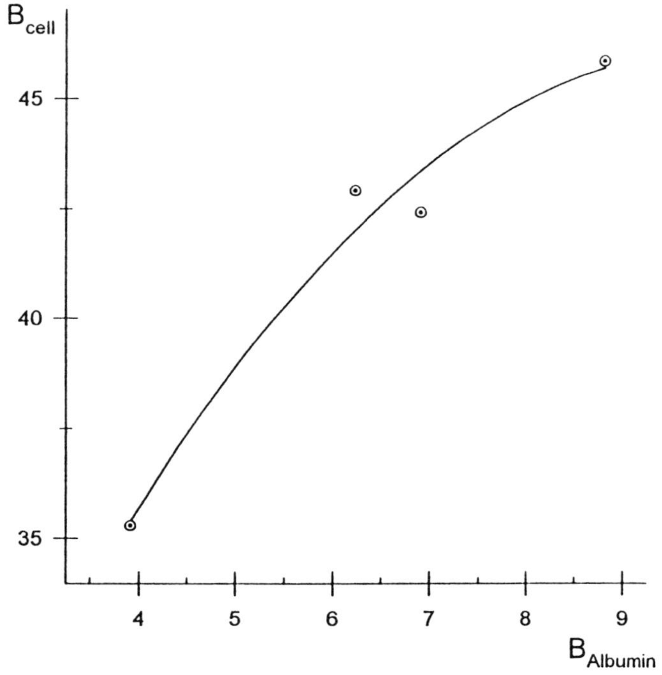

Figure 7.7. Relationship between the coefficients B in Equation 5.4 for erythrocytes and serum albumins from the same species.

(see, e.g., in [47,48]) but Equation 5.4 holds true for both [73]. It was found [52] that the coefficient B values for the serum albumins and those for erythrocytes from the same species are interrelated as shown in Figure 7.7.

Although statistically unreliable in view of the small number of species examined, the relationship presented in Fig. 7.7 supports the assumption that the effect of the composition of an aqueous medium upon the hydrophobic properties of biological solutes and particles is important for regulation of their biological functions.

Lectins offer an example of carbohydrate-binding proteins of different non-immune origin which can agglutinate selected carbohydrate-containing cell surfaces, e.g., red blood cells. The hemagglutinating activity of lectins is readily measurable in quantitative terms [74]. Hence, analysis of the relative hydrophobicity of lectins from different sources provides an opportunity to explore an existence of the biological activity - relative hydrophobicity relationship for the proteins. The relative hydrophobicity of various lectins was

Table 7.12

Relative Hydrophobicity of Lectins from Different Sources as Measured by Partitioning in Aqueous Dex—Ficoll Two-Phase Systems.

LECTIN	α_i [a]	β_i [a]	$n(CH_2)$ [b]	$n(CH_2)$ [c]
Soy bean agglutinin	175.0	-383.0	107.5 ± 1.1	64.6 ± 0.9
Lens culinaris lectin	109.3	-242.8	66.5 ± 0.5	39.3 ± 0.5
Peanut agglutinin	74.8	-58.9	64.4 ± 1.8	57.8 ± 0.5
Pea seed agglutinin	-	-	58.3 ± 0.7	-
Potato lectin	-	-	35.8 ± 1.2	-
Wheat germ agglutinin	41.6	-76.8	28.1 ± 1.0	19.5 ± 0.4
Concanavalin A	31.7	-54.5	22.1 ± 1.0	16.0 ± 0.6
Phytohemagglutinin	21.7	-50.9	12.7 ± 0.7	7.0 ± 0.4
Ricinus communis agglutinin	-	-	11.7 ± 1.3	-
Helix pomatia lectin	-141.1	370.5	-75.8 ± 1.2	-34.3 ± 0.4

[a] α_i and β_i are coefficients in Equation 7.1; [b] relative hydrophobicity of a protein in the aqueous medium containing 0.15 mole/kg NaCl in 0.01 mole/kg sodium phosphate buffer, pH 7.40; [c] relative hydrophobicity of a protein in the aqueous medium containing 0.11 mol/kg sodium phosphate buffer, pH 7.40.

examined by partitioning in the aqueous Dex-Ficoll two-phase systems containing 0.15 mole/kg NaCl in 0.01 mole/kg sodium phosphate buffer, pH 7.4 and 0.11 mole/kg sodium phosphate buffer, pH 7.4 [56]. The results obtained [56] and treated according to Equation 7.1 are presented in Table 7.12 as the relative hydrophobicities of lectins at the two salt compositions and the corresponding α_i and β_i values.

The data in Table 7.12 indicate that the carbohydrate specificity is not related to the hydrophobic properties of lectins. No general trend between the

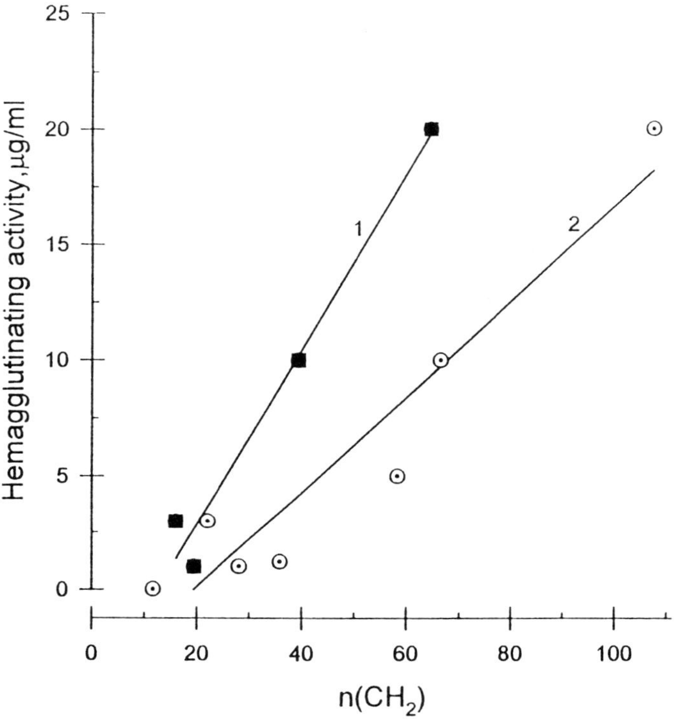

Figure 7.8. Relationships between the hemagglutinating activity and relative hydrophobicity for various lectins at two different salt compositions: 1 - 0.11 mole/kg sodium phosphate buffer, pH 7.4; 2- 0.15 mole/kg NaCl in 0.01 mole/kg sodium phosphate buffer, pH 7.4.

characteristics of the lectin-aqueous medium interactions and the glycoprotein nature of lectins could be found as well.

The hemagglutinating activity of the lectins toward rabbit red blood cells determined by Lakhtin [74] is plotted against the relative hydrophobicity of the lectins in the aqueous medium containing 0.15 mole/kg NaCl in 0.01 mole/kg sodium phosphate buffer, pH 7.4 and 0.11 mole/kg sodium phosphate buffer, pH 7.4 in Figure 7.8.

The relationships presented in Fig. 7.8 may be described as:

$$C_{aggl} = -4.0 + 0.21 \cdot n(CH_2)^{*} \quad (7.13a)$$
$$N = 7;\ r^2 = 0.955$$

and

$$C_{aggl} = -4.7 + 0.38 \cdot n(CH_2)^{**} \quad (7.13b)$$
$$N = 4;\ r^2 = 0.987$$

where C_{aggl} is the limit concentration (in μg/ml) of lectin producing agglutination of rabbit erythrocytes; $n(CH_2)$ is the lectin's relative hydrophobicity; and superscripts * and ** denote the salt composition of the aqueous medium - 0.15 mole/kg NaCl in 0.01 mole/kg sodium phosphate buffer, pH 7.4 and 0.11 mole/kg sodium phosphate buffer, pH 7.4, respectively.

It should be noted that protein activity is used in the above relationships directly instead of generally-accepted expression of the biological potency in terms of log(1/C) units. The reason is that the experimental data fit the relationships 7.13a and 7.13b much better when expressed as shown above.

The relationships described by Equations 7.13a and 7.13b are, to my knowledge, the first example of quantitative structure-activity relationship for proteins using the relative hydrophobicity as a general structural descriptor of a protein.

An existence of the relationships in question supports the above assertion that the relative hydrophobicity of a biological solute is the potency-related needed to alter the solute therapeutic potency, toxicity, etc. Some other implications of this conclusion will be discussed below.

Since the protein relative hydrophobicity is the surface property of a macromolecule, the changes in the protein conformation accompanied by alterations in the type, relative amount, and topography of the residues exposed to the solvent should be detected by the partition technique. Such changes may be induced in a protein by the ligand binding. Results of the aqueous two-phase partition studies of protein-ligand complexes are considered below.

7.5. PROTEIN-LIGAND COMPLEXES

The relative hydrophobicity of concanavalin A 1:2 complexes with glucose, mannose, and their α-methyl-D-glycopyranosides was measured by partitioning in aqueous Dex-Ficoll two-phase systems containing 0.15 mole/kg NaCl in 0.01 mole/kg sodium phosphate buffer, pH 7.4 and 0.11 mole/kg sodium phosphate buffer, pH 7.4 [56]. The data obtained [56] are given in Table 7.13. These data indicate that the 1:2 complexes of concanavalin A with the carbohydrates display the relative hydrophobicity noticeably lower than that of ligand-free concanavalin A at both salt compositions used. Ligand binding changes also the dependence of the lectin relative hydrophobicity on the salt composition of the aqueous medium as represented by the β_i values.

As indicated by the β_i values in Table 7.13, the relative hydrophobicity of the lectin complexes with mannose and mannopyranoside decreases with increasing sodium phosphate buffer concentration qualitatively as that of ligand-free concanavalin.

The relative hydrophobicity of the lectin complexes with glucose and glucopyranoside under the same conditions increases. The results obtained [56] imply that the carbohydrate binding produces significant changes in the concanavalin A - aqueous medium interactions, presumably because of conformational changes in the protein macromolecule. These changes clearly depend on the specific structure of a carbohydrate in agreement with the literature data (see, e.g., in [75]).

The possibility of examining the conformational changes produced in a protein macromolecule by a ligand binding with the aqueous two-phase partition technique was also explored in the study of the albumin-drug complexes [76].

The relative hydrophobicities of the 1:1 and 1:2 complexes of human serum albumin with three series of drugs including a group of 9 penicilline derivatives, a group of 10 sulfonamides, and a series of 7 cephalosporin antibiotics were studied [76] by the technique of partitioning in the aqueous Dex—Ficoll two-phase systems containing 0.15 mole/kg NaCl in 0.01 mole/kg sodium phosphate buffer, pH 7.4. As the drug binding to serum albumin is known to be an important determining factor in the drug overall pharmacology and pharmacokinetics [77], the relative hydrophobicity of the albumin-drug complexes is of both theoretical and practical interest due to the likely implications for the drug distribution throughout the body tissues, blood circulation time, etc. Information provided by the study [76] may also be useful from the viewpoint of the sensitivity of the aqueous two-phase partition technique to changes in a protein macromolecule induced by the presence of small amounts of low molecular weight contaminants (see below).

Results obtained in [66] are presented in Table 7.14. It may be noticed, first, that the relative hydrophobicity of the ligand-free human serum albumin used in the study [76] is much lower than that of the albumin samples examined in [51,52] under the same conditions (see in Tables 7.9 and 7.10). The likely reason is that both human serum albumin samples 1 and 2 analyzed in [51,52] contained lipids. The data given in Table 7.14 indicate in particular that the relative hydrophobicity of all the albumin-ligand 1:1 and 1:2 complexes studied so far [76] exceeds that of the ligand-free albumin.

Binding of different drugs to human serum albumin is often accompanied by conformational changes in the protein. These changes were observed

Table 7.13

Relative Hydrophobicity of Concanavalin A and its 1:2 Complexes with Different Carbohydrates as Measured by Partitioning in Aqueous Dex—Ficoll Two-Phase Systems.

Carbohydrate	α_i [a]	β_i [a]	$n(CH_2)$ [b]	$n(CH_2)$ [c]
-	31.7	-54.5	22.1 ± 1.0	16.0 ± 0.6
α-D-glucose	9.0	6.2	10.1 ± 1.3	10.8 ± 0.6
α-D-mannose	15.8	-11.6	13.8 ± 1.7	12.5 ± 0.8
α-methyl-D-glucopyra-noside	7.1	28.6	12.1 ± 1.1	15.3 ± 0.6
α-methyl-D-mannopyranoside	21.1	-20.5	17.5 ± 1.5	15.2 ± 0.3

[a] α_i and β_i are coefficients in Equation 7.1;
[b] relative hydrophobicity of a protein complex in the aqueous medium containing 0.15 mole/kg NaCl in 0.01 mole/kg sodium phosphate buffer, pH 7.40;
[c] relative hydrophobicity of a protein complex in the aqueous medium containing 0.11 mole/kg sodium phosphate buffer, pH 7.40.

by several spectroscopic techniques, including u.v. difference, fluorescence and infrared spectroscopy, circular dichroism measurements, volumetry, etc. [78, 79]. All available information indicates the large conformational flexibility of the human serum albumin molecule, the structure of which may be strongly affected by different ligands [78].

It must be indicated that the partition experiments [76] were performed using a stock albumin solution containing an appropriate amount of a drug in the 1:1 or 2:1 molar ratio with the protein. Varied amounts of such a solution were added to the aqueous two-phase system, and the protein concentrations in the two phases were determined. The possibility that the protein may partition in the system as a mixture of the drug-free protein and protein-drug complexes of different molar ratios was disregarded [76]. Once the molar ratio of the amount of the drug added to the protein had reached 1:1 or 2:1, it was assumed [76] that the protein concentrations in the two phases corresponded to those of the respective protein-drug complexes. This assumption may be true for some of the drugs examined but not for others, as the drugs are known to bind to human serum albumin differently [78, 79]. It seems quite possible that in some cases drug-free albumin may coexist in the solution with the 1:1 and 1:2 albumin-drug complexes. Since the partition coefficient of the total protein is measured, this value may sometimes represent the overall partition coefficient for, e.g., 1:1 and 1:2 albumin-drug complexes and drug-free protein. To a first approximation, however, the assumption [76] that the relative hydrophobicity values listed in Table 7.14 characterize the 1:1 and 1:2 albumin-drug complexes may be accepted.

The process of binding of a drug molecule to protein may be considered as including the following hypothetical steps: (i) removal of a drug from its aqueous surrounding; (ii) formation of protein-drug bonds (hydrophobic, van-der-Waals, electrostatic, etc.); (iii) conformational changes in a protein macromolecule; and (iv) rearrangement of protein-water interactions. Other steps, e.g., conformational changes in a drug molecule are to be considered in some cases, but those listed above seem to be the most typical. It seems reasonable to suggest that the rearrangement of protein-water interactions (step iv) may result from both formation of protein-drug bonds and protein conformational changes (steps ii and iii, respectively).

It is currently recognized that protein-ligand interactions may be accompanied by significant changes in the protein-solvent interactions. The most illustrative example is the binding of 60 extra water molecules to hemoglobin during the transition from the fully deoxygenated state to the fully oxygenated state [80]. It is reasonable to suggest that not only the number of water molecules participating in the protein-solvent interactions, but also the relative intensity of these interactions may change as the result of the protein conformational changes induced by the ligand binding. The data presented in Table 7.14 clearly support this suggestion.

The difference between the relative hydrophobicity of a given albumin-drug complex and that of the drug-free albumin characterizes the changes in the protein-water interactions resulting from the protein-drug complex formation and the presumable protein conformational changes. The drug affinity for albumin is usually correlated with drug lipophilicity [81-83]. Hence the albumin conformational changes represented by the difference in the relative hydrophobicities of albumin-drug complex and drug-free albumin, $\Delta n(CH_2)$, was correlated [76] with the drug lipophilicity measured as the logarithm of the drug partition coefficient in octanol/water system [84]. The relationships observed are given in Figure 7.9.

The relationships presented in Fig. 7.9 fit two different though similar parabolic curves described as:

$$\Delta n(CH_2)_j = 22.22 + 1.44 \cdot \log P_j - 1.21 \cdot [\log P_j]^2 \qquad (7.14a)$$

Table 7.14

Relative Hydrophobicity of 1:1 and 1:2 Complexes of Human Serum Albumin with Different Drugs as Measured by Partitioning in Aqueous Dex—Ficoll Two-Phase Systems.

Drug	logP [a]	$n(CH_2)_{1:1}$ [b]	$n(CH_2)_{1:2}$ [c]
Ligand-free albumin		-52.6	
Sulfonamides			
Sulfisoxazole	1.01	-28.1	-30.0
Sulfamethoxazole	0.89	-27.4	-33.0
Sulfamoxole	0.58	-30.4	1.1
Sulfamethizole *	0.54	-40.8	-46.8
Sulfapyridine **	0.35	-20.4	-43.0
Sulfamethazin *	0.28	-40.5	-31.9
Sulfamerazine	0.14	-30.8	-23.1
Sulfathiazole *	0.05	-41.1	-29.8
Sulfadiazine *	-0.09	-38.8	-39.4
Sulfanilamide *	-0.62	-42.2	-43.4
Penicillin Derivatives			
Naphcillin	3.46	-40.4	-33.6
Dicloxacillin *	2.91	-47.4	-23.2
Cloxacillin *	2.43	-43.9	-18.6
Oxacillin	2.38	-36.2	-41.3
Phenoxymethylcillin	2.09	-32.7	-20.2
Benzylpenicillin *	1.83	-41.0	-37.0
Methicillin *	1.22	-40.5	-24.7
Carbenicillin	1.13	-30.6	-25.0
Ampicillin *	-1.13	-43.8	-41.1
Cephalosporin Antibiotics			
Cephalexin *	0.65	-38.9	-37.7
Cephalothin	0.53	-30.5	-16.9
Cefamandole	0.50	-30.3	-24.9
Cefuroxime	-0.16	-32.9	-25.0
Cefazolin	-0.24	-30.9	-27.9
Cefachlor	-1.79	-36.5	-23.7
Cephradine	-1.99	-37.6	-24.8

Table 7.14 Continued.

Warfarin	2.52	-32.2	-34.0
Na-salicilate **	-0.85	-29.2	-50.1

[a] P is the partition coefficient of a drug in the octanol-water two-phase system; values of logP are listed as kindly provided by Dr. A. Leo (Pomona College, CA, USA); note that the values listed have been determined for non-ionic drugs derivatives, while complexes with albumin have been studied [76] using sodium salts of the corresponding drugs;

[b] relative hydrophobicity of the albumin-drug 1:1 complex in the aqueous medium containing 0.15 mole/kg NaCl in 0.01 mole/kg Na-phosphate buffer, pH 7.40;

[c] relative hydrophobicity of the albumin-drug 1:2 complex in the aqueous medium containing 0.15 mole/kg NaCl in 0.01 mole/kg Na-phosphate buffer, pH 7.40.

* denotes drugs fitting curve B in Figure 7.9 and Equation 7.13b;

** denotes the drugs not fitting any of the two relationships.

$$N = 15;\ r^2 = 0.9191$$

where $\Delta n(CH_2)_j$ is the difference between the relative hydrophobicities of the drug-free albumin and the albumin-drug 1:1 complex calculated as $\Delta n(CH_2) = n(CH_2)_{Alb\text{-}drug} - n(CH_2)_{Alb}$; P is the partition coefficient of a drug in the octanol-water system; subscript "j" denotes the j-th drug; r^2 is the correlation coefficient; and N is the number of points (curve A, Figure 7.9), and

$$\Delta n(CH_2)_j = 12.32 + 1.72 \cdot \log P_j - 1.38 \cdot [\log P_j]^2 \qquad (7.14b)$$

$$N = 11;\ r^2 = 0.9428$$

for curve B (Fig. 7.9).

The so-called "optimal" lipophilicity $\log P_o$ values for both sets of drugs may be calculated from the above equations. Such a value represents the lipophilicity of a drug displaying the maximum effect achievable for a given series of compounds, which in the present case is the maximum conformational change induced by a drug in the albumin molecule. The $\log P_o$ values corresponding to the maxima of the relationships described by curves A and B (Fig. 7.9) amount to essentially the same value of 0.6 ± 0.1 logP units.

It follows from the relationships observed (Equations 7.14a and 7.14b) that the drugs examined may be divided into two groups according to their effects on the albumin conformation. It seems rather unexpected that the difference between the two groups is not related to differences between the antibiotics classes, e.g., sulfonamides and cephalosporins or penicillins and sulfonamides. Analysis of structures of the drugs tested [76] seems to imply that the difference

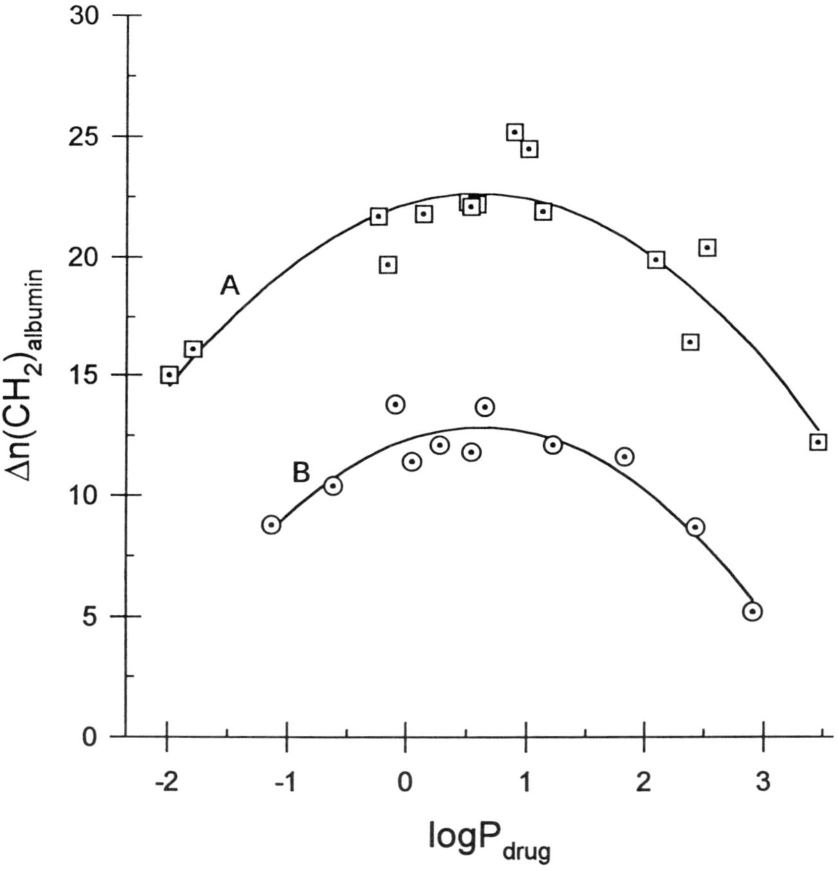

Figure 7.9. Change in the relative hydrophobicity of human serum albumin induced by the albumin binding of one drug molecule versus the logarithm of the drug octanol-water partition coefficient.

in question is based on some slight differences between the closely similar structures of, e.g., cephradine and cephalexin, or sulfisoxazole and sulfamoxole. It seems hard to explain why drugs of different classes, e.g., carbenicillin, sulfanilamide, and cefazolin fit the same relationship (curve A, Fig.7.9). It is also unclear at present why a given compound fits one or the other relationship. To get a better insight into this issue an examination of larger series of drugs

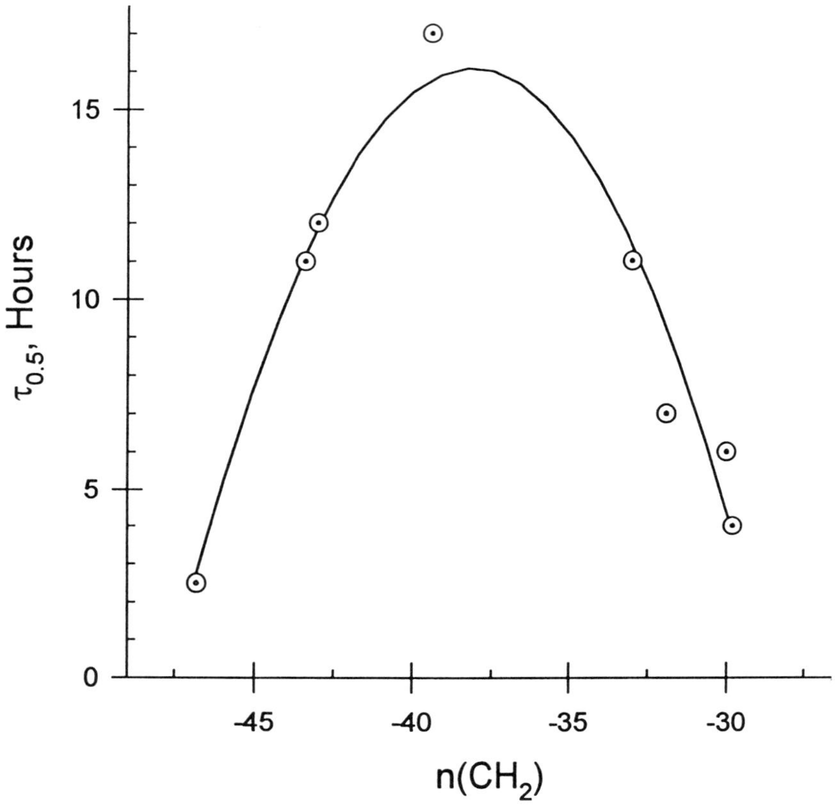

Figure 7.10. Relationship between the half-life time of sulfonamides in humans, $\tau_{0.5}$, and the relative hydrophobicity of the albumin-sulfonamide 1:2 complex, $n(CH_2)_{1:2}$.

and 3-D structure analysis of the drugs fitting the different curves in Fig. 7.9 seems to be necessary.

An additional relationship established [76] between the relative hydrophobicity of the 1:2 albumin-drug complexes and the drug half-life in humans for sulfonamides [85] is shown in Figure 7.10.

The relationship given in Fig. 7.10 may be described as:

$$\tau_{0.5} = -241.1 - 13.5 \cdot n(CH_2)^* - 0.18 \cdot [n(CH_2)^*]^2 \quad (7.15)$$

$$N = 8; r^2 = 0.9707$$

where $n(CH_2)^*$ is the relative hydrophobicity of the albumin-sulfonamide 1:2 complex [76]; and $\tau_{0.5}$ is the sulfonamide half-life time in humans [85]. The sulfonamides fitting Equation 7.15 are listed in Fig. 7.10.

According to Equation 7.15, the "optimal" relative hydrophobicity of the albumin-sulfonamide 1:2 complex amounts to -37.5 ± 5 equivalent CH_2 groups. This relative hydrophobicity of the albumin-drug 1:2 complex presumably provides the largest possible duration of the sulfonamide drug in a human body for a given series of sulfonamides. The practical value of the relationship 7.15 is limited, since it does not fit all sulfonamides examined. This result may indicate that the relative hydrophobicity of a protein-drug complex may be important for the drug persistence in a body, but this is not the only decisive factor. This issue must be explored in much more detail and on a larger scale before any definite conclusion may be drawn.

The results [56,76] presented in Tables 7.13 and 7.14 clearly indicate the possibility to use the aqueous two-phase partition technique for studying the ligand-induced conformational changes in a biological macromolecule. These results also imply the high sensitivity of the technique to the presence of small amounts of conformational perturbants in a protein. That indicates the possibility to apply the technique for analysis of purity of proteins and other biological materials. This and some other bioanalytical applications of the partition technique are considered in the next chapter.

7.6. SUMMARY

In summing up the foregoing, it must be stressed, first, that the experimental data discussed in this chapter pose more questions then provide the answers. The reason is self-evident. The aqueous two-phase partition technique provides a fundamentally new experimental information unavailable by any other physical or physicochemical method. This complicates the interpretation of experimental data obtained with the technique. The usual way out of such complication would be to accumulate a large body of experimental information studying as simple objects as possible. Unfortunately, none of biological solutes may be regarded as sufficiently "simple". The experimental data obtained in the studies of biological solutes considered above are consistent with those discussed in previous chapters for structurally much more simple organic solutes. These data are in agreement with the aforementioned view that the logarithm of the solute partition coefficient in an aqueous Dex-Ficoll or Dex-PEG two-phase system may serve as a measure of the solute relative hydrophobicity. That leads to the conclusion that:

(i) The partition coefficient of a solute in an aqueous polymer two-phase system is the solute biological potency-related characteristic of the solute.

Results of several quantitative structure-activity relationship (QSAR) studies for biological solutes considered above indicate an additional important conclusion:

(ii) The relative hydrophobicity of a biological solute (peptides, nucleotides, proteins, glycoproteins, etc.) as measured by the aqueous two-phase partition technique may be used as a general descriptor of a structure of a conformationally flexible solute. This descriptor may be used in the QSAR analysis of biological (and synthetic) solutes.

An additional observation with important biological implications is that the influence of the chemical (ionic, in particular) composition of an aqueous medium on the relative hydrophobicity of a biological solute may regulate biological potency of the solute. This simple physicochemical way of regulating functions and transport of solutes in a living body, to my knowledge, has never been considered in biochemical/biophysical literature. This issue, however, seems important enough to warrant additional experimental study.

As follows from the experimental data discussed in this chapter, the solute partition coefficient is highly sensitive to the solute molecular structure and conformation. Changes in the conformation of a biological macromolecule due to the presence of small amounts of ligands is readily detected by the technique. This observation complies with the known ability of aqueous two-phase systems to separate closely related biological molecules. The same high sensitivity of the aqueous two-phase partition technique that allows separation of biological materials can be exploited for the analysis of these materials as discussed in the next chapter.

REFERENCES:

1. B. Y. Zaslavsky, A. A. Borovskaya, N. D. Gulaeva, L. M. Miheeva, Biotechnol.Bioeng., 40, 1 (1992).
2. A. Leo, C. Hansch, D. Elkins, Chem.Rev., 71, 525 (1971).
3. L. G. Shchyukina, B. Y. Zaslavsky, 1984, unpublished results.
4. S. A. Galema, H. Hoiland, J.Phys.Chem., 95, 5321 (1991).
5. F. Franks, Cryobiology, 20, 335 (1983).
6. J. M. Harvey, M. C. Symons, J.Solution Chem., 7, 571 (1978).
7. R. D. Poretz, I. J. Goldstein, Arch.Biochem.Biophys., 125, 1034 (1968).

8. B. Y. Zaslavsky, L. M. Miheeva, S. V. Rogozhin, J.Chromatogr., 216, 103 (1981).
9. E. P. Baranov, V. Kh. Brikenshtein, B. Y. Zaslavsky, G. V. Karasyk, L. M. Miheeva, A. A. Ovchinnikov, I. L. Shamovsky, Bioorganicheskaya Khimia (Rus.), 14, 103 (1988).
10. C. Hansch, A. Leo, Substituent Constants for Correlation Analysis in Chemistry and Biology, New York, Academic Press, 1979.
11. B. Y. Zaslavsky, L. G. Shchyukina, S. V. Rogozhin, Mol. Biologia (Rus), 15, 1315 (1981).
12. L. G. Shchyukina, B. Y. Zaslavsky, S. V. Rogozhin, V. L. Florentiev, Mol.Biologia (Rus.), 18, 1128 (1984).
13. B. Y. Zaslavsky, L. M. Miheeva, N. M. Mestechkina, V. Pogorelov, S. V. Rogozhin, FEBS Letts., 94, 77 (1978).
14. H. Walter, J. L. Anderson, FEBS Lett., 131, 73 (1981).
15. J. F. Tinker, G. B. Brown, J.Biol.Chem., 173, 585 (1948).
16. L. Y. Tychinskaya, B. P. Gottih, A. M. Kricyn, Y. P. Lysov, A. K. Schelkina, V. L. Florentiev, Mol.Biologia (Rus.), 14, 1159 (1980).
17. Z. E. Rassi, C. Horvath, In: Practical Aspects of Modern HPLC (I. Molnar, ed.), New York, Berlin, Walter de Gruyter, 1983, pp.1-14.
18. J. Jacobson, Z. E. Rassi, C. Horvath, In: Practical Aspects of Modern HPLC (I. Molnar, ed.), New York, Berlin, Walter de Gruyter, 1983, pp.15-23.
19. B. Y. Zaslavsky, N. M. Mestechkina, L. M. Miheeva, S. V. Rogozhin, J.Chromatogr., 256, 49 (1983).
20. B. Y. Zaslavsky, N. M. Mestechkina, L. M. Miheeva, S. V. Rogozhin, G. Y. Bakalkin, G. G. Rjazhsky, E. V. Chetverina, A. A. Asmuko, J. D. Bespalova, N. V. Korobov, O. N. Chichenkov, Biochem. Pharmacol., 31, 3757 (1982).
21. B. Y. Zaslavsky, N. M. Mestechkina, L. M. Miheeva, S. V. Rogozhin, J.Chromatogr., 240, 21 (1982).
22. A. Diamond, K. Yu, J. T. Hsu, In: "Protein Purification: From Molecular Mechanisms to Large-Scale Processes", ACS Symposium Series, Vol. 427 (M. R. Ladisch, R. C. Willson, C. C. Painton, S. E. Builder, eds.), pp.52-65, American Chemical Society, Washington, D.C., 1990.
23. M. Akamatsu, T. Fujita, J.Pharm.Sci., 81, 164 (1992).
24. T. Sasagawa, T. Okuyama, D. C. Teller, J.Chromatogr., 240, 329 (1982).
25. Y. Audigier, H. Mazarguil, R. Gout, J. Cros, Eur.J.Pharmacol., 63, 35 (1980).
26. C. Yamashiro, C. H. Li, J.Chromatogr., 215, 255 (1981).

27. T. W. Hearn, S. E. Rewgnier, C. T. Wehr, Int.Lab., 1, 16 (1983).
28. R. Schwyzer, Biochemistry, 25, 6335 (1986).
29. J. D. Bespalova, N. V. Korobov, M. I. Titov, O. N. Chichenkov, Pharmacologia i Toxicologia (Rus), 45, 39 (1982).
30. O. N. Chichenkov, N. V. Korobov, Pharmacologia i Toxicologia (Rus), 47, 23 (1984).
31. J. A. H. Lord, A. A. Waterfield, J. Hughes, H. W. Kosterlitz, Nature, 267, 495 (1977).
32. S. Lemaire, A. Berube, G. Derome, I. Lemaire, J. Magnan, D. Regoli, S. St-Pierre, J.Med.Chem., 21, 1232 (1978).
33. T. M. Cho, P. Y. Law, H. H. Loh, In: Advances in Biochemical Psychopharmacology (H. H. Loh, D. H. Ross, eds.), Vol.20, Raven Press, New York, 1979, pp.69-101.
34. L. Terenius, Acta Pharmac.Tox., 37, 211 (1975).
35. R. Schwyzer, Biopolymers, 31, 785 (1991).
36. C. Tanford, Science, 200, 1012 (1978).
37. B. Y. Zaslavsky, E. A. Masimov, S. V. Rogozhin, Doklady Akademii Nauk SSSR (Rus), 277, 1163 (1984).
38. B. Y. Zaslavsky, A. V. Baevskii, S. V. Rogozhin, A. V. Gedrovich, A. V. Shishkov, A. A. Gasanov, E. A. Masimiov, J.Chromatogr., 285, 63 (1984).
39. B. Y. Zaslavsky, L. M. Miheeva, S. V. Rogozhin, Y. A. Davidovich, A. V. Gedrovich, A. V. Shishkov, A. A. Gasanov, E. A. Masimiov, J.Chromatogr., 291, 203 (1984).
40. A. I. Kurmaeva, L. I. Vedikhina, V. P. Barabanov, B. Y. Zaslavsky, N. D. Gulaeva, Kolloidnyi Zhurnal (Colloid Journal) (Rus), 52, 136 (1990).
41. E. A. Masimov, B. Y. Zaslavsky, A. A. Gasanov, S. V. Rogozhin, J.Chromatogr., 284, 337 (19847).
42. E. A. Masimov, B. Y. Zaslavsky, A. V. Baevskii. S. V. Rogozhin, A. V. Gedrovich, A. V. Shishkov, V. D. Shcherbukhin, Prikladnaya Biokhimia i Mikrobiologia (Appl.Biochem.Microbiol., Rus.), 20, 733 (1984).
43. B. Y.Zaslavsky, L. M.Miheeva, M. N.Rodnikova, G. V.Spivak, V. S. Harkin, A. U. Mahmudov, J.Chem.Soc., Faraday Trans.I, 85, 2857 (1989).
44. B. Y. Zaslavsky, L. M. Miheeva, E. A. Masimov, S. F. Djafarov, C. Reichardt, J.Chem.Soc. Faraday Trans., 86, 519 (1990).
45. B. Y. Zaslavsky, E. A. Masimov, A. A. Gasanov, S. V. Rogozhin, J.Chromatogr., 294, 261 (1984).

46. P. H. von Hippel, T. Schleich, In: Structure and Stability of Biological Macromolecules (S. N. Timasheff, G. D. Fasman, eds.), Marcel Dekker, New York, 1969, pp.412-574.
47. P. Å. Albertsson, Partition of Cell Particles and Macromolecules, 3rd.ed., Wiley, New York, 1986.
48. Partitioning in Aqueous Two-Phase Systems: Theory, Methods, Uses, and Applications to Biotechnology (H. Walter, D. E. Brooks, D. Fisher, eds.), Academic Press, Orlando, Fla, 1985.
49. B. Y. Zaslavsky, Anal.Chem., 64, 765A (1992).
50. B. Desbuquois, G. D. Aurbach, Biochem.J., 143, 83 (1974).
51. B. Y. Zaslavsky, N. M. Mestechkina, S. V. Rogozhin, Biochim. Biophys.Acta, 579, 463 (1979).
52. B. Y. Zaslavsky, N. M. Mestechkina, S. V. Rogozhin, J.Chromatogr., 260, 329 (1983).
53. R. Mothes, B. Y. Zaslavsky, N. M. Mestechkina, L. M. Miheeva, S. V. Rogozhin, K. D. Schwenke, Nahrung, 30, 1043 (1986).
54. V. V. Mozhaev, V. A. Siksnis, N. S. Melik-Nubarov, N. Z. Galkantaite, G. J. Denis, E. P. Butkus, B. Y. Zaslavsky, N. M. Mestechkina, K. Martinek, Eur.J.Biochem., 173, 147 (1988).
55. A. L. Unguryte, V. N. Kasho, L. M. Miheeva, B. Y. Zaslavsky, Biologicheskie Membrany (Biol.Membranes, Rus.), 7, 580 (1990).
56. N. D. Gulaeva, I. A. Yamskov, B. Y. Zaslavsky, 1990, unpublished data.
57. R. Kuboi, H. Tanaka, I. Komasawa, Kagaku Kogaku Ronbunshu, 16, 755 (1990).
58. R. Kuboi, H. Tanaka, I. Komasawa, In: Solvent Extraction 1990 (Proceedings of International Solvent Extraction Conference (ISEC-90) (T. Sekine, ed.), Elsevier, Amsterdam, 1991, pp.1857-1862.
59. R. Kuboi, H. Tanaka, I. Komasawa, Advances in Bioseparation Engineering 1991, Japan, 51 (1991).
60. H. Tanaka, R. Kuboi, I. Komasawa, J.Chem.Eng.Japan, 24, 661 (1991).
61. R. Kuboi, H. Tanaka, I. Komasawa, Kagaku Kogaku Ronbunshu, 17, 67; 75 (1991).
62. R. Kuboi, K. Yano, H. Tanaka, I. Komasawa, J.Chem.Eng.Japan, 26, 286 (1993).
63. K. Yano, A. Wakayama, R. Kuboi, I. Komasawa, Bunseki Kagaku, 42, 673 (1993).
64. W. H. Wang, R. Kuboi, I. Komasawa, Chem.Eng.Sci., 47, 113 (1992).
65. Y. Nozaki, C. Tanford, J.Biol.Chem., 246, 2211 (1971).
66. M. A. Eiteman, J.L. Gainer, Biotechnol.Prog., 6, 479 (1990).
67. K. Zakrewski, H. Goch, Biochemistry, 7, 1835 (1968).

68. R. F. Chen, J.Biol.Chem., 242, 173 (1967).
69. W. W. Sukow, J. Healey, P. Witkewicz, Physiol.Chem.Phys., 10, 133 (1978).
70. G. V. Iyengar, W. E. Kollmer, H. J. M. Bowen, The Elemental Composition of Human Tissues and Body Fluids, New York, Verlag Chemie, 1978.
71. V. M. Schlimak (Central Institute of Hematology and Blood Transfusion, Moscow, Russia), 1989, personal communication.
72. J. B. Matthew, G. I. H. Hanania, F. R. N. Gurd, Biochemistry, 18, 1928 (1979).
73. B. Y. Zaslavsky, L. M. Miheeva, S. V. Rogozhin, Biochim. Biophys.Acta, 588, 89 (1979).
74. V. Lakhtin, Ph.D. Thesis, Institute of Biochemistry, USSR Academy of Sciences, Moscow, 1989.
75. J. L. Ochoa, J.Chromatogr., 215, 351 (1981).
76. N. D. Gulaeva, B. Y. Zaslavsky, 1990, unpublished data.
77. T. A. G. Noctor, M. J. Diaz-Perez, I. W. Wainer, J.Pharm.Sci., 82, 675 (1993).
78. W. E. Muller, K. J. Fehske and S. A. C.Schlafer, In: Drug-Protein Binding (M. M. Reidenberg and S.Erill, eds.), Praeger Publishers, New York, 1986, pp.7-23.
79. U. Kragh-Hansen, Pharmacol.Rev., 33, 17 (1981).
80. M. F. Colombo, D. C. Rau, V. A. Parsegian, Science, 256, 655 (1992).
81. T. Fujita, J.Med.Chem., 15, 1049 (1972).
82. A. E. Bird and A. C. Marshall, Biochem.Pharmacol, 16, 2275 (1967).
83. W. J. Dunn, III, J.Med.Chem., 16, 484 (1973).
84. A. Leo (Pomona College, CA), personal communication, 1991.
85. T. Struller, Prog. Drug Res., 12, 389 (1968).

CHAPTER 8.

ANALYSIS OF INDIVIDUAL BIOPOLYMERS AND THEIR MIXTURES

As mentioned above, the method of aqueous two-phase partitioning is widely recognized as a cost-effective separation technique capable of fractionation, purification, and isolation of a wide variety of biological materials, ranging from proteins and nucleic acids to viruses, cells, and subcellular particles [1-5]. Separation or purification procedures based on this technique rely on differences between the partition behavior of a given target product and that of other solutes (or particles) present in an initial sample. The procedure is fundamentally similar to that of liquid-liquid partition chromatography [4]. The aqueous two-phase partition technique may be regarded as a variant of liquid partition chromatography, and even the instruments developed for the countercurrent partition chromatography may be used with aqueous two-phase systems (see, e.g., in [6,7]).

The liquid partition chromatography method is commonly used as both a separation and an analytical tool in a variety of different areas of research. However, the aqueous two-phase partition technique is currently used in biochemistry, cell biology, and biotechnology solely as a separation method on the laboratory and industrial scales [8]. Analytical applications of the technique (see, e.g., in [2-5]) are usually limited to analytical separations. It is clear from the practice of liquid chromatography as well as from the experimental data considered in the preceding chapters that the technique under consideration may also be used successfully as an analytical tool.

The most obvious merits of the aqueous two-phase partition technique as an analytical method are as follows.

Partition behavior of a solute reflects the difference in the intensity of the solute-solvent interactions in two aqueous media of different chemical composition. The partition technique is unique in the sense that no other current method can provide the same information. As shown in previous chapters the information obtained from monitoring the solute partitioning is related to the spatial arrangement of the solvent-accessible groups in the solute molecule and to the biological potency of the solute.

The solvent in both phases of an aqueous two-phase system is of the same aqueous nature. Therefore, the difference between the solvent features of the two phases is small. As an example, it may be repeated that the free energy of transfer of a CH_2- group between the two phases usually varies from about 10 to 200 cal/mole CH_2, depending on the polymer and salt composition of the system being used. The same characteristic for common water-organic solvent systems varies from about 400 to 1000 cal/mole CH_2, depending on the organic solvent used. The small difference between the solvent properties of the two aqueous phases is a disadvantage since the difference between the partition coefficient values of, for example, two similar compounds is clearly the dominant factor. This difference, however, is not the only important factor.

As mentioned, partitioning of a solute in a two-phase system is governed by the difference between the intensity of the solute-solvent interactions in the two phases. When the phases are of different solvent nature, the incorporation of a new chemical group into the structure of a solute, or the replacement of one group with another, affects the solute partitioning much more than merely a spatial rearrangement of the existing groups in the same molecular structure. Transfer of a solute from one solvent phase into the other results in the replacement of, for example, solute-water interactions with solute-octanol interactions. The energy of this replacement depends mostly on the chemical nature of the groups in the solute molecular structure. The exact spatial arrangement of the groups in the structure is of secondary importance.

When a polar solute is partitioned in an aqueous two-phase system, the effect of the solute spatial structure on the solute partitioning increases significantly. The likely reason is the direction-dependent nature of solute-solvent interactions in the aqueous media. The process of solute partitioning between the two aqueous phases may be viewed as the entropy-driven process. As the result, the aqueous two-phase partition technique may differentiate between very closely related molecular structures. The foregoing implies the possibility of an application of the aqueous two-phase partition technique as an analytical tool in virtually all fields dealing with biological materials. These applications, although experimentally explored to a very limited degree as yet, are considered in this chapter.

8.1. ANALYSIS OF INDIVIDUAL BIOPOLYMERS

Advancement of recombinant DNA technology during the past decade, and application of large-scale cell cultures for protein expression, has generated a new class of 'biological' therapeutic agents. The medicinal products classed as "biologicals" have been defined as "substances whose potency, purity and identity cannot be adequately defined by physicochemical means alone" [9,10]. Thus, analysis, characterization, and quality control of a "biological" medicinal agent depends directly on the use of biological test procedures. The use of animals in these procedures (in-vivo bioassays) provokes increasing ethical and political pressures [10]. Even in-vitro bioassays too often are costly, labor-consuming, and imprecise, if not inappropriate. Physicochemical techniques usually are much more precise and (being automated) less time- and labor-consuming. It must also be recognized, however, that depending on the product to be analyzed, these techniques may be as costly or even more expensive than bioassays.

The difficulties encountered when physicochemical methods are used for analysis of "biologicals" are well illustrated by Spellman [11] for glycoproteins. The carbohydrate moieties of recombinant glycoproteins, such as tissue plasminogen activator, erythropoietin, ß-, and γ-interferons, are known to be critically important for biological function and potency of a glycoprotein, as well as clearance from circulation, tissue targeting, stability, etc. (see references cited in [11]). The expression of a glycoprotein in a heterologous cell line may result in the production of recombinant glycoproteins carrying carbohydrate heterophile antigenic determinants [12]. The possibility of expression of glycoproteins with 'wrong' carbohydrate structures in nonmammalian cells is greater, but even mammalian cell lines, e.g., cultured human adenocarcinoma cells [13], may produce antigenic carbohydrate structures.

The carbohydrate structures of a glycoprotein are determined in part by the glycosylation apparatus of the host cell and in part by the tertiary structure of the particular glycoprotein being produced. Thus, heterologous expression of a particular glycoprotein would be predicted to yield different populations of oligosaccharides in each of the different expression host cells [14]. Additionally, the processing of glycoprotein carbohydrates may result in heterogeneous populations of structurally related oligosaccharides (microheterogeneity). Different glycosylation sites within a glycoprotein may also have different populations of attached carbohydrate structures (site heterogeneity). All these factors combine to complicate consistent lot-to-lot manufacturing of a recombinant medicinal product [11].

Similar microheterogeneity problems often arise in manufacturing recombinant proteins. We may not ask for perfection when a host cell culture is made to produce (express) a product (protein, glycoprotein, etc.) alien to this cell. The purity desired of the protein being genetically engineered or isolated from a natural source is usually greater than 99.9% (sometimes even ≥ 99.99%) [15]. To monitor this level of purity (even when it is achieved) is extremely difficult, particularly in view of the fact that the impurities may be structurally related to the product.

Problems arising in analysis and characterization of chemically modified proteins should also be mentioned. Covalent conjugation of poly(ethylene glycol) (PEG) to proteins has recently been shown to dramatically alter the protein's pharmacology and immunogenicity. PEG modification of a number of enzymes provided longer in-vivo half-lives of the modified enzymes. Similar modification of interleukin-2 increased the protein solubility and antitumor potency in mice and reduced plasma clearance and immunogenicity in rabbits and mice (for example, see [16] and references cited therein). Numerous physicochemical methods used in the attempt to characterize different PEG-protein preparations have been reviewed by Kunitani et al. [16]. According to these authors three dimensions of heterogeneity (length of each PEG fragment, number of PEG fragments attached to a given protein molecule, and locations of the fragments on the molecule) of the PEG-protein conjugate, complicates its characterization and quality control immensely. Possibility to use the aqueous two-phase partition technique for characterization of PEGylated proteins clearly follows from the results reported by Delgado et al.[17].

The problem of quality control, characterization, and analysis of lot-to-lot consistency of recombinant products and 'biologicals' isolated from natural sources is currently one of the major challenges in bioanalytical chemistry.

As may be clear from the foregoing, there is a strong necessity for new approaches to characterization and quality control of biological products. From a physicochemical point of view, an "ideal" physicochemical method for analysis of recombinant products and other "biologicals" should meet the following requirements:

The "ideal" technique should be capable of detecting extremely small changes in the structure of large macromolecules, such as changes in the sequence of a few saccharide residues in a glycoprotein, or replacement of a single amino acid residue with another residue in a protein. It is desirable and perhaps necessary for the technique to be especially sensitive to the structural changes affecting the biological function and/or potency of a biological macromolecule. In other words, the technique should provide the potency- and function-related information. The "ideal" technique should also be sensitive to the presence of impurities in the product in the amounts as small as 0.1 to 0.01 % wt. Additionally, the technique should be simple, precise and time- and cost-effective so as not to increase the overall cost of the product.

The method meeting all these requirements might be viewed as panacea. Needless to say, no such method is likely to be developed in reality. Although the technique of partitioning in aqueous two-phase systems is not an "ideal" method, it does appear to meet at least some of the above requirements.

As shown above, the partition coefficient of a biological solute in an aqueous (Dex-PEG or Dex-Ficoll) two-phase system may serve as a general descriptor of the solute structure and is related to the solute biological potency. Furthermore, because of the entropic nature of the partition process in aqueous two-phase polymer systems the solute partition coefficient value is highly sensitive to the three-dimensional structure of the solute.

As indicated above, the partition coefficient value for a given biopolymer (e.g., protein) depends on the specific features and purity of the biopolymer. From this viewpoint, the solute partition coefficient value is similar to other analytical parameters, such as the use of the melting point as a simple test of the purity of a synthetic compound. The partition coefficient, K, for a given solute in an aqueous two-phase system with a fixed polymer and salt composition is a constant feature of the solute and similar to other variables such as, for example, the specific absorption coefficient, dielectric constant or retention index.

The K values shown in Table 8.1 are quite different for closely related compounds - even chiral pairs. The aforementioned differences between the K values for reversal dipeptides [18], hexapeptides differing in enantiomers of the Arg residue [19], and isomers of dinucleosidephosphates [20] are large enough not just to be detected but, for these compounds, to be separated by the countercurrent chromatography (CCC) technique [6,7].

One of the most recent examples includes partitioning of ecdysone and 20-hydroxyecdysone in the aqueous Dex-UCON (ethylene oxide-propylene oxide random copolymer) two-phase system [21]. Ecdysone and 20-hydroxyecdysone are steroids of identical structure except for substitution of a hydroxyl group for a hydrogen at position 20 in the 20-hydroxyecdysone molecule. The partition coefficients of the two steroids have been reported in [21] to differ significantly, amounting to 1.28 for ecdysone and 1.04 for 20-hydroxyecdysone in the system containing 5% ethanol and 0.012 M sodium phosphate buffer, pH 7.0.

The other illustrative example comes from the results by Raymond et al. [22]. It was shown that the two membrane-bound alkaline phosphatase

Table 8.1
Partition Coefficients of Closely Related Biological Solutes in Different Aqueous Polymer Two-Phase Systems.

Compound	K	Phase system*
Gly-Leu	0.592 ± 0.018	System 1
Leu-Gly	0.687 ± 0.021	-"-
Ile-Gly	0.714 ± 0.021	-"-
Val-Gly	0.511 ± 0.015	-"-
Gly-Asp	0.247 ± 0.010	-"-
Asp-Gly	0.387 ± 0.012	-"-
Tyr-D-Ala-Gly-Phe-D-Leu-L-Arg	1.121 ± 0.021	System 2
Tyr-D-Ala-Gly-Phe-D-Leu-D-Arg	1.053 ± 0.018	-"-
Tyr-D-Ala-Gly-Phe-Leu-Arg-Lys-Arg	0.900 ± 0.019	System 2a
Tyr-Gly-Gly-Phe-D-Leu-L-Arg	0.814 ± 0.020	-"-
5'-ApA	1.418 ± 0.027	System 2a
2'-ApA	0.940 ± 0.022	-"-
5'-UpU	1.362 ± 0.025	-"-
2'-UpU	1.008 ± 0.015	-"-
Insulin (intact)	1.00	System 3**
Insulin (iodinated, 1:1)	0.38	-"-
Glucagon (intact)	1.15	-"-
Glucagon (iodinated, 1:1)	0.45	-"-
Glucagon (iodinated, 1:2.1)	0.24	-"-
Albumin bovine (with lipid traces)	0.486 ± 0.022	System 2
Albumin bovine (lipid-free)	0.558 ± 0.020	-"-
Albumin human (HSA)	0.388 ± 0.009	System 4
HSA-warfarin (1:1) complex	0.589 ± 0.018	-"-
HSA-cloxacillin (1:1) complex	0.486 ± 0.017	-"-
HSA-cloxacillin (1:2) complex	0.715 ± 0.020	-"-
HSA-sulfamoxole (1:1) complex	0.549 ± 0.018	-"-
HSA-sulfamoxole (1:2) complex	1.019 ± 0.015	-"-
HSA-sulfisoxazole (1:1) complex	0.641 ± 0.012	-"-
Concanavalin A	0.450 ± 0.018	-"-
Concanavalin A complexes (1:1) with		-"-
α-D-glucopyranoside	0.230 ± 0.020	-"-
α-D-mannopyranoside	0.298 ± 0.024	-"-

* The polymer and salt compositions of the aqueous two-phase systems used are given as indicated in the references shown:

Table 8.1 (Footnotes continued.)

System 1: ca.8.67% (w/w) PEG-3400; 13.36% (w/w) potassium phosphate (a salt mixture with the ratio of 306.9 g K_2HPO_4 to 168.6 g KH_2PO4) - no exact total composition of the system employed is given in [18];
System 2: 10.8% (w/w) Dex-70; 12.5% (w/w) Ficoll-400; 0.15 mole/kg NaCl; 0.01 mole/kg sodium phosphate buffer, pH 7.40 [19];
System 2a: polymer composition as in system 2 but different salt composition - 0.11 mole/kg sodium phosphate buffer, pH 7.40;
System 3: 12.0% (w/w) Dex-500; 6.0% (w/w) PEG-6000; 0.02 M citrate- phosphate buffer, pH 5.0 [23];
System 4: 11.6% (w/w) Dex-70; 13.5% (w/w) Ficoll-400; 0.15 mole/kg NaCl; 0.01 mole/kg sodium phosphate buffer, pH 7.4.
** Partition coefficients are taken from [23] (presented in form of a plot with no experimental error indications).

isoforms released from the cell surface by treatments with inositol-specific phospholipases C and D demonstrate significantly different partition behavior in the aqueous Dex-PEG two-phase systems of varied polymer concentration containing 0.15 M NaCl in 0.01 M sodium phosphate buffer, pH 7.5. Thus the partition coefficient values are clearly different for the two molecular forms of alkaline phosphatase differing by a single phosphate group. It was also reported [22] that the commonly used analytical methods, such as electrophoresis on gradient gels, could not distinguish between the isoforms that are readily differentiated using aqueous two-phase partition techniques.

Differences in partition coefficients of peptides, proteins, and glycoproteins following rather slight chemical modifications [22,23] or conformational changes induced by formation of 1:1 or 1:2 complexes with low molecular weight compounds (see Table 8.1) support the generally accepted opinion [1-5] that the partition coefficient value is a highly sensitive characteristic of a given solute. Typically, as mentioned above, albumin samples from different manufacturers that contain traces of different impurities but are indistinguishable by standard analytical methods (e.g., HPLC and electrophoresis) are easily differentiated using partition technique [24]. The presence of 0.1-0.05% wt. of a low molecular weight compound in an albumin sample seems to be easily detected with this technique, an impressive outcome.

In reality, the effect of a compound present as a contaminant in a given biopolymer sample on the biopolymer conformation may be more important for the biopolymer partition coefficient value than the amount of the compound itself. That means that the solute partition coefficient value may often be a highly sensitive measure of the biopolymer purity, although not in all cases. The protein partition coefficients examined so far showed quite high sensitivity to the protein purity. An additional typical example is offered by Mothes et al.[25].

Table 8.2

Partition Coefficients of Proteins from Rapeseed.

Protein	K [a]
12 S Globulin	1.157 ± 0.023
Albumin (globulin-containing)	0.828 ± 0.011
Albumin (globulin-free)	0.785 ± 0.014

[a] aqueous Dex-Ficoll two-phase system containing 0.15 mole/kg NaCl in 0.01 mole/kg sodium phosphate buffer, pH 7.4 was used in the partition experiments.

Partitioning of 12S globulin, globulin-containing albumin sample, and chromatographically purified globulin-free albumin (all isolated from rapeseed) was examined in the aqueous Dex-Ficoll two-phase system containing 0.15 mole/kg NaCl in 0.01 mole/kg sodium phosphate buffer, pH 7.4 [25]. The data [25] given in Table 8.2 indicate that the K-value for the globulin-containing albumin sample differs significantly from that of pure globulin-free albumin. That result is clearly due to the presence of small amount of globulin in the sample (the K-value lies between those for the two pure proteins).

It should be emphasized that the partition coefficient value has been found to be constant, for example, for human (and bovine) serum albumin samples of the highest commercially available purity from different manufacturers. Significant differences between the K values for the commercial protein samples of lower purity from different manufactrurers have been observed for all the proteins studied so far.

An example of the application of the partition technique to analysis of the lot-to-lot consistency of production of recombinant products is offered by the data obtained by Gulaeva et al.[26]. Partitioning of 10 different lots of the therapeutic recombinant human growth hormone preparation produced by Scientific-Industrial Association "Biotechnologia" (Moscow, Russia) was examined [26] in the aqueous Dex-polyvinylpyrrolidone (PVP) two-phase system containing an additive of sodium sulfate. The partition coefficient values of the samples were compared to that of the reference sample (2.58 ± 0.12) [26]. The partition coefficient values for different samples varied from 2.48 to 4.07, and no correlation between the K values determined and electrophoretic purity, monomer content or results of the RIA (radioimmunoassay) analysis could be found. Only about half of the samples analyzed displayed the partition behavior identical (within the experimental error limits) with that of the reference sample [26]. A relationship between the partition coefficient value of a hormone sample and its biological potency was observed. The partition technique was found [26] to be much more sensitive and more time- and cost-effective than other methodologies currently in use for analysis of the therapeutic recombinant human growth hormone preparations [10].

It may be concluded that the aqueous two-phase partition technique can be used as a simple, inexpensive, and highly sensitive method for analysis and quality control of production of recombinant products and other biological materials.

The use of aqueous two-phase partition techniques for separation or purification of biological solutes is based on the observation that the components of an initial multicomponent mixture of the solutes (peptides, proteins, nucleic acids, etc.) distribute between the two phases independently of each other, provided the components do not interact with each other. That means that a change in the partition behavior of a single given component of a mixture may be monitored without any special sample preparation or clean-up procedure, a distinct and impressive advantage. For example, chemical modification of a given protein resulting in a change of the protein partition coefficient value may be observed by partitioning of an original mixture of the protein with other proteins, peptides, etc., provided the target protein-specific method for the concentration measurements is used. The protein concentration may be measured directly in the phases by any appropriate assay. The partitioning procedure is similar to that of extraction and may be readily automated. Hence, it seems that combining the aqueous two-phase partitioning technique with immunochemical and other selective methods of concentration measurements may lead to development of more efficient procedures for clinical diagnostics and other biomedical applications.

Finally, using an analogy with electrophoretic profile of serum proteins, it was suggested [27] that the partition behavior of a multicomponent protein mixture, described by an overall partition coefficient value (see below), may be viewed as a general quantitative characteristic of the total protein mixture. Therefore, analysis of multicomponent protein mixtures by the aqueous two-phase partition technique is discussed below.

8.2. ANALYSIS OF MULTICOMPONENT PROTEIN MIXTURES

An overall partition coefficient of a multicomponent protein mixture in an aqueous two-phase system, $K(\Sigma)$, may be defined [27] as the ratio between the total protein concentrations in the two phases:

$$K(\Sigma) = (\Sigma c_i)_1/(\Sigma c_i)_2 \qquad (8.1)$$

where Σc_i is the total protein concentration in a given phase; subscript "i" denotes the i-th protein component of the mixture being analyzed; and subscripts "1" and "2" denote the two phases.

To determine the $K(\Sigma)$ value in accordance with Equation 8.1, however, it is necessary to determine the concentrations of each protein component of the mixture in the two phases, and that is not often achieved in practice. Therefore an overall partition coefficient of a protein mixture in practice is defined as:

$$K(\Sigma)^j = (\Sigma a_i^j \cdot c_i)_1 / (\Sigma a_i^j \cdot c_i)_2 \quad (8.1a)$$

where a_i^j is a proportionality constant for the strength of an analytical signal specific for a given i-th protein component of the mixture and the analytical detection technique; superscript "j" denotes the analytical technique in use.

Equation 8.1a implies that the overall partition coefficient value for a given protein mixture should depend on the particular analytical method employed. Commonly used total protein detection methods compatible with the aqueous two-phase partition technique include direct spectrophotometry, dye binding reactions with Coomassie G-250 [28] or fluorescamine [29]. Every component of a protein mixture being examined may contribute to the total protein concentration measurement differently ($a_i^j \neq a_{i+1}^j$). The contribution in question depends on both specific features of the component and particular analytical method employed. Hence the overall partition coefficient value should depend on the analytical method used, and it has been found to be so in practice (see below) [27]. This situation is similar to the one encountered in the liquid chromatography analysis of complex mixtures when different chromatographic profiles may be observed with different detectors.

Total serum or plasma proteins may serve as an example of a unified multicomponent dynamic extracellular protein system [30]. Following the current trends in biomedical and biochemical research, efforts in the studies of plasma proteins are focused mostly on analysis of the individual proteins. Analysis of the levels and functions of minor individual plasma components is clearly important. The possibility to characterize the total plasma proteins as a unified protein system seems, however, to be equally important (even though not as fashionable) from the theoretical and practical viewpoint.

Partitioning of total plasma proteins from apparently healthy people, rabbits, rats, and sheep has been studied [31] in the aqueous Dex-Ficoll two-phase systems containing varied amounts of NaCl and sodium phosphate buffer, pH 7.4. The overall partition coefficient $K(\Sigma)^j$ values were determined [31] by measuring the total protein concentrations in the phases with the Coomassie G-250 technique [28]. It was found [31], first, that the overall partition coefficient $K(\Sigma)^j$ value for the total plasma proteins varies with the ionic

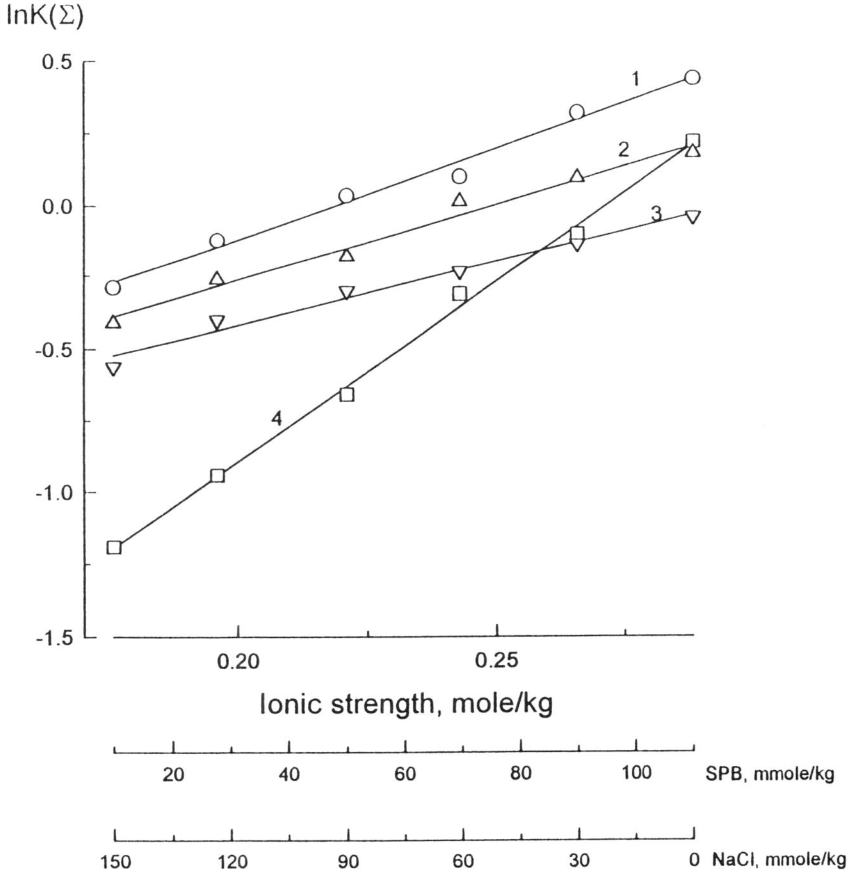

Figure 8.1. Logarithm of the overall partition coefficient for total plasma proteins from different species as a function of the salt composition of the aqueous Dex—Ficoll two-phase system containing varied amounts of NaCl and sodium phosphate buffer, pH 7.4. Plasma from 1 - human; 2 - rat; 3 - rabbit; 4 - sheep.

Table 8.4

Coefficients C and B Characterizing Partition Behavior of Total Plasma Proteins and Serum Albumins from Different Species in the Aqueous Dex—Ficoll Two-Phase Systems.*

Species **	Total plasma proteins		Serum albumin	
	C	B	C	B
Human (25)	-1.64 ± 0.06	6.20 ± 0.20	-2.28 ± 0.09	8.81 ± 0.08
Rabbit (3)	-1.28 ± 0.06	4.31 ± 0.24	-1.98 ± 0.13	5.91 ± 0.49
Sheep (4)	-3.48 ± 0.23	12.93 ± 0.98	-1.38 ± 0.05	4.34 ± 0.06
Rat (3)	-1.14 ± 0.05	4.30 ± 0.23	-1.65 ± 0.16	6.23 ± 0.56

* coefficients C and B determined from experimental data [31,32] according to Equation 5.4; salt composition of the two-phase system varied from 0.11 molekg sodium phosphate buffer, pH 7.4 to 0.15 molekg NaCl in 0.01 molekg sodium phosphate buffer, pH 7.4;

** number of individual donors of plasma indicated in parentheses.

composition of the system as shown in Figure 8.1 following the relation established for individual solutes (see in Chapter 5):

$$\ln K(\Sigma)^j = C + B \cdot I \qquad (5.4)$$

where I is the ionic strength value used as a quantitative index of the system salt composition; C and B are constants.

The coefficients C and B values for the total plasma proteins from different species are presented in Table 8.4 together with those reported [32] for albumins from the same species.

The data given in Table 8.4 indicate no relationship between the partition behavior, i.e., coefficients C and B values, of the total plasma proteins and serum albumins (which are the major components of the total plasma proteins).

It should also be noticed that, as reported by Mestechkina [31], the overall partition coefficient values for total plasma proteins from different sources fit the Collander-type relationship (see Equation 5.10a) established for individual solutes in aqueous two-phase systems of different polymer composition. This observation [31] supports the assumption [27] that the overall

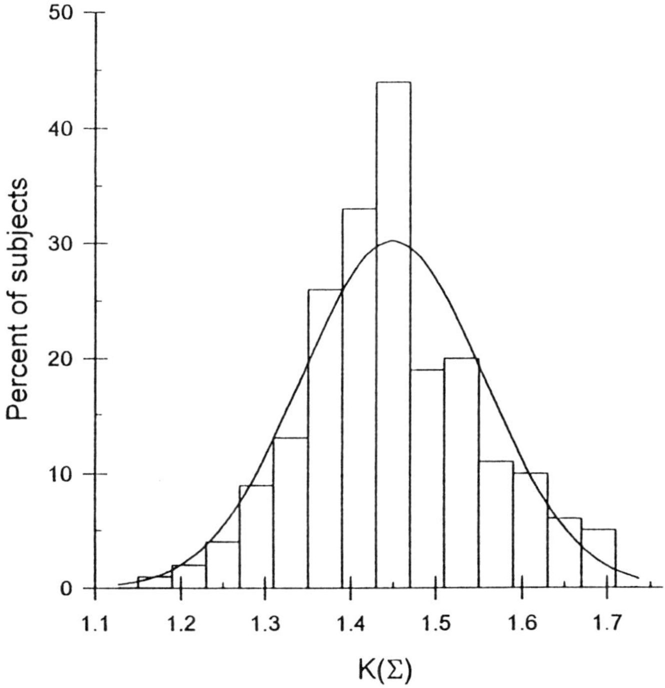

Figure 8.2. Histogram and Gaussian distribution of overall total plasma protein partition coefficient values in 103 apparently healthy people. Aqueous Dex—Ficoll two-phase system containing 0.11 mole/kg sodium phosphate buffer, pH 7.4.

partition coefficient value for total plasma proteins may be considered as a quantitative measure of the overall state of the blood protein system as a whole.

Partitioning of total plasma proteins from 103 apparently healthy people in the aqueous Dex-Ficoll two-phase system with fixed polymer and salt composition was found [27,31] to be independent of blood type, gender, age, smoking habit, etc. The results obtained [27,31] are presented in Figure 8.2 as a histogram within the corresponding normal (Gaussian) distribution curve. It should be mentioned that the average K(Σ) values for total plasma proteins from apparently healthy people in Fig. 8.2 (1.45 ± 0.11) and in Table 8.4 (1.16 ± 0.07 at the same salt composition) differ significantly. The reason for this difference is that the aqueous Dex-Ficoll two-phase systems employed different polymer concentrations in the experiments under discussion.

Table 8.5

Coefficients C and B Characterizing Partition Behavior of Total Plasma Proteins from Patients with Different Liver Pathologies in the Aqueous Dex—Ficoll Two-Phase Systems. *

Pathology **	C	B
Healthy people (25)	-1.64 ± 0.06	6.20 ± 0.20
Aggressive chronic hepatitis (11)	-1.46 ± 0.02	5.90 ± 0.40
Post-hepatitic and alcoholic cirrhosis (17)	-2.14 ± 0.03	8.80 ± 0.10
Secondary biliary cirrhosis (9)	-1.74 ± 0.07	7.50 ± 0.02

* coefficients C and B determined from experimental data [31,32] according to Equation 5.4; salt composition of the two-phase system varied from 0.11 mole/kg sodium phosphate buffer, pH 7.4 to 0.15 mole/kg NaCl in 0.01 mole/kg sodium phosphate buffer, pH 7.4

** number of plasma donors indicated in parentheses.

It is known that the biochemical and physicochemical characteristics of blood in healthy people may vary within definite narrow limits, and deviation from this limited range usually indicates pathology. Therefore the observed narrow normal distribution curve (standard deviation does not exceed 7.6%) supports the above view of the validity of the overall partition coefficient value as a measure of the state of the blood protein system. Accepting this view, it is reasonable to expect that the overall partition coefficient values for the total plasma proteins from patients with various pathologies will differ from those for the total plasma proteins from apparently healthy people.

This assumption was examined [31] using the total plasma proteins from patients with several liver disorders. The results obtained [31] in the aqueous Dex-Ficoll two-phase systems with varied amounts of NaCl and sodium phosphate buffer, pH 7.4, are shown in Table 8.5.

The data presented in Table 8.5 were obtained for a limited number of patients and may not be viewed as statistically reliable. These data, however, indicate definitely that the overall partition coefficient value for the total plasma proteins from patients with certain disorders deviate from the "normal" value. The possibility to differentiate between aggressive chronic hepatitis and secondary biliary cirrhosis using such simple procedure as partitioning should be particularly noticed.

Development of various techniques for early detection of malignancies, the likely aggressiveness or responsiveness of individual cancers, etc., is one of the most challenging problems in clinical biochemistry. One of the approaches in current use is based on detection and monitoring the so-called tumor markers in a patient's blood or in tissue samples. Tumor markers are compounds (enzymes, hormones, antigens, etc.) whose appearance or concentration in a biological liquid or tissue is related to the occurrence and/or stage of development of a given malignant tumor (see [33]). These compounds usually comprise minor components of a sample (especially at the early stage of the cancer development).

It has been assumed [27] that occurrence of a tumor marker(s) in the patient blood may affect the state of a plasma protein system as monitored by the aqueous two-phase partition technique. Analysis of plasma samples from patients with clinically established breast cancer and various mastopathies was performed by Blokhina et al.[27]. The results obtained using the aqueous Dex-Ficoll two-phase system containing 0.11 mol/kg sodium phosphate buffer, pH 7.4 are presented as the corresponding histograms in Figure 8.3.

The data presented in Figure 8.3 may be described by Gaussian distribution curves with the following average values of the overall partition coefficient $K(\Sigma)$: 1.61 ± 0.16 for patients with breast cancer (29 cases), 1.56 ± 0.12 for patients with various forms of mastopathy (26 cases), and 1.45 ± 0.11 for apparently healthy people (103 cases). The $K(\Sigma)$ values in both groups of patients exceed those for healthy people, but the changes in the partitioning of total plasma proteins observed in patients are too small to be of diagnostic value.

The reason for this shortcoming was attributed [27] to the possibility that the aqueous Dex-Ficoll two-phase system did not provide the conditions needed for differentiating between the properties of the total plasma proteins being examined. The aqueous Dex-PEG two-phase system was chosen [27] on the assumption that the difference between the solvent properties of the two phases in this system exceeded that of the Dex-Ficoll system. The difference between the solvent properties of the two phases in the aqueous Dex-PEG system is known to increase with increasing salt concentration (see above). Hence, the aqueous Dex-500-PEG-6000 two-phase system containing 1.8 mol/kg KCl in 0.01 mol/kg sodium phosphate buffer, pH 7.4 was used [27]. An additional advantage of this system in reference to the Dex-Ficoll system is that the time of the phase settling in the former system is much shorter than in the latter one.

Results of the study [27] of total plasma proteins from 103 apparently healthy people and 109 patients with clinically established breast cancer are

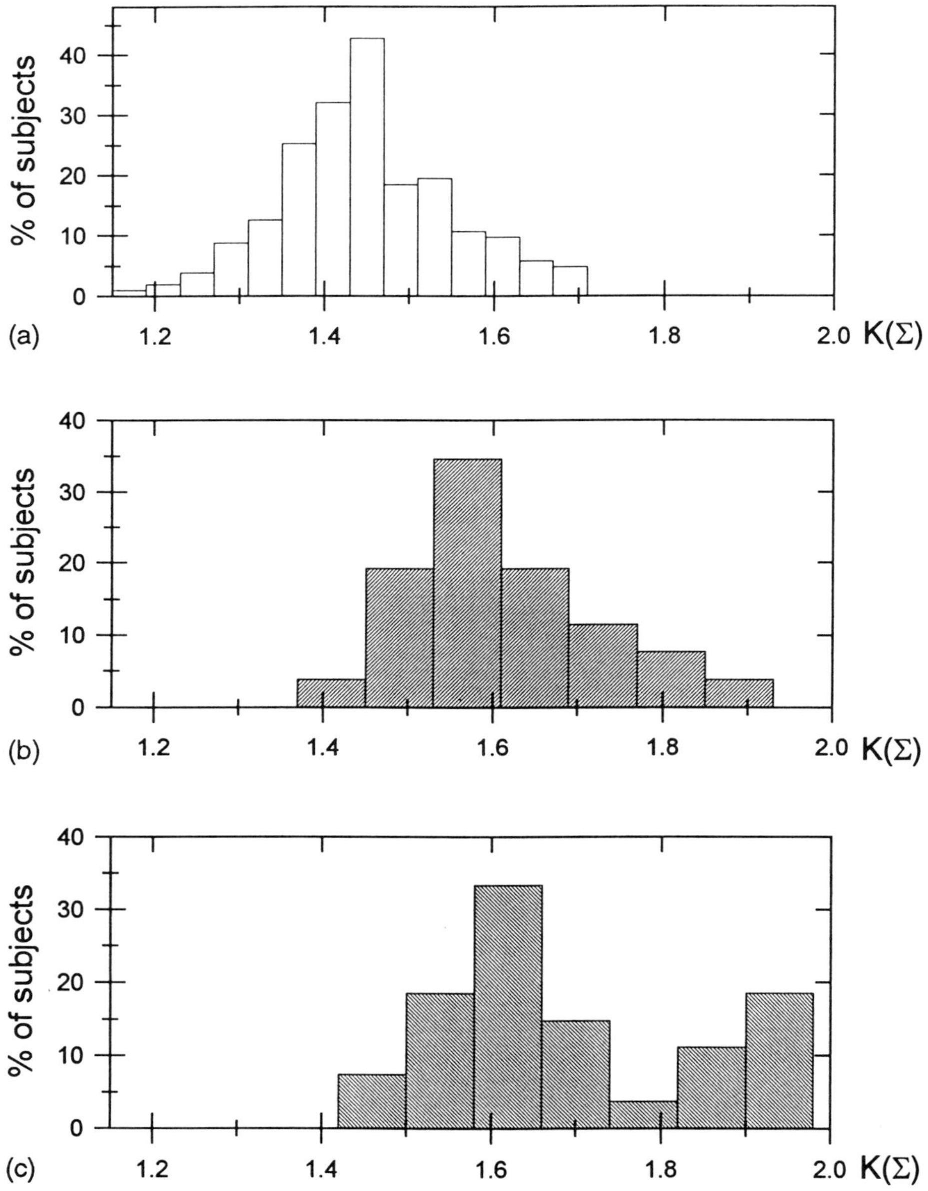

Figure 8.3. Histograms of overall total plasma protein partition coefficient values in (a) 103 apparently healthy people; (b) 26 patients with various mastopathies; and (c) 27 patients with malignant breast tumor. Aqueous Dex-Ficoll two-phase system containing 0.11 mole/kg sodium phosphate buffer, pH 7.4.

presented in Figure 8.4 as the histograms and the corresponding Gaussian distribution curves

The data given in Figure 8.4 are characterized by the average values of $K(\Sigma)$ 0.75 ± 0.07 for apparently healthy people and 0.92 ± 0.11 for patients with malignant breast tumor [27]. The results reported [27] indicate the sensitivity of the partition test (i.e., the likelihood of the result being "abnormal" in a patient with a malignant breast tumor) of 77%, the specificity (i.e., the likelihood of a test being normal in patients who do not have the malignant breast tumor) of 84.5%, and the predictive value of the test of 84%, which means that a positive test is significant for a breast tumor malignancy diagnosis. It may be mentioned that the specificity and predictive value of the partition test [27] are similar to the values obtained, for example, by radioimmunoassay of the serum concentration of arylsulfatase A [34] and, moreover, the sensitivity of the partition test exceeded that of arylsulfatase A. Additionally, certain differences between the partition behavior of total plasma proteins from patients with the localized and disseminated forms of breast cancer have been noticed [27]. The results [27] presented as the corresponding histograms in Figure 8.5 indicate, however, that although a trend for the $K(\Sigma)$ to increase with tumor-spread may be observed, but that the differences are not significant enough for diagnostic purposes, probably because of non-optimal conditions used in the partition test.

The $K(\Sigma)$ values for the total plasma proteins from 104 patients with various mastopathies are presented in Figure 8.6 as a histogram, together with those from healthy people and patients with clinically established malignant breast tumor. The histogram in question may be described by the Gaussian distribution curve with the average $K(\Sigma)$ value of 0.85 ± 0.10. According to the $K(\Sigma)$ values determined for this group of patients [27], 56% patients had no breast cancer and 79% from this subgroup had the $K(\Sigma)$ values corresponding to that of apparently healthy people. 44% patients with mastopathies have shown false(?)-positive results, i.e. the $K(\Sigma)$ values for these patients have been in the range typical for those with malignant breast tumor. The reason for this observation might be that this subgroup included patients with clinically and histopathologically undetected cancer. Consequent monitoring of the patients from this subgroup of 46 patients has shown that about 30% patients have developed the clinically established malignant breast tumor during a period of about 2 years following the abnormal $K(\Sigma)$ value detection (D.F. Schirin, 1989, personal communication).

Attempts to use the partition test for early detection of stomach cancer and lymphogranulomatosis under the aforementioned conditions have failed [31]. The changes in the $K(\Sigma)$ values observed for the total plasma proteins from the patients with these disorders in reference to the "normal" $K(\Sigma)$ value were not large enough to be used as a tumor indicator. To increase the useful-

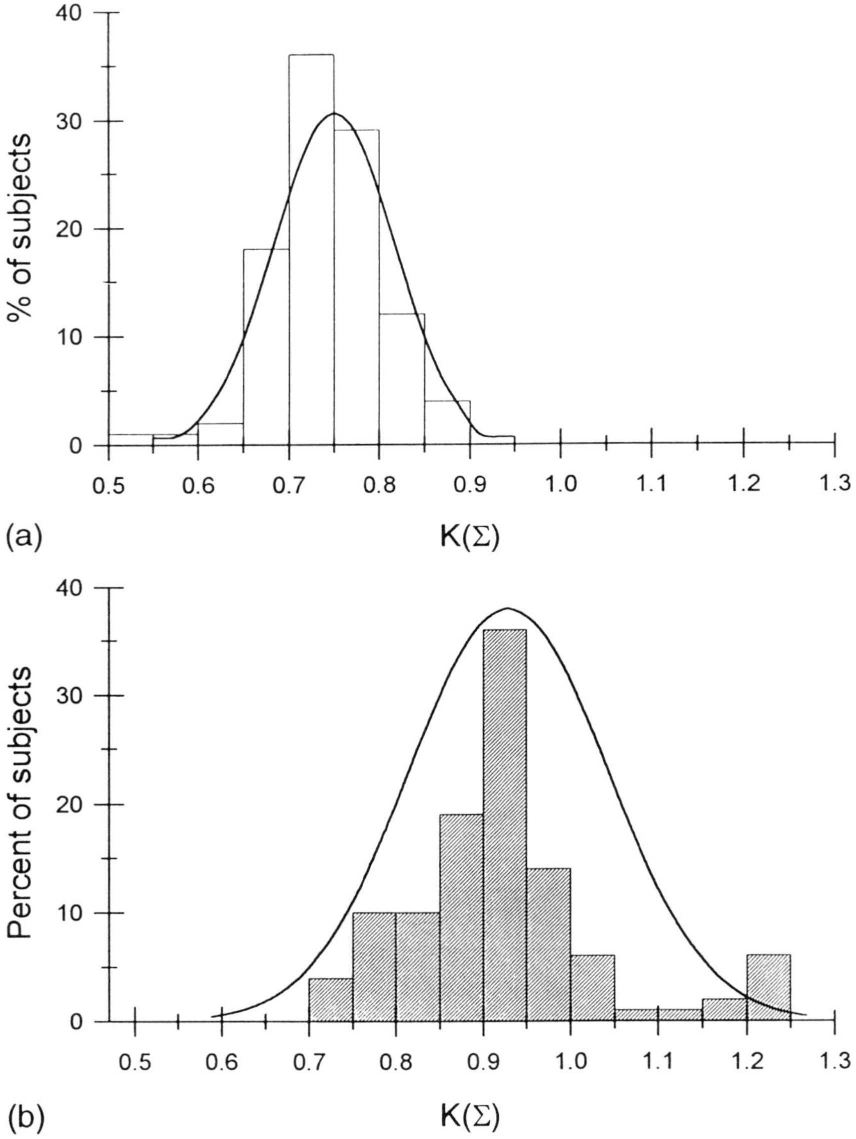

Figure 8.4. Histogram and Guassian distribution of overall total plasma protein partition coefficient values (a) in 103 apparently healthy people; (b) in 109 patients with breast cancer. Aqueous Dex—PEG two-phase system containing 1.8 mole/kg KCl in 0.01 mole/kg sodium phosphate buffer, pH 7.4.

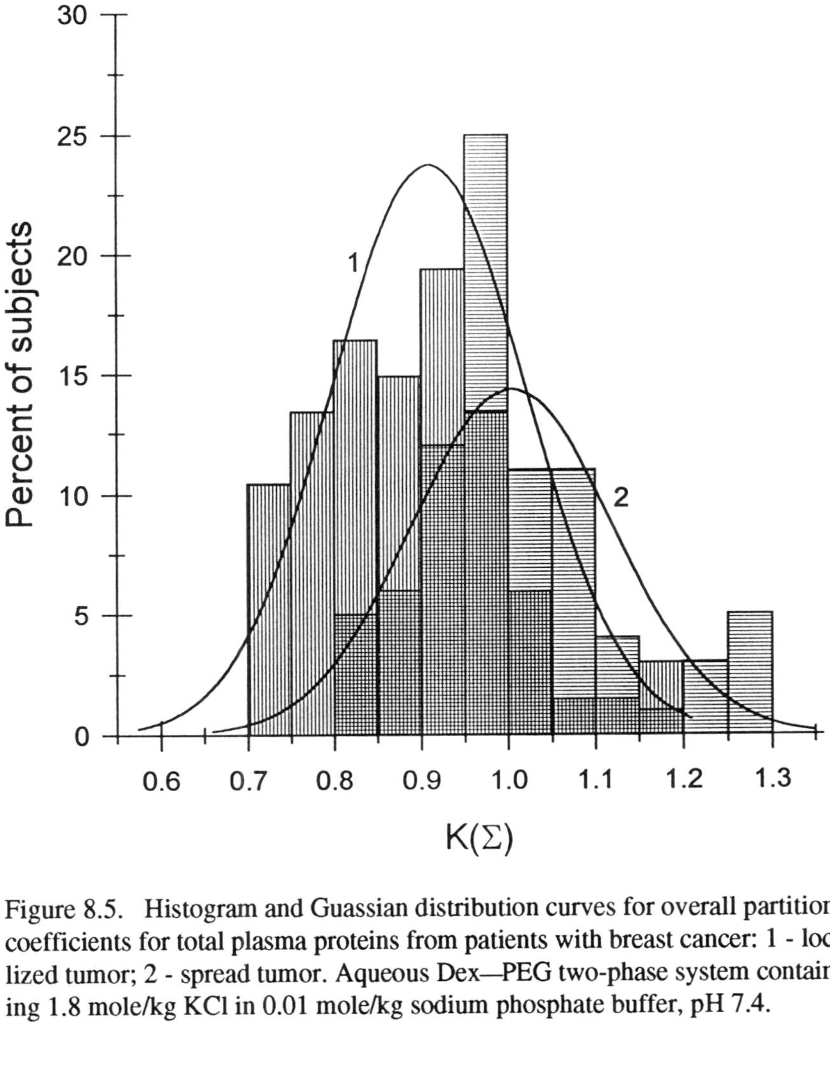

Figure 8.5. Histogram and Guassian distribution curves for overall partition coefficients for total plasma proteins from patients with breast cancer: 1 - localized tumor; 2 - spread tumor. Aqueous Dex—PEG two-phase system containing 1.8 mole/kg KCl in 0.01 mole/kg sodium phosphate buffer, pH 7.4.

ness of these changes the partitioning conditions may be modified and/or, for instance, the fluorescamine concentration measurements technique may be used instead of the Coomassie G-250 method. The diagnostic sensitivity of the test may also be increased if the partitioning procedure is applied not to the total plasma proteins but to a certain protein fraction isolated, for example, by fractional precipitation, chromatography, or electrophoresis. All these options should be explored to develop a new highly sensitive, simple, and inexpensive diagnostic procedure.

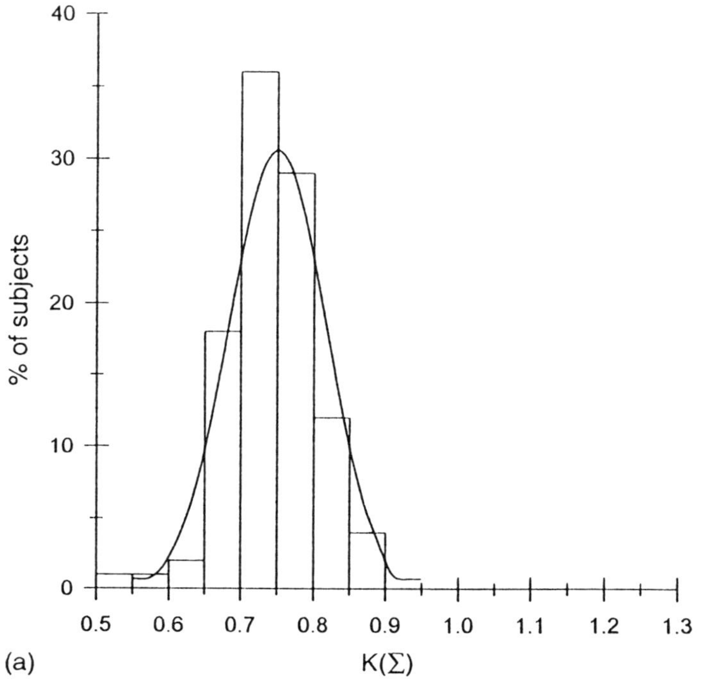

Figure 8.6. Histogram and Guassian distribution of overall total plasma protein partition coefficient values (a) in 103 apparently healthy people; (b) in 104 patients with various mastopathies; (c) in 109 patients with breast cancer. Aqueous Dex-PEG two-phase system containing 1.8 mole/kg KCl in 0.01 mole/kg sodium phosphate buffer, pH 7.4. Data from [27] with permission.

One additional example of the diagnostic use of the partition technique should be mentioned. It was found in experiments with rats kept under constant stressful conditions for 12-14 days, that while the absolute majority of the animals (ca.19 out of each 20) developed the desired depression syndrome, a few of the animals turned out to be exceptionally tolerant and capable of withstanding the stressful conditions used (N. Abdulov, 1987, unpublished data). Analysis of total plasma proteins by the partition test has shown the K(Σ) values for the "tolerant" animals to be equal to those of the animals not subjected to the stressful conditions ("normal" values), while those for the "depressed" animals have changed significantly. These results were taken as an indication that the partition procedure could be used as a "stress test" for analysis of plasma from humans.

Analysis of plasma samples from about 30 professional athletes (members of the professional alpine ski team and skating team of 'Dynamo'

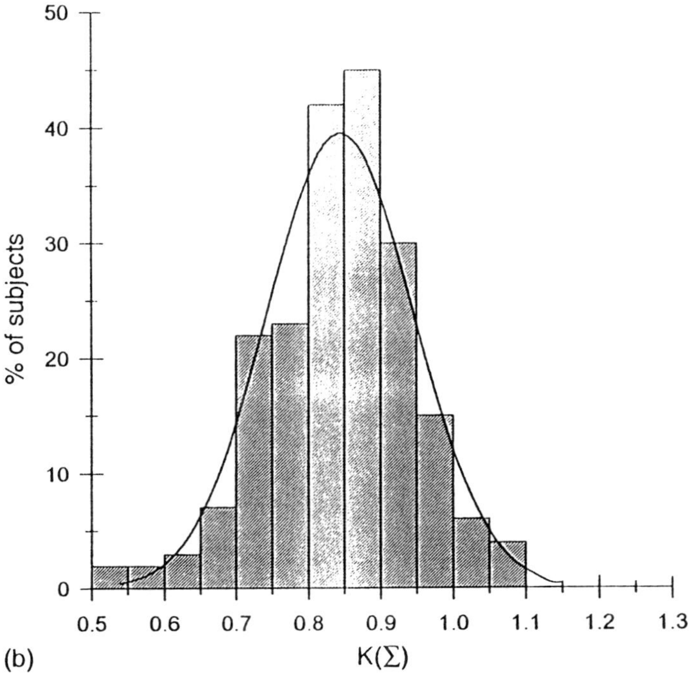

50
40
30
20
10
0
% of subjects
0.5
0.6
0.7
0.8
0.9
1.0
1.1
1.2
1.3
K(Σ)

(b)

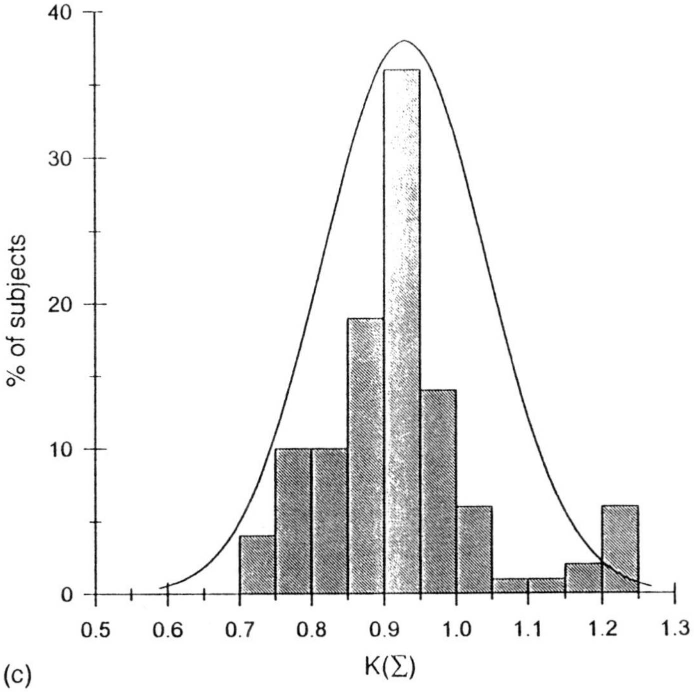

40
30
20
10
0
% of subjects
0.5
0.6
0.7
0.8
0.9
1.0
1.1
1.2
1.3
K(Σ)

(c)

club) has been performed [31]. The overall partition coefficient K(Σ) values for essentially all the athletes examined are located on the right-side part of the "normal" values distribution curve shown in Fig. 8.4. For several athletes (ca.2 out of each 10) the K(Σ) values were determined to be far to the right, and out of the "normal" range (i.e., increased in a similar way to the K(Σ) values for the "depressed" animals). It was found that individuals with increased K(Σ) values have periodically complained about sleep disorders, tiredness, persistent annoyance and other stress symptoms. It may be suggested that the aqueous two-phase partition technique may be promising as a diagnostic test for stress, although further studies are clearly necessary.

Use of the aqueous two-phase partition technique for analysis of protein-peptide extracts from different tissues has been examined by Zaslavsky et al.[35]. This possibility was assumed on the basis of the data reported by Irwin et al.[36] indicating that the SDS-PAGE electrophoregrams of the protein extracts from different mice tissues have been highly tissue-specific. The protein-peptide extracts obtained from different rat tissues (homogenized in 0.15 M NaCl in 0.01 M sodium phosphate buffer, pH 7.4) have been subjected to partitioning in the aqueous Dex-Ficoll two-phase systems of two different salt compositions, namely, 0.11 mole/kg sodium phosphate buffer, pH 7.4, and 0.15 mole/kg NaCl in 0.01 mole/kg sodium phosphate buffer, pH 7.4 [35]. The concentrations of the total protein-peptides extracts were determined by the Coomassie G-250 technique [28]. The overall partition coefficient K(Σ) was determined by the Coomassie G-250 technique [28]. The K(Σ) values for the extracts examined are presented in Table 8.6. It should be emphasized that the K(Σ) values listed in Table 8.6 have been averaged over all the samples of a given tissue extract (from 5 to 16) examined. It has been found that the K(Σ) value for the tissue extract is independent of an individual animal [35].

The K(Σ) values given in Table 8.6 may be seen to be clearly tissue-specific, and it has been verified experimentally that the K(Σ) values obtained in different aqueous two-phase systems fit the Collander-type relationship (see Equation 5-10a) established for individual solutes in aqueous two-phase systems of different polymer composition.

It should be mentioned that the overall partition coefficient K(Σ) values of the extracts from the rat brain have been found [31] to vary, depending on the extraction medium and the method used to determine concentrations in the phases.

The overall partition coefficient of the protein-peptide extracts, obtained from the brain with 0.05 M phosphate buffer (pH 7.4), and determined using the Coomassie G-250 technique K(Σ) was 2.70 ± 0.10 [31]. When the extraction medium was replaced with Tris HCl buffer (pH 7.4), the K(Σ) value determined with the same Coomassie G-250 method was found to be different,

Table 8.6

Partition Behavior of Protein-Peptide Extracts from Different Rat Tissues in the Aqueous Dex—Ficoll Two-Phase Systems of Different Salt Composition.

Tissue	$K(\Sigma)_I$ *	$K(\Sigma)_{II}$ *
Blood	0.739 ± 0.007	1.195 ± 0.008
Brain	0.375 ± 0.009	2.790 ± 0.010
Liver	0.455 ± 0.005	2.073 ± 0.009
Spleen	0.439 ± 0.007	1.648 ± 0.008
Lung	0.674 ± 0.009	1.261 ± 0.006
Heart	0.653 ± 0.006	1.132 ± 0.008

* subscripts I and II denote the overall partition coefficient values obtained in the systems containing 0.15 mole/kg NaCl in 0.01 mole/kg sodium phosphate buffer, pH 7.4, and 0.11 mole/kg sodium phosphate buffer, pH 7.4, respectively.

$K(\Sigma) = 2.96 \pm 0.09$. The overall partition coefficient $K(\Sigma)$ value for the same extracts determined with the fluorescamine binding method [29] was 1.18 ± 0.09 [31]. This result supports the aforementioned suggestion that the results of partitioning of complex protein mixtures may vary depending on the particular assay method employed.

It should be stressed that the partition coefficient of an individual solute is independent of the analytical method used for the solute concentration measurements in the two phases. The dependence of the overall partition coefficient $K(\Sigma)$ of a multicomponent protein-peptide mixture on the analytical method employed is likely due to different contributions of different components into the analytical signals detected by various methods. Short peptides and proteins with relatively high lysine content and other solutes with primary amino groups, for example, would contribute strongly to the concentration measurements with the fluorescamine technique. The contributions of the same components may be much lower or absent when the Coomassie G-250 method is used. Therefore, when the partition behavior of a multicomponent mixtures of solutes is studied it is always desirable to test different analytical methods for concentration measurements. Different contributions of various components of a mixture into the mixture partitioning, as observed with a given analytical method, should also be taken into consideration.

The partition behavior of a multicomponent mixture of biological solutes characterized by the overall partition coefficient value, K(Σ), is clearly governed not only by the composition and relative concentrations of individual components of the mixture but also by the properties of the complexes formed between the components. Acceptance of the overall partition coefficient value as a measure of the protein-peptide composition in a given tissue raises, however, the obvious question about the physicochemical meaning of this measure. The possible answer to this question is discussed below.

8.3. RELATIVE HYDROPHOBIC CHARACTER OF BIOLOGICAL LIQUIDS AND TISSUES

As shown above, the overall partition coefficient value for a given tissue extract depends on the type of extraction medium as well as the particular analytical method used for the concentration measurements in the two phases. Hence, the K(Σ) value may hardly be considered as a true measure of the chemical composition or physicochemical properties of the tissue. It has been suggested by Zaslavsky et al.[35], however, that the overall partition coefficient value may provide an information of theoretical and practical interest.

The concept developed in [35] is as follows. The relative hydrophobicity of drugs is known to influence their biological potency (see Chapter 6). Non-linear relationships between biological potency and the relative hydrophobicity are generally observed in homologous series of drugs if a fairly wide range of hydrophobicity is considered (see Equations 6.9, 6.10). It is generally accepted that the relative hydrophobicity of a drug governs its distribution throughout the body "compartments" or "biophases".

According to Martin [37], however, the term "compartment" (or "biophase") has no clear physical meaning. One "compartment" as defined [37] consists of all regions within the biological system which are of similar physical properties, and in which the drug is evenly dispersed within the time scale of the experiment. For example, in an equilibrium model [38] all nonaqueous phases of the same hydrophobicity form a single compartment [37]. According to Tanford [39] an organized biological structural framework may be considered as being essentially at equilibrium. Therefore, it is possible to view the observed biological response as proportional to the number of the receptor sites occupied by the drug [37,38], i.e. to the concentration of the drug in the "receptor compartment", and to ignore the way by which the observed distribution of the drug is achieved. The aforementioned nonlinear relationships therefore imply that the concentration of a drug in the target tissue "compartment" is governed by the compliance between the relative hydrophobicity of the drug and the relative hydrophobic character of the tissue.

Thus, it has been suggested [35] that it is necessary to understand the rather obscure term "compartment" and to define it more clearly as a biological fluid or tissue. It has also been suggested that the physical meaning of the term "hydrophobic character of a tissue or biological fluid" should be defined. From the purely physicochemical point of view a given biological liquid (or tissue), to a first approximation, may be regarded as a complex multicomponent aqueous medium of a particular chemical composition. This definitely rough approximation [35] neglects the structural organization of the tissue. According to this view, the relative hydrophobic character of the tissue has been defined as the relative affinity of the tissue's aqueous medium for an apolar CH_2 group [40]. An ever-growing number of researchers currently share the view that the structure and/or thermodynamic state and solvent features of cell water are different from those of pure water and depend on the type of the cell, its physiological state, and its macromolecular composition (see, e.g., in [41,42]). Hence, it has been suggested [35] that the relative affinity of the aqueous medium for a CH_2 group (i.e., the relative hydrophobic character of the medium) would be different in different cells and in different tissues and biological fluids. According to the hypothesis suggested in [40], the differences in the hydrophobic character of different biological fluids and tissues are supposed to be mainly due to different effects of their macromolecular components on the aqueous media.

As shown in Chapter 2, the relative hydrophobic character of aqueous solutions of different macromolecular composition may be estimated from the free energy of hypothetical transfer of a CH_2 group from pure water to the solution in question. It was also mentioned in Chapter 7 that there is a linear correlation between the relative hydrophobic character of the macromolecular solution and the relative hydrophobicity of the macromolecule [43] (see Equation 7.7 and Fig. 7.5).

The relative affinity of the aqueous solution of the extracted tissue components for a CH_2 group (the relative hydrophobic character of the tissue) has been calculated [35] from the $K(\Sigma)$ values for the protein-peptide extracts listed in Table 8.6 according to:

$$\lim[\Delta g(CH_2)] = 64.12 + 18.81 \cdot \ln K(\Sigma) \qquad (8.2)$$

where $\lim[\Delta g(CH_2)]$ is the maximum value (in cal/mole CH_2) of the relative hydrophobic character attainable for the aqueous solution of a given extract (see in Chapters 2 and 7).

The $\lim[\Delta g(CH_2)]$ values reported in [35] are listed in Table 8.7. Utilizing the free energy - additivity principle it is possible to calculate the difference in the relative hydrophobic character of the aqueous saline solutions of the extracts examined. The $\Delta g(CH_2)_{tr}$ values obtained [35] represent the free

Table 8.7

Relative Hydrophobic Character of Different Rat Tissues.

Tissue	$\lim[\Delta g(CH_2)]_I$ *	$\lim[\Delta g(CH_2)]_{II}$ *	$[\Delta g(CH_2)_{tr}]_I$ **	$[\Delta g(CH_2)_{tr}]_{II}$ **
Blood	-58.4	-67.5	0 ***	0 ***
Brain	-45.6	-83.5	12.8	-16.0
Liver	-49.2	-77.9	9.2	-10.4
Spleen	-48.6	-73.6	9.8	-6.1
Lung	-56.7	-68.5	1.7	-1.0
Heart	-56.1	-66.5	2.3	1.0

* $\lim[\Delta g(CH_2)]$ is the maximum value (in cal/mole CH_2) of the relative hydrophobic character attainable for the aqueous solution of a given extract (see Equation 8.2); subscripts I and II denote the overall partition coefficient values obtained in the systems containing 0.15 mole/kg NaCl in 0.01 mole/kg sodium phosphate buffer, pH 7.4, and 0.11 mole/kg sodium phosphate buffer, pH 7.4, respectively;

** $[\Delta g(CH_2)]_{tr}$ is the free energy of transfer of a CH_2 group from blood plasma to the tissue indicated; subscripts I and II as indicated above;

*** according to definition [35].

energy of transfer of a methylene group from the blood plasma medium to the aqueous medium of a given tissue. The $\Delta g(CH_2)_{tr}$ values reported in [35] are presented in Table 8.7.

All the considerations used by the authors [35] are based on the assumption that the relative hydrophobic character of a given tissue may be estimated as that of the aqueous saline solution of the proteins extractable from the tissue. This assumption [35], if correct, is clearly no more than a very rough approximation.

With the $\Delta g(CH_2)_{tr}$ values given in Table 8.7 the tissues can be arranged in a definite order with respect to their hydrophobic character which appears to depend upon the ionic composition of the medium. It follows from the data in Table 8.7 that in the presence of 0.11 M sodium phosphate buffer, pH 7.4 (isotonic medium) the relative hydrophobic character of the rat tissues decreases as shown: brain > liver > spleen > lung ≥ blood plasma > heart. In the presence of 0.15 M NaCl in 0.01 M sodium phosphate buffer, pH 7.4 (isotonic saline) the above order is changed: blood plasma, heart, lung > liver > spleen > brain. It should be noted that both ionic compositions used [35] are usually supposed to be rather close to those of the physiological medium. It is known, however, that the ionic composition of different tissues is significantly different [44]. The data given in Table 8.7, therefore, imply that an alteration in the ionic composition of a given tissue may induce a change in the relative hydrophobic character of the tissues. It seems only reasonable to assume that the latter change may lead to a redistribution of at least some endogenous solutes between the tissues and fluids of the organism, perhaps resulting in a pathological disorder.

It should be mentioned that the effect of the ionic composition of the medium on partitioning of solutes is observed in aqueous two-phase systems as well as in water-organic solvent systems (see in Chapters 4-6). Therefore, the results [35] under discussion appear to be in line with the theoretical considerations of the partition processes, and particularly with the equilibrium model of the drug distribution in the living organism describing process similar to those observed in an equilibrium multi-phase system [37-39].

It should be noticed also that the small $\Delta g(CH_2)_{tr}$ values listed in Table 8.7 are quite reasonable since the solutes providing the changes in the thermodynamic state of the aqueous medium in the biological fluids and tissues examined are of the same nature.

If the model of the equilibrium distribution of a solute between the organism tissues is adopted it is possible to describe the distribution of the solute between blood plasma and two different tissues utilizing the Collander-type relationship (see Equation 5.10a):

$$\ln K_i = \alpha_{ij} \cdot \ln K_j + \beta_{ij} \qquad (8.3)$$

where K_i is the partition coefficient of a solute between blood plasma and the i-th tissue; K_j is the partition coefficient of the same solute between blood plasma and the j-th tissue; α_{ij} and β_{ij} are constants. (Both K values may be determined as the ratios between the solute concentrations in blood plasma and in the corresponding tissue.)

As shown in Chapter 5, the slope (parameter α_{ij}) in the Collander-type equation is related to the free energies of transfer of a methylene group between the two phases of the systems when compared according to Equation 5.12:

$$a_{ij} = \Delta g(CH_2)_{tr}^{i} / \Delta g(CH_2)_{tr}^{j} \qquad (8.4)$$

where $\Delta g(CH_2)_{tr}^{i}$ is the free energy of transfer of a CH_2 group from blood plasma to the i-th tissue; and $\Delta g(CH_2)_{tr}^{j}$ - the free energy of transfer of a CH_2 group from blood plasma to the j-th tissue.

The intercept (coefficient β_{ij}) of the Collander-type equation 8.3 (see Chapter 5) represents the difference between the specific interactions of the solute with the solvents in the phases of the two-phase systems under consideration. It has been suggested [35] that the physical meaning of parameter β_{ij} may be extended to cover not only the specific solute-solvent interactions in the phases but any specific interactions of the solute with the components of the phases of the systems under comparison.

If the nature of the tissue distribution of a drug in a living organism is known (i.e., the K_i and K_j values are determined as the ratios of the drug concentrations in the blood and the i-th and j-th tissues, respectively) the α_{ij} value may be estimated, and the β_{ij} value may be calculated using Equation 8.3. Thus, the relative intensity of the drug-tissue specific interactions may be evaluated.

In this context, the results reported by Clausen and Bickel [45] on the distribution of various drugs between blood and several rat tissues are of particular interest. The data reported [45] have been obtained by the so-called distribution dialysis technique. This technique is similar to conventional equilibrium dialysis except that instead of one chamber filled with the buffer solution of a drug being distributed and the other chamber with a "binder", both chambers are filled with different "binders". In the experiments under discussion [45] one chamber contained a rat tissue homogenate, and the other contained human citrate-treated blood, both diluted 1:10 with 0.15 M NaCl in 0.01 M sodium phosphate buffer, pH 7.4. According to the authors [45], rat and human blood yielded identical results so the latter was used in the experiments. The tissue/blood drug concentration ratios reported by the Clausen and Bickel [45] may be considered as the partition coefficients, K_{app}, for the drugs in the corresponding blood-tissue two-phase systems. The logarithms of the K_{app} values [45] characterizing the drugs distribution between the blood and a given tissue are plotted in Figure 8.7 versus the logarithms of the K_{app} values for the same drugs distributed between the blood and the brain tissue homogenate.

As may be seen from Fig. 8.7, the data reported by Clausen and Bickel [45] fit the Collander-type relationships. The relationships observed may be described as:

$$\ln K_{app}(\text{drug j})_{\text{blood/tissue i}} = a_i \cdot \ln K_{app}(\text{drug j})_{\text{blood/brain}} + b_i \qquad (8.5)$$

where $K_{app}(\text{drug j})_{\text{blood/tissue i}}$ is the ratio of the concentration of the j-th drug in the dialysis chamber containing blood to the concentration of the drug in the chamber containing the homogenate of the i-th tissue. Subscript "blood/ brain" denotes the case when the brain homogenate is used as the i-th tissue homogenate; a_i and b_i are constants.

The a_i and b_i values calculated from the data [45] treated according to Equation 8.5 are listed in Table 8.8. It is rather surprising that different drugs

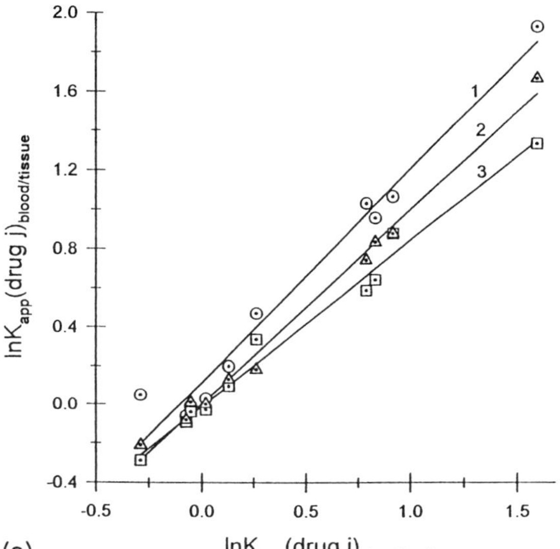

(a)

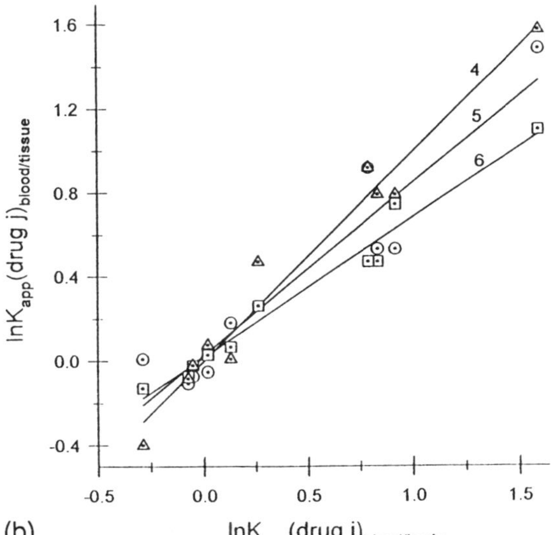

(b)

Figure 8.7. Logarithm of the tissue/blood drug concentration ratio versus logarithm of the brain/blood drug concentration ratio. The following rat tissues are compared: (a) 1 - liver; 2 - kidneys; 3 - lungs; (b) 4 - small intestine; 5 - adipose tissue; 6 - skeletal muscle. (Calculated from the data reported in [45].)

Table 8.8

Characteristics [a] of the Collander-Type Relationships Between the Data [44] for the Distribution of Ten Different Drugs Between Blood [b] and Various Rat Tissues' Homogenates.

Tissue	a_i	b_i	r^2
Liver	1.097 ± 0.067	0.113 ± 0.046	0.9855
Lungs	0.857 ± 0.039	-0.012 ± 0.027	0.9919
Kidneys	0.999 ± 0.032	0.002 ± 0.023	0.9958
Adipose tissue [c]	0.820 ± 0.105	0.028 ± 0.077	0.9468
Small intestine	0.997 ± 0.066	-0.0001 ± 0.046	0.9829
Skeletal muscle	0.667 ± 0.040	0.016 ± 0.028	0.9862

[a] a_i and b_i are coefficients of Equation 8.5; r^2 is the correlation coefficient;
[b] citrate-treated human (not rat) blood has been used in the experiments [45];
[c] nine drugs fit the relationship (the data for thiopental has been excluded).

such as methadone, imipramine, morphine, pentobarbital, salicylic acid, etc. [45], fit the same linear curve. It was expected that different binding of different drugs to various components of the tissues' homogenates examined [45] would lead to noticeable deviations from the curves as observed for several drugs such as, e.g., phenylbutazone in liver tissue, thiopental in adipose tissue, etc. (see Fig. 8.7). Hence, it seems reasonable to assume that the results obtained in the distribution dialysis experiments [45] represent not just the differences between the drugs' binding to the components of the different tissues. These results [45], as follows from the relationships shown in Fig. 8.7, represent the partitioning of the drugs between the different media in the two dialysis chambers affected by the binding in some particular cases. Assuming conditions to be at equilibrium, it is possible to consider the relationships shown in Fig. 8.7 as the typical Collander-type relationships with the physical meaning of coefficients a_i and b_i as discussed above (see Chapter 5, Equations 5.12-5.14, and Eqn. 8.4).

It follows from the a_i values presented in Table 8.8 that the hydrophobic character of the tissues' homogenates examined [45], relative to that of blood, decreases in the following order: liver > brain = kidneys = small intestine > lungs ≥ adipose tissue > skeletal muscle (salt composition of the media: 0.15 M NaCl in 0.01 M sodium phosphate buffer, pH 7.4).

The relative hydrophobic character of the brain, liver, and lungs tissues (as derived from the distribution dialysis data [45]) changes in an order different from the one established with the aqueous two-phase partition technique. The likely reason may be that while only the proteins (and peptides) extracted from a given tissue have been examined in the partition experiments [35], the total tissue homogenates have been used in the distribution dialysis studies [45]. Additionally, the cellulose membrane with a 12000 dalton cutoff used in the dialysis experiments [45] might not separate the low molecular weight components of a given tissue and blood affecting the relative hydrophobic character of both as detected by the aqueous two-phase partition technique [35].

One implication of the above results is worthy of particular notice. It follows from all the partition data obtained in the aqueous Dex-PEG and Dex-Ficoll two-phase systems that the dialysis technique should be used in studies of the macromolecule-small solute interactions with extreme caution. A given macromolecule may affect the small solute distribution in the dialysis experiments, not through the solute binding, but via its effect on the solvent features of the aqueous medium, and this effect must be taken into consideration.

Returning to the analysis of the relative hydrophobic character of biological liquids and tissues by the aqueous two-phase partition technique, it should be emphasized once more that the model suggested and the considerations used [35] are obviously based on concepts that are too simple since, for example, the complex structural organization of the tissues is neglected. Hence, the estimates of the relative hydrophobic character of tissues presented in Table 8.7 may be viewed only as a very crude approximation.

It may be suggested, nevertheless, that an estimation of the relative hydrophobic character of the biological liquids and tissues of a living organism may provide new information about the possible changes in the internal media of the organism induced by different factors.

This suggestion was investigated by Zaslavsky et al.(1988, unpublished data) using experimental animals (rats) treated with two different anabolic steroids commonly used as body-building drugs. The protein-peptide extracts from various tissues of the animals subjected to the treatments were examined by the partition technique in comparison to those from "control" animals treated with placebo. It was found that treatment with both anabolic steroids changed the partition behavior of total plasma proteins as well as those of protein-peptide extracts from liver and brain tissues. The partition behavior of the extracts from other tissues was found not to be affected by the treatments.

Two implications of the results obtained should be noticed. First, the relative hydrophobic character of different tissues as measured by the partition technique [35] may vary depending on the effects of chemical, physical, or biological factors. That result means that the drug's distribution throughout the body tissues and liquids may vary depending on the particular disorder and physiological peculiarities of an individual under treatment (see, e.g., in [46]). Second, the partition behavior of the protein-peptide extract from a given tissue may indicate changes in the features of the tissue undetectable with standard diagnostic procedures. It should be stressed that the partition test for the protein extract from a given tissue may hardly be expected to replace existent procedures. The partition test may indicate that the composition and/or features of the extractable components of the tissue under examination are different from those in a "normal" healthy tissue. In certain cases, the differences observed may be of diagnostic value. In other cases the difference detected may serve as an indication of a "disorder" which should be looked into by the other more specific experimental techniques.

The history of biomedical research indicates that the current knowledge of the mechanisms of various pathologies, toxic effects of different factors, etc., originates for the most part from the histopathological, morphological, and symptomatic observations. The functional and chemical links between different organs and tissues, and their role in pathological processes, are not understood very well. The use of the partition test for analysis of different tissues may help to develop better insight, at least in terms of what tissues should be examined in more detail. From this point of view the aqueous two-phase partition technique may be useful to obtain better understanding of the mechanisms of fundamental biological and medical importance. On the purely practical side, the partition test may be used for the biopsy analysis, in the studies of drug side-effects, toxic effect of chemicals, etc. [47].

Finally, the implications of the results obtained with the aqueous two-phase partition technique for organization of biological systems should be considered.

8.4. AQUEOUS TWO-PHASE SYSTEMS AS A MODEL OF BIOLOGICAL SYSTEMS

One of the most far-reaching applications of aqueous two-phase systems to be discussed finally is its application to the study of organization of biological systems. Before considering this application, current views on the properties of water in biological systems, and the role of water in the organization and function of metabolic processes should be briefly outlined.

Organization of a biological system is realized through spatial separation but functional integration of different components of the system. That is achieved for the most part by occurrence of biological membranes formed by water-insoluble and sparingly-soluble compounds. Biological membranes, however, are not the only means by which "compartments" can be generated in cells. Recently it has been realized that a number of metabolic pathways previously thought to occur in solution by random thermal motion of enzymes and substrates are, instead, spatially constrained in the intracellular aqueous compartments' undivided by membranes. Furthermore, transport of solutes in a multicellular organism is not always governed solely by the presence of cellular membranes. For examples, consider intercellular skin permeation of chemicals through the stratum corneum and the other layers (see, e.g., in [48] and references cited therein), and the passive diffusion of small peptides between the intestinal cells [49,50], or intercellular diffusion of solutes permeating the blood-nerve barrier from nerve blood vessels in the endoneurium [51]. The implication of these observations is that there should be a general principle of spatial organization of aqueous media into different compartments without a "mechanical" insoluble barrier between any two compartments. It seems reasonable to suggest that at least one mechanism involved in this "microcompartmentation" may be that of phase separation.

It is known that proteins and polysaccharides may phase-separate in aqueous mixtures (see, e.g., in [52]). Typical phase diagrams reported by Tolstoguzov and his co-workers [52] are reproduced in Figure 8.8. These phase diagrams appear to be similar to those of aqueous two-phase systems formed by two non-ionic polymers considered above (see in Chapter 3).

Phase separation in aqueous protein-protein, polysaccharide-polysaccharide and protein-polysaccharide mixtures has been studied by Tolstoguzov and his associates (see, e.g., in [52] and references therein). It has been concluded from the study of about 100 pairs of biological macromolecules, that [52] "phase separation is one of the most characteristic properties of (aqueous) solution mixtures, containing structurally dissimilar polysaccharides". Phase separation in about 20 aqueous protein-protein systems has been observed and judged [52] to be a general phenomenon particularly for proteins belonging to different classes within the Osborne classification (i.e., albumins, globulins, glutelins, and prolamines). Aqueous mixtures of native and denatured forms of the same protein, e.g., ovalbumin, have been found also to undergo phase separation, with different protein forms concentrated in different phases.

According to Tolstoguzov [52], the feature specific for aqueous two-protein, two-phase systems is the asymmetry of the phase diagram displayed as a large (up to an order of magnitude and more) difference in the concentrations of the proteins in the phases. The other "specific" feature of the systems is claimed [52] to be a fairly high value of phase separation threshold, e.g., exceeding 12% wt. for mixtures of globular proteins. Both these features,

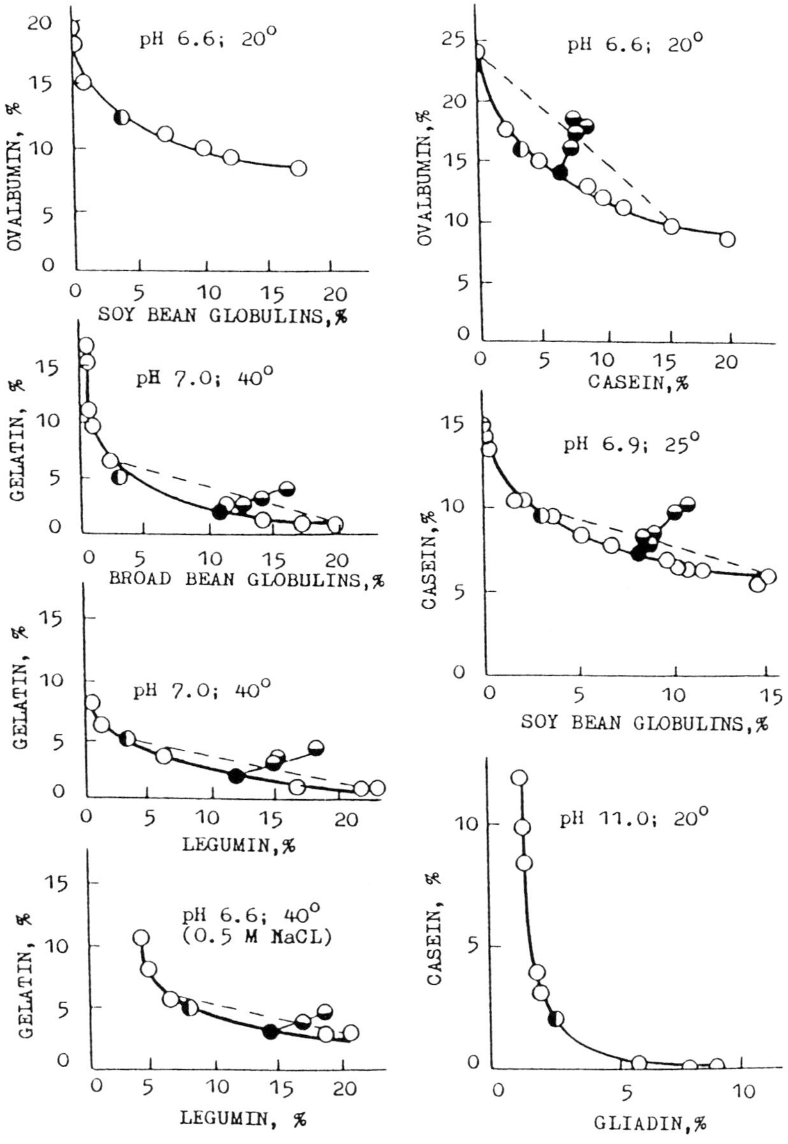

Figure 8.8. Phase diagrams for aqueous mixtures of proteins.
(From V. B. Tolstoguzov, Food Hydrocolloids, 2, 339 (1988). Reprinted by permission of Oxford University Press.)

however, are not as different from other aqueous polymer two-phase systems as suggested by Tolstoguzov [52] (see in Chapter 3). Phase diagrams of aqueous protein—salt two-phase systems presented in [52] appear also to be similar to those of aqueous PEG—salt systems considered above. The temperature and salt effects on phase separation in aqueous two-protein systems and protein—salt systems [52] appear to be rather moderate and not exceeding those in aqueous two-phase systems formed by nonionic polymers. However, effects of pH [52] are clearly greater than these effects as might be expected for the systems formed by polyelectrolytes.

Studies of aqueous two-phase systems formed by proteins [52] may be considered as too limited as yet to warrant any generalization in regard to their differences from the systems formed by nonionic polymers. The only certain conclusion that may be drawn from these studies is that proteins, polysaccharides, and other biological macromolecules are capable of phase separation in an aqueous medium. Additionally, temperature and low molecular nonionic and ionic solutes present in the aqueous intracellular environment may affect phase separation in the systems under discussion. There are many technological [52] as well as biological implications. Phase separation in concentrated aqueous solutions of proteins γ-crystallins is supposed to be one of the mechanisms for lens opacification likely to be related to cataract occurrence (for review see [53]).

In order to discuss the biological implications of the above phenomenon it is first necessary to outline the current views on the problem of structural organization of metabolic processes in the aqueous cell cytoplasm. The experimental evidence reviewed by Clegg (see, e.g., in [42,54-56] and references cited therein) indicates that (a) the aqueous cytoplasm, nucleoplasm, and the interiors of subcellular organelles may not be viewed as concentrated solutions of macromolecules, metabolites, ions, etc., randomly dispersed and freely diffusible in the aqueous medium; and (b) the solvent properties of the intracellular aqueous medium may well be different from those of pure water (or dilute solutions). Hence, an alternative to the traditional view of the intracellular architecture of eukaryotes as a suspension of subcellular organelles in the highly concentrated "cytosol" (often regarded as ca. 20% protein solution) was suggested [54]. According to Clegg and Drost-Hansen [57], the alternative view may be that of a structural network (surrounded by a dilute aqueous phase) observed as the microtrabecular lattice (MTL) [58], an extensive highly branched network of protein strands that connect with almost all cytoplasmic ultrastructures. Clegg suggested [42, 54-56] that this lattice houses most of the enzymes of aqueous cytoplasm, the aqueous medium between trabeculae containing relatively low concentrations of proteins and other macromolecules. The solvent properties of the aqueous medium have been suggested [57] to differ from those of pure water due to the influence of macromolecular surfaces of the subcellular structures on water in close proximity to the surfaces. According to the estimates [59] based on the image analysis of high voltage electron microscopy (HVEM) photographs, at least 50% of the total water occurs within 50 Å from some surface. Properties of water in the vicinity of macromolecular surface defined as vicinal water were outlined above (see Table 2.1).

Here are the consequences of the model suggested by Clegg and Drost-Hansen [57]:

> "1. Much of what is known about macromolecular function in cells is based on data obtained *in vitro*, almost always in highly diluted solutions. That approach is very convenient, but if intracellular water differs from that in test tubes, as we believe it does, then information obtained *in vitro* may not allow us to construct (or better "reconstruct") an accurate description of these molecules and their activities when they operate within cells...
> 2. It is widely accepted that direct interactions between macromolecules and their surrounding water of hydration play critical roles in their structure and function. There is no debate about this issue, and it seems very likely that water plays subtle but important roles in metabolism through water-enzyme interactions. However, to understand those roles we must know the details of the aqueous microenvironment in which this activity occurs.
> 3. Available evidence suggests, to us at least, that the solvent properties of at least a large fraction of the total cell water, notably in cytoplasm, differ from those of ordinary aqueous solutions. On this basis, some contribution to the uneven distribution of certain solutes across the plasma membrane, as well across membranes within cells (organelles), could arise from such solvent differences. In addition, small metabolites might "partition" between various intracellular aqueous phases [60]. Even protein distribution within cells could be influenced in this fashion. A speculative "model" on the organization of enzymes in the aqueous cytoplasm includes the possibility that a loose association of enzymes with the cytomatrix may be driven by water interactions involving their respective surfaces, similar to those involved in association through hydrophobic interactions [61].
> 4. Assembly-disassembly processes are influenced by the properties of the aqueous phase within which they occur. Such mechanisms could be critical to enzyme-enzyme associations and the dynamic turnover of the cytomatrix, and possibly other cell structures...
> 5. Many molecular interactions in cells involve electrostatic interactions which are, of course, very sensitive to the dielectric properties of the

aqueous phase in which they occur. Thus, the possibility that the dielectric permittivity of cell water is reduced [62] relative to dilute solutions, may be of some importance.

6. A reasonably good correlation exists between modifications of the cytomatrix and changes in the amount and properties of cell water, both of which commonly, although not always, accompany cell transformation by viruses or carcinogens. While that may be fortuitous, it is notable that the usual observation is a reduction in cytomatrix surface area and an increase in the amount of cell water that has "bulk-like" properties (see [42, 63]). That is consistent with the proposed relationship between the cytomatrix and its effects on the properties of the surrounding aqueous environment, and vice versa. It has also not escaped our attention [54, 56,61] that many of the metabolic changes accompanying the transformation process are associated with "soluble" enzymes which, in the view of some of us, are not really "soluble" at all but are instead part of the water-cytomatrix system.

7. Without the concept of vicinal water, and its characteristic thermal anomalies at several different temperatures of physiological interest, it is difficult to see how a large body of usual thermal responses of intact organisms can be explained completely. On the other hand, accepting the thermal transitions in the vicinal water structures allows for relatively facile explanations of (sometimes dramatic) complex thermal responses of organisms including some very abrupt thermal death limits, selection of body temperatures and multiple temperature growth optima."

The concept that the aqueous cytoplasm is highly organized as suggested by Clegg (see, e.g., in [54]) is consistent with a large body of experimental evidence. The model of a network of macromolecular surfaces in contact with "dilute aqueous solution" [54-57] is open to arguments, however. There is no doubt, in my view, that the macromolecular surfaces of subcellular structures do influence the solvent properties of the aqueous medium in their vicinity, and that the occurrence of vicinal water does play an important role in regulation of metabolic processes in the aqueous cytoplasm. We should be reminded, however, that in accordance with the so-called "paradoxical effect" (see, e.g., in [57]) the vicinal water is induced by any kind of macromolecular surface, and its (water) structure and solvent properties are independent of the specific chemical nature of the surface. If correct (and that remains to be proven) then the solvent properties of vicinal water must be the same throughout the cytoplasm, independent of the specific macromolecular surface with which the water is in contact. The implication is that there are two different aqueous phases in the aqueous cytoplasm of any cell - one phase composed of vicinal water, and the other phase composed of water not affected by the surfaces and possessing solvent properties similar to those of a highly dilute aqueous solution (i.e., "normal" water). Additionally, any physiological or pathological change in a given cell can be expected to alter only the relative volumes of these two phases but not their properties.

It is hard to imagine how these rather limited alteration may affect regulation or organization of the metabolic pathways in the cytoplasm. Finally, the model [54-57] under consideration does not allow for occurrence of "various intracellular aqueous phases" [60] mentioned in the above "consequence 3". Therefore the model [54-57] is certainly an oversimplification even though it is clearly much more realistic than the one based on the "concentrated solution" concept.

In view of the aforementioned experimental data on phase separation as a common phenomenon in aqueous protein mixtures, the likely principle of organization of aqueous cytoplasm may be suggested to include phase separation as a general mechanism readily controlled by chemical effectors and fitting most of the experimental evidence available at present [54-57]. What should be emphasized is that phase separation has very rarely even been considered in cell biology.

Aqueous mixtures of several, e.g., six different polymers are known to phase separate into as many as eighteen phases [1, pp.13-15]. Based on the experimental data discussed above, the solvent properties of the aqueous media in these phases, as well as their composition, may be assumed to be readily altered by relatively small overall or local changes in the composition. The aqueous cytoplasm hence may conceivably be composed of numerous coexisting aqueous phases of different solvent properties. The influence of the macromolecular surfaces of subcellular structures (including the MTL network) on the solvent features of adjacent (vicinal) water may be involved in "fine tuning" the solvent properties of the presumed different aqueous phases. Temperature effects known to affect phase separation [1] may eliminate some of the phases or generate new ones as well as change the solvent properties of the phases depending on the requirements for cell function and even survival.

The suggested model does not contradict the experimental data on the ability of a variety of cells to survive up to over 50% dehydration (see, e.g., in [54]). Dehydration may be viewed as a decrease in the solvent (water) concentration or an increase in the solute concentration. In an aqueous polymer two-phase system an increase in the polymers concentrations is known to alter the difference in the solvent properties of the phases but by no means to dispose of phase separation itself meaning that the suggested aqueous cytoplasm organization principle should not be destroyed by dehydration.

The experimental data obtained by the centrifugal stratification of viable cells (see, e.g., in [54]) should also be mentioned. In a typical experiment, intact *Euglena* cells (a unicellular eukaryote) have been subjected to centrifugation (100,000 g for 1 h) and examined by quick-freeze and several cytochemical methods [65,66]. The "soluble" layer of the stratified cells was distinguished by the lack of organized ultrastructure and was considered to represent the soluble part of cytoplasm. No macromolecules were found in that layer, consistent with the idea that the aqueous cytoplasm is diluted with respect to macromolecules. Similar results were obtained in stratification experiments with a few other cell types. Attractive as this interpretation appears there are difficulties about accepting it since emphasizing the reversible character of the changes in the cytoplasm structure produced by stratification ignores the possibility of dramatic effects of strong shear forces on initial organization of the aqueous cytoplasm. There is no experimental information, to my knowledge, about behavior of aqueous multiphase systems in a strong centrifugal field. It would be valuable to obtain such data. It seems possible, however, that the small "drops" of different aqueous phases presumably existing in the aqueous cytoplasm might settle together with "particulate" layers including, e.g., mitochondria, lysosomes, etc. This possibility should depend on the size and density of the phases, interfacial tension values, etc. There are presently no experimental data, however, to support or discard the above assumption vs. the interpretation by Clegg [54].

If the assumption that the aqueous cytoplasm is an aqueous multiphase system [61,67] is accepted as a working hypothesis, then organization of metabolic pathways in the aqueous cytoplasm is readily explainable. An illustrative example is offered by applications of aqueous two-phase systems for the so-called extractive bioconversion (see, e.g., [1,pp.221-225]). It has been repeatedly shown that, under appropriate partition conditions, an enzyme and its substrate may be concentrated in one aqueous phase while the product distributes into the other aqueous phase. This process could play a very important role in cell structure and function. Multienzyme systems, subcellular organelles, or cells may be used instead of individual enzymes, and the examples of extractive bioconversion include, e.g., hydrolysis of starch by amylase, saccharification of cellulose by cellulase and ß-glucosidase, deacetylation of penicillin G to 6-aminopenicillanic acid by penicillinacylase, etc. (see, e.g., in [1] and references cited therein).

It may be speculated (remaining to be explored experimentally) that the product of a given enzymatic reaction in the local phase in the aqueous cytoplasm may initiate an additional phase separation or affect the solvent properties of the aqueous medium in the phase it distributes into, providing a feedback for the reaction. Aqueous two- and multi-phase systems should be explored as the media for multienzyme reactions not only from the practical technological view [1] but also in terms of simulating the suggested principle of organization of metabolic pathways in the aqueous cytoplasm. This would be a difficult and time-consuming effort but it may be helpful for better understanding of the fundamental principles of organization of biological systems.

On a much higher organization level a living organism may also be considered as a multiphase system. This approximation is successfully used in pharmacology (see, e.g., in [46]). Using this approximation, the distribution of a solute (exo- or endogenous) between the biological liquids and tissues may be treated as that between the coexisting phases of a predominantly aqueous nature.

As shown above, a change in the solvent properties of a given phase is followed by re-distribution of the solutes. The solvent properties of an aqueous phase may be induced by different factors (see in Chapters 4 and 5), polymer additives being among the most effective ones.

It should also be mentioned that among various water-soluble polymers examined in regard to their influence on the relative hydrophobic character and other solvent features of an aqueous medium [43,68-71] (see in Chapter 2) polyvinylpyrrolidone (PVP), polyvinyl alcohol (PVA) and arabinogalactan have been found [40] to be the most effective. These polymers of different chemical structures and different molecular weights displayed the largest influence (e.g., in terms of lim[$\Delta g(CH_2)$] values, see Fig. 2.7) among the polymers examined in this regard so far [43,68-71]. It should also be pointed out that PVP and PVA are known to be commonly used (in Europe) as intravenously administered detoxicating agents. Arabinogalactan [71] has also shown significant detoxicating effects in animal studies [72].

A common explanation of the detoxication influence of PVP is that this polymer forms complexes with toxins in the blood stream, which are then eliminated as the body clears it of the toxins. The structural features of PVP allow one to provide a reasonable explanation for the high binding capacity of this polymer numerously shown toward a large variety of chemicals *in vitro*. The question of how PVP may differentiate between binding toxic xenobiotics and harmless endogenous solutes seems not to have any plausible explanation, however. Additionally, the structures of PVA and arabinogalactan (different from those of PVP) do not leave any room for a common explanation of the detoxicating effects of these two polymers.

The hypothesis suggested in [40] is that all three polymers (PVP, PVA, and arabinogalactan) may affect the solvent features of the aqueous medium of the blood plasma. Depending on the local concentration of a polymer in the blood stream, these effects may result in re-distribution of chemicals increasing their total blood content. This may lead to the binding of these chemicals by serum albumin or other proteins, or PVP (when it is used) followed by their clearance from the body. This hypothesis [40] has not been subjected to an experimental test as yet. The implication of the model worthy of particular notice is that the intravenous administration of any of the above polymers may affect drug distribution throughout the body liquids and tissues. The reason is that the drug behavior in a living organism should be controlled by the physicochemical mechanisms governing that of toxins [73]. Data [74] on the enhancement of the analgesic effect of morphine when administered intravenously together with PVP seem to indirectly support the above hypothesis [40]. If proved correct, the model [40] may lead to development of a new approach to more efficient drug targeting.

It should be mentioned that the above considerations are in agreement with the thermodynamic analysis of biological organization suggested by Tanford about 15 years ago [39] to be viewed as consisting of two stages: biosynthesis and assembly. The assembly process has been suggested [39] to be, for the most part, under thermodynamic control, meaning ".. as a first approximation it represents a search by each structural molecule for its state of lowest chemical potential". According to Tanford [39] "The thermodynamics of biological organization. ... focuses solely on where molecules prefer to go after they have been synthesized" (or administered in the organism). It may be suggested that analysis of partition behavior of biological solutes in aqueous two-phase systems may help us to gain much better understanding of "where and why" molecules prefer to go in cells, tissues and extracellular fluids.

8.5. SUMMARY

Summing up the experimental data and semi-theoretical considerations presented above, it can be concluded that the partition coefficient of an individual biological solute may be used as a simple, highly sensitive and cost-effective relative measure of the solute identity and/or purity, provided the value for a standard reference solute is known. The partition coefficient value is considered similar to the chromatographic retention index or melting point of a substance which is widely used as a simple control of the purity of a synthetic product. The information provided by the technique about a given biological solute is unique, quantitative, and in certain cases of paramount importance. The technique can be of particular value in assessing lot-to-lot consistency of production (of recombinant proteins, for example).

As the information provided by the technique about biomolecules is unique, it is likely that it may be helpful in elucidating the pathophysiological mechanisms of certain diseases and in suggesting lines for further investigation. A mixture of natural products can be characterized by the overall partition coefficient value. If the specific overall partition coefficient of a standard reference mixture is known, it is possible with the partition technique to check the identity of a given mixture with the reference one. The technique makes it possible to assess the lot-to-lot consistency of raw biological materials, e.g., a given plasma protein fraction, tissue or cell extract as well as individual proteins, glycoproteins, etc. The partition test may also be used on medical diagnostics. Fractionation of plasma proteins before the partition test would indicate the disease-related protein fraction, or single protein, and increase the test diagnostic predictive value, selectivity and specificity. Being tissue-specific, the partitioning test may also be of clinical value for analysis of plasma proteins as well as tissue biopsies.

As is usually the case, the advantages and drawbacks of the partition technique are closely interrelated. The advantages of the technique are that it is informationally unique, highly sensitive, simple, cheap, and time- and labor-effective. The accuracy of the technique is about 2-3% for a given partition coefficient K. The accuracy is determined by maintenance of a given fixed polymer and salt composition of the two-phase system, by the procedure employed to determine the K-value, and by the analytical method used to assay the concentrations of the sample being partitioned. Simple additional procedures are needed to increase the accuracy. First, the partition coefficient of a given solute in the one-step procedure should be determined as a slope of the linear function for several separate partition experiments with varied total concentration of the solute in the system of a given composition. Second, the material balance for the solute should be checked in each partition experiment.

The serious limitation of the partition technique is the generality of information obtained about the sample under examination. The difference in the partition coefficient values for two different samples of, e.g., glycoprotein, indicates only that the interactions of the two samples with water are different, but does not provide any specific information about detailed differences between their structures. It should be noted, however, that other analytical techniques are similarly limited (e.g., electrophoretic and chromatographic methods) though it does not decrease their utility. It is also possible to "calibrate" the likely differences in the partition coefficient values for a given biopolymer by studying partitioning of chemically and/or enzymatically modified structures.

There are many analytical applications for the technique which have been explored partially, or not at all. These include, for example, (i) study of hydrophobicity of biopharmaceuticals for exploring their corresponding quantitative structure-activity relationships (QSARs) and targeted modifications of the molecules; (ii) study of hydrophobicity of drugs and their presumed targets to explain the known structure-activity relationships; (iii) clinical biochemistry for diagnosis and clinical treatment monitoring; (iv) recombinant protein production monitoring; (v) analysis of chemical modifications of proteins and other biopolymers (e.g., PEG-conjugation and biochemical engineering); and (vi) simulation of transport of biological solutes and organization of biological systems.

The analytical potential of the technique as applied for characterization of surface properties of cells and cell organelles seems to be as great as for soluble materials. In these and other fields the technique may clearly provide similarly unique information. The main use of the technique of partitioning in aqueous polymer two-phase systems in biotechnology at present is in the downstream processing, large-scale recovery and purification of fermentation products. Applications of the technique for analytical separations are considered in the next chapter. It can be concluded, however, that the partition technique also provides great analytical potential and is a very promising analytical method for biotechnology, pharmacology, biochemistry and medicine.

REFERENCES:

1. P. Å. Albertsson, Partition of Cell Particles and Macromolecules, 3rd. ed., Wiley, New York, 1986.
2. Partitioning in Aqueous Two-Phase Systems: Theory, Methods, Uses, and Applications to Biotechnology (H. Walter, D. E. Brooks, D. Fisher, eds.), Academic Press, Orlando, Fla, 1985.
3. Separations Using Aqueous Phase Systems: Applications in Cell Biology and Biotechnology (D. Fisher and I. A. Sutherland, eds.), Plenum Press, New York, 1989.
4. W. Muller, Liquid-Liquid Partition Chromatography of Biopolymers, GIT Verlag, Darmstadt, 1988.
5. H. Walter, G. Johansson, D. E. Brooks, Anal.Biochem., 197, 1 (1991).
6. W. D. Conway, Countercurrent Chromatography: Apparatus, Theory, and Applications, VCH Publishers, New York, 1990.
7. A. Foucault, K. Nakanishi, J.Liquid Chromatogr., 13, 2421 (1990).
8. M. R. Kula, Bioseparation, 1, 181 (1990).
9. S. L. Jeffcoate, Biologicals, 19, 139 (1991).
10. A. F. Bristow, S. L. Jeffcoate, Biologicals, 20, 221 (1992).
11. M. W. Spellman, Anal.Chem., 62, 1714 (1990).
12. T. Feizi, R. A. Childs, Biochem.J., 245, 1 (1987).
13. Y. Kagawa, S. Takasaki, J. Utsumi, K. Honsoi, H. Shimizu, N. Kochibe, A. Kobata, J.Biol.Chem., 263, 17508 (1988).
14. T. W. Rademacher, R. B. Parekh, R. A. Dwek, Annu.Rev. Biochem., 57, 785 (1988).
15. M. Roman, P.R. Brown, J.Chromatogr., 592, 3 (1992).
16. M. Kunitani, G. Dollinger, D. Johnson, L. Kresin, J.Chromatogr., 588, 125 (1991).

17. C. Delgado, F. Malik, B. Selisko, D. Fisher, G. E. Francis, Abstracts of papers, 207th ACS National Meeting, San Diego 1994, Part 1, Abstract # 88 (I&EC), American Chemical Society, 1994.
18. A. D. Diamond, X. Lei, J. T. Hsu, Biotechnol. Techniques, 3, 271 (1989).
19. B. Y. Zaslavsky, N. M. Mestechkina, L. M. Miheeva, S. V. Rogozhin, G. Y. Bakalkin, G. G. Rjazhsky, E. V. Chetverina, A. A. Asmuko, J. D. Bespalova, N. V. Korobov, O. N. Chichenkov, Biochem. Pharmacol., 31, 3757 (1982).
20. L. G. Shchyukina, B. Y. Zaslavsky, S. V. Rogozhin, V. L. Florentiev, Mol.Biologia (Rus.), 18, 1128 (1984).
21. P. A. Alred, F. Tjerneld, R. F. Modlin, J.Chromatogr., 628, 205 (1993).
22. F. D. Raymond, D. W. Moss, D. Fisher, Biochim.Biophys. Acta, 1156, 117 (1993).
23. B. Desbuquois, G. D. Aurbach, Biochem.J., 143, 83 (1974).
24. B. Y. Zaslavsky, N. M. Mestechkina, S. V. Rogozhin, Biochim. Biophys. Acta, 579, 463 (1979).
25. R. Mothes, B. Y. Zaslavsky, N. M. Mestechkina, L. M. Miheeva, S. V. Rogozhin, K. D. Schwenke, Nahrung, 30, 1043 (1986).
26. N. D. Gulaeva, M. A. Chlenov, B. Y. Zaslavsky, 1989, unpublished data.
27. N. G. Blokhina, D. F. Schirin, V. A. Hailenko, B. Y. Zaslavsky, N. M. Mestechkina, S. V. Rogozhin, Vestnik Acad.Med.Nauk SSSR (Rus), 1984, 35 (1984).
28. M. M. Bradford, Anal.Biochem., 72, 248 (1976).
29. S. Udenfriend, S. Stein, P. Bohlen, W. Dairman, Science, 178, 871 (1972).
30. F. W. Putnam, In: The Proteins (H. Neurath, ed.), Vol.3, New York, Academic Press, 1965, pp.153-267.
31. N. M. Mestechkina, Ph.D. Thesis, Institute of Elementoorganic Compounds, USSR Academy of Sciences, Moscow, USSR, 1984.
32. B. Y. Zaslavsky, N. M. Mestechkina, S. V. Rogozhin, J.Chromatogr., 260, 329 (1983).
33. M. K. Schwartz, Clin.Chim.Acta, 206, 77 (1992).
34. P. Laidler, D. Kowalski, J. Silberring, Clin.Chim.Acta, 204, 69 (1991).
35. B. Y. Zaslavsky, N. D. Gulaeva, S. V. Rogozhin, A. A. Gasanov, E. A. Masimov, Mol.Cell.Biochem., 65, 125 (1984).
36. D. Irwin, I. D. Dauphinais, Anal.Biochem., 92, 193 (1979).
37. Y. C. Martin, In: Drug Design (E. J. Ariens, ed.), Vol.8, Academic Press, New York, pp.1-72 (1979).

38. T. Higuchi, S. S. Davis, J.Pharm.Sci., 59, 1376 (1970).
39. C. Tanford, Science, 200, 1012 (1978).
40. B. Y. Zaslavsky, E. A. Masimov, S. V. Rogozhin, Doklady Acad. Nauk USSR (Rus), 277, 1163 (1984).
41. P. T. Beall, Cryobiology, 20, 324 (1983).
42. J. Clegg, In: Microcompartmentalization (D. P. Jones, ed.), CRC Press, Boca Raton, 1988, pp.1-16.
43. B. Y. Zaslavsky, E. A. Masimov, A. A. Gasanov, S. V. Rogozhin, J.Chromatogr., 294, 261 (1984).
44. G. V. Iyengar, W. E. Kollmer, H. J. M. Bowen, The Elemental Composition of Human Tissues and Body Fluids, New York, Verlag Chemie, 1978.
45. J. Clausen, M. H. Bickel, J.Pharm.Sci., 82, 345 (1993).
46. A. J. Atkinson, Jr., T. I. Ruo, M. C. Frederiksen. Trends Pharm.Sci., 12, 96 (1991).
47. B. Zaslavsky, Anal.Chem., 64, 765A (1992).
48. N. E. Tayar, R. S. Tsai, B. Testa, P. A. Carrupt, C. Hansch, A. Leo, J.Pharm.Sci., 80, 744 (1991).
49. V. H. Lee, A. Yamamoto, Adv.Drug Delivery Rev., 4, 171 (1990).
50. S. Davis, Trends Pharm.Sci., 11, 353 (1990).
51. E. Rechthand, Q. R. Smith, S. I. Rapoport, Am.J.Physiol., 252, H1175 (1987).
52. V. B. Tolstoguzov, Food Hydrocolloids, 2, 339 (1988).
53. J. I. Clark, In: Methods in Enzymology, Vol.228 (H. Walter, G. Johansson, eds.), Academic Press, San Diego, 1994, pp.525-537.
54. J. S. Clegg, Amer.J.Physiol., 246, R133 (1984).
55. J. S. Clegg, In: The Organization of Cell Metabolism (G. R. Welch, J. S. Clegg, eds.), Plenum, New York, 1986, pp.41-55.
56. J. S. Clegg, M. B. Barrios, In: Cell Function and Disease (L. E. Canedo, L. E. Todd, L. Packer, J. Jaz, eds.), Plenum, New York, Plenum, 1988, pp.159-170.
57. J. S. Clegg, W. Drost Hansen, In: The Biochemistry and Molecular Biology of Fishes (P. W. Hochachka, M. T. P. Mommsen, eds.), Vol.1, Elsevier Science Publishers, Amsterdam, 1991, pp. 1-23.
58. K. R. Porter, In: The Organization of Cell Metabolism (G. R.Welch, J. S. Clegg, eds.), Plenum, New York, 1986, pp.9-26.
59. N. D. Gershon, K. R. Porter, B. L. Trus, Proc. Natl.Acad.Sci. USA, 82, 5030 (1985).
60. K. D. Garlid, In: Cell-Associated Water (W. Drost-Hansen, J. S. Clegg, eds.), Academic Press, New York, 1979, pp.293.

61. J. S. Clegg, In: Cell-Associated Water (W. Drost-Hansen, J. S. Clegg, eds.), Academic Press, New York, 1979, pp.363-413.
62. J. S. Clegg, S. Szwarnowski, V. E. R. McClean, R. J. Sheppard, E. H. Grant, Biochim.Biophys.Acta, 721, 458 (1982).
63. C. F. Hazlewood, In: Cell-Associated Water (W. Drost-Hansen, J. S. Clegg, eds.), Academic Press, New York, 1979, pp.165-260.
64. W. Drost-Hansen, In: Microstructure of Fine-Grained Sediments: From Mud to Shales (R. H. Bennett, W. R. Bryant, M. H. Hulbert, eds.), Springer Verlag, Berlin, 1990, pp.259-266.
65. E. S. Kempner, J. H. Miller, Exp.Cell Res., 51, 141 (1968).
66. E. S. Kempner, J. H. Miller, Exp.Cell Res., 51, 151 (1968).
67. J. S. Clegg, Proceedings of 6th International Conference on Partitioning in Aqueous Two-Phase Systems, Assmannshausen, 1989, p. 30.
68. B. Y. Zaslavsky, E. A. Masimov, L. M. Miheeva, S. V. Rogozhin, D. P. Hasaev, Doklady Acad.Nauk USSR (Rus), 261, 669 (1981).
69. E. A. Masimov, B. Y. Zaslavsky, A. A. Gasanov, S. V. Rogozhin, J.Chromatogr., 284, 337 (1984).
70. E. A. Masimov, B. Y. Zaslavsky, A. A. Gasanov, Y. A. Davidovich, S. V. Rogozhin, J.Chromatogr., 284, 349 (1984).
71. B. Y. Zaslavsky, L. M. Miheeva, N. D. Gulaeva, A. A. Borovskaya, M. I. Rubtsov, L. L. Lukatskaya, N. O. Mchedlov-Petrossyan, J.Chem.Soc. Faraday Trans., 87, 931 (1991).
72. V. M. Schlimak (Central Institute of Hematology and Blood Transfusion, Moscow, Russia), 1984, personal communication.
73. E. J. Lien, SAR: Side Effects and Drug Design, Marcel Dekker, New York, 1987.
74. Y. Kirsch (Institute of Technology of Hormones and Plasma Substitutes, Moscow, Russia), 1988, personal communication.

CHAPTER 9.
SEPARATION OF BIOMOLECULES

Mechanisms of phase separation and solute partitioning in aqueous polymer systems have been discussed in previous parts of the book with regard to the information about a solute provided by the solute partition behavior. The purpose of this chapter is to consider how the concepts developed above may be applied to the use of aqueous two-phase partition technique as a separation tool.

Partitioning in aqueous two-phase systems is well known as a highly efficient technique for separation and purification of various biological materials, ranging from proteins and nucleic acids to cells and viruses. Recently the technique has gained attention as a method of separation of inorganic materials as well (see below). Several books [1-5] and numerous reviews and book chapters describing these applications are available in the literature. Analysis of the original publications in the field implies that, with a certain amount of luck, an experienced investigator may separate essentially anything with this method.

A certain amount of luck is important because a particular separation procedure using aqueous two-phase partitioning is usually developed empirically by trial-and-error iteration. This approach is often a long and arduous task, and one that does not always lead to optimum results. The highly empirical character of the partition technique originates from our inadequate understanding of the basic principles and mechanisms of the partition process, and physicochemical features of biomolecules governing their partition behavior. The concepts developed above at least allow one to design certain ground rules to improve the situation. Implementation of these rules may also help to accumulate the information most important for further development of basic "rules of thumb" similar to those used, for example, in the use of HPLC. It should also be mentioned that development of general ground rules unavoidably includes a certain degree of generalization of experimental data. An attempt to generalize highly diverse experimental information is rarely successful, even for relatively simple organic compounds. For complex biological solutes such as proteins and nucleic acids, essentially any generalization turns out to be wrong at one point or another. Hence, all of the generalizations noted below should be viewed as hypothetical.

It is important to emphasize that only analytical separations are discussed below. Applications of the method on the industrial scale are described in excellent reviews by Kula et al.[6], Hustedt et al.[7,8] and others [9,10]. Furthermore, the analytical separations to be discussed refer to partition procedures aimed at separation of biomolecules performed with the amounts of soluble material not exceeding 0.1 %wt. of the total weight of the two-phase system used. It was shown (see, for example, in [11]) that when the amount of the material being partitioned exceeds this value, the material may affect the properties of the system. The likely explanation is that the material being separated may play a direct role as a phase-forming constituent of the system, resulting in unpredictable changes in partition behavior. Sometimes, however, the separation conditions developed for analytical procedures can be scaled up without significant changes in the procedure employed. Some of these examples will be considered.

Separation of two or more solutes by the partition technique obviously depends on the differences in their partition behavior: the more the difference, the better separation. There are essentially two possibilities to increase the difference in the solute partition behavior: (a) changing the properties of the phases; and (b) changing the properties of the solutes being separated. One may also improve the separation by increasing the efficiency of the procedure by using, for example, the liquid chromatography mode instead of a single extraction procedure.

In order to discuss the strategy for optimal design of separation conditions, it is necessary to first consider what technical procedures may be used, and how different the partition behavior of two solutes must be to achieve good separation.

9.1. SEPARATION PROCEDURES

Three procedures are generally used for separation of soluble materials in aqueous two-phase systems. These procedures are similar to those used with any solvent two-phase system. They include (1) single-step or multi-step extraction; (2) liquid-liquid partition column chromatography; and (3) countercurrent chromatography.

Extraction.

The technically easiest procedure is the single-step extraction. An aqueous two-phase system of a certain composition is prepared and the mixture to be separated is added to the system. After vigorous mixing the system is allowed to settle or is centrifuged to speed phase settling. The phases are separated and analyzed or used for recovery of the separated components of the initial mixture. This procedure may be highly efficient if the target product concentrates in one phase and the materials it is separated from concentrate in the other phase. In terms of the partition coefficient, K, the partition coefficient of a target product (K_{target}) must be, for example >> 1, while those for the other components of the mixture (K_i) are << 1 (i denotes i-th component of the mixture being separated). An equally "good" case is when K_{target} << 1 and K_i >> 1.

As surprising as it may seem, there are quite a few examples when this situation is realized in practice. One spectacular example is offered by the process described by Lowlis and Heinsohn [12] for the recovery of calf chymosin, also known as rennin, directly from the fermentation medium. It was found [12] that the partition coefficient of rennin is high (about 100) in the aqueous PEG—salt two-phase system (no particular details of the system used were reported). As a result, highly purified enzyme can be extracted directly from acidified fermentation broth simply by addition of polyethylene glycol [12].

Essentially similar procedures were successfully used by Hart et al.[13] to isolate recombinant insulin-like growth factor I (IGP-I) expressed with a signal sequence in *E.coli*. The product was found to be about 97% pure as the result of direct extraction from the fermentation broth in the aqueous PEG—salt two-phase system [13].

The high sensitivity of the aqueous two-phase partition technique to small differences in protein structure may be illustrated, for example, by different partitioning of β-lactoglobulins A and B in the aqueous PEG-3400—potassium phosphate two-phase system [14]. β-lactoglobulin A has aspartic acid and valine, while β-lactoglobulin B has glycine and alanine at residues positions 64 and 118, respectively. The proteins have an isoelectric point difference of only 0.1 pH units. The partition coefficients for β-lactoglobulins A and B in the aforementioned system amount to 0.08 and 0.04, respectively [14]. That means that the separation factor α, defined as the ratio of the partition coefficients for the two solutes, amounts to 2.0 for two proteins that differ by only two amino acid residues. A separation factor of about 1.25 is usually considered sufficient for base-line resolution of two solutes in liquid chromatography (see, for example, in [15]). Partition coefficients of horse and pig insulins in the same system were also found to be different: 21.2 and 19.4, respectively [14]. The separation factor for these two proteins, which differ only at one residue (position 9) amounts to 1.09 and possibly may be increased under appropriate conditions.

Complete separation of particular biological solutes is not always necessary for analytical purposes. Numerous analytical procedures used in clinical chemistry, toxicological and forensic analysis, studies of metabolism or pharmacological research are often based on monitoring the concentration or physicochemical features of a given component of a multicomponent biological sample. Technically simple, highly sensitive and selective, and readily automated extraction procedures may be used to advantage in many of these cases.

Extraction in an aqueous two-phase system may also be used to concentrate a solute of interest, to separate it from some other components interfering with the solute measurements, or to detect changes in the structure and/or physicochemical features of the solute related to its biological function(s) and/or potency.

Development of an efficient, sensitive and robust analytical procedure based on the aqueous two-phase partition technique in every particular case depends on our ability to manipulate the partition behavior of the analyte of interest and those of the other components of the biological sample being examined.

As emphasized in a previous part of the book (Chapters 4 and 5), partition behavior of a solute in an aqueous two-phase system depends on many factors. Hence, there are many possible ways to steer partitioning of different solutes to achieve their separation. All the currently known factors influencing partition behavior of a solute may be divided into two categories - factors affecting the solvent features of the two phases, and factors affecting the properties of a solute important for the solute interactions with the aqueous media in the phases. These factors are listed in Table 9.1. Some of the factors belong to both categories since they affect the solute and the system. That complicates predictibility of the influence of these factors.

Among the factors listed in Table 9.1 under category of those affecting the solvent features of the phases, the most commonly used in current practice are concentration and molecular weight of phase-forming polymers, and type and concentration of additives (usually inorganic salts). These factors are generally viewed as the most important to manipulate partitioning of biomolecules to achieve better separation.

Table 9.1
Factors Capable of Steering Solute Partitioning in Aqueous Two-Phase Systems.

Factors affecting solvent features of the two phases	Factors affecting properties of a solute
Type of phase-forming polymers [a]	pH
Molecular weight of phase-forming polymers [a]	Type of additive [b]
Concentrations of phase-forming polymers [a]	Concentration of additive [b]
Type of additive [b]	Presence of complex-forming additives[c]
Concentration of additive [b]	Structural modification [d]
Temperature	

[a] in aqueous single polymer-salt systems type and concentration of phase-forming salt is the factor equal to those of phase-forming polymer in two-polymer systems;
[b] additive of low molecular weight, such as inorganic salts, saccharides, urea, etc., with no specific affinity for the solute;
[c] affinity ligands, such as drugs, triazine dyes, organic complexons, fatty acids, etc.;
[d] modification by chemical, enzymatic, etc. treatment resulting in elimination, incorporation, or alteration of topography of solvent-accessible moieties in the solute molecule.

Table 9.2 illustrates effects of the phase-forming polymers molecular weights on partitioning of proteins in aqueous two-polymer and single polymer-salt two-phase systems [16-18]. The effects in question appear to increase in the range of relatively low molecular weights of PEG when it not only increases or decreases the solute affinity for a given phase, but seems to determine the direction of the solute partitioning, i.e. whether the solute will partition into one or the other phase. The possible reasons will be discussed below.

Concentrations of phase-forming polymers (in two—polymer systems) or phase-forming polymers and salts (in single polymer—salt system) will influence solute partitioning, as shown in Figure 9.1. Increasing polymer

Table 9.2

Partition Coefficients, K, of Proteins in Aqueous Two-Phase Systems Formed by Polymers of Different Molecular Weights.

PEG		Polymer 2 or Salt		ΔPEG	K_{Tran} [a]	K_{Lys} [a]	K_{Gal} [a]	Ref.
Mol.wt.	%wt.	Mol.wt.	%wt.	%wt.				
		Dextran						
4,000	8.74	10,000	13.23	12.7	0.014 [b]	1.70 [c]	-	[16]
20,000	7.63	10,000	10.29	12.4	0.029 [b]	1.85 [c]	-	-"-
4,000	7.87	500,000	11.50	12.4	0.038 [b]	1.12 [c]	-	-"-
20,000	7.65	500,000	10.29	12.4	0.018 [b]	1.58 [c]	-	-"-
		$(NH_4)_2SO_4$						
300	21.0		19.5		> 50 [d]	-	-	[17]
400	19.0		17.0		2.80 [d]	-	-	-"-
600	17.5		15.5		0.196 [d]	-	-	-"-
1,000	15.0		13.0		0.027 [d]	-	-	-"-
6,000	10.8		9.3		<<0.01 [d]	-	-	-"-
		Aquaphase PPT [a]						
3,600	7.0	35,000	19.0		-	-	45.0 [e]	[18]
20,000	7.0	35,000	19.0		-	-	0.26 [e]	-"-

[a] K_{Tran} - partition coefficient of human transferrin; K_{Lys} partition coefficient of lysozyme; K_{Gal} partition coefficient of galactosidase from an extract of *E.colil*; Aquaphase PPT - hydroxypropyl starch, M_w 35,000; [b] pH 5.2; [c] pH 5.6; [d] pH 5.0; [e] pH 7.0.

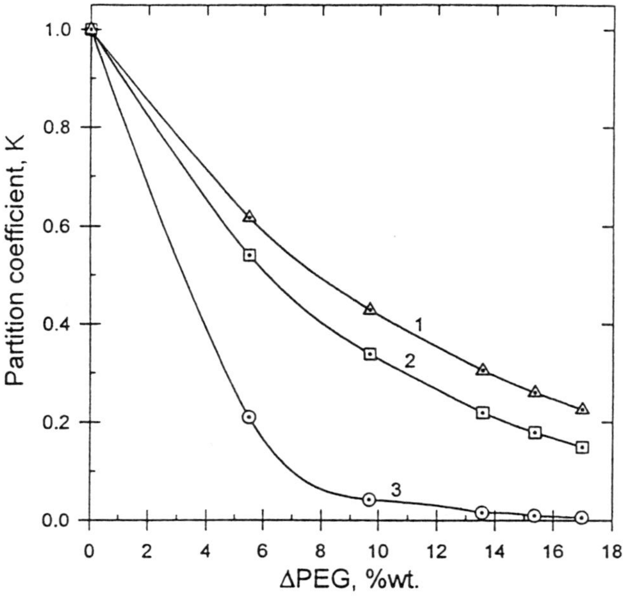

Figure 9.1. Influence of concentrations of phase-forming polymers expressed as the difference between PEG concentration in the two phases, ΔPEG, on partitioning of horse myoglobin (1), cytochrome c (2), and human serum albumin (3) in the aqueous Dex-70—PEG-6000 two-phase system containing 0.15 mole/kg NaCl in 0.01 mole/kg sodium phosphate buffer, pH 7.4.

(and/or salt) concentrations generally increases the differences in the partition behavior of two solutes. The effect in question seems to be the most readily predictable. Increasing the concentrations of phase polymers (and/or salt), however, also increases the viscosities of the phases. This effect also complicates handling the systems and aggravates the solute mass transfer (see below). Therefore, increasing concentrations of phase polymers is not always the best way to optimize separation.

Temperature may influence protein partitioning as illustrated by the data by Forciniti et al.[19] partially shown in Table 9.3. Studies of the temperature effects [1,19] are still too limited to allow any generalization. However, it follows from theoretical considerations (see below) that differences in solute partitioning (i.e., separation) should increase with decreasing temperature; however, it must be highly solute-specific and also depends on the phase-forming polymers used. No general conclusion, however hypothetical, may be drawn and the temperature effects are hardly likely to become predictible in the near future.

Table 9.3
Partition Coefficients of Lysozyme and Catalase in Four Different Aqueous Dex—PEG Two-Phase System, pH 9.2 at Different Temperatures.

Temperature, °C	$K_{Lysozyme}$	$K_{Catalase}$	Separation factor α
4.0 [a]	0.54	0.046	11.7
25.0 [a]	0.44	0.063	7.0
40.0 [a]	0.85	0.174	4.9
4.0 [b]	0.35	0.014	25
25.0 [b]	0.31	0.016	19.4
40.0 [b]	0.53	0.020	26.5
4.0 [c]	0.65	0.22	3.0
25.0 [c]	0.47	0.21	2.2
40.0 [c]	0.98	0.43	2.3
4.0 [d]	0.58	0.052	11.2
25.0 [d]	0.36	0.042	8.6
40.0 [d]	0.72	0.068	10.6

[a] System composition: 12.2% wt. Dex-10, 8.4% wt. PEG-4000;
[b] System composition: 10.0% wt. Dex-10, 5.6% wt. PEG-20,000;
[c] System composition: 11.3-11.5% wt. Dex-500, 7.9% wt. PEG-4000;
[d] System composition: 10.3% wt. Dex-500, 7.65% wt. PEG-20,000.
(Data from D. Forciniti, C. K. Hall, M. R. Kula, Bioseparation, 2, 115 (1991). Reprinted by permission of Kluwer Academic Publishers.)

Changing pH seems to be an obvious way to manipulate partitioning of ionic solutes in a readily predictible manner. That is why manipulations of the pH are usually viewed to be of primary importance in partitioning studies of proteins [20]. If the pH change would affect merely the net charge on the solute molecule, the pH effect might be predicted, for example, as described by Equation 5.5 (see above):

$$\ln K = \ln K(o) + \gamma \cdot Z \tag{5.5}$$

where K(o) is the partition coefficient of the solute at the pH value corresponding to the solute isoelectric point; Z is the net charge of the protein; and γ is the protein-specific factor dependent on all external parameters, such as polymer and salt composition of the two-phase system, temperature, etc.

Even taking into account an extremely high specificity of the factor γ, Equation 5.5 predicts that the logarithm of the protein partition coefficient should be a monotonic function of the protein net charge. That feature could be useful, unfortunately, it is not generally true (see Chapter 5). That is so because the net charge of a macromolecule is not the feature of primary importance for its partition behavior. The balance of all the interactions of charged and uncharged groups with the aqueous media in the two phases is the major factor. This balance certainly changes with variation in the ratio of the charged or uncharged groups. Hydration interactions of the other groups located in the vicinity of the ionizable ones may also be affected in the process, and the total outcome is hardly predictible at the current level of our knowledge of solute-solvent interactions. Additionally, a pH shift may induce structural changes in macromolecules, such as aggregation, dissociation into subunits, conformational changes, etc. Hence, the pH effect is much easier to study experimentally than to predict theoretically. Finally, because changes of pH are usually produced by changing the type of buffer used, the salt composition of the two-phase system is also altered. The latter complicates interpretation of results, and hinders the possibility to extrapolate the observed trend out of the examined pH range.

Partition coefficients of bovine serum albumin and immunoglobulin G in aqueous Dex—PEG and Aquaphase PPT (hydroxypropyl starch)—PEG at different salt composition and pH are shown in Table 9.4, illustrating typical pH effects. These data indicate that pH can be useful to steer the protein partitioning as needed for better separation. For example, the best separation conditions for bovine serum albumin and immunoglobulin G are achieved in the aqueous Aquaphase—PEG two-phase system containing 0.01 M sodium phosphate buffer, pH 8.0. The type of the buffer used should always be taken into consideration, however, as it may influence the solute partitioning to a degree similar or even exceeding that for the pH effect. Compare partition coefficients values (Table 9.4) observed for the proteins at the same pH of 8.0 in sodium phosphate buffer and in Tris-HCl buffer. Clearly, the type and concentration of salt additives strongly affects the biomolecules partitioning in aqueous two-phase systems.

Table 9.4

Partition Coefficients of Bovine Serum Albumin (BSA) and Immunoglobulin G (IgG).

pH	Buffer	Partition coefficient K *		
		Dex—PEG **	Aquaphase—PEG ***	
		BSA	BSA	IgG
5.0	Na-acetate	0.085	0.34	0.005
6.0	Na-phosphate	0.18	0.56	0.02
7.0	Na-phosphate	0.41	1.50	0.11
8.0	Na-phosphate	0.73	2.14	0.12
7.0	Na-phosphate + 0.1 M NaCl	0.06	0.21	0.24
7.2	Tris-HCl	0.08	1.10	<0.005
8.0	Tris-HCl	0.10	1.47	<0.005

* Partition coefficient is defined as the ratio of the concentration of a protein in the PEG-rich phase to the protein concentration in the Dex-rich (or Aquaphase-rich) phase; protein concentration in the system 5 g/l; temperature 23°C;

** System composition: 8% wt. Dex-500; 6% wt. PEG-8000; 0.01 M buffer;

*** System composition: 14% wt. Aquaphase PPT; 5% wt. PEG-8000; 0.01 M buffer (Aquaphase PPT - hydroxypropyl starch, M_w 35,000).

(Data from S. Sturesson, F. Tjerneld, G. Johansson, Appl.Biochem.Biotechnol., 26, 281 (1990). Reprinted by permission of The Humana Press, Inc.)

Salt composition (type and concentration of salt additive) is the factor probably most commonly used additionally to the pH-value to manipulate partitioning of biological solutes. However, the influence of this factor on partitioning of proteins or nucleic acids is also not as predictible as it may seem. The reason is that salt additives may influence both solvent features of

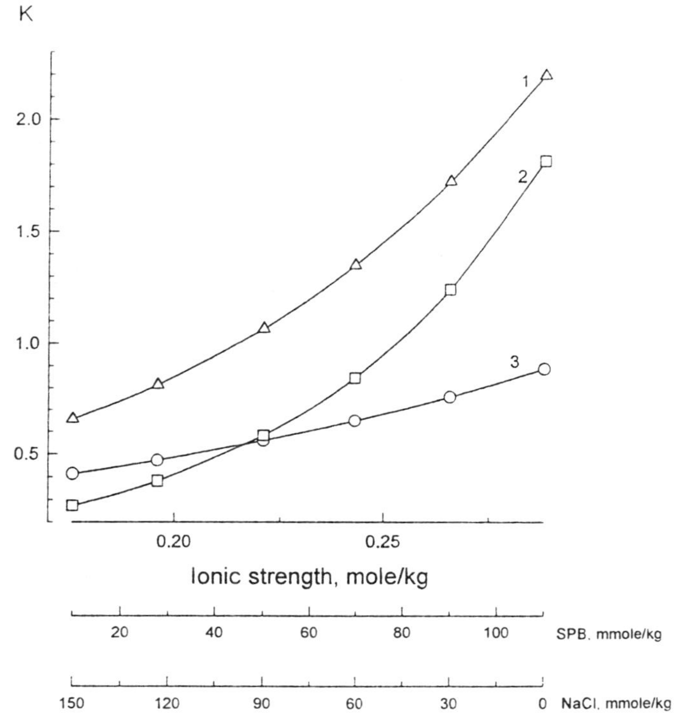

Figure 9.2. Partitioning of β-1,4-glucomannans in the aqueous Dex—Ficoll two-phase system containing varied amounts of NaCl and sodium phosphate buffer, pH 7.4 [21]: 1 - β-1,4-glucomannan from *E. hissaricus* (M_w 360,000; mannose/glucose ratio of 1.5; 2 - β-1,4-glucomannan from *E. fuscus* (M_w 158,000; mannose/glucose ratio of 2.6; 3 - β-1,4-glucomannan from *E. comosus* (M_w 60,000; mannose/glucose ratio of 3.2).

the phases and properties of biomolecules governing their partition behavior. Partitioning of nonionic polysaccharides in the aqueous Dex—Ficoll two-phase systems of varied salt composition [21] shown in Figure 9.2 illustrates this point. It is hardly predictible that the best separation conditions (among those tested [21] would be provided by the salt composition of 0.15 mole/kg NaCl in 0.01 mole/kg sodium phosphate buffer, pH 7.4 for β-1,4-glucomannans from *E. hissaricus* and *E. fuscus*, and 0.11 mole/kg sodium phosphate buffer, pH 7.4 for β-1,4-glucomannans from *E. hissaricus* and *E. comosus*. There are some general trends, however; as indicated in Chapter 5, these trends come down to the opposite effects of water-structure-making and water-structure-breaking salts on the solute partitioning.

The affinity of bovine serum albumin for the PEG-rich phase in the aqueous Aquaphase—PEG system in sodium phosphate buffer, pH 7.0 decreases significantly upon addition of 0.1 M NaCl (partition coefficient K changes from

Table 9.5

Salt Effects on Protein Partitioning in Aqueous Two-Phase Systems.

Salt (mole/kg)	Partition coefficient, K			
	BSA	Hemoglobin	Cytochrome C	Transferrin
8.5% Dex-70 — 5.3 % PEG-6000 — 0.01 mole/kg universal buffer, pH 5.0				
-	0.179	0.242	0.440	
NaCl (0.1)	0.194	0.477	0.477	
NaCl (0.5)	0.0.085	0.399	0.577	
NaSCN (0.1)	0.094	0.533	0.803	
NaSCN (0.5)	0.141	0.387	0.756	
Na_2SO_4 (0.05)	0.210	0.177	0.340	
Na_2SO_4 (0.25)	0.034	0.058	0.183	
PEG — $(NH_4)_2SO_4$ two-phase system [a]				
	PEG Molecular weight		pH	
-	400		6.0	3.22
NaSCN (0.1)	-"-		-"-	0.186
-	600		-"-	0.196
NaSCN (0.05)	-"-		-"-	0.031

[a] From L. M. Miheeva, J. Fitzpatrick, 1992, unpublished data. By permission of the authors.

1.50 to 0.21), see Table 9.4. Other salts may provide even more significant effects (Table 9.5).

Partition coefficients of human transferrin in aqueous PEG-$(NH_4)_2SO_4$ two-phase systems formed by PEGs of different molecular weights were studied [17] as functions of pH and addition of NaSCN. The results [17] partially shown in Table 9.5 indicate that NaSCN affects the protein partitioning significantly, changing the protein partition coefficient from 1.87 to 0.138 at pH 8.2 and from 3.22 to 0.186 at pH 6.0 in the systems formed by PEG of the low molecular weights of 600 and 400, respectively. Similar additions of NaSCN in the above systems formed by PEGs with molecular weights exceeding 1000 do not affect partitioning of transferrin.

Salt additives affect partitioning of biomacromolecules in aqueous two—polymer systems as well as in the PEG—salt systems. According to the data reported by Kuboi et al.[22], addition of 0.1mole/kg Na_2SO_4 increases the partition coefficient of β-galactosidase in the aqueous Dex-100000—PEG-4000 two-phase system containing 0.005 M Tris-HCl buffer, pH 7.6 from about 0.03 to about 80, while addition of 0.1 mole/kg NaSCN reduced the partition coefficient of the protein to about 0.0002. Effects of these salts on the solvent features of aqueous media in the two phases of Dex–PEG systems were discussed above. These effects while clearly different are hardly likely to provide the changes in the protein partition behavior as drastic as those observed for β-galactosidase [22]. Therefore it is reasonable to suggest that the effects in question are mostly due to specific changes in the protein conformation induced by the protein-salt interactions.

The most dramatically salt-dependent partition behavior in aqueous two-phase systems is usually observed for nucleic acids. Typical effects of salt composition on the partition behavior of nucleic acids are illustrated by the data [1,18] shown in Table 9.6.

Small changes in the salt composition hardly change the polymer composition or solvent features of the aqueous media in the two phases. Partitioning of proteins is generally invariable under such minor changes. Therefore, the dramatic changes in the partition behavior of nucleic acids are likely to be caused by salt-induced modifications of the nucleic acid-aqueous media interactions because of altered properties of functional groups and/or conformation of the biopolymer. The aforementioned unusually large effects of salt composition (and ionic strength) of the medium on the partition behavior of mononucleotides and dinucleosidephosphates (see Table 7.4) imply that the altered hydration interactions of the functional groups are the main cause. Given the significant differences in the partition coefficients of polynucleotides shown in Table 9.6, it is expected that the aqueous two-phase partition technique is highly efficient in separation of nucleic acids of different composition (see, for example, in [1,4]). It may be concluded that salt additives affect the partition behavior of nucleic acids mostly due to their influence on the properties of

Table 9.6
Partition of Nucleic Acids in Aqueous Polymer Two-Phase Systems of Different Salt Composition.

Nucleic acid	Polymer system and Salt composition	Partition coefficient K
	Dex-500—PEG-6000 [a]	
Poly A	5 mM Na_2HPO_4 + 5 mM Na_3PO_4	114 [b]
Poly A	7 mM Na_2HPO_4 + 3mM Na_3PO_4	0.08 [b]
Poly A	30 mM Na_2HPO_4	< 0.02 [b]
Poly H	5 mM Na_2HPO_4 + 5 mM Na_3PO_4	0.115 [b]
Poly U	5 mM Na_2HPO_4 + 5 mM Na_3PO_4	28 [b]
Poly I	30 mM Na_2HPO_4	0.1 [b]
Poly A + Poly I (1:1)	30 mM Na_2HPO_4	1.0 [b]
Poly A + Poly I (1:2)	30 mM Na_2HPO_4	18.3 [b]
	Aquaphase PPT—PEG-8000 [c]	
DNA [d]	0.01 M Na-phosphate buffer, pH 7.0	250
DNA [d]	0.01 M Na-phosphate buffer, pH 7.0 + 0.1 M NaCl	17
DNA [d]	0.01 M Li-phosphate buffer, pH 7.0	100
DNA [d]	0.01 M Li-phosphate buffer, pH 7.0 + 0.1 M NaCl	73
RNA [e]	0.01 M Na-phosphate buffer, pH 7.0	2.8
RNA [e]	0.01 M Na-phosphate buffer, pH 7.0 + 0.1 M NaCl	0.82
RNA [e]	0.01 M Na-acetate buffer, pH 5.0	2.5
RNA [e]	0.01 M Na-phosphate buffer, pH 5.0 + 0.1 M NaCl	0.73

[a] System composition: 5% wt. Dex-500; 4% wt. PEG-6000; 4°C [1];
[b] approximate K values determined from graphic representation in [1];
[c] System composition: 13% wt. Aquaphase PPT (M_w 35,000); 5% wt. PEG-8000; 23°C [18];
[d] Deoxyribonucleic acid from calf thymus [18];
[e] Ribonucleic acid from bovine liver [18].

biopolymers rather than on the properties of the phases. The role of counterions in the partitioning of ionic solutes is likely to be very important (see below) although still poorly understood.

An obvious question is why salt composition does not affect the partition behavior of proteins as dramatically as it does that of nucleic acids. The likely reason seems to be related to the specific features of phosphate group. As mentioned above, two molecular forms of alkaline phosphatase differing in structure by a single phosphate group were reported by Raymond et al.[23] to partition in the opposite phases of an aqueous Dex-500—PEG-6000 system containing 0.15 M NaCl in 0.01 M sodium phosphate buffer, pH 7.5. Charges on phosphate groups are unlikely to be the most important factor as implied by the data presented in Table 9.7.

Partition coefficients of β-galactosidase fused to the carboxy terminus with polyaspartic acid peptides of the general structure -Gly-Asp-Pro-Met-Ala-$(Asp)_n$-Tyr with n varied from 4 to 15 [24] are shown in Table 9.7. The data obtained by Luther and Glatz [24] indicate that while the partition behavior of the modified proteins is different from that of the original protein, the changes observed are much less spectacular than those reported by Raymond et al.[23]. It should be mentioned that if the above assumption in regard to the specific features of phosphate groups is correct, partitioning of phosphorylated proteins should be very different from the dephosphorylated ones. No investigation of this issue has been performed as yet, to my knowledge, but is certainly worth doing.

Development of recombinant DNA technology allows one to produce proteins with additional residues genetically fused to the original protein. This type of "genetic" modification of proteins may be used to great advantage for protein separation. β-Galactosidase was found [25,26] to display strong affinity for the PEG-rich phase in aqueous PEG—salt and Dex—PEG two-phase systems. It has been suggested [25] that the high content of hydrophobic tryptophan residues in the protein is the reason for the high affinity of β–galactosidase for the relatively hydrophobic PEG-rich phase. The effect of insertion of a tetrapeptide sequence, -Ala-Trp-Trp-Pro-, in the protein on its partitioning in the aqueous PEG-4000—potassium phosphate two-phase system demonstrated by Kohler et al.[28] supports this suggestion. The results reported by Eiteman et al.[30] indicate, particularly, that the partition coefficient of a staphylococcal protein A derivative may be increased from about 0.2 to about 5.3 as the result of the insertion of octapeptide fragment into the protein. This approach of "genetic" structural modifications of biomolecules seems to be very promising, though rather expensive for purely theoretical work. It should be mentioned also that the effect of the fused "handle" on the protein partitioning may be altered by the conformational changes induced by the "handle" in the protein molecule.

Table 9.7

Partition Coefficients of β-Galactosidase Forms Modified by Polyaspartic Tail Fusion. *

Number of additional Asp residues	Partition coefficient K**
-	20.0
5	30.0
11	15.8
16	46.7

* polyaspartic acid peptides of the general structure -Gly-Asp-Pro-Met-Ala-$(Asp)_n$-Tyr with n varied from 4 to 15 were fused to the carboxy terminus of the protein;
** system composition: 8.0% wt. Dex-40; 8.5% wt. PEG-3350; 0.01 M potassium phosphate buffer, pH 7.5; 0.1 mM $MgCl_2$.
(Determined from the data presented in graphic form in J. R. Luther, C. E. Glatz, Biotechnol.Bioeng., 44, 147 (1994).)

It should be noted also that the "fused" handles of relatively small size may affect partitioning of large proteins rather dramatically. That confirms the aforementioned conclusion that the size or molecular weight of a solute has essentially no effect on the solute partitioning. This conclusion follwos even more unambigously from the studies of partitioning of inorganic salts (for review see [31]). The logarithms of the partition coefficients of inorganic salts of cations of first and second groups (Na^+, Rb^+, and Cs^+, and Ca^{2+}, Sr^{2+}, and Ba^{2+}, respectively) appear to be linearly related to the cation enthalpy of hydration [32]. Typical partition coefficient values for several closely related inorganic ions under different extraction conditions in aqueous PEG—salt two-phase systems are presented in Table 9.8.

Partition behavior of a given metal ion depends on the type of phase-forming salt and type of additional salt present, i.e. type of counterion [31]. As an example, the partition coefficient of Fe^{3+} in the aqueous two-phase system formed by 18.8% wt. PEG-2000 and 17.2% wt. $(NH_4)_2SO_4$ containing 1.0 M H_2SO_4 amounts to 0.2 in the presence of 2.0 M NH_4Cl, and to 60 in the presence of 0.96 M NH_4SCN. The effect of the type of anion on partitioning of cations is in line with that ions may distribute between the two phases only as electrically neutral combinations with corresponding counterions. Significant

Table 9.8

Partition Coefficients of Metal Ions * in Aqueous PEG-2000—$(NH_4)_2SO_4$ Two-Phase Systems Containing 18.8% wt. PEG-2000; 21.2% wt. $(NH_4)_2SO_4$ and Extractant in the Concentration Indicated.

Ion	Extractant (concentration)	Partition coefficient
Th^{4+}	-	0.040
UO_2^{2+}	-	0.082
Pu^{4+}	-	0.023
Am^{3+}	-	0.013
Th^{4+}	10^{-3} M Alizarin Complexone	610
UO_2^{2+}	10^{-3} M Alizarin Complexone	0.30
Pu^{4+}	10^{-3} M Alizarin Complexone	280
Am^{3+}	10^{-3} M Alizarin Complexone	0.10
Th^{4+}	10^{-3} M Xylenol Orange	510
UO_2^{2+}	10^{-3} M Xylenol Orange	0.14
Pu^{4+}	10^{-3} M Xylenol Orange	28
Am^{3+}	10^{-3} M Xylenol Orange	0.02
Th^{4+}	$5*10^{-4}$ Arsenazo III	870
UO_2^{2+}	$5*10^{-4}$ Arsenazo III	250
Pu^{4+}	$5*10^{-4}$ Arsenazo III	440
Am^{3+}	$5*10^{-4}$ Arsenazo III	0.38

* As indicated above, ions distribute between the two phases only as electrically neutral combination with corresponding counterions, probably, as sulfates.

(From R. D. Rogers, A. H. Bond, C. B. Bauer, Sep.Sci.Technol., 28, 1091 (1993). Reprinted by permission of Marcel Dekker, Inc.)

pH-dependence of the partition behavior of actinide and lanthanide elements may be used for separation of these elements by extraction in aqueous two-phase systems [31].

Extraction of metal ions in water—organic solvent systems is usually manipulated by chelating agents capable of forming complexes with certain ions and displaying high affinity for the organic phase. Partitioning of metal ions in aqueous two-phase systems may also be manipulated by the addition of a water-soluble extractant which coordinates the metal ion. The difference between extraction in water—organic solvent systems and in aqueous two-phase systems is that the extractants have to be water-soluble [31].

An important additional requirement for a given extractant to be efficient in aqueous two-phase systms is that the extractant-metal complex must be stable in both phases. This requirement is met readily enough in common solvent two-phase systems due to the different nature of solvent media in the two phases. Dissociation of the metal ion-extractant complex in a nonaqueous phase, for example, is strongly hindered by the low dielectric constant of the organic solvent. The same (aqueous) nature of both phases in aqueous two-phase systems changes this situation rather drastically. If the stability of a complex is not high enough, for example, in the upper phase, the efficiency of the extraction procedure may decrease significantly. It may be possible that the selectivity of extraction observed with certain complexons [31] (see Table 9.8) is actually caused by the differences in stability of complexes with different ions and not by different affinities of the complexes for the polymer-rich phase. No studies of this issue have been done as yet, to my knowledge.

Further work is clearly needed in order to understand specific mechanisms of salt partitioning in aqueous two-phase systems. It seems very likely, however, that these mechanisms are similar to those for biological solutes.

Partition behavior of biological solutes in aqueous two-phase systems may also be affected by complex formation. Any additive changing partitioning of a target solute due to the complex formation may be viewed as playing the role of an extractant.

Addition of ligands capable of specific binding to a target protein may affect the protein partition behavior to a very significant degree. Thus, addition of a triazine dye Blue MX-R changes the partition coefficient of lactate dehydrogenase from rabbit muscle in the aqueous PEG—potassium phosphate two-phase system (12% wt. PEG-1540; 12% wt. potassium phosphate; pH 7.5; 4°C) from 0.005 to 0.15 at the free dye concentration of 5 mM [33]. The partition coefficient of alcohol dehydrogenase from baker's yeast was reported [34] to increase from about 0.7 to about 2 in the aqueous Dex—PEG two-phase system as the result of adding 5 mM surfactant Triton X-405. Similar, though protein-specific, effects were observed for several other proteins [34]. Addition of 5 mM Triton X-405 and 0.1 mM EDTA increased the separation factor for α-lactalbumin and β-lactoglobulin in the aqueous Dex-150—PEG-4000 two-phase system from 4 to 13 [34].

Essentially similar results were reported by Guiliano [35] indicating that the partition coefficient of egg white lysozyme in the aqueous malto-dextrin—polyvinylpyrrolidone (PVP) two-phase system is dramatically changed by addition of different textile dyes serving as affinity ligands for lysozyme. The partition coefficient of the protein was increased from about 2 up to about 55 depending on the structure of the particular dye used as a protein ligand. Dye structure and concentration were found [35] to affect the protein partition coefficient as well as ionic strength and pH of the medium in the system. An attempt to explain the effect of the ligand on the protein partition behavior as originated from the protein-ligand-PVP interactions [35] is open to argument. The important conclusion, however, is very straightforward: partitioning of protein-ligand complexes may be very different from those of free protein macromolecules. Thus, judicious use of ligands may prove to be of great analytical value.

In spite of tremendous research efforts, our knowledge of physico-chemical properties of biological macromolecules and the ways to modify them selectively remains very limited and highly empirical. What we do know is that biological macromolecules are generally capable of specific recognition. That means that there usually are ligands specific for a given biopolymer, and binding of these ligands may change the biopolymer properties in a useful way. Taking into account the high selectivity of specific ligands and the common practice of affinity chromatography, these ligands offer an obvious means to steer partitioning of a given biopolymer in an aqueous two-phase system.

<u>Affinity Partition.</u>

The so-called affinity partition technique is based on steering partitioning of a target biological macromolecule by selective binding of a ligand specific for this macromolecule. The efficiency of this procedure depends on the difference between the partition behavior of free target product and that of the product-ligand complex. It is generally believed that in order to shift the complex to a particular phase the ligand must be covalently coupled to the phase polymer concentrated in this phase [1,2,5,36-40]. The basic principle of the technique [41] is schematically represented in Figure 9.3.

The obvious advantage of this variant of partition technique compared to the usual procedures is the supposed predictability and selectivity of the effect of the polymer-bound affinity ligand on the partition behavior of a given biomolecule. The advantages of the affinity partition technique over liquid-solid affinity chromatographic systems are evident. The ligand binding to a target

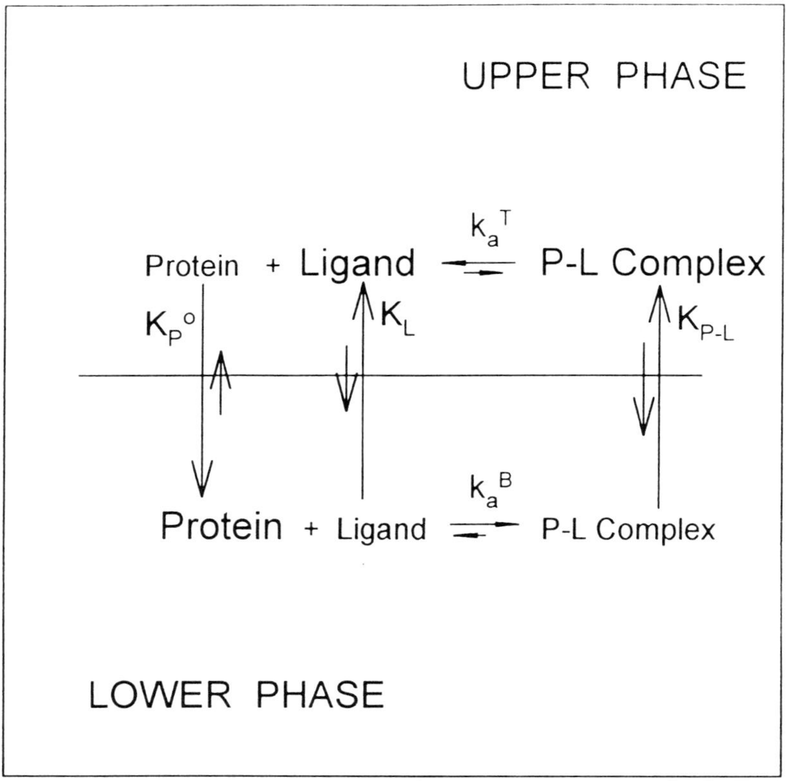

Figure 9.3. Schematic presentation of the affinity partitioning principle. Sizes of the symbols correspond to the relative concentrations of the solutes and those of the arrows indicate the directions of partitioning.

macromolecule is increased as it occurs in the homogeneous phase rather than only at the solid-liquid interface. The absence of the solid matrix prevents the loss of product due to nonspecific sorption. Additionally, stability of biological products in the aqueous media of polymer two-phase systems is usually increased. Also there is no decrease in the ligand capacity due to the loss of the polymer-bound ligand frequently observed in the column affinity chromatography as the result of the rupture of the solid matrix. The polymer-bound ligand dissolved in an aqueous two-phase system may be quantitatively recovered and returned to the process.

Regretably, all these advantages come with a price: the high cost of biospecific ligands and coupling of these ligands to phase polymers. The synthetic procedures developed for activation of phase polymers, usually PEG and dextran, and coupling of ligands to these polymers have been recently reviewed in the literature (see, for example, in [5,42,43]) and are beyond the present scope.

It is clear that the highly specific ligand, while being the most selective, usually is the most expensive. The unavoidable loss of some amount of the ligand in the process of covalent coupling to the polymer increases the price of the resulting polymer-bound ligand. Hence, the commonly used ligands are fatty acids (see in [1,36]) and triazine dyes (see, for example, in [38-40]). These ligands usually bound to PEG are effective in steering partitioning of certain proteins into the PEG-rich phase (see below).

Simple charged groups covalently attached to PEG are sometimes also highly effective. In the recent example, replacement of 8 % of PEG with PEG-trimethylamine in the aqueous PEG-potassium phosphate system was reported [44] to result in purification of penicillin acylase from *E.coli* with the enzyme yield in the PEG-rich phase up to 98 % and purification factor of 25.7.

Metallated iminodiacetic acid derivative of PEG, Cu(II)IDA-PEG was shown [45,46] to be an effective ligand with affinity for proteins rich in surface histidines. Partition coefficients of horse and human hemoglobins in the aqueous PEG—Na_2SO_4 two-phase system have been reported [45] to increase from 0.10 to more than 33 and 95, respectively, as the result of replacement of 1% of PEG with the Cu(II)IDA-PEG in the system formed by 14% wt. PEG-5000 and 8% wt. sodium sulfate. Complete separation of human hemoglobin from serum albumin in the model 1:1 protein mixture was achieved in a single extraction procedure with the additive of this ligand [45]. It has been shown [46] that partitioning of heme-containing proteins such as cytochrome C and myoglobins of different origins in aqueous Dex—PEG two-phase systems are affected by the above ligand to a very significant degree.

Monoclonal antibodies against horseradish peroxidase and porcine lactate dehydrogenase isoenzyme 5, respectively, were coupled to PEG [47] to produce the affinity ligands for selective extraction of the corresponding antigen in the so-called immunoaffinity partitioning procedure developed [48] for analysis of antigen-antibody affinity constants. Partitioning of antigen was changed significantly by the ligands [47]. Free horseradish peroxidase in the aqueous Dex-70—PEG-6000 two-phase system, pH 7.5 distributes between the two phases with the partition coefficient K about 1.0 to 1.05, and the complex of this antigen with unmodified antibody has K about 0.1 to 0.15 [47]. Addition of the PEG-conjugated antibody as the affinity ligand to the system increased the partition coefficient of the complex to 2.0 to 4.2 depending on the particular ligand structure (degree of PEGylation) and concentration [47].

PEG-bound triazine dyes are among the most favorite ligands for use in the affinity partitioning [38-40]. These dyes with polysulfonated aromatic structures are commercially available as low-cost chemicals, are readily coupled to PEG, display moderate to high specificity and binding capacity toward many proteins, and are resistant to biological and chemical degradation. To give just a few examples out of many, human serum albumin may be separated from α-fetoprotein by partitioning in the aqueous Dex-500—PEG-6000 two-phase system containing 0.5% wt. of PEG-bound Cibacron Blue F3G-A [49]. Partition coefficient of albumin in the presence of the indicated amount of the affinity ligand increases from 0.006 (in the ligand-free system) to 4.88, while that of α-fetoprotein changes less significantly - from 0.047 to 0.25, presumably because of weaker binding to the ligand [49].

Phosphofructokinase from rat erythrocyte haemolysate can be purified tenfold with a yield of about 95% by a single extraction under appropriate conditions in the aqueous Dex-500—PEG-6000 two-phase system containing about 0.2% wt. PEG-bound Cibacron Blue F3G-A [50]. Partition coefficient of the enzyme under influence of the above amount of the affinity ligand increases from less than 0.1 (in the ligand-free system) to 20, while the partition coefficient of total proteins remains at about 0.1.

Separation of isoenzymes of lactate dehydrogenase from rabbit tissues has been performed using the affinity partition technique with the PEG-bound Procion Blue H-5R [51]. The results reported by Kirchberger et al.[52] imply the possibility to separate some of human alkaline phosphatase isoenzymes using the PEG-bound Procion Navy MX-RB as an affinity ligand.

Alred et al.[53] have shown that the affinity partitioning in aqueous Dex-40—UCON 50-HB-5100 (monobutyl ether of ethylene oxide-propylene oxide random copolymer with the weight ratio of two monomers 1:1 and molecular weight of about 4000) allows one to isolate glucose-6-phosphate dehydrogenase from the bakers' yeast extract with the yield of about 79 % and purification factor of 4 using UCON-bound dye Procion Yellow HE-3G as an affinity ligand.

PEG esters of fatty acids are also often used as the affinity ligands in the so-called hydrophobic affinity partition technique [1,36,54,55]. The basic concept of this variant is that the effect of the affinity ligand on the protein partitioning is related to the ability of the protein to participate in hydrophobic interactions with the ligand. First, the effect in question may be used to separate proteins with different relative intensity of hydrophobic interactions with the ligand, and, second, the effect may be used as a measure of the protein ability to participate in hydrophobic interactions. The hydrophobic affinity partition technique while modestly successful in separation of certain proteins (see, for example, in [1,36]), usually serves as an analytical tool.

Comparative analysis of two proteins very similar in many aspects (human α_2-macroglobulin and pregnancy zone protein) by Birkenmeier et al. [56] may serve as an illustration of the analytical use of the technique in question. Partition coefficients of these two proteins in the aqueous Dex-70—PEG-8000 two-phase system containing 0.01 M NaCl in 0.02 M sodium phosphate buffer, pH 7.0 at 0°C are 0.16 for α_2-macroglobulin and 0.30 for pregnancy zone protein, indicating α_2-macroglobulin to be more hydrophobic. Treatment with chymotrypsin or methylamine does not affect the relative hydrophobicity of the proteins within the limits of experimental error. Partitioning of the proteins, however, changes noticeably and differently in the presence of PEG-palmitate. These changes are commonly expressed in terms of $\Delta \log K_{max}$, i.e., the maximum value of $\Delta \log K = \log(K_{P\text{-}L})_{max} - \log K_P$, where K_P is the protein partition coefficient in the ligand-free system; $K_{P\text{-}L}$ is the protein partition coefficient in the presence of the affinity ligand; and subscript "max" denotes the maximum value of the partition coefficient attainable in the presence of excessive amount of the ligand (for more detail, see below).

The $\Delta \log K_{max}$ values reported in [56] are 1.0 for α_2-macroglobulin and 1.42 for pregnancy zone protein supposedly implying that the pregnancy zone protein participates in hydrophobic interaction with the ligand with the intensity exceeding that characteristic for α_2-macroglobulin. Similar data were obtained in the systems with additives of PEG-bound fatty acids of various alkyl chain length [56].

Despite the great potential for large-scale continuous processing, and quite obvious success in laboratory separations, the use of the affinity partition technique remains rather limited. The application of the technique, analytical ones in particular, clearly depends on how predictable the effects of affinity ligands are on the partition behavior of macromolecules.

A straightforward thermodynamic analysis of affinity partitioning was given by Flanagan and Barondes [41], in more general form by Brooks et al. [57] and extended to include effects of more than one type of binding sites by Cordes et al.[58]. For the simple case of N identical and independent binding sites on a protein molecule this analysis leads to the following equation

$$K_{P\text{-}L} = K_P^{o} \cdot (1 + k_a^{T} \cdot [L]^{T})^{N} / (1 + k_a^{B} \cdot [L]^{B})^{N} \quad (9.1)$$

where $K_{P\text{-}L}$ is the partition coefficient of the protein in the presence of ligand; K_P^{o} is the partition coefficient of the ligand-free protein; k_a^{T} and k_a^{B} are the association constants for binding of ligand to protein sites in the top and bottom phases, respectively; $[L]^{T}$ and $[L]^{B}$ are the equilibrium concentrations of ligand in the top and bottom phases, respectively; N is the number of equivalent independent ligand binding sites in the protein.

If the ligand concentration in both phases is very high, all macro-

Table 9.9

Partition Coefficients of Procion Yellow HF-3G Ligand-Coupled Polymers of Different Structure in Various Aqueous Two-Phase Systems.

Two-phase system	Ligand carrier	K_L
7% wt. Dex-500; 5% wt. PEG-8000; 0.025 M sodium phosphate buffer, pH 7.5	PEG-8,000	31.6
	Ficoll-70,000	7.9
	Dex-70,000	39.8
	EHEC [a]	42.7
	HPS [a]	2.1
4.2% wt. Dex-500; 6% wt. PVA-14000; 0.025 M sodium phosphate buffer, pH 7.5	PVA-14,000	55.0
	PEG-8,000	9.1
	Ficoll-70,000	0.5
	Dex-70,000	107
	HPS [a]	4.4
4% wt. Dex-500; 11.5% wt. Ficoll-400 0.025 M sodium phosphate buffer, pH 7.5	Ficoll-70,000	0.50
	PEG-8,000	0.52
	Dex-70,000	0.138
	HPS [a]	0.34
15% wt. HPS [a]; 5.8% wt. PEG-8000; 0.025 M sodium phosphate buffer, pH 7.5	HPS [a]	0.043
	PEG-8,000	28.8
	Ficoll-70,000	3.0
	Dex-70,000	18.2
	EHEC [a]	> 200
10% wt. Dex-40; 7% wt. M-PEG-500 [b]	M-PEG-peptide [c]	3.0
0.025 M citrate buffer, pH 5.2	M-PEG-peptide [d]	3.1

[a] EHEC - ethylhydroxyethylcellulose; HPS - hydroxypropyl starch (Aquaphase PPT; Mol.wt. 35,000); [b] M-PEG - methoxypolyethylene glycol; [c] peptide structure: D-Ala-D-Ala; [d] peptide structure: D-Ala-D-Ala-D-Ala

(Data from G. Johansson, M. Joelsson, J.Chromatogr., 411, 161 (1987) and P. P. Godbole, R. S. Tsai, W. M. Clark, Biotechnol.Bioeng., 38, 535 (1991).)

molecules in both phases will be saturated with the ligand and Equation 9.1 reduces to:

$$K_{P\text{-}L} = K_P^o \cdot K_L^N (k_a^T / k_a^B)^N \tag{9.2}$$

where K_L is the partition coefficient of a ligand.

If the binding constants are the same in the two phases Equation 9.1 simplifies even more:

$$K_{P\text{-}L} = K_P^o \cdot K_L^N \tag{9.3}$$

or in the logarithmic form

$$\Delta \log K_{max} = \log(K_{P\text{-}L}/K_P^o) = N \cdot \log K_L \tag{9.4}$$

provided the protein is saturated with the ligand.

The whole concept of the affinity partition technique is based on the assumption that as the ligand is bound to one of the phase-forming polymers, it should distribute in a two-phase system in an extreme manner, either in favor of the upper or into the lower phase. In this regard the results obtained by Johansson and Joelsson [59,60] should be considered. Some of these data on the partition coefficients of polymer-bound dye Procion Yellow HE-3G in different aqueous two-phase systems are given in Tables 9.9. and 9.10.

The data shown in Table 9.9 indicate, first, that the extremely one-sided partitioning of an affinity ligand is observed in three cases. Hydroxypropyl starch (HPS)-bound dye distributing into the HPS-rich phase (K_L = 0.043) in the HPS-PEG system, ethylhydroxyethylcellulose (EHEC)-bound dye partitioning into the PEG-rich phase (K_L > 200) in the same system, and Dex-bound ligand with relatively high degree of substitution (8.3) partitioning into the PVA-rich phase in the Dex-PVA two-phase system [60]. In all the other cases partitioning of the polymer-bound ligands, while clearly one-sided, is not extreme (K varying from 0.138 to 55). Secondly, partition behavior of the polymer-bound ligands observed by Johansson and Joelsson [60] in several cases contradicts the other fundamental assumption of the affinity partition technique; namely that the polymer-bound ligand should distribute between the two phases in a manner similar to that of the parent phase-forming polymer. In contrast to this view, the Dex-bound ligand in the Dex-PEG system favors the PEG-rich phase (K_L = 39.8), and the Ficoll-bound ligand prefers Dex-rich phase (K_L = 0.5) in the Dex-Ficoll system.

The data on partitioning of Dex-70-bound dye Procion Yellow HF-3G with various number of dye molecules coupled to one dextran macromolecule [59] are shown in Table 9.10. These data indicate much larger effects of the dye fragment on the partition behavior of the ligand macromolecule than might be expected. As mentioned above, the whole concept of the affinity partition

Table 9.10

Partition Coefficients of Procion Yellow HF-3G Ligand-Coupled Dex-70 with Different Degree of Substitution in Various Aqueous Dex—PEG Two-Phase Systems.

Two-phase system	Ligand carrier	K_L
8% wt. Dex-40; 5% wt. PEG-8000	Dex-70 (n = 1.3) [a]	0.15
0.05 M sodium phosphate buffer, pH 7.9	Dex-70 (n = 2.3) [a]	0.30
	Dex-70 (n = 5.3) [a]	2.0
	Dex-70 (n = 8.3) [a]	17.5
5% wt. Dex-500; 3.8% wt. PEG-8000	Dex-70 (n = 1.3) [a]	0.97
0.05 M sodium phosphate buffer, pH 7.9	Dex-70 (n = 2.3) [a]	1.62
	Dex-70 (n = 5.3) [a]	6.2
	Dex-70 (n = 8.3) [a]	24.3
5% wt. Dex-2000; 4% wt. PEG-8000	Dex-70 (n = 1.3) [a]	1.03
0.05 M sodium phosphate buffer, pH 7.9	Dex-70 (n = 2.3) [a]	1.83
	Dex-70 (n = 5.3) [a]	7.5
	Dex-70 (n = 8.3) [a]	22.5

[a] n - degree of substitution representing an average number of dye molecules coupled to one dextran macromolecule.

(From G. Johansson, M. Joelsson, J.Chromatogr., 393, 195 (1987). Reprinted by permission of Elsevier Scientific Publishing Co.)

technique is based on the predictable behavior of the polymer-bound ligand. This concept appears to be in conflict with reality as implied by the K_L values in Table 9.10.

Increasing numbers of dye molecules coupled to the dextran macromolecule changes the ligand affinity for the parent polymer-rich phase from rather weak one initially, to an almost nonexistent affinity (K_L changes from 0.97 to 24.3 in the Dex-500-PEG-8000 two-phase system with the substitution degree increasing from 1.3 to 8.3). Partitioning of methoxy-PEG-bound di- and tri-D-Ala peptides [61] in the aqueous Dex—methoxy-PEG system (K_L = 3 and 3.1, respectively) is more in line with the aforementioned concept.

Partition behavior of the affinity ligands deviating so much from the theoretical concept may be one of the reasons why partitioning of proteins in some cases was reported (see, for example, in [36]) to be affected by the affinity ligands in a manner directly opposite to that expected from a theoretical viewpoint. These cases, while not very common, should be noticed as they may help us to get better insight into the mechanism of the affinity partition process.

As indicated above, the effect of an affinity ligand on the solute partitioning may be estimated by the $\Delta \log K_{max}$ value according to Equation 9.4. Any PEG-bound hydrophobic ligand in an aqueous Dex—PEG two-phase system is expected to favor the upper PEG-rich phase. That means that the ligand partition coefficient $K_L > 1$, i.e. $\log K_L > 0$, hence, the $\Delta \log K_{max}$ value may be zero (no ligand binding) or positive. That result is usually observed. Examination of an extensive set of data by Johansson (see in [36] and references cited therein) on the effects of various PEG-bound ligands on partitioning of different proteins in aqueous Dex-PEG two-phase systems reveals rather unexpected deviations from the general trend. Some of the data published in [36] are presented in Table 9.11 as $\Delta \log K$ values. These values were calculated as the difference between the logarithms of the protein partition coefficients in the ligand-containing and ligand-free systems.

The data shown in Table 9.11 indicate, first, that the partition coefficient of a protein may be decreased rather than increased under influence of the PEG-bound hydrophobic ligand ($\Delta \log K < 0$). Second, this effect depends not only on the structure of the ligand, but also on the particular protein being partitioned. As an example, PEG-capronate slightly increases the partition coefficient of ovalbumin from about 0.36 to about 0.4, while clearly decreases that of human serum albumin (from 0.148 to 0.098). The observed dependence of the affinity ligand effects upon the particular protein being partitioned does not allow one to explain the effects in question as being due merely to the specific partition behavior of a given ligand.

Discrepancies between the theoretical predictions and actual partitioning of biological macromolecules in aqueous two-phase systems containing a polymer-bound affinity ligand are usually attributed to (a) different solute-ligand binding constants in the two phases and/or (b) existence of multiple and interdependent binding sites in the solute molecule.

Table 9.11

Effects of PEG-Bound Ligands on Partitioning of Different Proteins. [a]

PEG	ΔlogK				
derivative	HSA	BSA	Lysozyme	Ovalbumin	RNase
Acetate	-0.05	0	-0.03	+0.03	-0.01
Butyrate	-0.12	+0.03	-0.04	0	-0.02
Capronate	-0.18	-0.06	-0.04	+0.05	+0.02
Caprylate	-0.04	+0.09	-0.01	+0.05	0
Myristate	+0.92	+1.29	-0.02	+0.05	-0.01
Palmitate	+1.13	-0.02	-0.02	+0.01	+0.02
Succinate	-0.07	+0.01	-0.02	+0.01	-0.01
Benzoate	-0.13	+0.08	-0.02	+0.10	+0.05

[a] Two-phase system composition: 7% wt. Dex-500; 7% wt. PEG-8000 including 1/10 of indicated PEG-8000 ester; 0.125 M K_2SO_4, pH 7.0; HSA - human serum albumin; BSA - bovine serum albumin; RNase - ribonuclease.

(From G. Johansson, Biochim. Biophys. Acta, 451, 517 (1976). Reprinted by permission of Elsevier Scientific Publishing Co.)

The aforementioned assumption of binding constants equality in the two phases being an obvious oversimplification is hardly likely. The reason follows from the supposedly large difference between the affinity of the ligand for the two phases. A biological macromolecule being partitioned may be considered as a separate microphase in the medium of a given macrophase of an aqueous two-phase system. In that case, the ligand binding may be viewed as involving thermodynamically governed transfer of the ligand molecule from the medium of the macrophase into the microphase of the macromolecule. It follows from this simplified consideration that the transfer in question will be much less favorable in the phase with higher affinity for the ligand. The higher affinity of a ligand for a given phase, the less strong will be the binding to the

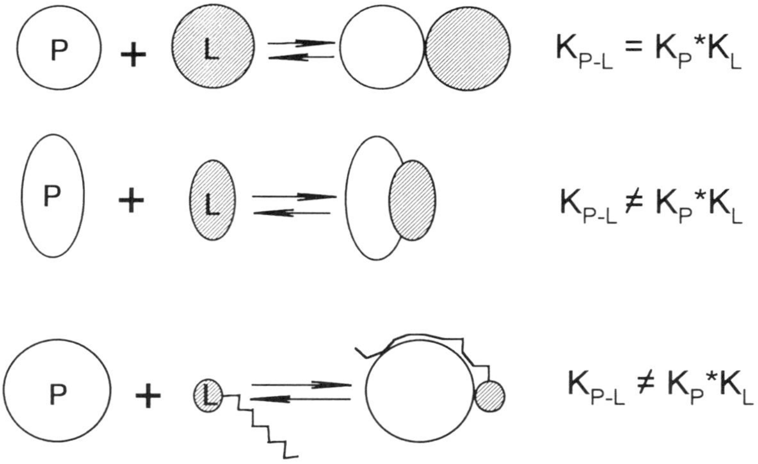

Figure 9.4. Schematic representation of complex formation between the biological macromolecule and polymer-bound affinity ligand. (After P.Å. Albertsson, Partition of Cell Particles and Macromolecules, 3rd.ed., Wiley, New York, 1986. Reprinted by permission of John Wiley & Sons, Inc.)

product in this phase. A similar conclusion was drawn by Baskir et al.[62] from consideration of polymer-bound ligand—protein interactions in the similar polymer-rich phase, in terms of a polymer lattice solution model.

It follows from the model by Baskir et al.[62], as well as from more simplistic considerations described above, that an increase in the polymer concentrations in the phases - while leading to more one-sided partitioning of a ligand (observed experimentally [59]) - will result in weaker product-ligand binding in the parent polymer-rich phase. This conclusion was experimentally confirmed by the results reported by Alred et al.[63]. The PEG-bound dye Procion Yellow HE-3G - lactate dehydrogenase binding constants were examined in the phases of aqueous Dex—PEG two-phase system by fluorescence titration [63]. They found that the strength of binding of the PEG-bound ligand was slightly lower in the PEG-rich phase compared to that in the Dex-rich phase at the system composition (5.2% wt. Dex-500; 3.8% wt. PEG-8000; 0.01 M sodium phosphate buffer, pH 7.0) close to the critical point, and twice lower at the system composition (7% wt. Dex-500; 5% wt. PEG-8000, same salt composition) further away from the critical point. Hence, it may be concluded that the affinity partitioning process is the result of two oppositely directed effects: the one-sidedness of the ligand partitioning, and the strength of the ligand-product binding in the phases. If the ligand in complex with the product distributes into the upper phase, but the complex in this phase is highly unstable, it will dissociate and the product will return into the lower phase, i.e., the efficiency of the extraction will decrease.

The above considerations may explain why the theoretically predicted level of extraction is not reached but does not provide reasons for the ΔlogK value being negative.

An explanation was suggested by Godbole et al.[61]. The authors [61] examined experimentally all the parameters of the basic affinity partitioning model as described by Equation 9.2 for the case of one-to-one binding interaction between the solute and polymer-bound ligand. Godbole et al.[61] suggested that the theoretical model does not fit the experimental results because the ligand-solute complex cannot be viewed as the constituent free molecules joined at a single point (or at several points in the case of multiple binding). This explanation is in line with the model for complex formation suggested by Albertsson [1,64].

The model for the formation of a complex between a biological macromolecule and a polymer-bound affinity ligand is shown in Figure 9.4. The surface of contact between the ligand and macromolecule may be small or large depending on the specific features of the macromolecule surface. If it is small, the partition coefficient of a complex may be treated in terms of Equation 9.1. If it is relatively large, the model as described by this Equation is unsuitable. If the polymer "tail" of the ligand "covers" the relatively large fraction of the solvent-accessible groups at the protein surface, the partition behavior of the complex may hardly be predictible. It will take time and a lot of research effort to bring this method to the stage at which its undoubtedly great separation and analytical potential will be completely utilized.

To improve the efficiency of an extraction in aqueous two-phase systems the multi-step extraction procedure is often used. This procedure is usually performed in the so-called thin layer countercurrent distribution apparatus or centrifugal countercurrent distribution devices (see, for example, in [1,65]).

The basic original instrument for countercurrent distribution in aqueous two-phase systems consists of two stacked discs mounted on a device capable of shaking both disks and rotating the upper disc step-wise over the immobile lower disk. Each disk has a concentric row of 60 to 120 cavities of the height usually not exceeding 2-3 mm (that is the reason for the name "thin-layer countercurrent distribution apparatus"). When the cavities of the top disk are aligned with those of the bottom disc they form a set of separate chambers which may be filled with a two-phase system. The volume ratio of the two phases is usually chosen in such a way that the lower phases fill the cavities in the bottom disk, and the upper phases fill those in the top disk. Thus, we have initially a concentric row of of two-phase systems. The sample is loaded in the "first" chamber. Then both disks are shaken to mix the sample with the phases and stopped for a time needed for phase settling. After that the top disk is slided over the bottom one step-wise to align its cavities with the neighboring ones. The upper phase in each chamber is replaced with that from the adjacent chamber, resulting in the original sample being located in two chambers, both containing one sample-free (upper or lower) phase, and one sample-containing (lower or upper) phase. The whole procedure is then repeated. After a given series of steps usually called transfers (i.e. extractions) the phases from all the chambers are withdrawn and analyzed for the components of the original sample.

In the more recent design developed by Åkerlund [66] phase settling is enhanced by centrifugation. The two stacked disks in the operating unit are replaced with the two concentric ring plates. The outer ring plate has 60 cavities for the lower phases, and the inner plate contains the upper phases. The upper phases are transfered by step-wise rotation of the inner plate relative to the outer plate, and rotation of both plates together provides the centrifugal force for phase settling. The procedure involving 50 to 60 transfers takes less than 2 hours [1].

The countercurrent distribution procedure is simple and straight-forward. It may be used with any aqueous two-phase system, with or without an affinity ligand, and it increases the separation efficiency significantly. Numerous examples of protein separation with this procedure are described in detail in the literature (see, for example, [1-3,5,9]). To give just one more recent example out of many, extraction of xanthine oxidase from total dairy milk in the aqueous Dex-500—PEG-8000 two-phase system containing 0.2 mM EDTA and 0.01 M sodium phosphate buffer, pH 7 in the centrifugal counter-current distribution mode, included 57 transfers at 0-4°C and resulted in isolation of the enzyme of the purity clearly exceeding that of the commercially available enzyme preparation [67].

Separation efficiency of an extraction, even when performed in the multiple-step mode, is obviously inferior to the use of liquid chromatography.

Chromatographic Procedures.

When the difference between the partition coefficients of the products to be separated is relatively small, the obvious way to separate the products is to perform a successive series of single-step extractions. The material to be separated may be repeatedly extracted from one phase with the other phase under variable or unvariable extraction conditions. This procedure is called by Albertsson [1] gradient extraction, as the composition of the two-phase system is usually changed during the procedure. This procedure may be very efficient, for example, for separation of nucleic acids from proteins, or separation of DNA from RNA (see, for example, in [1]). Much more efficient techniques for separation of solutes with different partition coefficients in a pair of immiscible liquids are based on a very large number of extractions, usually performed as a liquid chromatographic procedure, i.e. as the procedure in which one phase is moved relative to the other. Two different modes of chromatography are currently used with aqueous two-phase systems: column chromatography and countercurrent chromatography.

Column Chromatography.

Any separation method based on different distribution of solutes between two immiscible phases may be realized in the column chromatography mode once the suitable solid support is found. If one liquid phase serves as a stationary phase and the second phase plays the role of the mobile phase, the speed of migration of a solute in reference to the solvent front is governed by the relative amounts of the two phases in the column and the partition coefficient of the solute. The solid supports capable of immobilizing one of the two aqueous phases were designed by W.Muller [4,68,69].

Two not readily fulfilled requirements for the support material for chromatography in aqueous two-phase systems must be met [68]. The support material has to be selective for one of the phases and to be inert enough not to affect the partition behavior of solutes by nonspecific adsorption [69]. Taking into consideration the very small difference between the properties of the immiscible aqueous phases, the first requirement is particularly hard to meet. Cellulose powder and cross-linked polyacrylamide were found to be capable of retaining considerable amounts of the Dex-rich phase in aqueous Dex—PEG systems [4]. Much better materials were developed by grafting polyacrylamide chains on surfaces carrying primary or secondary hydroxyl groups [68]. The support materials coated with polyacrylamide were found to be capable of retaining the Dex-rich phase in different aqueous Dex—polymer i (polymer i: PEG, polyvinyl alcohol, polyvinylpyrrolidone) two-phase systems. A diol-modified silica was found to bind the PEG-rich phase in aqueous PEG—salt systems. The support materials developed by Muller [4,68,69] are currently commercially available from E. Merck (Germany) under trade-name LiParGel. Chromatographic procedures using these materials are described in detail in the book by Muller [4].

High selectivity of the polymer-coated surface in regard to retaining aqueous media of different composition may be of primary biological importance as one of the factors providing laminar flow in blood stream, influencing coagulation process, regulating transport of biomolecules in various biological fluids, etc. According to the "vicinal water" theory (see in Chapter 2), the support material may affect the solvent features of the stationary phase and, consequently, the separation capability of the two-phase system. This possibility, while particularly important for column chromatography in aqueous two-phase systems, has not been explored as yet.

LiParGel 650 and 750 have been examined as supports for the protein partition chromatography in aqueous Dex-PEG two-phase systems by Walsdorf and Kula [70]. It has been found that the partition behavior of proteins in a given two-phase system is not the only factor governing their elution pattern. All effects typical for polymer matrices such as size exclusion, hydrophobic and electrostatic interactions of proteins with the matrix functional groups appear to affect the chromatographic process. Experimental results obtained by Walsdorf and Kula [70] plotted together with the "ideal" elution curve for the column used are reproduced in Figure 9.5. The "ideal" elution curve is described as:

$$V_{el} = V_s/K + V_m \tag{9.5}$$

where V_{el} is the volume of mobile phase required to elute the center of the band of a given compound; V_s and V_m are the volumes of the stationary and mobile phases, respectively; K is the partition coefficient of a given compound. [It should be mentioned that the partition coefficient K in the case under consideration is defined as the ratio of the compound concentration in the PEG-rich (mobile) phase to the compound concentration in the Dex-rich (stationary) phase. Hence Equation 9.5 is slightly different from the generally used expression.]

Different support materials: commercially available LiParGel 650 and LiParGel 750, and an experimental sample of polyacrylated porous spherical silica LiChrospher Diol 1000/10 were compared in the study by Walsdorf and Kula [70]. It was reported [70] that the large pore size of 1000-Å of the latter material seems to be sufficient to avoid size exclusion effects observed with the other support materials. Interactions of proteins with hydrophobic and/or charged groups on the support material were concluded [70] to affect their elution pattern.

Actually, the observed deviations from the theoretical elution behavior may originate not from the direct matrix-protein interactions but from the effect of the matrix on the solvent features of the stationary Dex-rich phase. The "ideal" elution curve was calculated [70] from the partition coefficients of solutes determined in the batch, i.e. extraction-like, partition procedures. If the solvent features of the stationary phase are affected by the support matrix due,

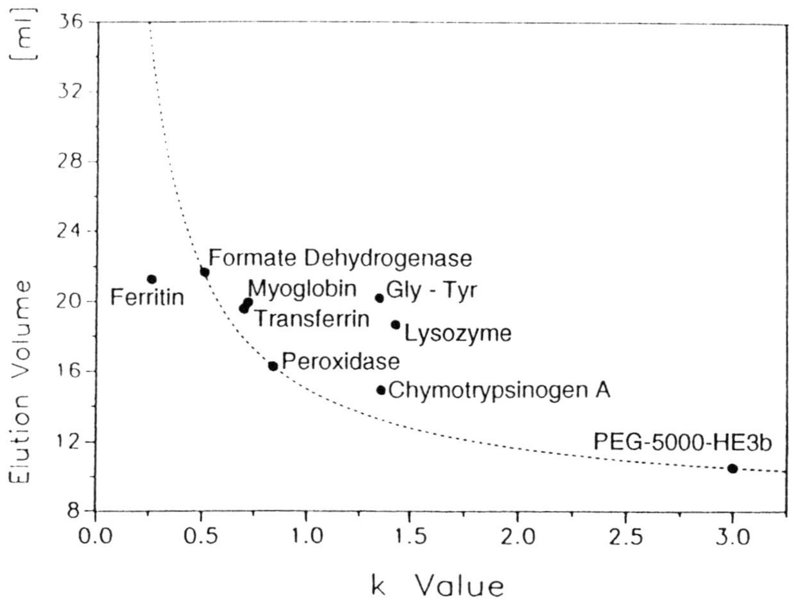

Figure 9.5. Relationship between partition coefficient and elution volume for a set of standard proteins, dipeptide glycyl-L-tyrosine and a PEG-5000 triazine dye derivative in a LiParGel 750 column (30 x 1 cm I.D.). The aqueous two-phase system composition: 2.7% wt. PEG-20000; 4.5% wt. Dex-500; 0.075 M KBr in 0.01 M phosphate buffer, pH 7.0. T = 30°C; flow-rate 1 ml/min. Dashed line is the computer-derived hyperbola illustrating the "ideal" elution curve of the column (V_s = 7.2 ml; V_m = 8.37 ml). The values for glycyl-L-tyrosine and lysozyme were omitted in calculating V_s and V_m. (Reprinted from A. Walsdorf, M. R. Kula, J.Chromatogr., 542, 55 (1991) by permission of Elsevier Scientific Publishing Co.).

for example, to the "vicinal water"-effect, the partition coefficients of the same solutes would change and that would be observed as deviations from the theoretical elution curve. The cause of the deviations observed [70] is practically important as the first possibility (matrix-protein interactions) calls for an improvement of the matrix, while the second one implies the need for changing merely the calibration procedure. To resolve the issue it may be necessary to calbrate the column with a homologous series of solutes and compare the results with those obtained in batch experiments.

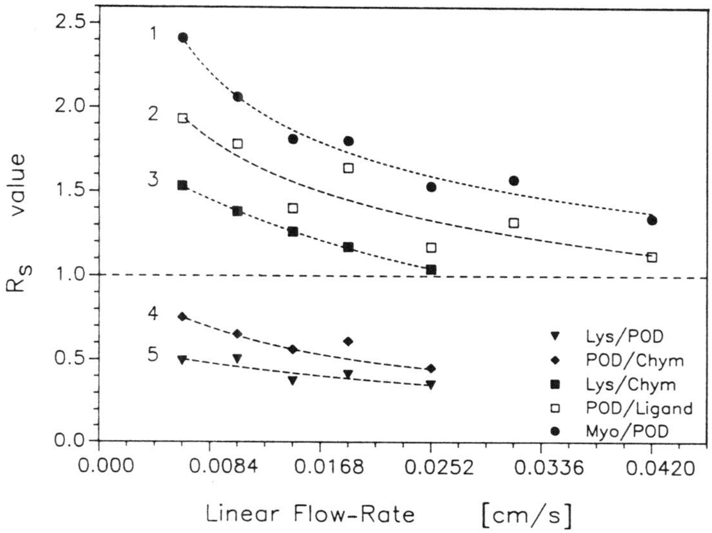

Figure 9.6. Influence of the linear flow-rate on peak resolution of different proteins. Lys - lysozyme; POD - horseradish peroxidase; Chym - chymotrypsin; myo - myoglobin; ligand - PEG-5000-bound triazine dye Procion Red HE3b. The aqueous two-phase system composition: 5.4% wt. PEG-6000; 9.0% wt. Dex-40; 0.1 M NaCl in 0.05 M sodium phosphate buffer, pH 7.5. LiParGel 650 column (30 x 0.5 cm I.D.); sample volume 100 µl. (Reprinted from M. R. Kula, L. Elling, A. Walsodrf, J.Chromatogr., 548, 3 (1991) by permission of Elsevier Scientific Publishing Co.).

The column resolution was examined by Kula et al.[71] using LiParGel 750 and 650 columns under conditions denoted in the caption to Figure 9.6. Figure 9.6 shows the influence of linear flow-rate on the resolution (R_s value) of different proteins. For curves 1-3 in Fig. 9.6 $R_s > 1$ indicating good resolution of the indicated pairs of proteins. Resolution of horseradish peroxidase and chymotrypsin peaks (curve 4) and lysozyme and peroxidase peaks (curve 5) was not attainable under the conditions applied. To improve the resolution in these two cases the system composition must be changed [71].

Figure 9.7 from the work by Kula et al.[71] illustrates the effect of the sample volume on the number of theoretical plates (N) for the LiParGel 750 column. In total agreement with liquid partition chromatography theory [72] the number of theoretical plates is inversely proportional to the sample volume [71]. The number of theoretical plates decreases with increasing protein molecular weight as shown by the results reported by Kula et al.[71] for bovine serum albumin.

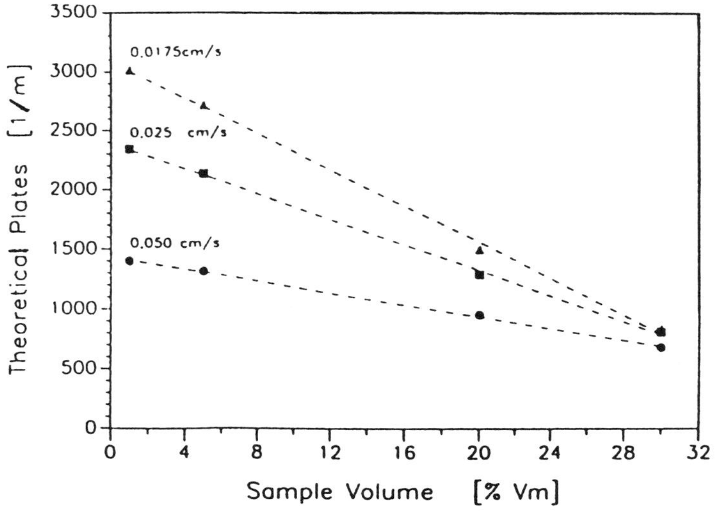

Figure 9.7. Number of theoretical plates as a function of sample volume of myoglobin (17,600 dalton; 1 mg/ml). The aqueous two-phase system composition: 2.7% wt. PEG-20000; 4.5% wt. Dex-500; 0.05 M $NaClO_3$ in 0.0126 M potassium phosphate buffer, pH 7.0. LiParGel 750 column (30 x 1 cm I.D.); mobile phase volume V_m 10.6 ml; stationary phase volume V_s 5.34 ml. (Reprinted from M. R. Kula, L. Elling, A. Walsdorf, J.Chromatogr., 548, 3 (1991) by permission of Elsevier Scientific Publishing Co.).

Thus, liquid-liquid partition chromatography in aqueous two-phase systems is a highly efficient method for separation of biological macromolecules. Numerous examples of separation of various proteins and nucleic acids, including separations of DNA restiction fragments, ribosomal RNAs, isolation of highly purified enzymes from crude cellular extracts, etc., may be found in the books by Muller [4], Albertsson [1] and several reviews [69,73]. Among the more recent examples, a LiParGel 750 column and aqueous Dex—PEG two-phase system was used by Walsdorf et al.[74] to isolate formate dehydrogenase from the crude extract of *Candida boidini* using affinity partitioning with PEG-bound Procion Red HE3b as the affinity ligand.

The work by Andersson et al.[75] should be particularly mentioned in the context of this book. Following earlier observations [76] that the partition behaviour of antibodies appeared to be related to their specificity, Andersson et al.[75] used LiParGel 650 column and the aqueous Dex-500—PEG-8000 two-phase systems containing 0.1 M NaCl, 0.01 or 0.05 M Na_2HPO_4 and 0.1 or 0.2 M glycine, pH 7.0 to examine partitioning of several antigens, antibodies, and antigen-antibody complexes. First, the elution pattern of intact human IgG was found [75] to be different from that of IgG acylated with β-propiolactone and IgG labelled with ^{125}I. This observation agrees with the high sensitivity of the partition technique toward chemical modifications of proteins as discussed above (Chapters 7 and 8). The assumption by Andersson et al.[75] that the altered elution pattern implies conformational changes of the protein macromolecule is likely to be true for acylated IgG as it is supported by size-exclusion chromatography data. It should be emphasized, however, that the change in the protein partition behavior alone does not imply that the protein conformation is changed. Iodination of IgG, particularly, changes the chemical nature of some of the solvent-accessible groups at the protein surface and that may be the only reason for the changes observed [75] in the protein partitioning.

Partition coefficients of proteins were estimated from elution volume measurements [75] according to Equation 9.5. It is particularly stressed by the authors [75] that the partition coefficients of human transferrin (K = 0.5), IgG (K = 0.71), and albumin (K = 0.56) are different from those of the corresponding rabbit antibodies (partition coefficients: 1.67, 1.25, and 1.0, respectively) and those of the antigen-antibody 2:1 complexes (partition coefficients: 0.4, 0.38, and 0.45, respectively). However, the conclusion by Andersson et al.[75] that the dominating surface properties of complex differ from those of the unbound antigen and antibody, based on the observation that the partition coefficients of each component of the complex exceeds that of the complex itself, may hardly be viewed as completely valid, unfortunately. The reason is that different systems were used for chromatography of free antigens and antibodies (0.1 M glycine-containing system) and for antigen-antibody complexes (0.2 M glycine-containing system) [75].

The most important observation by Andersson et al.[75] is that the elution pattern (i.e., partition behavior) of murine IgG 1 monoclonal antibodies against three different epitopes on human albumin or IgG is different. In contrast, no difference in the properties of these antibodies could be detected by high performance size-exclusion chromatography, SDS-PAGE, and isoelectric focusing on PhastGel. Two murine monoclonal antibodies of different IgG subclasses (IgG 1 and IgG 2a) with pI of 5.8 and 6.8 values, with specificity to the same epitope on human IgG Fc, displayed identical partition behavior. That observation supports the conclusion [75] that the surface properties of immunoglobulins detected by LLPC (liquid-liquid partition chromatography in aqueous two-phase systems) seem to be dominated by those determining the specificity of the molecules, i.e. the conformation of the antigen binding site.

Two implications of the results reported by Andersson et al.[75] should be emphasized. First, the partition behavior of proteins monitored by the column chromatography procedure is related to the protein's biological function, in agreement with the results obtained in the single-step extraction procedure as discussed above (Chapter 7). Secondly, the biological function and/or potency-related features of biopolymers detected by the aqueous two-phase partition technique could not be observed with commonly used analytical techniques.

Clearly, liquid-liquid column partition chromatography in aqueous two-phase systems is the separation technique of greatest potential, even though it is at the developmental stage. Effects of polymer supports on protein separation are observed and remain to be understood, although they must be dealt with. These effects are nonexistent in the countercurrent chromatography mode.

Countercurrent Chromatography.

Countercurrent chromatography is a form of liquid-liquid partition chromatography. Its distinctive feature is that to restrain a stationary liquid phase centrifugal or gravitational forces are used instead of porous supporting matrix. These forces maintain a bed of stationary phase in a coil or train of chambers (playing the role of chromatographic column), while a stream of a mobile phase is passed through the system in contact with the stationary phase [15]. There is no adsorption in this case, so that solute retention depends only on the phase-volume ratio, and the partition coefficient of the solute. Sometimes the so-called true countercurrent chromatography mode, in which both phases are pumped in opposite directions to produce an actual countercurrent flow, is also used. Instruments specifically designed for countercurrent chromatography in aqueous two-phase systems are described in the literature [1,15,77] and some of them are currently commercially available. Basic principles of countercurrent chromatography are described in numerous books and reviews (see, for example, [15,78,79]).

One advantage of the countercurrent chromatography mode over high performance liquid chromatography (HPLC) is the lack of solid support, preventing any loss of separated material due to nonspecific adsorption. The other advantage is that the volume ratio of the stationary phase is much higher than that in HPLC column. Hence, the capacity of a column in countercurrent chromatography is much higher than that of an HPLC column for the same total volume.

The resolution for two adjacent peaks is described as [15]:

$$R_S = 0.25(\alpha - 1)\sqrt{N} \cdot \left[\frac{K_1}{K_1((\alpha+1)/2) + ((1-S_F)/S_F)} \right] \tag{9.6}$$

where K_1 is the partition coefficient of a given solute 1; α is the separation factor (the ratio of the partition coefficients for the two solutes always expressed

Table 9.12

Viscosities (in mPa·sec) of the Phases of Aqueous Polymer and Solvent Two-Phase Systems (at 22-25°C).

System composition *	Upper Phase [a]	Lower Phase [b]	Ref.
10% Dex-500 — 7% PEG-8000 [c]	7.00	332	[82]
10% Dex-500 — 7% PEG-3350 [c]	3.18	248	[82]
10% Dex-200 — 7% PEG-3350 [c]	3.48	99	[82]
14% Aquaphase PPT — 5% PEG-8000 [d]	6.87	42	[83]
23.79% MDX — 3.75% PEG-8000 [e]	7.66	17.3	[84]
14% PEG-1550 — 12% Potassium phosphate, pH 7.0	9	2.0	[7]
15% PEG-4000 — 8% Sodium sulfate [f]	20.33	1.35	[85]
1% Water in MeOH—Hexane (1:1) [g]	0.34	0.37	[80]
CCl_4—MeOH—Water (10:9:1) [g]	0.67	1.22	[80]
BuOH—Water (1:1) [g]	3.06	1.38	[80]

* System composition for aqueous polymer systems is given in weight percent;

[a] PEG-rich phase in the aqueous two-phase systems listed;

[b] Polysacharide-rich or salt-rich phase in the aqueous two-phase systems listed;

[c] Salt composition: 0.1 M potassium phosphate buffer, pH 7.0;

[d] Aquaphase PPT - hydroxypropyl starch (Mol.wt. 35,000);

[e] MDX - maltodextrin (Mol. wt. 1800), salt composition: potassium phosphate buffer, pH 6.9-7.1;

[f] Additional salt composition: potassium phosphate buffer, pH 4.3;

[g] System composition is given in volume ratios for the solvents indicated; MeOH - methanol; BuOH - butanol.

as a value greater than unity); N is the column efficiency or number of theoretical plates; and S_F is the retained stationary phase volume expressed as a fraction of total column volume.

Mutually interdependent influences of different factors in Equation 9.6 on the resolution of two solutes was analyzed in detail by Conway [15]. It was concluded [15, p.212] that a separation factor (α) of about 1.25 is needed usually to obtain base-line resolution in a common preparative countercurrent chromatography apparatus. An α value of 1.1 may be sufficient for good resolution in current analytical devices using small-bore columns and operating at very slow flow rates [15].

According to Oka et al.[80], the following requirements must be met for a given two-phase system to be useful in countercurrent chromatography: (a) two phases should be of approximately equal volumes; (b) the phases should yield a reasonable short settling time (less than 30 s for satisfactory retention of the stationary phase in the column for the high-speed countercurrent chromatography to be performed); and (c) the partition coefficient of the compound to be isolated should be close to one (that provides the retntion volume close to the total column capacity). The first and the last conditions are readily fulfilled in aqueous two-phase systems. However, the second condition clearly hinders the use of these systems in countercurrent chromatography. It should be noted that settling time increases with increasing viscosity of the phases. Also, phase viscosity is particularly important for the separation efficiency of the countercurrent chromatographic procedure. Efficiency generally increases with lowering phase viscosities, likely due to improved mass transfer between the two phases [81].

Typical viscosities of the phases of aqueous polymer systems are shown in Table 9.12 together with those for solvent systems commonly used in countercurrent chromatography. These data indicate that viscosities of the phases of aqueous two-phase systems exceed those of typical solvent phases by one to two orders of magnitude. These data imply that: (a) countercurrent chromatography may be performed in aqueous polymer systems at low flow rate; and (b) a search for new phase-forming polymers of low molecular weight is in order to design aqueous two-phase systems with higher separation efficiency.

Experimental observations (see, for example, in [86-89] and references cited therein) support these recommendations. Increasing flow rate of a mobile phase above 1 ml/min. usually decreases retention of a stationary phase [88,89] consequently reducing the separation efficiency. Increasing rotational speed, i.e., centrifugal force retaining the stationary phase, was reported [89] to increase peak resolution at low flow rates. Analysis of the factors influencing solute separation in aqueous two-phase systems by countercurrent chromato-

graphy in a commercially available device led Foucault and Nakanishi to the following conclusion [86]:
"The very poor mass transfer causes the loss of efficiency as the droplets of mobile phase travel too fast in a channel, compared to the kinetic of transfer of the protein between mobile and stationary phases. Three steps are required for highly efficient separation:

1 - Mixing the two phases until equilibrium is reached, i.e., when the ratio of the concentrations for a given protein is equal or very close to the partition coefficient - this step requires a long time (slow diffusivity for proteins in a viscous solution) and/or a very large interface area compared to the bulk volumes of the two phases. The same problem has been encountered in classical chromatography, and it has been solved first by using pellicular solid beads, then, later, small particles of silica with a high surface per mass unit.

Adequate shaking to assure complete mixing between phases and material to be partitioned and to increase the area of interface is critical.
2 - Settling the phases which is readily achieved with a strong centrifugal field.
3 - Transfer of the mobile phase to the next cell.

Steps 2 and 3 are easily obtained in CPC (centrifugal partition chromatography) with aqueous two-phase systems, while step 1 is not reached because so far we cannot use sufficiently high flow rate to obtain good mixing in each cell."

The technical procedures used for separation of biomolecules in aqueous two-phase systems are currently at an early stage of development. Successful separation methods based on the aqueous two-phase partition technique are the result of an extremely high selectivity of the technique (i.e., large difference in partitioning of structurally closely related biomolecules) rather than provided by technical means capable of separating solutes with slightly different partition behavior employed, for example, in HPLC and countercurrent chromatography in common solvent two-phase systems. Hence, the development of an aqueous two-phase separation method must be focused on the design of the conditions providing as large a difference between the partition coefficients of the solutes of interest as is possible.

For the following discussion of the strategy for the method development it is useful to consider the relative importance of the factors steering distribution of solutes between immiscible aqueous phases.

9.2. RELATIVE IMPORTANCE OF FACTORS INFLUENCING THE SOLUTE PARTITION BEHAVIOR

Understanding the relative importance of all the factors affecting the solute partition behavior is obviously critical for an effective separation method. These factors were considered previously from the viewpoint of the mechanism of solute partitioning in aqueous two-phase systems. Based on the foregoing, we may discuss the relative importance of these factors and how they should be varied, in an economical time-saving manner, to provide an efficient, satisfactory separation. Unavoidably, many of the issues considered above will be briefly mentioned here again.

As the first approximation, let us assume that there is no direct solute-phase polymer interaction in a given aqueous two-phase system (as shown above for Dex—PEG and Dex—Ficoll systems). The solute partitioning in this case is governed by the difference in the solute-solvent interactions in the two phases. The solute-solvent interactions depend on the two categories of factors listed in Table 9.1. One category, listed in the left column, includes the factors governing the solvent features of the aqueous media in the two phases. The other category (right column) covers the factors affecting the solute and its ability to participate in the interactions with the solvent. For consideration of the relative importance of the "solvent-relevant" factors of the first category it is necessary to describe distribution of a solute between two immiscible aqueous phases in very general and simple terms.

When a solute distributes between the phases of an water—organic solvent system, the differences between the interactions experienced by the solute in the phases are very significant because of the different chemical nature of the solvents. That is why the partition coefficients for different solutes in these systems vary over several orders of magnitude. In aqueous two-phase systems the nature of the solvent in both phases is the same; hence, the partition coefficient, K, usually varies over a range of 0.1 to 10. Using the water structure concept, the solute distribution between two immiscible aqueous phases may be viewed in terms of fitting the solute molecule into two different water structures specific for the aqueous media in these phases.

Macromolecules of a phase-forming polymer alter the original water structure, causing changes in the orientation and strength of the water-water H-bonds, the ratio of unbound water molecules, and so on. That results in formation of a specific structure of the water H-bond network, and particular solvent features. When the difference between the water structures induced by two different polymers exceeds a critical level, they become immiscible and phase separation occurs.

Biomolecules distribute between immiscible aqueous phases according to the difference between their relative affinity for the phases, i.e. for the different water structures. Since the nature of the interactions is the same in both phases, the affinity in question may be regarded in terms of "structural fitness". The better a given biomolecule fits into the structure of the water network in a given phase, the higher the affinity it has for this phase.

As mentioned above (see Chapter 1), the energy of transfer of a solute molecule into an aqueous medium includes the energy of formation of a cavity to accommodate the molecule, and the energy of the solute-water interactions. The energy of the cavity formation is likely to depend not only on the size but also on the shape of the cavity, i.e., size and shape of the biomolecule. Additionally, since the strength of the water-water H-bonds is orientation-dependent, the nature and topology of the solvent-accessible groups at the biomolecule surface must affect the total interaction energy. This total energy may be considered in terms of rearrangement of the water structure in the phase as a result of the solute transfer into the phase. The less rearrangement is needed, the better the structural fit.

It is currently impossible to describe all the parameters governing the structural fitness of biomolecules and aqueous media of varied composition. It seems possible, however, to estimate what factors among those listed in Table 9.1 may affect it most.

According to the data discussed in the first part of this book, the water structure in an aqueous polymer solution is determined mostly by the chemical structure, concentration, and molecular weight of the polymer. Low molecular weight additives also affect the solvent features and structure of the aqueous medium, but to a lesser degree. That condition follows, for example, from inorganic salts, such as KSCN and K_2SO_4 which, while very different in regard to the water-structure-effects, do not phase-separate in an aqueous mixture. Hence, polymers may be concluded to be the main components of a system responsible for the specific water structures formed in their aqueous solutions.

Molecular weight and concentrations of a polymer are clearly important for the water structure formed in the polymer solution. Once a certain critical polymerization degree is exceeded, however, increasing polymer concentration may compensate for decreasing molecular weight (for example, see Figs. 2.1 and 2.3). The data discussed in Chapter 2 imply that both concentration and molecular weight are of secondary importance relative to the specific polymer chemical structure.

Experimental data on protein partitioning support this point of view. According to the data reported by Albertsson et al.[90], the affinity of a protein for a given phase in an aqueous two-phase system may usually be affected by changes in the molecular weights of the phase polymers, but will not be reversed (i.e. the protein will not favor the other phase). Partitioning of nine different proteins was examined [90] in the aqueous Dex—PEG two-phase systems of the same polymer concentrations (6% wt. PEG and 8% wt. Dex) and varied polymer molecular weights. Variations of the Dex molecular weight from $4 \cdot 10^4$ to $2 \cdot 10^6$ in the systems formed by Dex and PEG-6000 do change the protein partition coefficient K values, but only for catalase and β-galactosidase, out of nine proteins, the K value was changed from < 1 to > 1. Ovalbumin is the only protein that responded to decreasing the PEG molecular weight from

$4 \cdot 10^4$ to $4 \cdot 10^3$ in the systems formed by PEG and Dex-500 by the change of the K value from less than 1 to 1.25 in the Dex-500-PEG-4000 system [1, p.62-63]. Much more detailed data reported by Forciniti et al.[20] do not agree with the aforementioned results [1] for catalase, and do not show any exception for what likely may be a general rule: namely, that a change in the phase polymer molecular weight may not reverse the protein affinity for a given phase. (Note that all the generalizations given here, including the above one, are to be viewed as hypothetical ones only!) This rule, even if correct, covers only a limited range of the polymer molecular weights. For example, as follows from data in Table 9.5, this rule covers only PEGs with molecular weights above 4,000.

Thus, the chemical structure of phase-forming polymers appears to be the most important factor governing the partition behavior of solutes in aqueous two-phase systems. Unfortunately, this factor is currently the least explored one. Partition behavior of solutes is studied usually in Dex—PEG systems. Other systems, such as hydroxypropyl starch (trade name "Aquaphase PPT")—PEG, Dex—Ficoll, Dex—polyvinylpyrrolidone (PVP), Dex—polyvinyl alcohol (PVA), maltodextrin—PEG, pullulan—PEG, etc., were explored to a much more limited extent. Studying partitioning of solutes in systems formed with only a fixed pair of polymer types seems hardly to be promising, as would be an attempt to develop HPLC techniques using merely one type of solid matrix.

One of the two phase-forming polymers is replaced with a phase-forming inorganic salt in aqueous single polymer—salt two-phase systems. The role of the salt is probably not as decisive as that of the polymer, but it is clearly an important one. Both effects of concentration and molecular weight of the phase-forming polymer, and type and concentration of salt additives, appear to be displayed in aqueous single polymer—salt systems much more strongly than in two—polymer systems. The likely reason for this seems to be that both polymer and salt compositions of the two phases are much more dissimilar in the latter systems, compared to those in two—polymer ones.

Similarly important function appears to be performed by a salt additive in an aqueous two-polymer system. From the viewpoint of the "structural fitness" concept, an analogy between the solute distribution between immiscible aqueous phases and selection of an adequate housing seems to be appropriate. The type of phase polymers, in this analogy, may be viewed as the "location and architecture of the house", while salt additives would play the role of "interior design" - not a small factor, when making the decision to buy or not to buy.

As discussed above, salt additives may steer the solute partitioning from one phase to the other, providing the means to manipulate distribution and hence separation of biomolecules. The only serious difficulty encountered in using salt additives as the separation-effective factor is currently the rather low predictability of the salt effects. The reason is that the presence of a salt additive affects not only the properties of the phases but also those of the biomolecules. The salt effect on the solvent features of the phases may be experimentally explored (see in Part 2) and sometimes predicted. Effects of salts on the physicochemical properties of biomolecules are, however, poorly understood. Before discussing this issue in more detail the "structural fitness" concept in regard to the effects of different phase polymers and salt additives should be described.

The concept in question is schematically presented in Figure 9.8. Different water structures may be characterized by a set of solvent features. Two parameters may be used to describe the solvent features of different water structures formed in the presence of phase-forming polymers. The first one is the hydrophobic character of the aqueous media in reference to that of pure water (see in Chapters 2 and 4). The other factor may be parameter C, the solvent solvatochromic polarity E_T^N, or some other factor characterizing the ability of water to participate in ionic and other polar interactions with a solute. Additional important factors may include the dielectric constant, thermodynamic activity coefficient of water, etc.; however, such details would overcomplicate the description attempted here.

The chemical structure of a given pair of two polymers (polymers 1 and 2, for example) determines the areas of different water structures and solvent features achievable in the aqueous phases formed in mixtures of these polymers, under varied polymer molecular weights and concentrations, and different salt compositions. The other pair of chemically different polymers (for example, polymers 3 and 4) form aqueous phases of different properties. The total areas corresponding to different polymer-induced water structures are likely to overlap to a greater extent than is shown in Fig. 9.8. When a solute is partitioned in a given two-phase system, it distributes into one or the other phase of the system according to the fitness of its structure into that of the aqueous medium of the phase. Since biological molecules are usually conformationally flexible, it seems reasonable to suggest that there should be a variety of different water structures corresponding to the perfect (or suitable) structural fit. These different structures of aqueous media are shown in Fig. 9.8 as the separate area of "optimal" water structures. Depending on the particular composition of a two-phase system the "optimal" water structure may be formed in the lower or upper phase (as shown in Fig. 9.8 for the polymer 1 - polymer 2 system) or only in the lower (or only upper) phase (as shown in Fig. 9.8 for the polymer 3 - polymer 4 system). It should be repeated that salt additives are as important in providing the "perfect fit" situation (in a given two-phase system) as the phase-forming polymers themselves.

Salt effects on the properties of biomolecules, including conformation and hydration, are poorly understood, and are usually unpredictable. From the "structural fitness" viewpoint, these effects may lead to a change in the position

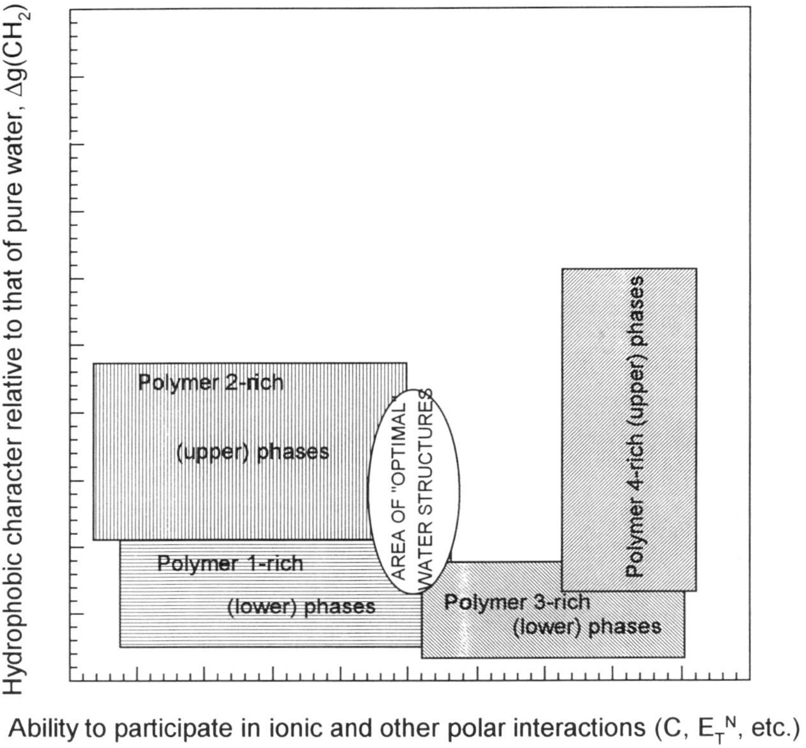

Figure 9.8. Schematic representation of the "structural fitness" concept. For explanation see text.

of the area of the "optimal" water structures. That is not easily predictable.

More readily predictable are effects of pH. Physicochemical features of biomolecules are routinely studied as functions of pH. Therefore, there usually is at least some background information in regard to previously described and characterized proteins. When dealing with protein extracts from new natural sources, such pH effects are not as predictable. The range of possible variations of pH is, however, much more limited than those of salt composition, and hence pH-effects are easier to study. It should be kept in mind that the purpose of pH manipulations is to modify the solute of interest in such a way as to steer its partitioning into one or the other phase. Generally, the more charged groups on the biomolecule, the less affinity it should display for the more hydrophobic phase.

Use of complex-forming ligands and chemical modification of biomolecules share a common goal - to impart new properties to the molecule and, therefore, steer its partitioning. This approach is straightforward, except for one problem: to guide the molecule partitioning by chemical modification we need to know exactly what effect would be expected from a specific kind of modification. Work on this aspect seems currently to be in progress. The task is complicated in that the topography of the solvent-accessible groups in a given molecule may be as important as the chemical nature of the groups. Ligands capable of forming complexes with biomolecules are usually considered as providing the molecules with additional features required for partitioning into one or the other phase. The polymer-bound affinity ligands offer an especially illustrative example. These ligands are generally viewed as providing a polymer handle capable of pulling the biomolecule out into the parent polymer-rich phase. In one of the older reviews, the principle of affinity partitioning was illustrated by a cartoon of a fisherman using the polymer-bound ligand as a fishing rod with the ligand as a bait. As any oversimplified view, this one is not only incorrect but possibly confusing. The reason is that while the polymer-bound ligand endows the protein with additional features, the protein in turn gives new properties to the polymer-bound ligand. The total outcome is hardly predictable as it is governed by the poorly understood protein-ligand—polymer interactions, and protein-ligand—polymer—water interactions. Empirical information on the influence of affinity ligands on partitioning of various biomolecules is accumulating fast, and it seems reasonable to hope that the predictability of these effects will increase in the near future.

Finally, the influence of temperature on solute partitioning should be commented upon. As mentioned above, temperature effects are among the least explored variables. Increasing the temperature should decrease the difference between the water structures of the aqueous media in the two immiscible phases. That effect is partially compensated for by an increase in the polymer concentrations required for phase separation under these conditions. It seems possible, however, that the structural fit may be achieved easier at an elevated temperature. In that case, the selectivity of a given two-phase system, with regard to structurally closely related compounds, would likely decrease. Decreasing temperature may have an opposite effect, but that remains to be examined experimentally.

9.3. SEPARATION METHOD DEVELOPMENT

Next I consider some practical recommendations for those interested in applying the aqueous two-phase partition technique for bioseparations. The first and most important step is selection of two phase-forming polymers, or a single polymer and a salt. Not only must the polymer chemical structure be selected (for example, PEG and Dex or polyvinyl alcohol (PVA) and polyvinylpyrrolidone (PVP)) the molecular weight fractions of the polymers to be used must also be selected. This selection is hard to make because the experimental information is rather limited, and somewhat confusing. To stress this point, the data [60] shown above in Table 9.9, and presented in Table 9.13 in a slightly modified form, will be discussed.

The PEG-bound Procion Yellow distributes in a predictable manner into more hydrophobic PEG-rich or PVA-rich phases in three out of four systems. In the Dex—Ficoll system the same solute does not favor the more hydrophobic Ficoll-rich phase but prefers the lower Dex-rich phase, which it tends to avoid in other systems. The same dye bound to Ficoll-70 distributes in an even more unpredictable manner. It favors the more hydrophilic Dex-rich phase in the Dex—Ficoll (!!) and Dex—PVA systems, while it distributes into the more hydrophobic PEG-rich phases in the Dex—PEG and HPS—PEG systems. Dex-bound dye distributes into the Dex-rich phase only in the Dex—Ficoll system; in the Dex—PEG and Dex—PVA systems it clearly favors more hydrophobic phases. It should be stressed that the polymer-bound dyes examined by Johansson et al.[60] are much less structurally complex than proteins. However, the partition behavior of these solutes in the systems formed by different pairs of polymers is hardly predictable.

Comparative analysis of the partition behavior of different biomolecules in aqueous two-phase systems of the same salt composition, but formed by chemically different polymers, is presently one of the most important issues to be investigated.

From the viewpoint of bioseparation method development, selection of the system to be used should be based on results obtained experimentally in the aqueous Dex—PEG and PEG—salt systems formed with two different molecular weight fractions of PEG [for example, PEG-8000 and PEG-1000 in two-polymer Dex—PEG systems, and PEG-8000 and PEG-300 or PEG-600 in the PEG—salt systems].

For biomolecules capable of withstanding relatively high salt concentrations the aqueous PEG—salt [particularly, PEG-(potassium, sodium, or ammonium) phosphate or sulfate] may be a reasonable first choice. Two systems formed by the salt and PEG of low molecular weight (for example, PEG-300 or PEG-600) and salt and more commonly used PEG-8000 are recommended to be investigated because of the possible large effect of PEG molecular weight. Two different salts, such as chloride and thiocyanate of the cation com-mon with that of the phase-forming salt, should be tested at different concentrations if possible. Additionally, pH variations, with reasonably large steps (for example, 3-4, 6-7, and 8-9) may be a factor capable of noticeable influence on solute partition behavior.

Table 9.13

Partition Coefficients of Procion Yellow HF-3G-Coupled Polymers of Different Structure in Various Aqueous Two-Phase Systems.

Solute [a]	Partition coefficient K in the indicated two-phase system [b]			
	Dex—PEG [c]	Dex—PVA [c]	Dex—Ficoll [c]	HPS—PEG [c]
PEG-dye	31.6	9.1	0.52	18.2
Ficoll-dye	7.9	0.5	0.50	28.8
Dex-dye	39.8	107	0.138	3.0
HPS-dye	42.7	4.4	0.34	0.043

[a] Dye Procion Yellow HF-3G bound to different polymers - PEG-8000; Ficoll-70,000; Dex-70,000, and HPS - hydroxypropyl starch-35,000; [b] parition coefficient K in the systems was defined as the ratio of the solute concentration in the upper (second polymer-rich) phase to the solute concentration in the lower (first polymer-rich) phase; EHEC - ethylhydroxyethylcellulose; HPS - hydroxypropyl starch (Aquaphase PPT; Mol.wt. 35,000); [c] system composition: 7% wt. Dex-500 — 5% wt. PEG-8000; 4.2% wt. Dex-500 — 6% wt. PVA-14000; 4% wt. Dex-500 — 11.5% wt. Ficoll-400; 15% wt. HPS — 5.8% wt. PEG-8000; salt composition the same in all the systems: 0.025 M sodium phosphate buffer, pH 7.5.
(From G. Johansson, M. Joelsson, J.Chromatogr., 411, 161 (1987). Reprinted by permission of Elsevier Scientific Publishing Co.)

The advantages of the PEG—salt systems are that the effects of all the above factors on the solute partitioning are generally stronger in these systems compared with those observed in aqueous two—polymer systems. Phase diagrams for many of these systems are known (see in Chapter 10) providing at least some guidance for choosing phase polymer and salt concentrations needed to obtain a two-phase system with 1:1, or any other volume ratio, as needed. Fine-tuning of the separation conditions in these systems may be achieved by adjusting the concentrations of the phase-forming components, changing the type of the cation in the phase-forming salt, increasing or decreasing concentration of the salt additive, adjusting the pH value, etc. One of the practical advantages of these systems is the short settling time and relatively low viscosity of the phases, compared to those in aqueous two-polymer systems.

If the aqueous PEG—salt systems cannot be used or are found to be unsuitable for separation, the Dex—PEG systems seem to be a reasonable second choice. Once again, two different systems formed by PEG of the molecular weight of about 1000 to 3000 and PEG-8000 and Dex of any molecular weight from 40,000 to 500,000 should be used. Molecular weight of dextran seems to be of secondary importance at this stage of development. Effects of two different salt additives should be examined. Sodium (potassium, ammonium) sulfate, or phosphate and sodium (potassium, ammonium) thiocyanate may be recommended as the salt additives most likely to provide an opposite influence on solute partitioning. These additives may be used at concentrations from about 0.05 M (for sulfate/phosphate) or 0.1 M (for thiocyanates) to as high as 1 to 2 M depending on the particular properties of the biomolecules to be separated. pH effects over the range dependent on the biomolecules under study should be examined. As shown above, increasing the concentrations of the phase-forming polymers generally increases the affinity of a given solute for the phase it favors initially. Hence, once the partition coefficients K of the solutes to be separated are known at a given polymer concentration and salt composition, the linear lnK—ΔPEG relationships may be plotted and extrapolated to the larger polymer concentrations to find the trends and make a reasonably intelligent decision on how to further manipulate solute partitioning.

If the solute of interest favors the Dex-rich phase in the Dex—PEG systems under all the aforementioned conditions, PEG may be replaced with a less hydrophobic polymer, such as PVA or Ficoll, for example. The Dex—Ficoll system is usually not an optimal choice for separation because phase settling takes too long (21 to 24 hours) and cannot be accelerated by low-speed centrifugation. Aquaphase—PEG system seems to be very appropriate in some cases.

If the target analyte in the aqueous Dex—PEG systems still remains in the PEG-rich phase, PEG may be replaced with PVP in the Dex—PVP system or PEG—PVP system may be a good choice.

Information about physicochemical features of the biomolecules to be separated is critically important for method development. Development of separation methods for small organic compounds by liquid chromatography is based on the view that the general chemical structure predetermines selection of the mode of the liquid chromatography (for example, normal or reversed-phase mode). That, in turn, governs the choice of the solid matrix and commonly used mobile phase composition, with further adjustments and fine-tuning as needed. Spatial structure and physicochemical properties of conformationally flexible biomolecules vary significantly with solvent composition, pH, temperature, and so on, and are understood much less than those of small organic compounds. Extraction procedures to separate proteins from nucleic acids, or proteins from polysaccharides (see, for example, in [1-5]) are usually readily developed. It is much more difficult to separate a subpopulation of proteins coexisting in the same biological liquid, such as serum γ-globulins for example. Even the specific physicochemical information about these proteins remains inadequate at present. An important advantage of the systematic method development for bioseparation by aqueous two-phase partitioning is that the fundamentally important information is obtained in the process. That makes the process of method development in the field of aqueous two-phase partitioning challenging, intellectually rewarding, and fundamentally important for biomedical research.

9.4. SUMMARY

It clearly follows from the foregoing that the aqueous two-phase partitioning technique is an efficient separation method, even though in its infancy. The technique is widely recognized as a very powerful separation tool for biological particles such as cells, viruses, and subcellular organelles. This area of technique application is extremely important, especially in view of the limited number of alternative separation methods currently available for biological particles. There are, however, so many highly successful and well established techniques for separation of biomolecules that one may ask why we need an additional separation technique.

The answer provided above is pretty clear: separations available with aqueous two-phase techniques often cannot be achieved by any other method. The reason is that the basic principle of separation by extraction in aqueous two-phase systems is completely different from those realized in other commonly used separation techniques such as electrophoresis, ion-exchange, or size-exclusion chromatography, etc. That means, for example, that a heterogeneity in apparently homogeneous samples may be detected, and otherwise inseparable subpopulations may be isolated and their structure and function studied. Low cost and high efficiency of readily scaled up extraction procedures for separation and purification of biomolecules on an industrial scale are additional important factors, but beyond the scope of the present work.

All these reasons justify research efforts necessary for further development of the aqueous two-phase partition technique. Further design and advancement of other technical procedures (affinity partitioning, column and countercurrent chromatography, etc.) are undoubtedly important. Much more critical for the further development of the partition technique is the essentially neglected issue of new aqueous polymer systems. I believe that future research efforts should be focused on extending the variety of phase-forming polymers.

Comparative analysis of partitioning of biomolecules and small solutes (from inorganic ions to homologous series of organic compounds) in aqueous two-phase systems formed by polymers of various chemical nature is currently a critical need. That information will establish the "ground rules" needed for solute partitioning and selection of separation conditions. Equally important appears to be the study of the solvent features of aqueous media in the phases, as functions of their polymer (and salt) composition. This research may appear to be very time-consuming and tedious, but without it the great potential of the partition technique may never be completely realized. The positive and extremely important and rewarding aspect of research along these lines, is that the information obtained in the process would be of great fundamental interest and importance, not only from the partition technique development viewpoint but also with regard to better insight into the function and potency relevant properties of biomolecules in solution.

REFERENCES:

1. P. Å. Albertsson, Partition of Cell Particles and Macromolecules, 3rd. ed., Wiley, New York, 1986.
2. Partitioning in Aqueous Two-Phase Systems: Theory, Methods, Uses, and Applications to Biotechnology (H. Walter, D. E. Brooks, D. Fisher, eds.), Academic Press, Orlando, Fla, 1985.
3. Separations Using Aqueous Phase Systems: Applications in Cell Biology and Biotechnology (D. Fisher and I. A. Sutherland, eds.), Plenum Press, New York, 1989.
4. W. Muller, Liquid-Liquid Partition Chromatography of Biopolymers, GIT Verlag, Darmstadt, 1988.
5. Methods in Enzymology, Vol. 222 (H. Walter, G. Johansson, eds.), Academic Press, San Diego, 1994.
6. M. R. Kula, Bioseparation, 1, 181 (1990).
7. H. Hustedt, K. H. Kroner, M. R. Kula, In: Partitioning in Aqueous Two-Phase Systems: Theory, Methods, Uses, and Applications to Biotechnology (H. Walter, D. E. Brooks, D. Fisher, eds.), Academic Press, Orlando, Fla, 1985, pp.529-587.
8. H. Hustedt, K. H. Kroner, H. Schutte, M. R. Kula, In: Enzyme Technology (R.M. Lafferty, ed.), Springer-Verlag, Berlin, 1983, pp.135-145.
9. P. Å. Albertsson, G. Johansson, F. Tjerneld, In: Separation Processes in Biotechnology (J. A. Asenjo, ed.), Marcel Dekker, 1990, pp.287-327.

10. J. G. Huddleston, A. Lyddiatt, Appl. Biochem. Biotechnol., 26, 249 (1990).
11. C. A. Haynes, F. J. Benitez, H. W. Blanch, J. M. Prausnitz, AIChE J., 39, 1539 (1993).
12. V. B. Lowlis, H. Heinsohn, LC-GC, 11, 720 (1993).
13. R. A. Hart, P. M. Lester, D. H. Reifsnyder, J. R. Ogez, S. E. Builder, Book of Abstracts, 207th ACS National Meeting, San Diego 1994, American Chemical Society, San Diego, CA, 1994, Abstract # 89 (I&EC).
14. A. D. Diamond, K. Yu, J. T. Hsu, ACS Symposium Series, # 427, 52 (1990).
15. W. D. Conway, Countercurrent Chromatography. Apparatus, Theory, and Applications, VCH., New York, 1989.
16. D. Forciniti, C. K. Hall, M. R. Kula, Biotechnol.Bioeng., 38, 986 (1991).
17. L. M. Miheeva, J. Fitzpatrick, 1992, unpublished data.
18. S. Sturesson, F. Tjerneld, G. Johansson, Appl.Biochem.Biotechnol., 26, 281 (1990).
19. D. Forciniti, C. K. Hall, M. R. Kula, Bioseparation, 2, 115 (1991).
20. D. Forciniti, C. K. Hall, M. R. Kula, Chem.Eng.Sci., 47, 165 (1992).
21. E. A. Masimov, B. Y. Zaslavsky, A. V. Baevsky, S. V. Rogozhin, A. V. Gedrovich, A. V. Shishkov, V. D. Scherbukhin, Prikladnaya Biokhimia i Microbiologia (Rus.), 20, 733 (1984).
22. R. Kuboi, H. Tanaka, K. Yano, I. Komasawa, Advances in Bioseparation Engineering 1991, 51, (1991).
23. F. D. Raymond, D. W. Moss, D. Fisher, Biochim.Biophys.Acta, 1156, 117 (1993).
24. J. R. Luther, C. E. Glatz, Biotechnol.Bioeng., 44, 147 (1994).
25. S.O. Enfors, K. Kohler, A. Veide, Bioseparation, 1, 305 (1990).
26. A. Veide, L. Strandberg, H. Hellebust, S.O. Enfors, In: Separations Using Aqueous Phase Systems: Applications in Cell Biology and Biotechnology (D. Fisher, I. A. Sutherland, eds.), New York, Plenum Press, 1986, pp. 317-324
27. A. Veide, L. Strandberg, S.-O. Enfors, Enzyme Microb. Technol., 9, 730 (1987).
28. K. Kohler, C. Ljungquist, A. Kondo, A. Veide, B. Nilsson, Bio/technology, 9, 642 (1991).
29. R. J. Todd, M. E. Van Dam, D. Casimiro, B. L. Haymore, F. H. Arnold, Proteins: Structure, Function and Genetics, 10, 156 (1991).
30. M. A. Eiteman, C. Hassinen, A. Veide, 1994, in press.

31. R. D. Rogers, A. H. Bond, C. B. Bauer, Sep.Sci.Technol., 28, 1091 (1993).
32. R. D. Rogers, C. B. Bauer, A. H. Bond, presented on 8th Symposium on Separation Science and Technology for Energy Applications, Gattinburg, TN, October, 1993.
33. W. H. Wang, R. Kuboi, I. Komasawa, Chem.Eng.Sci., 47, 113 (1992).
34. K. Yano, A. Wakayama, R. Kuboi,I. Komasawa, Bunseki Kagaku, 42, 673 (1993).
35. K. A. Giuliano, Anal. Biochem., 197, 333 (1991).
36. G. Johansson, In: Partitioning in Aqueous Two-Phase Systems: Theory, Methods, Uses, and Applications to Biotechnology (H. Walter, D.E. Brooks, D. Fisher, eds.), Academic Press, Orlando, Fla, 1985, pp.161-226.
37. H. Walter, G. Johansson, D. E. Brooks, Anal.Biochem., 197, 1 (1991).
38. G. Kopperschlager, G. Birkenmeier, Bioseparation, 1, 235 (1990).
39. G. Kopperschlager, G. Birkenmeier, In: Molecular Interactions in Bioseparation (T. Ngo, ed.), Plenum, New York, 1992.
40. G. Kopperschlager, J. Kirchberger, In: Immobilized Macromolecules: Application Potentials (U. B. Sleytr, P. Messner, D. Pum, M. Sara, eds.), Springer-Verlag, Berlin, 1993, pp. 87-104.
41. S. D. Flanagan, S. Barondes, J.Biol.Chem., 250, 1484 (1975).
42. J. M. Harris, M. Yalpani, In: Partitioning in Aqueous Two-Phase Systems: Theory, Methods, Uses, and Applications to Biotechnology (H. Walter, D.E. Brooks, D. Fisher, eds.), Academic Press, Orlando, Fla, 1985, pp. 589-626.
43. J. M. Harris, ed., Poly(Ethylene Glycol) Chemistry: Biotechnical and Biomedical Applications, Plenum, New York, 1992.
44. Y. Guan, X.-Y. Wu, T. E. Treffry, T. H. Lilley, Biotechnol.Bioeng., 40, 517 (1992).
45. S. D. Plunkett, F. H. Arnold, Biotechnol.Techniques, 4, 45 (1990).
46. S. Suh, F. H. Arnold, Biotechnol.Bioeng., 35, 682 (1990).
47. L. Elling, M.-R. Kula, Biotechnol.Appl.Biochem., 13, 354 (1991).
48. L. Elling, M.-R. Kula E. Hadas, E. Katchalski-Katzir, Anal.Biochem., 192, 74 (1991).
49. G. Birkenmeier, E. Usbeck, G. Kopperschlager, Anal.Biochem., 136, 264 (1984).
50. M. C. Tejedor, C. Delgado, M. Grupeli, J. Luque, J.Chromatogr., 589, 127 (1992).
51. J. Kirchberger, G. Kopperschlager, M. A. Vijayalakshmi, J.Chromatogr., 557, 325 (1991).
52. J. Kirchberger, U. Domar, G. Kopperschlager, T. Stigbrand, J.Chromatogr., 574, 237 (1992).

53. P. A. Alred, F. Tjerneld, A. Kozlowski, J. M. Harris, Bioseparation, 2, 363 (1992).
54. V. P. Shanbhag, G. Johansson, Biochem.Biophys.Res.Comm., 61, 1141 (1974).
55. V. P. Shanbhag, C. G. Axelsson, Eur.J.Biochem., 60, 17 (1975).
56. G. Birkenmeier, L. Carlsson-Bostedt, V. Shanbhag, T. Kriegel, G. Kopperschlager, L. Sottrup-Jensen, T. Stigbrand, Eur.J.Biochem., 183, 239 (1989).
57. D. E. Brooks, K. A. Sharp, D. Fisher, In: Partitioning in Aqueous Two-Phase Systems: Theory, Methods, Uses, and Applications to Biotechnology (H. Walter, D.E. Brooks, D. Fisher, eds.), Academic Press, Orlando, Fla, 1985, pp. 11-84.
58. A. Cordes, J. Flossdorf, M.-R. Kula, Biotechnol.Bioeng., 30, 514 (1987).
59. G. Johansson, M. Joelsson, J.Chromatogr., 393, 195 (1987).
60. G. Johansson, M. Joelsson, J.Chromatogr., 411, 161 (1987).
61. P. P. Godbole, R. S. Tsai, W. M. Clark, Biotechnol.Bioeng., 38, 535 (1991).
62. J. N. Baskir, T. A. Hatton, U. W. Suter, J.Phys.Chem., 93, 969 (1989).
63. P. A. Alred, G. Johansson, F. Tjerneld, Anal.Biochem., 205, 351 (1992).
64. P. Å. Albertsson, Meth.Biochem.Anal., 29, 1 (1983).
65. T. E. Treffry, P. T. Sharpe, H. Walter, D. E. Brooks, In: Partitioning in Aqueous Two-Phase Systems: Theory, Methods, Uses, and Applications to Biotechnology (H. Walter, D.E. Brooks, D. Fisher, eds.), Academic Press, Orlando, Fla, 1985, pp. 132-148.
66. H. -E. Åkerlund, J. Biochem. Biophys. Methods, 9, 133 (1984).
67. A. Ortin, M. J. Lopez-Perez, M. T. Muino-Blanco, J. A. Cebrian-Perez, J.Chromatogr., 558, 357 (1991).
68. W. Muller, Eur.J.Biochem., 155, 213 (1986).
69. W. Muller, Bioseparation, 1, 265 (1990).
70. A. Walsdorf, M. R. Kula, J.Chromatogr., 542, 55 (1991).
71. M. R. Kula, L. Elling, A. Walsodrf, J.Chromatogr., 548, 3 (1991).
72. L. R. Snyder, J. J. Kirkland, Introduction to Modern Liquid Chromatography, Wiley-Interscience, New York, 2nd ed., 1979.
73. W. Muller, In: Partitioning in Aqueous Two-Phase Systems: Theory, Methods, Uses, and Applications to Biotechnology (H. Walter, D. E. Brooks, D. Fisher, eds.), Academic Press, Orlando, Fla, 1985, pp. 227-266.
74. A. Walsdorf, D. Forciniti, M. R. Kula, J.Chromatogr., 523, 103 (1990).

75. K. Andersson, C. Wingren, U. B. Hansson, Scand. J. Immunol., 38, 95 (1993).
76. U. B. Hansson, K. Andersson, Y. Liu, P. Å. Albertsson,, Anal. Biochem., 183, 305 (1989).
77. I. A. Sutherland, In: Partitioning in Aqueous Two-Phase Systems: Theory, Methods, Uses, and Applications to Biotechnology (H. Walter, D.E. Brooks, D. Fisher, eds.), Academic Press, Orlando, Fla, 1985, pp. 149-159.
78. Countercurrent Chromatography: Theory and Practice (N. B. Mandava, Y. Ito, eds.) Marcel Dekker, New York, 1988.
79. A. P. Foucault, Anal. Chem., 63, 569A (1991).
80. E. Oka, H. Oka, Y. Ito, J.Chromatogr., 538, 99 (1991).
81. P. W. Armstrong, Anal.Chem., 60, 2513 (1988).
82. S. B. Sawant, S. K. Sikdar, J. B. Joshi, Biotechnol.Bioeng., 36, 109 (1990).
83. S. J. Gluck, M. P. Wingeier, J.Chromatogr., 547, 69 (1991).
84. K. S. M. S. R. Raghava Rao, R. Stewart, P. Todd, Sep. Sci. Technol., 26, 257 (1991).
85. S. V. Save, V. G. Pangarkar, S. V. Kumar, Biotechnol.Bioeng., 41, 72 (1993).
86. A. Foucault, K. Nakanishi, J. Liquid Chromatogr., 13, 2421 (1990).
87. Y. Shibusawa, Y. Ito, J.Chromatogr., 550, 695 (1991).
88. Y. Shibusawa, Y. Ito, J. Liquid Chromatogr., 15, 2787 (1992).
89. X. Lei, J. T. Hsu, J. Liquid Chromatogr., 15, 2801 (1992).
90. P. Å. Albertsson, A. Cajarville, D. E. Brooks, F. Tjerneld, Biochim. Biophys. Acta, 926, 87 (1987).

CHAPTER 10.
PHASE DIAGRAMS

Until recently the monograph published by P.Å. Albertsson [1] was the only source of phase diagrams of various aqueous polymer systems. The last edition of the book was published in 1986, and many new phase diagrams for different aqueous polymer systems have been reported in the literature since. Composition of the coexisting phases is of the primary importance for consideration of phase separation in aqueous polymer systems, properties of the phases, partition behavior of solutes or particles in the two-phase systems, etc. Hence an attempt to compile the phase diagrams for aqueous two-phase systems available in the literature seems to be worthwile.

The only criterion used for selection of the diagrams presented below was that the composition of the phases was reported in an original publication in numerical form. It was presumed that any analysis of diagrams reported in graphic form might increase an original (usually unspecified) experimental error too much to be of any reasonable value. In some cases only compositions of the phases and not of the total system were reported in the literature and they are presented the same way below. In many cases the authors did not report composition of the critical point of phase diagram, and hence the central part of the binodial curve is somewhat arbitrary.

Methods of analysis of the polymer and salt composition of phases have been extensively covered in the literature [1,2], and are not discussed here. Compositions of the phases are given for polymers and salts (where known), concentration of water is not given as it may be readily calculated by subtracting the concentrations of the components given from 100%. Questionable data are denoted in the tables with short explanations why they are considered as such. Otherwise no comments are given. I believe that those interested enough to look into the data presented below have read the previous chapters and may make their own conclusions.

The data are presented below in the following order:

Systems:	Table and Figure Numbers
Dex—PEG—(salt)—water	10.1-102
Dex—Polyvinylpyrrolidone—(salt)—water	103-121
Dex—Poly(vinyl alcohol)—(salt)—water	122-130
Dex—Ficoll—(salt)—water	131-133
PEG—Poly(vinyl methyl ether)—water	134-136
PEG—salt—water	137-163

REFERENCES:

1. P. Å. Albertsson, Partition of Cell Particles and Macromolecules, 3rd ed., Wiley, New York, 1986.
2. S. Bamberger, D. E. Brooks, K. A. Sharp, J. M. Van Alstine, T. J. Webber, In: Partitioning in Aqueous Two-Phase Systems: Theory, Methods, Uses, and Applications to Biotechnology (H. Walter, D. E. Brooks, D. Fisher, eds.), Academic Press, Orlando, FL, 1985, pp.85-130.

Table 10.1. Phase Diagram and Phase Composition of the Dextran—Poly(ethylene glycol) System Dex-40—PEG-3400 at 22°C.
(From A. D. Diamond, J. T. Hsu, Biotechnol. Bioeng., 34, 1000 (1989) with permission of John Wiley & Sons, Inc.)

PEG - M_w ~3400; Manufacturer: Aldrich (Milwaukee, WI, USA); Lot 00917PT
Dextran T-40 - M_w 40,200; M_n 24,400; Manufacturer: Pharmacia Fine Chemicals (Piscataway, NJ, USA); Lot 01852

Total system		Bottom phase		Top phase		STL*
PEG % w/w	Dex % w/w	PEG % w/w	Dex % w/w	PEG % w/w	Dex % w/w	
6.21	9.59	5.24	11.68	8.11	6.16	-0.520
6.59	10.01	3.74	15.98	9.93	3.41	-0.492
7.32	9.91	2.78	18.63	11.11	2.35	-0.512
8.22	9.91	2.20	21.25	12.48	1.56	-0.522
					av.:	-0.512 ±0.014

* - STL - Tie-Line Slope defined as the ratio STL = (ΔPEG)/(ΔDex) where Δ is the difference between the concentrations of a given polymer in the two coexisting phases.

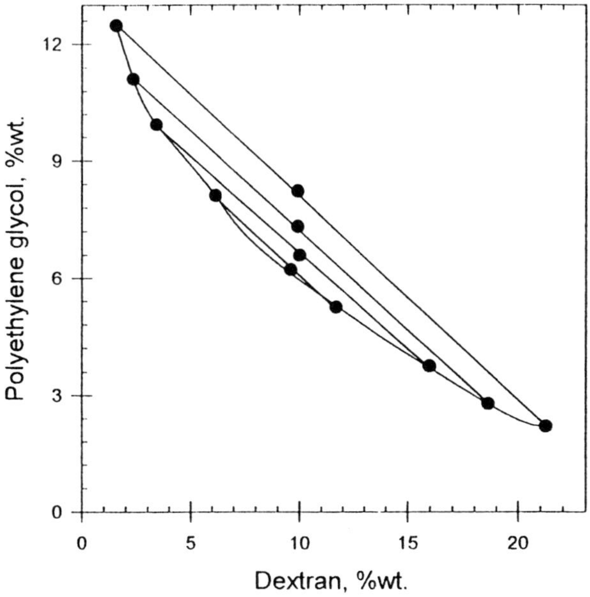

Table 10.2. Phase Diagram and Phase Composition of the Dextran—Poly(ethylene glycol) System Dex-40—PEG-3400 at 4°C.
(From A. D. Diamond, J. T. Hsu, Biotechnol. Bioeng., 34, 1000 (1989) with permission of John Wiley & Sons, Inc.)

PEG - M_w ~3400; Manufacturer: Aldrich (Milwaukee, WI, USA); Lot 00917PT
Dextran T-40 - M_w 38,8000; M_n 24,200; Manufacturer: Pharmacia Fine Chemicals (Piscataway, NJ, USA); Lot 03375

Total system		Bottom phase		Top phase		STL*
PEG % w/w	Dex % w/w	PEG % w/w	Dex % w/w	PEG % w/w	Dex % w/w	
6.50	8.80	3.28	15.83	8.82	3.70	-0.457
6.70	9.30	2.77	17.58	9.62	2.86	-0.465
6.90	10.00	2.39	19.57	10.71	2.13	-0.477
7.10	10.60	2.09	21.21	11.19	1.76	-0.468
					av.:	-0.467 ±0.008

* - STL - Tie-Line Slope defined as the ratio STL = (ΔPEG)/(ΔDex) where Δ is the difference between the concentrations of a given polymer in the two coexisting phases.

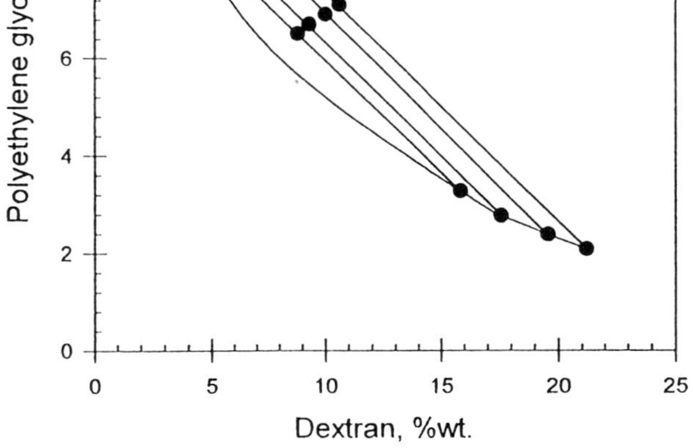

Table 10.3. Phase Diagram and Phase Composition of the Dextran—Poly(ethylene glycol) System Dex-70—PEG-3400 at 22°C.
(From A. D. Diamond, J. T. Hsu, Biotechnol. Bioeng., 34, 1000 (1989) with permission of John Wiley & Sons, Inc.)

PEG - M_w ~3400; Manufacturer: Aldrich (Milwaukee, WI, USA); Lot 00917PT
Dextran T-70 - M_w 72,200; M_n 38,400; Manufacturer: Pharmacia Fine Chemicals (Piscataway, NJ, USA); Lot 02377

Total system		Bottom phase		Top phase		STL*
PEG % w/w	Dex % w/w	PEG % w/w	Dex % w/w	PEG % w/w	Dex % w/w	
6.12	8.54	5.16	10.71	7.77	4.87	-0.447
6.76	8.16	3.73	14.55	9.46	2.55	-0.478
6.58	9.10	3.10	16.31	10.20	1.98	-0.495
7.12	9.07	2.44	18.15	11.09	1.29	-0.513
					av.:	-0.483 ±0.028

* - STL - Tie-Line Slope defined as the ratio STL = (ΔPEG)/(ΔDex) where Δ is the difference between the concentrations of a given polymer in the two coexisting phases.

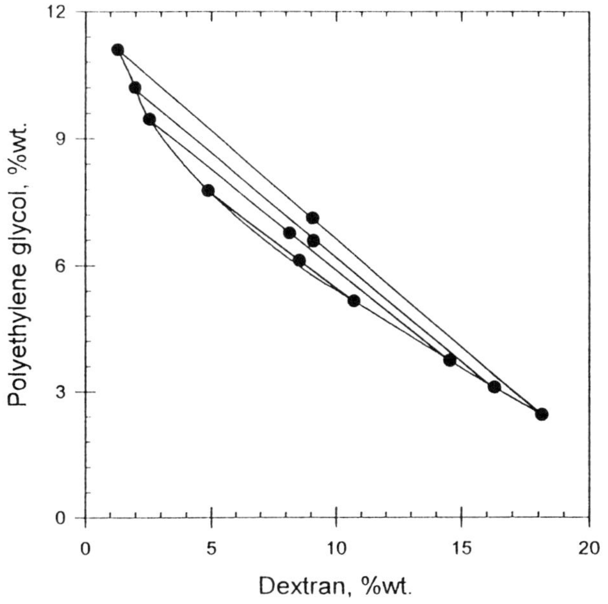

Table 10.4. Phase Diagram and Phase Composition of the Dextran—Poly(ethylene glycol) System Dex-70—PEG-3400 at 4°C.
(From A. D. Diamond, J. T. Hsu, Biotechnol. Bioeng., 34, 1000 (1989) with permission of John Wiley & Sons, Inc.)

PEG - M_w ~3400; Manufacturer: Aldrich (Milwaukee, WI, USA); Lot 00917PT
Dextran T-70 - M_w 72,200; M_n 38,400; Manufacturer: Pharmacia Fine Chemicals (Piscataway, NJ, USA); Lot 02377

Total system		Bottom phase		Top phase		STL*
PEG % w/w	Dex % w/w	PEG % w/w	Dex % w/w	PEG % w/w	Dex % w/w	
6.30	7.30	3.06	14.62	8.24	2.32	-0.421
6.45	7.80	2.82	15.89	8.96	1.82	-0.436
6.55	8.40	2.61	17.34	9.54	1.48	-0.437
6.70	9.00	2.42	18.93	10.07	1.18	-0.431
					av.:	-0.431 ±0.007

* - STL - Tie-Line Slope defined as the ratio STL = (ΔPEG)/(ΔDex) where Δ is the difference between the concentrations of a given polymer in the two coexisting phases.

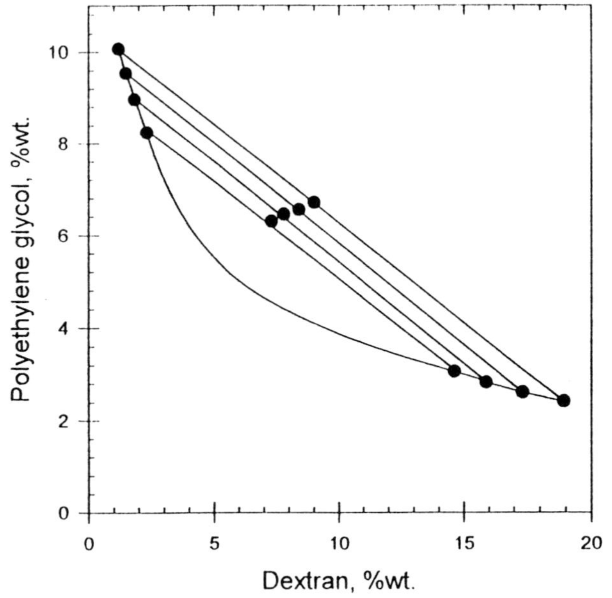

Table 10.5. Phase Diagram and Phase Composition of the Dextran—Poly(ethylene glycol) System Dex-500—PEG-3400 at 22°C.
(From A. D. Diamond, J. T. Hsu, Biotechnol. Bioeng., 34, 1000 (1989) with permission of John Wiley & Sons, Inc.)

PEG - M_w ~3400; Manufacturer: Aldrich (Milwaukee, WI, USA); Lot 00917PT
Dextran T-500 - M_w 507,000; M_n 234,200; Manufacturer: Pharmacia Fine Chemicals (Piscataway, NJ, USA); Lot 05163

Total system		Bottom phase		Top phase		STL*
PEG % w/w	Dex % w/w	PEG % w/w	Dex % w/w	PEG % w/w	Dex % w/w	
6.00	6.14	4.08	10.77	8.41	0.94	-0.440
6.50	6.50	3.56	13.44	9.11	0.43	-0.427
7.00	7.00	3.26	15.84	9.88	0.19	-0.423
8.00	8.00	2.57	20.03	11.59	0.04	-0.451
					av.:	-0.435 ±0.013

* - STL - Tie-Line Slope defined as the ratio STL = (ΔPEG)/(ΔDex) where Δ is the difference between the concentrations of a given polymer in the two coexisting phases.

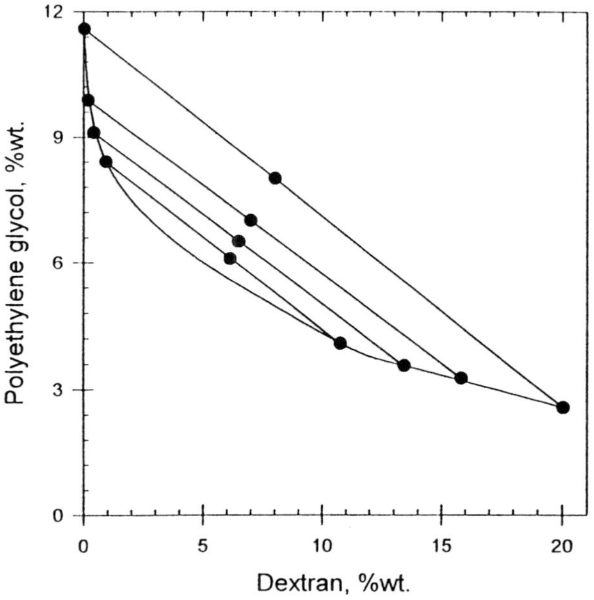

Table 10.6. Phase Diagram and Phase Composition of the Dextran—Poly(ethylene glycol) System Dex-500—PEG-3400 at 4°C.
(From A. D. Diamond, J. T. Hsu, Biotechnol. Bioeng., 34, 1000 (1989) with permission of John Wiley & Sons, Inc.)

PEG - M_w ~3400; Manufacturer: Aldrich (Milwaukee, WI, USA); Lot 00917PT
Dextran T-500 - M_w 507,000; M_n 234,200; Manufacturer: Pharmacia Fine Chemicals (Piscataway, NJ, USA); Lot 05163

Total system		Bottom phase		Top phase		STL*
PEG % w/w	Dex % w/w	PEG % w/w	Dex % w/w	PEG % w/w	Dex % w/w	
5.00	5.70	3.45	9.20	6.43	1.51	-0.388
6.50	7.00	2.10	17.48	9.39	0.12	-0.420
7.00	8.00	1.98	20.12	10.28	0.07	-0.414
8.00	8.00	1.50	22.77	11.51	0.04	-0.440
					av.:	-0.416 ±0.021

* - STL - Tie-Line Slope defined as the ratio STL = (ΔPEG)/(ΔDex) where Δ is the difference between the concentrations of a given polymer in the two coexisting phases.

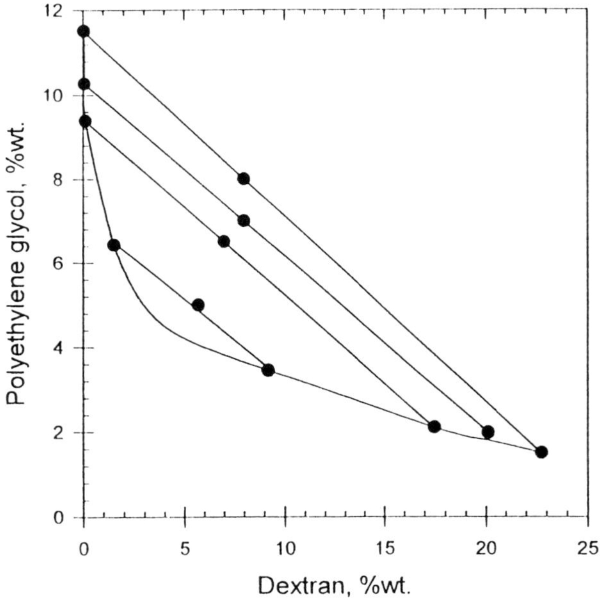

Table 10.7. Phase Diagram and Phase Composition of the Dextran—Poly(ethylene glycol) System Dex-10—PEG-4000 at 25°C.
(From D. Forciniti, C. K. Hall, M.-R. Kula, Fluid Phase Equilibria, 61, 243 (1991) and Bioseparation, 2, 115 (1991) by permission of Elsevier Scientific Publishing Co. and Kluwer Academic Publishers, respectively.)

PEG - M_w 4,100; M_n 3,800; Manufacturer: Fluka AG (Switzerland); Lot not indicated;
Dextran - M_w 19,300; M_n 13,200; Manufacturer: Fluka AG (Switzerland); Lot not indicated

Total system		Bottom phase		Top phase		STL*
PEG % w/w	Dex % w/w	PEG % w/w	Dex % w/w	PEG % w/w	Dex % w/w	
7.71	10.29	4.0	18.6	11.1	4.4	-0.500
7.81	11.23	3.5	20.8	12.3	3.4	-0.506
8.38	12.20	2.9	23.6	13.9	2.5	-0.521
8.74	13.23	2.6	26.3	15.3	1.9	-0.520
					av.:	-0.512 ±0.010

*- STL - Tie-Line Slope defined as the ratio STL = (ΔPEG)/(ΔDex) where Δ is the difference between the concentrations of a given polymer in the two coexisting phases.

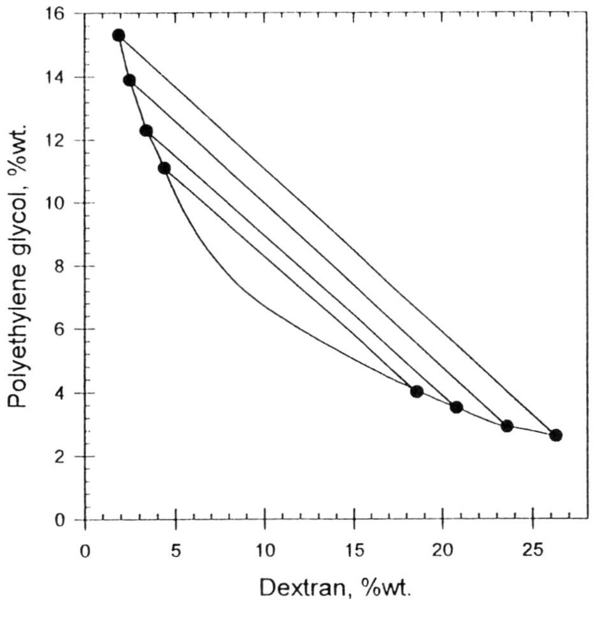

Table 10.8. Phase Diagram and Phase Composition of the Dextran—Poly(ethylene glycol) System Dex-10—PEG-4000 at 4°C.
(From D. Forciniti, C. K. Hall, M.-R. Kula, Fluid Phase Equilibria, 61, 243 (1991) and Bioseparation, 2, 115 (1991) by permission of Elsevier Scientific Publishing Co. and Kluwer Academic Publishers, respectively.)

PEG - M_w 4,100; M_n 3,800; Manufacturer: Fluka AG (Switzerland); Lot not indicated;
Dextran - M_w 19,300; M_n 13,200; Manufacturer: Fluka AG (Switzerland); Lot not indicated

Total system		Bottom phase		Top phase		STL*
PEG % w/w	Dex % w/w	PEG % w/w	Dex % w/w	PEG % w/w	Dex % w/w	
7.6	10.3	2.8	20.4	10.9	3.4	-0.476
7.9	11.3	2.6	22.6	12.0	2.7	-0.472
8.4	12.2	2.1	25.1	13.4	2.0	-0.489
8.7	13.2	1.5	27.0	14.0	1.6	-0.492
					av.:	-0.482 ± 0.010

*- STL - Tie-Line Slope defined as the ratio STL = (ΔPEG)/(ΔDex) where Δ is the difference between the concentrations of a given polymer in the two coexisting phases.

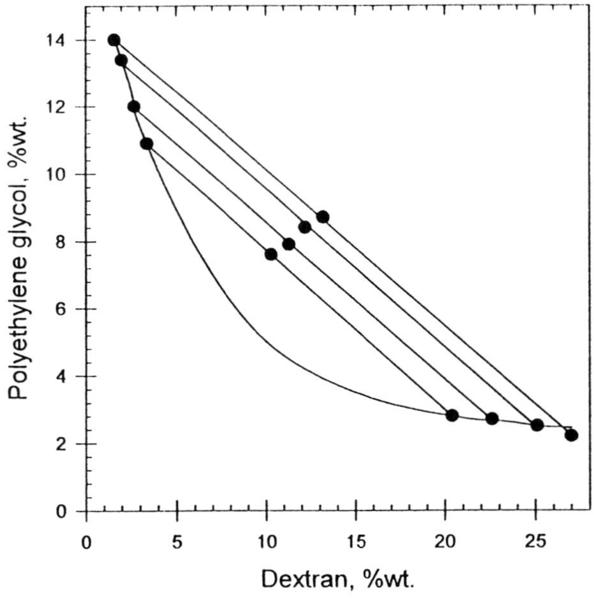

Table 10.9. Phase Diagram and Phase Composition of the Dextran—Poly(ethylene glycol) System Dex-10—PEG-4000 at 40°C.
(From D. Forciniti, C. K. Hall, M.-R. Kula, Fluid Phase Equilibria, 61, 243 (1991) and Bioseparation, 2, 115 (1991) by permission of Elsevier Scientific Publishing Co. and Kluwer Academic Publishers, respectively.)

PEG - M_w 4,100; M_n 3,800; Manufacturer: Fluka AG (Switzerland); Lot not indicated;
Dextran - M_w 19,300; M_n 13,200; Manufacturer: Fluka AG (Switzerland); Lot not indicated

Total system		Bottom phase		Top phase		STL*
PEG % w/w	Dex % w/w	PEG % w/w	Dex % w/w	PEG % w/w	Dex % w/w	
7.6	10.3	4.4**	15.4**	10.7**	5.3**	-0.624**
7.9	11.3	2.6	20.4	12.2	3.7	-0.575
8.4	12.2	2.3	22.6	14.2	2.6	-0.595
8.7	13.2	2.4	24.0	15.3	2.0	-0.586
					av.:	-0.585 ± 0.010

*- STL - Tie-Line Slope defined as the ratio STL = (ΔPEG)/(ΔDex) where Δ is the difference between the concentrations of a given polymer in the two coexisting phases;
** - Composition of the phases questionable as the STL value is inconsistent with the other values; was not used in calculations of the average STL value.

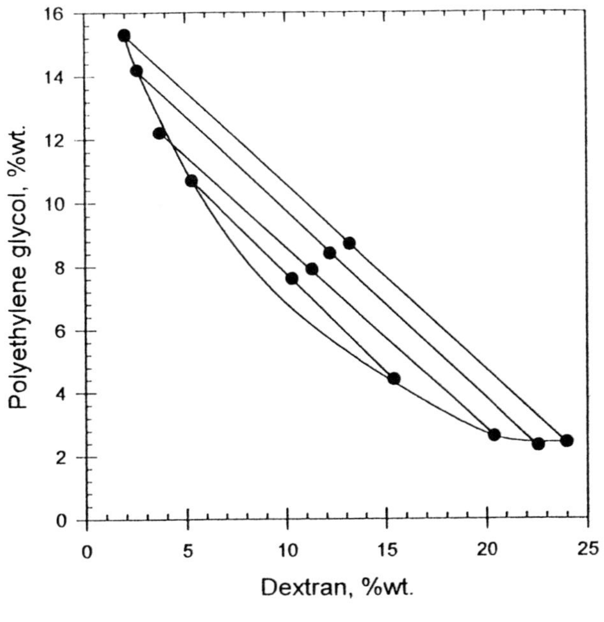

Table 10.10. Phase Diagram and Phase Composition of the Dextran—Poly(ethylene glycol) System Dex-40—PEG-4000 at 25°C.
(From D. Forciniti, C. K. Hall, M.-R. Kula, Fluid Phase Equilibria, 61, 243 (1991) and Bioseparation, 2, 115 (1991) by permission of Elsevier Scientific Publishing Co. and Kluwer Academic Publishers, respectively.)

PEG - M_w 4100; M_n 3800; Manufacturer: Fluka AG (Switzerland); Lot not indicated;
Dextran - M_w 37,000; M_n 27,700; Manufacturer: Pfeifer and Langen (Dormagen, Germany); Lot not indicated

Total system		Bottom phase		Top phase		STL*
PEG % w/w	Dex % w/w	PEG % w/w	Dex % w/w	PEG % w/w	Dex % w/w	
5.7	10.0	4.0**	13.2**	8.0**	4.8**	-0.476**
7.6	10.3	1.9	21.2	11.9	1.3	-0.503
7.9	11.3	1.7	23.1	12.9	1.0	-0.507
8.3	12.1	1.5	25.0	14.0	0.7	-0.514
					av.:	-0.508
						± 0.006

*- STL - Tie-Line Slope defined as the ratio STL = (ΔPEG)/(ΔDex) where Δ is the difference between the concentrations of a given polymer in the two coexisting phases;
** - Composition of the phases questionable as the STL value is inconsistent with the other values; was not used in calculations of the average STL value.

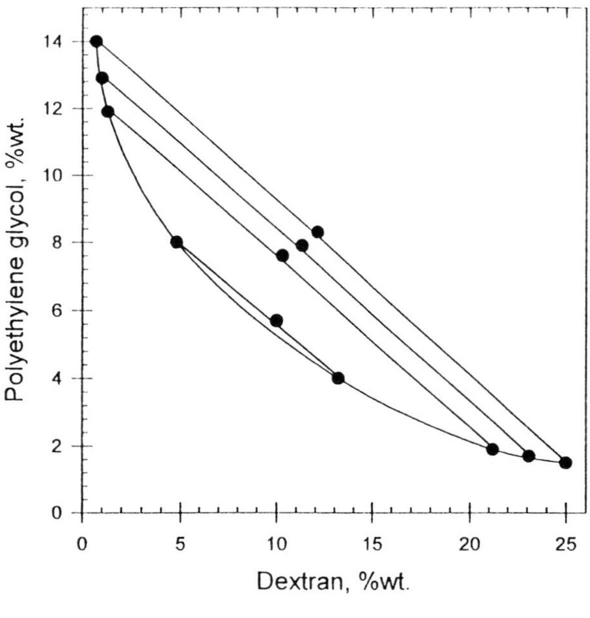

Table 10.11. Phase Diagram and Phase Composition of the Dextran—Poly(ethylene glycol) System Dex-40—PEG-4000 at 4°C.
(From D. Forciniti, C. K. Hall, M.-R. Kula, Fluid Phase Equilibria, 61, 243 (1991) and Bioseparation, 2, 115 (1991) by permission of Elsevier Scientific Publishing Co. and Kluwer Academic Publishers, respectively.)

PEG - M_w 4100; M_n 3800; Manufacturer: Fluka AG (Switzerland); Lot not indicated;
Dextran - M_w 37,000; M_n 27,700; Manufacturer: Pfeifer and Langen (Dormagen, Germany); Lot not indicated

Total system		Bottom phase		Top phase		STL*
PEG % w/w	Dex % w/w	PEG % w/w	Dex % w/w	PEG % w/w	Dex % w/w	
5.6	10.0	3.0**	15.8**	8.3**	3.4**	-0.427**
7.6	10.3	1.6	23.5	12.3	0.96	-0.475
7.9	11.3	1.0***	25.2***	12.7	0.69	-0.477
8.4	12.2	1.0	27.3	13.9	0.48	-0.481
					av.:	-0.478 ± 0.003

*- STL - Tie-Line Slope defined as the ratio STL = (ΔPEG)/(ΔDex) where Δ is the difference between the concentrations of a given polymer in the two coexisting phases;
**-Composition of the phases questionable as the STL value is inconsistent with the other values; was not used in calculations of the average STL value;
***-Composition of the phase questionable as it does not fit phase diagram.

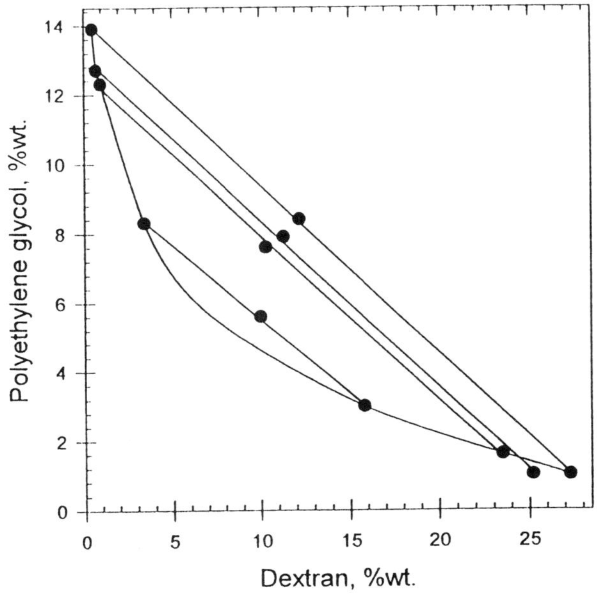

Table 10.12. Phase Diagram and Phase Composition of the Dextran—Poly(ethylene glycol) System Dex-40—PEG-4000 at 40°C.
(From D. Forciniti, C. K. Hall, M.-R. Kula, Fluid Phase Equilibria, 61, 243 (1991) and Bioseparation, 2, 115 (1991) by permission of Elsevier Scientific Publishing Co. and Kluwer Academic Publishers, respectively.)

PEG - M_w 4100; M_n 3800; Manufacturer: Fluka AG (Switzerland); Lot not indicated;
Dextran - M_w 37,000; M_n 27,700; Manufacturer: Pfeifer and Langen (Dormagen, Germany); Lot not indicated

Total system		Bottom phase		Top phase		STL*
PEG % w/w	Dex % w/w	PEG % w/w	Dex % w/w	PEG % w/w	Dex % w/w	
6.2	9.9	2.9	15.3	10.5	2.5	-0.594
7.6	10.3	1.9	21.3	12.3	1.3	-0.520
7.9	11.3	1.3	23.9	13.2	0.99	-0.519
8.4	12.2	0.6**	25.9**	14.4	0.69	-0.547
					av.:	-0.545 ±0.035

*- STL - Tie-Line Slope defined as the ratio STL = (ΔPEG)/(ΔDex) where Δ is the difference between the concentrations of a given polymer in the two coexisting phases;
** - Composition of the phase questionable as it does not fit phase diagram.

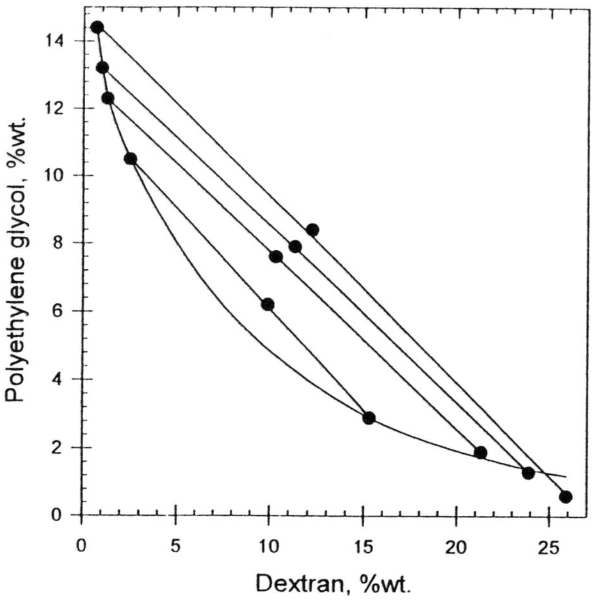

Table 10.13. Phase Diagram and Phase Composition of the Dextran—Poly(ethylene glycol) System Dex-110—PEG-4000 at 25°C.
(From D. Forciniti, C. K. Hall, M.-R. Kula, Fluid Phase Equilibria, 61, 243 (1991) and Bioseparation, 2, 115 (1991) by permission of Elsevier Scientific Publishing Co. and Kluwer Academic Publishers, respectively.)

PEG - M_w 4100; M_n 3800; Manufacturer: Fluka (Switzerland); Lot not indicated;
Dextran - M_w 86,200; M_n 52.100; Manufacturer: Fluka (Switzerland); Lot not indicated

Total system		Bottom phase		Top phase		STL*
PEG % w/w	Dex % w/w	PEG % w/w	Dex % w/w	PEG % w/w	Dex % w/w	
5.8	10.0	2.5	17.1	10.0	0.9	-0.463
7.6	10.3	2.1	22.3	12.5	0.3	-0.473
7.9	11.3	2.0	24.0	13.3	0.3	-0.477
8.4	12.2	2.2	25.8	14.6	0.2	-0.484
					av.:	-0.474 ± 0.009

* - STL - Tie-Line Slope defined as the ratio STL = (ΔPEG)/(ΔDex) where Δ is the difference between the concentrations of a given polymer in the two coexisting phases.

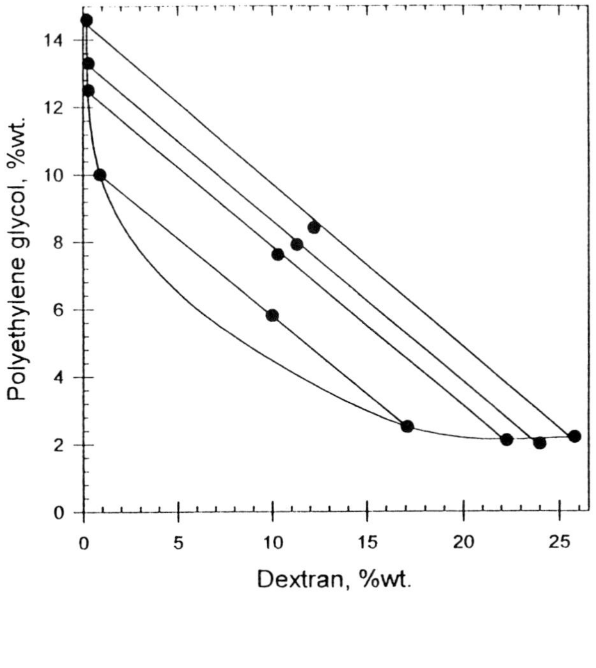

Table 10.14. Phase Diagram and Phase Composition of the Dextran—Poly(ethylene glycol) System Dex-110—PEG-4000 at 4°C.
(From D. Forciniti, C. K. Hall, M.-R. Kula, Fluid Phase Equilibria, 61, 243 (1991) and Bioseparation, 2, 115 (1991) by permission of Elsevier Scientific Publishing Co. and Kluwer Academic Publishers, respectively.)

PEG - M_w 4100; M_n 3800; Manufacturer: Fluka (Switzerland); Lot not indicated;
Dextran - M_w 86,200; M_n 52.100; Manufacturer: Fluka (Switzerland); Lot not indicated

Total system		Bottom phase		Top phase		STL*
PEG % w/w	Dex % w/w	PEG % w/w	Dex % w/w	PEG % w/w	Dex % w/w	
5.6	10.0	1.8**	20.7**	9.0**	0.65**	-0.359**
7.6	10.3	0.9	25.5	12.1	0.23	-0.443
7.9	11.3	0.9	26.4	12.9	0.18	-0.458
8.4	12.2	1.0	27.9	14.0	0.15	-0.468
					av.:	-0.456 ± 0.013

* - STL - Tie-Line Slope defined as the ratio STL = (ΔPEG)/(ΔDex) where Δ is the difference between the concentrations of a given polymer in the two coexisting phases;
** - Composition of the phases questionable as the STL value is inconsistent with the other values; was not used in calculations of the average STL value.

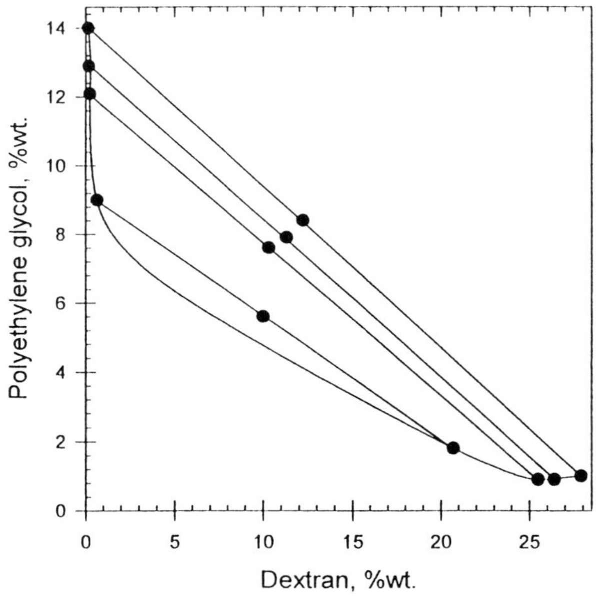

Table 10.15. Phase Diagram and Phase Composition of the Dextran—Poly(ethylene glycol) System Dex-110—PEG-4000 at 40°C.
(From D. Forciniti, C. K. Hall, M.-R. Kula, Fluid Phase Equilibria, 61, 243 (1991) and Bioseparation, 2, 115 (1991) by permission of Elsevier Scientific Publishing Co. and Kluwer Academic Publishers, respectively.)

PEG - M_w 4100; M_n 3800; Manufacturer: Fluka (Switzerland); Lot not indicated;
Dextran - M_w 86,200; M_n 52.100; Manufacturer: Fluka (Switzerland); Lot not indicated

Total system		Bottom phase		Top phase		STL*
PEG % w/w	Dex % w/w	PEG % w/w	Dex % w/w	PEG % w/w	Dex % w/w	
5.6	10.0	2.4**	16.6**	10.0**	1.0**	-0.487**
7.6	10.3	2.0	21.2	12.7	0.35	-0.513
7.9	11.3	1.2***	23.9***	13.5	0.25	-0.520
8.2	12.0	1.6	24.6	14.4	0.21	-0.525
					av.:	-0.519 ± 0.006

* - STL - Tie-Line Slope defined as the ratio STL = (ΔPEG)/(ΔDex) where Δ is the difference between the concentrations of a given polymer in the two coexisting phases;
** - Composition of the phases questionable as the STL value is inconsistent with the other values; was not used in calculations of the average STL value;
*** - Composition of the phase questionable as it does not fit phase diagram.

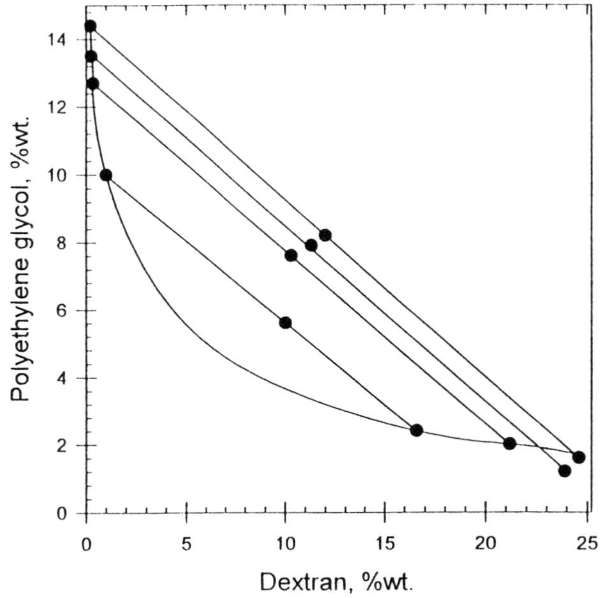

Table 10.16. Phase Diagram and Phase Composition of the Dextran—Poly(ethylene glycol) System Dex-500—PEG-4000 at 25°C.
(From D. Forciniti, C. K. Hall, M.-R. Kula, Fluid Phase Equilibria, 61, 243 (1991) and Bioseparation, 2, 115 (1991) by permission of Elsevier Scientific Publishing Co. and Kluwer Academic Publishers, respectively.)

PEG - M_w 4100; M_n 3800; Manufacturer: Fluka AG (Switzerland); Lot not indicated;
Dextran - M_w 215,000; M_n 88,200; Manufacturer: Pfeifer and Langen (Dormagen, Germany); Lot not indicated

Total system		Bottom phase		Top phase		STL*
PEG % w/w	Dex % w/w	PEG % w/w	Dex % w/w	PEG % w/w	Dex % w/w	
7.6	10.5	1.5	22.8	12.5	0.3	-0.489
5.6	10.2	1.6	17.8	10.0	0.6	-0.488
8.4	12.4	1.2	26.2	14.4	0.2	-0.508
7.9	11.5	1.3	24.9	13.5	0.2	-0.494
					av.:	-0.495 ± 0.009

* - STL - Tie-Line Slope defined as the ratio STL = (ΔPEG)/(ΔDex) where Δ is the difference between the concentrations of a given polymer in the two coexisting phases.

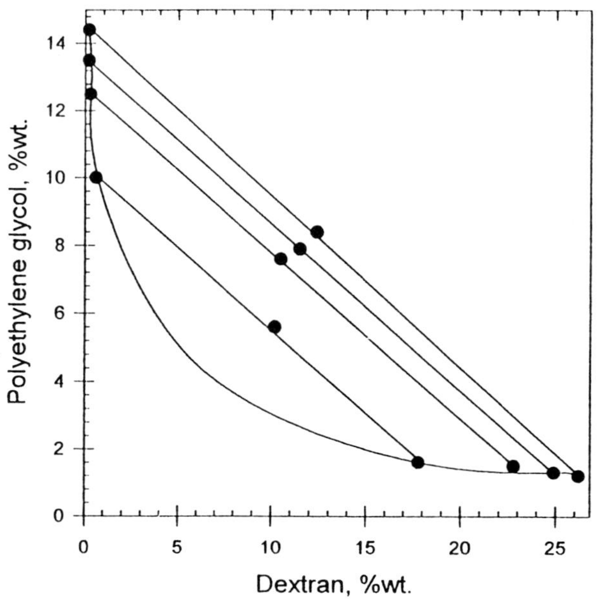

Table 10.17. Phase Diagram and Phase Composition of the Dextran—Poly(ethylene glycol) System Dex-500—PEG-4000 at 4°C.
(From D. Forciniti, C. K. Hall, M.-R. Kula, Fluid Phase Equilibria, 61, 243 (1991) and Bioseparation, 2, 115 (1991) by permission of Elsevier Scientific Publishing Co. and Kluwer Academic Publishers, respectively.)

PEG - M_w 4100; M_n 3800; Manufacturer: Fluka AG (Switzerland); Lot not indicated;
Dextran - M_w 215,000; M_n 88,200; Manufacturer: Pfeifer and Langen (Dormagen, Germany); Lot not indicated

Total system		Bottom phase		Top phase		STL*
PEG % w/w	Dex % w/w	PEG % w/w	Dex % w/w	PEG % w/w	Dex % w/w	
7.6	10.3	1.5	24.8	11.9	0.09	-0.421
5.6	10.0	1.9	19.5	9.4	0.32	-0.391
8.4	12.2	1.5	27.6	13.4	0.05	-0.432
7.9	11.3	1.4	26.5	12.4	0.07	-0.416
					av.:	-0.415 ±0.017

* - STL - Tie-Line Slope defined as the ratio STL = (ΔPEG)/(ΔDex) where Δ is the difference between the concentrations of a given polymer in the two coexisting phases.

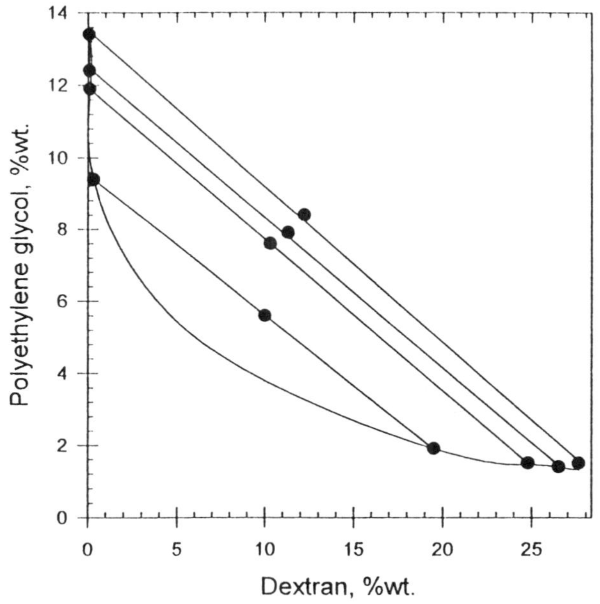

Table 10.18. Phase Diagram and Phase Composition of the Dextran—Poly(ethylene glycol) System Dex-500—PEG-4000 at 40°C.
(From D. Forciniti, C. K. Hall, M.-R. Kula, Fluid Phase Equilibria, 61, 243 (1991) and Bioseparation, 2, 115 (1991) by permission of Elsevier Scientific Publishing Co. and Kluwer Academic Publishers, respectively.)

PEG - M_w 4100; M_n 3800; Manufacturer: Fluka AG (Switzerland); Lot not indicated;
Dextran - M_w 215,000; M_n 88,200; Manufacturer: Pfeifer and Langen (Dormagen, Germany); Lot not indicated

Total system		Bottom phase		Top phase		STL*
PEG % w/w	Dex % w/w	PEG % w/w	Dex % w/w	PEG % w/w	Dex % w/w	
7.6	10.3	1.4	22.9	12.5	0.17	-0.488
5.6	10.0	2.1	17.7	10.0	0.59	-0.462
8.4	12.2	0.8	26.5	14.9	0.08	-0.534
7.9	11.3	1.4	24.2	13.8	0.12	-0.515
					av.:	-0.500 ±0.031

* - STL - Tie-Line Slope defined as the ratio STL = (ΔPEG)/(ΔDex) where Δ is the difference between the concentrations of a given polymer in the two coexisting phases.

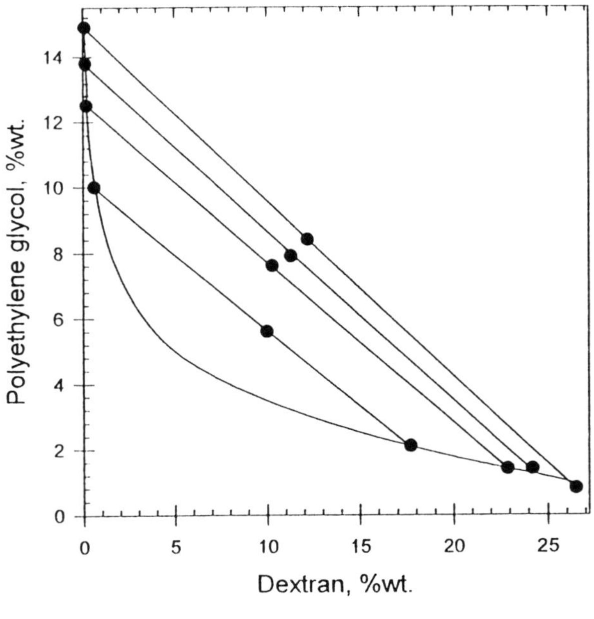

Table 10.19. Phase Diagram and Phase Composition of the Dextran—Poly(ethylene glycol) System Dex-10—PEG-6000 at 25°C.

(From D. Forciniti, C. K. Hall, M.-R. Kula, Fluid Phase Equilibria, 61, 243 (1991) and Bioseparation, 2, 115 (1991) by permission of Elsevier Scientific Publishing Co. and Kluwer Academic Publishers, respectively.)

PEG - M_w 5600; M_n 5300; Manufacturer: Fluka AG (Switzerland); Lot not indicated;

Dextran - M_w 19,300; M_n 13,200; Manufacturer: Fluka AG (Switzerland); Lot not indicated

Total system		Bottom phase		Top phase		STL*
PEG % w/w	Dex % w/w	PEG % w/w	Dex % w/w	PEG % w/w	Dex % w/w	
5.9	9.95	4.7	13.0	7.9	7.1	-0.542
6.3	8.9	4.7**	13**	7.6**	7.1**	-0.492**
7.75	10.3	2.1	21.4	12.2	2.5	-0.534
7.9	11.3	1.9	23.1	13.1	2.0	-0.531
					av.:	-0.536
						±0.006

*- STL - Tie-Line Slope defined as the ratio STL = (ΔPEG)/(ΔDex) where Δ is the difference between the concentrations of a given polymer in the two coexisting phases;
** - Composition of the phases questionable as the STL value is inconsistent with the other values; was not used in calculations of the average STL value; not shown on phase diagram.

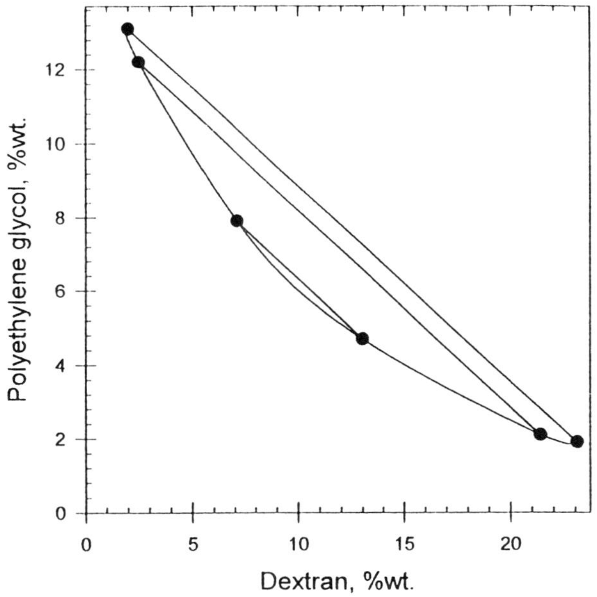

Table 10.20. Phase Diagram and Phase Composition of the Dextran—Poly(ethylene glycol) System Dex-10—PEG-6000 at 4°C.
(From D. Forciniti, C. K. Hall, M.-R. Kula, Fluid Phase Equilibria, 61, 243 (1991) and Bioseparation, 2, 115 (1991) by permission of Elsevier Scientific Publishing Co. and Kluwer Academic Publishers, respectively.)

PEG - M_w 5600; M_n 5300; Manufacturer: Fluka AG (Switzerland); Lot not indicated;
Dextran - M_w 19,300; M_n 13,200; Manufacturer: Fluka AG (Switzerland); Lot not indicated

Total system		Bottom phase		Top phase		STL*
PEG % w/w	Dex % w/w	PEG % w/w	Dex % w/w	PEG % w/w	Dex % w/w	
5.6	10.0	3.5**	15.5**	7.7**	5.6**	-0.424**
6.8	9.35	2.5	18.7	9.9	3.4	-0.484
7.6	10.3	1.3***	22.4***	11.5	2.2	-0.505
7.9	11.3	1.2	24.1	12.3	1.8	-0.498
					av.:	-0.496 ±0.011

*- STL - Tie-Line Slope defined as the ratio STL = (ΔPEG)/(ΔDex) where Δ is the difference between the concentrations of a given polymer in the two coexisting phases;
** - Composition of the phases questionable as the STL value is inconsistent with the other values; was not used in calculations of the average STL value;
*** - Composition of the phase questionable as it does not fit phase diagram.

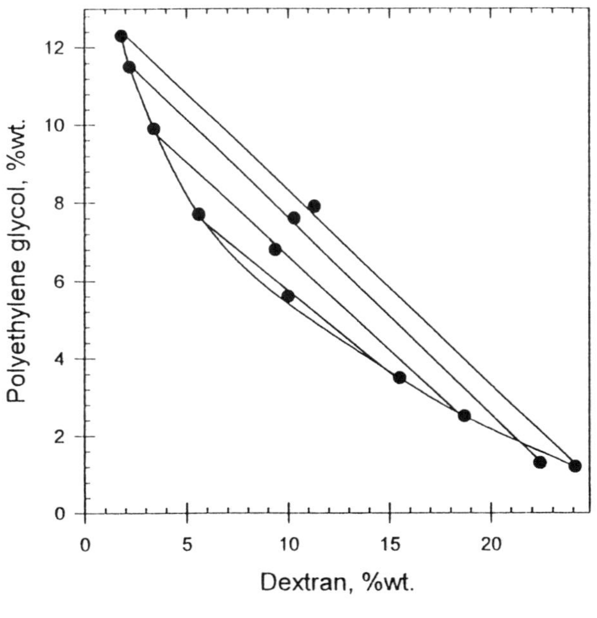

Table 10.21. Phase Diagram and Phase Composition of the Dextran—Poly(ethylene glycol) System Dex-10—PEG-6000 at 40°C.
(From D. Forciniti, C. K. Hall, M.-R. Kula, Fluid Phase Equilibria, 61, 243 (1991) and Bioseparation, 2, 115 (1991) by permission of Elsevier Scientific Publishing Co. and Kluwer Academic Publishers, respectively.)

PEG - M_w 5600; M_n 5300; Manufacturer: Fluka AG (Switzerland); Lot not indicated;
Dextran - M_w 19,300; M_n 13,200; Manufacturer: Fluka AG (Switzerland); Lot not indicated

Total system		Bottom phase		Bottom phase		STL*
PEG % w/w	Dex % w/w	PEG % w/w	Dex % w/w	PEG % w/w	Dex % w/w	
5.6	10.0	2.2**	12.7**	9.5**	4.9**	-0.936**
7.5	10.2	3.3**	19.3**	11.6**	2.9**	-0.506**
7.6	10.25	2.1	20.0	12.1	2.6	-0.575
7.9	11.3	1.8	21.9	13.1	2.1	-0.571
					av.:	-0.573 ±0.003

*- STL - Tie-Line Slope defined as the ratio STL = (ΔPEG)/(ΔDex) where Δ is the difference between the concentrations of a given polymer in the two coexisting phases;
** - Composition of the phases questionable as the STL value is inconsistent with the other values; was not used in calculations of the average STL value; not shown on phase diagram.

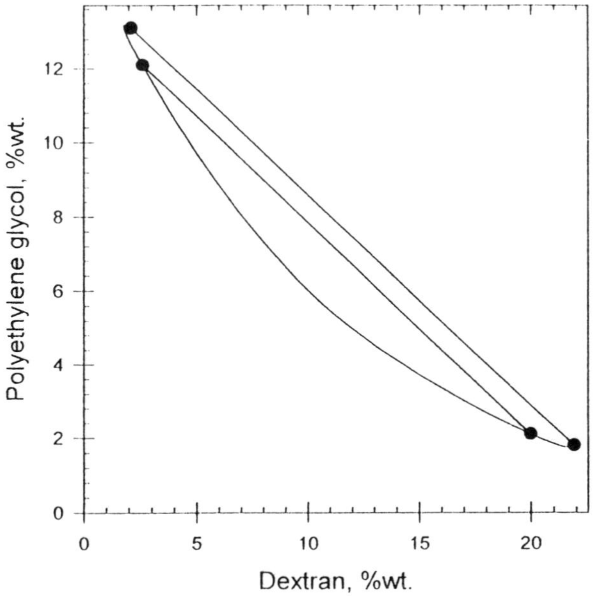

Table 10.22. Phase Diagram and Phase Composition of the Dextran—Poly(ethylene glycol) System Dex-40—PEG-6000 at 25°C.
(From D. Forciniti, C. K. Hall, M.-R. Kula, Fluid Phase Equilibria, 61, 243 (1991) and Bioseparation, 2, 115 (1991) by permission of Elsevier Scientific Publishing Co. and Kluwer Academic Publishers, respectively.)

PEG - M_w 5600; M_n 5300; Manufacturer: Fluka AG (Switzerland); Lot not indicated;
Dextran - M_w 37,000; M_n 27,700; Manufacturer: Pfeifer and Langen (Dormagen, Germany); Lot not indicated

Total system		Bottom phase		Top phase		STL*
PEG % w/w	Dex % w/w	PEG % w/w	Dex % w/w	PEG % w/w	Dex % w/w	
5.0	8.5	4.0**	13.2**	8.0**	4.8**	-0.476**
5.6	10.0	1.9	21.2	11.9	1.3	-0.503
7.6	10.3	1.7	23.1	12.9	1.0	-0.507
7.9	11.3	1.5	25.0	14.0	0.7	-0.514
					av.:	-0.508 ±0.006

*- STL - Tie-Line Slope defined as the ratio STL = (ΔPEG)/(ΔDex) where Δ is the difference between the concentrations of a given polymer in the two coexisting phases;
** - Composition of the phases questionable as the STL value is inconsistent with the other values; was not used in calculations of the average STL value.

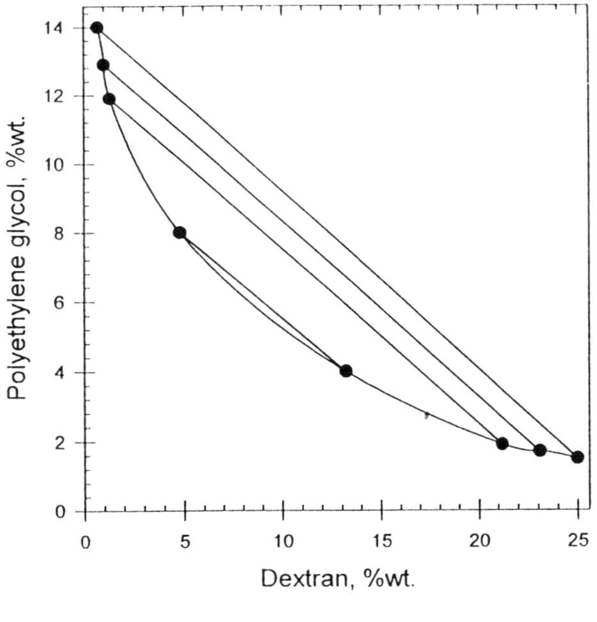

Table 10.23. Phase Diagram and Phase Composition of the Dextran—Poly(ethylene glycol) System Dex-40—PEG-6000 at 4°C.
(From D. Forciniti, C. K. Hall, M.-R. Kula, Fluid Phase Equilibria, 61, 243 (1991) and Bioseparation, 2, 115 (1991) by permission of Elsevier Scientific Publishing Co. and Kluwer Academic Publishers, respectively.)

PEG - M_w 5600; M_n 5300; Manufacturer: Fluka AG (Switzerland); Lot not indicated;
Dextran - M_w 37,000; M_n 27,700; Manufacturer: Pfeifer and Langen (Dormagen, Germany); Lot not indicated

Total system		Bottom phase		Top phase		STL*
PEG % w/w	Dex % w/w	PEG % w/w	Dex % w/w	PEG % w/w	Dex % w/w	
5.0	8.5	2.1**	14.5**	7.9**	3.2**	-0.513**
5.6	10.0	1.1	19.3	9.5	1.5	-0.472
7.6	10.3	0.9	24.2	12.2	0.59	-0.479
7.9	11.3	1.0	25.8	13.0	0.45	-0.473
					av.:	-0.475 ±0.004

*- STL - Tie-Line Slope defined as the ratio STL = (ΔPEG)/(ΔDex) where Δ is the difference between the concentrations of a given polymer in the two coexisting phases;
**-Composition of the phases questionable as the STL value is inconsistent with the other values; was not used in calculations of the average STL value.

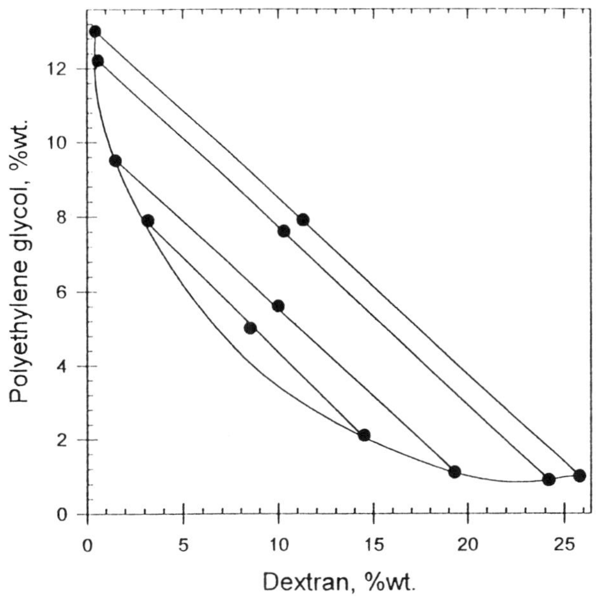

Table 10.24. Phase Diagram and Phase Composition of the Dextran—Poly(ethylene glycol) System Dex-40—PEG-6000 at 40°C.
(From D. Forciniti, C. K. Hall, M.-R. Kula, Fluid Phase Equilibria, 61, 243 (1991) and Bioseparation, 2, 115 (1991) by permission of Elsevier Scientific Publishing Co. and Kluwer Academic Publishers, respectively.)

PEG - M_w 5600; M_n 5300; Manufacturer: Fluka AG (Switzerland); Lot not indicated;
Dextran - M_w 37,000; M_n 27,700; Manufacturer: Pfeifer and Langen (Dormagen, Germany); Lot not indicated

Total system		Bottom phase		Top phase		STL*
PEG % w/w	Dex % w/w	PEG % w/w	Dex % w/w	PEG % w/w	Dex % w/w	
5.56	8.4	2.5	14.3	8.6	2.6	-0.521
5.6	10.0	1.2	18.6	9.8	2.0	-0.518
7.6	10.3	0.4	23.9	12.3	0.88	-0.517
7.9	11.3	0.5	25.2	13.5	0.6	-0.528
					av.:	-0.521 ±0.005

* - STL - Tie-Line Slope defined as the ratio STL = (ΔPEG)/(ΔDex) where Δ is the difference between the concentrations of a given polymer in the two coexisting phases.

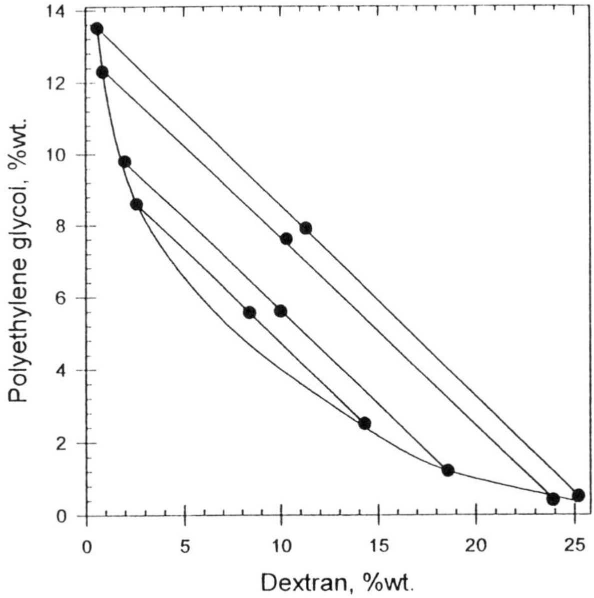

Table 10.25. Phase Diagram and Phase Composition of the Dextran-70—Poly(ethylene glycol)-6000 System at 23°C.

Dextran-70 - M_w 57,200; M_n 28,700; Manufacturer: Minmedprom (Moscow, Russia); Lot 680480;
PEG - M_w ~6000; Manufacturer: Serva Fine Biochemicals (Heidelberg, Germany); Lot 419-80

Total system		Bottom phase		Top phase		STL*
PEG % w/w	Dextran % w/w	PEG % w/w	Dextran % w/w	PEG % w/w	Dextran % w/w	
8.72	13.95	0.46	28.48	16.47	0.32	-0.569
7.92	12.88	0.57	26.14	14.81	0.44	-0.554
7.15	11.69	0.66	23.42	13.24	0.69	-0.553
6.74	10.82	0.92	21.44	12.20	0.86	-0.548
5.81	9.67	1.20	18.46	10.13	1.42	-0.524
5.28	8.52	1.65	15.38	8.68	2.09	-0.529
4.54	7.50	2.28	11.62	6.66	3.64	-0.549
4.10**	6.90**				av.:	-0.547 ±0.015

* - STL - Tie-Line Slope defined as the ratio STL = (ΔPEG)/(ΔDex) where Δ is the difference between the concentrations of a given polymer in the two coexisting phases;
** - Composition of critical point.

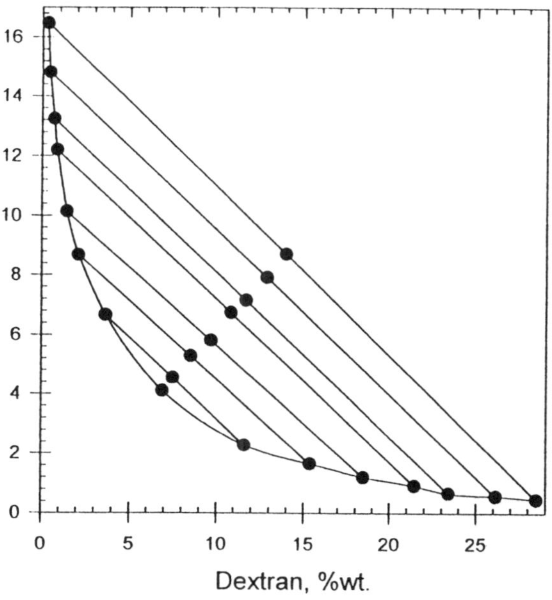

Table 10.26. Phase Diagram and Phase Composition of the Dextran-70—Poly(ethylene glycol)-6000 System at 8°C.

Dextran-70 - M_w 57,200; M_n 28,700; Manufacturer: Minmedprom (Moscow, Russia); Lot 680480;
PEG - M_w ~6000; Manufacturer: Serva Fine Biochemicals (Heidelberg, Germany); Lot 463-80

Total system		Bottom phase		Top phase		STL*
PEG % w/w	Dextran % w/w	PEG % w/w	Dextran % w/w	PEG % w/w	Dextran % w/w	
8.77	14.00	0.16	30.49	15.90	0.35	-0.522
8.00	12.86	0.22	27.77	14.45	0.50	-0.522
6.89	10.81	0.43	23.18	12.05	0.93	-0.522
6.06	9.58	0.64	19.96	10.30	1.45	-0.522
4.00**	8.00**				av.:	-0.522

*- STL - Tie-Line Slope defined as the ratio STL = (ΔPEG)/(ΔDex) where Δ is the difference between the concentrations of a given polymer in the two coexisting phases;

**- Composition of critical point.

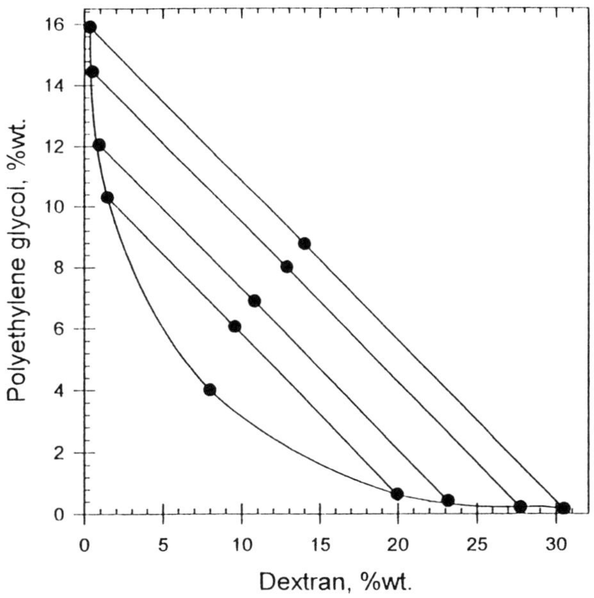

Table 10.27. Phase Diagram and Phase Composition of the Dextran-70—Poly(ethylene glycol)-6000 System at 23°C.

Dextran-70 - M_w 57,200; M_n 28,700; Manufacturer: Minmedprom (Moscow, Russia); Lot 680480;
PEG - M_w ~6000; Manufacturer: Serva Fine Biochemicals (Heidelberg, Germany); Lot 463-80

Total system		Bottom phase		Top phase		STL*
PEG % w/w	Dextran % w/w	PEG % w/w	Dextran % w/w	PEG % w/w	Dextran % w/w	
8.19	12.99	0.66	26.60	15.08	0.52	-0.553
6.78	10.78	0.75	21.69	12.14	1.08	-0.553
6.04	9.67	1.05	18.70	10.45	1.69	-0.553
5.34	8.52	1.63	15.23	8.55	2.72	-0.553
4.30**	7.90**				av.:	-0.553

* - STL - Tie-Line Slope defined as the ratio STL = (ΔPEG)/(ΔDex) where Δ is the difference between the concentrations of a given polymer in the two coexisting phases;

** - Composition of critical point.

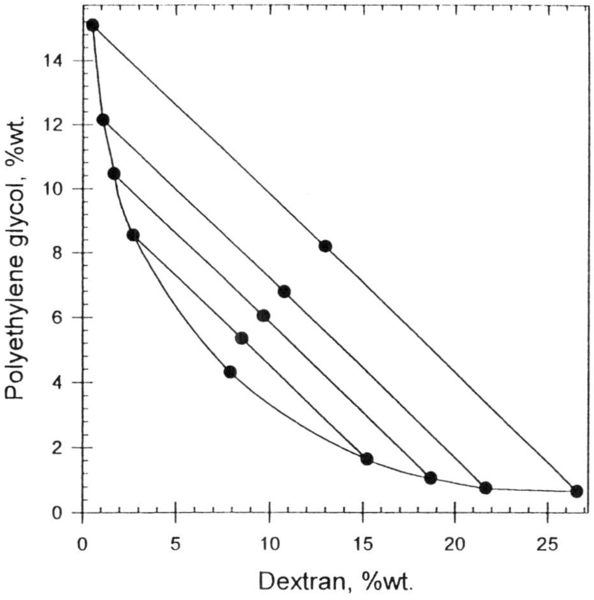

Table 10.28. Phase Diagram and Phase Composition of the Dextran-70—Poly(ethylene glycol)-6000 System at 38°C.

Dextran-70 - M_w 57,200; M_n 28,700; Manufacturer: Minmedprom (Moscow, Russia); Lot 680480;
PEG - M_w ~6000; Manufacturer: Serva Fine Biochemicals (Heidelberg, Germany); Lot 463-80

Total system		Bottom phase		Top phase		STL*
PEG % w/w	Dextran % w/w	PEG % w/w	Dextran % w/w	PEG % w/w	Dextran % w/w	
8.76	14.05	0.33	28.36	16.80	0.40	-0.589
7.99	12.77	0.49	25.52	15.20	0.59	-0.590
6.80	10.80	0.67	21.21	12.50	1.12	-0.589
6.04	9.62	0.99	18.19	10.67	1.76	-0.589
4.70**	7.90**				av.:	-0.589 ±0.001

* - STL - Tie-Line Slope defined as the ratio STL = (ΔPEG)/(ΔDex) where Δ is the difference between the concentrations of a given polymer in the two coexisting phases;
** - Composition of critical point.

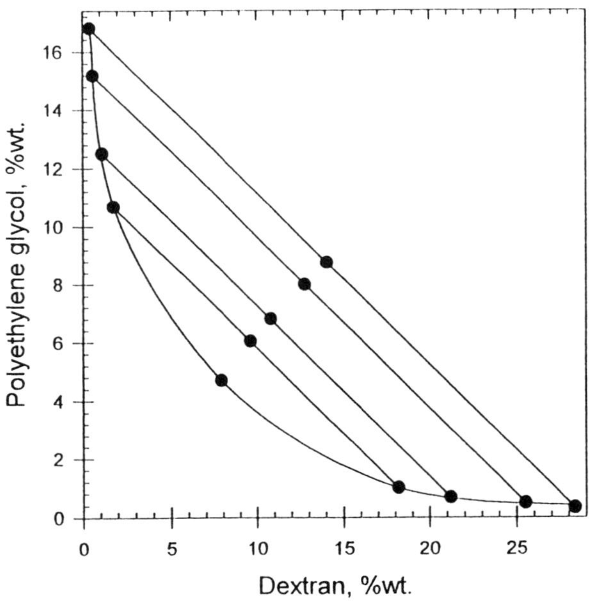

Table 10.29. Phase Diagram and Phase Composition of the Dextran-70—Poly(ethylene glycol)-6000 System at 50°C.

Dextran-70 - M_w 57,200; M_n 28,700; Manufacturer: Minmedprom (Moscow, Russia); Lot 680480;
PEG - M_w ~6000; Manufacturer: Serva Fine Biochemicals (Heidelberg, Germany); Lot 463-80

Total system		Bottom phase		Top phase		STL*
PEG % w/w	Dextran % w/w	PEG % w/w	Dextran % w/w	PEG % w/w	Dextran % w/w	
8.75	14.06	0.24	27.76	17.26	0.37	-0.621
8.03	12.77	0.44	25.00	15.62	0.55	-0.621
7.38	11.84	0.56	22.83	14.27	0.75	-0.621
6.74	10.70	0.66	20.49	12.71	1.08	-0.621
4.80**	7.80**				av.:	-0.621

* - STL - Tie-Line Slope defined as the ratio STL = (ΔPEG)/(ΔDex) where Δ is the difference between the concentrations of a given polymer in the two coexisting phases;

** - Composition of critical point.

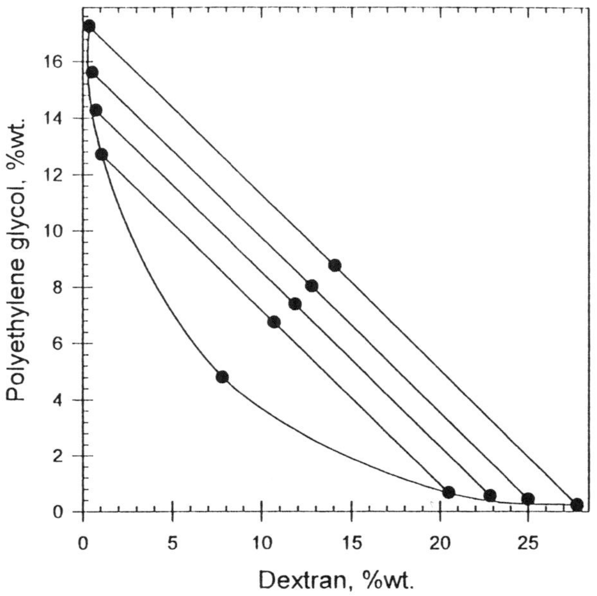

Table 10.30. Phase Diagram and Phase Composition of the Dextran-70—Poly(ethylene glycol)-6000 System Containing 2.0 mole/kg Urea at 23°C.

Dextran-70 - M_w 57,200; M_n 28,700; Manufacturer: Minmedprom (Moscow, Russia); Lot 680480;
PEG - M_w ~6000; Manufacturer: Serva Fine Biochemicals (Heidelberg, Germany); Lot 463-80

Total system		Bottom phase		Top phase		STL*
PEG % w/w	Dextran % w/w	PEG % w/w	Dextran % w/w	PEG % w/w	Dextran % w/w	
8.75	13.98	0.48	28.64	16.23	0.65	-0.563
7.96	12.79	0.62	26.24	14.47	0.93	-0.547
6.74	10.82	1.21	20.93	11.72	1.72	-0.547
6.13	9.65	1.68	17.78	9.84	2.86	-0.547
4.80**	8.60**				av.:	-0.551 ±0.008

* - STL - Tie-Line Slope defined as the ratio STL = (ΔPEG)/(ΔDex) where Δ is the difference between the concentrations of a given polymer in the two coexisting phases;
** - Composition of critical point.

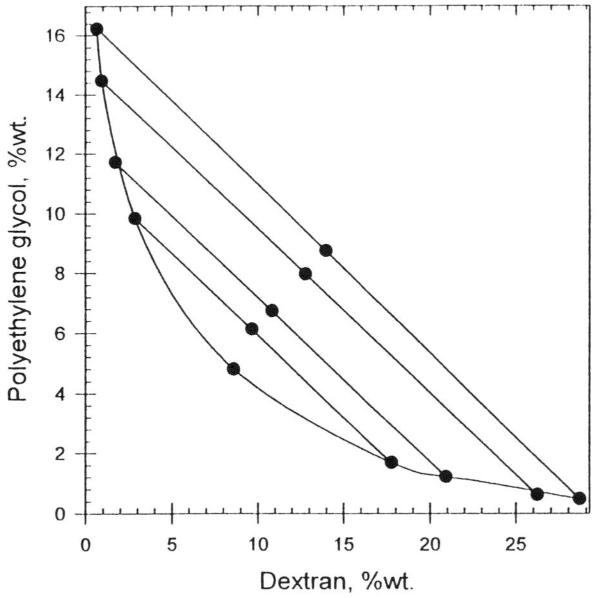

Table 10.31. Phase Diagram and Phase Composition of the Dextran-70—Poly(ethylene glycol)-6000 System Containing 4.0 mole/kg Urea at 23°C.

Dextran-70 - M_w 57,200; M_n 28,700; Manufacturer: Minmedprom (Moscow, Russia); Lot 680480;
PEG - M_w ~6000; Manufacturer: Serva Fine Biochemicals (Germany); Lot 463-80

Total system		Bottom phase		Top phase		STL*
PEG % w/w	Dextran % w/w	PEG % w/w	Dextran % w/w	PEG % w/w	Dextran % w/w	
8.70	13.92	0.89	27.44	16.06	1.08	-0.575
7.94	12.72	0.95	24.86	14.40	1.53	-0.577
6.80	10.70	1.80	19.37	11.33	2.85	-0.577
6.01	9.48	3.28	14.21	8.52	5.13	-0.577
5.62**	9.00**				av.:	-0.577 ±0.001

*- STL - Tie-Line Slope defined as the ratio STL = (ΔPEG)/(ΔDex) where Δ is the difference between the concentrations of a given polymer in the two coexisting phases;

**- Composition of critical point.

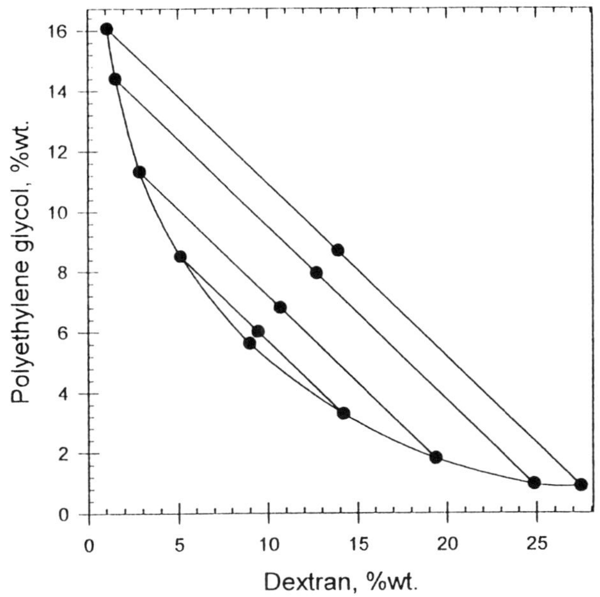

Table 10.32. Phase Diagram and Phase Composition of the Dextran-70—Poly(ethylene glycol)-6000 system Containing 0.1 mole/kg NH_4SCN at 23°C

Dextran-70 - M_w 57,200; M_n 28,700; Manufacturer: Minmedprom (Moscow, Russia); Lot 680480;
PEG - M_w ~6000; Manufacturer: Serva Fine Biochemicals (Germany); Lot 463-80

Total system		Top phase			Top phase			STL*
PEG % w/w	Dex % w/w	PEG % w/w	Dex % w/w	NH_4SCN % w/w	PEG % w/w	Dex % w/w	NH_4SCN % w/w	
8.74	13.98	0.26	29.20	0.738	16.32	0.37	0.783	-0.557
7.99	12.77	0.32	26.54	0.743	14.79	0.55	0.780	-0.557
6.77	10.82	0.64	21.82	0.744	12.17	1.05	0.775	-0.555
6.01	9.57	0.94	18.67	0.745	10.42	1.67	0.770	-0.558
4.20**	7.60**						av.:	-0.557

* - STL - Tie-Line Slope defined as the ratio STL = (ΔPEG)/(ΔDex) where Δ is the difference between the concentrations of a given polymer in the two coexisting phases;

** - Composition of critical point.

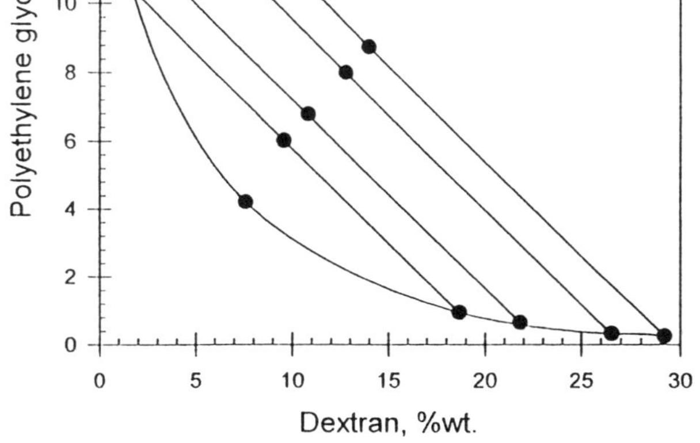

Table 10.33. Phase Diagram and Phase Composition of the Dextran-70—Poly(ethylene glycol)-6000 System Containing 0.1 mole/kg NaSCN at 23°C.

Dextran-70 - M_w 57,200; M_n 28,700; Manufacturer: Minmedprom (Moscow, Russia); Lot 680480;
PEG - M_w ~6000; Manufacturer: Serva Fine Biochemicals (Germany); Lot 463-80

Total system		Bottom phase			Top phase			STL*
PEG % w/w	Dextran % w/w	PEG % w/w	Dextran % w/w	KSCN % w/w	PEG % w/w	Dextran % w/w	KSCN % w/w	
8.75	14.01	0.21	29.09	0.69	16.48	0.38	0.92	-0.567
8.00	12.78	0.34	26.55	0.70	14.79	0.57	0.91	-0.556
6.76	10.77	0.55	21.94	0.72	12.14	1.10	0.89	-0.556
6.06	9.58	0.83	18.99	0.74	10.45	1.69	0.87	-0.556
5.39	8.68	1.26	16.10	0.75	8.54	3.00	0.86	-0.556
4.69	7.69	2.57	11.50	0.78	5.96	5.40	0.83	-0.556
4.10**	7.80**						av.:	-0.558 ±0.005

* - STL - Tie-Line Slope defined as the ratio STL = (ΔPEG)/(ΔDex) where Δ is the difference between the concentrations of a given polymer in the two coexisting phases;
** - Composition of critical point.

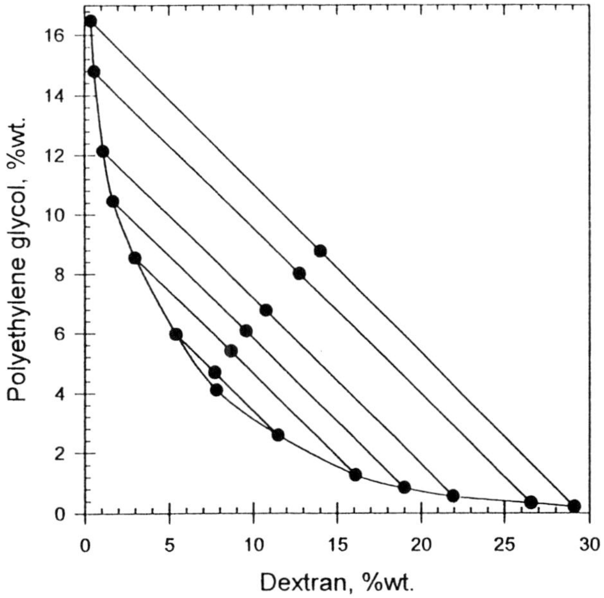

Table 10.34. Phase Diagram and Phase Composition of the Dextran-70—Poly(ethylene glycol)-6000 System Containing 0.1 mole/kg KSCN at 23°C.

Dextran-70 - M_w 57,200; M_n 28,700; Manufacturer: Minmedprom (Moscow, Russia); Lot 680480;
PEG - M_w ~6000; Manufacturer: Serva Fine Biochemicals (Germany); Lot # 463-80

Total system		Bottom phase			Top phase			STL*
PEG % w/w	Dextran % w/w	PEG % w/w	Dextran % w/w	KSCN % w/w	PEG % w/w	Dextran % w/w	KSCN % w/w	
8.04	12.86	0.73	25.84	0.884	15.01	0.48	1.060	-0.563
6.81	10.86	0.85	21.45	0.900	12.38	0.96	1.040	-0.563
5.44	8.66	1.70	15.30	0.926	8.95	2.41	1.013	-0.562
4.64	7.65	4.01	8.77	0.966	5.28	6.52	0.981	-0.564
4.60**	7.60**						av.:	-0.563 ±0.001

* - STL - Tie-Line Slope defined as the ratio STL = (ΔPEG)/(ΔDex) where Δ is the difference between the concentrations of a given polymer in the two coexisting phases;
** - Composition of critical point.

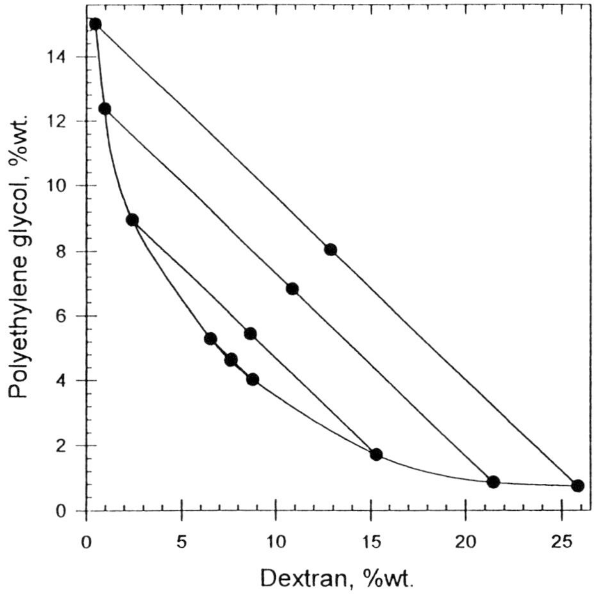

Table 10.35. Phase Diagram and Phase Composition of the Dextran-70—Poly(ethylene glycol)-6000 System Containing 0.5 mole/kg KSCN at 23°C.

Dextran-70 - M_w 57,200; M_n 28,700; Manufacturer: Minmedprom (Moscow, Russia); Lot 680480;
PEG - M_w ~6000; Manufacturer: Serva Fine Biochemicals (Germany); Lot 463-80

Total system		Bottom phase			Top phase			STL*
PEG % w/w	Dextran % w/w	PEG % w/w	Dextran % w/w	KSCN % w/w	PEG % w/w	Dextran % w/w	KSCN % w/w	
7.79	12.84	0.60	24.13	4.423	15.78	0.30	5.341	-0.637
6.90	10.92	0.69	20.67	4.478	13.46	0.63	5.253	-0.637
5.39	8.61	1.57	14.61	4.620	9.68	1.87	5.110	-0.637
4.76	7.57	2.93	10.48	4.787	7.24	3.71	5.050	-0.637
4.80**	6.75**						av.:	-0.637

* - STL - Tie-Line Slope defined as the ratio STL = (ΔPEG)/(ΔDex) where Δ is the difference between the concentrations of a given polymer in the two coexisting phases;

** - Composition of critical point.

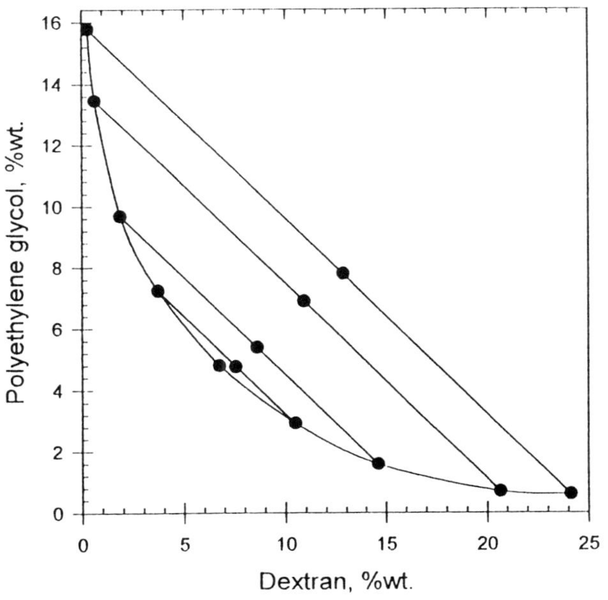

Table 10.36. Phase Diagram and Phase Composition of the Dextran-70—Poly(ethylene glycol)-6000 System Containing 0.75 mole/kg KSCN at 23°C.

Dextran-70 - M_w 57,200; M_n 28,700; Manufacturer: Minmedprom (Moscow, Russia); Lot 680480;
PEG - M_w ~6000; Manufacturer: Serva Fine Biochemicals (Germany); Lot 463-80

Total system		Bottom phase			Top phase			STL*
PEG % w/w	Dextran % w/w	PEG % w/w	Dextran % w/w	KSCN % w/w	PEG % w/w	Dextran % w/w	KSCN % w/w	
8.02	12.82	0.44	23.93	6.620	16.63	0.22	8.090	-0.683
6.82	10.87	0.50	19.70	6.727	13.89	0.52	7.945	-0.698
5.37	8.59	1.39	14.42	6.918	10.24	1.46	7.720	-0.683
4.60	7.64	2.59	10.58	7.110	7.78	2.98	7.586	-0.683
4.80**	6.20**						av.:	-0.687 ±0.008

* - STL - Tie-Line Slope defined as the ratio STL = (ΔPEG)/(ΔDex) where Δ is the difference between the concentrations of a given polymer in the two coexisting phases;

** - Composition of critical point.

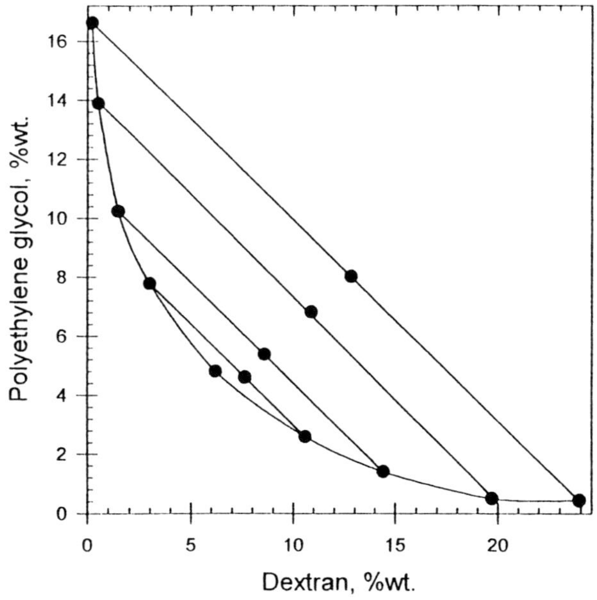

Table 10.37. Phase Diagram and Phase Composition of the Dextran-70—Poly(ethylene glycol)-6000 System Containing 0.1 mole/kg KCl at 23°C.

Dextran-70 - M_w 57,200; M_n 28,700; Manufacturer: Minmedprom (Moscow, Russia); Lot 680480;
PEG - M_w ~6000; Manufacturer: Serva Fine Biochemicals (Germany); Lot 463-80

Total system		Bottom phase			Top phase			STL*
PEG % w/w	Dextran % w/w	PEG % w/w	Dextran % w/w	KCl % w/w	PEG % w/w	Dextran % w/w	KCl % w/w	
8.04	12.95	0.30	26.68	0.760	15.10	0.44	0.730	-0.564
6.74	10.84	0.38	22.12	0.766	12.33	0.94	0.741	-0.564
5.37	8.56	1.08	16.17	0.756	8.95	2.21	0.740	-0.564
4.73	7.61	2012	12.24	0.750	6.82	3.91	0.740	-0.564
4.20**	7.50**						av.:	-0.564

* - STL - Tie-Line Slope defined as the ratio STL = (ΔPEG)/(ΔDex) where Δ is the difference between the concentrations of a given polymer in the two coexisting phases;
** - Composition of critical point.

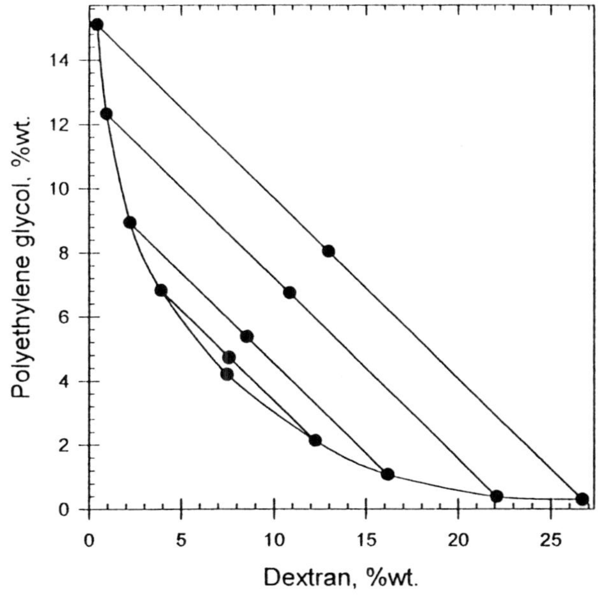

Table 10.38. Phase Diagram and Phase Composition of the Dextran-70—Poly(ethylene glycol)-6000 System Containing 0.5 mole/kg KCl at 23°C.

Dextran-70 - M_w 57,200; M_n 28,700; Manufacturer: Minmedprom (Moscow, Russia); Lot 680480;
PEG - M_w ~6000; Manufacturer: Serva Fine Biochemicals (Germany); Lot 463-80

Total system		Bottom phase			Top phase			STL*
PEG % w/w	Dextran % w/w	PEG % w/w	Dextran % w/w	KCl % w/w	PEG % w/w	Dextran % w/w	KCl % w/w	
7.96	12.86	0.51	25.53	3.85	15.33	0.33	3.61	-0.586
6.81	10.77	0.57	21.38	3.83	12.69	0.78	3.63	-0.588
5.38	8.64	1.27	15.64	3.80	9.33	1.93	3.66	-0.588
4.74	7.62	2.39	11.62	3.79	7.14	3.55	3.71	-0.589
4.45**	7.00**						av.:	-0.588 ±0.001

* - STL - Tie-Line Slope defined as the ratio STL = (ΔPEG)/(ΔDex) where Δ is the difference between the concentrations of a given polymer in the two coexisting phases;

** - Composition of critical point.

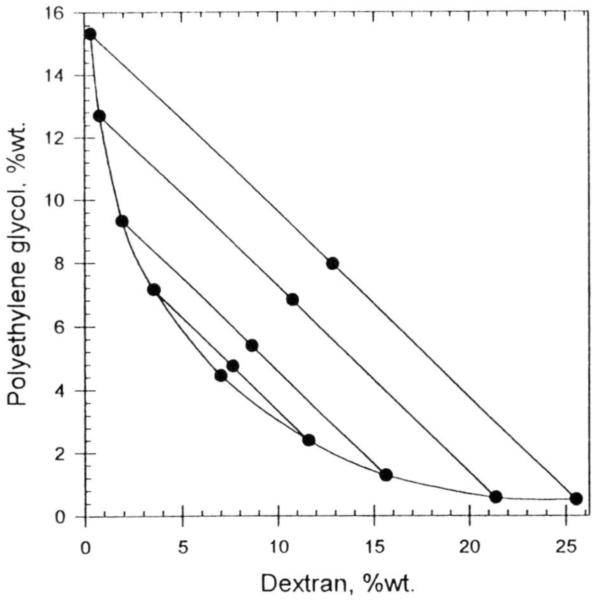

Table 10.39. Phase Diagram and Phase Composition of the Dextran-70—Poly(ethylene glycol)-6000 System Containing 0.75 mole/kg KCl at 23°C.

Dextran-70 - M_w 57,200; M_n 28,700; Manufacturer: Minmedprom (Moscow, Russia); Lot 680480;
PEG - M_w ~6000; Manufacturer: Serva Fine Biochemicals (Germany); Lot 463-80

Total system		Bottom phase			Top phase			TLS*
PEG % w/w	Dextran % w/w	PEG % w/w	Dextran % w/w	KCl % w/w	PEG % w/w	Dextran % w/w	KCl % w/w	
7.99	12.80	0.70	24.70	5.91	15.65	0.28	5.27	-0.612
6.72	10.79	0.69	20.64	5.84	12.90	0.70	5.35	-0.612
5.25	8.59	1.48	14.72	5.73	9.45	1.77	5.46	-0.615
4.71	7.61	2.29	11.55	5.65	7.44	3.15	5.53	-0.613
4.55**	6.65**						av.:	-0.613 ±0.002

*- STL - Tie-Line Slope defined as the ratio STL = (ΔPEG)/(ΔDex) where Δ is the difference between the concentrations of a given polymer in the two coexisting phases;

** - Composition of critical point.

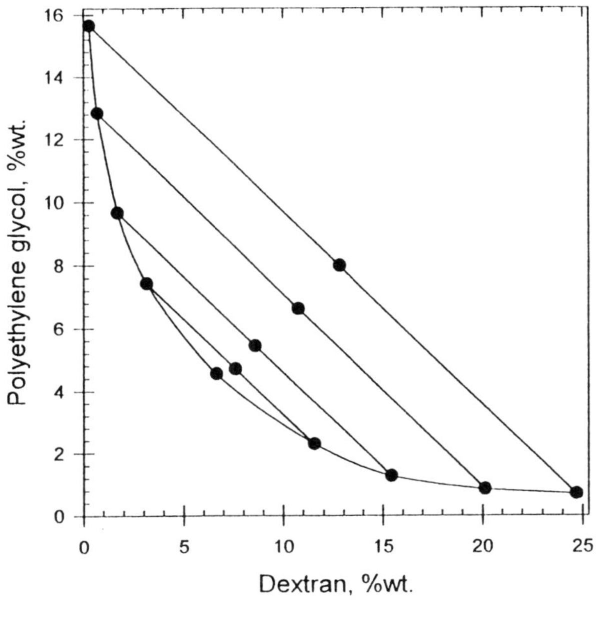

Table 10.40. Phase Diagram and Phase Composition of the Dextran-70—Poly(ethylene glycol)-6000 System Containing 0.1 mole/kg Na_2SO_4 at 23°C.

Dextran-70 - M_w 57,200; M_n 28,700; Manufacturer: Minmedprom (Moscow, Russia); Lot 680480;
PEG - M_w ~6000; Manufacturer: Serva Fine Biochemicals (Germany); Lot 463-80

Total system		Bottom phase			Top phase			STL*
PEG % w/w	Dextran % w/w	PEG % w/w	Dextran % w/w	Na_2SO_4 % w/w	PEG % w/w	Dextran % w/w	Na_2SO_4 % w/w	
8.77	14.03	0.21	26.53	1.73	18.23	0.20	1.08	-0.684
8.07	12.72	0.30	24.28	1.70	16.49	0.29	1.11	-0.675
6.82	10.75	0.40	20.18	1.65	13.64	0.74	1.17	-0.681
5.29	8.59	1.02	14.81	1.58	9.92	1.78	1.24	-0.683
4.74	7.70	1.95	11.80	1.52	7.95	2.99	1.31	-0.681
4.40**	7.03**						av.:	-0.681 ±0.003

*- STL - Tie-Line Slope defined as the ratio STL = (ΔPEG)/(ΔDex) where Δ is the difference between the concentrations of a given polymer in the two coexisting phases;
** - Composition of critical point.

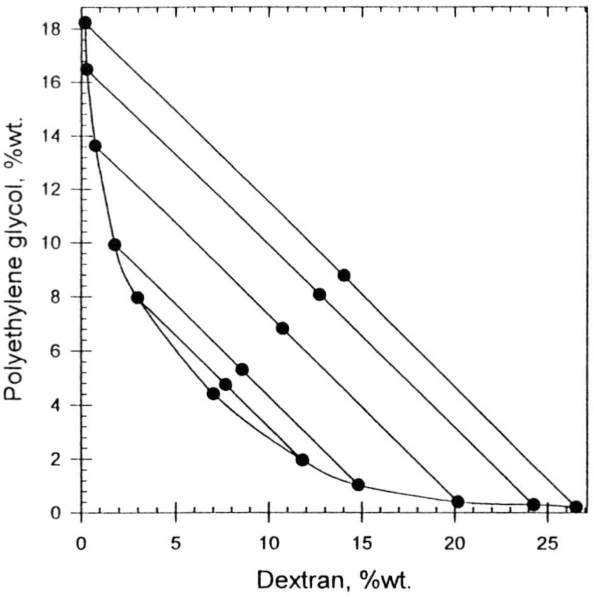

Table 10.41. Phase Diagram and Phase Composition of the Dextran-70—Poly(ethylene glycol)-6000 System Containing 0.05 mole/kg K_2SO_4 at 23°C.

Dextran-70 - M_w 57,200; M_n 28,700; Manufacturer: Minmedprom (Moscow, Russia); Lot 680480;
PEG - M_w ~6000; Manufacturer: Serva Fine Biochemicals (Germany); Lot 419-80

Total system		Bottom phase			Top phase			STL*
PEG % w/w	Dextran % w/w	PEG % w/w	Dextran % w/w	K_2SO_4 % w/w	PEG % w/w	Dextran % w/w	K_2SO_4 % w/w	
8.00	12.74	0.66	25.18	1.146	15.27	0.42	0.680	-0.590
6.67	10.57	0.69	20.70	1.092	12.39	0.88	0.719	-0.590
5.23	8.60	1.70	14.59	1.030	9.05	2.13	0.792	-0.590
4.59	7.57	2.64	10.88	0.966	6.59	4.18	0.839	-0.590
4.40**	7.20**						av.:	-0.590

* - STL - Tie-Line Slope defined as the ratio STL = (ΔPEG)/(ΔDex) where Δ is the difference between the concentrations of a given polymer in the two coexisting phases;

** - Composition of critical point.

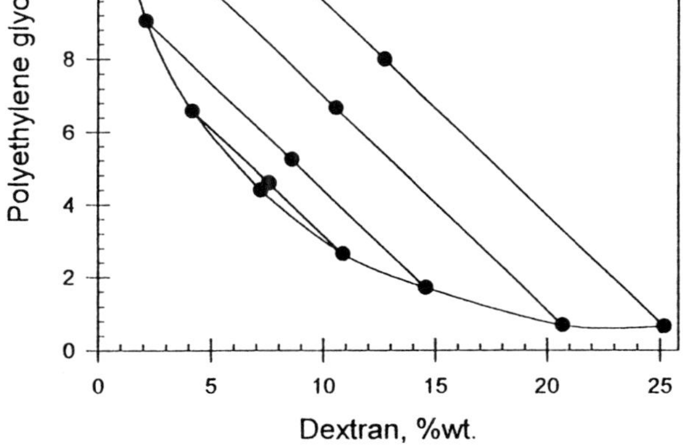

Table 10.42. Phase Diagram and Phase Composition of the Dextran-70—Poly(ethylene glycol)-6000 System Containing 0.1 mole/kg K_2SO_4 at 23°C.

Dextran-70 - M_w 57,200; M_n 28,700; Manufacturer: Minmedprom (Moscow, Russia); Lot 680480;
PEG - M_w ~6000; Manufacturer: Serva Fine Biochemicals (Germany); Lot 419-80

Total system		Bottom phase			Top phase			STL*
PEG % w/w	Dextran % w/w	PEG % w/w	Dextran % w/w	K_2SO_4 % w/w	PEG % w/w	Dextran % w/w	K_2SO_4 % w/w	
7.96	12.77	0.47	24.66	2.25	15.82	0.30	1.139	-0.630
6.70	10.78	0.62	20.43	2.13	13.10	0.62	1.225	-0.630
5.28	8.50	1.04	15.23	1.99	9.61	1.62	1.36	-0.630
4.67	7.45	1.98	11.72	1.85	7.57	2.85	1.44	-0.630
4.38	7.23	2.81	9.72	1.79	6.33	4.14	1.53	-0.631
4.40**	6.53**						av.:	-0.630 ±0.001

* - STL - Tie-Line Slope defined as the ratio STL = (ΔPEG)/(ΔDex) where Δ is the difference between the concentrations of a given polymer in the two coexisting phases;

** - Composition of critical point.

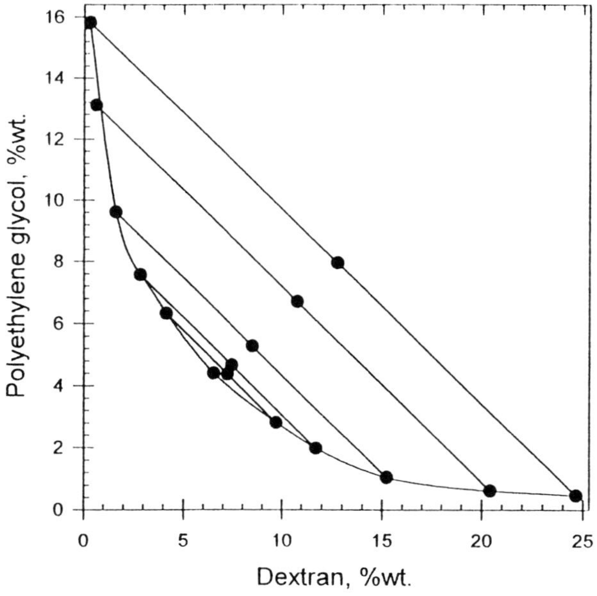

Table 10.43. Phase Diagram and Phase Composition of the Dextran-70—Poly(ethylene glycol)-6000 System Containing 0.25 mole/kg K_2SO_4 at 23°C.

Dextran-70 - M_w 57,200; M_n 28,700; Manufacturer: Minmedprom (Moscow, Russia); Lot 680480;
PEG - M_w ~6000; Manufacturer: Serva Fine Biochemicals (Germany); Lot 419-80

Total system		Bottom phase			Top phase			STL*
PEG % w/w	Dextran % w/w	PEG % w/w	Dextran % w/w	K_2SO_4 % w/w	PEG % w/w	Dextran % w/w	K_2SO_4 % w/w	
6.46	10.78	0.28	18.73	5.99	14.68	0.20	3.11	-0.777
5.34	8.53	0.62	14.65	5.71	11.51	0.55	3.47	-0.772
4.65	7.52	1.10	12.09	5.49	9.72	0.99	3.70	-0.777
4.43	6.69	1.40	10.59	5.345	8.50	1.45	3.87	-0.777
4.02	6.22	2.00	8.85	5.14	6.98	2.36	4.10	-0.767
4.10**	5.00**						av.:	-0.774 ±0.005

*- STL - Tie-Line Slope defined as the ratio STL = (ΔPEG)/(ΔDex) where Δ is the difference between the concentrations of a given polymer in the two coexisting phases;
** - Composition of critical point.

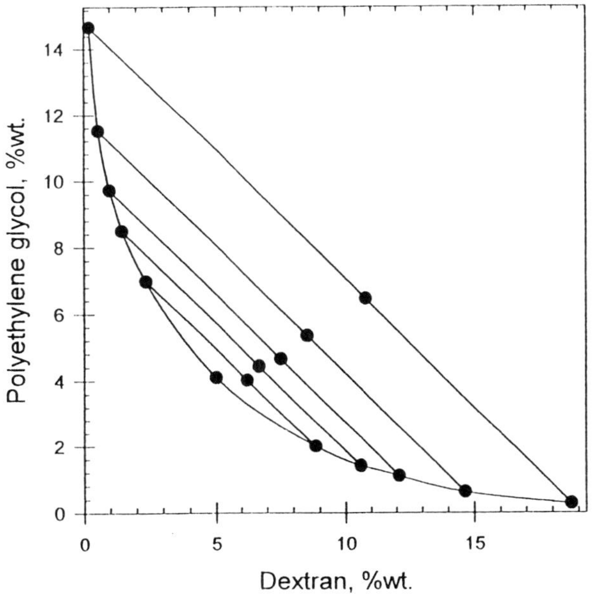

Table 10.44. Phase Diagram and Phase Composition of the Dextran-70—Poly(ethylene glycol)-6000 System Containing 0.1 mole/kg Cs_2SO_4 at 23°C.

Dextran-70 - M_w 57,200; M_n 28,700; Manufacturer: Minmedprom (Moscow, Russia); Lot 680480;
PEG - M_w ~6000; Manufacturer: Serva Fine Biochemicals (Germany); Lot 463-80

Total system		Bottom phase			Top phase			STL*
PEG % w/w	Dextran % w/w	PEG % w/w	Dextran % w/w	Cs_2SO_4 % w/w	PEG % w/w	Dextran % w/w	Cs_2SO_4 % w/w	
8.78	14.03	0.23	27.39	4.80	17.64	0.18	2.39	-0.640
8.05	12.78	0.34	24.82	4.71	16.07	0.25	2.49	-0.640
6.75	10.76	0.50	20.59	4.48	13.26	0.64	2.73	-0.640
5.37	8.56	1.02	15.36	4.26	9.78	1.67	2.95	-0.640
4.34**	6.95**						av.:	-0.640

* - STL - Tie-Line Slope defined as the ratio STL = (ΔPEG)/(ΔDex) where Δ is the difference between the concentrations of a given polymer in the two coexisting phases;

** - Composition of critical point.

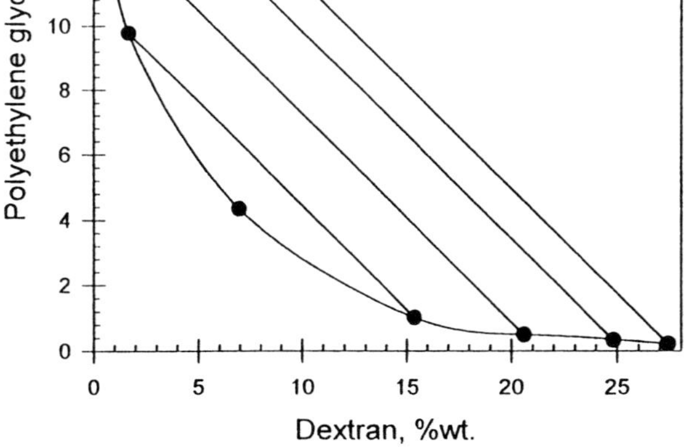

Table 10.45. Phase Diagram and Phase Composition of the Dextran-70—Poly(ethylene glycol)-6000 System Containing 0.1 mole/kg $(NH_4)_2SO_4$ at 23°C.

Dextran-70 - M_w 57,200; M_n 28,700; Manufacturer: Minmedprom (Moscow, Russia); Lot 680480;
PEG - M_w ~6000; Manufacturer: Serva Fine Biochemicals (Germany); Lot 463-80

Total system		Bottom phase			Top phase			STL*
PEG % w/w	Dex % w/w	PEG % w/w	Dex % w/w	$(NH_4)_2SO_4$ % w/w	PEG % w/w	Dex % w/w	$(NH_4)_2SO_4$ % w/w	
8.72	13.98	0.34	27.46	1.444	17.20	0.32	0.923	-0.621
7.98	12.76	0.36	25.02	1.440	15.60	0.47	0.961	-0.621
6.75	10.77	0.41	20.98	1.420	12.86	0.93	1.020	-0.621
6.02	9.61	0.70	18.18	1.415	11.09	1.45	1.074	-0.621
4.40**	7.70**						av.:	-0.621

* - STL - Tie-Line Slope defined as the ratio STL = (ΔPEG)/(ΔDex) where Δ is the difference between the concentrations of a given polymer in the two coexisting phases;

** - Composition of critical point.

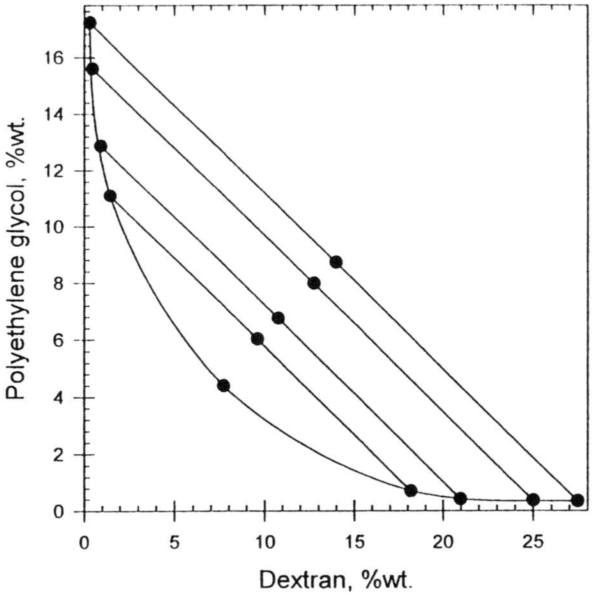

Table 10.46. Phase Diagram and Phase Composition of the Dextran-70—Poly(ethylene glycol)-6000 System Containing 0.15 mole/kg NaCl in 0.01 mole/kg Sodium Phosphate Buffer, pH 7.4 at 23°C.

Dextran-70 - M_w 57,200; M_n 28,700; Manufacturer: Minmedprom (Moscow, Russia); Lot 680480;
PEG - M_w ~6000; Manufacturer: Serva Fine Biochemicals (Germany); Lot 419-80

Total system		Bottom phase			Top phase			STL*
PEG % w/w	Dextran % w/w	PEG % w/w	Dextran % w/w	Na^+ ** % w/w	PEG % w/w	Dextran % w/w	Na^+ ** % w/w	
8.74	14.09	0.35	27.93	1.015	17.32	0.25	0.944	-0.613
8.16	12.76	0.35	25.81	0.995	15.71	0.37	0.931	-0.604
7.28	11.77	0.36	23.34	0.957	13.91	0.61	0.903	-0.596
5.94	9.56	0.92	18.61	0.928	10.60	1.23	0.890	-0.557
4.62	7.40	1.70	12.42	0.896	7.20	2.95	0.875	-0.581
3.90***	6.60***						av.:	-0.592 ±0.022

* - STL - Tie-Line Slope defined as the ratio STL = (ΔPEG)/(ΔDex) where Δ is the difference between the concentrations of a given polymer in the two coexisting phases;
** - Salt concentrations in the phases determined as the sodium concentrations;
*** - Composition of critical point.

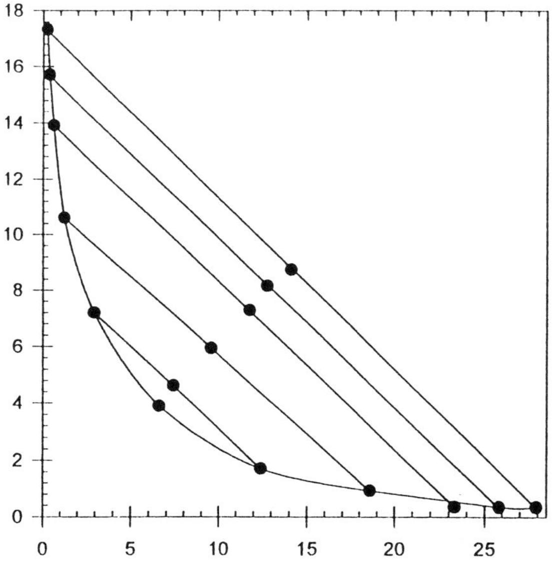

Table 10.47. Phase Diagram and Phase Composition of the Dextran-70—Poly(ethylene glycol)-6000 System Containing 0.11 mole/kg Sodium Phosphate Buffer, pH 7.4 at 23°C.

Dextran-70 - M_w 57,200; M_n 28,700; Manufacturer: Minmedprom (Moscow, Russia); Lot 680480;
PEG - M_w ~6000; Manufacturer: Serva Fine Biochemicals (Germany); Lot 463-80

Total system		Bottom phase			Top phase			STL *
PEG % w/w	Dextran % w/w	PEG % w/w	Dextran % w/w	Na^+ ** % w/w	PEG % w/w	Dextran % w/w	Na^+ ** % w/w	
8.88	13.88	0.34	27.60	0.605	17.41	0.16	0.317	-0.622
7.46	11.97	0.40	23.33	0.570	14.69	0.36	0.332	-0.622
6.20	10.01	0.49	19.18	0.546	11.94	0.79	0.354	-0.623
4.91	8.17	1.31	13.97	0.524	8.77	1.97	0.395	-0.622
4.20***	6.80***						av.:	-0.622 ±0.001

* - STL - Tie-Line Slope defined as the ratio STL = (ΔPEG)/(ΔDex) where Δ is the difference between the concentrations of a given polymer in the two coexisting phases;

** - Concentrations of sodium phosphate salts used as the components of the buffer, pH 7.4 were determined in the phases as those of Na^+;

*** - Composition of critical point.

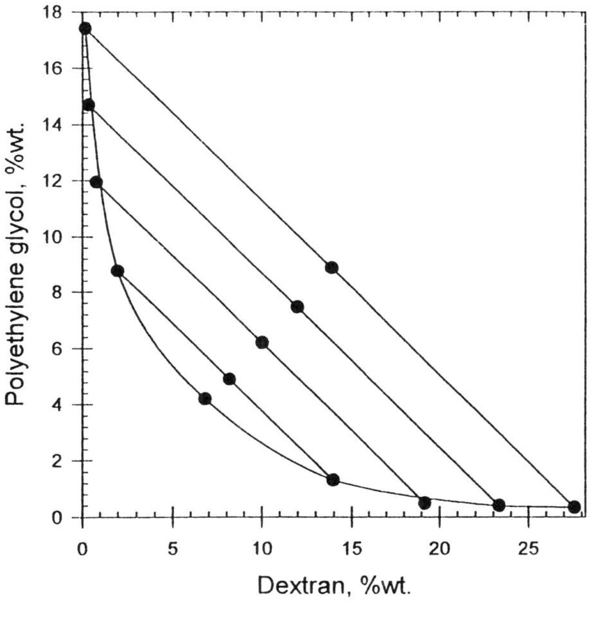

Table 10.48. Phase Diagram and Phase Composition of the Dextran-70—Poly(ethylene glycol)-6000 System Containing 0.01 mole/kg Universal Buffer, pH 7.5 at 23°C.

Dextran-70 - M_w 57,200; M_n 28,700; Manufacturer: Minmedprom (Moscow, Russia); Lot 680480;
PEG - M_w ~6000; Manufacturer: Serva Fine Biochemicals (Germany); Lot 463-80

Total system		Bottom phase			Top phase			STL*
PEG % w/w	Dextran % w/w	PEG % w/w	Dextran % w/w	Na^+ ** % w/w	PEG % w/w	Dextran % w/w	Na^+ ** % w/w	
8.17	12.93	0.65	26.40	0.147	15.01	0.50	0.115	-0.554
6.87	10.77	0.80	21.75	0.142	12.23	1.02	0.117	-0.551
6.10	9.73	0.96	19.05	0.141	10.61	1.57	0.120	-0.552
5.21	8.65	1.88	14.71	0.137	8.45	2.76	0.123	-0.550
4.64***	7.45***						av.:	-0.552 ±0.002

* - STL - Tie-Line Slope defined as the ratio STL = (ΔPEG)/(ΔDex) where Δ is the difference between the concentrations of a given polymer in the two coexisting phases;

** - Buffer salts concentrations in the phases determined as the sodium concentrations;

*** - Composition of critical point.

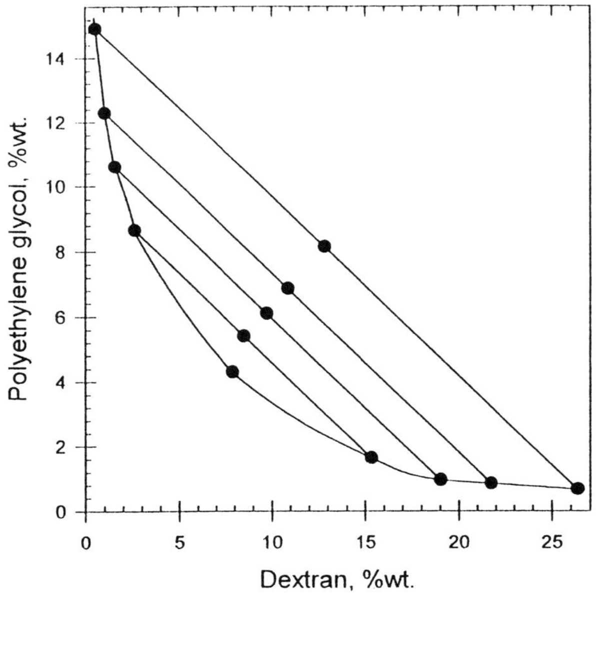

Table 10.49. Phase Diagram and Phase Composition of the Dextran-70—Poly(ethylene glycol)-6000 System Containing 0.1 mole/kg NaSCN in 0.01 mole/kg Universal Buffer, pH 7.5 at 23°C.

Dextran-70 - M_w 57,200; M_n 28,700; Manufacturer: Minmedprom (Moscow, Russia); Lot 680480;
PEG - M_w ~6000; Manufacturer: Serva Fine Biochemicals (Germany); Lot 463-80

Total system		Bottom phase			Top phase			STL*
PEG % w/w	Dextran % w/w	PEG % w/w	Dextran % w/w	Na^+ ** % w/w	PEG % w/w	Dextran % w/w	Na^+ ** % w/w	
8.06	12.91	0.32	26.28	0.528	15.28	0.43	0.572	-0.579
7.28	11.70	0.53	23.35	0.531	13.68	0.65	0.569	-0.579
6.04	9.65	0.90	18.74	0.536	11.00	1.30	0.565	-0.579
5.28	8.53	1.44	15.16	0.539	8.89	2.30	0.561	-0.579
4.50***	7.50***							

* - STL - Tie-Line Slope defined as the ratio STL = (ΔPEG)/(ΔDex) where Δ is the difference between the concentrations of a given polymer in the two coexisting phases;
** - Salt concentrations in the phases determined as the sodium concentrations;
*** - Composition of critical point.

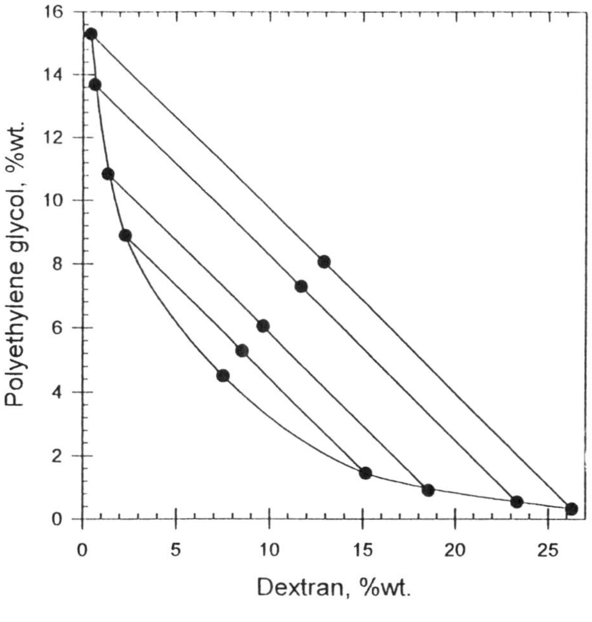

Table 10.50. Phase Diagram and Phase Composition of the Dextran-70—Poly(ethylene glycol)-6000 System Containing 0.5 mole/kg NaSCN in 0.01 mole/kg Universal Buffer, pH 7.5 at 23°C.

Dextran-70 - M_w 57,200; M_n 28,700; Manufacturer: Minmedprom (Moscow, Russia); Lot 680480;
PEG - M_w ~6000; Manufacturer: Serva Fine Biochemicals (Germany); Lot 463-80

Total system		Bottom phase			Top phase			STL*
PEG % w/w	Dextran % w/w	PEG % w/w	Dextran % w/w	Na^+ ** % w/w	PEG % w/w	Dextran % w/w	Na^+ ** % w/w	
8.11	12.90	0.37	25.45	2.055	15.85	0.35	2.406	-0.617
7.33	11.65	0.48	22.75	2.073	14.18	0.55	2.387	-0.617
6.63	10.55	0.56	20.30	2.093	12.70	0.80	2.368	-0.623
6.02	9.625	0.90	18.14	2.112	11.35	1.20	2.349	-0.617
5.34	8.47	1.37	14.91	2.140	9.32	2.02	2.320	-0.617
4.50***	7.20***						av.:	-0.618 ±0.003

* - STL - Tie-Line Slope defined as the ratio STL = (ΔPEG)/(ΔDex) where Δ is the difference between the concentrations of a given polymer in the two coexisting phases;
** - Salt concentrations in the phases determined as the sodium concentrations;
*** - Composition of critical point.

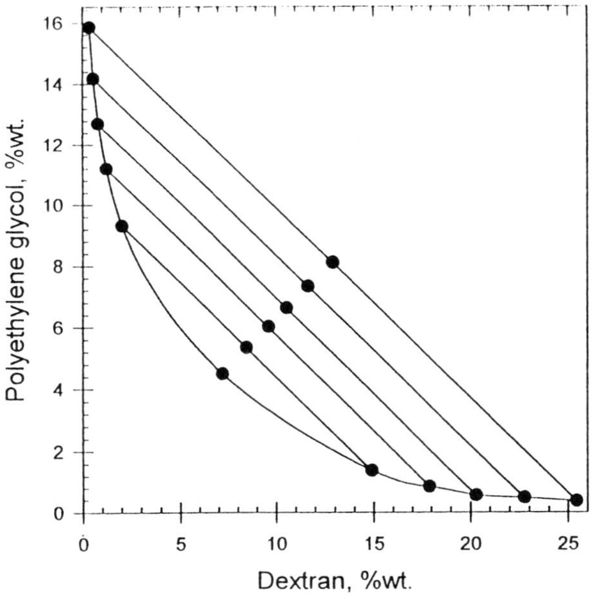

Table 10.51. Phase Diagram and Phase Composition of the Dextran-70—Poly(ethylene glycol)-6000 System Containing 0.1 mole/kg $NaClO_4$ in 0.01 mole/kg Universal Buffer, pH 7.5 at 23°C.

Dextran-70 - M_w 57,200; M_n 28,700; Manufacturer: Minmedprom (Moscow, Russia); Lot 680480;
PEG - M_w ~6000; Manufacturer: Serva Fine Biochemicals (Germany); Lot 463-80

Total system		Bottom phase			Top phase			STL*
PEG % w/w	Dextran % w/w	PEG % w/w	Dextran % w/w	Na^+ ** % w/w	PEG % w/w	Dextran % w/w	Na^+ ** % w/w	
8.03	12.94	0.26	26.27	0.555	15.34	0.40	0.621	-0.583
7.32	11.73	0.46	23.50	0.573	13.81	0.60	0.633	-0.583
5.99	9.61	0.80	18.80	0.598	11.06	1.290	0.646	-0.583
5.35	8.48	1.34	15.35	0.616	9.08	2.08	0.653	-0.583
4.76	7.61	2.32	11.78	0.645	7.045	3.69	0.668	-0.584
4.40***	7.25***						av.:	-0.583 ±0.001

* - STL - Tie-Line Slope defined as the ratio STL = (ΔPEG)/(ΔDex) where Δ is the difference between the concentrations of a given polymer in the two coexisting phases;
** - Salt concentrations in the phases determined as the sodium concentrations;
*** - Composition of critical point.

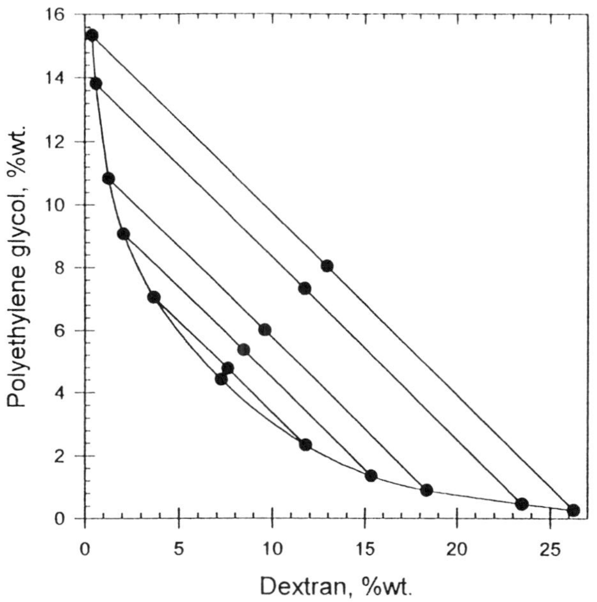

Table 10.52. Phase Diagram and Phase Composition of the Dextran-70—Poly(ethylene glycol)-6000 System Containing 0.5 mole/kg $NaClO_4$ in 0.01 mole/kg Universal Buffer, pH 7.5 at 23°C.

Dextran-70 - M_w 57,200; M_n 28,700; Manufacturer: Minmedprom (Moscow, Russia); Lot # 680480;
PEG - M_w ~6000; Manufacturer: Serva Fine Biochemicals (Germany); Lot # 463-80

Total system		Bottom phase			Top phase			STL *
PEG % w/w	Dextran % w/w	PEG % w/w	Dextran % w/w	Na^+ ** % w/w	PEG % w/w	Dextran % w/w	N^+ ** % w/w	
7.65	12.39	0.29	23.93	2.800	15.35	0.315	3.580	-0.638
7.27	11.66	0.37	22.48	2.830	14.45	0.40	3.555	-0.638
6.00	9.64	0.70	18.25	2.906	11.83	0.80	3.480	-0.638
5.30	8.47	1.08	15.09	2.972	9.71	1.56	3.415	-0.638
4.33***	6.65***						av.:	-0.638

* - STL - Tie-Line Slope defined as the ratio STL = (ΔPEG)/(ΔDex) where Δ is the difference between the concentrations of a given polymer in the two coexisting phases;
** - Salt concentrations in the phases determined as the sodium concentrations;
*** - Composition of critical point.

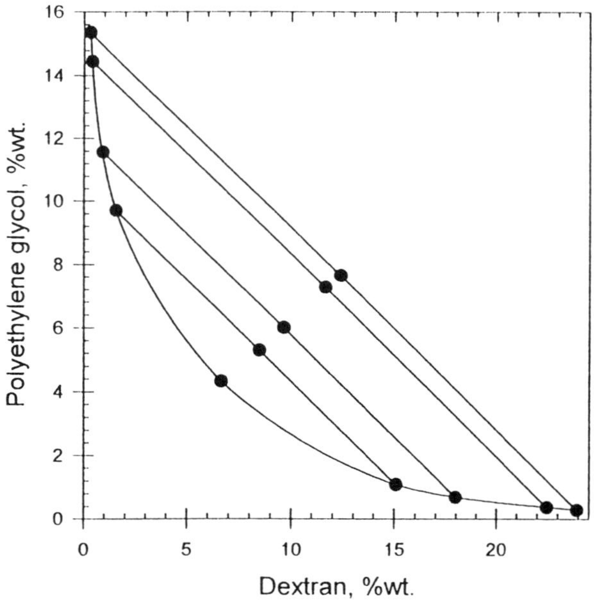

Table 10.53. Phase Diagram and Phase Composition of the Dextran-70—Poly(ethylene glycol)-6000 System Containing 0.1 mole/kg NaCl in 0.01 mole/kg Universal Buffer, pH 7.5 at 23°C.

Dextran-70 - M_w 57,200; M_n 28,700; Manufacturer: Minmedprom (Moscow, Russia); Lot # 680480;
PEG - M_w ~6000; Manufacturer: Serva Fine Biochemicals (Germany); Lot # 463-80

Total system		Bottom phase			Top phase			STL *
PEG % w/w	Dextran % w/w	PEG % w/w	Dextran % w/w	Na^+ ** % w/w	PEG % w/w	Dextran % w/w	Na^+ ** % w/w	
8.16	12.86	0.34	26.97	0.442	15.04	0.45	0.400	-0.554
6.76	10.84	1.00	21.30	0.435	12.26	0.92	0.403	-0.553
6.08	9.66	1.08	18.68	0.437	10.66	1.40	0.409	-0.554
5.29	8.47	1.49	15.33	0.436	8.64	2.42	0.415	-0.554
4.30***	7.70***						av.:	-0.554 ±0.001

* - STL - Tie-Line Slope defined as the ratio STL = (ΔPEG)/(ΔDex) where Δ is the difference between the concentrations of a given polymer in the two coexisting phases;

** - Salt concentrations in the phases determined as the sodium concentrations;

*** - Composition of critical point.

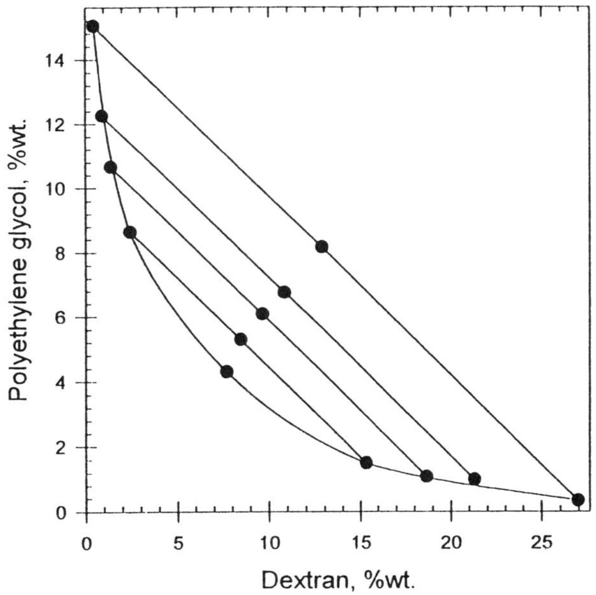

Table 10.54. Phase Diagram and Phase Composition of the Dextran-70—Poly(ethylene glycol)-6000 System Containing 0.5 mole/kg NaCl in 0.01 mole/kg Universal Buffer, pH 7.5 at 23°C.

Dextran-70 - M_w 57,200; M_n 28,700; Manufacturer: Minmedprom (Moscow, Russia); Lot # 680480;
PEG - M_w ~6000; Manufacturer: Serva Fine Biochemicals (Germany); Lot # 463-80

Total system		Bottom phase			Top phase			STL *
PEG % w/w	Dextran % w/w	PEG % w/w	Dextran % w/w	Na^+ **	PEG % w/w	Dextran % w/w	Na^+ **	
8.00	12.98	0.37	25.34	2.958	15.75	0.42	2.884	-0.617
7.36	11.68	0.55	22.72	2.951	14.18	0.62	2.892	-0.617
6.10	9.71	0.86	18.19	2.948	11.27	1.32	2.896	-0.617
5.31	8.56	1.48	14.77	2.940	9.18	2.28	2.902	-0.617
4.60***	7.40***						av.:	-0.617

* - STL - Tie-Line Slope defined as the ratio STL = (ΔPEG)/(ΔDex) where Δ is the difference between the concentrations of a given polymer in the two coexisting phases;
** - Salt concentrations in the phases determined as the sodium concentrations;
*** - Composition of critical point.

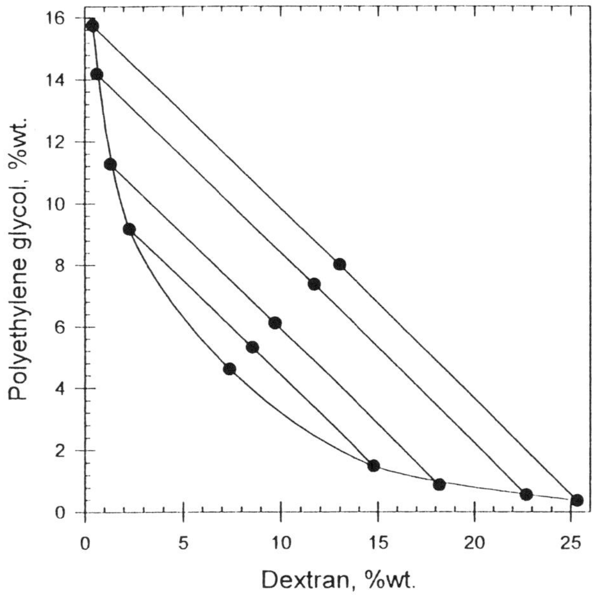

Table 10.55. Phase Diagram and Phase Composition of the Dextran-70—Poly(ethylene glycol)-6000 System Containing 0.05 mole/kg Na_2SO_4 in 0.01 mole/kg Universal Buffer, pH 7.5 at 23°C.

Dextran-70 - M_w 57,200; M_n 28,700; Manufacturer: Minmedprom (Moscow, Russia); Lot # 680480;
PEG - M_w ~6000; Manufacturer: Serva Fine Biochemicals (Germany); Lot # 463-80

Total system		Bottom phase			Top phase			STL *
PEG % w/w	Dextran % w/w	PEG % w/w	Dextran % w/w	Na^+ ** % w/w	PEG % w/w	Dextran % w/w	Na^+ ** % w/w	
8.04	12.90	0.33	25.75	0.593	15.55	0.38	0.376	-0.600
7.29	11.62	0.50	22.94	0.582	13.91	0.58	0.390	-0.600
6.02	9.60	0.80	18.55	0.560	11.15	1.30	0.411	-0.600
5.26	7.90	1.74	13.76	0.535	8.51	2.49	0.451	-0.601
4.50***	7.10***						av.:	-0.600 ±001

* - STL - Tie-Line Slope defined as the ratio STL = (ΔPEG)/(ΔDex) where Δ is the difference between the concentrations of a given polymer in the two coexisting phases;
** - Salt concentrations in the phases determined as the sodium concentrations;
*** - Composition of critical point.

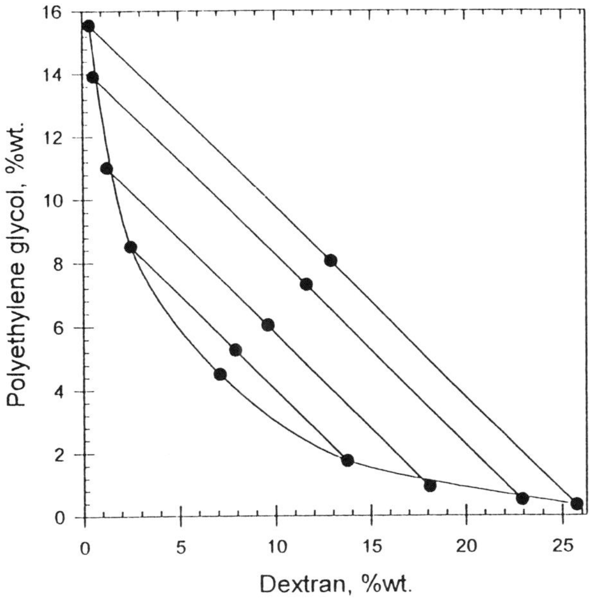

Table 10.56. Phase Diagram and Phase Composition of the Dextran-70—Poly(ethylene glycol)-6000 System Containing 0.25 mole/kg Na_2SO_4 in 0.01 mole/kg Universal Buffer, pH 7.5 at 23°C.

Dextran-70 - M_w 57,200; M_n 28,700; Manufacturer: Minmedprom (Moscow, Russia); Lot # 680480;
PEG - M_w ~6000; Manufacturer: Serva Fine Biochemicals (Germany); Lot # 463-80

Total system		Bottom phase			Top phase			STL *
PEG % w/w	Dextran % w/w	PEG % w/w	Dextran % w/w	Na^+ ** % w/w	PEG % w/w	Dextran % w/w	Na^+ ** % w/w	
7.31	11.72	0.21	20.40	2.508	16.80	0.12	1.298	-0.818
6.10	9.70	0.40	19.11	2.473	15.87	0.20	1.338	-0.818
6.04	9.67	0.46	16.49	2.396	13.71	0.29	1.416	-0.818
5.35	8.52	0.65	14.27	2.324	11.88	0.54	1.488	-0.818
4.42	7.03	1.23	10.93	2.200	9.06	1.36	1.612	-0.818
4.26***	5.10***						av.:	-0.818

* - STL - Tie-Line Slope defined as the ratio STL = (ΔPEG)/(ΔDex) where Δ is the difference between the concentrations of a given polymer in the two coexisting phases;

** - Salt concentrations in the phases determined as the sodium concentrations;

*** - Composition of critical point.

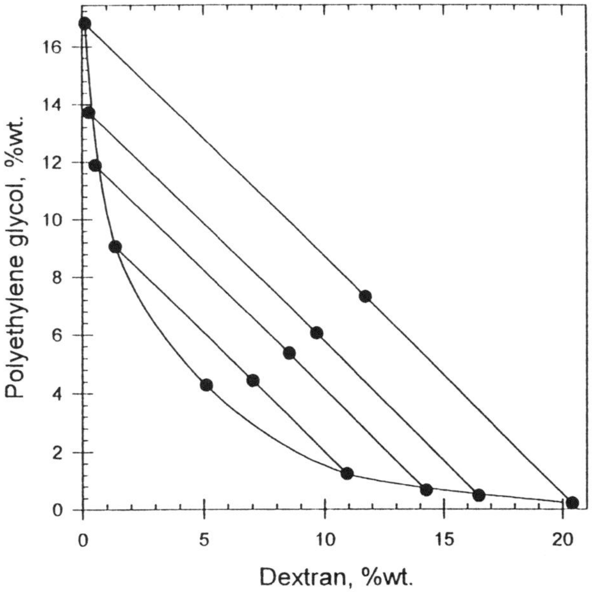

Table 10.57. Phase Diagram and Phase Composition of the Dextran-110—Poly(ethylene glycol)-6000 System at 25°C.
(From D. Forciniti, C. K. Hall, M.-R. Kula, Fluid Phase Equilibria, 61, 243 (1991) and Bioseparation, 2, 115 (1991) by permission of Elsevier Scientific Publishing Co. and Kluwer Academic Publishers, respectively.)

PEG - M_w 5600; M_n 5300; Manufacturer: Fluka (Switzerland); Lot not indicated
Dextran - M_w 86,200; M_n 52,100; Manufacturer: Fluka (Switzerland); Lot not indicated

Total system		Bottom phase		Top phase		
PEG % w/w	Dex % w/w	PEG % w/w	Dex % w/w	PEG % w/w	Dex % w/w	STL *
4.4	6.8	2.0	11.6	6.5	1.9	-0.464
5.0	8.0	1.4	14.1	8.2	0.66	-0.506
5.6	10.0	1.5	18.7	10.1	0.33	-0.468
7.6	10.2	1.0	23.4	12.6	0.15	-0.499
					av.:	-0.484 ±0.021

* - STL - Tie-Line Slope defined as the ratio STL = (ΔPEG)/(ΔDex) where Δ is the difference between the concentrations of a given polymer in the two coexisting phases.

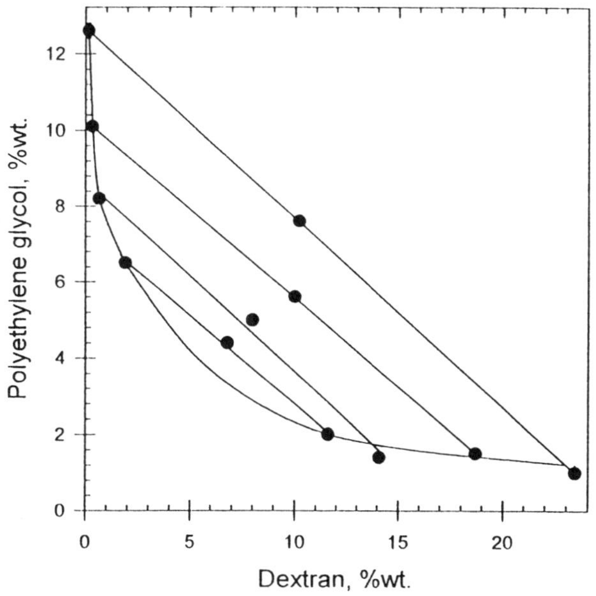

Table 10.58. Phase Diagram and Phase Composition of the Dextran-110—Poly(ethylene glycol)-6000 System at 4°C.
(From D. Forciniti, C. K. Hall, M.-R. Kula, Fluid Phase Equilibria, 61, 243 (1991) and Bioseparation, 2, 115 (1991) by permission of Elsevier Scientific Publishing Co. and Kluwer Academic Publishers, respectively.)

PEG - M_w 5600; M_n 5300; Manufacturer: Fluka (Switzerland); Lot not indicated;
Dextran - M_w 86,200; M_n 52,100; Manufacturer: Fluka (Switzerland); Lot not indicated

Total system		Bottom phase *		Top phase *		STL **
PEG % w/w	Dex % w/w	PEG % w/w	Dex % w/w	PEG % w/w	Dex % w/w	
4.4	6.75	2.8	12.0	5.9	1.9	-0.307
5.0	8.5	1.7	17.1	8.1	0.66	-0.389
5.6	10.0	0.9	21.1	9.8	0.33	-0.428
7.6	10.3	0.6	25.7	12.2	0.15	-0.454
					av.:	-0.395 ±0.064

* - Compositions of all the phases questionable as the STL values are inconsistent;
** - STL - Tie-Line Slope defined as the ratio STL = (ΔPEG)/(ΔDex) where Δ is the difference between the concentrations of a given polymer in the two coexisting phases.

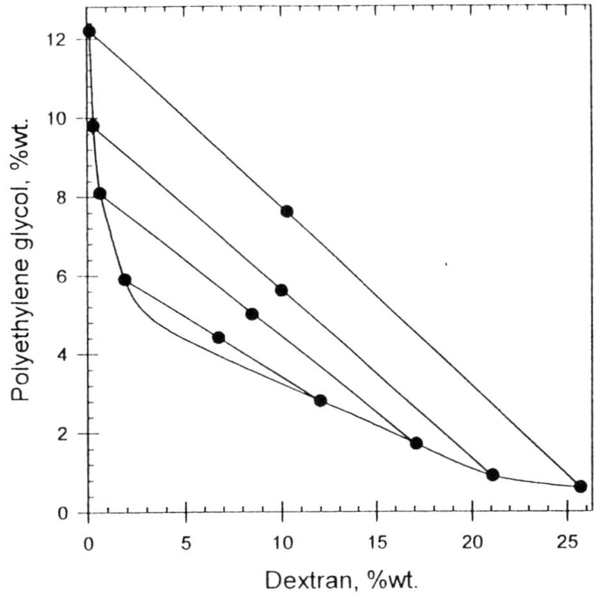

Table 10.59. Phase Diagram and Phase Composition of the Dextran-110—Poly(ethylene glycol)-6000 System at 40°C.

(From D. Forciniti, C. K. Hall, M.-R. Kula, Fluid Phase Equilibria, 61, 243 (1991) and Bioseparation, 2, 115 (1991) by permission of Elsevier Scientific Publishing Co. and Kluwer Academic Publishers, respectively.)

PEG - M_w 5600; M_n 5300; Manufacturer: Fluka (Switzerland); Lot not indicated;

Dextran - M_w 86,200; M_n 52,100; Manufacturer: Fluka (Switzerland); Lot not indicated

Total system		Bottom phase *		Top phase *		STL **
PEG % w/w	Dex % w/w	PEG % w/w	Dex % w/w	PEG % w/w	Dex % w/w	
5.0	8.5	4.9	8.9	8.2	1.5	-0.446
5.6	10.0	3.9	13.4	10.2	0.63	-0.493
7.6	10.3	3.5	17.4	13.1	0.28	-0.561
					av.:	-0.500 ±0.058

*- Compositions of all the phases questionable as the STL values are inconsistent;

** - STL - Tie-Line Slope defined as the ratio STL = (ΔPEG)/(ΔDex) where Δ is the difference between the concentrations of a given polymer in the two coexisting phases.

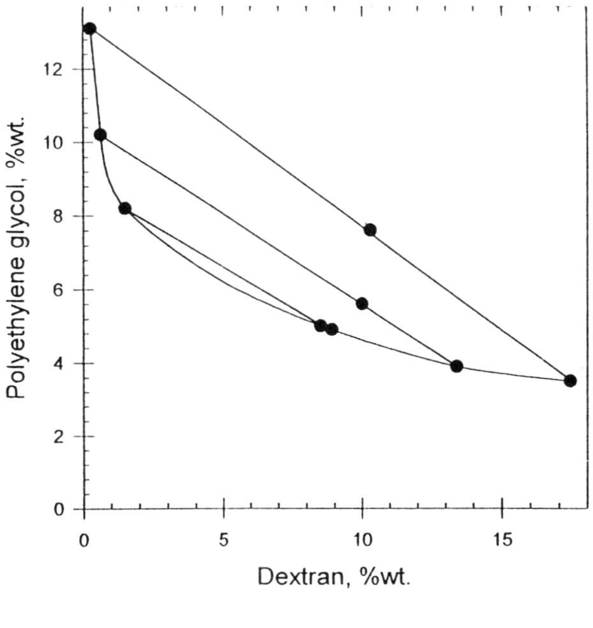

Table 10.60. Phase Diagram and Phase Composition of the Dextran-500—Poly(ethylene glycol)-6000 System at 25°C.
(From D. Forciniti, C. K. Hall, M.-R. Kula, Fluid Phase Equilibria, 61, 243 (1991) and Bioseparation, 2, 115 (1991) by permission of Elsevier Scientific Publishing Co. and Kluwer Academic Publishers, respectively.)

PEG - M_w 5600; M_n 5300; Manufacturer: Fluka (Switzerland); Lot not indicated
Dextran - M_w 215,000; M_n 88,200; Manufacturer: Pfeifer and Langen (Dormagen, Germany); Lot not indicated

Total system		Bottom phase		Top phase		STL *
PEG % w/w	Dex % w/w	PEG % w/w	Dex % w/w	PEG % w/w	Dex % w/w	
4.4	6.9	2.2	11.7	6.6	1.4	-0.427
5.0	8.6	1.3	16.5	8.6	0.8	-0.465
5.6	10.1	0.9	19.9	9.8	0.8	-0.466
7.6	10.0	0.8	23.4	12.3	0.2	-0.496
					av.:	-0.464 ±0.028

* - STL - Tie-Line Slope defined as the ratio STL = (ΔPEG)/(ΔDex) where Δ is the difference between the concentrations of a given polymer in the two coexisting phases.

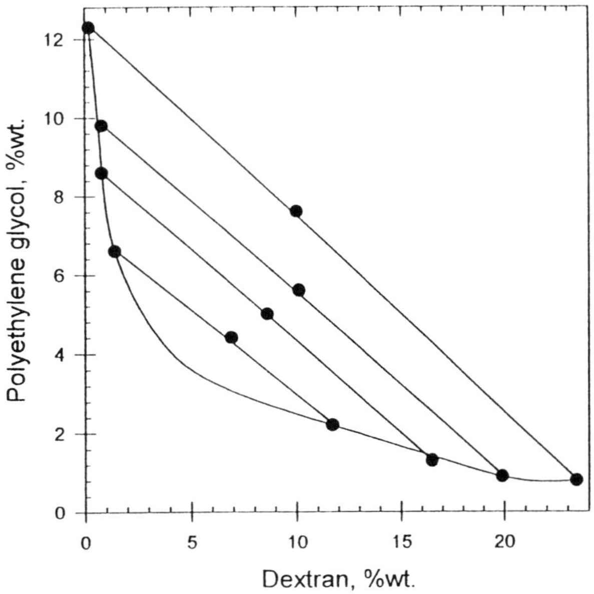

Table 10.61. Phase Diagram and Phase Composition of the Dextran-500—Poly(ethylene glycol)-6000 System at 4°C.
(From D. Forciniti, C. K. Hall, M.-R. Kula, Fluid Phase Equilibria, 61, 243 (1991) and Bioseparation, 2, 115 (1991) by permission of Elsevier Scientific Publishing Co. and Kluwer Academic Publishers, respectively.)

PEG - M_w 5600; M_n 5300; Manufacturer: Fluka (Switzerland); Lot not indicated;
Dextran - M_w 215,000; M_n 88,200; Manufacturer: Pfeifer and Langen (Dormagen, Germany); Lot not indicated

Total system		Bottom phase		Top phase		STL *
PEG % w/w	Dex % w/w	PEG % w/w	Dex % w/w	PEG % w/w	Dex % w/w	
4.4	6.75	1.7	13.4	6.4	1.0	-0.379
5.0	8.5	1.3	17.4	8.2	0.36	-0.405
5.6	10.0	1.4	20.3	9.9	0.18	-0.422
7.6	10.3	1.4	24.6	11.9	0.07	-0.428
					av.:	-0.409 ±0.022

* - STL - Tie-Line Slope defined as the ratio STL = (ΔPEG)/(ΔDex) where Δ is the difference between the concentrations of a given polymer in the two coexisting phases.

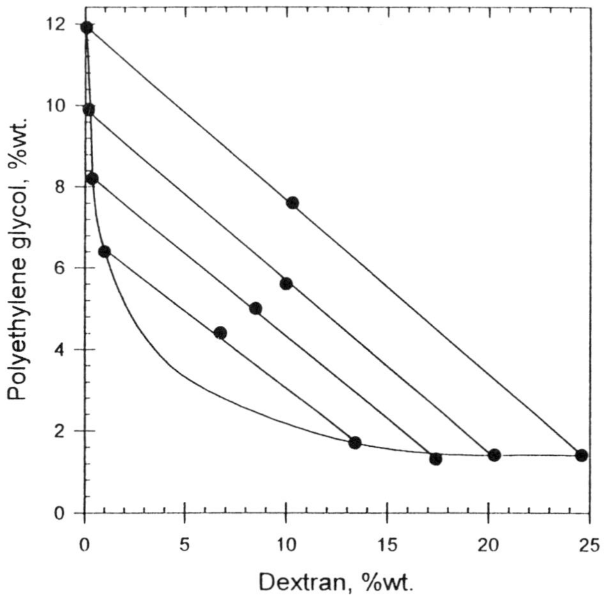

Table 10.62 Phase Diagram and Phase Composition of the Dextran-500—Poly(ethylene glycol)-6000 System at 40°C.
(From D. Forciniti, C. K. Hall, M.-R. Kula, Fluid Phase Equilibria, 61, 243 (1991) and Bioseparation, 2, 115 (1991) by permission of Elsevier Scientific Publishing Co. and Kluwer Academic Publishers, respectively.)

PEG - M_w 5600; M_n 5300; Manufacturer: Fluka (Switzerland); Lot not indicated;
Dextran - M_w 215,000; M_n 88,200; Manufacturer: Pfeifer and Langen (Dormagen, Germany); Lot not indicated

Total system		Bottom phase		Top phase		STL *
PEG % w/w	Dex % w/w	PEG % w/w	Dex % w/w	PEG % w/w	Dex % w/w	
4.4	6.75	2.8 **	27.8 **	6.2 **	2.0 **	-0.132 **
5.6	10.0	1.8	19.2	9.8	0.26	-0.422
7.6	10.3	1.2	23.7	12.7	0.1	-0.487
					av.:	-0.455
						±0.046

* - STL - Tie-Line Slope defined as the ratio STL = (ΔPEG)/(ΔDex) where Δ is the difference between the concentrations of a given polymer in the two coexisting phases;
** - Composition of the phases questionable as the STL value is inconsistent with the other values; was not used in calculations of the average STL value; not shown on phase diagram.

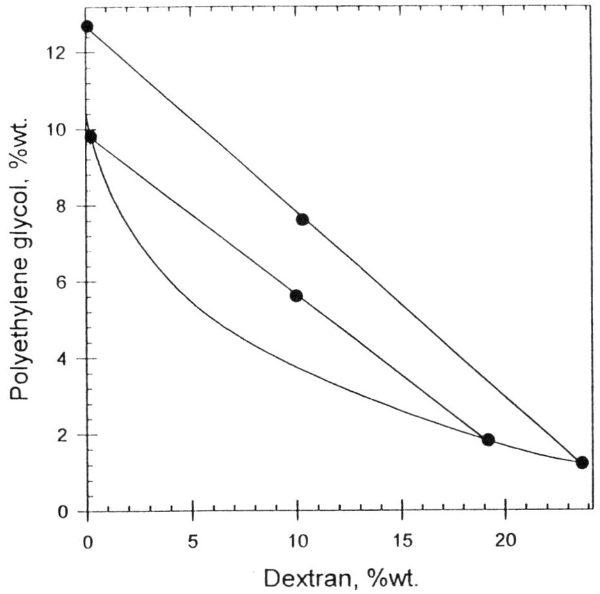

Table 10.63. Phase Diagram and Phase Composition of the Dextran-10—Poly(ethylene glycol)-8000 System at 22°C.
(From J. Hsu, personal communication, 1991, with permission of the author.)

PEG - M_w~8000; Manufacturer: Aldrich (Milwaukee, WI, USA); Lot 00917PT; Dextran T-10 - M_w 10,900; M_n 5,300; Manufacturer: Pharmacia Fine Chemicals (Piscataway, NJ, USA); Lot 00985

Total system		Bottom phase		Top phase		STL *
PEG % w/w	Dex % w/w	PEG % w/w	Dex % w/w	PEG % w/w	Dex % w/w	
5.47	12.18	2.50	16.96	7.82	8.42	-0.623
5.70	12.84	1.96	19.01	9.16	6.92	-0.596
5.81	14.00	1.25	21.56	11.11	5.44	-0.612
5.96	15.00	0.82	23.46	12.37	4.63	-0.613
					av.:	-0.611 ±0.011

* - STL - Tie-Line Slope defined as the ratio STL = (ΔPEG)/(ΔDex) where Δ is the difference between the concentrations of a given polymer in the two coexisting phases.

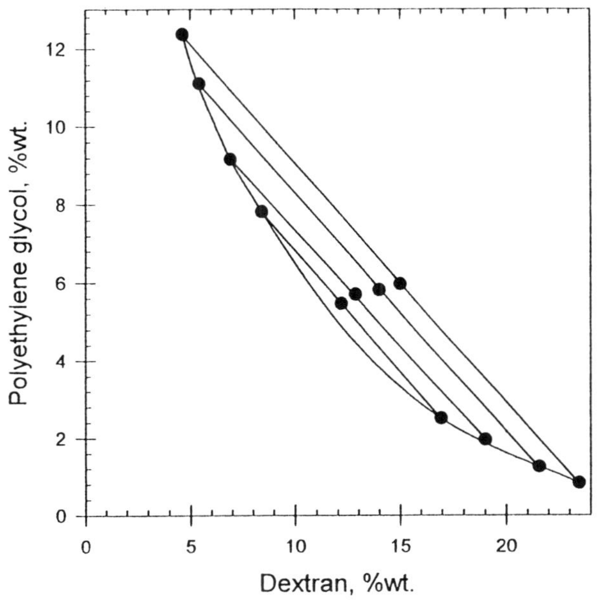

Table 10.64. Phase Diagram and Phase Composition of the Dextran-70—Poly(ethylene glycol)-8000 System at 10ºC.
(From J. Hsu, personal communication, 1991, with permission of the author.)

PEG - M_w~8000; Manufacturer: Aldrich (Milwaukee, WI, USA); Lot 02521LT; Dextran T-70 - M_w 72,200; M_n 38,400; Manufacturer: Pharmacia Fine Chemicals (Piscataway, NJ, USA); Lot 02377

Total system		Bottom phase		Top phase		STL *
PEG % w/w	Dex % w/w	PEG % w/w	Dex % w/w	PEG % w/w	Dex % w/w	
4.13	7.19	1.75	12.43	6.43	2.02	-0.450
3.89	9.19	1.29	14.98	7.45	1.32	-0.451
4.22	9.77	0.92	17.14	8.59	0.91	-0.473
4.35	10.47	0.79	17.95	9.03	0.75	-0.479
					av.:	-0.463 ±0.015

* - STL - Tie-Line Slope defined as the ratio STL = (ΔPEG)/(ΔDex) where Δ is the difference between the concentrations of a given polymer in the two coexisting phases.

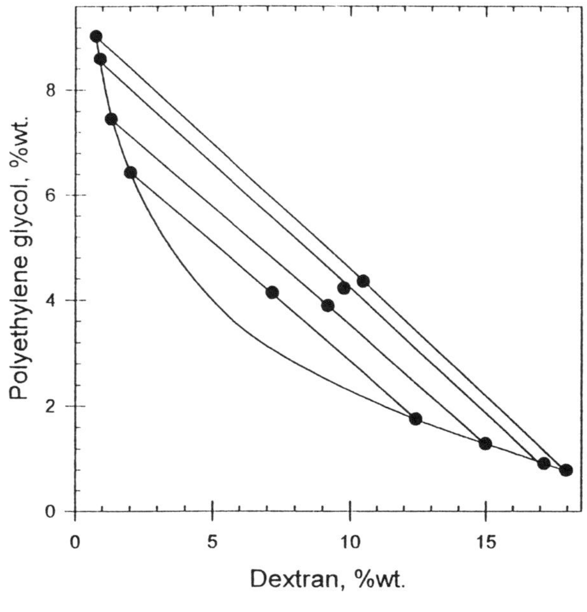

Table 10.65. Phase Diagram and Phase Composition of the Dextran-500—Poly(ethylene glycol)-8000 System at 10°C.
(From J. Hsu, personal communication, 1991, with permission of the author.)

PEG - M_w~8000; Manufacturer: Aldrich (Milwaukee, WI, USA); Lot 02521LT; Dextran T-500 - M_w 507,000; M_n 234,200; Manufacturer: Pharmacia Fine Chemicals (Piscataway, NJ, USA); Lot # 05163

Total system		Bottom phase		Top phase		STL *
PEG % w/w	Dex % w/w	PEG % w/w	Dex % w/w	PEG % w/w	Dex % w/w	
3.50	5.00	1.86	8.87	5.31	0.95	-0.436
4.00	7.00	0.94	13.74	7.07	0.25	-0.454
4.50	8.00	0.63	16.15	8.20	0.13	-0.472
5.80	8.40	0.49	19.27	9.73	0.09	-0.482
					av.:	-0.461 ±0.020

* - STL - Tie-Line Slope defined as the ratio STL = (ΔPEG)/(ΔDex) where Δ is the difference between the concentrations of a given polymer in the two coexisting phases.

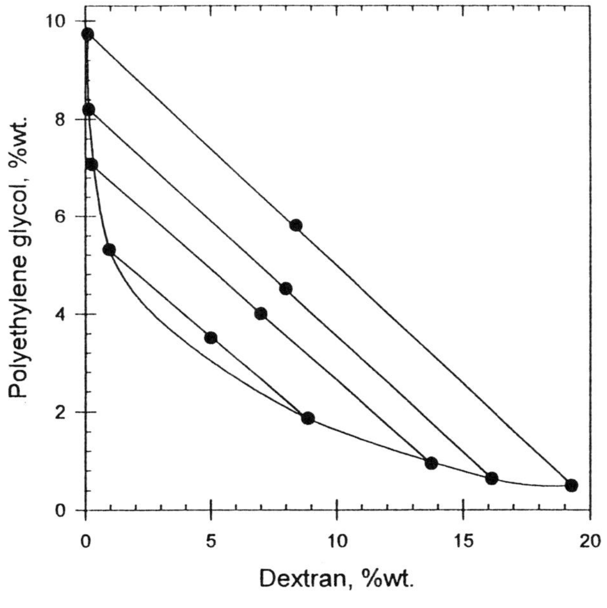

Table 10.66. Phase Diagram and Phase Composition of the Dextran-70—Poly(ethylene glycol)-8000 System at 22°C.
(From A. D. Diamond, J. Hsu, Biotechnol. Techniques, 3, 119 (1989) with permission of Eaton Publishing Co.)

PEG - M_w~8000; Manufacturer: Aldrich (Milwaukee, WI, USA); Lot 02521LT; Dextran T-70 - M_w 72,200; M_n 38,400; Manufacturer: Pharmacia Fine Chemicals (Piscataway, NJ, USA); Lot 02377

Total system		Bottom phase		Top phase		STL *
PEG % w/w	Dex % w/w	PEG % w/w	Dex % w/w	PEG % w/w	Dex % w/w	
4.07	6.39	3.33	7.81	4.57	5.08	-0.454
4.16	8.41	1.53	13.74	7.53	1.62	-0.495
4.84	8.44	1.16	15.87	8.67	1.08	-0.508
5.32	9.06	0.83	17.74	9.68	0.77	-0.522
3.89	14.30	0.44	20.52	11.15	0.46	-0.534
					av.:	-0.503 ±0.031

* - STL - Tie-Line Slope defined as the ratio STL = (ΔPEG)/(ΔDex) where Δ is the difference between the concentrations of a given polymer in the two coexisting phases.

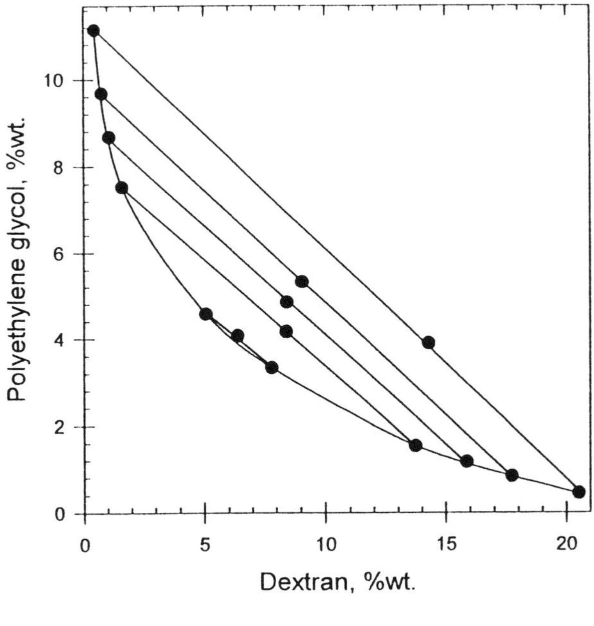

Table 10.67. Phase Diagram and Phase Composition of the Dextran-40—Poly(ethylene glycol)-8000 System at 22°C.
(From A. D. Diamond, J. Hsu, Biotechnol. Techniques, 3, 119 (1989) with permission of Eaton Publishing Co.)

PEG - M_w~8000; Manufacturer: Aldrich (Milwaukee, WI, USA); Lot 02521LT; Dextran T-40 - M_w 40,200; M_n 24,400; Manufacturer: Pharmacia Fine Chemicals (Piscataway, NJ, USA); Lot 01852

Total system		Bottom phase		Top phase		STL *
PEG % w/w	Dex % w/w	PEG % w/w	Dex % w/w	PEG % w/w	Dex % w/w	
4.18	7.89	2.70	10.77	5.83	4.61	-0.508
3.89	8.79	2.23	11.86	6.52	3.88	-0.538
4.43	8.27	1.65	13.52	7.24	2.99	-0.531
6.24	8.60	0.82	18.86	10.18	1.16	-0.529
					av.:	-0.527
						±0.013

* - STL - Tie-Line Slope defined as the ratio STL = (ΔPEG)/(ΔDex) where Δ is the difference between the concentrations of a given polymer in the two coexisting phases.

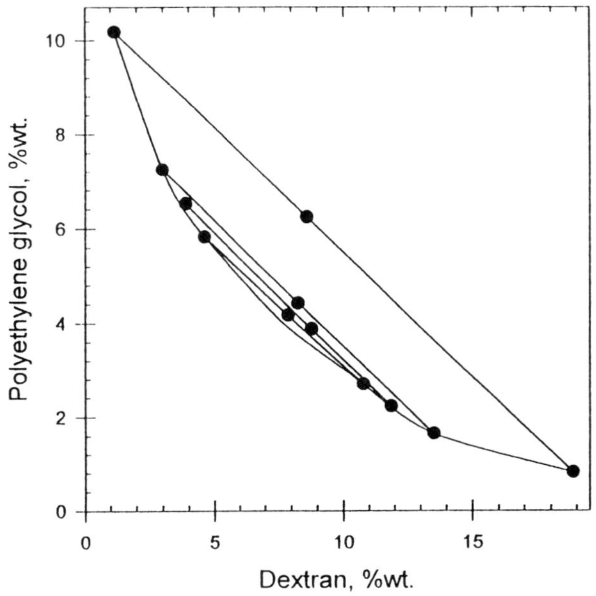

Table 10.68. Phase Diagram and Phase Composition of the Dextran-500—Poly(ethylene glycol)-8000 System at 22°C.
(From A. D. Diamond, J. Hsu, Biotechnol. Techniques, 3, 119 (1989) with permission of Eaton Publishing Co.)

PEG - M_w~8000; Manufacturer: Aldrich (Milwaukee, WI, USA); Lot 02521LT; Dextran T-500 - M_w 507,000; M_n 234,200; Manufacturer: Pharmacia Fine Chemicals (Piscataway, NJ, USA); Lot 05163

Total system		Bottom phase		Top phase		STL *
PEG	Dex	PEG	Dex	PEG	Dex	
% w/w	% w/w	% w/w	% w/w	% w/w	% w/w	
3.80	5.20	1.67	9.69	5.89	0.79	-0.474
4.40	6.20	0.92	13.19	7.15	0.20	-0.480
5.00	7.00	0.71	15.71	8.28	0.10	-0.485
5.80	8.40	0.47	18.92	10.01	0.04	-0.505
					av.:	-0.486
						±0.013

* - STL - Tie-Line Slope defined as the ratio STL = (ΔPEG)/(ΔDex) where Δ is the difference between the concentrations of a given polymer in the two coexisting phases.

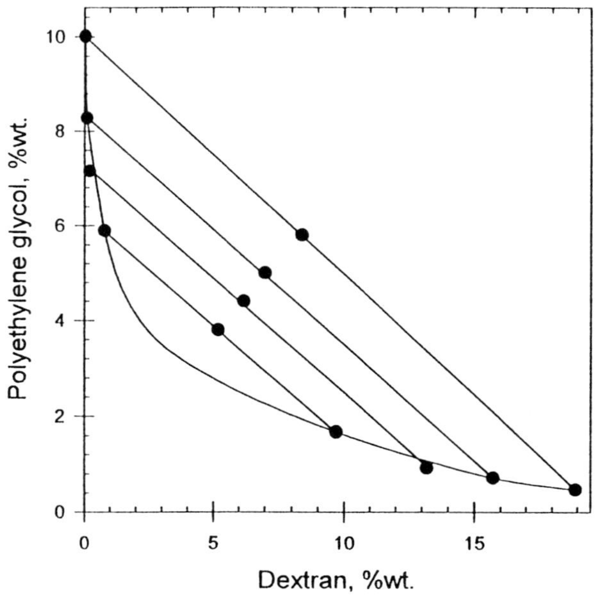

Table 10.69. Phase Diagram and Phase Composition of the Dextran-40—Poly(ethylene glycol)-8000 System at 4°C.
(From A. D. Diamond, J. T. Hsu, Biotechnol. Bioeng., 34, 1000 (1989) with permission of John Wiley & Sons, Inc.)

PEG - M_w ~6000; Manufacturer: Aldrich (Milwaukee, WI, USA); Lot 02521LT; Dextran T-40 - M_w 38,800; M_n 24,200; Manufacturer: Pharmacia Fine Chemicals (Piscataway, NJ, USA); Lot # 03375

Total system		Bottom phase		Top phase		STL *
PEG % w/w	Dex % w/w	PEG % w/w	Dex % w/w	PEG % w/w	Dex % w/w	
3.90	7.50	2.80	10.04	4.69	5.73	-0.438
4.20	8.20	1.59	13.26	6.72	2.89	-0.495
4.60	8.70	1.17	15.58	7.87	2.01	-0.494
4.90	9.20	0.72	17.42	8.55	1.62	-0.496
					av.:	-0.481
						±0.028

* - STL - Tie-Line Slope defined as the ratio STL = (ΔPEG)/(ΔDex) where Δ is the difference between the concentrations of a given polymer in the two coexisting phases.

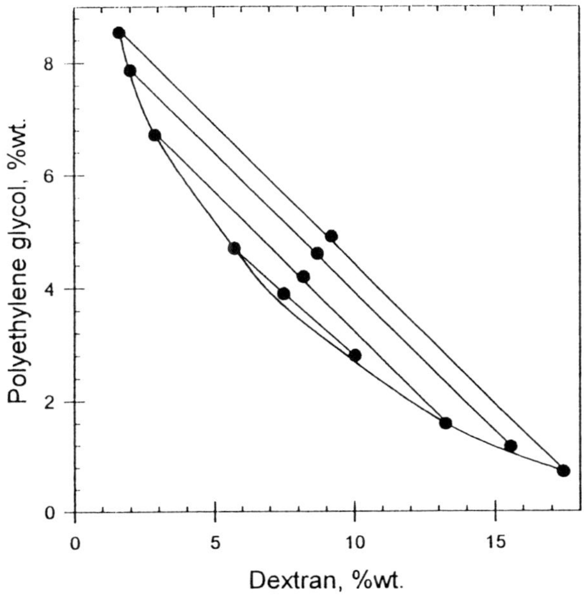

Table 10.70. Phase Diagram and Phase Composition of the Dextran-70—Poly(ethylene glycol)-8000 System at 4°C.
(From A. D. Diamond, J. T. Hsu, Biotechnol. Bioeng., 34, 1000 (1989) with permission of John Wiley & Sons, Inc.)

PEG - M_w ~6000; Manufacturer: Aldrich (Milwaukee, WI, USA); Lot 02521LT; Dextran T-70 - M_w 72,200; M_n 38,400; Manufacturer: Pharmacia Fine Chemicals (Piscataway, NJ, USA); Lot # 02377

Total system		Bottom phase		Top phase		STL *
PEG % w/w	Dex % w/w	PEG % w/w	Dex % w/w	PEG % w/w	Dex % w/w	
3.70	7.25	1.56	11.91	5.95	2.35	-0.459
5.00	8.00	0.76	17.17	8.29	0.88	-0.462
4.70	9.40	0.72	18.05	8.64	0.77	-0.458
5.10	10.50	0.58	20.25	9.71	0.55	-0.463
					av.:	-0.461 ±0.002

* - STL - Tie-Line Slope defined as the ratio STL = (ΔPEG)/(ΔDex) where Δ is the difference between the concentrations of a given polymer in the two coexisting phases.

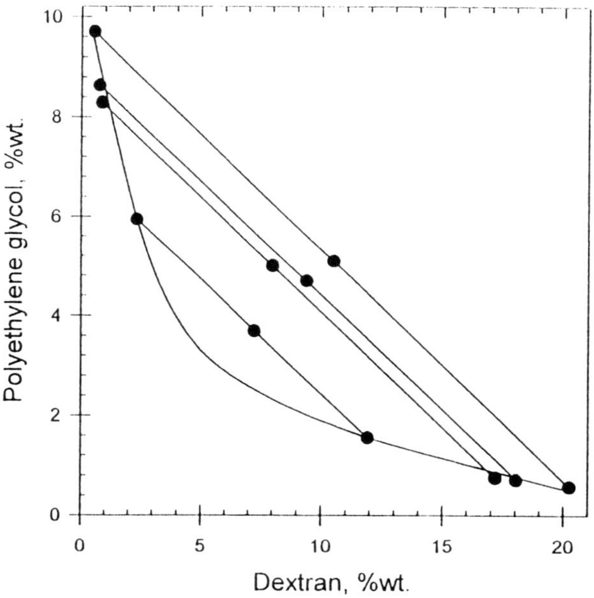

Table 10.71. Phase Diagram and Phase Composition of the Dextran-500—Poly(ethylene glycol)-8000 System at 4°C.
(From A. D. Diamond, J. T. Hsu, Biotechnol. Bioeng., 34, 1000 (1989) with permission of John Wiley & Sons, Inc.)

PEG - M_w ~6000; Manufacturer: Aldrich (Milwaukee, WI, USA); Lot 02521LT; Dextran T-500 - M_w 507,000; M_n 234,200; Manufacturer: Pharmacia Fine Chemicals (Piscataway, NJ, USA); Lot 05163

Total system		Bottom phase		Top phase		STL *
PEG % w/w	Dex % w/w	PEG % w/w	Dex % w/w	PEG % w/w	Dex % w/w	
3.27	4.84	1.63	8.83	4.91	0.86	-0.412
4.50	5.86	0.73	14.11	7.03	0.11	-0.450
5.76	7.50	0.43	18.73	9.27	0.03	-0.473
7.00	8.00	0.30	21.61	10.83	0.01	-0.488
					av.:	-0.456
						±0.033

* - STL - Tie-Line Slope defined as the ratio STL = (ΔPEG)/(ΔDex) where Δ is the difference between the concentrations of a given polymer in the two coexisting phases.

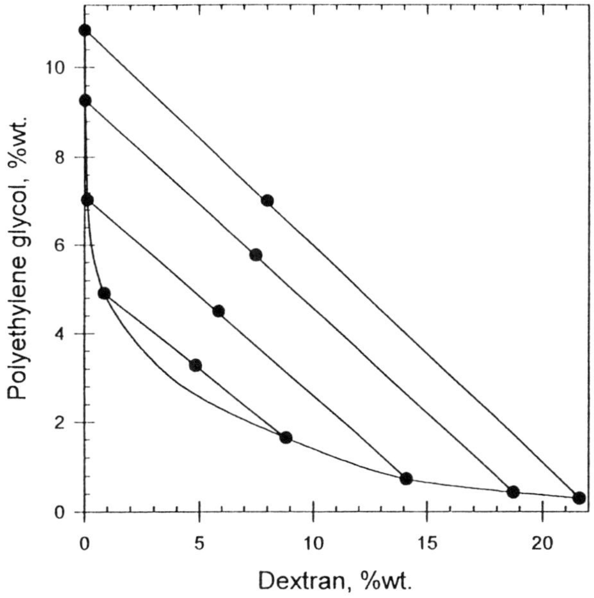

Table 10.72. Phase Diagram and Phase Composition of the Dextran-500—Poly(ethylene glycol)-8000 System Containing 0.1 mole/kg K_2SO_4 at 21°C. (From D. E. Brooks, K. A. Sharp, S. Bamberger, C. H. Tamblyn, G. V. F. Seaman, H. Walter, J.Colloid Interface Sci., 102, 1 (1984) with permission of Academic Press, Inc.)

PEG - M_w ~8000; Manufacturer: Union Carbide (New York, USA); Lot not indicated;
Dextran - M_w 511,000; M_n 191,600; Manufacturer: Pharmacia (Uppsala, Sweden); Lot 3447

Total system		Bottom phase		Top phase		STL *
PEG % w/w	Dex % w/w	PEG % w/w	Dex % w/w	PEG % w/w	Dex % w/w	
4.0.	4.0	1.39	8.68	5.31	0.73	-0.493
4.0	6.0	0.86	11.69	6.47	0.27	-0.491
4.0	7.0	0.66	13.02	7.15	0.20	-0.506
					av.:	-0.497 ±0.008

* - STL - Tie-Line Slope defined as the ratio STL = (ΔPEG)/(ΔDex) where Δ is the difference between the concentrations of a given polymer in the two coexisting phases.

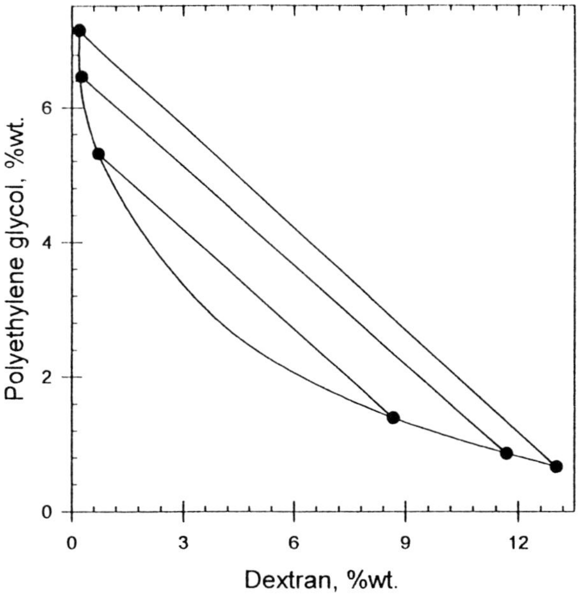

Table 10.73. Phase Diagram and Phase Composition of the Dextran-500—Poly(ethylene glycol)-8000 System Containing 0.4 mole/kg K_2SO_4 at 21°C. (From D. E. Brooks, K. A. Sharp, S. Bamberger, C. H. Tamblyn, G. V. F. Seaman, H. Walter, J.Colloid Interface Sci., 102, 1 (1984) with permission of Academic Press, Inc.)

PEG - M_w ~8000; Manufacturer: Union Carbide (New York, USA); Lot not indicated;
Dextran - M_w 511,000; M_n 191,600; Manufacturer: Pharmacia (Uppsala, Sweden); Lot 3447

Total system		Bottom phase		Top phase		STL *
PEG % w/w	Dex % w/w	PEG % w/w	Dex % w/w	PEG % w/w	Dex % w/w	
4.0.	4.0	1.01	8.33	6.56	0.05	-0.670
4.0	6.0	0.71	10.43	8.13	0.08	-0.717
4.0	7.0	0.73	11.42	8.89	0.08	-0.720
					av.:	-0.702 ±0.028

* - STL - Tie-Line Slope defined as the ratio STL = (ΔPEG)/(ΔDex) where Δ is the difference between the concentrations of a given polymer in the two coexisting phases.

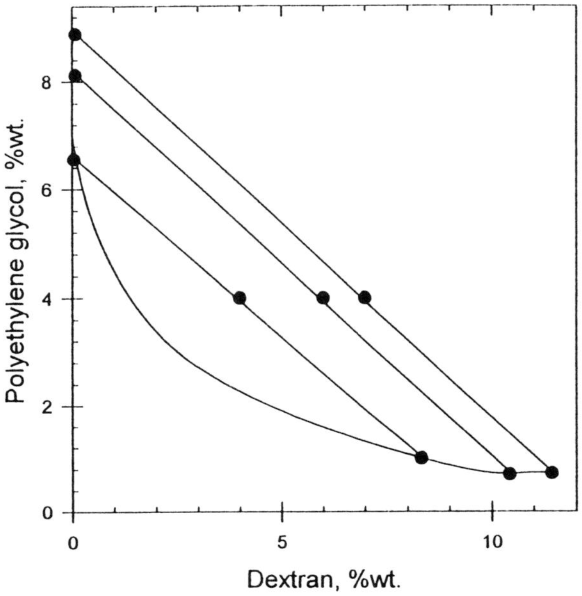

Table 10.74. Phase Diagram and Phase Composition of the Dextran-10—Poly(ethylene glycol)-10000 System at 25°C.
(From D. Forciniti, C. K. Hall, M.-R. Kula, Fluid Phase Equilibria, 61, 243 (1991) and Bioseparation, 2, 115 (1991) by permission of Elsevier Scientific Publishing Co. and Kluwer Academic Publishers, respectively.)

PEG - M_w 11,400; M_n 10,200; Manufacturer: Fluka AG (Switzerland); Lot not indicated;
Dextran - M_w 19,300; M_n 13,200; Manufacturer: Fluka AG (Switzerland); Lot not indicated

Total system		Bottom phase		Top phase		STL *
PEG % w/w	Dex % w/w	PEG % w/w	Dex % w/w	PEG % w/w	Dex % w/w	
5.6	6.7	2.5	13.0	6.4	5.2	-0.500
5.0	8.5	2.3	14.1	7.2	4.6	-0.516
5.6	10.0	1.1	18.3	9.8	2.6	-0.554
7.7	10.3	0.9	22.6	12.7	1.5	-0.559
					av.:	-0.532
						±0.029

*- STL - Tie-Line Slope defined as the ratio STL = (ΔPEG)/(ΔDex) where Δ is the difference between the concentrations of a given polymer in the two coexisting phases.

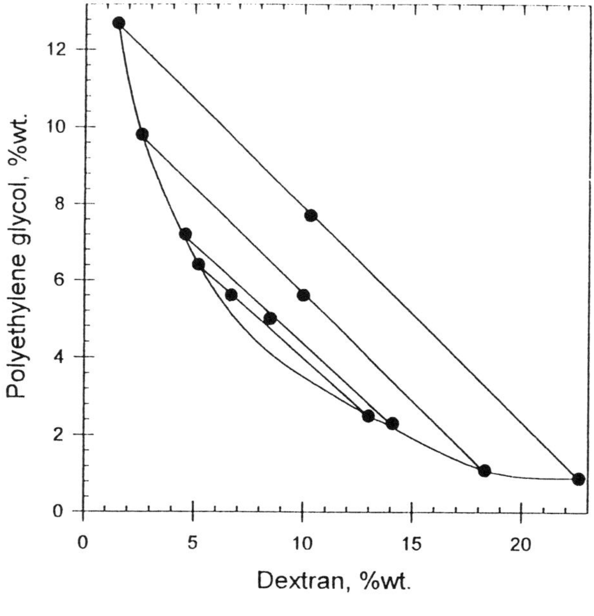

Table 10.75. Phase Diagram and Phase Composition of the Dextran-10—Poly(ethylene glycol)-10000 System at 4°C.
(From D. Forciniti, C. K. Hall, M.-R. Kula, Fluid Phase Equilibria, 61, 243 (1991) and Bioseparation, 2, 115 (1991) by permission of Elsevier Scientific Publishing Co. and Kluwer Academic Publishers, respectively.)

PEG - M_w 11,400; M_n 10,200; Manufacturer: Fluka AG (Switzerland); Lot not indicated;
Dextran - M_w 19,300; M_n 13,200; Manufacturer: Fluka AG (Switzerland); Lot not indicated

Total system		Bottom phase		Top phase		STL *
PEG % w/w	Dex % w/w	PEG % w/w	Dex % w/w	PEG % w/w	Dex % w/w	
5.0	7.4	2.4 **	13.8 **	6.4 **	4.9 **	-0.449
5.0	8.5	2.1	14.5	6.8	4.0	-0.448
5.6	10.0	1.4	18.9	9.1	2.4	-0.467
7.6	10.3	1.1	23.3	12.0	1.4	-0.498
					av.:	-0.466 ±0.023

*- STL - Tie-Line Slope defined as the ratio STL = (ΔPEG)/(ΔDex) where Δ is the difference between the concentrations of a given polymer in the two coexisting phases;
** - Composition of the phases questionable as it does not fit phase diagram.

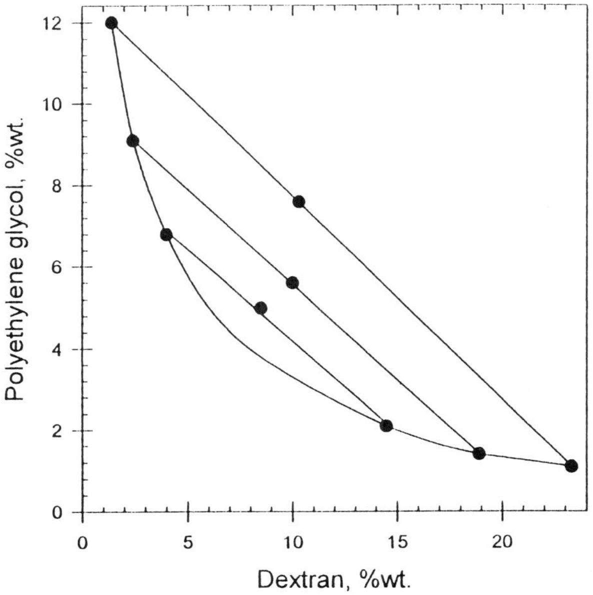

Table 10.76. Phase Diagram and Phase Composition of the Dextran-10—Poly(ethylene glycol)-10000 System at 40°C.

(From D. Forciniti, C. K. Hall, M.-R. Kula, Fluid Phase Equilibria, 61, 243 (1991) and Bioseparation, 2, 115 (1991) by permission of Elsevier Scientific Publishing Co. and Kluwer Academic Publishers, respectively.)

PEG - M_w 11,400; M_n 10,200; Manufacturer: Fluka AG (Switzerland); Lot not indicated;

Dextran - M_w 19,300; M_n 13,200; Manufacturer: Fluka AG (Switzerland); Lot not indicated

Total system		Bottom phase		Top phase		STL *
PEG % w/w	Dex % w/w	PEG % w/w	Dex % w/w	PEG % w/w	Dex % w/w	
5.35	7.8	1.5	13.3	7.8	4.1	-0.685
5.0	8.5	1.3	14.0	7.9	3.9	-0.653
5.6	10.0	1.3 **	17.2 **	10.2 **	2.3 **	-0.597 **
7.6	10.3	0.5	21.9	13.0	1.4	-0.610
					av.:	-0.649 ±0.038

*- STL - Tie-Line Slope defined as the ratio STL = (ΔPEG)/(ΔDex) where Δ is the difference between the concentrations of a given polymer in the two coexisting phases;

** - Composition of the phases questionable as the STL value is inconsistent with the other values; was not used in calculations of the average STL value; not shown on phase diagram.

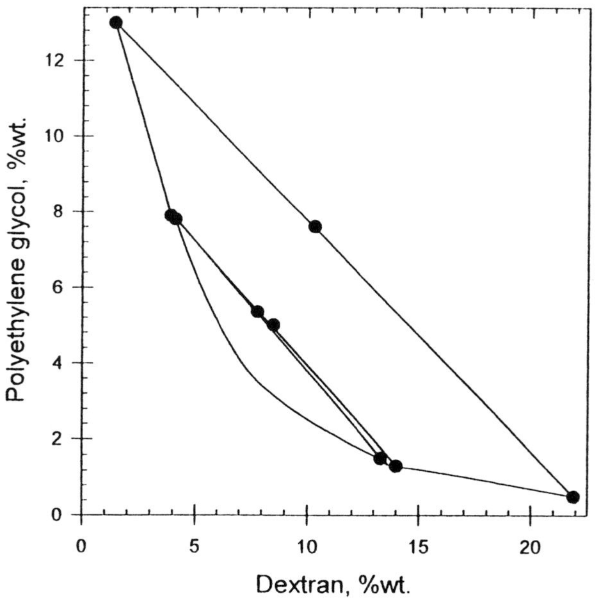

Table 10.77. Phase Diagram and Phase Composition of the Dextran-40—Poly(ethylene glycol)-10000 System at 25°C.
(From D. Forciniti, C. K. Hall, M.-R. Kula, Fluid Phase Equilibria, 61, 243 (1991) and Bioseparation, 2, 115 (1991) by permission of Elsevier Scientific Publishing Co. and Kluwer Academic Publishers, respectively.)

PEG - M_w 11,400; M_n 11,200; Manufacturer: Fluka AG (Switzerland); Lot not indicated;
Dextran - M_w 37,000; M_n 27,700; Manufacturer: Pfeifer and Langen (Dormagen, Germany); Lot not indicated

Total system		Bottom phase		Top phase		STL *
PEG % w/w	Dex % w/w	PEG % w/w	Dex % w/w	PEG % w/w	Dex % w/w	
4.4	6.75	1.7	11.7	5.7	3.3	-0.476
5.0	8.5	0.8	16.2	8.3	1.4	-0.507
5.6	10.0	0.6	19.2	10.2	0.8	-0.522
7.7	10.3	0.4	23.1	12.8	0.4	-0.546
					av.:	-0.513 ±0.029

* - STL - Tie-Line Slope defined as the ratio STL = (ΔPEG)/(ΔDex) where Δ is the difference between the concentrations of a given polymer in the two coexisting phases.

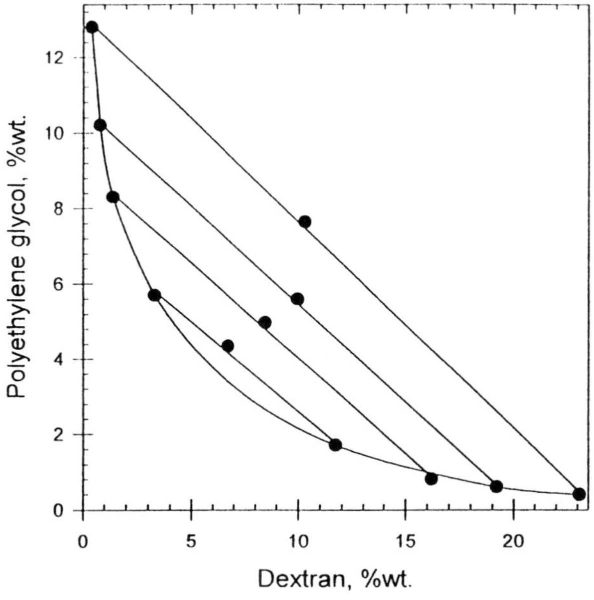

Table 10.78. Phase Diagram and Phase Composition of the Dextran-40—Poly(ethylene glycol)-10000 System at 4°C.
(From D. Forciniti, C. K. Hall, M.-R. Kula, Fluid Phase Equilibria, 61, 243 (1991) and Bioseparation, 2, 115 (1991) by permission of Elsevier Scientific Publishing Co. and Kluwer Academic Publishers, respectively.)

PEG - M_w 11,400; M_n 11,200; Manufacturer: Fluka AG (Switzerland); Lot not indicated;
Dextran - M_w 37,000; M_n 27,700; Manufacturer: Pfeifer and Langen (Dormagen, Germany); Lot not indicated

Total system		Bottom phase		Top phase		STL *
PEG % w/w	Dex % w/w	PEG % w/w	Dex % w/w	PEG % w/w	Dex % w/w	
4.4	6.75	1.4	13.2	6.3	2.6	-0.462
5.0	8.5	0.9	17.3	8.4	1.3	-0.469
5.6	10.0	0.9	20.3	9.9	0.74	-0.460
7.6	10.3	0.5 **	24.6 **	12.6 **	0.34 **	-0.499 **
					av.:	-0.464 ±0.005

* - STL - Tie-Line Slope defined as the ratio STL = (ΔPEG)/(ΔDex) where Δ is the difference between the concentrations of a given polymer in the two coexisting phases;

** - Composition of the phases questionable as the STL value is inconsistent with the other values; was not used in calculations of the average STL value.

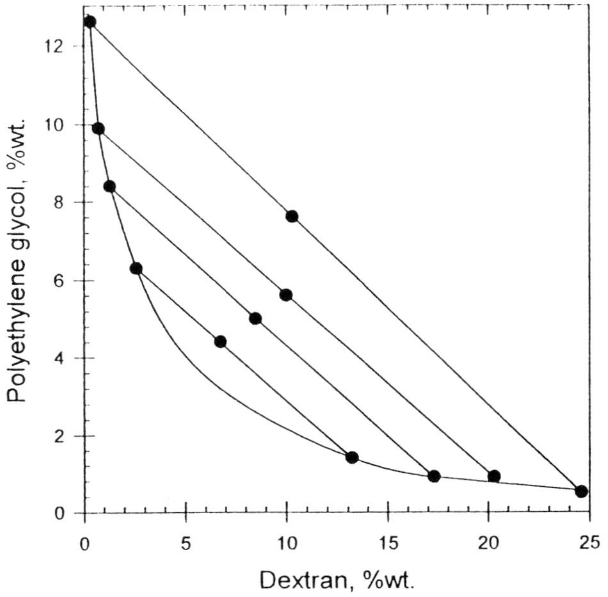

Table 10.79. Phase Diagram and Phase Composition of the Dextran-40—Poly(ethylene glycol)-10000 System at 40°C.
(From D. Forciniti, C. K. Hall, M.-R. Kula, Fluid Phase Equilibria, 61, 243 (1991) and Bioseparation, 2, 115 (1991) by permission of Elsevier Scientific Publishing Co. and Kluwer Academic Publishers, respectively.)

PEG - M_w 11,400; M_n 11,200; Manufacturer: Fluka AG (Switzerland); Lot not indicated;
Dextran - M_w 37,000; M_n 27,700; Manufacturer: Pfeifer and Langen (Dormagen, Germany); Lot not indicated

Total system		Bottom phase		Top phase		STL *
PEG % w/w	Dex % w/w	PEG % w/w	Dex % w/w	PEG % w/w	Dex % w/w	
4.4	6.75	1.4	12.5	6.2	3.3	-0.522
5.0	8.5	0.8	16.5	8.9	1.3	-0.533
5.6	10.0	0.4	20.2	10.4	0.86	-0.517
7.6	10.3	0.2 **	21.3 **	12.4 **	0.44 **	-0.585 **
					av.:	-0.524 ±0.008

*- STL - Tie-Line Slope defined as the ratio STL = (ΔPEG)/(ΔDex) where Δ is the difference between the concentrations of a given polymer in the two coexisting phases;
**- Composition of the phases questionable as the STL value is inconsistent with the other values; was not used in calculations of the average STL value.

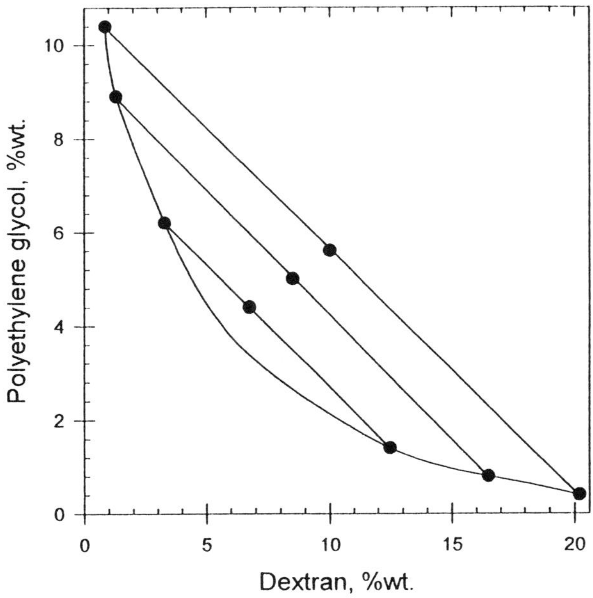

Table 10.80. Phase Diagram and Phase Composition of the Dextran-110—Poly(ethylene glycol)-10000 System at 25°C.
(From D. Forciniti, C. K. Hall, M.-R. Kula, Fluid Phase Equilibria, 61, 243 (1991) and Bioseparation, 2, 115 (1991) by permission of Elsevier Scientific Publishing Co. and Kluwer Academic Publishers, respectively.)

PEG - M_w 11,400; M_n 10,200; Manufacturer: Fluka (Switzerland); Lot not indicated;
Dextran - M_w 86,200; M_n 52.100; Manufacturer: Fluka (Switzerland); Lot not indicated

Total system		Bottom phase		Top phase		STL *
PEG % w/w	Dex % w/w	PEG % w/w	Dex % w/w	PEG % w/w	Dex % w/w	
4.4	6.8	0.8	13.6	7.2	0.7	-0.496
5.0	8.0	0.6	16.5	8.6	0.4	-0.497
5.6	10.0	0.5	19.9	10.6	0.2	-0.513
7.7	10.3	0.5	23.6	12.9	0.1	-0.528
					av.:	-0.509 ±0.015

* - STL - Tie-Line Slope defined as the ratio STL = (ΔPEG)/(ΔDex) where Δ is the difference between the concentrations of a given polymer in the two coexisting phases.

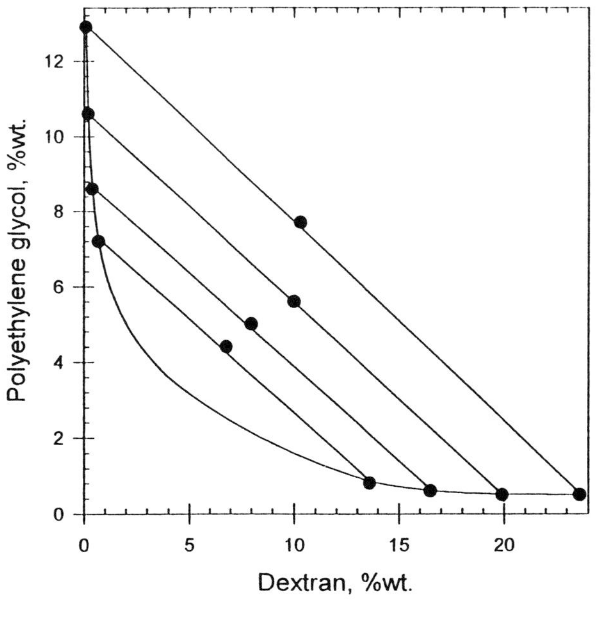

Table 10.81. Phase Diagram and Phase Composition of the Dextran-110—Poly(ethylene glycol)-10000 System at 4°C.
(From D. Forciniti, C. K. Hall, M.-R. Kula, Fluid Phase Equilibria, 61, 243 (1991) and Bioseparation, 2, 115 (1991) by permission of Elsevier Scientific Publishing Co. and Kluwer Academic Publishers, respectively.)

PEG - M_w 11,400; M_n 10,200; Manufacturer: Fluka (Switzerland); Lot not indicated;
Dextran - M_w 86,200; M_n 52.100; Manufacturer: Fluka (Switzerland); Lot not indicated

Total system		Bottom phase *		Top phase *		STL **
PEG % w/w	Dex % w/w	PEG % w/w	Dex % w/w	PEG % w/w	Dex % w/w	
4.4	6.75	1.1	15.5	6.9	0.55	-0.388
5.0	8.5	0.4	19.2	8.5	0.28	-0.428
5.6	10.0	0.5	21.6	10.0	0.18	-0.444
7.6	10.3	0.4	26.0	12.3	0.09	-0.459
					av.:	-0.430
						±0.031

* - Compositions of all the phases questionable as the STL values are inconsistent
** - STL - Tie-Line Slope defined as the ratio STL = (ΔPEG)/(ΔDex) where Δ is the difference between the concentrations of a given polymer in the two coexisting phases.

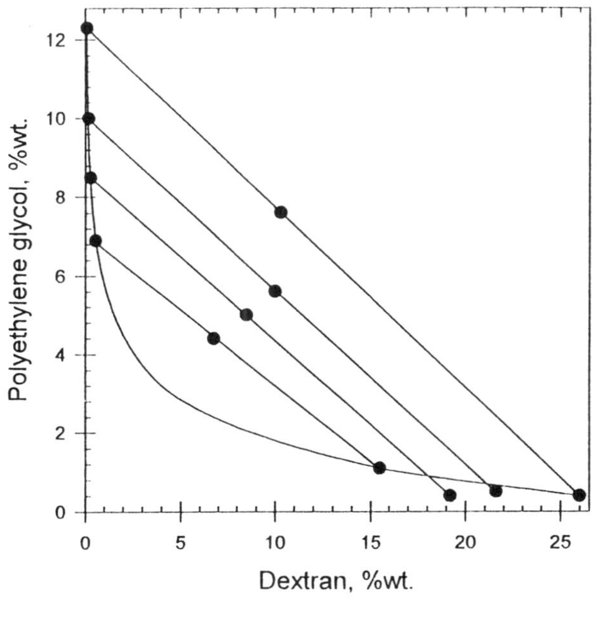

Table 10.82. Phase Diagram and Phase Composition of the Dextran-110—Poly(ethylene glycol)-10000 System at 40°C.
(From D. Forciniti, C. K. Hall, M.-R. Kula, Fluid Phase Equilibria, 61, 243 (1991) and Bioseparation, 2, 115 (1991) by permission of Elsevier Scientific Publishing Co. and Kluwer Academic Publishers, respectively.)

PEG - M_w 11,400; M_n 10,200; Manufacturer: Fluka (Switzerland); Lot not indicated;
Dextran - M_w 86,200; M_n 52.100; Manufacturer: Fluka (Switzerland); Lot not indicated

Total system		Bottom phase		Top phase		STL *
PEG % w/w	Dex % w/w	PEG % w/w	Dex % w/w	PEG % w/w	Dex % w/w	
4.4	6.75	0.7	13.7	7.5	0.69	-0.523
5.0	8.5	0.3	16.9	9.6	0.33	-0.561
5.6	10.0	0.2	19.7	11.0	0.2	-0.554
7.6	10.3	0.3	23.4	13.3	0.11	-0.558
					av.:	-0.549 ±0.018

* - STL - Tie-Line Slope defined as the ratio STL = (ΔPEG)/(ΔDex) where Δ is the difference between the concentrations of a given polymer in the two coexisting phases.

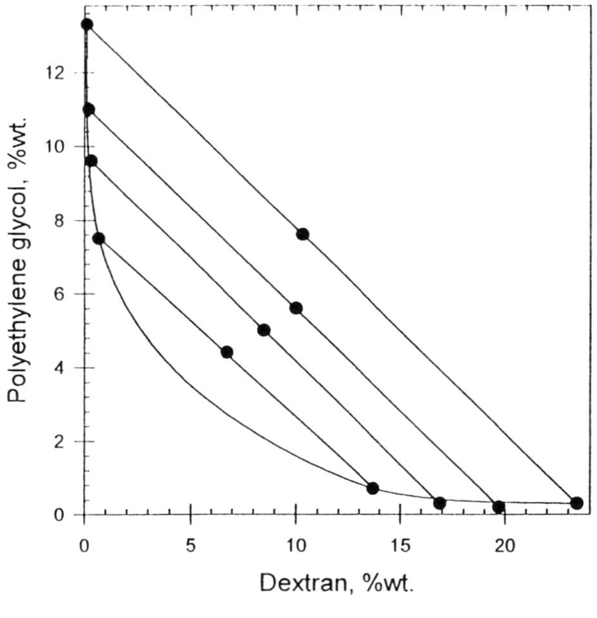

Table 10.83. Phase Diagram and Phase Composition of the Dextran-500—Poly(ethylene glycol)-10000 System at 25°C.
(From D. Forciniti, C. K. Hall, M.-R. Kula, Fluid Phase Equilibria, 61, 243 (1991) and Bioseparation, 2, 115 (1991) by permission of Elsevier Scientific Publishing Co. and Kluwer Academic Publishers, respectively.)

PEG - M_w 11,400; M_n 10,200; Manufacturer: Fluka (Switzerland); Lot not indicated;
Dextran - M_w 215,000; M_n 88,200; Manufacturer: Pfeifer and Langen (Dormagen, Germany); Lot not indicated

Total system		Bottom phase		Top phase		STL *
PEG % w/w	Dex % w/w	PEG % w/w	Dex % w/w	PEG % w/w	Dex % w/w	
4.4	6.75	0.7	14.1	7.0	0.4	-0.460
5.0	8.5	0.7	17.0	8.5	0.4	-0.470
5.6	9.95	0.4	19.2	9.8	0.3	-0.497
7.65	10.3	0.4	23.0	12.3	0.3	-0.524
					av.:	-0.488 ±0.029

* - STL - Tie-Line Slope defined as the ratio STL = (ΔPEG)/(ΔDex) where Δ is the difference between the concentrations of a given polymer in the two coexisting phases.

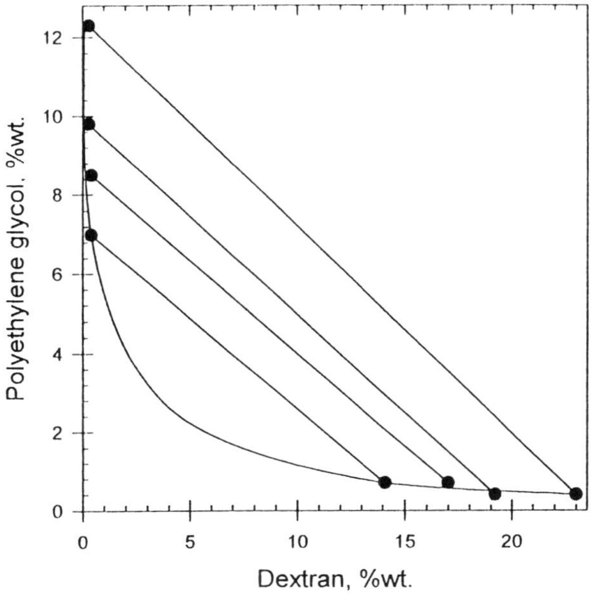

Table 10.84. Phase Diagram and Phase Composition of the Dextran-500—Poly(ethylene glycol)-10000 System at 4°C.
(From D. Forciniti, C. K. Hall, M.-R. Kula, Fluid Phase Equilibria, 61, 243 (1991) and Bioseparation, 2, 115 (1991) by permission of Elsevier Scientific Publishing Co. and Kluwer Academic Publishers, respectively.)

PEG - M_w 11,400; M_n 10,200; Manufacturer: Fluka (Switzerland); Lot not indicated;
Dextran - M_w 215,000; M_n 88,200; Manufacturer: Pfeifer and Langen (Dormagen, Germany); Lot not indicated

Total system		Bottom phase		Top phase		STL *
PEG % w/w	Dex % w/w	PEG % w/w	Dex % w/w	PEG % w/w	Dex % w/w	
4.4	6.75	0.6	15.9	7.2	0.34	-0.424
5.0	8.5	0.5	18.9	8.5	0.15	-0.427
5.6	10.0	0.6	21.3	10.0	0.08	-0.443
7.6	10.3	0.8	25.1	12.1	0.04	-0.451
					av.:	-0.436 ±0.013

* - STL - Tie-Line Slope defined as the ratio STL = (ΔPEG)/(ΔDex) where Δ is the difference between the concentrations of a given polymer in the two coexisting phases.

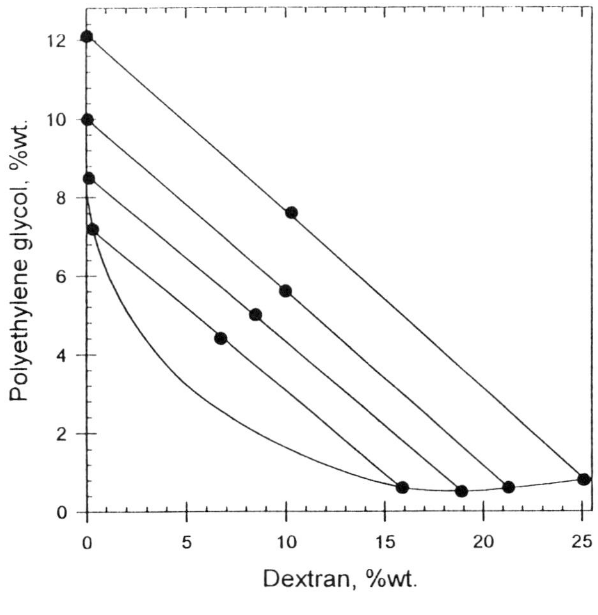

Table 10.85. Phase Diagram and Phase Composition of the Dextran-500—Poly(ethylene glycol)-10000 System at 40°C.
(From D. Forciniti, C. K. Hall, M.-R. Kula, Fluid Phase Equilibria, 61, 243 (1991) and Bioseparation, 2, 115 (1991) by permission of Elsevier Scientific Publishing Co. and Kluwer Academic Publishers, respectively.)

PEG - M_w 11,400; M_n 10,200; Manufacturer: Fluka (Switzerland); Lot not indicated;
Dextran - M_w 215,000; M_n 88,200; Manufacturer: Pfeifer and Langen (Dormagen, Germany); Lot not indicated

Total system		Bottom phase *		Top phase *		STL **
PEG % w/w	Dex % w/w	PEG % w/w	Dex % w/w	PEG % w/w	Dex % w/w	
4.4	6.75	0.8	15.4	7.2	0.27	-0.423
5.0	8.5	0.9	16.9	9.1	0.15	-0.490
5.6	10.0	0.7	19.8	10.7	0.1	-0.508
7.65	10.3	0.6	23.1	13.3	0.03	-0.550
					av.:	-0.493 ±0.053

* - Compositions of all the phases questionable as the STL values are inconsistent;
** - STL - Tie-Line Slope defined as the ratio STL = (ΔPEG)/(ΔDex) where Δ is the difference between the concentrations of a given polymer in the two coexisting phases.

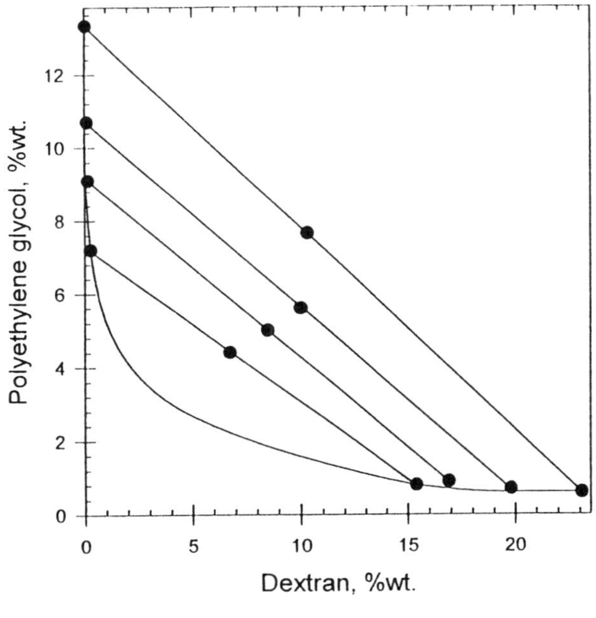

Table 10.86. Phase Diagram and Phase Composition of the Dextran-40—Poly(ethylene glycol)-20000 System at 22°C.
(From A. D. Diamond, J. Hsu, Biotechnol. Techniques, 3, 119 (1989) with permission of Eaton Publishing Co.)

PEG - M_w ~20,000; Manufacturer: Union Carbide (New York, NY, USA); Lot not indicated;
Dextran T-40 - M_w 38,800; M_n 24,200; Manufacturer: Pharmacia Fine Chemicals (Piscataway, NJ, USA); Lot 03375

Total system		Bottom phase		Top phase		STL *
PEG % w/w	Dex % w/w	PEG % w/w	Dex % w/w	PEG % w/w	Dex % w/w	
4.24	6.50	2.10	9.57	7.52	1.79	-0.697
5.20	7.51	1.28	13.66	9.19	1.23	-0.636
6.03	8.36	0.80	16.75	10.75	0.82	-0.625
6.77	9.54	0.50	20.05	12.41	0.52	-0.610
					av.:	-0.642 ±0.038

* - STL - Tie-Line Slope defined as the ratio STL = (ΔPEG)/(ΔDex) where Δ is the difference between the concentrations of a given polymer in the two coexisting phases.

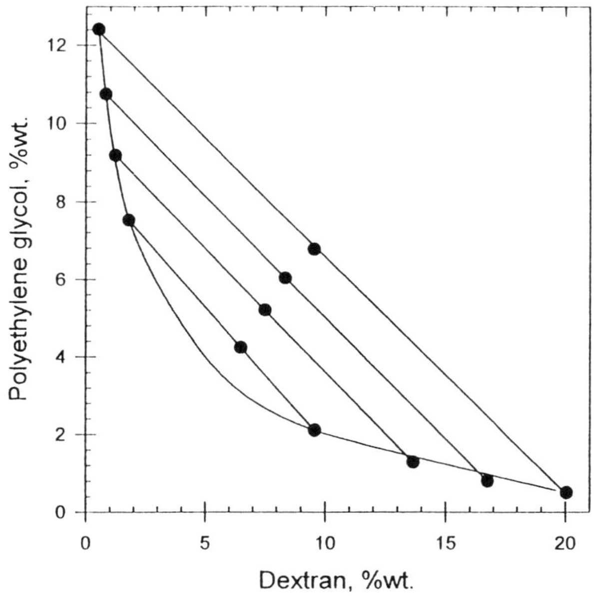

Table 10.87. Phase Diagram and Phase Composition of the Dextran-40—Poly(ethylene glycol)-20000 System at 4°C.
(From A. D. Diamond, J. T. Hsu, Biotechnol. Bioeng., 34, 1000 (1989) with permission of John Wiley & Sons, Inc.)

PEG - M_w ~20,000; Manufacturer: Union Carbide (New York, NY, USA); Lot not indicated;
Dextran T-40 - M_w 38,800; M_n 24,200; Manufacturer: Pharmacia Fine Chemicals (Piscataway, NJ, USA); Lot 03375

Total system		Bottom phase *		Top phase *		STL **
PEG % w/w	Dex % w/w	PEG % w/w	Dex % w/w	PEG % w/w	Dex % w/w	
4.20	4.20	3.02	5.64	4.72	3.13	-0.677
4.50	5.00	2.34	7.99	6.52	1.92	-0.689
5.00	6.50	1.41	12.57	7.93	1.42	-0.585
5.50	8.00	0.82	16.22	9.37	0.90	-0.558
					av.:	-0.627 ±0.065

* - Compositions of all the phases questionable as the STL values are inconsistent;
** - STL - Tie-Line Slope defined as the ratio STL = (ΔPEG)/(ΔDex) where Δ is the difference between the concentrations of a given polymer in the two coexisting phases.

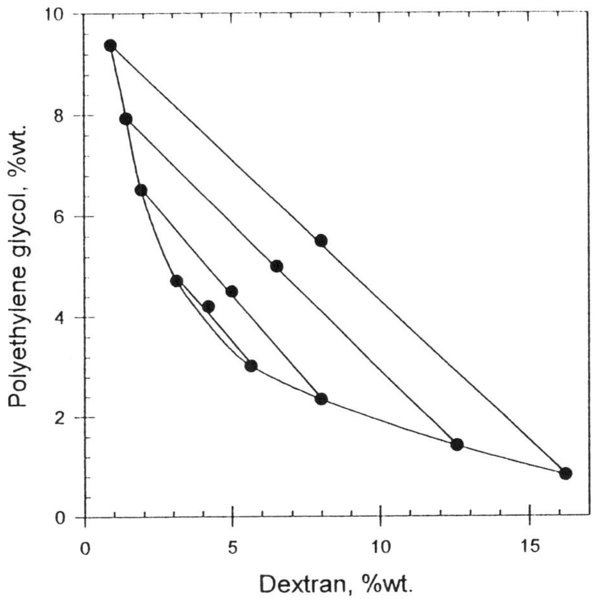

Table 10.88. Phase Diagram and Phase Composition of the Dextran-70—Poly(ethylene glycol)-20000 System at 22°C.
(From A. D. Diamond, J. Hsu, Biotechnol. Techniques, 3, 119 (1989) with permission of Eaton Publishing Co.)

PEG - M_w ~20,000; Manufacturer: Union Carbide (New York, NY, USA); Lot not indicated;
Dextran T-70 - M_w 72,200; M_n 38,400; Manufacturer: Pharmacia Fine Chemicals (Piscataway, NJ, USA); Lot 02377

Total system		Bottom phase *		Top phase *		STL **
PEG % w/w	Dex % w/w	PEG % w/w	Dex % w/w	PEG % w/w	Dex % w/w	
4.01	3.67	3.00	5.20	5.30	1.90	-0.697
4.13	4.11	2.57	6.50	6.03	1.56	-0.700
4.39	4.63	2.03	8.17	6.66	1.28	-0.672
4.27	5.23	1.88	8.95	7.05	1.16	-0.664
5.10	6.29	1.22	12.76	8.77	0.74	-0.628
5.58	7.28	1.02	15.27	9.81	0.54	-0.597
6.26	8.78	0.84	18.53	11.44	0.34	-0.583
					av.:	-0.649 ±0.047

* - Compositions of all the phases questionable as the STL values are inconsistent;
** - STL - Tie-Line Slope defined as the ratio STL = (ΔPEG)/(ΔDex) where Δ is the difference between the concentrations of a given polymer in the two coexisting phases.

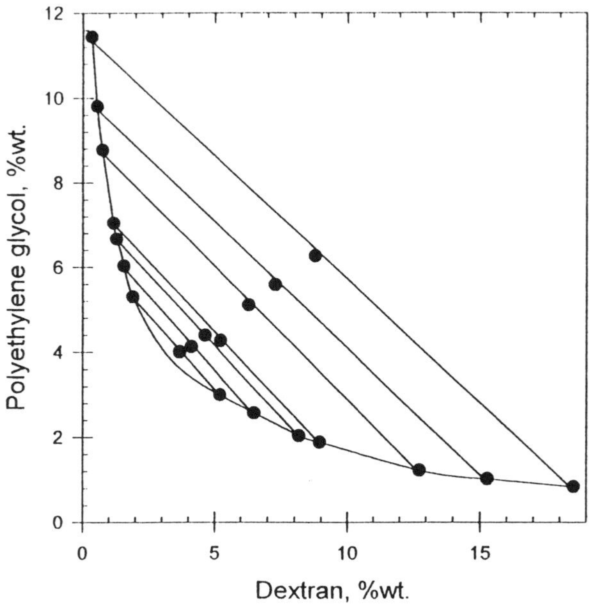

Table 10.89. Phase Diagram and Phase Composition of the Dextran-70—Poly(ethylene glycol)-20000 System at 4°C.
(From A. D. Diamond, J. T. Hsu, Biotechnol. Bioeng., 34, 1000 (1989) with permission of John Wiley & Sons, Inc.)

PEG - M_w ~20,000; Manufacturer: Union Carbide (New York, NY, USA); Lot not indicated;
Dextran T-70 - M_w 72,200; M_n 38,400; Manufacturer: Pharmacia Fine Chemicals (Piscataway, NJ, USA); Lot 02377

Total system		Bottom phase		Top phase		STL *
PEG % w/w	Dex % w/w	PEG % w/w	Dex % w/w	PEG % w/w	Dex % w/w	
4.15	4.10	2.33 **	7.03 **	5.92 **	1.29 **	-0.625 **
4.65	5.15	1.52	10.52	7.01	0.95	-0.574
5.10	6.30	1.07	13.68	8.04	0.71	-0.537
5.80	7.70	0.57	17.18	9.65	0.48	-0.544
					av.:	-0.552 ±0.020

*- STL - Tie-Line Slope defined as the ratio STL = (ΔPEG)/(ΔDex) where Δ is the difference between the concentrations of a given polymer in the two coexisting phases;
**- Composition of the phases questionable as the STL value is inconsistent with the other values; was not used in calculations of the average STL value.

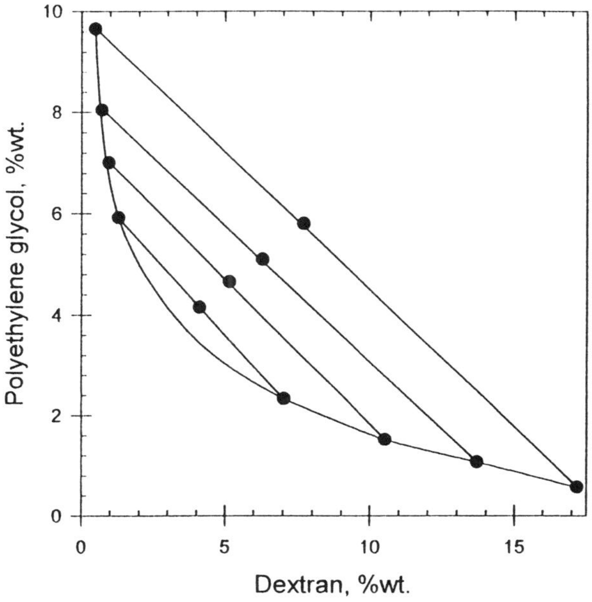

Table 10.90. Phase Diagram and Phase Composition of the Dextran-500—Poly(ethylene glycol)-20000 System at 22°C.
(From A. D. Diamond, J. Hsu, Biotechnol. Techniques, 3, 119 (1989) with permission of Eaton Publishing Co.)

PEG - M_w ~20,000; Manufacturer: Union Carbide (New York, NY, USA); Lot not indicated;
Dextran T-500 - M_w 507,000; M_n 234,200; Manufacturer: Pharmacia Fine Chemicals (Piscataway, NJ, USA); Lot 05163

Total system		Bottom phase		Top phase		STL *
PEG % w/w	Dex % w/w	PEG % w/w	Dex % w/w	PEG % w/w	Dex % w/w	
2.38	3.57	2.09 **	3.36 **	4.21 **	0.79 **	-0.825 **
2.98	4.64	1.69	6.13	5.40	0.33	-0.640
3.81	6.19	1.09	10.97	7.29	0.07	-0.569
4.76	8.33	0.55	15.54	9.51	0.05	-0.578
					av.:	-0.596 ±0.039

* - STL - Tie-Line Slope defined as the ratio STL = (ΔPEG)/(ΔDex) where Δ is the difference between the concentrations of a given polymer in the two coexisting phases;

** - Composition of the phases questionable as the STL value is inconsistent with the other values; does not fit phase diagram; was not used in calculations of the average STL value.

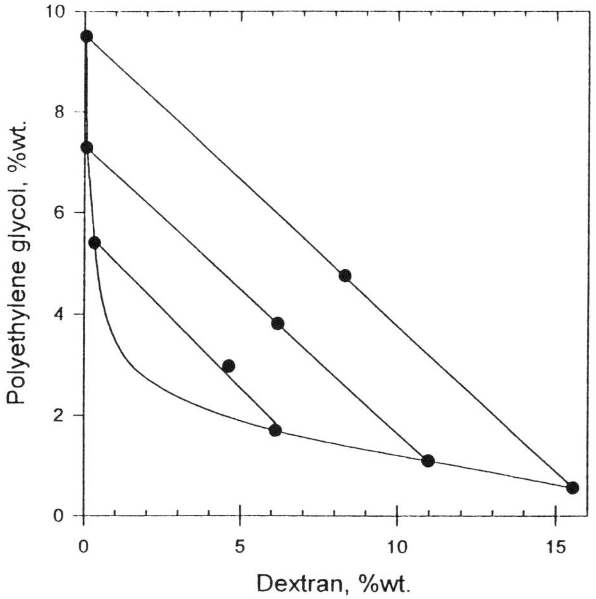

Table 10.91. Phase Diagram and Phase Composition of the Dextran-500—Poly(ethylene glycol)-20000 System at 4°C.
(From A. D. Diamond, J. T. Hsu, Biotechnol. Bioeng., 34, 1000 (1989) with permission of John Wiley & Sons, Inc.)

PEG - M_w ~20,000; Manufacturer: Union Carbide (New York, NY, USA); Lot not indicated;
Dextran T-500 - M_w 507,000; M_n 234,200; Manufacturer: Pharmacia Fine Chemicals (Piscataway, NJ, USA); Lot 05163

Total system		Bottom phase		Top phase		STL *
PEG % w/w	Dex % w/w	PEG % w/w	Dex % w/w	PEG % w/w	Dex % w/w	
3.10	2.00	1.87	3.88	3.65	0.78	-0.574
2.60	3.93	1.50	5.78	4.86	0.28	-0.611
3.35	5.84	1.04	9.53	6.29	0.11	-0.557
3.95	7.14	0.64**	13.73**	7.37**	0.04**	-0.492**
					av.:	-0.581 ±0.028

* - STL - Tie-Line Slope defined as the ratio STL = (ΔPEG)/(ΔDex) where Δ is the difference between the concentrations of a given polymer in the two coexisting phases;
** - Composition of the phases questionable as the STL value is inconsistent with the other values; does not fit phase diagram; was not used in calculations of the average STL value.

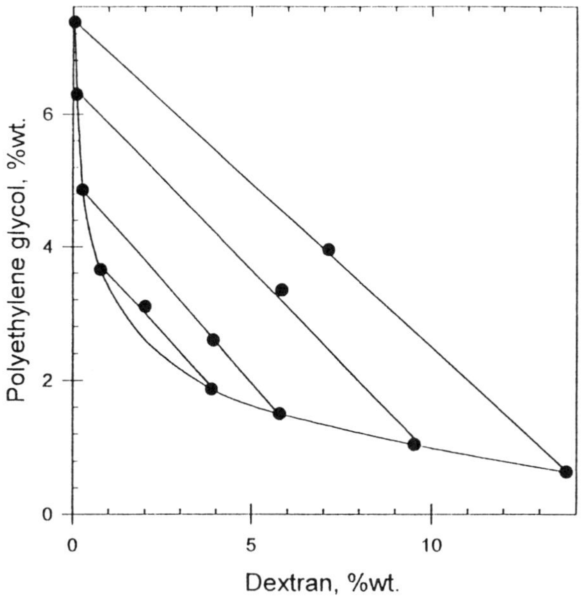

Table 10.92. Phase Diagram and Phase Composition of the Dextran-10—Poly(ethylene glycol)-20000 System at 25°C.
(From D. Forciniti, C. K. Hall, M.-R. Kula, Fluid Phase Equilibria, 61, 243 (1991) and Bioseparation, 2, 115 (1991) by permission of Elsevier Scientific Publishing Co. and Kluwer Academic Publishers, respectively.)

PEG - M_w 21,000; M_n 19,100; Manufacturer: Merck (Dormstadt, Germany); Lot not indicated;
Dextran - M_w 19,300; M_n 13,200; Manufacturer: Fluka AG (Switzerland); Lot not indicated

Total system		Bottom phase		Top phase		STL *
PEG % w/w	Dex % w/w	PEG % w/w	Dex % w/w	PEG % w/w	Dex % w/w	
5.6	10.0	0.8	18.8	10.2	1.8	-0.553
7.6	10.3	0.6	22.5	13.0	1.1	-0.579
8.2	11.3	0.6	24.9	14.2	0.8	-0.564
8.4	12.2	0.8	25.3	15.3	0.8	-0.592
					av.:	-0.572 ±0.017

*- STL - Tie-Line Slope defined as the ratio STL = (ΔPEG)/(ΔDex) where Δ is the difference between the concentrations of a given polymer in the two coexisting phases.

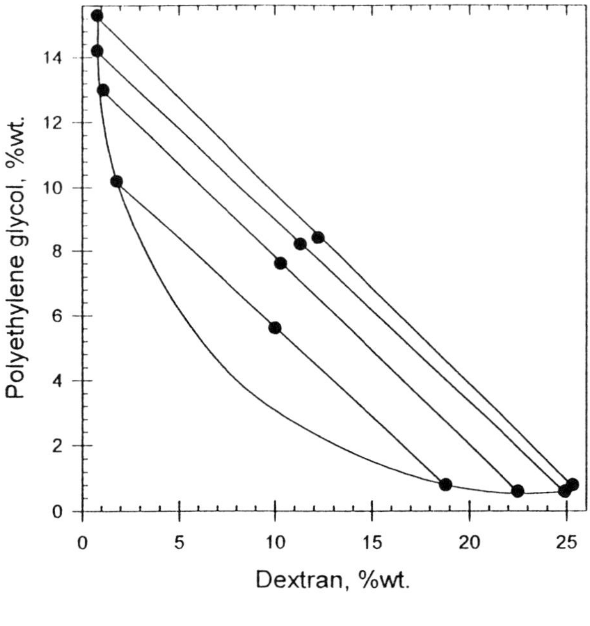

Table 10.93. Phase Diagram and Phase Composition of the Dextran-10—Poly(ethylene glycol)-20000 System at 4°C.

(From D. Forciniti, C. K. Hall, M.-R. Kula, Fluid Phase Equilibria, 61, 243 (1991) and Bioseparation, 2, 115 (1991) by permission of Elsevier Scientific Publishing Co. and Kluwer Academic Publishers, respectively.)

PEG - M_w 21,000; M_n 19,100; Manufacturer: Merck (Dormstadt, Germany); Lot not indicated;
Dextran - M_w 19,300; M_n 13,200; Manufacturer: Fluka AG (Switzerland); Lot not indicated

Total system		Bottom phase		Top phase		STL *
PEG % w/w	Dex % w/w	PEG % w/w	Dex % w/w	PEG % w/w	Dex % w/w	
4.4	6.75	2.0	12.7	5.7	4.4	-0.446
5.0	8.5	0.9**	16.7**	7.7	2.5	-0.479
5.6	10.0	1.1	19.4	9.5	1.8	-0.477
7.6	10.3	0.97	23.1	12.2	1.1	-0.510
					av.:	-0.478 ±0.026

*- STL - Tie-Line Slope defined as the ratio STL = (ΔPEG)/(ΔDex) where Δ is the difference between the concentrations of a given polymer in the two coexisting phases;

** - Composition of the phase questionable as it does not fit phase diagram.

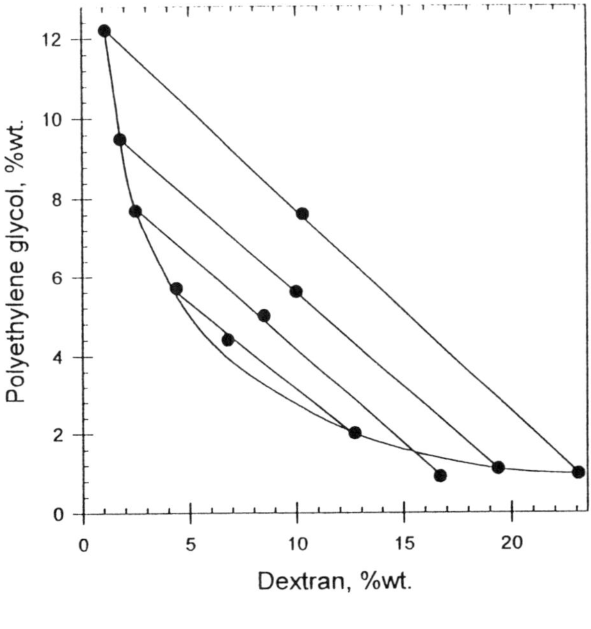

Table 10.94. Phase Diagram and Phase Composition of the Dextran-10—Poly(ethylene glycol)-20000 System at 40°C.
(From D. Forciniti, C. K. Hall, M.-R. Kula, Fluid Phase Equilibria, 61, 243 (1991) and Bioseparation, 2, 115 (1991) by permission of Elsevier Scientific Publishing Co. and Kluwer Academic Publishers, respectively.)

PEG - M_w 21,000; M_n 19,100; Manufacturer: Merck (Dormstadt, Germany); Lot not indicated;
Dextran - M_w 19,300; M_n 13,200; Manufacturer: Fluka AG (Switzerland); Lot not indicated

Total system		Bottom phase		Top phase		STL *
PEG % w/w	Dex % w/w	PEG % w/w	Dex % w/w	PEG % w/w	Dex % w/w	
4.4	6.75	2.3	10.0	5.7	4.6	-0.630
5.0	8.5	1.1	15.1	8.8	2.4	-0.606
5.6	10.0	0.6	17.8	10.9	1.7	-0.640
7.6	10.3	0.4	21.4	13.7	1.0	-0.652
					av.:	-0.632 ±0.020

*- STL - Tie-Line Slope defined as the ratio STL = (ΔPEG)/(ΔDex) where Δ is the difference between the concentrations of a given polymer in the two coexisting phases.

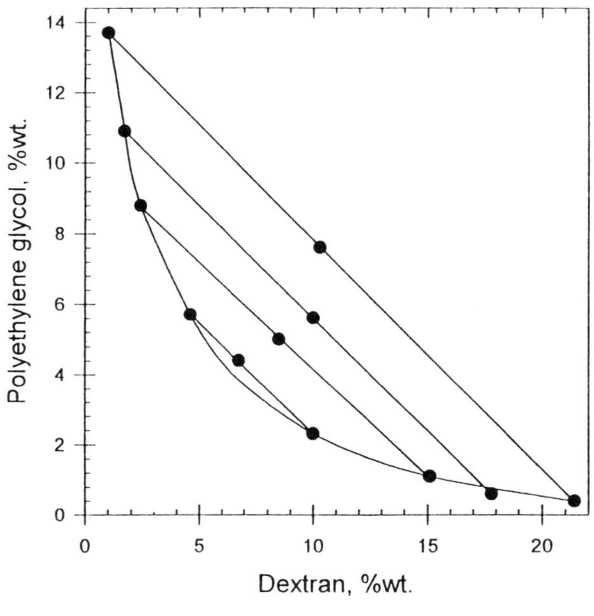

Table 10.95. Phase Diagram and Phase Composition of the Dextran-40—Poly(ethylene glycol)-20000 System at 25°C.
(From D. Forciniti, C. K. Hall, M.-R. Kula, Fluid Phase Equilibria, 61, 243 (1991) and Bioseparation, 2, 115 (1991) by permission of Elsevier Scientific Publishing Co. and Kluwer Academic Publishers, respectively.)

PEG - M_w 21,000; M_n 19,100; Manufacturer: Merck (Dormstadt, Germany; Lot not indicated;
Dextran - M_w 37,000; M_n 27,700; Manufacturer: Pfeifer and Langen (Dormagen, Germany); Lot not indicated

Total system		Bottom phase		Top phase		STL *
PEG % w/w	Dex % w/w	PEG % w/w	Dex % w/w	PEG % w/w	Dex % w/w	
4.3	6.7	0.5	13.5	6.5	1.7	-0.508
5.05	8.5	0.4	16.9	8.8	0.9	-0.525
5.6	10.0	0.4	19.2	10.4	0.5	-0.535
7.6	10.3	0.3	22.8	13.0	0.3	-0.564
					av.:	-0.533
						±0.023

* - STL - Tie-Line Slope defined as the ratio STL = (ΔPEG)/(ΔDex) where Δ is the difference between the concentrations of a given polymer in the two coexisting phases.

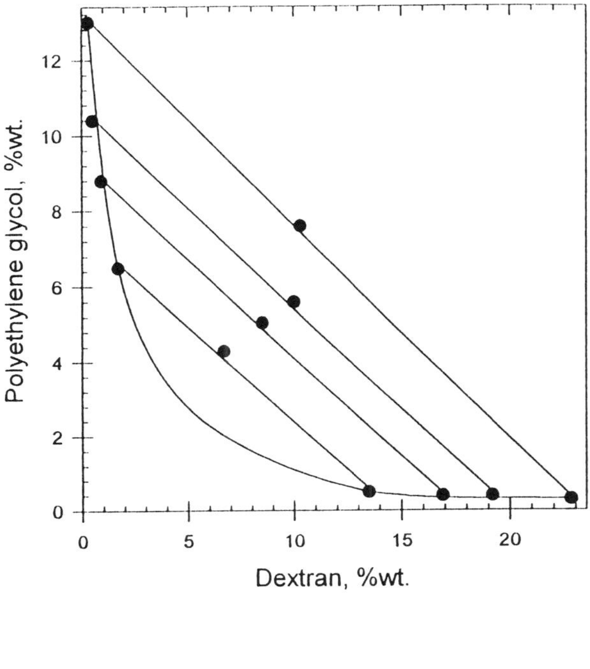

Table 10.96. Phase Diagram and Phase Composition of the Dextran-40—Poly(ethylene glycol)-20000 System at 4°C.
(From D. Forciniti, C. K. Hall, M.-R. Kula, Fluid Phase Equilibria, 61, 243 (1991) and Bioseparation, 2, 115 (1991) by permission of Elsevier Scientific Publishing Co. and Kluwer Academic Publishers, respectively.)

PEG - M_w 21,000; M_n 19,100; Manufacturer: Merck (Dormstadt, Germany; Lot not indicated;
Dextran - M_w 37,000; M_n 27,700; Manufacturer: Pfeifer and Langen (Dormagen, Germany); Lot not indicated

Total system		Bottom phase		Top phase		STL *
PEG	Dex	PEG	Dex	PEG	Dex	
% w/w	% w/w	% w/w	% w/w	% w/w	% w/w	
4.4	6.75	0.7	15.0	6.8	1.4	-0.449
5.0	8.5	0.8**	18.2**	8.4	0.78	-0.436
5.6	10.0	0.5	20.6	10.1	0.49	-0.477
7.6	10.3	0.4***	23.7***	13.1***	0.3***	-0.543***
					av.:	-0.454 ±0.021

* - STL - Tie-Line Slope defined as the ratio STL = (ΔPEG)/(ΔDex) where Δ is the difference between the concentrations of a given polymer in the two coexisting phases;
** - Composition of the phase questionable as it does not fit phase diagram;
*** - Composition of the phases questionable as the STL value is inconsistent with the other values; was not used in calculations of the average STL value; is not shown on phase diagram.

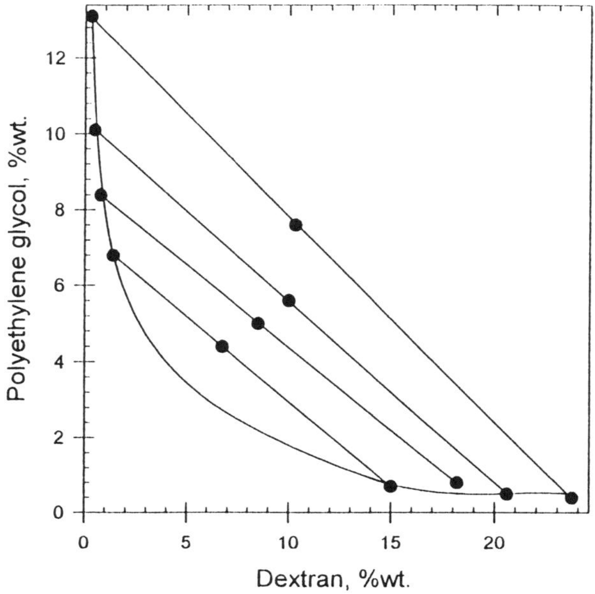

Table 10.97. Phase Diagram and Phase Composition of the Dextran-40—Poly(ethylene glycol)-20000 System at 40°C.
(From D. Forciniti, C. K. Hall, M.-R. Kula, Fluid Phase Equilibria, 61, 243 (1991) and Bioseparation, 2, 115 (1991) by permission of Elsevier Scientific Publishing Co. and Kluwer Academic Publishers, respectively.)

PEG - M_w 21,000; M_n 19,100; Manufacturer: Merck (Dormstadt, Germany; Lot not indicated;
Dextran - M_w 37,000; M_n 27,700; Manufacturer: Pfeifer and Langen (Dormagen, Germany); Lot not indicated

Total system		Bottom phase		Top phase		STL *
PEG % w/w	Dex % w/w	PEG % w/w	Dex % w/w	PEG % w/w	Dex % w/w	
4.4	6.75	0.7	12.9	7.5	1.4	-0.591
4.8	8.2	0.6	15.6	9.2	0.81	-0.581
5.65	10.0	0.1	19.2	11.7	0.46	-0.619
7.6	10.3	0.1	22.9	13.6	0.24	-0.596
					av.:	-0.597
						±0.016

* - STL - Tie-Line Slope defined as the ratio STL = (ΔPEG)/(ΔDex) where Δ is the difference between the concentrations of a given polymer in the two coexisting phases.

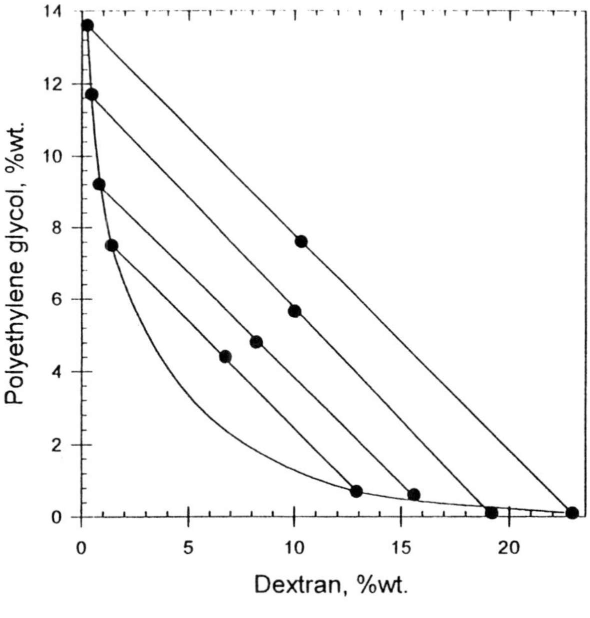

Table 10.98. Phase Diagram and Phase Composition of the Dextran-110—Poly(ethylene glycol)-20000 System at 25°C.
(From D. Forciniti, C. K. Hall, M.-R. Kula, Fluid Phase Equilibria, 61, 243 (1991) and Bioseparation, 2, 115 (1991) by permission of Elsevier Scientific Publishing Co. and Kluwer Academic Publishers, respectively.)

PEG - M_w 21,000; M_n 19,100; Manufacturer: Merck (Dormstadt, Germany); Lot not indicated;
Dextran - M_w 86,200; M_n 52.100; Manufacturer: Fluka (Switzerland); Lot not indicated

Total system		Bottom phase		Top phase		STL *
PEG % w/w	Dex % w/w	PEG % w/w	Dex % w/w	PEG % w/w	Dex % w/w	
4.4	6.8	0.9**	14.5**	7.2**	0.4**	-0.447**
5.0	8.5	0.5	17.1	9.1	0.2	-0.509
5.6	10.0	0.5	19.9	10.7	0.2	-0.518
7.6	10.3	0.3	23.3	13.1	0.1	-0.552
					av.:	-0.526 ±0.023

* - STL - Tie-Line Slope defined as the ratio STL = (ΔPEG)/(ΔDex) where Δ is the difference between the concentrations of a given polymer in the two coexisting phases;

** - Composition of the phases questionable as the STL value is inconsistent with the other values; was not used in calculations of the average STL value.

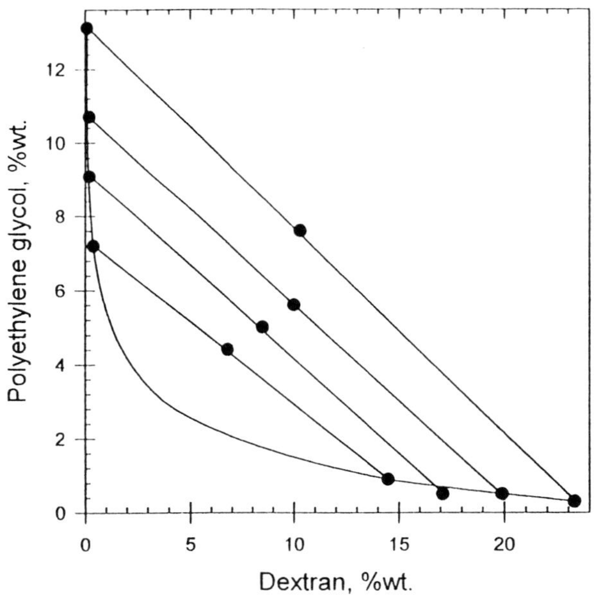

Table 10.99. Phase Diagram and Phase Composition of the Dextran-110—Poly(ethylene glycol)-20000 System at 4°C.
(From D. Forciniti, C. K. Hall, M.-R. Kula, Fluid Phase Equilibria, 61, 243 (1991) and Bioseparation, 2, 115 (1991) by permission of Elsevier Scientific Publishing Co. and Kluwer Academic Publishers, respectively.)

PEG - M_w 21,000; M_n 19,100; Manufacturer: Merck (Dormstadt, Germany); Lot not indicated;
Dextran - M_w 86,200; M_n 52.100; Manufacturer: Fluka (Switzerland); Lot not indicated

Total system		Bottom phase		Top phase		STL *
PEG % w/w	Dex % w/w	PEG % w/w	Dex % w/w	PEG % w/w	Dex % w/w	
4.4	6.75	0.5	15.5	7.1	0.35	-0.436
5.0	8.5	0.4	20.0	8.7	0.21	-0.419
5.6	10.0	0.2	21.5	10.5	0.15	-0.482
7.6	10.3	0.2	25.9	12.6	0.1	-0.481
					av.:	-0.455 ±0.032

* - STL - Tie-Line Slope defined as the ratio STL = (ΔPEG)/(ΔDex) where Δ is the difference between the concentrations of a given polymer in the two coexisting phases.

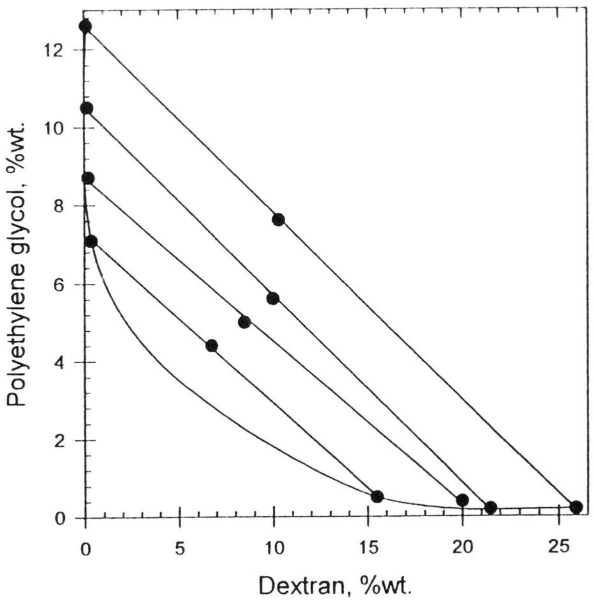

Table 10.100. Phase Diagram and Phase Composition of the Dextran-110—Poly(ethylene glycol)-20000 System at 40°C.
(From D. Forciniti, C. K. Hall, M.-R. Kula, Fluid Phase Equilibria, 61, 243 (1991) and Bioseparation, 2, 115 (1991) by permission of Elsevier Scientific Publishing Co. and Kluwer Academic Publishers, respectively.)

PEG - M_w 21,000; M_n 19,100; Manufacturer: Merck (Dormstadt, Germany); Lot not indicated;
Dextran - M_w 86,200; M_n 52.100; Manufacturer: Fluka (Switzerland); Lot not indicated

Total system		Bottom phase		Top phase		STL *
PEG	Dex	PEG	Dex	PEG	Dex	
% w/w	% w/w	% w/w	% w/w	% w/w	% w/w	
4.4	6.75	0.5	14.5	7.7	0.36	-0.509
5.0	8.5	0.3	17.46	9.4	0.21	-0.528
5.6	10.0	0.4	20.0	11.0	0.15	-0.534
7.6	10.3	0.4	23.6	13.4	0.11	-0.553
					av.:	-0.531
						±0.018

* - STL - Tie-Line Slope defined as the ratio STL = (ΔPEG)/(ΔDex) where Δ is the difference between the concentrations of a given polymer in the two coexisting phases.

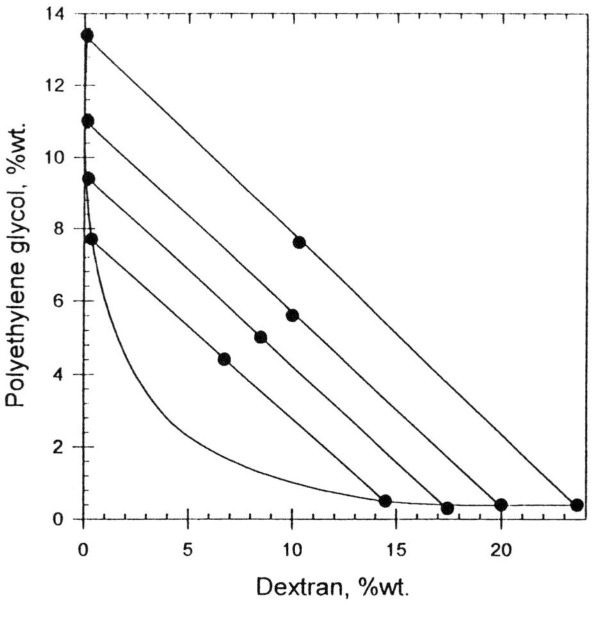

Table 10.101. Phase Diagram and Phase Composition of the Dextran-500—Poly(ethylene glycol)-20000 System at 25°C.
(From D. Forciniti, C. K. Hall, M.-R. Kula, Fluid Phase Equilibria, 61, 243 (1991) and Bioseparation, 2, 115 (1991) by permission of Elsevier Scientific Publishing Co. and Kluwer Academic Publishers, respectively.)

PEG - M_w 21,000; M_n 19,100; Manufacturer: Merck (Dormstadt, Germany); Lot not indicated;
Dextran - M_w 215,000; M_n 88,200; Manufacturer: Pfeifer and Langen (Dormagen, Germany); Lot not indicated

Total system		Bottom phase		Top phase		STL *
PEG	Dex	PEG	Dex	PEG	Dex	
% w/w	% w/w	% w/w	% w/w	% w/w	% w/w	
4.4	6.75	0.4**	14.6**	7.2**	0.3**	-0.476**
5.0	8.5	0.3	17.3	9.0	0.2	-0.509
5.6	10.0	0.5	20.0	10.8	0.07	-0.517
7.65	10.3	0.7	23.3	13.1	0.04	-0.533
					av.:	-0.520
						±0.012

* - STL - Tie-Line Slope defined as the ratio STL = (ΔPEG)/(ΔDex) where Δ is the difference between the concentrations of a given polymer in the two coexisting phases;

** - Composition of the phases questionable as the STL value is inconsistent with the other values; was not used in calculations of the average STL value.

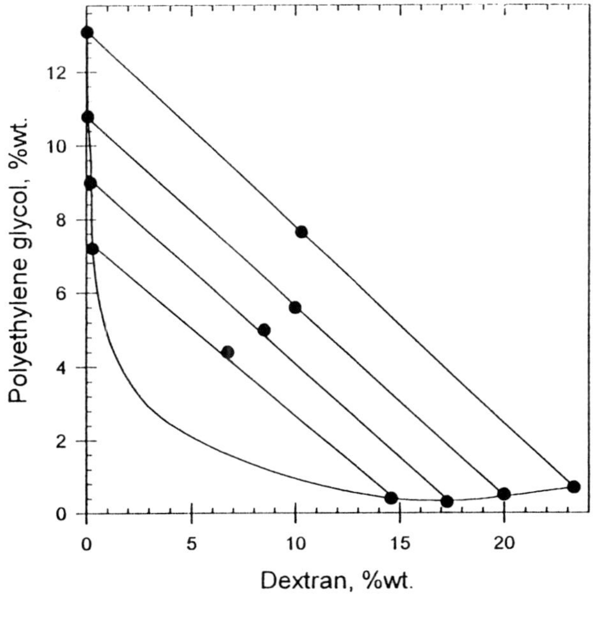

Table 10.102. Phase Diagram and Phase Composition of the Dextran-500—Poly(ethylene glycol)-20000 System at 4°C.
(From D. Forciniti, C. K. Hall, M.-R. Kula, Fluid Phase Equilibria, 61, 243 (1991) and Bioseparation, 2, 115 (1991) by permission of Elsevier Scientific Publishing Co. and Kluwer Academic Publishers, respectively.)

PEG - M_w 21,000; M_n 19,100; Manufacturer: Merck (Dormstadt, Germany); Lot not indicated;
Dextran - M_w 215,000; M_n 88,200; Manufacturer: Pfeifer and Langen (Dormagen, Germany); Lot not indicated

Total system		Bottom phase *		Top phase *		STL **
PEG % w/w	Dex % w/w	PEG % w/w	Dex % w/w	PEG % w/w	Dex % w/w	
4.4	6.75	0.8	15.7	6.9	0.2	-0.394
5.0	8.5	0.7	18.6	8.6	0.1	-0.427
5.6	9.95	0.7	20.8	9.8	0.06	-0.439
7.65	10.3	0.4	25.27	12.4	0.04	-0.476
					av.:	-0.434
						±0.034

*- Compositions of all the phases questionable as the STL values are inconsistent;
**- STL - Tie-Line Slope defined as the ratio STL = (ΔPEG)/(ΔDex) where Δ is the difference between the concentrations of a given polymer in the two coexisting phases.

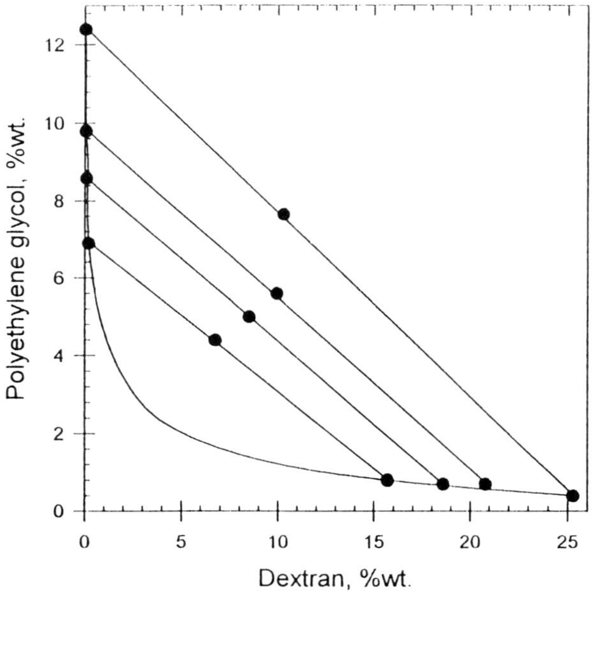

Table 10.103. Phase Diagram and Phase Composition of the Dextran-70—Polyvinylpyrrolidone System at 23°C.

Dextran-70 - M_w 57,200; M_n 28,700; Manufacturer: Minmedprom (Moscow, Russia); Lot 680480;
Polyvinylpyrrolidone (PVP) - M_w 12,700; Manufacturer: Minmedprom (Moscow, Russia); Lot 135-84

Total system		Bottom phase		Top phase		STL *
PVP % w/w	Dextran % w/w	PVP % w/w	Dextran % w/w	PVP % w/w	Dextran % w/w	
18.32	16.42	4.62	31.37	32.18	1.29	-0.916
17.43	15.61	5.01	29.17	30.38	1.47	-0.916
17.60	12.36	6.03	24.99	27.02	2.07	-0.916
15.58	11.00	6.45	20.97	23.17	2.72	-0.916
14.03	9.97	7.00	17.64	19.23	4.29	-0.916
13.20	9.40	9.19	13.78	16.24	6.08	-0.916
12.00**	9.60**				av.:	-0.916

* - STL - Tie-Line Slope defined as the ratio STL = (ΔPVP)/(ΔDex) where Δ is the difference between the concentrations of a given polymer in the two coexisting phases;
** - Composition of critical point.

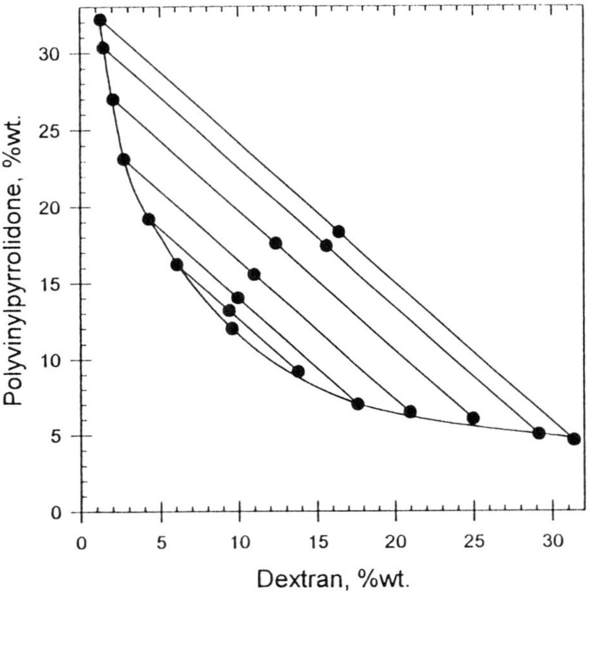

Table 10.104. Phase Diagram and Phase Composition of the Dextran-70—Polyvinylpyrrolidone System at 8°C.

Dextran-70 - M_w 57,200; M_n 28,700; Manufacturer: Minmedprom (Moscow, Russia); Lot 680480;
Polyvinylpyrrolidone (PVP) - M_w 12,700; Manufacturer: Minmedprom (Moscow, Russia); Lot 135-84

Total system		Bottom phase		Top phase		STL *
PVP % w/w	Dextran % w/w	PVP % w/w	Dextran % w/w	PVP % w/w	Dextran % w/w	
16.55	13.92	3.19	28.86	27.57	1.41	-0.888
14.67	12.42	3.80	24.98	24.23	1.97	-0.888
13.57	11.44	4.65	21.63	21.46	2.70	-0.888
10.80**	9.80**				av.:	-0.888

* - STL - Tie-Line Slope defined as the ratio STL = (ΔPVP)/(ΔDex) where Δ is the difference between the concentrations of a given polymer in the two coexisting phases;
** - Composition of critical point.

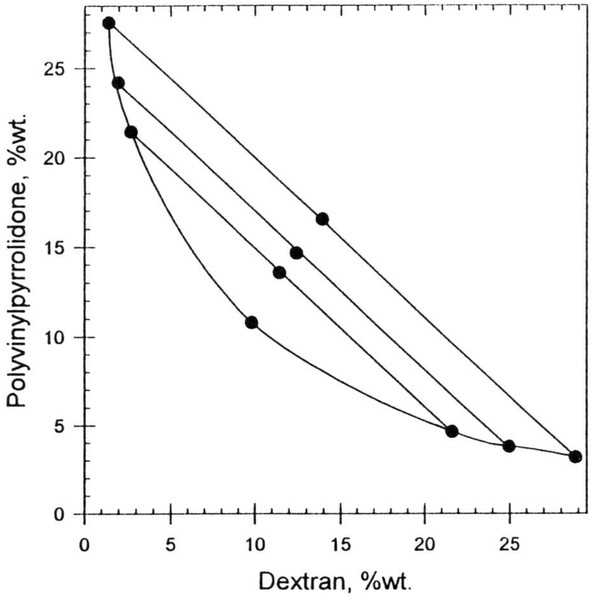

Table 10.105. Phase Diagram and Phase Composition of the Dextran-70—Polyvinylpyrrolidone System at 38°C.

Dextran-70 - M_w 57,200; M_n 28,700; Manufacturer: Minmedprom (Moscow, Russia); Lot 680480;
Polyvinylpyrrolidone (PVP) - M_w 12,700; Manufacturer: Minmedprom (Moscow, Russia); Lot 135-84

Total system		Bottom phase		Top phase		STL *
PVP % w/w	Dextran % w/w	PVP % w/w	Dextran % w/w	PVP % w/w	Dextran % w/w	
16.31	13.81	5.53	24.92	26.69	3.11	-0.970
14.47	12.26	7.32	19.65	22.53	3.97	-0.970
13.50	11.34	8.71	16.28	19.48	5.17	-0.969
13.30**	9.70**				av.:	-0.970 ±0.001

* - STL - Tie-Line Slope defined as the ratio STL = (ΔPVP)/(ΔDex) where Δ is the difference between the concentrations of a given polymer in the two coexisting phases;
** - Composition of critical point.

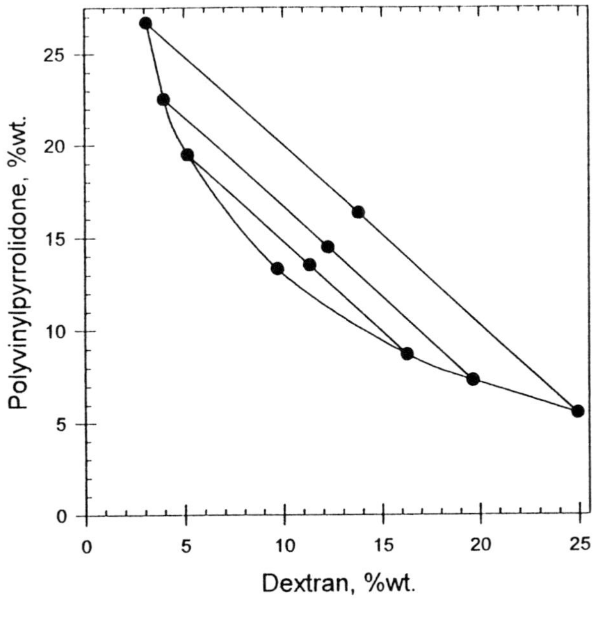

Table 10.106. Phase Diagram and Phase Composition of the Dextran-70—Polyvinylpyrrolidone System at 50°C.

Dextran-70 - M_w 57,200; M_n 28,700; Manufacturer: Minmedprom (Moscow, Russia); Lot 680480;
Polyvinylpyrrolidone (PVP) - M_w 12,700; Manufacturer: Minmedprom (Moscow, Russia); Lot 135-84

Total system		Bottom phase		Top phase		STL *
PVP % w/w	Dextran % w/w	PVP % w/w	Dextran % w/w	PVP % w/w	Dextran % w/w	
16.33	13.81	6.12	24.21	27.12	2.82	-0.982
14.47	12.30	7.05	19.68	22.83	3.98	-1.006
13.46	11.38	8.46	16.37	19.60	5.32	-1.008
13.00**	10.00**				av.:	-0.999 ±0.015

* - STL - Tie-Line Slope defined as the ratio STL = (ΔPVP)/(ΔDex) where Δ is the difference between the concentrations of a given polymer in the two coexisting phases;
** - Composition of critical point.

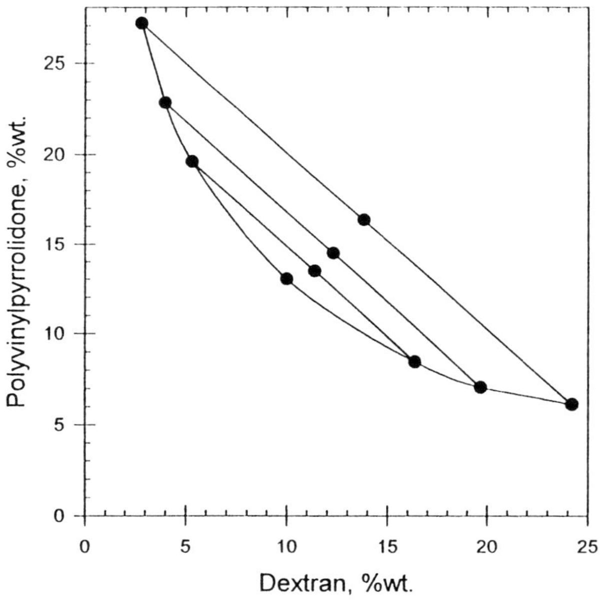

Table 10.107. Phase Diagram and Phase Composition of the Dextran-70—Polyvinylpyrrolidone System Containing 0.5 mole/kg Urea at 23°C.

Dextran-70 - M_w 57,200; M_n 28,700; Manufacturer: Minmedprom (Moscow, Russia); Lot 680480;
Polyvinylpyrrolidone (PVP) - M_w 12,700; Manufacturer: Minmedprom (Moscow, Russia); Lot 135-84

Total system		Bottom phase		Top phase		STL *
PVP % w/w	Dextran % w/w	PVP % w/w	Dextran % w/w	PVP % w/w	Dextran % w/w	
18.28	16.40	4.75	32.36	30.95	1.62	-0.852
17.30	15.58	5.49	29.50	28.87	1.93	-0.848
16.00	14.27	7.12	25.99	26.17	2.34	-0.805
13.00**	9.80**				av.:	-0.835 ±0.026

* - STL - Tie-Line Slope defined as the ratio STL = (ΔPVP)/(ΔDex) where Δ is the difference between the concentrations of a given polymer in the two coexisting phases;

** - Composition of critical point.

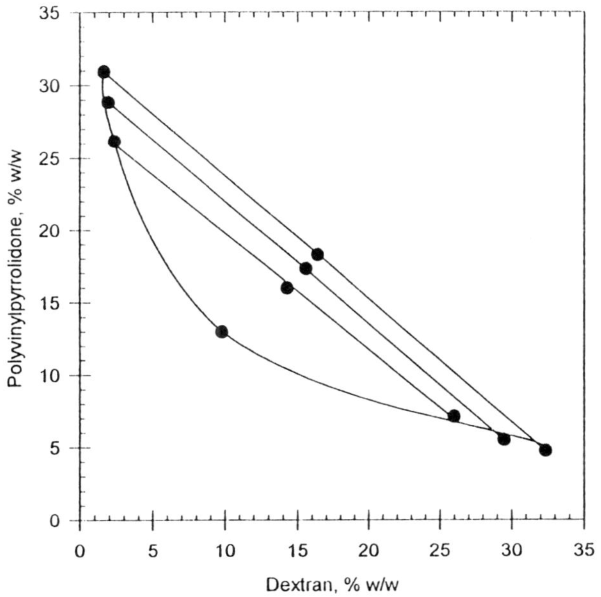

Table 10.108. Phase Diagram and Phase Composition of the Dextran-70—Polyvinylpyrrolidone System Containing 2.0 mole/kg Urea at 23°C.

Dextran-70 - M_w 57,200; M_n 28,700; Manufacturer: Minmedprom (Moscow, Russia); Lot 680480;
Polyvinylpyrrolidone (PVP) - M_w 12,700; Manufacturer: Minmedprom (Moscow, Russia); Lot 135-84

Total system		Bottom phase		Top phase		STL *
PVP % w/w	Dextran % w/w	PVP % w/w	Dextran % w/w	PVP % w/w	Dextran % w/w	
18.24	16.35	5.79	29.62	31.44	2.28	-0.938
17.39	15.59	6.34	27.37	29.65	2.54	-0.939
15.95	14.33	6.93	23.95	26.48	3.10	-0.938
13.80**	10.30**				av.:	-0.938 ±0.001

* - STL - Tie-Line Slope defined as the ratio STL = (ΔPVP)/(ΔDex) where Δ is the difference between the concentrations of a given polymer in the two coexisting phases;
** - Composition of critical point.

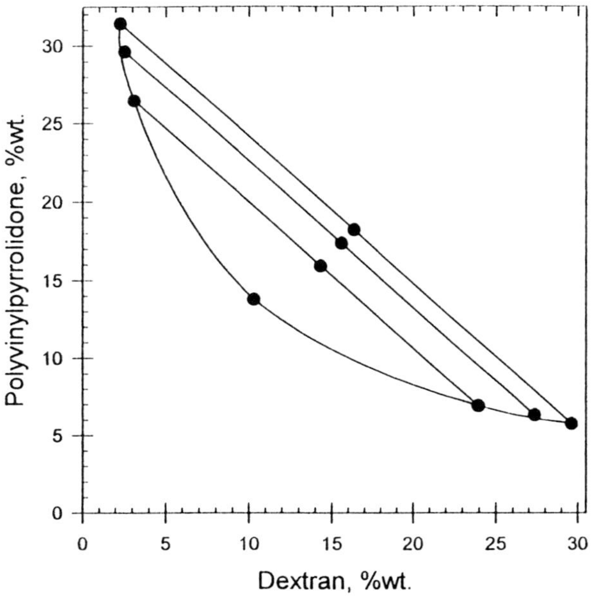

Table 10.109. Phase Diagram and Phase Composition of the Dextran-70—Polyvinylpyrrolidone System Containing 0.1 mole/kg KSCN at 23°C.

Dextran-70 - M_w 57,200; M_n 28,700; Manufacturer: Minmedprom (Moscow, Russia); Lot 680480;
Polyvinylpyrrolidone (PVP) - M_w 12,700; Manufacturer: Minmedprom (Moscow, Russia); Lot 135-84

Total system		Bottom phase			Top phase			STL *
PVP % w/w	Dextran % w/w	PVP % w/w	Dextran % w/w	KSCN % w/w	PVP % w/w	Dextran % w/w	KSCN % w/w	
18.69	16.70	4.79	29.99	0.895	31.39	4.56	1.031	-1.046
17.60	15.69	6.60	26.20	0.916	28.16	5.59	1.027	-1.046
16.48	14.70	9.59	21.29	0.932	24.21	7.31	1.010	-1.046
15.74	14.04	13.77	15.92	0.957	18.80	11.11	0.984	-1.046
16.20**	13.30**						av.:	-1.046

* - STL - Tie-Line Slope defined as the ratio STL = (ΔPVP)/(ΔDex) where Δ is the difference between the concentrations of a given polymer in the two coexisting phases;
** - Composition of critical point.

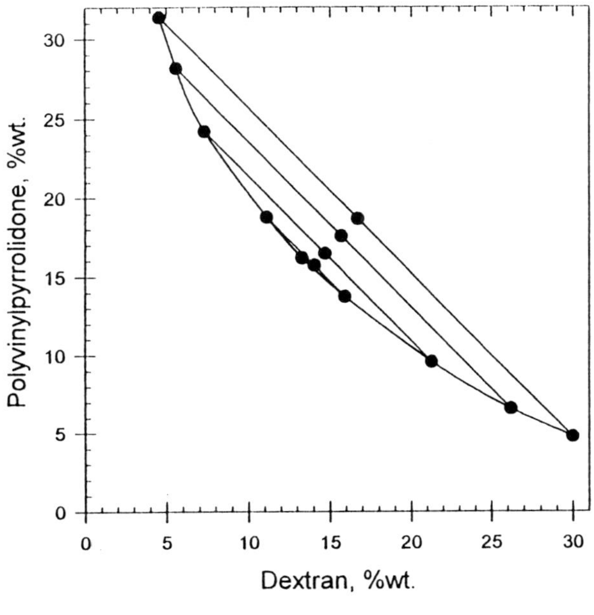

Table 10.110. Phase Diagram and Phase Composition of the Dextran-70—Polyvinylpyrrolidone System Containing 0.1 mole/kg NaSCN at 23°C.

Dextran-70 - M_w 57,200; M_n 28,700; Manufacturer: Minmedprom (Moscow, Russia); Lot 680480;
Polyvinylpyrrolidone (PVP) - M_w 12,700; Manufacturer: Minmedprom (Moscow, Russia); Lot 135-84

Total system		Bottom phase			Top phase			STL *
PVP % w/w	Dextran % w/w	PVP % w/w	Dextran % w/w	NaSCN % w/w	PVP % w/w	Dextran % w/w	NaSCN % w/w	
18.28	16.43	4.67	30.32	0.629	31.36	3.14	0.972	-0.982
17.34	15.65	5.45	27.78	0.648	29.05	3.70	0.952	-0.980
15.95	14.36	7.26	23.25	0.691	24.99	5.14	0.921	-0.979
14.70**	12.50**						av.:	-0.980 ±0.002

* - STL - Tie-Line Slope defined as the ratio STL = (ΔPVP)/(ΔDex) where Δ is the difference between the concentrations of a given polymer in the two coexisting phases;
** - Composition of critical point.

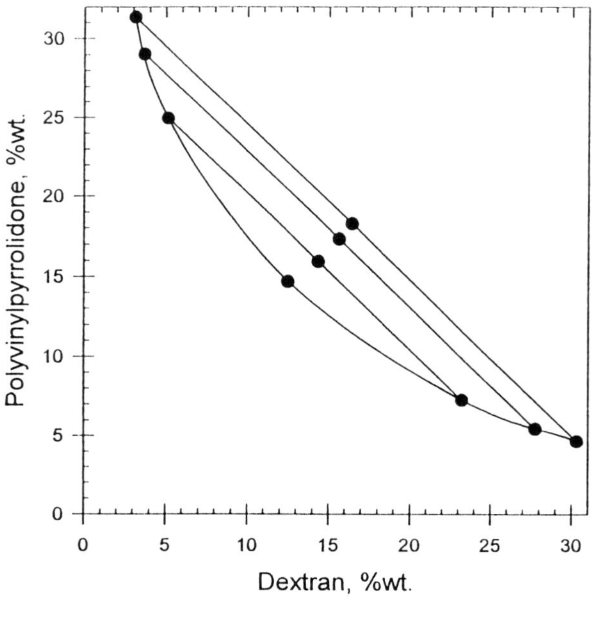

Table 10.111. Phase Diagram and Phase Composition of the Dextran-70—Polyvinylpyrrolidone System Containing 0.1 mole/kg NH_4SCN at 23°C.

Dextran-70 - M_w 57,200; M_n 28,700; Manufacturer: Minmedprom (Moscow, Russia); Lot 680480;
Polyvinylpyrrolidone (PVP) - M_w 12,700; Manufacturer: Minmedprom (Moscow, Russia); Lot 135-84

Total system		Bottom phase			Top phase			STL *
PVP % w/w	Dex % w/w	PVP % w/w	Dex % w/w	NH_4SCN % w/w	PVP % w/w	Dex % w/w	NH_4SCN % w/w	
18.29	16.43	3.56	31.10	0.688	32.45	2.33	0.830	-1.004
17.43	15.64	4.56	28.46	0.700	30.42	2.70	0.828	-1.004
15.99	14.33	5.78	24.50	0.707	26.69	3.67	0.810	-1.004
13.44**	11.98**						av.:	-1.004

* - STL - Tie-Line Slope defined as the ratio STL = (ΔPVP)/(ΔDex) where Δ is the difference between the concentrations of a given polymer in the two coexisting phases;

** - Composition of critical point.

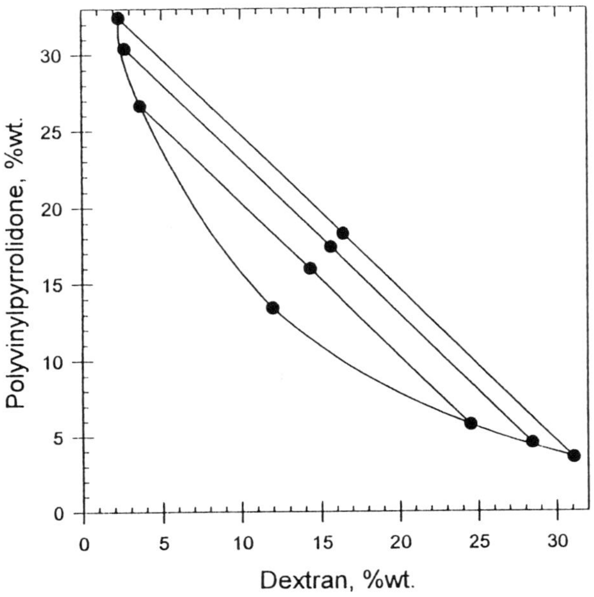

Table 10.112. Phase Diagram and Phase Composition of the Dextran-70—Polyvinylpyrrolidone System Containing 0.05 mole/kg $KClO_4$ at 23°C.

Dextran-70 - M_w 57,200; M_n 28,700; Manufacturer: Minmedprom (Moscow, Russia); Lot 680480;
Polyvinylpyrrolidone (PVP) - M_w 12,700; Manufacturer: Minmedprom (Moscow, Russia); Lot 135-84

Total system		Bottom phase			Top phase			
PVP % w/w	Dex % w/w	PVP % w/w	Dex % w/w	$KClO_4$ % w/w	PVP % w/w	Dex % w/w	$KClO_4$ % w/w	STL*
18.26	16.40	6.11	30.43	0.662	31.07	2.17	0.730	-0.883
17.40	15.60	6.47	27.84	0.660	28.78	2.84	0.720	-0.892
16.00	14.36	7.05	24.18	0.668	25.70	3.62	0.718	-0.907
14.00**	11.60**						av.:	-0.894 ±0.012

* - STL - Tie-Line Slope defined as the ratio STL = (ΔPVP)/(ΔDex) where Δ is the difference between the concentrations of a given polymer in the two coexisting phases;
** - Composition of critical point.

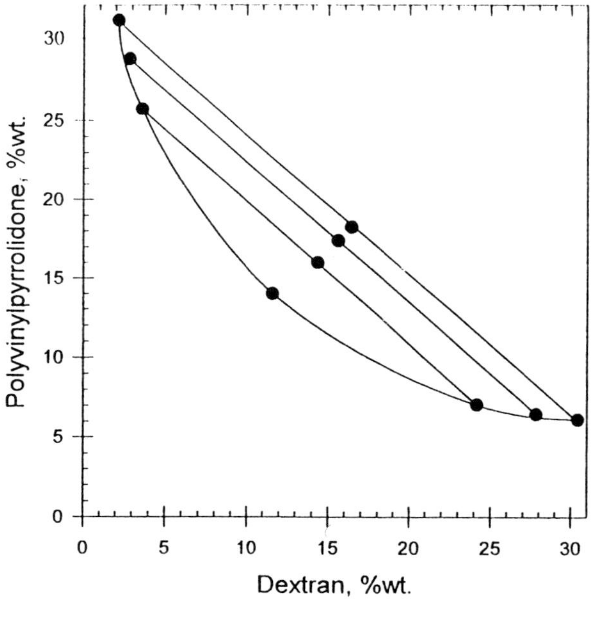

Table 10.113. Phase Diagram and Phase Composition of the Dextran-70—Polyvinylpyrrolidone System Containing 0.1 mole/kg KBr at 23°C.

Dextran-70 - M_w 57,200; M_n 28,700; Manufacturer: Minmedprom (Moscow, Russia); Lot 680480;
Polyvinylpyrrolidone (PVP) - M_w 12,700; Manufacturer: Minmedprom (Moscow, Russia); Lot 135-84

Total system		Bottom phase			Top phase			STL *
PVP	Dextran	PVP	Dextran	KBr	PVP	Dextran	KBr	
% w/w	% w/w	% w/w	% w/w	% w/w	% w/w	% w/w	% w/w	
18.23	16.60	6.00	30.16	1.319	30.26	3.19	1.070	-0.900
17.36	15.78	6.58	27.64	1.304	28.35	3.65	1.080	-0.907
16.06	14.50	8.12	23.19	1.273	24.83	4.78	1.101	-0.908
14.40**	12.00**						av.:	-0.905 ±0.004

* - STL - Tie-Line Slope defined as the ratio STL = (ΔPVP)/(ΔDex) where Δ is the difference between the concentrations of a given polymer in the two coexisting phases;
** - Composition of critical point.

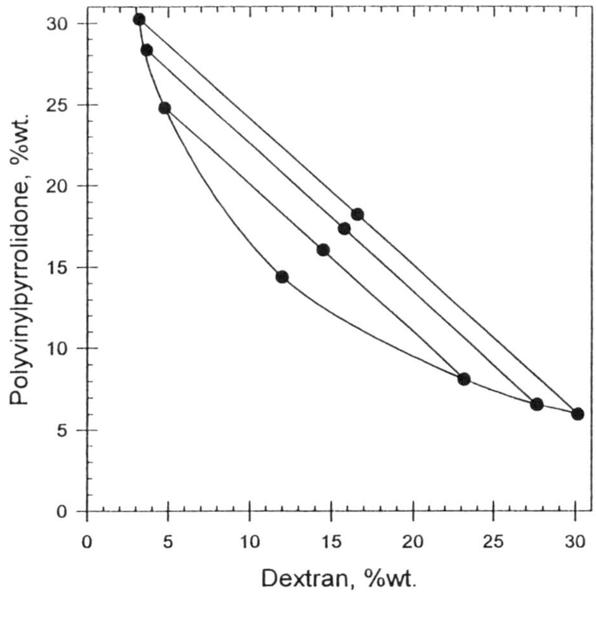

Table 10.114. Phase Diagram and Phase Composition of the Dextran-70—Polyvinylpyrrolidone System Containing 0.1 mole/kg KCl at 23°C.

Dextran-70 - M_w 57,200; M_n 28,700; Manufacturer: Minmedprom (Moscow, Russia); Lot 680480;
Polyvinylpyrrolidone (PVP) - M_w 12,700; Manufacturer: Minmedprom (Moscow, Russia); Lot 135-84

Total system		Bottom phase			Top phase			STL *
PVP % w/w	Dextran % w/w	PVP % w/w	Dextran % w/w	KCl % w/w	PVP % w/w	Dextran % w/w	KCl % w/w	
18.29	16.61	5.00	31.04	0.840	31.67	2.08	0.650	-0.921
17.40	15.79	5.45	28.78	0.837	29.73	2.39	0.663	-0.920
16.48	13.49	5.67	25.24	0.826	26.04	3.10	0.680	-0.917
15.00	12.01	7.37	20.30	0.808	22.31	4.06	0.700	-0.920
12.80**	10.50**						av.:	-0.920 ±0.002

* - STL - Tie-Line Slope defined as the ratio STL = (ΔPVP)/(ΔDex) where Δ is the difference between the concentrations of a given polymer in the two coexisting phases;
** - Composition of critical point.

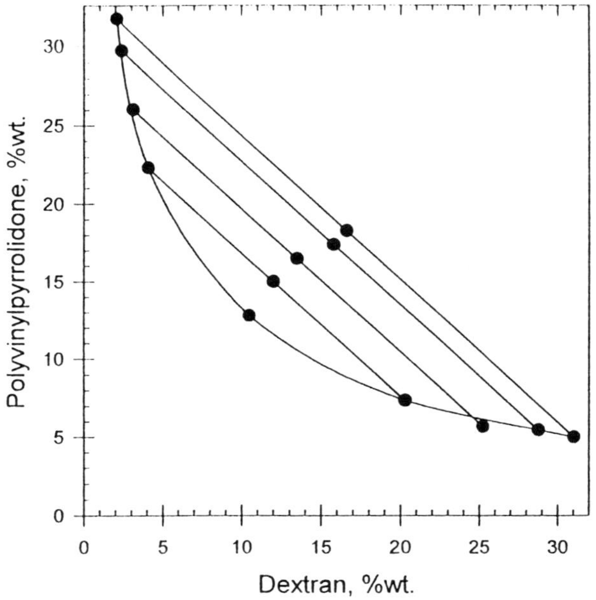

Table 10.115. Phase Diagram and Phase Composition of the Dextran-70—Polyvinylpyrrolidone System Containing 0.1 mole/kg KF at 23°C.

Dextran-70 - M_w 57,200; M_n 28,700; Manufacturer: Minmedprom (Moscow, Russia); Lot 680480;
Polyvinylpyrrolidone (PVP) - M_w 12,700; Manufacturer: Minmedprom (Moscow, Russia); Lot 135-84

Total system		Bottom phase			Top phase			STL *
PVP % w/w	Dextran % w/w	PVP % w/w	Dextran % w/w	KF % w/w	PVP % w/w	Dextran % w/w	KF % w/w	
18.28	16.43	2.83	33.10	0.743	32.61	0.97	0.431	-0.927
17.42	15.61	2.88	31.08	0.723	30.59	1.19	0.439	-0.927
15.95	14.33	3.06	27.70	0.717	27.78	1.68	0.456	-0.950
14.00	10.40	6.50	18.49	0.654	20.70	3.17	0.504	-0.927
10.90**	9.50**						av.:	-0.933 ±0.012

* - STL - Tie-Line Slope defined as the ratio STL = (ΔPVP)/(ΔDex) where Δ is the difference between the concentrations of a given polymer in the two coexisting phases;
** - Composition of critical point.

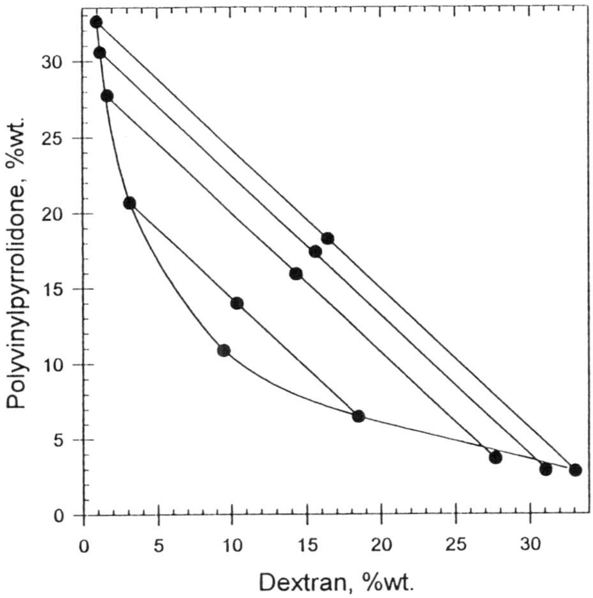

Table 10.116. Phase Diagram and Phase Composition of the Dextran-70—Polyvinylpyrrolidone System Containing 0.1 mole/kg Na_2SO_4 at 23°C.

Dextran-70 - M_w 57,200; M_n 28,700; Manufacturer: Minmedprom (Moscow, Russia); Lot 680480;
Polyvinylpyrrolidone (PVP) - M_w 12,700; Manufacturer: Minmedprom (Moscow, Russia); Lot 135-84

Total system		Bottom phase			Top phase			STL *
PVP % w/w	Dex % w/w	PVP % w/w	Dex % w/w	Na_2SO_4 % w/w	PVP % w/w	Dex % w/w	Na_2SO_4 % w/w	
18.23	16.38	3.21	31.86	1.956	33.60	0.53	0.872	-0.970
17.32	15.67	3.50	29.91	1.853	32.01	0.58	0.906	-0.972
15.95	14.37	3.88	26.87	1.850	29.03	0.82	0.984	-0.965
13.62	12.40	4.40	21.90	1.760	24.14	1.56	1.072	-0.971
12.03	11.03	5.52	17.74	1.660	20.25	2.55	1.147	-0.970
11.30**	7.90**						av.:	-0.970 ±0.003

* - STL - Tie-Line Slope defined as the ratio STL = (ΔPVP)/(ΔDex) where Δ is the difference between the concentrations of a given polymer in the two coexisting phases;
** - Composition of critical point.

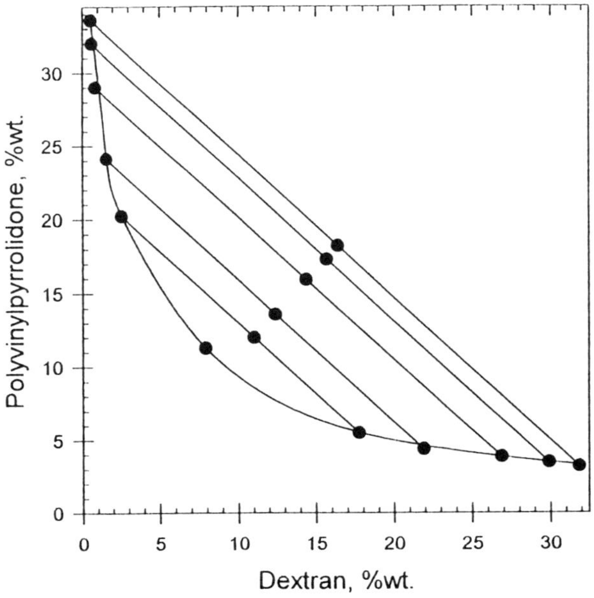

Table 10.117. Phase Diagram and Phase Composition of the Dextran-70—Polyvinylpyrrolidone System Containing 0.1 mole/kg K_2SO_4 at 23°C.

Dextran-70 - M_w 57,200; M_n 28,700; Manufacturer: Minmedprom (Moscow, Russia); Lot 680480;
Polyvinylpyrrolidone (PVP) - M_w 12,700; Manufacturer: Minmedprom (Moscow, Russia); Lot 135-84

Total system		Bottom phase			Top phase			STL *
PVP % w/w	Dex % w/w	PVP % w/w	Dex % w/w	K_2SO_4 % w/w	PVP % w/w	Dex % w/w	K_2SO_4 % w/w	
18.27	16.39	3.24	32.05	2.694	33.60	0.39	0.924	-0.959
17.41	15.61	3.42	30.18	2.616	31.94	0.47	0.960	-0.960
15.99	14.33	3.62	27.21	2.410	29.11	0.66	0.981	-0.960
12.52	11.01	4.46	19.40	2.263	21.53	1.62	1.240	-0.960
10.99	9.50	5.85	14.85	2.209	17.44	2.78	1.469	-0.960
10.48**	7.77**						av.:	-0.960 ±0.0004

*- STL - Tie-Line Slope defined as the ratio STL = (ΔPVP)/(ΔDex) where Δ is the difference between the concentrations of a given polymer in the two coexisting phases;
** - Composition of critical point.

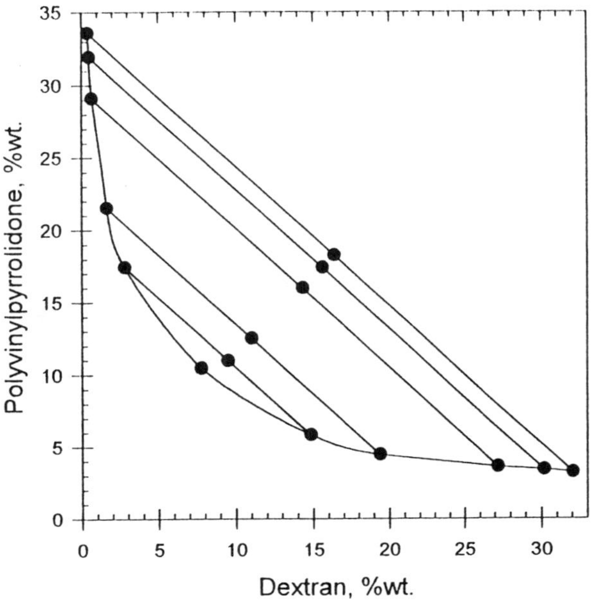

Table 10.118. Phase Diagram and Phase Composition of the Dextran-70—Polyvinylpyrrolidone System Containing 0.1 mole/kg Cs_2SO_4 at 23°C.

Dextran-70 - M_w 57,200; M_n 28,700; Manufacturer: Minmedprom (Moscow, Russia); Lot 680480;
Polyvinylpyrrolidone (PVP) - M_w 12,700; Manufacturer: Minmedprom (Moscow, Russia); Lot 135-84

Total system		Bottom phase			Top phase			STL *
PVP % w/w	Dex % w/w	PVP % w/w	Dex % w/w	Cs_2SO_4 % w/w	PVP % w/w	Dex % w/w	Cs_2SO_4 % w/w	
18.32	16.36	2.97	31.83	5.497	34.30	0.38	1.891	-0.996
17.41	15.62	3.20	29.88	5.333	32.50	0.48	1.968	-0.997
16.00	14.36	3.43	26.96	5.154	29.62	0.69	2.113	-0.997
13.58	12.40	3.38	23.79	4.770	24.77	1.16	2.304	-0.945
12.05	11.00	4.34	20.79	4.461	21.14	1.87	2.520	-0.888
10.26**	9.26**						av.:	-0.965 ±0.048

* - STL - Tie-Line Slope defined as the ratio STL = (ΔPVP)/(ΔDex) where Δ is the difference between the concentrations of a given polymer in the two coexisting phases;
** - Composition of critical point.

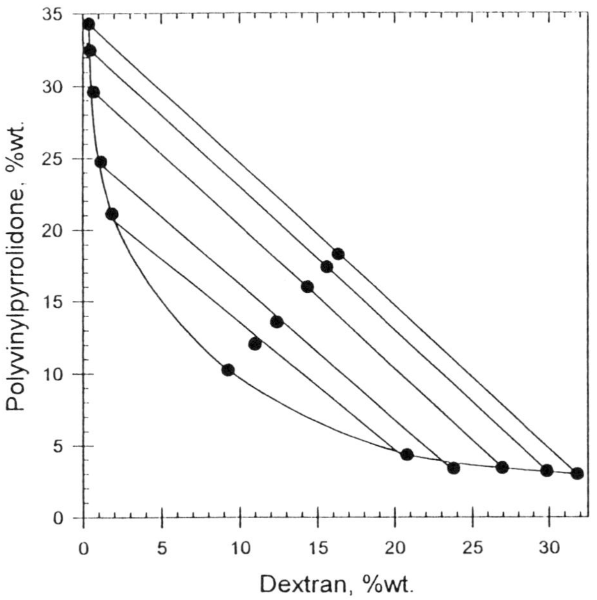

Table 10.119. Phase Diagram and Phase Composition of the Dextran-70—Polyvinylpyrrolidone System Containing 0.1 mole/kg $(NH_4)_2SO_4$ at 23°C.

Dextran-70 - M_w 57,200; M_n 28,700; Manufacturer: Minmedprom (Moscow, Russia); Lot 680480;
Polyvinylpyrrolidone (PVP) - M_w 12,700; Manufacturer: Minmedprom (Moscow, Russia); Lot 135-84

Total system		Bottom phase			Top phase			STL*
PVP % w/w	Dextran % w/w	PVP % w/w	Dextran % w/w	Salt % w/w	PVP % w/w	Dextran % w/w	Salt % w/w	
18.36	16.41	3.58	31.80	1.869	33.32	0.83	0.770	-0.960
17.46	15.82	3.91	29.78	1.812	31.56	0.98	0.794	-0.960
15.96	14.36	4.47	26.30	1.730	28.16	1.62	0.853	-0.960
14.45	13.00	5.00	22.90	1.620	25.00	1.90	0.892	-0.952
12.20	11.00	7.40	16.10	1.520	19.20	3.70	1.050	-0.952
12.10**	8.90**						av:	-0.957 ±0.004

* - STL - Tie-Line Slope defined as the ratio STL = (ΔPVP)/(ΔDex) where Δ is the difference between the concentrations of a given polymer in the two coexisting phases;
** - Composition of critical point.

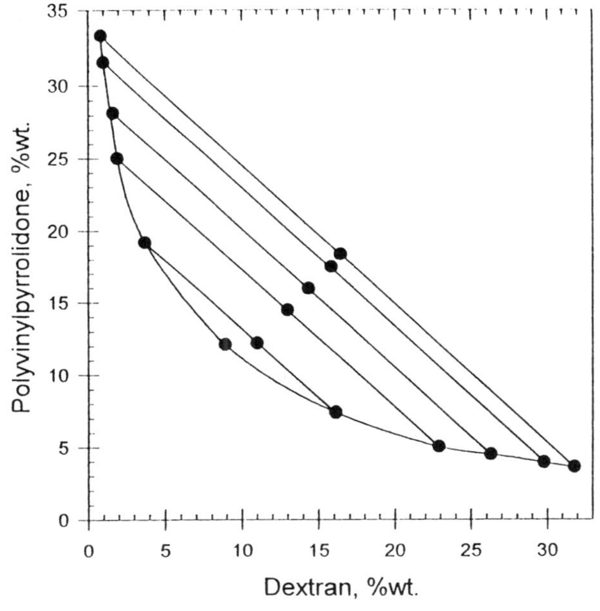

Table 10.120. Phase Diagram and Phase Composition of the Dextran-70—Polyvinylpyrrolidone System Containing 0.15 mole/kg NaCl in 0.01 mole/kg Sodium Phosphate Buffer, pH 7.4 at 23°C.

Dextran-70 - M_w 57,200; M_n 28,700; Manufacturer: Minmedprom (Moscow, Russia); Lot 680480;
Polyvinylpyrrolidone (PVP) - M_w 12,700; Manufacturer: Minmedprom (Moscow, Russia); Lot 135-84

Total system		Bottom phase			Top phase			STL **
PVP % w/w	Dextran % w/w	PVP % w/w	Dextran % w/w	Na^+ * % w/w	PVP % w/w	Dextran % w/w	Na^+ * % w/w	
18.22	16.37	4.87	32.00	0.489	30.49	2.00	0.431	-0.854
17.58	15.52	5.66	29.47	0.485	28.57	2.65	0.434	-0.854
16.24	14.37	6.86	25.35	0.486	25.81	3.16	0.443	-0.854
15.24	13.48	7.22	22.87	0.478	23.47	3.84	0.441	-0.854
14.59	13.00	8.46	20.18	0.480	22.06	4.25	0.449	-0.854
14.01	12.39	9.22	17.93	0.478	20.15	5.20	0.453	-0.859
13.29	11.81	9.89	15.80	0.474	18.26	5.99	0.456	-0.853
13.55***	9.75***						av.:	-0.855 ±0.002

* - Salt concentrations in the phases determined as the sodium concentrations;
** - STL - Tie-Line Slope defined as the ratio STL = (ΔPVP)/(ΔDex) where Δ is the difference between the concentrations of a given polymer in the two coexisting phases;
*** - Composition of critical point.

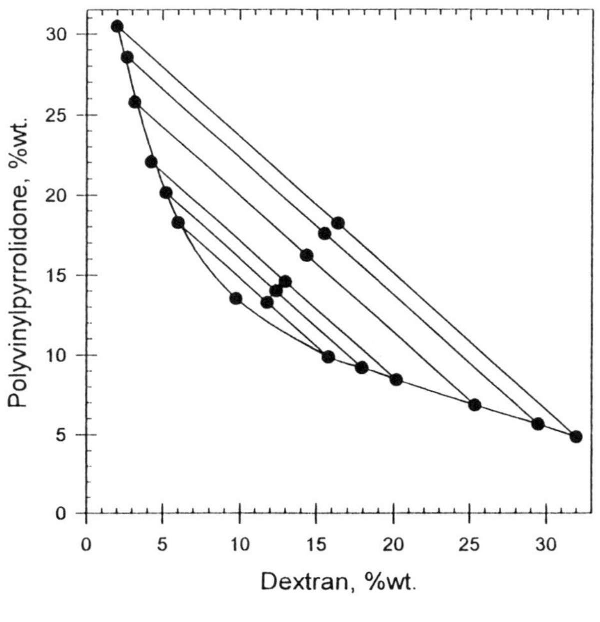

Table 10.121. Phase Diagram and Phase Composition of the Dextran-70—Polyvinylpyrrolidone System Containing 0.11 mole/kg Sodium Phosphate Buffer, pH 7.4 at 23°C.

Dextran-70 - M_w 57,200; M_n 28,700; Manufacturer: Minmedprom (Moscow, Russia); Lot 680480;
Polyvinylpyrrolidone (PVP) - M_w 12,700; Manufacturer: Minmedprom (Moscow, Russia); Lot 135-84

Total system		Bottom phase			Top phase			STL *
PVP	Dextran	PVP	Dextran	Na^+ **	PVP	Dextran	Na^+ **	
% w/w	% w/w	% w/w	% w/w	% w/w	% w/w	% w/w	% w/w	
16.25	14.41	2.52	28.08	0.63	29.76	0.59	0.29	-0.991
15.31	13.50	2.64	26.19	0.61	27.76	1.02	0.30	-0.998
12.50	11.00	3.82	19.70	0.56	21.91	1.57	0.33	-0.998
11.08	9.82	5.19	15.72	0.54	18.42	2.46	0.36	-0.998
10.00	9.00	7.02	11.98	0.50	15.01	3.98	0.39	-0.999
10.30***	7.20***						av.:	–0.997 ±0.003

*- STL - Tie-Line Slope defined as the ratio STL = (ΔPVP)/(ΔDex) where Δ is the difference between the concentrations of a given polymer in the two coexisting phases;
**- Concentrations of sodium phosphate salts used as the components of the buffer, pH 7.4 were determined in the phases as those of Na^+;
***- Composition of critical point.

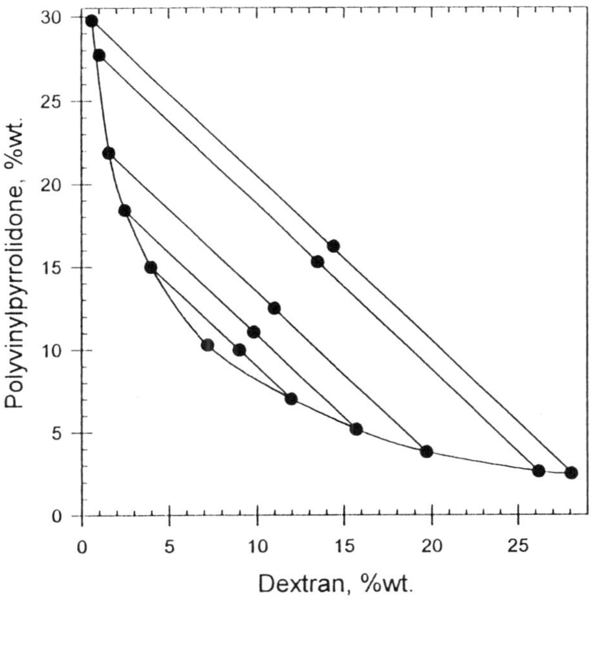

Table 10.122. Phase Diagram and Phase Composition of the Dextran-70—Polyvinyl Alcohol System at 22°C.

Dextran-70 - M_w 57,200; M_n 28,700; Manufacturer: Minmedprom (Moscow, Russia); Lot 680480;
Polyvinyl alcohol (PVA) - M_w 55,000; 1.3% of acetate groups; Manufacturer: Minchimprom (Moscow, Russia); Lot 1246-83

Total system		Bottom phase		Top phase		STL *
PVA % w/w	Dextran % w/w	PVA % w/w	Dextran % w/w	PVA % w/w	Dextran % w/w	
4.49	4.45	1.16	7.54	8.24	0.97	-1.078
4.00	3.99	1.23	6.56	7.13	1.09	-1.079
3.50	3.51	1.48	5.38	5.86	1.32	-1.079
3.00	2.98	1.90	4.00	4.30	1.77	-1.076
2.85	2.88	2.18	3.50	3.78	2.02	-1.081
2.85**	2.65**				av.:	-1.079 ±0.002

* - STL - Tie-Line Slope defined as the ratio STL = (ΔPVA)/(ΔDex) where Δ is the difference between the concentrations of a given polymer in the two coexisting phases;

** - Composition of critical point.

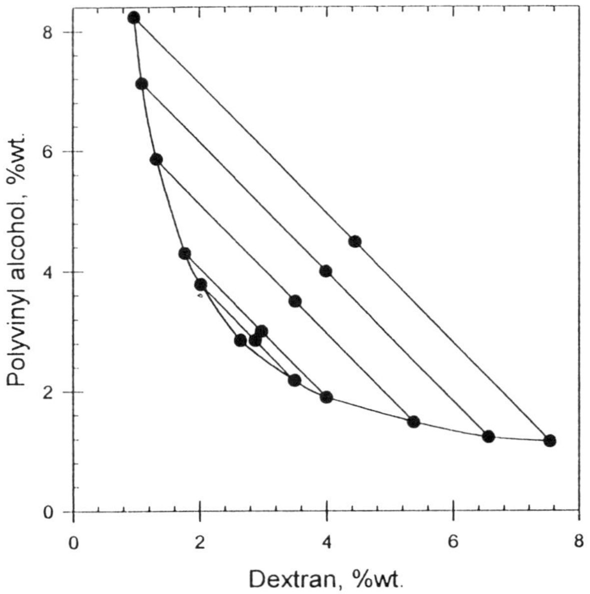

Table 10.123. Phase Diagram and Phase Composition of the Dextran-70—Polyvinyl Alcohol System at 38.5°C.

Dextran-70 - M_w 57,200; M_n 28,700; Manufacturer: Minmedprom (Moscow, Russia); Lot 680480;
Polyvinyl alcohol (PVA) - M_w 55,000; 1.3% of acetate groups; Manufacturer: Minchimprom (Moscow, Russia); Lot 1246-83

Total system		Bottom phase		Top phase		STL *
PVA % w/w	Dextran % w/w	PVA % w/w	Dextran % w/w	PVA % w/w	Dextran % w/w	
3.66	4.50	1.17	6.49	7.92	1.09	-1.250
3.25	4.01	1.31	5.56	6.78	1.19	-1.252
2.83	3.62	1.56	4.64	5.59	1.41	-1.248
3.05**	2.37**				av.:	-1.250 ±0.002

* - STL - Tie-Line Slope defined as the ratio STL = (ΔPVA)/(ΔDex) where Δ is the difference between the concentrations of a given polymer in the two coexisting phases;
** - Composition of critical point.

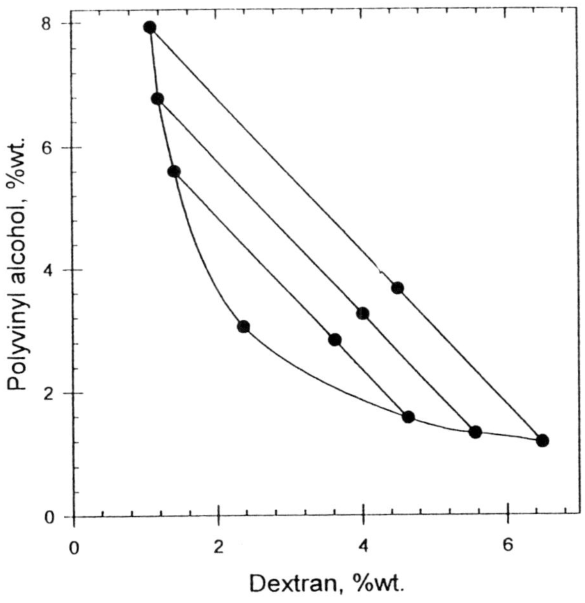

Table 10.124. Phase Diagram and Phase Composition of the Dextran-70—Polyvinyl Alcohol System at 50°C.

Dextran-70 - M_w 57,200; M_n 28,700; Manufacturer: Minmedprom (Moscow, Russia); Lot 680480;
Polyvinyl alcohol (PVA) - M_w 55,000; 1.3% of acetate groups; Manufacturer: Minchimprom (Moscow, Russia); Lot 1246-83

Total system		Bottom phase		Top phase		STL *
PVA % w/w	Dextran % w/w	PVA % w/w	Dextran % w/w	PVA % w/w	Dextran % w/w	
4.24	4.00	1.29	6.26	8.06	1.08	-1.307
3.70	3.45	1.49	5.14	6.64	1.20	-1.307
3.19	3.02	1.83	4.06	5.31	1.40	-1.308
3.05**	2.37**				av.:	-1.307 ±0.001

* - STL - Tie-Line Slope defined as the ratio STL = (ΔPVA)/(ΔDex) where Δ is the difference between the concentrations of a given polymer in the two coexisting phases;
** - Composition of critical point.

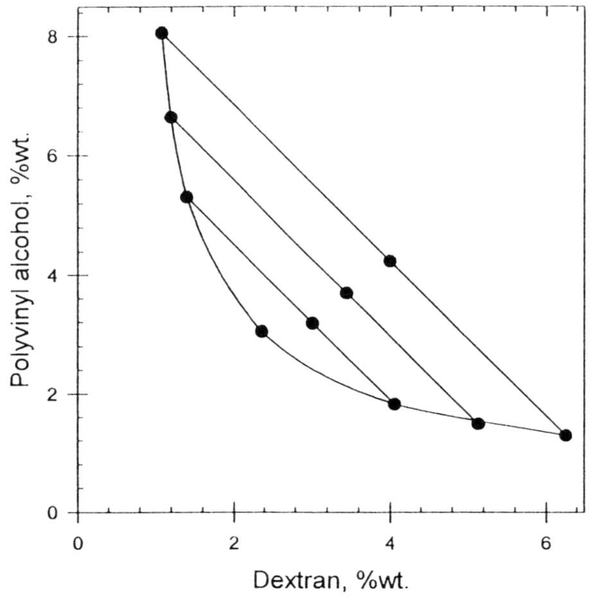

Table 10.125. Phase Diagram and Phase Composition of the Dextran-70—Polyvinyl Alcohol System Containing 0.5 mole/kg Urea at 23°C.

Dextran-70 - M_w 57,200; M_n 28,700; Manufacturer: Minmedprom (Moscow, Russia); Lot 680480;
Polyvinyl alcohol (PVA) - M_w 55,000; 1.3% of acetate groups; Manufacturer: Minchimprom (Moscow, Russia); Lot 1246-83

Total system		Bottom phase		Top phase		STL *
PVA % w/w	Dextran % w/w	PVA % w/w	Dextran % w/w	PVA % w/w	Dextran % w/w	
4.46	4.49	0.27	7.45	9.10	1.19	-1.411
3.97	4.00	0.46	6.47	7.68	1.36	-1.413
3.52	3.51	0.70	5.51	6.19	1.62	-1.411
2.95**	2.65**				av.:	-1.412 ±0.001

* - STL - Tie-Line Slope defined as the ratio STL = (ΔPVA)/(ΔDex) where Δ is the difference between the concentrations of a given polymer in the two coexisting phases;
** - Composition of critical point.

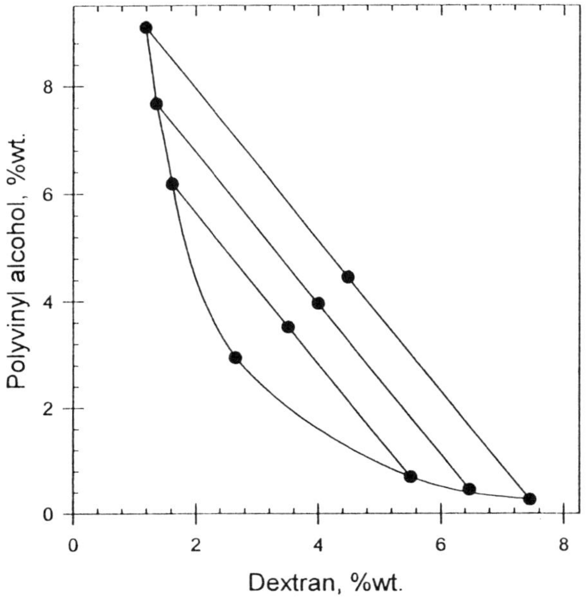

Table 10.126. Phase Diagram and Phase Composition of the Dextran-70—Polyvinyl Alcohol System Containing 2.0 mole/kg Urea at 23°C.

Dextran-70 - M_w 57,200; M_n 28,700; Manufacturer: Minmedprom (Moscow, Russia); Lot 680480;
Polyvinyl alcohol (PVA) - M_w 55,000; 1.3% of acetate groups; Manufacturer: Minchimprom (Moscow, Russia); Lot 1246-83

Total system		Bottom phase		Top phase		STL *
PVA % w/w	Dextran % w/w	PVA % w/w	Dextran % w/w	PVA % w/w	Dextran % w/w	
4.43	4.50	0.16	7.76	8.45	1.43	-1.310
3.93	4.01	0.32	6.77	6.99	1.67	-1.308
3.42	3.56	0.60	5.71	5.61	1.89	-1.312
2.60**	3.40**				av.:	-1.310 ±0.002

*- STL - Tie-Line Slope defined as the ratio STL = (ΔPVA)/(ΔDex) where Δ is the difference between the concentrations of a given polymer in the two coexisting phases;

**- Composition of critical point.

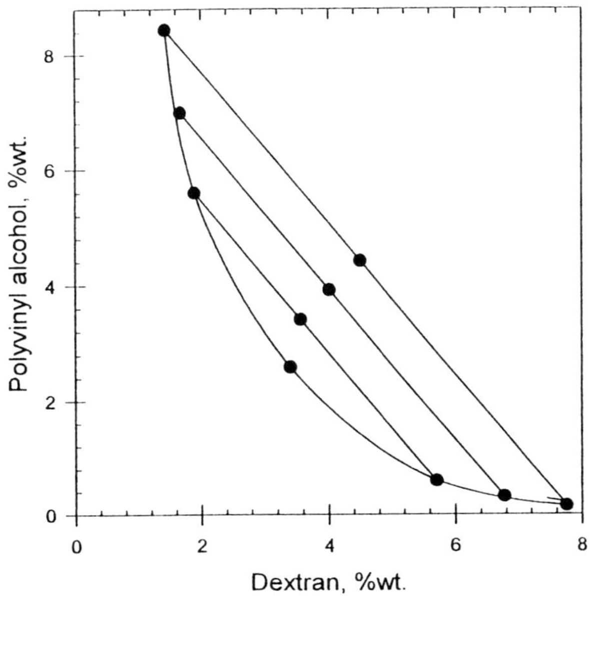

Table 10.127. Phase Diagram and Phase Composition of the Dextran-70—Polyvinyl Alcohol System Containing 0.15 mole/kg NaCl in 0.01 mole/kg Sodium Phosphate Buffer, pH 7.4 at 23°C.

Dextran-70 - M_w 57,200; M_n 28,700; Manufacturer: Minmedprom (Moscow, Russia); Lot 680480;
Polyvinyl alcohol (PVA) - M_w 55,000; 1.3% of acetate groups; Manufacturer: Minchimprom (Moscow, Russia); Lot 1246-83

Total system		Bottom phase			Top phase			STL *
PVA % w/w	Dextran % w/w	PVA % w/w	Dextran % w/w	Na^+ ** % w/w	PVA % w/w	Dextran % w/w	Na^+ ** % w/w	
4.99	4.36	0.15	7.82	0.988	9.87	0.86	0.963	-1.397
4.42	3.85	0.29	6.81	0.986	8.39	1.00	0.966	-1.394
3.58	3.04	0.84	5.00	0.983	5.90	1.38	0.970	-1.398
3.01	2.67	1.59	3.69	0.978	4.11	1.88	0.972	-1.392
2.67***	2.65***						av.:	-1.395 ±0.003

* - STL - Tie-Line Slope defined as the ratio STL = (ΔPVA)/(ΔDex) where Δ is the difference between the concentrations of a given polymer in the two coexisting phases;
** - Salt concentrations in the phases determined as the sodium concentrations;
*** - Composition of critical point.

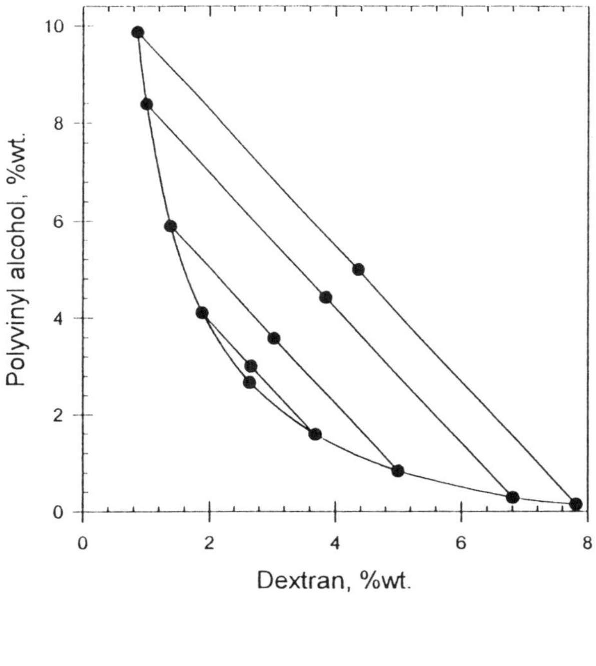

Table 10.128. Phase Diagram and Phase Composition of the Dextran-70—Polyvinyl Alcohol System Containing 0.11 mole/kg Sodium Phosphate Buffer, pH 7.4 at 23°C.

Dextran-70 - M_w 57,200; M_n 28,700; Manufacturer: Minmedprom (Moscow, Russia); Lot 680480;
Polyvinyl alcohol (PVA) - M_w 55,000; 1.3% of acetate groups; Manufacturer: Minchimprom (Moscow, Russia); Lot 1246-83

Total system		Bottom phase			Top phase			STL *
PVA % w/w	Dextran % w/w	PVA % w/w	Dextran % w/w	Na^+ ** % w/w	PVA % w/w	Dextran % w/w	Na^+ ** % w/w	
6.00	3.60	0.05	6.95	0.506	11.45	0.53	0.424	-1.776
5.50	3.30	0.09	6.35	0.501	10.38	0.55	0.429	-1.774
5.00	3.00	0.25	5.68	0.497	9.20	0.63	0.433	-1.772
4.50	2.75	0.30	5.12	0.493	8.17	0.68	0.437	-1.773
3.50	2.15	0.59	3.79	0.484	5.75	0.88	0.447	-1.773
3.00	1.85	2.18	2.31	0.469	3.13	1.77	0.462	-1.759
2.57***	2.00***						av.:	-1.771 ±0.006

* - STL - Tie-Line Slope defined as the ratio STL = (ΔPVA)/(ΔDex) where Δ is the difference between the concentrations of a given polymer in the two coexisting phases;
** - Salt concentrations in the phases determined as the sodium concentrations;
*** - Composition of critical point.

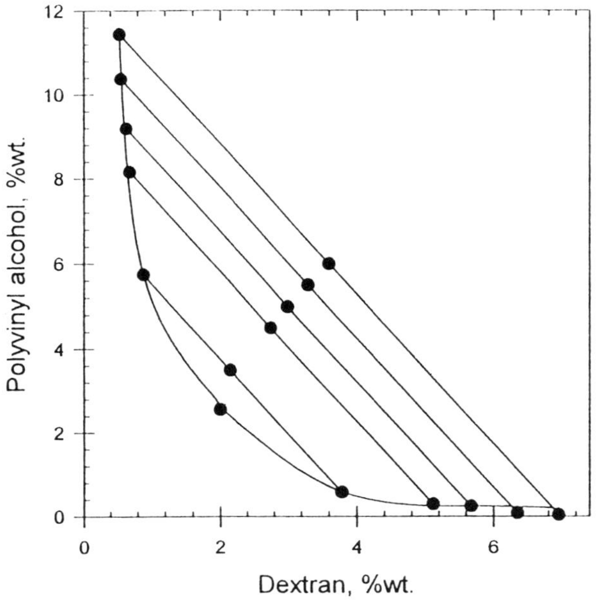

Table 10.129. Phase Diagram and Phase Composition of the Dextran-70—Polyvinyl Alcohol System Containing 0.1 mole/kg K_2SO_4 at 23°C.

Dextran-70 - M_w 57,200; M_n 28,700; Manufacturer: Minmedprom (Moscow, Russia); Lot 680480;
Polyvinyl alcohol (PVA) - M_w 55,000; 1.3% of acetate groups; Manufacturer: Minchimprom (Moscow, Russia); Lot 1246-83

Total system		Bottom phase			Top phase			STL *
PVA % w/w	Dextran % w/w	PVA % w/w	Dextran % w/w	K_2SO_4 % w/w	PVA % w/w	Dextran % w/w	K_2SO_4 % w/w	
3.50	2.15	1.01	3.46	1.90	5.65	1.02	1.58	-1.902
3.00	1.90	1.67	2.60	1.85	3.95	1.40	1.69	-1.900
2.68**	1.75**						av.:	-1.901 ±0.001

* - STL - Tie-Line Slope defined as the ratio STL = (ΔPVA)/(ΔDex) where Δ is the difference between the concentrations of a given polymer in the two coexisting phases;
** - Composition of critical point.

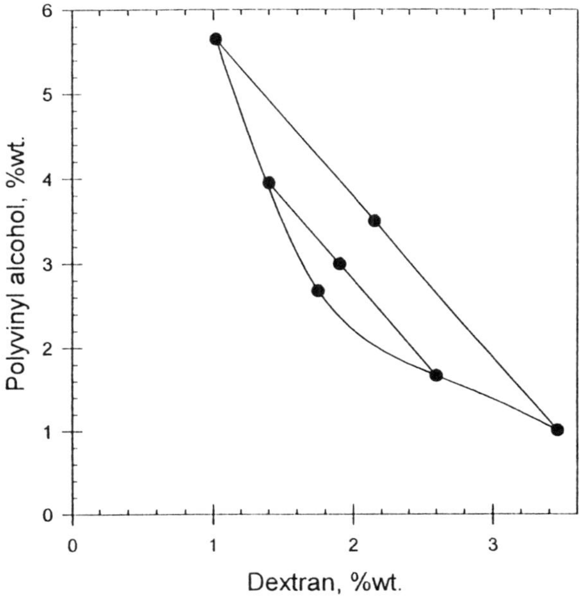

Table 10.130. Phase Diagram and Phase Composition of the Dextran-70—Polyvinyl Alcohol System Containing 0.5 mole/kg KSCN at 23°C.

Dextran-70 - M_w 57,200; M_n 28,700; Manufacturer: Minmedprom (Moscow, Russia); Lot 680480;
Polyvinyl alcohol (PVA) - M_w 55,000; 1.3% of acetate groups; Manufacturer: Minchimprom (Moscow, Russia); Lot 1246-83

Total system		Bottom phase			Top phase			STL *
PVA % w/w	Dextran % w/w	PVA % w/w	Dextran % w/w	KSCN % w/w	PVA % w/w	Dextran % w/w	KSCN % w/w	
3.65	4.55	0.53	7.31	5.09	6.79	1.77	4.63	-1.130
3.20	3.95	1.14	5.77	5.01	5.26	2.13	4.70	-1.132
2.70**	3.32**						av.:	-1.131 ±0.001

* - STL - Tie-Line Slope defined as the ratio STL = (ΔPVA)/(ΔDex) where Δ is the difference between the concentrations of a given polymer in the two coexisting phases;
** - Composition of critical point.

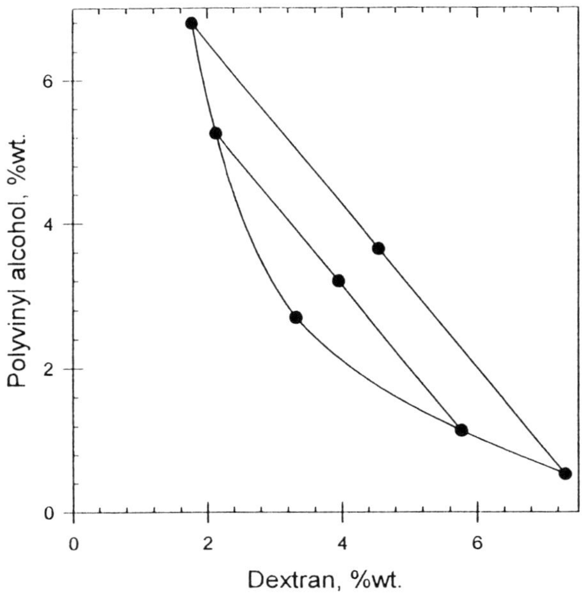

Table 10.131. Phase Diagram and Phase Composition of the Dextran-70—Ficoll-400 System at 23°C.

Dextran-70 - M_w 57,200; M_n 28,700; Manufacturer: Minmedprom (Moscow, Russia); Lot 680480;
Ficoll-400 - M_w ~400,000; Manufacturer: Pharmacia Fine Chemicals (Uppsala, Sweden); Lot HH-2637

Total system		Bottom phase		Top phase		STL *
Ficoll % w/w	Dextran % w/w	Ficoll % w/w	Dextran % w/w	Ficoll % w/w	Dextran % w/w	
18.16	15.60	4.02	28.32	31.70	3.43	-1.112
16.28	13.98	4.92	24.19	27.10	4.25	-1.112
13.50	11.60	6.15	18.21	20.53	5.28	-1.112
12.30	10.55	7.25	15.09	17.11	6.22	-1.112
11.67	10.02	9.08	12.35	14.14	7.80	-1.112
11.35**	9.80**				av.:	-1.112 ±0.001

* - STL - Tie-Line Slope defined as the ratio STL = (ΔPVA)/(ΔDex) where Δ is the difference between the concentrations of a given polymer in the two coexisting phases;
** - Composition of critical point.

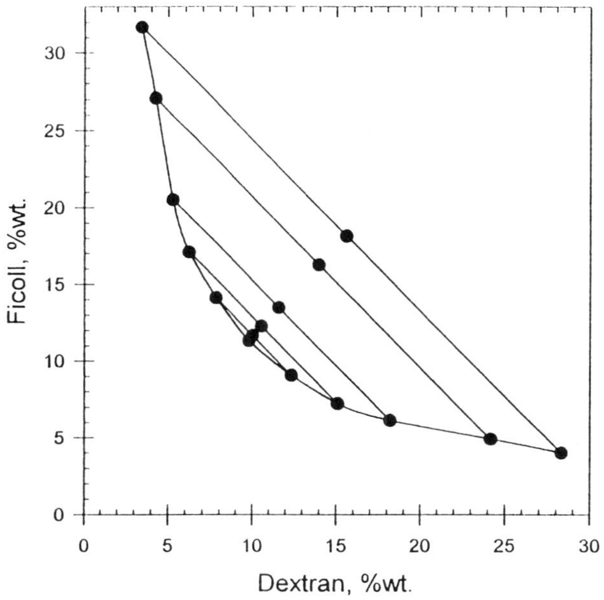

Table 10.132. Phase Diagram and Phase Composition of the Dextran-70—Ficoll-400 System Containing 0.15 mole/kg NaCl in 0.01 mole/kg Sodium Phosphate Buffer, pH 7.4 at 23°C.

Dextran-70 - M_w 57,200; M_n 28,700; Manufacturer: Minmedprom (Moscow, Russia); Lot 680480;
Ficoll-400 - M_w ~400,000; Manufacturer: Pharmacia Fine Chemicals (Uppsala, Sweden); Lot HH-26371

Total system		Bottom phase			Top phase			STL *
Ficoll % w/w	Dextran % w/w	Ficoll % w/w	Dextran % w/w	Na^+ ** % w/w	Ficoll % w/w	Dextran % w/w	Na^+ ** % w/w	
16.33	14.01	3.87	25.14	1.025	28.30	3.32	0.984	-1.120
14.84	12.77	4.26	22.21	1.023	25.04	3.66	0.987	-1.120
13.53	11.59	5.40	18.85	1.018	21.34	4.61	0.991	-1.119
12.28	10.55	6.65	15.57	1.014	17.69	5.72	0.996	-1.121
11.20***	9.80***						av.:	-1.120 ±0.001

* - STL - Tie-Line Slope defined as the ratio STL = (ΔFicoll)/(ΔDex) where Δ is the difference between the concentrations of a given polymer in the two coexisting phases;
** - Salt concentrations in the phases determined as the sodium concentrations;
*** - Composition of critical point.

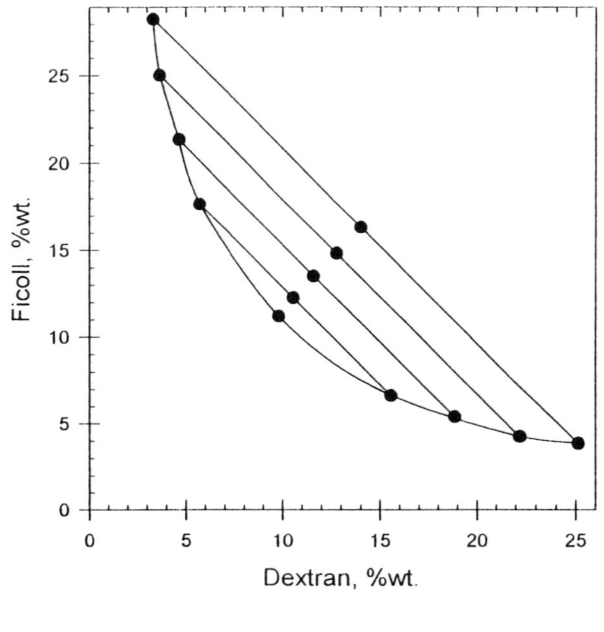

Table 10.133. Phase Diagram and Phase Composition of the Dextran-70—Ficoll-400 System Containing 0.11 mole/kg Sodium Phosphate Buffer, pH 7.4 at 23°C.

Dextran-70 - M_w 57,200; M_n 28,700; Manufacturer: Minmedprom (Moscow, Russia); Lot 680480;
Ficoll-400 - M_w ~400,000; Manufacturer: Pharmacia Fine Chemicals (Uppsala, Sweden); Lot HH-26371

Total system		Bottom phase			Top phase			STL *
Ficoll % w/w	Dextran % w/w	Ficoll % w/w	Dextran % w/w	Na^+ ** % w/w	Ficoll % w/w	Dextran % w/w	Na^+ ** % w/w	
14.86	12.74	2.88	22.29	0.526	27.73	2.48	0.394	-1.254
13.50	11.62	3.61	19.50	0.514	24.18	3.11	0.405	-1.255
12.28	10.57	4.55	16.74	0.502	20.63	3.93	0.417	-1.255
11.66	9.97	5.30	15.04	0.495	18.50	4.53	0.425	-1.256
11.07	9.39	6.74	12.84	0.484	15.68	5.72	0.436	-1.256
10.55***	8.80***						av.:	-1.255
								±0.001

* - STL - Tie-Line Slope defined as the ratio STL = (ΔFicoll)/(ΔDex) where Δ is the difference between the concentrations of a given polymer in the two coexisting phases;
** - Salt concentrations in the phases determined as the sodium concentrations;
*** - Composition of critical point.

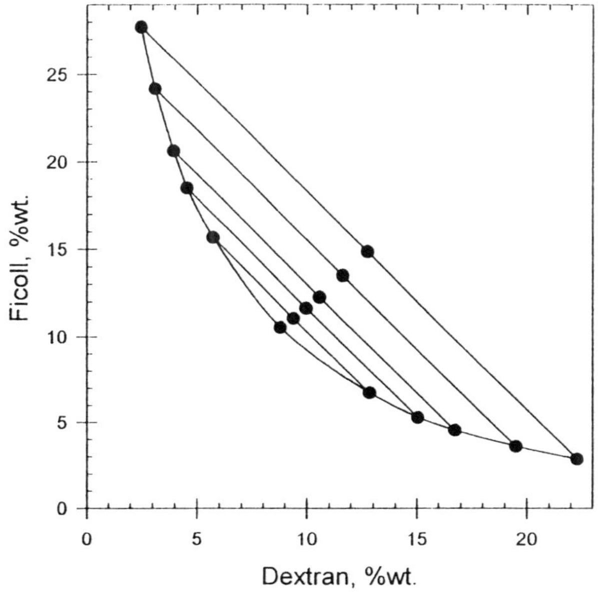

Table 10.134. Phase Diagram and Phase Composition of the Poly(ethylene glycol)-4000—Poly(vinyl methyl ether)-100 System at 20°C.
(From J. N. Baskir, T. A. Hatton, U. W. Suter, J.Phys.Chem., 93, 2111 (1989) by permission of the American Chemical Society.)

PEG - M_w 3720; M_n 3520; Manufacturer: Polysciences (Warrington, PA); Lot 45971;
PVME - M_w 110,000; M_n 51,000; Manufacturer: Scientific Polymer Products (Ontario, NY); Lot 8

Total system		Bottom phase		Top phase		STL *
PEG % w/w	PVME % w/w	PEG % w/w	PVME % w/w	PEG % w/w	PVME % w/w	
6.0	% w/w	7.8	3.6	3.4	16.8	-0.333
7.0	9.0	9.4	2.8	3.0	19.4	-0.386
8.0	9.0	10.6	2.2	2.8	22.8	-0.379
					av.:	-0.366 ±0.029

* - STL - Tie-Line Slope defined as the ratio STL = (ΔPEG)/(ΔPVME) where Δ is the difference between the concentrations of a given polymer in the two coexisting phases.

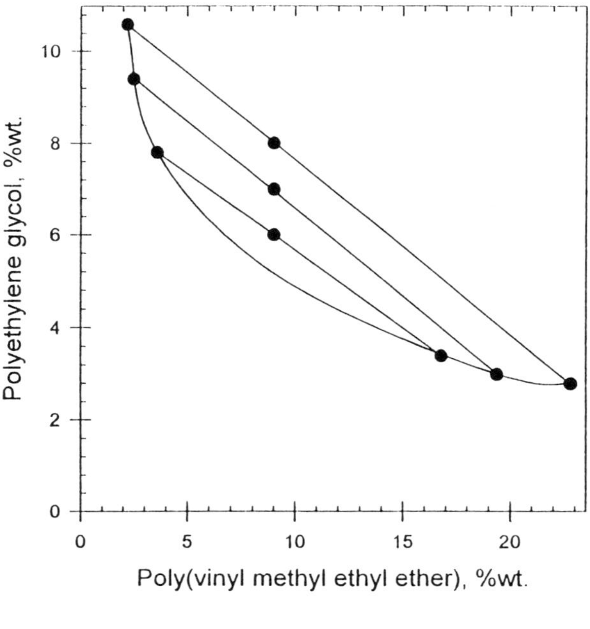

Table 10.135. Phase Diagram and Phase Composition of the Poly(ethylene glycol)-7500—Poly(vinyl methyl ether)-100 System at 20°C.
(From J. N. Baskir, T. A. Hatton, U. W. Suter, J.Phys.Chem., 93, 2111 (1989) by permission of the American Chemical Society.)

PEG - M_w 10,800; M_n 10,100; Manufacturer: Polysciences (Warrington, PA); Lot 62891;
PVME - M_w 110,000; M_n 51,000; Manufacturer: Scientific Polymer Products (Ontario, NY); Lot 8

Total system		Bottom phase		Top phase		STL *
PEG % w/w	PVME % w/w	PEG % w/w	PVME % w/w	PEG % w/w	PVME % w/w	
4.0	6.0	5.2	2.8	1.2	13.2	-0.385
3.0	10.0	6.2**	2.2**	1.0**	15.4**	-0.288**
3.0	12.0	7.3	1.7	0.6	18.4	-0.401
6.0	8.0	8.7	1.3	0.2	22.6	-0.399
					av.:	-0.395 ±0.009

*- STL - Tie-Line Slope defined as the ratio STL = (ΔPEG)/(ΔPVME) where Δ is the difference between the concentrations of a given polymer in the two coexisting phases;

** - Composition of the phases questionable as the STL value is inconsistent with the other values; was not used in calculations of the average STL value.

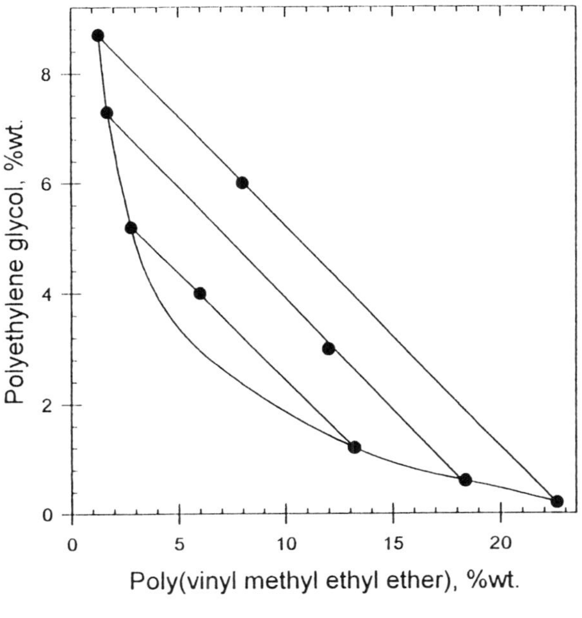

Table 10.136. Phase Diagram and Phase Composition of the Poly(ethylene glycol)-35000—Poly(vinyl methyl ether) System at 20°C.
(From J. N. Baskir, T. A. Hatton, U. W. Suter, J.Phys.Chem., 93, 2111 (1989) by permission of the American Chemical Society.)

PEG - M_w 34,300; M_n 29,400; Manufacturer: Fluka AG; Lot not indicated; PVME - M_w 110,000; M_n 51,000; Manufacturer: Scientific Polymer Products (Ontario, NY); Lot 8

Total system		Bottom phase		Top phase		STL *
PEG % w/w	PVME % w/w	PEG % w/w	PVME % w/w	PEG % w/w	PVME % w/w	
2.0	7.0	4.9	1.2	0.2	10.8	-0.490
2.0	9.0	6.0	1.0	0.1	13.0	-0.492
2.0	11.0	7.2	0.7	0.1	14.9	-0.500
					av.:	-0.494
						±0.005

* - STL - Tie-Line Slope defined as the ratio STL = (ΔPEG)/(ΔPVME) where Δ is the difference between the concentrations of a given polymer in the two coexisting phases.

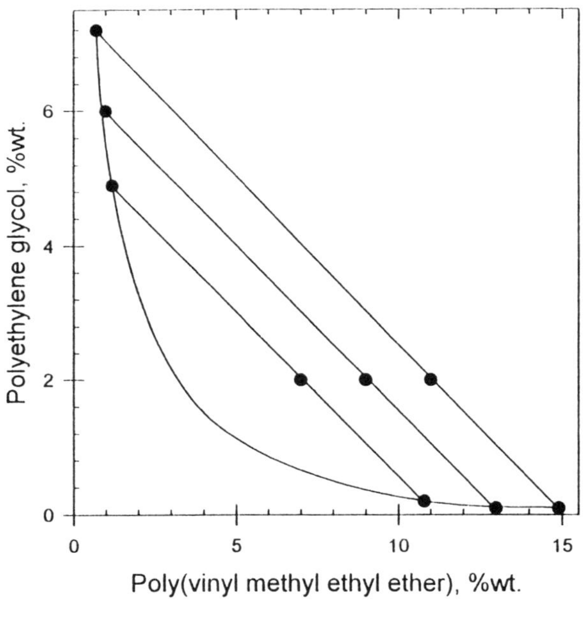

Table 10.137. Phase Diagram and Phase Composition of the Poly(ethylene glycol)-300—Ammonium Sulfate system at 23°C.

PEG - M_w ~ 300; Manufacturer: Merck (Germany); Lot not indicated

Total system		Bottom phase		Top phase		STL *
PEG % w/w	$(NH_4)_2SO_4$ % w/w	PEG % w/w	$(NH_4)_2SO_4$ % w/w	PEG % w/w	$(NH_4)_2SO_4$ % w/w	
20.59	19.55	9.34	27.32	32.20	11.53	-1.448
20.92	20.185	6.29	30.30	36.38	9.51	-1.447
22.195	21.24	3.34	34.26	42.42	7.27	-1.448
22.97	22.025	2.47	36.18	46.24	5.95	-1.448
24.02	22.82	1.64	38.28	49.42	5.28	-1.448
19.97**	19.21**				av.:	-1.448 ±0.0001

*- STL - Tie-Line Slope defined as the ratio STL = (ΔPEG)/(ΔSalt) where ΔPEG is the difference between the concentrations of PEG and ΔSalt the difference between the concentrations of $(NH_4)_2SO_4$ in the two coexisting phases;

**- Composition of critical point.

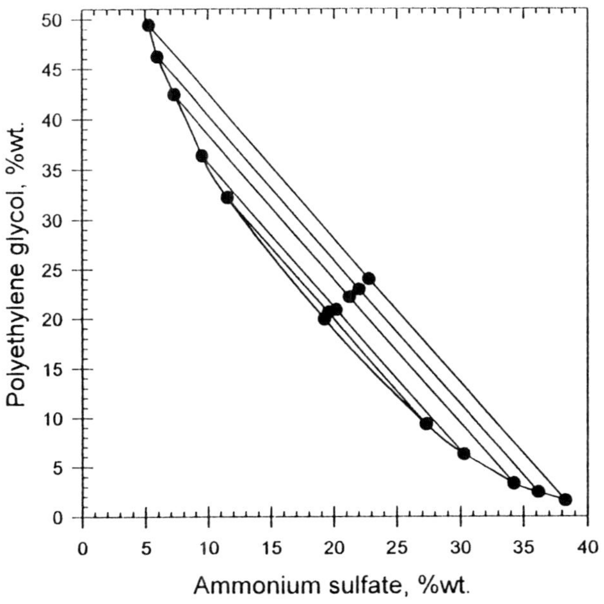

Table 10.138. Phase Diagram and Phase Composition of the Poly(ethylene glycol)-600—Ammonium Sulfate system at 23°C.

PEG - M_w ~ 600; Manufacturer: Loba Chemie (Austria); Lot not indicated

Total system		Bottom phase		Top phase		STL *
PEG % w/w	$(NH_4)_2SO_4$ % w/w	PEG % w/w	$(NH_4)_2SO_4$ % w/w	PEG % w/w	$(NH_4)_2SO_4$ % w/w	
18.69	14.74	6.32	22.36	29.08	8.56	-1.649
19.41	15.20	4.78	24.24	32.89	7.21	-1.651
20.58	16.32	2.94	27.00	38.57	5.40	-1.650
22.55	17.93	1.67	30.60	45.71	3.88	-1.648
24.07	18.80	0.62	33.10	50.10	3.09	-1.649
16.47**	14.92**				av.:	-1.649 ±0.001

*- STL - Tie-Line Slope defined as the ratio STL = (ΔPEG)/(ΔSalt) where ΔPEG is the difference between the concentrations of PEG and ΔSalt the difference between the concentrations of $(NH_4)_2SO_4$ in the two coexisting phases;

**- Composition of critical point.

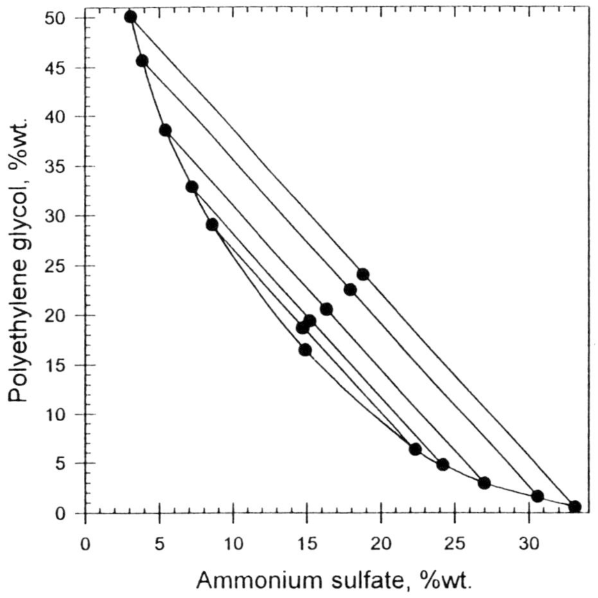

Table 10.139. Phase Diagram and Phase Composition of the Poly(ethylene glycol)-1000—Ammonium Sulfate System at 25°C.
(From S. M. Snyder, K. D. Cole, D. C. Szlag, Chem.Eng.Data, 37, 268 (1992) by permission of the American Chemical Society.)

PEG - M_w 1125; M_n 1075; Manufacturer and Lot not indicated

Total system		Bottom phase		Top phase		STL *
PEG % w/w	$(NH_4)_2SO_4$ % w/w	PEG % w/w	$(NH_4)_2SO_4$ % w/w	PEG % w/w	$(NH_4)_2SO_4$ % w/w	
15.0	14.0	3.1	19.8	29.6	7.2	-2.103
18.5	16.0	0.5	25.0	41.7	4.6	-2.020
21.0	20.0	0.1	34.1	53.6	2.9	-1.715
24.0	24.0	0.0	40.5	61.2	1.7	-1.577
					av.:	-1.854
						±0.001

*- STL - Tie-Line Slope defined as the ratio STL = (ΔPEG)/(ΔSalt) where ΔPEG is the difference between the concentrations of PEG and ΔSalt the difference between the concentrations of $(NH_4)_2SO_4$ in the two coexisting phases.

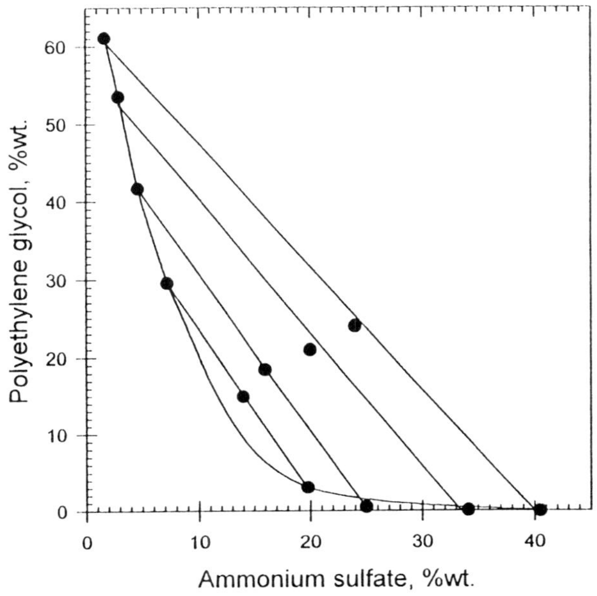

Table 10.140. Phase Diagram and Phase Composition of the Poly(ethylene glycol)-6000—Ammonium Sulfate system at 23°C.

PEG - M_w ~ 6000; Manufacturer: Serva Fine Biochemicals (Germany); Lot not indicated

Total system		Bottom phase		Top phase		STL *
PEG % w/w	$(NH_4)_2SO_4$ % w/w	PEG % w/w	$(NH_4)_2SO_4$ % w/w	PEG % w/w	$(NH_4)_2SO_4$ % w/w	
10.04	9.44	2.69	12.42	16.17	6.95	-2.464
12.70	10.87	0.27	16.01	29.00	4.25	-2.443
15.30	12.46	0.082	19.20	38.09	3.21	-2.376
17.88	14.26	0.044	22.10	46.77	2.53	-2.388
9.43**	9.21**				av.:	-2.418 ±0.042

*- STL - Tie-Line Slope defined as the ratio STL = (ΔPEG)/(ΔSalt) where ΔPEG is the difference between the concentrations of PEG and ΔSalt the difference between the concentrations of $(NH_4)_2SO_4$ in the two coexisting phases;

**- Composition of critical point.

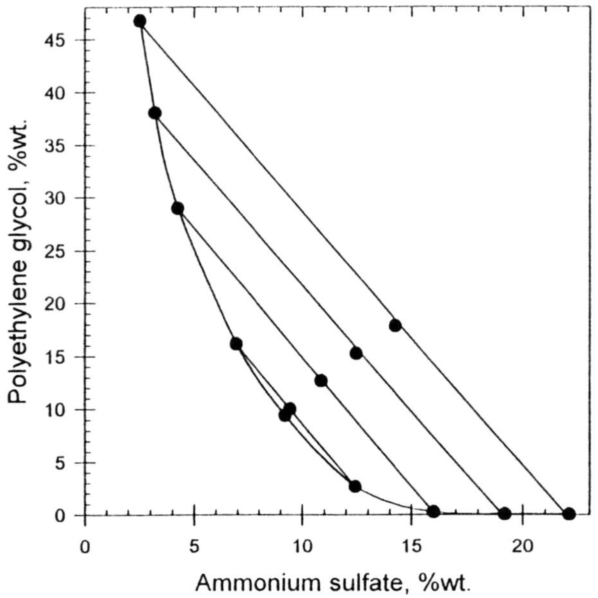

Table 10.141. Phase Diagram and Phase Composition of the Poly(ethylene glycol)-8000—Ammonium Sulfate system at 25°C.
(From S. M. Snyder, K. D. Cole, D. C. Szlag, Chem.Eng.Data, 37, 268 (1992) by permission of the American Chemical Society.)

PEG - M_w 9700; M_n 8070; Manufacturer and Lot not indicated

Total system		Bottom phase		Top phase		STL *
PEG % w/w	$(NH_4)_2SO_4$ % w/w	PEG % w/w	$(NH_4)_2SO_4$ % w/w	PEG % w/w	$(NH_4)_2SO_4$ % w/w	
10.0	14.0	0.0	18.6	36.4	3.6	-2.427
12.0	16.0	0.0	21.8	41.9	2.9	-2.217
14.0	20.0	0.0	27.7	52.5	2.1	-2.051
16.0	24.0	0.0**	32.6**	55.2**	1.4**	-1.769**
					av.:	-2.232 ±0.188

*- STL - Tie-Line Slope defined as the ratio STL = (ΔPEG)/(ΔSalt) where ΔPEG is the difference between the concentrations of PEG and ΔSalt the difference between the concentrations of $(NH_4)_2SO_4$ in the two coexisting phases;

**- Composition of the phases questionable as the STL value is incosistent with the other values; was not used in calculations of the average STL value.

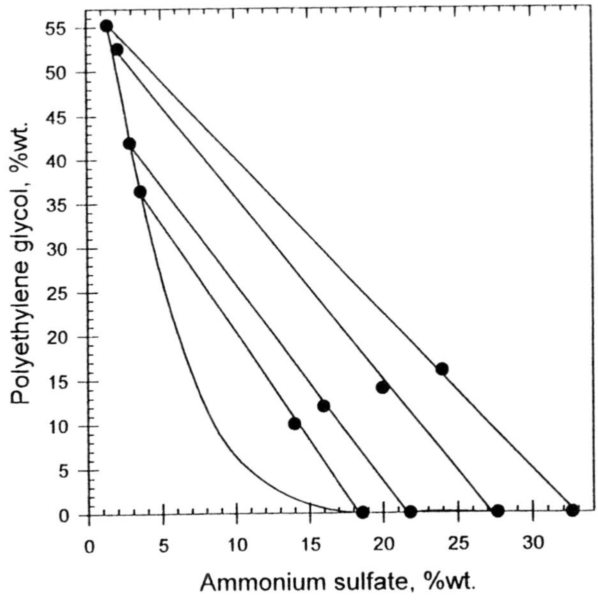

Table 10.142. Phase Diagram and Phase Composition of the Poly(ethylene glycol)-20000—Ammonium Sulfate system at 23°C.

PEG - M_w ~ 20,000; Manufacturer: Loba Chemie (Austria); Lot not indicated

Total system		Bottom phase		Top phase		STL *
PEG % w/w	$(NH_4)_2SO_4$ % w/w	PEG % w/w	$(NH_4)_2SO_4$ % w/w	PEG % w/w	$(NH_4)_2SO_4$ % w/w	
9.18	9.25	1.72	12.24	20.78	4.99	-2.630
11.47	9.65	0.62	14.00	26.92	3.86	-2.594
14.65	10.07	0.45	15.76	31.63	3.67	-2.579
17.26	10.94	0.08	17.82	37.42	2.86	-2.496
6.50**	8.80**				av.:	-2.575 ±0.057

*- STL - Tie-Line Slope defined as the ratio STL = (ΔPEG)/(ΔSalt) where ΔPEG is the difference between the concentrations of PEG and ΔSalt the difference between the concentrations of $(NH_4)_2SO_4$ in the two coexisting phases;

**- Composition of critical point.

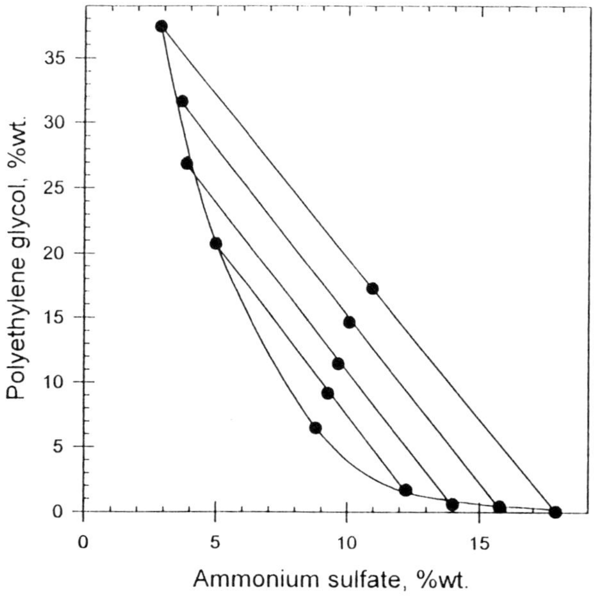

Table 10.143. Phase Diagram and Phase Composition of the Poly(ethylene glycol)-1000—Sodium Sulfate system at 25°C.
(From S. M. Snyder, K. D. Cole, D. C. Szlag, Chem.Eng.Data, 37, 268 (1992) by permission of the American Chemical Society.)

PEG- M_w 1125; M_n 1075 (Manufacturer and Lot not indicated)

Total system		Bottom phase		Top phase		TLS *
PEG % w/w	Na_2SO_4 % w/w	PEG % w/w	Na_2SO_4 % w/w	PEG % w/w	Na_2SO_4 % w/w	
12.8	9.6	5.2	13.7	26.6	4.5	-2.326
10.0	13.0	2.3	16.2	34.4	2.3	-2.309
11.0	14.0	1.4	18.3	35.2	2.3	-2.113
12.5	14.5	0.8	20.3	36.9	1.9	-1.962
14.0	15.0	0.6	21.6	41.5	1.8	-2.066
					av.:	-2.155 ±0.158

* - STL - Tie-Line Slope defined as the ratio STL = (ΔPEG)/(ΔSalt) where ΔPEG is the difference between the concentrations of PEG and ΔSalt the difference between the concentrations of Na_2SO_4 in the two coexisting phases.

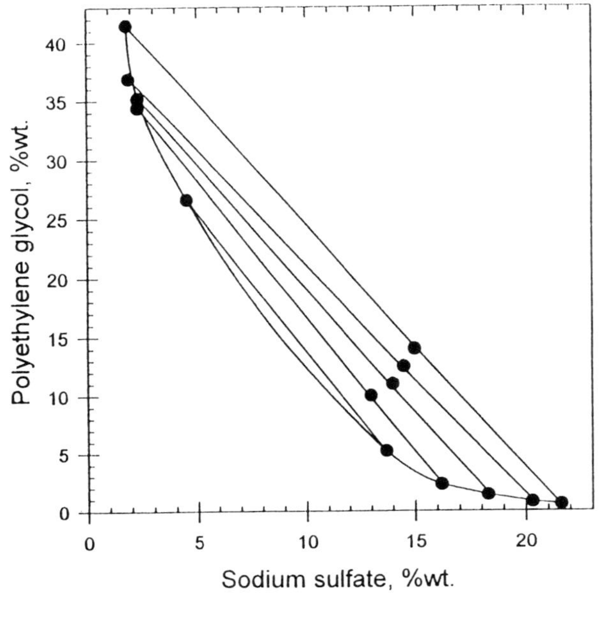

Table 10.144. Phase Diagram and Phase Composition of the Polyethylene glycol-3350—Sodium Sulfate system at 25°C.
(From S. M. Snyder, K. D. Cole, D. C. Szlag, Chem.Eng.Data, 37, 268 (1992) by permission of the American Chemical Society.)

PEG- M_w 3400; M_n 3200 (Manufacturer and Lot not indicated)

Total system		Bottom phase		Top phase		TLS *
PEG % w/w	Na_2SO_4 % w/w	PEG % w/w	Na_2SO_4 % w/w	PEG % w/w	Na_2SO_4 % w/w	
7.6	9.7	1.3**	12.1**	23.6**	4.9**	-3.097**
9.9	10.9	0.6	14.6	30.8	3.3	-2.673
14.0	10.8	0.4	16.4	34.9	3.4	-2.654
17.9	12.9	0.3	21.5	45.1	1.4	-2.229
					av.:	-2.519 ±0.251

*- STL - Tie-Line Slope defined as the ratio STL = (ΔPEG)/(ΔSalt) where ΔPEG is the difference between the concentrations of PEG and ΔSalt the difference between the concentrations of Na_2SO_4 in the two coexisting phases;
**- Composition of the phases questionable as the STL value is inconsistent with the other values; was not used in calculations of the average STL value.

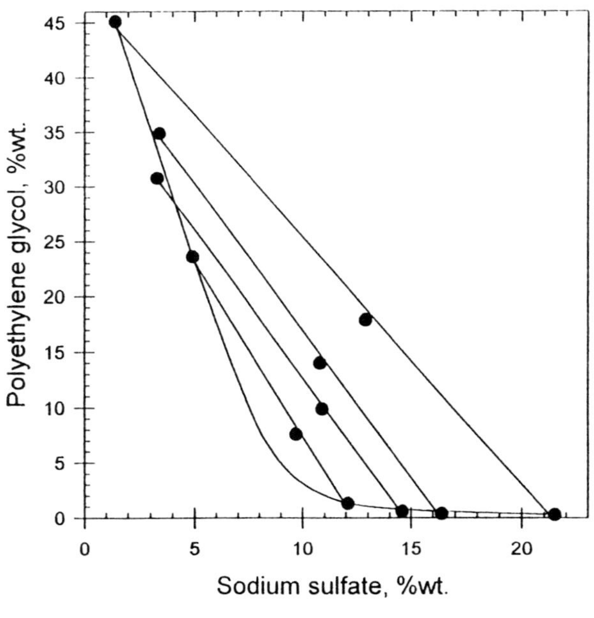

Table 10.145. Phase Diagram and Phase Composition of the Polyethylene glycol-8000—Sodium Sulfate system at 25°C.
(From S. M. Snyder, K. D. Cole, D. C. Szlag, Chem.Eng.Data, 37, 268 (1992) by permission of the American Chemical Society.)

PEG- M_w 9700; M_n 8070 (Manufacturer and Lot not indicated)

Total system		Bottom phase		Top phase		TLS *
PEG % w/w	Na_2SO_4 % w/w	PEG % w/w	Na_2SO_4 % w/w	PEG % w/w	Na_2SO_4 % w/w	
13.0	8.0	0.5**	11.9**	25.8**	4.3**	-3.329**
10.0	13.0	1.4	16.0	36.5	3.2	-2.742
11.0	14.0	1.1	17.7	38.7	3.1	-2.575
12.5	14.5	1.1	19.0	40.4	2.9	-2.441
14.0	15.0	1.2	20.3	41.7	3.2	-2.368
					av.:	-2.532 ±0.164

*- STL - Tie-Line Slope defined as the ratio STL = (ΔPEG)/(ΔSalt) where ΔPEG is the difference between the concentrations of PEG and ΔSalt the difference between the concentrations of Na_2SO_4 in the two coexisting phases;

**- Composition of the phases questionable as the STL value is inconsistent with the other values; was not used in calculations of the average STL value.

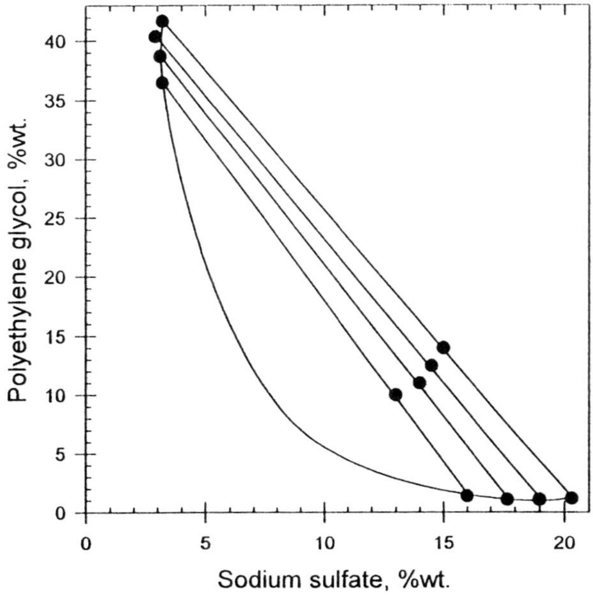

Table 10.146. Phase Diagram and Phase Composition of the Polyethylene glycol-1000—Magnesium Sulfate system at 25°C.
(From S. M. Snyder, K. D. Cole, D. C. Szlag, Chem.Eng.Data, 37, 268 (1992) by permission of the American Chemical Society.)

PEG- M_w 1125; M_n 1075 (Manufacturer and Lot not indicated)

Total system		Bottom phase		Top phase		TLS *
PEG % w/w	$MgSO_4$ % w/w	PEG % w/w	$MgSO_4$ % w/w	PEG % w/w	$MgSO_4$ % w/w	
14.0	9.5	6.3	13.2	30.6	3.3	-2.455
14.5	10.0	3.8	14.6	33.5	3.1	-2.583
15.0	10.5	2.6	15.7	36.1	3.1	-2.913
15.5	11.0	2.0	16.5	37.8	3.1	-2.672
					av.:	-2.656 ±0.193

*- STL - Tie-Line Slope defined as the ratio STL = (ΔPEG)/(ΔSalt) where ΔPEG is the difference between the concentrations of PEG and ΔSalt the difference between the concentrations of $MgSO_4$ in the two coexisting phases.

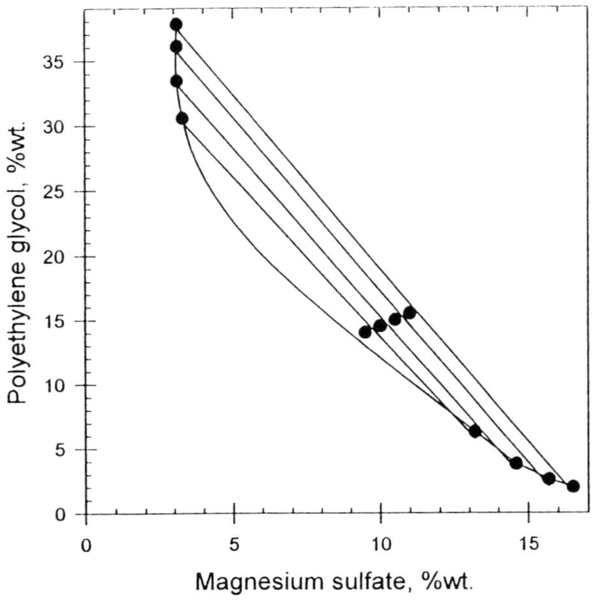

Table 10.147. Phase Diagram and Phase Composition of the Polyethylene glycol-3350—Magnesium Sulfate system at 25°C.
(From S. M. Snyder, K. D. Cole, D. C. Szlag, Chem.Eng.Data, 37, 268 (1992) by permission of the American Chemical Society.)

PEG- M_w 3400; M_n 3200 (Manufacturer and Lot not indicated)

Total system		Bottom phase		Top phase		TLS *
PEG % w/w	$MgSO_4$ % w/w	PEG % w/w	$MgSO_4$ % w/w	PEG % w/w	$MgSO_4$ % w/w	
12.3	7.9	4.8	12.2	25.4	3.1	-2.264
13.1	8.9	4.5	13.9	29.2	2.6	-2.186
14.0	9.8	5.1	14.2	32.8	2.5	-2.368
14.7	10.5	4.4	15.2	34.4	2.4	-2.344
					av.:	-2.291 ±0.083

*- STL - Tie-Line Slope defined as the ratio STL = (ΔPEG)/(ΔSalt) where ΔPEG is the difference between the concentrations of PEG and ΔSalt the difference between the concentrations of $MgSO_4$ in the two coexisting phases.

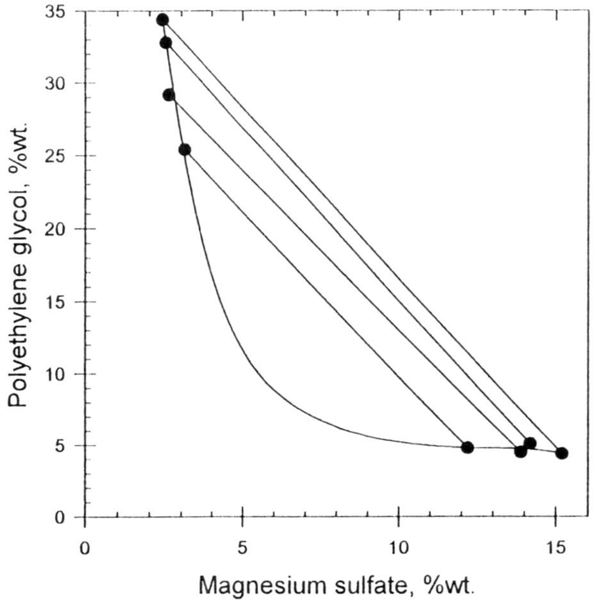

Table 10.148. Phase Diagram and Phase Composition of the Polyethylene glycol-8000—Magnesium Sulfate system at 25°C.
(From S. M. Snyder, K. D. Cole, D. C. Szlag, Chem.Eng.Data, 37, 268 (1992) by permission of the American Chemical Society.)

PEG- M_w 9700; M_n 8070 (Manufacturer and Lot not indicated)

Total system		Bottom phase		Top phase		TLS *
PEG % w/w	$MgSO_4$ % w/w	PEG % w/w	$MgSO_4$ % w/w	PEG % w/w	$MgSO_4$ % w/w	
12.0	8.0	3.8	12.5	25.2	3.8	-2.460
14.0	9.5	0.8	14.5	32.3	2.6	-2.647
14.5	10.0	0.7	15.1	33.5	2.3	-2.563
15.0	10.5	0.8	15.9	35.2	2.2	-2.511
15.5	11.0	1.2	16.7	38.2	1.7	-2.467
					av.:	-2.530 ±0.077

* - STL - Tie-Line Slope defined as the ratio STL = (ΔPEG)/(ΔSalt) where ΔPEG is the difference between the concentrations of PEG and ΔSalt the difference between the concentrations of $MgSO_4$ in the two coexisting phases.

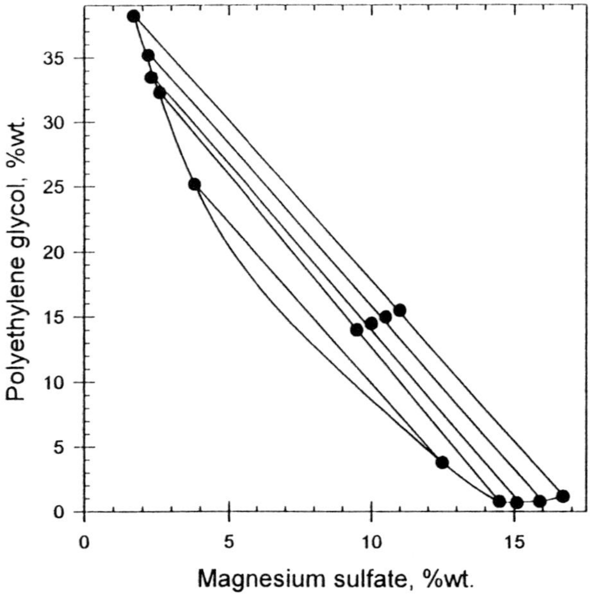

Table 10.149. Phase Diagram and Phase Composition of the Poly(ethylene glycol)-400—Potassium Phosphate, pH 7.0 System at 4°C.
(From X. Lei, A. D. Diamond, J. T. Hsu, J.Chem.Eng.Data, 35, 420 (1990) with permission of the American Chemical Society.)

PEG - M_w~ 400; Manufacturer: Aldrich (Milwaukee, WI, USA); Lot 02166TP;
Potassium Phosphate - mixture of K_2HPO_4 and KH_2PO_4 in the 1.82 : 1.0 ratio

Total system		Bottom phase		Top phase		STL *
PEG % w/w	Salt % w/w	PEG % w/w	Salt % w/w	PEG % w/w	Salt % w/w	
19.40	16.00	3.05	31.45	28.75	6.85	-1.045
20.60	16.80	2.54	33.51	32.02	5.78	-1.063
23.20	17.40	2.24	37.61	36.79	4.69	-1.050
25.10	17.90	2.12	39.73	39.73	3.95	-1.051
					av.:	-1.052 ±0.043

* - STL - Tie-Line Slope defined as the ratio STL = (ΔPEG)/(ΔSalt) where ΔPEG is the difference between the concentrations of PEG and ΔSalt the difference between the concentrations of potassium phosphate in the two coexisting phases.

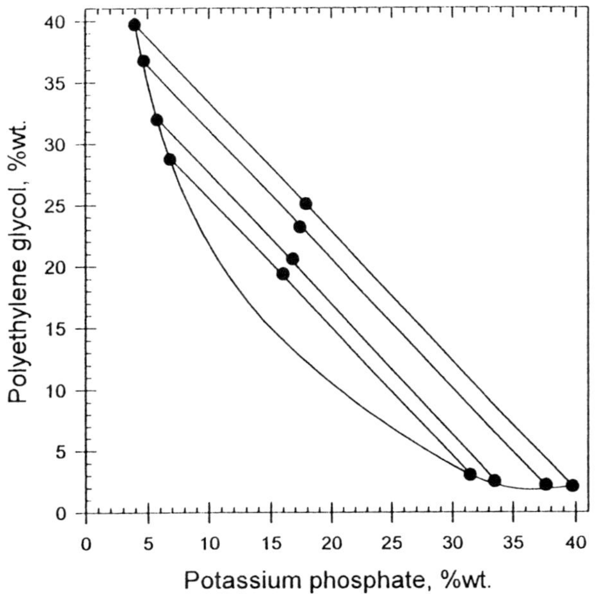

Table 10.150. Phase Diagram and Phase Composition of the Poly(ethylene glycol)-600—Potassium Phosphate, pH 7.0 System at 4°C.
(From X. Lei, A. D. Diamond, J. T. Hsu, J.Chem.Eng.Data, 35, 420 (1990) with permission of the American Chemical Society.)

PEG - $M_w \sim 600$; Manufacturer: Aldrich (Milwaukee, WI, USA); Lot 03403HM; Potassium Phosphate - mixture of K_2HPO_4 and KH_2PO_4 in the 1.82 : 1.0 ratio

Total system		Bottom phase		Top phase		STL *
PEG % w/w	Salt % w/w	PEG % w/w	Salt % w/w	PEG % w/w	Salt % w/w	
14.00	15.50	5.41	22.06	23.01	9.04	-1.352
15.51	16.43	3.47	26.05	28.73	6.43	-1.287
17.00	16.90	2.66	28.29	31.88	5.38	-1.275
18.30	17.40	2.00	30.59	34.48	4.78	-1.258
					av.:	-1.293 ±0.041

* - STL - Tie-Line Slope defined as the ratio STL = (ΔPEG)/(ΔSalt) where ΔPEG is the difference between the concentrations of PEG and ΔSalt the difference between the concentrations of potassium phosphate in the two coexisting phases.

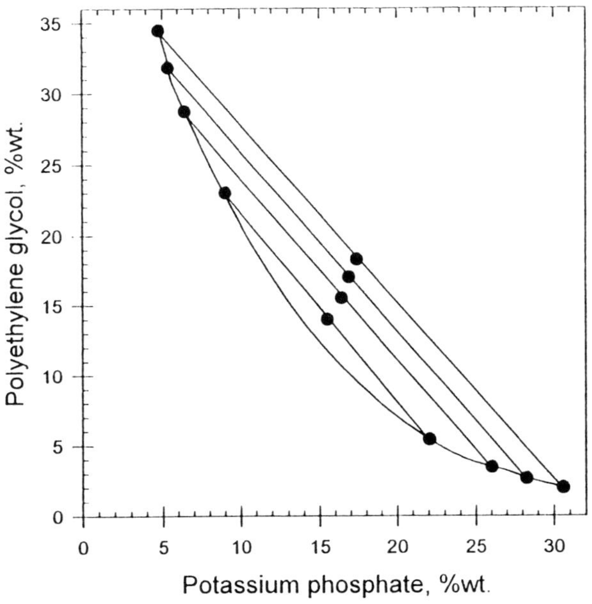

Table 10.151. Phase Diagram and Phase Composition of the Poly(ethylene glycol)-1000—Potassium Phosphate, pH 7.0 System at 4°C.
(From X. Lei, A. D. Diamond, J. T. Hsu, J.Chem.Eng.Data, 35, 420 (1990) with permission of the American Chemical Society.)

PEG - M_w ~ 1000; Manufacturer: Aldrich (Milwaukee, WI, USA); Lot 00404HM; Potassium Phosphate - mixture of K_2HPO_4 and KH_2PO_4 in the 1.82 : 1.0 ratio

Total system		Bottom phase		Top phase		STL *
PEG % w/w	Salt % w/w	PEG % w/w	Salt % w/w	PEG % w/w	Salt % w/w	
13.00	15.00	3.08	21.56	25.02	6.64	-1.470
15.00	15.70	1.08	25.56	29.02	5.53	-1.395
17.00	16.20	0.71	28.08	32.56	4.60	-1.356
19.00	17.00	0.56	30.81	36.37	3.70	-1.321
					av.:	-1.386
						±0.064

* - STL - Tie-Line Slope defined as the ratio STL = (ΔPEG)/(ΔSalt) where ΔPEG is the difference between the concentrations of PEG and ΔSalt the difference between the concentrations of potassium phosphate in the two coexisting phases.

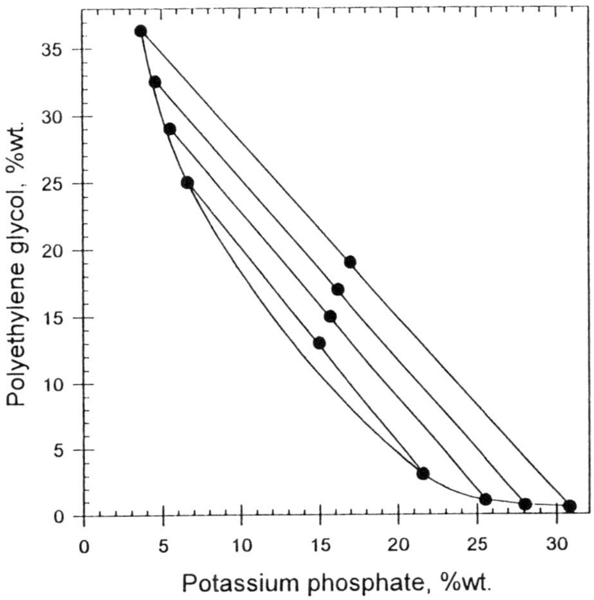

Table 10.152. Phase Diagram and Phase Composition of the Poly(ethylene glycol)-1500—Potassium Phosphate, pH 7.0 System at 4°C.
(From X. Lei, A. D. Diamond, J. T. Hsu, J.Chem.Eng.Data, 35, 420 (1990) with permission of the American Chemical Society.)

PEG - M_w ~ 1500; Manufacturer: Aldrich (Milwaukee, WI, USA); Lot not indicated;
Potassium Phosphate - mixture of K_2HPO_4 and KH_2PO_4 in the 1.82 : 1.0 ratio

Total system		Bottom phase		Top phase		STL *
PEG % w/w	Salt % w/w	PEG % w/w	Salt % w/w	PEG % w/w	Salt % w/w	
12.40	12.83	3.79	18.33	22.22	6.64	-1.577
13.66	13.12	2.50	20.37	25.30	5.69	-1.553
15.74	13.90	1.34	23.48	29.95	4.44	-1.503
18.64	15.17	0.98	27.71	35.13	3.46	-1.408
					av.:	-1.510 ±0.075

* - STL - Tie-Line Slope defined as the ratio STL = (ΔPEG)/(ΔSalt) where ΔPEG is the difference between the concentrations of PEG and ΔSalt the difference between the concentrations of potassium phosphate in the two coexisting phases.

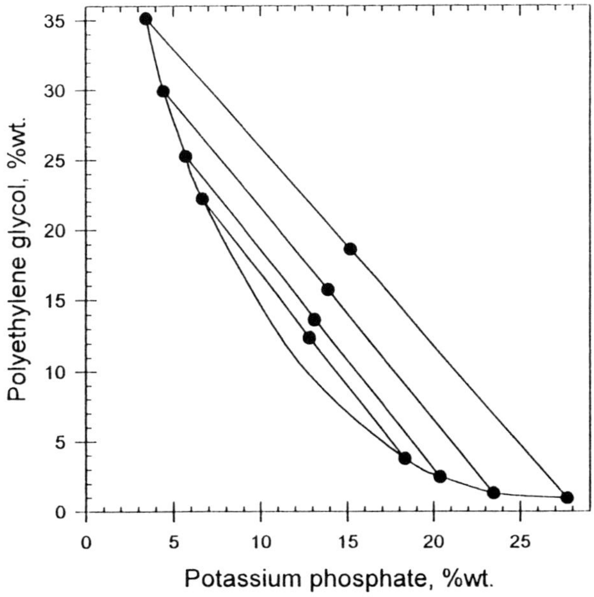

Table 10.153. Phase Diagram and Phase Composition of the Poly(ethylene glycol)-3400—Potassium Phosphate, pH 6.0 System at 4°C.
(From X. Lei, A. D. Diamond, J. T. Hsu, J.Chem.Eng.Data, 35, 420 (1990) with permission of the American Chemical Society.)

PEG - M_w ~ 3400; Manufacturer: Aldrich (Milwaukee, WI, USA); Lot 02607HV; Potassium Phosphate - mixture of K_2HPO_4 and KH_2PO_4 in the 1.0 : 2.0 ratio

Total system		Bottom phase		Top phase		STL *
PEG % w/w	Salt % w/w	PEG % w/w	Salt % w/w	PEG % w/w	Salt % w/w	
10.50	12.50	2.49	17.50	18.11	8.04	-1.651
11.00	13.00	1.88	18.68	20.81	7.10	-1.635
12.00	14.00	1.00	20.79	24.54	6.15	-1.608
13.00	15.00	0.60**	23.27**	27.14**	5.65**	-1.506**
					av.:	-1.631 ±0.022

* - STL - Tie-Line Slope defined as the ratio STL = (ΔPEG)/(ΔSalt) where ΔPEG is the difference between the concentrations of PEG and ΔSalt the difference between the concentrations of potassium phosphate in the two coexisting phases;

** - Composition of the phases questionable as the STL value is inconsistent with the other values; was not used in calculations of the average STL value.

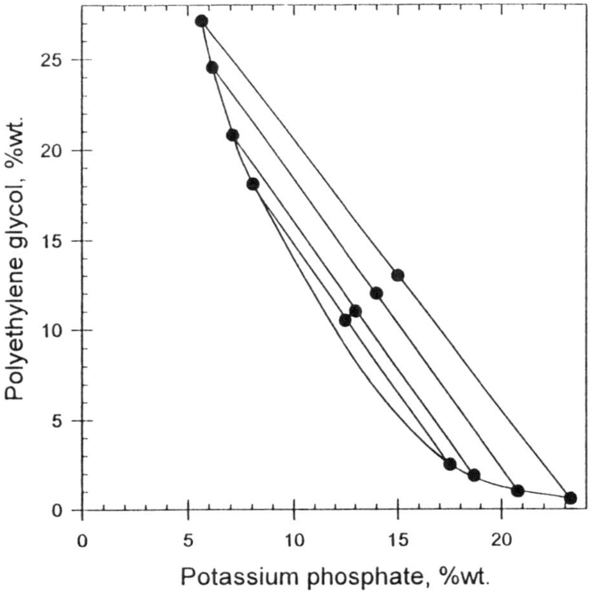

Table 10.154. Phase Diagram and Phase Composition of the Poly(ethylene glycol)-3400—Potassium Phosphate, pH 7.0 System at 4°C.
(From X. Lei, A. D. Diamond, J. T. Hsu, J.Chem.Eng.Data, 35, 420 (1990) with permission of the American Chemical Society.)

PEG - M_w ~ 3400; Manufacturer: Aldrich (Milwaukee, WI, USA); Lot 02607HV; Potassium Phosphate - mixture of K_2HPO_4 and KH_2PO_4 in the 1.82 : 1.0 ratio

Total system		Bottom phase		Top phase		STL *
PEG % w/w	Salt % w/w	PEG % w/w	Salt % w/w	PEG % w/w	Salt % w/w	
10.10	10.90	2.76	14.80	18.55	6.60	-1.926
11.00	11.40	1.61	16.48	22.14	5.51	-1.871
12.20	11.80	1.01	17.92	24.58	4.88	-1.808
13.70	12.30	0.78	19.85	27.66	4.21	-1.719
					av.:	-1.831
						±0.089

* - STL - Tie-Line Slope defined as the ratio STL = (ΔPEG)/(ΔSalt) where ΔPEG is the difference between the concentrations of PEG and ΔSalt the difference between the concentrations of potassium phosphate in the two coexisting phases.

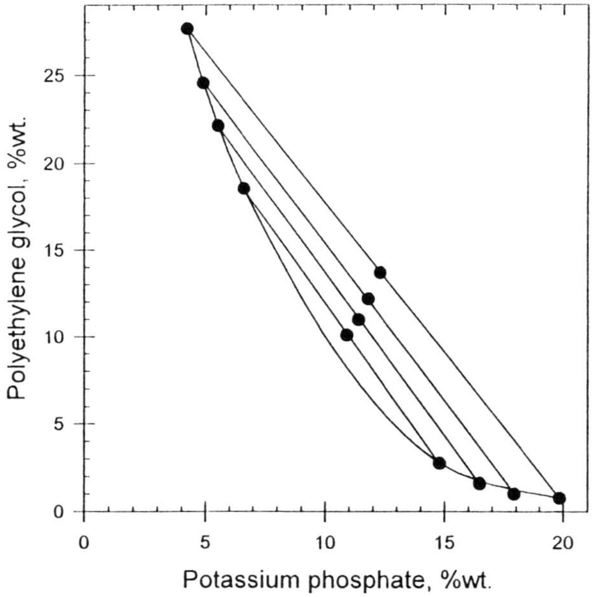

Table 10.155. Phase Diagram and Phase Composition of the Poly(ethylene glycol)-1000—Potassium Phosphate, pH 8.0 System at 25°C.
(From S. M. Snyder, K. D. Cole, D. C. Szlag, Chem.Eng.Data, 37, 268 (1992) by permission of the American Chemical Society.)

PEG - M_w 1125; M_n 1075; Manufacturer and Lot not indicated;
Potassium Phosphate - mixture of K_2HPO_4 and KH_2PO_4 providing pH 8.0

Total system		Bottom phase		Top phase		STL *
PEG % w/w	Salt % w/w	PEG % w/w	Salt % w/w	PEG % w/w	Salt % w/w	
16.1	10.0	5.0	16.0	22.7	6.8	-1.924
17.9	10.5	2.8	18.7	28.9	5.0	-1.905
20.0	11.4	2.1	21.6	36.1	3.5	-1.878
22.0	12.3	1.6	24.0	39.1	3.1	-1.794
					av.:	-1.875 ±0.057

* - STL - Tie-Line Slope defined as the ratio STL = (ΔPEG)/(ΔSalt) where ΔPEG is the difference between the concentrations of PEG and ΔSalt the difference between the concentrations of potassium phosphate in the two coexisting phases.

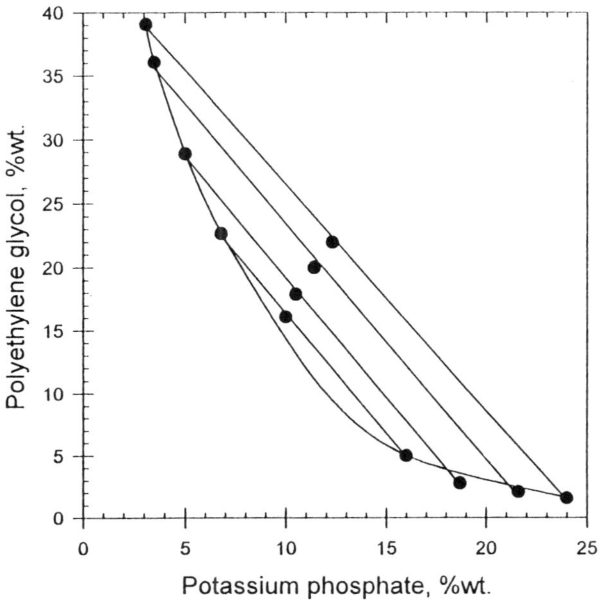

Table 10.156. Phase Diagram and Phase Composition of the Poly(ethylene glycol)-3400—Potassium Phosphate, pH 8.0 System at 4°C.
(From X. Lei, A. D. Diamond, J. T. Hsu, J.Chem.Eng.Data, 35, 420 (1990) with permission of the American Chemical Society.)

PEG - M_w ~ 3400; Manufacturer: Aldrich (Milwaukee, WI, USA); Lot 02607HV; Potassium Phosphate - mixture of K_2HPO_4 and KH_2PO_4 in the 15 : 1.0 weight ratio

Total system		Bottom phase		Top phase		STL *
PEG % w/w	Salt % w/w	PEG % w/w	Salt % w/w	PEG % w/w	Salt % w/w	
13.80	9.20	2.04	14.75	23.11	4.59	-2.074
14.80	10.30	0.78	17.47	27.33	3.74	-1.934
16.00	11.00	0.64	19.04	30.03	3.30	-1.867
17.00	12.00	0.52	21.18	32.89	2.93	-1.774
					av.:	-1.912 ±0.126

* - STL - Tie-Line Slope defined as the ratio STL = (ΔPEG)/(ΔSalt) where ΔPEG is the difference between the concentrations of PEG and ΔSalt the difference between the concentrations of potassium phosphate in the two coexisting phases.

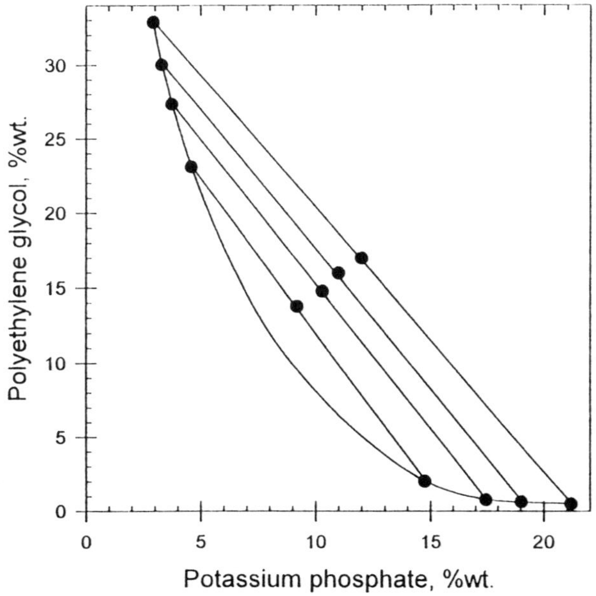

Table 10.157. Phase Diagram and Phase Composition of the Poly(ethylene glycol)-3400—Potassium Phosphate, pH 9.2 System at 4°C.
(From X. Lei, A. D. Diamond, J. T. Hsu, J.Chem.Eng.Data, 35, 420 (1990) with permission of the American Chemical Society.)

PEG - $M_w \sim 3400$; Manufacturer: Aldrich (Milwaukee, WI, USA); Lot 02607HV;
Potassium Phosphate - K_2HPO_4

Total system		Bottom phase		Top phase		STL *
PEG % w/w	Salt % w/w	PEG % w/w	Salt % w/w	PEG % w/w	Salt % w/w	
13.80	9.20	2.26	14.63	23.95	4.40	-2.120
14.80	10.30	0.73	17.53	27.91	3.75	-1.972
16.00	11.00	0.39	19.49	30.82	3.22	-1.870
17.00	12.00	0.21	21.48	33.91	2.79	-1.803
					av.:	-1.941 ±0.138

* - STL - Tie-Line Slope defined as the ratio STL = (ΔPEG)/(ΔSalt) where ΔPEG is the difference between the concentrations of PEG and ΔSalt the difference between the concentrations of potassium phosphate in the two coexisting phases.

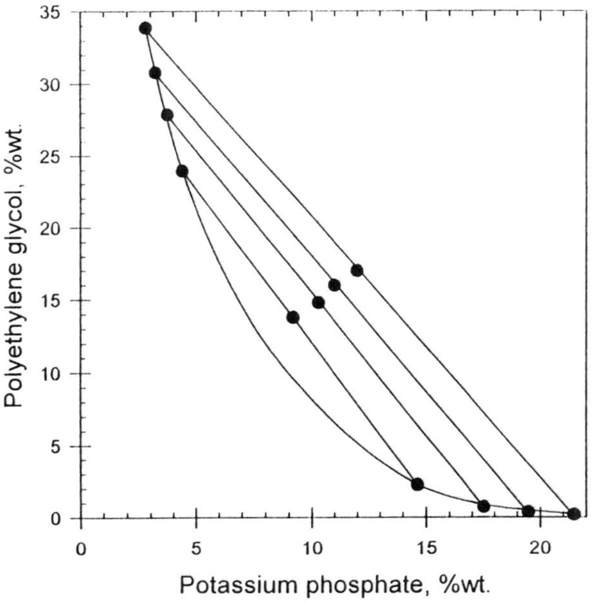

Table 10.158. Phase Diagram and Phase Composition of the Poly(ethylene glycol)-8000—Potassium Phosphate, pH 8.0 System at 25°C.
(From S. M. Snyder, K. D. Cole, D. C. Szlag, Chem.Eng.Data, 37, 268 (1992) by permission of the American Chemical Society.)

PEG - M_w 9700; M_n 8070; Manufacturer and Lot not indicated;
Potassium Phosphate - mixture of K_2HPO_4 and KH_2PO_4 providing pH 8.0

Total system		Bottom phase		Top phase		STL *
PEG % w/w	Salt % w/w	PEG % w/w	Salt % w/w	PEG % w/w	Salt % w/w	
12.1	7.7	1.9	11.5	21.7	4.4	-2.789
13.9	7.7	1.6	12.4	24.6	3.9	-2.706
16.1	10.0	1.6	16.3	34.6	2.6	-2.409
17.9	10.8	2.3	18.3	38.1	2.2	-2.224
20.0	11.6	3.0**	20.6**	41.2	1.8	-2.032
21.9	12.3	2.0	23.1	44.4	1.6	-1.972
					av.:	-2.355 ±0.341

*- STL - Tie-Line Slope defined as the ratio STL = (ΔPEG)/(ΔSalt) where ΔPEG is the difference between the concentrations of PEG and ΔSalt the difference between the concentrations of potassium phosphate in the two coexisting phases;
**- Composition of the phase questionable as it does not fit the phase diagram.

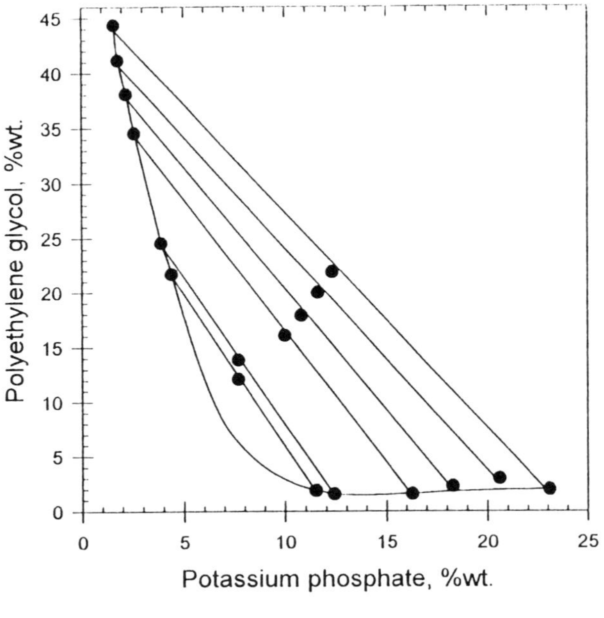

Table 10.159. Phase Diagram and Phase Composition of the Poly(ethylene glycol)-8000—Potassium Phosphate, pH 7.0 System at 4°C.
(From X. Lei, A. D. Diamond, J. T. Hsu, J.Chem.Eng.Data, 35, 420 (1990) with permission of the American Chemical Society.)

PEG - M_w ~ 8000; Manufacturer: Aldrich (Milwaukee, WI, USA); Lot 1722BW;
Potassium Phosphate - mixture of K_2HPO_4 and KH_2PO_4 in the 1.82 : 1.0 ratio

Total system		Bottom phase		Top phase		STL *
PEG % w/w	Salt % w/w	PEG % w/w	Salt % w/w	PEG % w/w	Salt % w/w	
12.20	9.70	2.00	14.77	22.19	4.68	-2.001
13.20	10.30	1.60	16.43	24.85	4.13	-1.890
14.30	10.90	1.35	17.98	27.14	3.84	-1.824
15.50	11.70	1.19	19.96	29.82	3.34	-1.723
					av.:	-1.860 ±0.117

* - STL - Tie-Line Slope defined as the ratio STL = (ΔPEG)/(ΔSalt) where ΔPEG is the difference between the concentrations of PEG and ΔSalt the difference between the concentrations of potassium phosphate in the two coexisting phases.

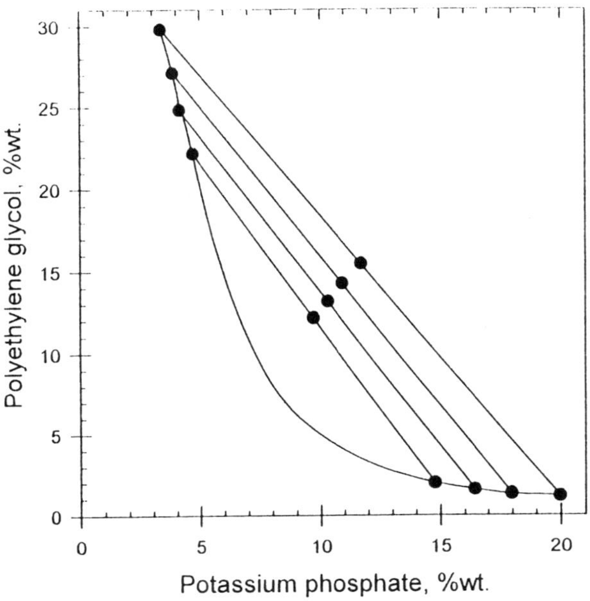

Table 10.160. Phase Diagram and Phase Composition of the Poly(ethylene glycol)-20000—Potassium Phosphate, pH 7.0 System at 4°C.
(From X. Lei, A. D. Diamond, J. T. Hsu, J.Chem.Eng.Data, 35, 420 (1990) with permission of the American Chemical Society.)

PEG - M_w ~ 20000; Manufacturer: Aldrich (Milwaukee, WI, USA); Lot not indicated;
Potassium Phosphate - mixture of K_2HPO_4 and KH_2PO_4 in the 1.82 : 1.0 ratio

Total system		Bottom phase		Top phase		STL *
PEG % w/w	Salt % w/w	PEG % w/w	Salt % w/w	PEG % w/w	Salt % w/w	
9.00	9.00	2.10	11.99	17.55	5.36	-2.330
10.00	9.60	1.20	13.56	20.06	4.70	-2.129
11.40	9.90	0.90	14.99	23.39	4.01	-2.04/8
13.00	10.40	0.88	16.36	26.21	3.69	-1.999
					av.:	-2.127 ±0.146

* - STL - Tie-Line Slope defined as the ratio STL = (ΔPEG)/(ΔSalt) where ΔPEG is the difference between the concentrations of PEG and ΔSalt the difference between the concentrations of potassium phosphate in the two coexisting phases.

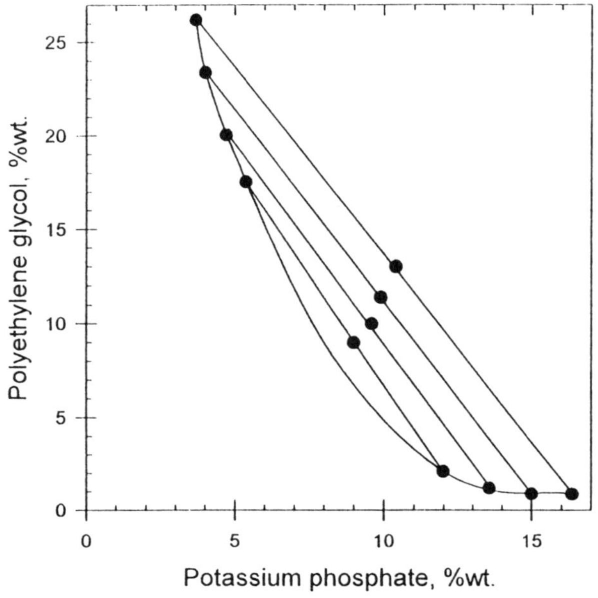

Table 10.161. Phase Diagram and Phase Composition of the Poly(ethylene glycol)-1000—Sodium Carbonate System at 25°C.
(From S. M. Snyder, K. D. Cole, D. C. Szlag, Chem.Eng.Data, 37, 268 (1992) by permission of the American Chemical Society.)

PEG- M_w 1125; M_n 1075 (Manufacturer and Lot not indicated)

Total system		Bottom phase		Top phase		STL *
PEG % w/w	Na_2CO_3 % w/w	PEG % w/w	Na_2CO_3 % w/w	PEG % w/w	Na_2CO_3 % w/w	
12.7	7.0	7.6	10.1	17.9	5.8	-2.395
9.9	10.9	2.0	13.7	34.2	2.1	-2.776
13.6	10.5	1.5	14.5	38.5	1.3	-2.803
11.9	11.9	1.1	16.3	39.4**	1.8**	-2.641
14.9	12.9	1.9	17.8	45.9	1.0	-2.619
					av.:	-2.647
						±0.162

*- STL - Tie-Line Slope defined as the ratio STL = (ΔPEG)/(ΔSalt) where ΔPEG is the difference between the concentrations of PEG and ΔSalt the difference between the concentrations of Na_2CO_3 in the two coexisting phases;
**- Composition of the phase questionable as it does not fit the phase diagram.

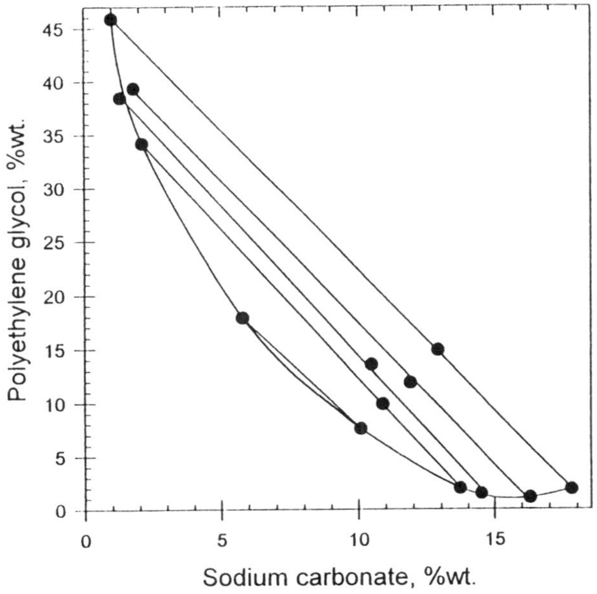

Table 10.162. Phase Diagram and Phase Composition of the Poly(ethylene glycol)-3350—Sodium Carbonate System at 25°C.
(From S. M. Snyder, K. D. Cole, D. C. Szlag, Chem.Eng.Data, 37, 268 (1992) by permission of the American Chemical Society.)

PEG- M_w 3400; M_n 3200 (Manufacturer and Lot not indicated)

Total system		Bottom phase		Top phase		STL *
PEG % w/w	Na_2CO_3 % w/w	PEG % w/w	Na_2CO_3 % w/w	PEG % w/w	Na_2CO_3 % w/w	
14.0	10.9	0.8	14.4	37.5	4.5	-3.707
16.1	11.9	0.5	16.5	43.2**	4.3**	-3.500
20.0	14.0	1.5***	20.3***	50.5***	3.4***	-2.899***
					av.:	-3.604 ±0.146

* - STL - Tie-Line Slope defined as the ratio STL = (ΔPEG)/(ΔSalt) where ΔPEG is the difference between the concentrations of PEG and ΔSalt the difference between the concentrations of Na_2CO_3 in the two coexisting phases;
** - Composition of the phase questionable as it does not fit phase diagram;
*** - Composition of the phases questionable as the STL value is inconsistent with the other values; was not used in calculations of the average STL value.

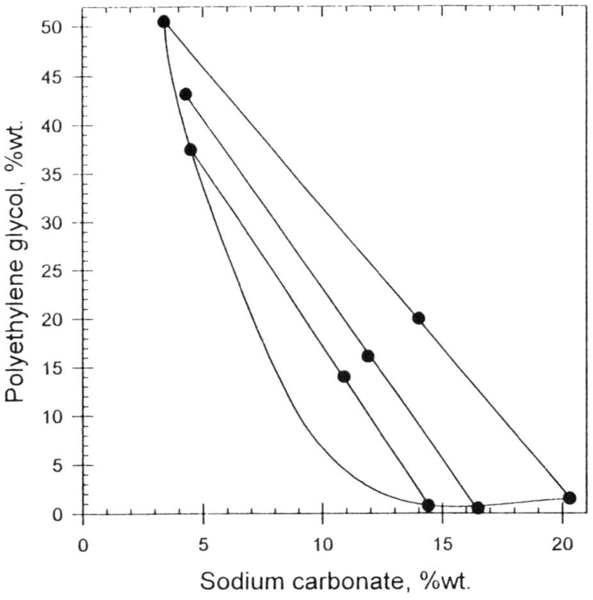

Table 10.163. Phase Diagram and Phase Composition of the Poly(ethylene glycol)-8000—Sodium Carbonate System at 25°C.
(From S. M. Snyder, K. D. Cole, D. C. Szlag, Chem.Eng.Data, 37, 268 (1992) by permission of the American Chemical Society.)

PEG- M_w 9700; M_n 8070 (Manufacturer and Lot not indicated)

Total system		Bottom phase *		Top phase *		STL **
PEG % w/w	Na_2CO_3 % w/w	PEG % w/w	Na_2CO_3 % w/w	PEG % w/w	Na_2CO_3 % w/w	
11.7	5.6	0.55	8.5	25.9	2.2	-4.032
19	6.0	0.5	11.7	34.9***	2.1***	-3.583
8.8	12.1	0.5	15.0	42.9	0.8	-2.986
10.3	14.3	0.3	18.1	48.0	0.1	-2.650
					av.:	-3.604 ±0.146

* - Composition of the phases questionable as the STL values are highly inconsistent;
** - STL - Tie-Line Slope defined as the ratio STL = (ΔPEG)/(ΔSalt) where ΔPEG is the difference between the concentrations of PEG and ΔSalt the difference between the concentrations of Na_2CO_3 in the two coexisting phases;
*** -Composition of the phase questionable as it does not fit phase diagram.

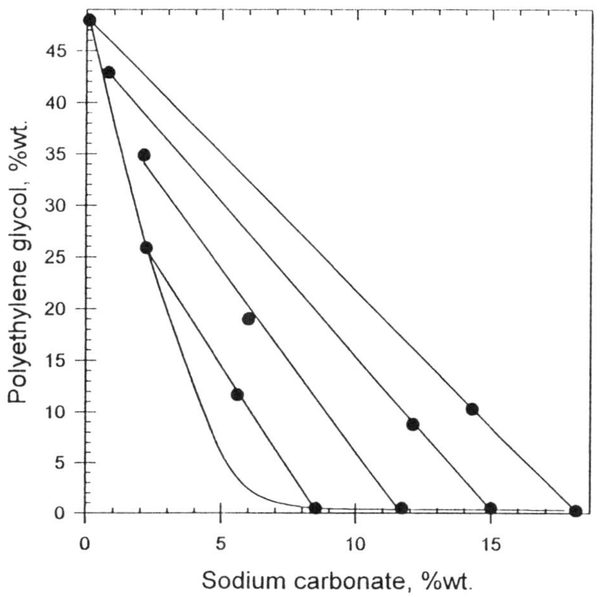

INDEX